U.S. Department of the Interior
U.S. Geological Survey

MINERALS YEARBOOK

Area Reports: International 2012

Europe and Central Eurasia

Volume III

Vincennes University
Shake Learning Resources Center
Vincennes, In 47591-9986

338.2
U58m
2012
Vol.3

U.S. DEPARTMENT OF THE INTERIOR
SALLY JEWELL, Secretary

U.S. GEOLOGICAL SURVEY
Suzette M. Kimball, Acting Director

UNITED STATES GOVERNMENT PRINTING OFFICE, WASHINGTON: 2015

Manuscript approved for publication February 27, 2015.

For more information on the USGS—the Federal source for science about the Earth, its natural and living resources, natural hazards, and the environment:
Internet: http://www.usgs.gov
Telephone: 1–888–ASK–USGS

Any use of trade, product, or firm names in this publication is for descriptive purposes only and does not imply endorsement by the U.S. Government.

Although this report is in the public domain, permission must be secured from the individual copyright owners to reproduce any copyrighted materials contained within this report.

Suggested citation:
U.S. Geological Survey, 2015, Area reports—International—Europe and Central Eurasia: U.S. Geological Survey Minerals Yearbook 2012, v. III, 352 p.

For sale by the Superintendent of Documents, U.S. Government Printing Office
Internet: bookstore.gpo.gov Phone: toll free (866) 512-1800; DC area (202) 512-1800
Fax: (202) 512-2104 Mail: Stop IDCC; Washington, DC 20402-0001

ISSN 0076-8952
ISBN 978 1 4113 3679 7

Foreword

This edition of the U.S. Geological Survey (USGS) Minerals Yearbook discusses the performance of the worldwide minerals and materials industries during 2012 and provides background information to assist in interpreting that performance. Content of the individual Minerals Yearbook volumes follows:
- Volume I, Metals and Minerals, contains chapters about virtually all metallic and industrial mineral commodities important to the U.S. economy. Chapters on survey methods, summary statistics for domestic nonfuel minerals, and trends in mining and quarrying in the metals and industrial mineral industries in the United States are also included.
- Volume II, Area Reports: Domestic, contains a chapter on the mineral industry of each of the 50 States and Puerto Rico and the Administered Islands. This volume also has chapters on survey methods and summary statistics of domestic nonfuel minerals.
- Volume III, Area Reports: International, is published as four separate reports. These regional reports contain the latest available minerals data on more than 190 foreign countries and discuss the importance of minerals to the economies of these nations and the United States. Each report begins with an overview of the region's mineral industries during the year. It continues with individual country chapters that examine the mining, refining, processing, and use of minerals in each country of the region and how each country's mineral industry relates to U.S. industry. Most chapters include production tables and industry structure tables, information about Government policies and programs that affect the country's mineral industry, and an outlook section.

The USGS continually strives to improve the value of its publications to users. Constructive comments and suggestions by readers of the Minerals Yearbook are welcomed.

Suzette M. Kimball, Acting Director

Contacts

Information about the U.S. Geological Survey, its programs, staff, and products may be accessed on the Internet at http://www.usgs.gov or by contacting the Earth Science Information Center at 1–888–ASK–USGS. For specific information about this publication, contact the Center secretary at (703) 648–4961. Additional minerals information may be accessed on the Internet at http://minerals.usgs.gov/minerals.

Acknowledgments

The Country Specialists in the National Minerals Information Center, U.S. Geological Survey, in preparing the International Review regional books of volume III of the Minerals Yearbook, extensively use statistics and data on mineral production, consumption, and trade provided by various foreign Government minerals and statistical agencies through various official publications. The cooperation and assistance of these organizations are gratefully acknowledged. Statistical and informational material was also obtained from reports of the U.S. Department of State, United Nations publications, and the domestic and foreign technical and trade press.

Of particular assistance were reports submitted by the Resource Reporting Officers and other officers of the Department of State in U.S. Embassies worldwide. Their contributions are sincerely appreciated. Internal statistical support was provided by the staff of the Data Collection and Coordination Section of the National Minerals Information Center.

The regimes of some countries reviewed in this volume may not be recognized by the U.S. Government. The information contained herein is technical and statistical in nature and is not to be construed as conflicting with or being contradictory of U.S. foreign policy.

The staff of the National Minerals Information Center gratefully acknowledges the invaluable contributions that John H. DeYoung, Jr., has made to the quality of the Minerals Yearbook during his tenure as the Director of the National Minerals Information Center from January 1996 until June 2013.

Steven M. Fortier
Director, National Minerals Information Center

Contents

Foreword, by Suzette M. Kimball iii
Contacts iv
Acknowledgments, by Steven M. Fortier v
Europe and Central Eurasia, by Alberto Alexander Perez, Elena Safirova, Steven T. Anderson, Alfredo C. Gurmendi,
    Harold R. Newman, Yadira Soto-Viruet, Glenn J. Wallace, and David R. Wilburn 1.1
Albania, by Mark Brininstool 2.1
Armenia, by Elena Safirova 3.1
Austria, by Steven T. Anderson 4.1
Azerbaijan, by Elena Safirova 5.1
Belarus, by Elena Safirova 6.1
Belgium and Luxembourg, by Alberto Alexander Perez 7.1
Bosnia and Herzegovina, by Yadira Soto-Viruet 8.1
Bulgaria, by Yadira Soto-Viruet 9.1
Croatia, by Harold R. Newman 10.1
Cyprus, by Harold R. Newman 11.1
Czech Republic, by Steven T. Anderson 12.1
Denmark, the Faroe Islands, and Greenland, by Harold R. Newman 13.1
Estonia, by Alberto Alexander Perez 14.1
Finland, by Alberto Alexander Perez 15.1
France, by Alberto Alexander Perez 16.1
Georgia, by Elena Safirova 17.1
Germany, by Steven T. Anderson 18.1
Greece, by Harold R. Newman 19.1
Hungary, by Steven T. Anderson 20.1
Iceland, by Harold R. Newman 21.1
Ireland, by Alberto Alexander Perez 22.1
Italy, by Alberto Alexander Perez 23.1
Kazakhstan, by Elena Safirova 24.1
Kosovo, by Mark Brininstool 25.1
Kyrgyzstan, by Elena Safirova 26.1
Latvia, by Alberto Alexander Perez 27.1
Lithuania, by Alberto Alexander Perez 28.1
Macedonia, by Yadira Soto-Viruet 29.1
Malta, by Harold R. Newman 30.1
Moldova, by Elena Safirova 31.1
Montenegro, by Harold R. Newman 32.1
Netherlands, by Alberto Alexander Perez 33.1
Norway, by Harold R. Newman 34.1
Poland, by Yadira Soto-Viruet 35.1
Portugal, by Alfredo C. Gurmendi 36.1
Romania, by Alberto Alexander Perez 37.1
Russia, by Elena Safirova 38.1
Serbia, by Yadira Soto-Viruet 39.1
Slovakia, by Harold R. Newman 40.1
Slovenia, by Harold R. Newman 41.1
Spain, by Alfredo C. Gurmendi 42.1
Sweden, by Alberto Alexander Perez 43.1
Switzerland, by Harold R. Newman 44.1
Tajikistan, by Elena Safirova 45.1
Turkmenistan, by Elena Safirova 46.1
Ukraine, by Elena Safirova 47.1
United Kingdom, by Alberto Alexander Perez 48.1
Uzbekistan, by Elena Safirova 49.1

THE MINERAL INDUSTRIES OF EUROPE AND CENTRAL EURASIA

By Alberto Alexander Perez, Elena Safirova, Steven T. Anderson, Alfredo C. Gurmendi, Harold R. Newman, Yadira Soto-Viruet, Glenn J. Wallace, and David R. Wilburn

The region of Europe and Central Eurasia as defined in this volume encompasses territory that extends from the Atlantic coast of Europe to the Pacific coast of the Russian Federation. It includes the British Isles, Iceland, and Greenland (a self-governing part of the Kingdom of Denmark).

The European Union (EU) is a supranational entity that at yearend 2012 comprised the following 27 countries: Austria, Belgium, Bulgaria, Cyprus, the Czech Republic, Denmark, Estonia, Finland, France, Germany, Greece, Hungary, Ireland, Italy, Latvia, Lithuania, Luxembourg, Malta, the Netherlands, Poland, Portugal, Romania, Slovakia, Slovenia, Spain, Sweden, and the United Kingdom. The euro (€) operates as a single currency for countries within the EU that have fulfilled the stated requirements of the European Central Bank (located in Frankfurt, Germany) for inclusion in the euro area. As of January 1, 2013, the EU countries that were part of the euro area were Austria, Belgium, Cyprus, Estonia, Finland, France, Germany, Greece, Ireland, Italy, Luxembourg, Malta, the Netherlands, Portugal, Slovakia, Slovenia, and Spain. Kosovo and Montenegro officially adopted the euro as their sole currency without an agreement with the euro area and therefore they did not have euro issuing rights in 2012 (European Commission, 2014a, b).

Other countries that were candidates to join the EU were Iceland, Macedonia, Montenegro, Serbia, and Turkey (although no date was given for expected accession, as they were all still in the negotiation stage). Albania, Bosnia and Herzegovina, and Kosovo (under UN Security Council Resolution 1244) were considered potential candidate countries and were expected to start negotiations for EU candidate country status (European Commission, 2014a).

The Commonwealth of Independent States (CIS) was founded in 1991 by several Republics of the former Soviet Union and later was extended to include all the former Soviet Republics except the Baltic States of Estonia, Latvia, and Lithuania. The countries that made up the CIS in 2012 were Armenia, Azerbaijan, Belarus, Kazakhstan, Kyrgyzstan, Moldova, Russia, Tajikistan, Turkmenistan, Ukraine, and Uzbekistan. Georgia withdrew from the CIS in 2008. The CIS does not have supranational powers, and all member countries have equal standing under international law. Although the member countries had pledged to work on economic integration, few actual measures have been taken to make the CIS a functioning integrated economic bloc similar to that of the EU. Some member states of the CIS, however, established the Eurasian Economic Community with the goal of creating a full-fledged common market (Korrespondent.net, 2008).

A Customs Union agreement among Belarus, Kazakhstan, and Russia went into effect on January 1, 2010. According to this agreement, the countries form a joint customs territory where no customs duties or other economic restrictions on the movement of goods between the three countries apply. Each of the members of the Customs Union applies the same customs rates and trade regulations for goods traded with countries outside of the Customs Union. The members of the Customs Union were projected to save more than $400 billion by 2015 owing to reduced shipping times. Armenia, Kyrgyzstan, and Tajikistan expressed their interest in joining the Customs Union in the future but, as of the end of 2012, no decisions had been made (International Centre for Trade and Sustainable Development, 2010).

Starting on January 1, 2012, the Customs Union among the three countries was transformed into a Common Economic Space (CES), which was the next step in the Eurasian integration process. The CES agreement removed barriers to the movement of goods, capital, and labor between the three countries. It also included coordinated principles of business regulation and coordination of macroeconomic and monetary policies, although it did not imply the introduction of a common currency. The Eurasian Economic Commission, which was a new supranational body, was expected to govern the integration processes within the CES framework and had the right to make decisions that would become mandatory for all three states. In 2012, the CES court in Minsk, Belarus, which was to resolve economic disputes between member states as well as between individual economic agents, started operations. The complete package of CES integration documents included 17 international treaties and was signed in November 2011 in Moscow. The ultimate goal of the integration among the CES members is creation of a Eurasian Economic Union (an organization similar to the EU), which was planned for 2015 (Utro.ru, 2012).

The European Free Trade Association (EFTA), which is an alternative entity to the EU in Western Europe, comprised Iceland, Liechtenstein, Norway, and Switzerland. The agreement on the European Economic Area (EEA), which had been in force since 1994, brings all 27 EU members and 3 of the EFTA members (Iceland, Liechtenstein, and Norway) into a single internal market. The EEA provides for the free movement of goods, services, persons, and capital among the 30 EEA states. Switzerland was not part of the EEA but had a bilateral agreement with the EU that addresses the same issues covered by the EEA (European Free Trade Association, 2014).

The 49 countries in the Europe and Central Eurasia region encompass an area of 29.4 million square kilometers, which is about three times larger than that of the United States; 17.1 million square kilometers of the area is accounted for by Russia. In 2012, the 49 countries had a total population of 822 million. The EU population as of January 1, 2013, was

505.7 million, which was about 60% larger than that of the United States. The total gross domestic product (GDP) based on purchasing power parity of the 49 countries in the region was about $20.6 trillion, and the weighted average per capita GDP was $25,000; the per capita GDP ranged from $2,219 in Tajikistan to $84,450 in Luxembourg (tables 1, 2).

Acknowledgments

The U.S. Geological Survey (USGS) acknowledges and expresses its sincere appreciation to the following foreign Government agencies, international institutions, and private research organizations for providing mineral-production statistics, basic economic data, and other mineral-related information:

- Armenia—National Statistical Service;
- Austria—Bundesministerium für Wirtschaft, Familie und Jugend;
- Belgium—Statistics Belgium (StatBel);
- Bosnia and Herzegovina—Agency for Statistics of Bosnia and Herzegovina;
- Cyprus—The Mine Service; Ministry of Agriculture, Natural Resources and Environment;
- Estonia—Geological Survey of Estonia;
- Finland—Geological Survey of Finland;
- Georgia—National Statistics Office of Georgia (GEOSTAT);
- Germany—Bundesanstalt für Geowissenschaften und Rohstoffe;
- Greece—the Government of Greece;
- Hungary—Magyar Köztársaság Gazdasági És Közlekedési Minisztérium Magyar Geológiai Szogálat (Hungarian Geological Survey);
- Iceland—Statistics Iceland;
- Latvia—Central Statistical Bureau of Latvia;
- Lithuania—Statistics Lithuania;
- Moldova—National Bureau of Statistics of the Republic of Moldova;
- Montenegro—Statistical Office of the Republic of Montenegro;
- Poland—Central Statistical Office;
- Portugal—Instituto Geológico Minero (IGM), Division of Statistical Studies;
- Romania—National Institute of Statistics;
- Serbia—Serbian Government;
- Slovenia—Slovenian Government;
- Tajikistan—Agency on Statistics under the President of the Republic of Tajikistan; and
- Ukraine—State Statistics Committee.

General Economic Conditions

Growth in the Europe and Central Eurasia region decelerated considerably in 2012 after relatively strong growth in 2011. All economies in the region had to deal with challenging external conditions, including the euro area's recession and debt problems, volatile global financial markets, and a slowing global economy. Although the economy of the region as a whole had average growth of 1.1%, the economies of a number of countries in the region grew at a much faster rate in 2012, including that of Turkmenistan (which expanded by 11.0%), Uzbekistan (8.0%), Tajikistan (7.5%), Armenia (7.2%), Georgia (6.5%), Latvia (5.8%), and Kazakhstan (5.0%) (International Monetary Fund, 2013; World Bank, The, 2013).

Uranium production in the region of Europe and Central Eurasia accounted for 51.6% of the world's production of this mineral commodity (measured in U_3O_8 content); lignite coal, 49.9%; and titanium metal, 48.2%. The region's output of potash (K_2O equivalent) accounted for 43.7% of world production; secondary aluminum, 41.6%; refined palladium, 38.8%; nickel metal, 36.3%; refined platinum, 32.1%; ammonia (N content), 24.4%; secondary lead, 24.1%; and secondary copper, 23.5%. The region also produced 20.4% of the world's output of primary aluminum, 19.4% of the world's output of primary copper, and 18.3% of the world's output of crude steel. The region was practically self-sufficient in the production of construction materials and remained among the world's leading producers of natural gas (34.0% of world production). Russia accounted for 26.8% of total natural diamond (gemstone and industrial) production in the world (table 4). The region was a leading crude oil producer and had significant coal reserves.

The EU countries were substantial participants in the world mineral economy and occupied an important role mostly as processors and consumers of most major mineral commodities. In Central Eurasia, however, mining of several mineral commodities remained economically important and made significant contributions to the GDPs and export revenues of the countries that produced them. In 2012, Central Eurasia remained a major world supplier of mined and processed minerals, and the consumption of these commodities in the region had increased in the past few years. The countries of Central and Eastern Europe and the CIS produced mineral commodities mainly for export, and the output of mineral commodities in these countries was significantly influenced by economic conditions in the rest of the world. China and the EU were especially significant markets for mineral products from Central and Eastern Europe and the CIS. As economies began to show signs of recovering from the global economic crisis that began in 2008, consumption of mineral commodities increased and drove the recovery of production in the region.

In the CIS, Russia and Kazakhstan were the two leading producers of mineral commodities. In Russia, mining and quarrying contributed $179 billion (10.9%) to the total value added in the economy in 2012. Mineral products made up 71.4% of the total value of Russian exports, and crude oil alone contributed 34.4% to the total value of exports. Petroleum products, natural gas, and ferrous metals accounted for 19.7%, 11.8%, and 4.8%, respectively, of the total value of Russia's exports (Federal State Statistics Service of the Russian Federation, 2013a, b).

Russia, which accounted for about 80% of the territory of the CIS, was by far the largest country in the CIS in terms of both population and territory, and it was the leading mineral producer. Many other CIS countries also were significant producers and processors of minerals. In 2012, Russia ranked among the top world producers or was a significant regional producer of such mineral commodities as aluminum, antimony, arsenic, asbestos, barite, bauxite, boron, cadmium, cement, coal, cobalt, copper, diamond, diatomite, fluorspar, gallium, gemstones, germanium, gold, graphite, gypsum, indium, iodine, iron ore, lead, lime, magnesium compounds and metals, mica (flake, scrap, and sheet), molybdenum, natural gas, nickel, nitrogen, palladium, peat, petroleum, phosphate rock, pig iron, platinum, potash, rhenium, selenium, silicon, steel, sulfur, tellurium, titanium sponge, tungsten, uranium, vanadium, and vermiculite.

In Kazakhstan, total industrial production was valued at $113 billion, of which $68.7 billion (60.8%) was from mining and $13.1 billion (12%) was from metallurgy (including $4.8 billion from ferrous metallurgy). Extraction of crude petroleum alone contributed $57.9 billion to the country's GDP. Overall, extractive industries contributed 34% to the country's GDP and metallurgy contributed another 6.4% to the GDP. Kazakhstan was a leading producer of uranium (37% of the world's production), the second-ranked producer of chromium (16%), the fourth-ranked producer of titanium sponge (11%) and magnesium metal (3%), and the fifth-ranked producer of rhenium (6%). The country was also a significant producer of bauxite, cadmium, copper, gallium, and zinc (Agency of Statistics of the Republic of Kazakhstan, 2013).

Ukraine was a significant producer of such mineral products as ferroalloys, iron ore, manganese ore, pig iron, steel, and titanium raw materials. Other CIS countries were significant world or regional producers of one or more mineral commodities, including Armenia (molybdenum), Azerbaijan (petroleum), Belarus (potash), Kyrgyzstan (antimony metal and mercury metal), Tajikistan (antimony ore), Turkmenistan (natural gas), and Uzbekistan (gold and uranium). All the CIS countries produced a number of other mineral commodities.

The EU was mostly dependent on imported mineral raw materials for metals, industrial minerals, and fuel minerals. The import dependence for many metal ores was 100% [including for antimony, cobalt, ilmenite, molybdenum, niobium, platinum-group metals (PGMs), rare-earth metals, rutile, tantalum, and vanadium], and the EU was from 70% to 90% import dependent for most other metallic ores. The EU's dependence on imports of metallic mineral raw materials (such as concentrates, ores, and scrap) and obtaining sources of energy for its metal refining and processing industries were key concerns for the EU's mineral industry (European Commission, 2008).

As a major world mineral processing and consuming area, the EU remained a significant determinant of world demand for nearly all mineral commodities. Its mineral processing and manufacturing industries accounted for a significant share of the world production of semimanufactured and fabricated ferrous and nonferrous metals. In 2012, Germany was still the EU's dominant smelter and refiner of most metals. With a high per capita income and standard of living, the EU was one of the world's major consumers of mineral fuels and of mineral products in consumer goods.

Legislation

In September, the President of Kyrgyzstan signed into law a new mining code that was adopted by the country's Parliament in June. The new code is intended to improve the investment climate in the country and it sets fees that mining companies are obligated to pay to the central and local governments for their use of the country's natural resources. The main principles of the new code are protection of investment as a form of private property, noninvolvement of local and national authorities in management decisions of private enterprises, and provision of exclusive rights for the transition of licenses from exploration to mining. The new code also describes how mining contributes to socioeconomic development of localities, outlines the "social packet" to be included in the application for exploration and mining licenses, and establishes renewal fees for extending licenses (Knews.kg, 2012).

In July, Romania's Zeta Petroleum plc announced that a new energy and gas law (law No.123/2012) had been published in the Romanian official Gazette No. 485, dated July 16, 2012. The main objective of this law is to put into Romanian law the provisions of the European Commission's third energy package concerning rules for the internal market in natural gas. The new law provides a calendar for the elimination of regulated prices for end users. These regulated prices ended on December 1, 2012, for nonresidential customers and was to end on July 1, 2013, for residential customers (Zeta Petroleum plc, 2012).

Several changes in the law focused on the provisions concerning exploration. In Greenland, legislation setting the framework for foreign exploration and mining in the country was passed and became law in 2012. The new law defines what is to be classified as a large-scale project and regulates the minimum salary levels for foreign workers (Creamer Media's Mining Weekly, 2012). According to Statistics Greenland, the number of exploration licenses granted in Greenland had increased to 75 in 2011 from 33 in 2005 (North of 60 Mining News, 2012).

A new tax law was ratified in Poland in 2012 that would tax copper and silver based on the mass of the extracted commodity and would be evaluated monthly. The maximum tax rate for the extraction of copper was 16,000 Polish zlotys (PLN), or US$5,000 per metric ton, and the maximum tax rate for the extraction of silver was PLN2,100 (US$660) per kilogram (U.S. Library of Congress, 2012).

In 2012, new export duties were imposed in Kazakhstan and Russia (Deloitte Development LLC, 2012). According to the March 2013 Fraser Institute survey, the top 10 destinations for mineral exploration based on favorable Government mineral policies in 2012 included Finland, Ireland, Norway, and Sweden. Finland, Greenland, and Sweden were included in the Fraser Institute 2012 survey of the top 10 destinations for mineral exploration based on their prospecting potential (assuming the regulations and land use restrictions that were in place in 2012) (Wilson, McMahon, and Cervantes, 2013).

Exploration

Information on exploration activities for Europe and Central Eurasia based on data provided by USGS compilations and economic data reported in U.S. nominal dollars by the SNL Metals Economics Group (SNL-MEG) has been included in a grouping of data for projects in mainland Asia, the CIS,

Europe, and the Middle East. As reported by SNL-MEG, the mineral exploration budget for this composite region increased by about 28% to about $3.1 billion in 2012 from the $2.4 billion in 2011. The exploration budget for Russia was reported to have increased to about $610 million in 2012 from about $337 million in 2011, and to account for about 3 percent of the world's exploration budget in 2012. These figures for exploration activity in Russia do not include activity conducted by Government-controlled entities (SNL Metals Economics Group, 2012).

In terms of the number of exploration sites, the greatest amount of exploration in Europe and Central Eurasia took place primarily in Kazakhstan, Russia, and Scandinavia (particularly Finland and Sweden). On the basis of exploration site data compiled by the USGS, Russia accounted for about 24% of the sites actively being explored in this regional grouping in 2012, Sweden accounted for about 9%, Finland accounted for about 8%, and Greenland and Kazakhstan each accounted for about 7%. The remaining 45% took place in 22 other countries located in the CIS and the EU. Exploration activity in the CIS focused primarily on gold (65%), copper (12%), iron and silver (6% each), potash (4%), rare earths (3%), and other minerals (4%). European mineral exploration focused primarily on gold (38%), nonferrous base metals (32%), iron ore (6%), rare-earth elements (4%), and uranium (3%); the remaining 17% of the exploration activity focused on 11 other mineral commodities.

Commodity Overview

This report includes mineral commodity outlook tables. In tables 5 through 20, estimates for the production of major mineral commodities for 2015 and beyond have been based upon supply-side assumptions, such as announced plans for increased production/new capacity construction and bankable feasibility studies. The outlook tables in this summary chapter show historic and projected production trends; therefore, no indication is made about whether the data are estimated or reported, and revisions are not identified. Data on individual mineral commodities in the tables in the individual country chapters are labeled to indicate estimates and revisions. The outlook segments of the mineral commodity tables are based on projected trends that could affect current (2012) producing facilities and on planned new facilities that operating companies, consortia, or Governments have projected to come online within the indicated timeframes. Forward-looking information, which includes estimates of future production, exploration and mine development, cost of capital projects, and timing of the start of operations, is subject to a variety of risks and uncertainties that could cause actual events or results to differ significantly from expected outcomes. Projects listed in the following section are presented as an indication of industry plans and are not a USGS prediction of what will take place.

Metals

Bauxite and Alumina and Aluminum.—In 2012, Russia and Kazakhstan produced the majority of bauxite output in the region, accounting for 5.7 million metric tons (Mt) and 5.2 Mt, respectively. By 2019, bauxite production was likely to increase slightly in both Russia and Kazakhstan to projected output levels of 6.0 million metric tons per year (Mt/yr) and 5.5 Mt/yr, respectively. In 2012, Russia was the leading source of alumina in Europe and Central Eurasia with annual production of 2.7 Mt. Ireland ranked second with 1.9 Mt and was followed by Kazakhstan and Spain (1.5 Mt each), and Ukraine (1.4 Mt) (tables 4, 5).

In 2012, Russia, which was the leading individual producer of primary and secondary aluminum in Europe and Central Eurasia, produced 3.9 Mt. The next-ranked producers in the region were Norway (2.05 Mt), Italy (1.11 Mt), and Germany (1.05 Mt). The projected output of primary and secondary aluminum in Russia was expected to remain unchanged through 2019. Production capacity was expected to be increased in Italy to 1.4 Mt/yr by 2019, but it was not expected to change significantly in Germany and Norway (tables 4, 6).

United Company RUSAL (RUSAL) of Russia was the world's leading producer of aluminum. RUSAL operated 14 smelters in Russia and Europe (12 in Russia, 1 in Ukraine, and 1 in Sweden). In 2012, RUSAL produced 12.37 Mt of bauxite, 7.48 Mt of alumina, and 4.17 Mt of aluminum at its facilities worldwide. Because of low aluminum prices in 2012, RUSAL was involved in cost-cutting activities and was devising a plan to divert aluminum production to its Eastern Division (where energy prices were lower) from its Western Division (where energy was more expensive). Instead, the plants in the Western Division were to focus on production of aluminum ferroalloys and flat ingots, which were much less energy-intensive to produce.

Cobalt.—The Europe and Central Eurasia region produced 26.2% of the world's total production of refined cobalt in 2012. According to the Cobalt Development Institute, Finland produced 10,547 metric tons (t) and was the second-ranked producer of refined cobalt in the world after China. The only producer in Finland was the OM Group of the United States. Umicore N.V. of Belgium (4,200 t) was the world's fourth-ranked producer and Xstrata of Switzerland at its operations in Norway (2,969 t) was the world's sixth-ranked producer. Other cobalt producers in the region were OAO GMK Norilsk Nickel (Nornickel) in Russia, which produced 2,186 t, and Eramet S.A. of France, which produced 326 t (table 7; Cobalt Development Institute, 2013, p. 3).

Copper.—In 2012, Russia was the region's leading producer of both mined copper and refined copper. Russia's mine production of copper was projected to increase to 1,000,000 metric tons per year (t/yr) by 2019 from 883,000 t in 2012. Other leading producers of mined copper in the region in 2012 were Poland (479,000 t), Kazakhstan (419,000 t), Bulgaria (108,000 t), and Uzbekistan (96,000 t). Russia's production of refined copper was 875,000 in 2012 and was projected to stay at about the same level through 2019. Other leading producers of refined copper in the region in 2012 were Germany (686,000 t), Poland (566,000 t), Belgium (380,000 t), and Kazakhstan (367,000 t). Production in Kazakhstan was projected to increase to 410,000 t/yr by 2019, and production in Belgium, Germany, and Poland was expected to remain at about the same level (tables 8, 9).

Russia had three leading vertically integrated copper producing companies— Nornickel, OAO Ural'skaya

Gorno-Metallurgicheskaya Kompaniya (UGMK), and ZAO Russkaya Mednaya Kompaniya (RMK). In 2012, RMK was building two new mines in Chelyabinskaya Oblast' in the South Urals and expected to commission them in late 2013 and 2014. When at full capacity, the two new mines were expected to produce 44 Mt/yr of copper ore. Metalloinvest Holding and State Corporation Gostechnologii continued to build a mine at the Udokan deposit in Zabaikal'skiy Kray. The mine was expected to start producing in 2014 and to reach projected capacity of 36 Mt/yr of copper ore by 2016.

Kazakhmys plc was the dominant producer of copper ore and metals in Kazakhstan. The company produced 306,100 t of copper contained in concentrate and 294,400 t of refined copper cathodes, which accounted for about 73% of the copper in concentrate and 89% of the refined copper produced in Kazakhstan in 2011, respectively. The average copper grade of crude ore produced by Kazakhmys decreased to 1.01% from 1.09% in 2010, resulting in a 6% decrease in the copper content of ore production despite a 1.5% increase in crude ore production. Ore grades were expected to continue to decrease, but Kazakhmys planned to partially offset this decrease by increasing crude ore production volumes. New mines at the Aktogai and the Bozshakol deposits were expected to open in 2015 and to produce, together, about 200,000 t/yr of copper in concentrate during the first 10 years of the life of the mines (Kazakhmys plc, 2013).

KGHM Polska Miedz S.A. was the only producer of mined copper and primary copper metal in Poland. In 2012, KGHM produced about 427,000 t of copper in concentrate and 566,000 t of electrolytically refined copper. The average copper content at the company's mines had been decreasing in recent years and was 1.59% in 2012. KGHM tried to maintain refined copper production by increasing the amounts of mined ore and by using increasing amounts of purchased copper scrap and imported copper concentrates.

Uzbekistan produced an estimated 96,000 t of refined copper in 2012. The only copper producer in Uzbekistan was the Almalyk mining and metallurgical complex (Almalyk GMK), which was located in Toshkent Voliyati. The company had mining, beneficiation, and metallurgical facilities. Copper ore was mined from the Kalmakyr and the Sary-Cheku deposits; a new deposit, Dal'nee, which is similar in ore structure to Kalmakyr, was to serve as a replacement as the first two deposits become depleted. In 2012, Almalyk completed reconstruction and expansion of the Kalmakyr Mine, which was expected to increase the mine capacity to 31.5 Mt/yr of ore. The Almalyk GMK was also planning to start developing the Dal'nee deposit in 2014.

Gold.—In 2012, Europe and Central Eurasia accounted for about 15% of world gold production; the majority of gold produced in the region came from Central Eurasia. The principal producers, by volume, were Russia, which produced about 217,800 kilograms (kg) of primary gold, followed by Uzbekistan (93,000 kg), Kazakhstan (about 39,900 kg), Finland (about 10,800 kg), and Kyrgyzstan (about 10,300 kg). Russia's production of gold is projected to increase to 240,000 kilograms per year (kg/yr) by 2019, and that of Uzbekistan and Kazakhstan is projected to increase to 110,000 kg/yr and 54,000 kg/yr, respectively. Russia, Uzbekistan, and Kazakhstan are projected to remain the principal producers of gold in the Europe and Central Eurasia region for the foreseeable future (table 10).

In 2012, Russia produced a total 226.3 t of gold, which included 217.8 t of primary mine production. The primary mine production in 2012 constituted a 9.1% increase compared with the output in 2011. As of 2012, Russia had 26 large gold mining companies which together produced about 80% of all the gold produced in the country, and the rest of gold was mined by about 400 smaller scale producers. In 2012, the leading gold-producing regions in Russia were Krasnoyarskiy Kray (44.0 t), Amurskaya Oblast' (29.3 t), Sakha Republic (Yakutiya) (21.2 t), Magadanskaya Oblast' (19.7 t), Irkutskaya Oblast' (19.0 t), and Chukotskiy Avtonomnyy Okrug (18.0 t). The leading producers of gold in 2012 were OAO Polyus Zoloto (48.8 t), Petropavlovsk plc (22.1 t), OAO Polymetall (15.2 t), and Kinross Gold Corp. (14.5 t). By the end of 2013, two new mining and beneficiation plants in Magadanskaya Oblast'—one at the Natalkinskoe deposit and one at the Pavlik deposit—were expected to become operational. The combined gold production of the two plants at full capacity was projected to reach 55 t/yr.

In 2012, Uzbekistan produced an estimated 96 t of gold. The main gold producer in Uzbekistan was the Navoi mining and metallurgical complex (Navoi GMK), which was responsible for more than 80% of Uzbekistan's gold production. The resources of the Navoi GMK included 13 deposits that made up about 85% of all gold resources of Uzbekistan. The largest deposit, Muruntau (located in the central Qizilqum region), contains gold quartz ores and was mined by an open pit method. The Navoi GMK included four metallurgical plants in Navoiy, Uchkuduk, Zarafzhan, and Zarmitan. In 2012, the Navoi GMK was planning to complete construction of a new gold mining complex that would use bioleaching technology and produce 20 t/yr of gold when it reaches its full capacity. Other renovation and expansion projects at the Navoi GMK in 2012 included modernization of heap-leaching facilities at the Muruntau Mine and a series of improvements at the Zarmitan plant. Other gold producers in Uzbekistan included the Almalyk GMK and Amantaytau Goldfields, which was a joint venture of the Uzbekistan's State Committee for Geology, the Navoi GMK, and Oxus Gold plc. of the United Kingdom. In 2012, the Almalyk GMK started construction of three new mines—the Samarchuk Mine, the Kairagach Mine, and the Uzun Mine, all located in Toshkent Viloyati.

The leading producers of gold in Kazakhstan were Kazzinc JSC and Kazakhmys plc, which accounted for 17,400 kg and about 4,000 kg, respectively, or 43.6% and 10%, respectively, of Kazakhstan's total gold production in 2012. Kazakhmys' substantial Bozshakol copper development project was reported to contain gold and could be a significant new source of production.

In 2012, Kyrgyzstan produced 10,333 kg of gold, which was a 44.6% decrease compared with the 2011 production level. The largest of the operating mines was the Kumtor gold mine, which is located about 350 kilometers (km) southeast of Bishkek. The Kumtor Mine was operated by Centerra Gold Inc. of Canada. In 2012, Centerra produced only 9.8 t of gold content compared with 18.1 t in 2011. The reason for the sharp reduction in output was an unexpected movement of ice from a glacier into the mine's open pit. Centerra expected production in 2013 to return to the level of about 18,000 kg/yr. In 2012, the Kumtor Mine contributed 5.5% to the GDP of Kyrgyzstan and 18.9% to the total industrial production of the country.

Dragon Mining Ltd. of Australia and Elgin Mining Inc. of Canada, which merged with Gold-Ore Resources Ltd. in May, had gold mines located in the Skelleftea mining district of Sweden. This district had been the focus of exploration for gold-rich polymetallic deposits since the mid-1920s. Dragon Mining's Svartliden Mine is located 700 km north of Stockholm, and Elgin's Bjorkdal Mine is located 750 km north of Stockholm. An updated measured and indicated mineral resource estimate for the Bjorkdal Mine's open pit and underground mine of 30,295 kg of gold was released in February (Elgin Mining, 2012; Gold-Ore Resources Ltd., 2012; Dragon Mining Ltd., 2013).

Boliden is the other main producer of gold in Sweden. Its polymetallic mines have an estimated capacity of about 2,000 kg/yr of gold. Its major operations were the Aitik Mine, which was principally a copper-producing mine, and the operations at the Boliden and Garpenberg sites (table 10; Boliden AB, 2013, p. 19).

In Romania, Gabriel Resources Ltd. reported proven reserves of 5.9 million troy ounces (180 t) of gold and 32.6 million troy ounces (1,010 t) of silver, and probable reserves of 4.2 million troy ounces (130 t) of gold and 15 million troy ounces (470 t) of silver at its Rosia Montana project. The company estimated that the project could produce an average of 511,000 troy ounces per year of gold during a 16-year mine life and could make Romania a significant European gold producer. The company also reported that 62.45% of the people consulted in a referendum vote were in favor of resuming the mining operations at the Rosia Montana project. The referendum had been initiated by 35 local mayors and conducted on December 9, 2012. The referendum was advisory in nature and did not have the power to enforce or bind the Government to any particular action (Gabriel Resources Ltd., 2012a, b).

Carpathian Gold Inc. of Canada (a junior mine developer) reported that it would account for about 7.2 million troy ounces (220 t) of gold and 635,000 t of copper in contained metal in its final prefeasibility study of the Rovina Valley project. Rovina's measured and indicated resources were estimated to be 406 Mt at grades of 0.55 gram per metric ton gold and 0.16% copper. Carpathian stated that, compared to a 2008 resource estimate, the copper grade decreased by 11%, and the gold grade, by 12% (Keen, 2012).

Iron and Steel.—Europe and Central Eurasia produced about 18.3% of the world's crude steel output, which was a slight decrease compared with that of 2011, and it produced 15.8% of the pig iron and direct-reduced iron output in 2012, which was a slightly higher share of the world production than in 2011. Russia was the leading producer of crude steel in the region; its output in 2012 was 70.4 Mt, which was a 3.4% increase compared with that of 2011. Germany was the second-ranked producer, by volume, with production of 42.7 Mt (a decrease of 3.6%) followed by Ukraine, 32.4 Mt (a decrease of 8.2%), and Italy, 27.3 Mt (a decrease of 22.3%).

Russia's production capacity was projected to increase to 74 Mt/yr by 2019. The production volume in Ukraine was expected to increase to 34 Mt/yr by 2019, and in Italy, to 32 Mt/yr. Germany's production was projected to decrease to 43 Mt/yr by 2019 (tables 4, 12).

According to the World Steel Association, in 2012, Ukraine was the third-ranked net exporter of steel in the world; it exported about 22.3 Mt of steel, which was about 69% of the country's total steel production. Ukraine was the 10th-ranked producer of steel, by volume, in the world. Metinvest Holding was the leading producer of crude steel in Ukraine and accounted for 41% of production. The iron and steel industry in Ukraine had the advantage of large domestic sources of iron ore, but it was dependent on export markets for product sales, and it operated inefficiently owing to a need for technical modernization.

Russia was the world's fifth-ranked producer of steel and the fourth-ranked net exporter; it exported 19.8 Mt in 2012, which was 28.1% of its total production. Germany was the 7th-ranked producer of steel in the world and the 11th-ranked net exporter in the world; it exported 3.1 Mt, or 7.2% of its total production. Italy was the 11th-ranked crude steel producer in the world in 2012, and it was the 8th-ranked net exporter; it exported 4.3 Mt, or 15.6% of its total production. Russia's apparent consumption of steel in 2012 increased by 2.2% to 41.8 Mt, and that of Italy, Ukraine, and Germany decreased by 18.1%, 9%, and 7.6%, respectively (World Steel Association, 2014).

Russia produced 52.9 Mt of pig iron in 2012, which was 29.2% of the total produced in Europe and Central Eurasia and 4.6% of the total world production. Germany produced 28.3 Mt, which was 15.6% of the total produced in Europe and Central Eurasia and 2.4% of total world production, and Ukraine produced 28.9 Mt, which was 16% of the total produced in Europe and Central Eurasia and 2.43% of total world production (World Steel Association, 2014).

Iron Ore.—Europe and Central Eurasia produced 10.3% of the world's iron ore in 2012; Russia produced 61.4 Mt (measured in Fe content); Ukraine produced 45.1 Mt; Sweden, 17.2 Mt; and Kazakhstan, 14.3 Mt. Russia's production was expected to increase to 64 Mt/yr by 2019; Ukraine's, to 46 Mt/yr; Sweden's, to 18,000 Mt/yr; and Kazakhstan's, to 14.7 Mt/yr (table 11).

Sweden's LKAB's Kiruna Mine was the world's largest underground iron ore mine in terms of volume; it has an ore body that is 4 km long and 80 meters wide and reaches to a depth of about 2 km. LKAB announced that it had been granted an environmental permit for a new open pit mine located at Gruvberget. This would be LKAB's first new iron ore mine in 50 years. Production at the new Gruvberget Mine was expected to be 2 Mt/yr. The ore body contains both hematite and magnetite (table 11; Luossavaara-Kiirunavaara AB, 2012a, b).

Lead and Zinc.—Europe and Central Eurasia produced about 10.2% of the world's production of mine output of zinc and about 21.8% of the world's zinc metal output in 2012. Kazakhstan and Ireland were the leading producers of zinc ore (zinc content) and produced 369,700 t and 337,500 t, respectively. Other significant zinc ore (zinc content) producers were Sweden (188,300 t); Russia (179,800 t); and Poland (89,000 t).

The principal producers of primary and secondary zinc in Europe and Central Eurasia in 2012 were Spain, which produced 490,000 t; Kazakhstan, 319,000 t; Finland, 314,742 t; Belgium, 290,000 t; Russia, 260,000 t; and the Netherlands, 257,000 t (table 4).

Europe and Central Eurasia produced about 7.2% of the world's production of mine output of lead and about 12.4% of primary lead metal production. Russia, Sweden, and Poland were the principal producers of mined lead, accounting for 93,000 t, 64,000 t and 58,000 t, respectively. Other producers of note were Ireland (47,000 t) and Kazakhstan (38,000 t). The United Kingdom was the principal producer of primary lead metal in the Europe and Central Eurasia region with an estimated production volume of 157,000 t, followed by Germany (134,000 t), Kazakhstan (88,000 t), Russia (85,000 t), and Sweden (62,000 t) (table 4).

Boliden Tara Mine's operation in Navan, Co. Meath, Ireland, which was the leading zinc mine in Europe, produced about 170,000 t of zinc and about 35,000 t of lead in 2012. Since the mine began its operations in 1977, production had totaled 80.7 Mt grading an average of 8.2% zinc and 1.9% lead. The mine's Joint Ore Reserves Committee (JORC)-classified ore reserves (proven and probable) were 14 Mt grading 7.2% zinc and 1.7% lead. The mine employed 718 people in 2012 (Boliden AB, 2013, p. 89, 117; Department of Communications, Energy and Natural Resources, 2013, p. 1). Galmoy Mines Ltd. (a subsidiary of Lundin Mining Corp. of Canada) ceased its underground mining operations at its mine in Galmoy in October 2012. The total mine production for 2012 amounted to 142,000 t of ore grading 14% zinc and 2.4% lead. The ore was processed at the Lisheen Mine operations located in Co. Kilkenny (Department of Communications, Energy and Natural Resources, 2013, p. 1; Lundin Mining Corp., 2013). Vedanta Resources Plc. (the owner of the Lisheen Mine) reported that it had produced 1.4 Mt of ore grading 11% zinc and 1.95% lead. The company also produced 321,000 t of zinc concentrates containing 53.3% zinc and 41,000 t of lead concentrates containing 60.3% lead. Since mining operations were started at Lisheen in 1999, a total of 19.72 Mt of ore grading an average of 11.8% zinc and 2% lead had been mined (Department of Communications, Energy and Natural Resources, 2013, p. 1).

In September 2011, Poland's Ministry of the Treasury had posted an invitation for bids for the purchase of shares of Zaklady Gorniczo-Hutnicze (ZGH) "Boleslaw" S.A. (ZGH Boleslaw), which was Poland's only producer of lead and zinc ore and the country's leading producer of refined zinc. The plant is located in the Bukowno region of Poland and had an estimated capacity of about 110,000 t/yr of zinc and 30,000 t/yr of lead. ArcelorMittal held 33.77% share of Stalprodukt S.A., and in November, Stalprodukt acquired an 86.92% share of ZGH Boleslaw (Ministry of the Treasury of the Republic of Poland, 2011; Thomson Reuters, 2012; Stalprodukt S.A., 2013a, p.14, 2013b).

In 2012, Russia had two large zinc deposits (the Kholodninskoe and the Ozyornoe), which are located in the Republic of Buryatiya. In Russia, more than 60% of the zinc produced was used for the production of galvanized steel, mainly for the automobile and construction industries.

Nickel.—In 2012, Europe and Central Eurasia accounted for 10.8% of the world's mined nickel and 36.3% of the world's refined nickel production. Production of the region's mine output of nickel was largely the result of Russian mining activity, and refined nickel production took place mainly in Russia and Western Europe. Russia accounted for about 70.2% of nickel mine output and 52.0% of nickel refinery production in the region in 2012. Other countries, most notably Finland and Greece, also mined nickel ore, but in smaller amounts. Production of refined nickel was more diversified across countries. Russia produced 258,000 t in 2012, and Norway, the United Kingdom, and Finland produced 92,000 t, 46,000 t, and 34,000 t, respectively (tables 4, 13).

Russia was one of the world's leading nickel mining countries in 2012, accounting for about 8% of the world's mined nickel. Nornickel was Russia's leading nickel producer and the world's leading nickel mining company; the company produced 8.9% of the world's mined nickel from its worldwide operations. Nornickel's operations in Russia were located on the Kola Peninsula in the northwest of the country and in the Norilsk region on the Taymyr Peninsula in East Siberia. Nornickel also owned assets in other countries; in particular, mines in Australia and Botswana and the Harjavalta smelter in Finland. In 2012, because of low prices on the world market, the company was considering reducing its investments and even halting the processing of ores mined on the Kola Peninsula by 2015, if the nickel prices remained low.

Finland's Talvivaara nickel mine was the largest nickel mine in Europe; it operated at two polymetallic deposits—the Kolmisoppi and the Kuusilampi—which are located about 30 km from Sotkamo. Based on estimated proven reserves, the deposit could produce about 2.5% of the world's nickel during its projected 24-year operating life. Talvivaara's bioheap-leach project was planned to produce nickel from an open pit operation with cobalt, copper, and zinc as byproducts. In 2012, Talvivaara reported production of 12,916 t of nickel, which was a 20% decrease compared with output in 2011 and much lower than the capacity of between 25,000 t/yr and 30,000 t/yr that was expected to be reached in 2012 (Mining Technology, 2008; Talvivaara Mining Co., 2012, p. 8).

Platinum-Group Metals.—Within the region of Europe and Central Eurasia, almost all mining for platinum-group metals (PGMs) took place in Russia, although small amounts of PGMs were also mined in Finland, Poland, and Serbia. Russia and South Africa were the world's leading PGM ore producers; Russia was the world's leading producer of palladium, accounting for 82,400 kg, or 38.8% of the world's production. It also produced 30,200 kg of platinum (tables 4, 14, 15).

The leading PGM producer in Russia was Nornickel, whose Zapolyarnyi division was mining three large PGM deposits in Krasnoyarskiy Kray—the Norilsk-1, the Oktyabr'skoye, and the Talnakhskoye deposits. Another division within Nornickel, Kol'skaya GMK, was mining several deposits in Murmanskaya Oblast'—the Kotsel'vaara-Kammikivi, the Semiletka, the Zapolyarnoye, and the Zhdanovskoye deposits. Altogether, Nornickel produced almost all the palladium and about 75% of the platinum output in Russia. Another platinum producer in Russia, Chernogorskaya Gornorudnaya Kompaniya (ChGRK), was planning to start mining the Chernogorskoye deposit of nonferrous and precious metals in 2013. The company was expected to reach its production capacity of about 15,000 t/yr of copper, 8,000 t/yr of nickel, 12.5 t/yr of palladium, and 6 t/yr of platinum by 2016.

Silver.—Europe and Central Eurasia produced about 4.6% of the world's production of mine output of silver in 2012. Russia was the principal silver producer, with an output of 1,679 t of silver content, which was about 40% of the total production of silver in Europe and Central Eurasia. Poland, which produced 1,149 t, was the second-ranked producer of silver, by volume, followed by Kazakhstan, which a produced 963 t. Other relevant producers were Sweden, which produced 309 t, and Finland, which produced 128 t (table 4).

Tin.—Europe and Central Eurasia produced only 0.1% of the total world production of mined tin and only 0.2% of the world production of tin metal, respectively. Russia and Portugal were the only producers of mined tin, and Russia was the only producer of tin metal in the region. Russia was at the lowest tin production point since 2010 and was trying to revive its tin industry. It was expected that, by 2019, mined tin production would increase to 3,000 t/yr from 100 t in 2012 and tin metal production would increase to 3,000 t/yr from 700 t in 2012. Data on tin mine and metal production and projections for future production are in tables 4, 16, and 17.

Titanium.—Europe and Central Eurasia produced about 4.2% of the world's production of ilmenite (which is an ore mineral of titanium) and about 48.2% of world titanium metal sponge output in 2012. Ukraine was the leading producer of ilmenite; it produced about 146,000 t of TiO_2 in 2012. Kazakhstan was a distant second, having produced 15,000 t of TiO_2. Russia was the leading producer of titanium sponge, with production of 42,000 t; Kazakhstan and Ukraine produced 21,000 t and an estimated 8,500 t, respectively (table 4).

The titanium industry in Ukraine consisted of ilmenite and rutile concentrate production, titanium sponge production at the Government-owned Zaporozhye Titanium & Magnesium Complex (ZTMK), and titanium ingot production by a small number of producers that had a combined capacity of 12,000 t/yr of titanium ingots. The dominant producers of ilmenite and rutile ores and concentrate were the Irshansk mining and beneficiation complex (GOK) and the Volnogorsk State Mining-Metals Complex. In addition, there was considerable activity in new projects involving production of mined titanium. Velta LLC began production of ilmenite from the Birzulovskoye deposit in Kirovograd Oblast' in December 2011 and was planning to open the second phase of mine production in the end of 2012. In January, the Government decided to privatize ZTMK, and, as of December, Tolexis Trading Ltd., which was part of the DF Group of Ukraine, had won the tender and acquired 49% of the shares of the titanium sponge producer.

The main producer of titanium sponge in Russia was OAO VSMPO-Avisma, which produced titanium sponge at its titanium and magnesium complex in Permskiy Kray. The raw material for the titanium production was imported, mostly from Ukraine. VSMPO-Avisma supplied titanium mill products to the world's leading aircraft manufacturers. In 2012, VSMPO-Avisma acquired 75% of the shares of Limpeza Ltd.

of Cyprus (owner of the Demurinskiy GOK, which mined the Volchanskiy deposit containing alluvial titanium and zirconium in Ukraine). It was reported that VSMPO-Avisma was planning to build a beneficiation plant next to the Demurinskiy GOK.

Industrial Minerals

Cement.—Europe and Central Eurasia produced about 7.7% of the world's production of hydraulic cement. Russia was the largest producer, in terms of volume, at 61.7 Mt, followed by Germany (32.4 Mt), Italy (26.2 Mt), and Spain (20 Mt). According to Cembureau, production in the 27 country members of the European Union decreased, on average, by 20%. Production in Spain and Italy (the countries with the most significant reduction in output) decreased by 39.5% and 20.8%, respectively. In Italy, the decrease was owing mostly to a slowdown in construction, including a decrease in residential construction of 6.3% and a decrease in civil engineering projects and nonresidential construction of 19.7%. In Spain, the decrease in the construction sector continued a 4-year trend. In 2012, the estimated decrease was 34% compared with that of the previous year and included decreased production in all construction sectors. The value of civil engineering projects decreased by 56%; nonresidential construction, by 23%; and residential construction, by 21%. Because of Spain's economic conditions, the construction sector is highly unlikely to register any increases in production in the coming year, and production in the construction sector was also likely to decrease further. The production decreases were observed in all areas of the EU, including in Eastern Europe, where production in the Czech Republic decreased by 16%, and that in Poland, by 10% (table 4; Cembureau, 2013).

Diamond.—Russia was the world's leading diamond producer and the only significant diamond mining country in Europe and Central Eurasia. Almost all Russia's output of diamond was mined by the Joint Stock Company ALROSA (ALROSA) of Russia, which had its main operation in Sakha Republic (Yakutiya) in Eastern Siberia. ALROSA was one of the world's leading companies in the field of diamond exploration, diamond mining, sales of rough diamond, and diamond processing, and the company accounted for 97% of Russia's diamond production. Russia's share of global natural, gemstone, and industrial diamond production was 26.8% in 2012. Data on historic and projected diamond production are in table 18.

According to the Antwerp World Diamond Centre, the city of Antwerp was the center of the world's open rough diamond market. The city of Antwerp has 1,850 diamond companies and 4,500 diamond dealers, and about 10,000 people work in the industry in the city. In 2012, Belgium's exports of polished diamond decreased sharply—by 18.91% in terms of volume, and by 17.63% in terms of value, to $997,878,263. The average price per carat of exported diamond was $1,866 in December 2012 (Antwerp World Diamond Centre, 2013a, b).

Lithium.—Portugal was the only lithium producer in the region. In 2012, lithium production in Portugal decreased to 20,700 t from 40,110 t in 2010. It was expected that, by 2015, lithium production would return to its 2010 level, and, by 2019, would increase to about 44,000 t/yr (table 19).

Potash.—In 2012, Europe and Central Eurasia produced 15.4 Mt of potash (in K_2O equivalent), or 43.7% of the world's production. Russia was the leading regional potash producer in 2012, with output of 5.6 Mt; it was followed by Belarus (4.8 Mt) and Germany (3.8 Mt) (table 4). In 2012, world prices of potash decreased and led to sluggish sales as well as to a slowdown in expansions in the industry.

OAO Uralkali (Uralkali) of Russia was the world's second-ranked producer of potash. In 2012, Uralkali reduced its production of potash by 14.4% to 5.6 Mt in potassium dioxide equivalent. On average, only about 80% of the capacity of Uralkali's mines was used. After completing the expansion of the Berezniki-4 Mine, Uralkali increased its annual capacity to 13 Mt/yr of potassium chloride. In addition to Uralkali, several other potash projects were underway. OOO Verkhnekamskaya Potash Co. was preparing the Talitskiy sector of the Verkhnekamskoye potash deposit for production, which was expected to begin in 2016. OAO MHK Eurokhim was continuing to build a mine at the Gremyachinskoye deposit, which was to start producing potassium chloride in 2014.

OAO Belaruskali of Belarus was one of the world's leading producers of potash mineral fertilizers, and it had a 15% share of the world market. In 2012, the production of potash in Belarus decreased to about 4.8 Mt, or by 9.8%. In 2012, only 55.7% of the produced potash was exported compared with the 88.5% in 2011. Despite the slowdown, Belaruskali continued with expansion of its production facilities. In July, Belaruskali started operations in the Beryozovskiy section of the deposit and increased its annual capacity to 10.3 Mt/yr of potassium chloride. By 2015, the company was planning to increase its production capacity to 11 Mt/yr of potassium chloride.

Mineral Fuels and Related Materials

Coal.—In 2012, Europe and Central Eurasia accounted for 49.9% of the world's lignite production, 9.8% of the world's bituminous coal production, and 5.4% of the world's anthracite production. In Central Eurasia, Kazakhstan, Russia, and Ukraine were the leading coal producers, and within the EU, Germany and Poland were the leading coal producers. A number of other countries throughout the region also mined coal (tables 4, 20).

The dynamics of coal consumption and production among the EU member countries and the CIS countries demonstrates the different priorities of those countries. Countries in the CIS (for example, Kyrgyzstan, Tajikistan, and Ukraine) made concerted efforts to switch their energy-intensive enterprises to coal from natural gas, in part because natural gas is more costly and entails regular conflicts with Russia, which was the main supplier of the natural gas in the region. The EU countries, on the other hand, had been trying to reduce coal production and consumption during the previous decade. By 2012, however, this trend was reversed. Coal consumption in Europe was 707 Mt in 2010 and increased to 762 Mt in 2011 and 784 Mt in 2012. The reason for the shift was that recent developments in the shale gas industry have led to record low prices for natural gas in the United States and, at the same time, reduced the demand for American coal. The oversupply of coal on world markets, in turn, put downward pressure on coal prices in Europe. One more important factor is that the prices of carbon permits issued by the EU Emissions Trading Scheme (EU ETS) had also dropped and made the combined price (the sum of the price of coal and the price of a carbon permit) of using coal for electricity generation affordable. The drop of the EU ETS emissions prices came about because, during the Great Recession (the common term for the general economic decline observed in the world markets in the end of the first decade of the 21st century), production declined and the companies were able to accumulate significant amounts of emissiond permits that they did not need when the production was low. Many analysts think that the effect of the 2012 substitution from cleaner natural gas to dirtier coal is a random deviation from the general course toward greater use of renewable energy. They expect that, in the near future, coal consumption will continue to decrease compared with consumption of natural gas and renewable energy (Birnbaum, 2013).

Russia ranked sixth in the world in the total amount of coal mined following China, the United States, India, Indonesia, and Australia, and produced 366 Mt/yr. In 2011, the Russian Federation adopted a new program for development of the coal industry through 2030. The goal of the program was to increase Russia's coal production to 450 Mt/yr in 2030 from 334.8 Mt in 2011. The more detailed goals included an increase in the share of Eastern Siberia in coal production, a 100% increase in the production of coking coal, and a 150% increase in coal exports (tables 4, 20; U.S. Energy Information Administration, 2013).

Ukraine was among the world's leading coal mining countries. According to the BP Statistical Review of World Energy, proven resources of coal in Ukraine were 33,900 Mt, or about 4% of the world's reserves (BP p.l.c., 2013, p. 30). In 2012, Ukraine produced 85.7 Mt of coal, which was a 4.6% increase compared with production in 2011. The country was the fourth-ranked coal producer in Europe after Russia, Germany, and Poland. Despite the increases in production, about 80% of all coal mines in Ukraine operated at a loss. Ukrainian coal was unable to compete with the coal from Germany either on price or quality. In 2012, the Government mandated that heating plants in the country switch from natural gas, which was imported from Russia, to domestic coal as their energy source. Many residents and environmental activists were concerned that the switch could significantly worsen the environmental situation in the country.

Natural Gas.—In 2012, Europe and Central Eurasia contributed 34.0% of the world's production of natural gas. Russia was the leading producer in the region and the second-ranked producer in the world with 2012 production of 655 billion cubic meters; it was followed by Norway (106.7 billion cubic meters), the Netherlands (80.8 billion cubic meters), Turkmenistan (69 billion cubic meters), Uzbekistan (62.9 billion cubic meters), and the United Kingdom (an estimated 57 billion cubic meters). Russia was the world's second-ranked (after the United States) natural gas producer and the leading exporter; it had the world's largest natural gas reserves (47.6 trillion cubic meters), which was about 24% of the world's total natural gas reserves. Many countries in the Europe and Central Eurasia region produced natural gas, but generally not in large volumes. Norway, the Netherlands, and the United Kingdom, in order of volume, were significant regional producers of natural gas in Europe; Turkmenistan and Uzbekistan were notable regional natural gas producers in the CIS (table 4; U.S. Energy Information Administration, 2012).

Almost 90% of Russia's natural gas was produced in the Ndym-Pur-Taz (NPT) region in northern West Siberia (the region's name was derived from the names of three rivers that border it). The NPT region hosts three massive Russian gasfields (the Medvezh'ye, the Urengoy, and the Yamburg), which had been the country's main producers and had supplied about 70% of the country's gas production. These three fields were in decline, however, as reserves were being depleted. To keep up with the growth in the Russian economy and the country's long-term export commitments to Europe to increase gas output, Russia was expected to have to incur greater costs to develop fields further north and to the east in an even more difficult physical environment than in the NPT region. A main target for future development would be the Yamal Peninsula, where large reserves were discovered in several fields. The newly developed Zapolyarnoye field on the Yamal Peninsula was a major contributor to replacing decreasing production from large older fields where reserves were more than 50% depleted.

OAO Gazprom, which was Russia's leading gas producer, projected that between 2008 and 2030, Russia would increase natural gas output to between 885 billion and 940 billion cubic meters per year. Most of the increases in natural gas output were projected to come from independent gas companies in Russia, such as Itera, Northgaz, and Novatek, which, although blocked from the export market, had found a niche supplying the domestic market.

Petroleum.—In 2012, Europe and Central Eurasia produced 5.9 billion barrels (Gbbl), or 20.7% of total world production of petroleum. Russia was the leading oil producer in the region and a top exporting nation; in 2012, it produced 3.6 Gbbl of crude oil, or 61% of the total regional output. Other significant producers were Norway, which produced 694 million barrels (Mbbl); Kazakhstan (576 Mbbl); the United Kingdom (368 Mbbl), and Azerbaijan (321 Mbbl). Azerbaijan was engaged in major oil development projects offshore in the Caspian Sea, and Kazakhstan was engaged in major projects both onshore and offshore.

In May in the United Kingdom, Maersk Oil UK Ltd. (Maersk) signed an agreement with Noble Energy Inc. for the purchase of 30% of its interest in the Maersk-operated Dumbarton and Lochranza fields. Maersk reported that this agreement also included the control of the Global Producer III floating production, storage, and offloading installation in the United Kingdom's central North Sea. With this investment, Maersk would hold a 100% interest in the Dumbarton and the Lochranza fields. Maersk stated that it had paid Noble Energy $127 million for the assets. The Dumbarton and the Lochranza fields produced a combined output of about 20,000 barrels per day of oil equivalent in 2012(Maersk Oil Ltd., 2012).

According to the BP Statistical Review of World Energy, at the end of 2012, Russia's proven reserves of petroleum were 87,200 Mbbl (BP p.l.c., 2013, p. 6). For the coming decade, Russian oil production was projected to increase at an annual rate of between 1.5% and 2.5% owing in part to increased oil output from Sakhalin Island. This increase would be coupled with a slowdown in growth from the major mature oilfields in West Siberia, a number of which had passed peak production. New fields that were under development were expected to account for almost all Russia's increase in annual oil output in the next 5 years and would probably produce more than one-half of the country's oil in 2020.

In 2012, crude petroleum was produced in Russia by nine vertically integrated oil and gas companies, the largest four of which were LUKOIL Group, OAO NK Rosneft', OAO Surgutneftegaz, and TNK–BP Holding. The country had 28 large refineries and more than 200 small ones; the total refining capacity of the country was 290 Mt/yr of petroleum. More than 90% of the refining capacity belonged to the vertically integrated companies. Beginning in 2013, Russian law bans automotive gasoline for which the environmental requirements are below the Euro-3 standard.

Uranium.—In 2012, Central Eurasia accounted for 51.6% of the world's uranium production. Kazakhstan was the leading uranium producer in the world, and its production volume amounted to 24,648 t (U_3O_8 content). The next-ranked producers in the region were Russia (3,348 t); Uzbekistan (3,190 t); Kyrgyzstan (an estimated 2,150 t), and Ukraine (1,132 t). Uranium mining took place in several other countries in the region (the Czech Republic, Germany, and Romania), but in smaller quantities (table 4).

In 2012, Kazakhstan remained the leading producer of mined uranium, having produced 36.5% of the world's output. Kazakhstan had no nuclear powerplants, and all uranium production was exported. Within the past 9 years, Kazakhstan rapidly increased investment in its uranium industry, and the country's production of uranium oxide increased to 24,648 t in 2012 from 3,300 t in 2003. AO NAK Kazatomprom (the leading Government-owned producer) mined 11,900 t of uranium, or about 20% of the world's production. The leadership of Kazatomprom stated that the country could increase its uranium production to 30,000 t/yr within the next 3 years (World Nuclear Association, 2013).

References Cited

Agency of Statistics of the Republic of Kazakhstan, 2013, Statistical yearbook—Kazakhstan in 2011: Astana, Kazakhstan, Agency of Statistics of the Republic of Kazakhstan, November, 218 p.

Antwerp World Diamond Centre, 2013a, History: Antwerp World Diamond Centre. (Accessed July 14, 2013, at http://www.awdc.be/en/history.)

Antwerp World Diamond Centre, 2013b, Rough import and export figures show increase compared to December 2011: Antwerp World Diamond Centre. (Accessed November 30, 2013, at http://www.awdc.be/en/news?type=6.)

Birnbaum, Michael, 2013, Europe consuming more coal: The Washington Post, February 7. (Accessed June 9, 2014, at http://www.washingtonpost.com/world/europe-consuming-more-coal/2013/02/07/ec21026a-6bfe-11e2-bd36-c0fe61a205f6_story.html.)

Boliden AB, 2013, Annual report 2012: Boliden AB, 129 p. (Accessed November 14, 2013, at http://www.boliden.com/Documents/Press/Publications/Boliden_AR12_ENG.pdf.)

BP p.l.c., 2013, BP statistical review of world energy: BP p.l.c., June, 48 p. (Accessed April 22, 2014, at http://www.bp.com/assets/bp_internet/globalbp/globalbp_uk_english/reports_and_publications/statistical_energy_review_2012/STAGING/local_assets/pdf/statistical_review_of_world_energy_full_report_2013.pdf.)

Cembureau, 2013, Activity report 2011: Brussels, Belgium, Cembureau, 46 p.

Creamer Media's Mining Weekly, 2012, Greenland passes mining projects bill, opens for cheap labor: Creamer Media's Mining Weekly, December 8. (Accessed December 8, 2012, at http://www.miningweekly.com/print-version/greenland-passes-mining-projects-bill-opens-for-cheap-labour-2012-12-08.)

Cobalt Development Institute, 2013, Cobalt news—Production: Guildford, United Kingdom, Cobalt Development Institute, April, 6 p.

Deloitte Development LLC, 2012, Tracking the trends—The top 10 trends mining companies may face in the coming year: Deloitte Development LLC, 33 p. (Accessed February 2, 2013, at http://www.deloitte.com/assets/Dcom-SouthAfrica/Local Assets/Documents/Industries/Mining/Tracking the trends 2012.pdf.)

Department of Communications, Energy and Natural Resources, 2013, Ireland—Exploration and mining news: Dublin, Ireland, Department of Communications, Energy and Natural Resources, May 1, 6 p. (Accessed August 15, 2013, at http://www.dcenr.gov.ie/NR/rdonlyres/5695C011-D2A6-407A-A104-DE9A4909187A/0/IndustryNews_1stMay2013.pdf.)

Dragon Mining Ltd., 2013, Annual report 2012: Dragon Mining Ltd., 102 p. (Accessed November 19, 2013, at http://www.dragon-mining.com.au/sites/default/files/dragon_ar2012_final_web_version_0.pdf.)

Elgin Mining, 2012, Elgin Mining and Gold-Ore Resources announce closing of merger creating growth oriented international gold producer, developer, and explorer: Elgin Mining, May 2, 2 p. (Accessed November 19, 2013, at http://www.elginmining.com/s/NewsReleases.asp?ReportID=522555&_Type=News-Releases&_Title=Elgin-Mining-and-Gold-Ore-Resources-Announce-Closing-of-Merger-Creating-Gro... .)

European Commission, 2008, On the competitiveness of the metals industries: Brussels, Belgium, Communication from the Commission to the Council and the European Parliament, no. 108, February 22, 11 p.

European Commission, 2014a, Economic and financial affairs: Brussels, Belgium, European Commission. (Accessed May 30, 2014, at http://ec.europa.eu/economy_finance/euro/index_en.htm.)

European Commission, 2014b, European Union—Countries: Brussels, Belgium, European Commission. (Accessed May 30, 2014, at http://europa.eu/about-eu/countries/index_en.htm.)

European Free Trade Association, 2014, About EFTA: Geneva, Switzerland, European Free Trade Association. (Accessed May 30, 2014, at http://www.efta.int/about-efta/european-free-trade-association.)

Federal State Statistics Service of the Russian Federation, 2013a, Rossijskij statisticheskij ezhegodnik [Russian statistical yearbook]—Schet proizvodstva po vidam jekonomicheskoj dejatelnosti [Production account by economic activity]: Federal State Statistics Service of the Russian Federation. (Accessed June 5, 2014, at http://www.gks.ru/bgd/regl/b11_13/IssWWW.exe/Stg/d3/11-09-02.htm.)

Federal State Statistics Service of the Russian Federation, 2013b, Rossiyskiy statisticheskiy ezhegodnik [Russian statistical yearbook]—Tovarnaya struktura eksporta rossiyskoy federatsii [Commodity structure of exports of the Russian Federation]: Federal State Statistics Service of the Russian Federation. (Accessed June 5, 2014, at http://www.gks.ru/bgd/regl/b11_13/IssWWW.exe/Stg/d6/25-08.htm.)

Gabriel Resources Ltd., 2012a, Projects: Toronto, Ontario, Canada, Gabriel Resources Ltd. (Accessed November 9, 2013, at http://www.gabrielresources.com/site/projects.aspx.)

Gabriel Resources Ltd., 2012b, Referendum voting confirmed as overwhelmingly in favour of mining in Rosia Montana: Gabriel Resources Ltd. press release, December 12, 2 p. (Accessed August 9, 2013, at http://www.gabrielresources.com/documents/GBU_Referendum_Results_000.pdf.)

Gold-Ore Resources Ltd., 2012, Gold-Ore reports increase in gold resource estimates at Bjorkdal gold mine: Gold-Ore Resources Ltd. (Accessed November 22, 2013, at http://www.businesswire.com/news/home/20120305005916/en/Gold-Ore-Reports-Increase-Gold-Resource-Estimates-Bjorkdal.)

International Centre for Trade and Sustainable Development, 2010, Vstupil v silu Edinyi kodeks Tamozhennogo Soyuza [Unified customs code of the Customs Union went into force]: International Centre for Trade and Sustainable Development, July 6. (Accessed June 5, 2014, at http://ictsd.org/i/news/bridgesrussiandigest/79510/.)

International Monetary Fund, 2013, Regional economic outlook—Faster, higher, stronger—Raising growth potential of CESEE: International Monetary Fund, October, 45 p. (Accessed June 5, 2014, at https://www.imf.org/external/pubs/ft/reo/2013/eur/eng/pdf/ereo1013.pdf.)

Kazakhmys plc, 2013, Annual report and accounts 2012: London, United Kingdom, Kazakhmys plc, April 10, 183 p.

Keen, Kip, 2012, Carpathian doubles gold-copper resource bound for Rovina prefeasibilty study: Mineweb.com, July 17. (Accessed August 13, 2013, at http://www.mineweb.com/mineweb/content/en/mineweb-europe-and-middle-east?oid=155250&sn=Detail.)

Knews.kg, 2012, V Kyrgyzstane vstupaet v silu novyi zakon o nedrakh [A new mining code goes into effect in Kyrgyzstan]: Knews.kg, August 14. (Accessed June 5, 2014, at http://www.knews.kg/econom/20287_v_kyirgyizstane_vstupaet_v_silu_novyiy_zakon_o_nedrah/.)

Korrespondent.net, 2008, SNG ofitsial'no isklyuchilo Gruziyu iz svoih ryadov [SIC officially removed Georgia from its members]: Korrespondent.net, October 9. (Accessed June 5, 2014, at http://korrespondent.net/world/609809-sng-oficialno-isklyuchilo-gruziyu-iz-svoih-ryadov.)

Lundin Mining Corp., 2013, Production statistics: Lundin Mining Corp. (Accessed August 15, 2013, at http://www.lundinmining.com/s/ProductionStats.asp.)

Luossavaara-Kiirunavaara AB, 2012a, Gruvberget: Luossavaara-Kiirunavaara AB. (Accessed November 15, 2013, at http://www.lkab.com/en/Future/Urban-Transformations/Why/What-is-Iron-Ore/The-Ore-in-Svappavaara1/Gruvberget/.)

Luossavaara-Kiirunavaara AB, 2012b, LKAB gets the go-ahead for new iron ore mine in Sweden: Luossavaara-Kiirunavaara AB. (Accessed November 15, 2013, at http://www.lkab.com/en/About-us/Overview/Operations-Areas/Kiruna/.)

Maersk Oil Ltd., 2012, Maersk Oil buys 30% interest in Dumbarton and Lochranza fields in UK North Sea: Maersk Oil Ltd. (Accessed January 5, 2014, at http://www.maerskoil.com/Media/Newsroom/Pages/MaerskOilbuys30interestinDumbartonandLochranzafieldsinUKNorthSea.aspx.)

Mining Technology, 2008, Talvivaara bioheapleach, Finland: Mining Technology. (Accessed June 5, 2014, at http://www.mining-technology.com/projects/talvivaara/.)

Ministry of the Treasury of the Republic of Poland, 2011, Invitation to negotiations regarding the purchase of shares of Zaklady Gorniczo-Hutnicze "Boleslaw" S.A. with its registered office in Bukowno: Ministry of the Treasury of the Republic of Poland, September 27. (Accessed December 6, 2013, at http://msp.gov.pl/portal/en/43/2677/Invitation_to_negotiations_regarding_the_purchase_of_shares_of_Zaklady_GorniczoH.html.)

North of 60 Mining News, 2012, Greenland—The next mining destination in the North: Petroleum Newspapers of Alaska, LLC, v. 6, no. 38, September 20. (Accessed September 27, 2012, at http://www.petroleumnews.com/mnarch/06-38-4.html.)

SNL Metals Economics Group, 2012, Trends in worldwide exploration budgets: Strategic Report, v. 25, no. 6, November/December, p. 1–10.

Stalprodukt S.A., 2013a, Report of the Stalprodukt S.A., supervisory board for the period from 1 January 2012 to 31 December 2012: Bochnia, Poland, Stalprodukt S.A., 20 p. (Accessed December 6, 2013, at http://www.stalprodukt.com.pl/pub/File/lad_korporacyjny_EN/Sprawozdanie Rady Nadzorczej_2012_English.pdf.)

Stalprodukt S.A., 2013b, Shares and shareholders: Bochnia, Poland, Stalprodukt S.A. (Accessed December 6, 2013, at http://www.stalprodukt.com.pl/akcje_akcjonariusze.)

Talvivaara Mining Co., 2012, Annual report: Talvivaara Mining Co., 178 p. (Accessed May 30, 2014, at http://www.talvivaara.com/files/talvivaara/CG2014/Annual Report 2013.pdf.)

Thomson Reuters, 2012, Stalprodukt SA updates on acquisition of Zaklady Gorniczo-Hutnicze Boleslaw SA: Thomson Reuters, November 7. (Accessed December 6, 2013, at http://www.reuters.com/finance/stocks/STPEUR.PAp/key-developments/article/2638729.)

U.S. Energy Information Administration, 2012, International energy statistics: Washington, DC, U.S. Energy Information Administration, 342 p.

U.S. Energy Information Administration, 2013, Russia: U.S. Energy Information Administration country analysis brief, September 18. (Accessed June 5, 2014, at http://www.eia.gov/countries/cab.cfm?fips=RS.)

U.S. Library of Congress, 2012, Poland—Tax on metal exploitation introduced: U.S. Library of Congress, Law Library of Congress, April 12. (Accessed February 5, 2013, at http://www.loc.gov/lawweb/servlet/lloc_news?disp3_l205403087_text.)

Utro.ru, 2012, Tamozhennyi Soyuz pereros v Edinoe Prostranstvo [Customs Union graduated into Common Space]: Utro.ru, January 1. (Accessed June 5, 2014, at http://www.utro.ru/articles/2012/01/01/1020668.shtml.)

Wilson, Alana, McMahon, Fred, and Cervantes, Miguel, 2013, Fraser Institute annual survey of mining companies 2012/2013: Vancouver, British Columbia, Canada, Fraser Institute, February, 134 p. (Accessed March 1, 2013, at http://www.fraserinstitute.org/uploadedFiles/fraser-ca/Content/research-news/research/publications/mining-survey-2012-2013.pdf.)

World Bank, The, 2013, Global economic prospects: The World Bank, January, 178 p. (Accessed June 5, 2014, at http://siteresources.worldbank.org/INTPROSPECTS/ Resources/334934-1322593305595/8287139-1358278153255/ GEP13AFinalFullReport_.pdf .)

World Nuclear Association, 2013, World uranium mining production 2012: World Nuclear Association, May. (Accessed June 5, 2014, at http://www.world-nuclear.org/info/Nuclear-Fuel-Cycle/Mining-of-Uranium/ World-Uranium-Mining-Production/.)

World Steel Association, 2014, World steel in figures 2013: Brussels, Belgium, World Steel Association, 17 p.

Zeta Petroleum plc, 2012, New energy and gas law; liberalises Romanian domestic gas market: Zeta Petroleum plc., July 26. (Accessed August 9, 2013, at http://www.zetapetroleum.com/files/files/70_120726_ RomanianOfficialGazette_1_3.pdf.)

TABLE 1
EUROPE AND CENTRAL EURASIA: AREA AND POPULATION IN 2012

Country	Area[1] (square kilometers)	Estimated population[2] (thousands)
Albania	28,748	3,200
Armenia	29,743	3,000
Austria	83,871	8,400
Azerbaijan	86,600	9,300
Belarus	207,600	9,500
Belgium	30,528	11,100
Bosnia and Herzegovina	51,197	3,800
Bulgaria	110,879	7,300
Croatia	56,594	4,300
Cyprus	9,251	1,100
Czech Republic	78,867	10,500
Denmark, including Greenland	2,209,180	5,700
Estonia	45,228	1,300
Finland	338,145	5,400
France	551,500	65,700
Georgia	69,700	4,500
Germany	357,022	80,400
Greece	131,957	11,100
Hungary	93,028	9,900
Iceland	103,000	300
Ireland	70,273	4,600
Italy	301,340	59,500
Kazakhstan	2,724,900	16,800
Kosovo	10,887	1,800
Kyrgyzstan	199,951	5,600
Latvia	64,589	2,000
Lithuania	65,300	3,000
Luxembourg	2,586	500
Macedonia	25,713	2,100
Malta	316	400
Moldova	33,851	3,600
Montenegro	13,812	600
Netherlands	41,543	16,800
Norway	323,802	5,000
Poland	312,685	38,500
Portugal	92,090	10,500
Romania	238,391	20,100
Russia	17,098,242	143,500
Serbia	77,474	7,200
Slovakia	49,035	5,400
Slovenia	20,273	2,100
Spain	505,370	46,800
Sweden	450,295	9,500
Switzerland	41,277	8,000
Tajikistan	143,100	8,000
Turkmenistan	488,100	5,200
Ukraine	603,550	45,600
United Kingdom	243,610	63,600
Uzbekistan	447,400	29,800
Regional total	29,362,393	821,900
World total	510,072,000	7,046,368

[1]Source: U.S. Central Intelligence Agency, The World Factbook 2013.
[2]Source: The World Bank, 2013 World Development Indicators Database.

TABLE 2
EUROPE AND CENTRAL EURASIA: GROSS DOMESTIC PRODUCT[1, 2]

Country	Gross domestic product in 2012 based on purchasing power parity		Real gross domestic product growth rate (percentage)		
	Gross value (million dollars)	Per capita (dollars)	2010	2011	2012
Albania	26,110	8,159	3.5	2.0	1.3
Armenia	19,649	6,550	2.1	4.4	7.2
Austria	359,021	42,741	2.1	3.1	0.8
Azerbaijan	96,768	10,405	5.0	0.0	2.2
Belarus	146,745	15,447	7.6	5.3	1.5
Belgium	420,307	37,865	2.1	1.9	-0.2
Bosnia and Herzegovina	31,909	8,397	0.7	1.7	-0.7
Bulgaria	103,816	14,221	0.2	1.7	0.8
Croatia	78,400	18,233	-1.2	0.0	-0.2
Cyprus	23,613	21,466	1.0	0.5	-2.4
Czech Republic	286,952	27,329	2.3	1.7	-1.2
Denmark, including Greenland	210,147	36,868	1.7	1.1	-1.6
Estonia	29,088	22,375	3.1	7.6	3.2
Finland	197,476	36,570	3.6	2.9	-0.2
France	2,254,067	34,308	1.4	1.7	0.0
Georgia	26,670	5,927	6.4	6.7	6.5
Germany	3,197,069	39,765	3.6	3.1	0.9
Greece	276,879	24,944	-4.4	-6.7	-6.4
Hungary	195,630	19,761	1.2	1.7	-1.7
Iceland	12,831	42,770	-3.5	3.1	1.6
Ireland	192,223	41,788	-0.4	0.7	0.9
Italy	1,832,916	30,805	1.3	0.4	-2.4
Kazakhstan	231,787	13,797	7.3	7.5	5.0
Kosovo	13,369	7,427	4.0	5.0	2.1
Kyrgyzstan	13,279	2,371	-1.4	5.7	-0.9
Latvia	37,272	18,636	-0.3	5.5	5.8
Lithuania	65,014	21,671	1.3	5.9	3.6
Luxembourg	42,225	84,450	3.5	1.0	0.1
Macedonia	21,861	10,410	1.8	3.0	-0.3
Malta	11,260	28,150	3.1	2.1	0.8
Moldova	12,156	3,377	6.9	6.4	-0.8
Montenegro	7,340	12,233	1.1	2.5	0.0
Netherlands	706,955	42,081	1.6	1.3	-0.9
Norway	277,152	55,430	0.3	1.7	3.0
Poland	800,934	20,803	3.8	4.4	2.0
Portugal	246,523	23,478	1.3	-1.5	-3.1
Romania	273,411	13,603	-1.3	2.5	0.3
Russia	2,513,299	17,514	4.0	4.3	3.4
Serbia	78,721	10,933	1.0	1.8	-1.8
Slovakia	131,893	24,425	4.0	3.3	2.0
Slovenia	57,955	27,598	1.2	-0.2	-2.3
Spain	1,410,628	30,142	-0.1	0.7	-1.4
Sweden	392,956	41,364	5.7	4.0	1.2
Switzerland	363,421	45,428	2.7	1.9	1.0
Tajikistan	17,749	2,219	6.5	7.4	7.5
Turkmenistan	48,948	9,413	9.2	14.7	11.0
Ukraine	335,172	7,350	4.2	5.2	0.2
United Kingdom	2,336,295	36,734	1.4	0.7	0.2
Uzbekistan	104,694	3,513	8.5	8.3	8.0
Regional total	20,570,555	25,028[3]	2.3[3]	2.2[3]	1.1[3]
World total	83,193,418	12,154[3]	5.1[3]	3.9[3]	3.2[3]

[1]Source: International Monetary Fund, World Economic Outlook Database, April 2013.

[2]Gross domestic product (GDP) listed may differ from that reported in individual country chapters owing to differences in source or date of reporting.

[3]Weighted average.

TABLE 3
SELECTED SIGNIFICANT EXPLORATION ACTIVITIES IN EUROPE AND CENTRAL EURASIA IN 2012

Country	Type[1]	Site	Commodity[2]	Company	Resource notes[3,4]
Azerbaijan	P	Gedabek	Au, Cu, Ag	Anglo Asian Mining plc.	744,000 oz Au, 60,000 t Cu, 6.2 Moz Ag (R)
Portugal	P	Neves-Corvo	Cu, Zn, Pb, Ag	Lundin Mining Corp.	815,000 t Cu, 1.9 Mt Zn, 454 kt Pb, 82 Moz Ag (R)

[1]Exploration at producing (P) site.
[2]Abbreviations used in this table for commodities are as follows: Ag, silver; Au, gold; Cu, copper; Pb, lead; Zn, zinc.
[3]Abbreviations used in this table for units of measurement are as follows: oz, troy ounces; Moz, million troy ounces; Mt, million metric tons; t, metric tons.
[4]Based on 2012 data reported from various sources. R—proven + probable. Resource data not verified by the U.S. Geological Survey.

TABLE 4
EUROPE AND CENTRAL EURASIA: PRODUCTION OF SELECTED MINERAL COMMODITIES IN 2012[1, 2]

(Thousand metric tons unless otherwise specified)

Country	Aluminum				Metals Antimony, mine output, metal content (metric tons)	Chromite, mine output, gross weight	Mine output, metal content	Copper Metal, refined	
	Alumina	Bauxite	Metal Primary[3]	Secondary				Primary[3]	Secondary
Albania[e]	--	--	--	--	--	330	5	--	--
Armenia	--	--	--	--	--	--	41	--	--
Austria	--	--	--	373 [e]	--	--	1	--	114 [p]
Azerbaijan	15 [e]	--	55 [e]	--	--	--	1	--	--
Belarus	--	--	--	NA	--	--	--	--	--
Belgium	--	--	--	--	--	--	--	380 [e]	--
Bosnia and Herzegovina	202	800	126 [e]	--	--	--	--	--	--
Bulgaria	--	--	--	--	--	--	108 [e]	226	--
Croatia	--	--	--	--	--	--	--	--	--
Cyprus	--	--	--	--	--	--	--	4	--
Czech Republic	--	--	--	50 [e]	--	--	--	--	--
Denmark-Greenland	--	--	--	25 [e]	--	--	--	--	--
Estonia	--	--	--	--	--	--	--	--	--
Finland	430 [e]	69	--	20	--	(4)	NA	148 [e]	--
France	--	--	349	184	--	--	--	--	--
Georgia	--	--	--	--	--	--	7 [e]	--	--
Germany	967	--	410	635	--	--	--	390 [p]	296 [p]
Greece	784	1,816	165	--	--	1 [e]	--	--	--
Hungary	150 [e]	255	--	150	--	--	--	--	--
Iceland	--	--	803	--	--	--	--	--	--
Ireland	1,927	--	--	--	--	--	--	--	--
Italy	--	--	110	1,003	--	--	--	8 [e]	--
Kazakhstan	1,510	5,170	249	--	865	3,590	419	367	--
Kosovo	--	--	--	--	--	2	--	--	--
Kyrgyzstan	--	--	--	--	1,200 [e]	--	--	--	--
Latvia	--	--	--	--	--	--	--	--	--
Lithuania	--	--	--	--	--	--	--	--	--
Luxembourg	--	--	--	--	--	--	--	--	--
Macedonia	--	--	--	--	--	--	10 [e]	2	--
Malta	--	--	--	--	--	--	--	--	--
Moldova	--	--	--	--	--	--	--	--	--
Montenegro	--	--	90 [e]	--	--	--	--	--	--
Netherlands	--	--	110	--	--	--	--	--	--
Norway[e]	--	--	1,800	250	--	--	--	36	--
Poland	--	--	--	11 [e]	--	--	479	566	--
Portugal[p]	--	--	--	18 [e]	--	--	74	--	--

See footnotes at end of table.

TABLE 4—Continued
EUROPE AND CENTRAL EURASIA: PRODUCTION OF SELECTED MINERAL COMMODITIES IN 2012[1,2]

(Thousand metric tons unless otherwise specified)

		Aluminum			Metals	Chromite, mine output, gross weight	Copper			
			Metal		Antimony, mine output, metal content (metric tons)			Metal, refined		
Country	Alumina	Bauxite	Primary[3]	Secondary			Mine output, metal content	Primary[3]	Secondary	
Romania	450	--	252	13	--	--	8 [e]	--	--	
Russia	2,719	5,700	3,924	--	7,300 [e]	670 [e]	883 [e]	621	254	
Serbia[e]	--	--	--	--	--	--	42	32	2	
Slovakia	--	--	181	--	--	--	--	--	--	
Slovenia[e]	--	--	40	15	--	--	--	--	--	
Spain[e]	1,500	--	408	243	--	--	75	255 [e]	35 [e]	
Sweden	--	--	129	30 [e]	--	--	82	179	40 [e]	
Switzerland[e]	--	--	--	(4)	--	--	--	--	--	
Tajikistan	--	--	273	--	4,700 [e]	--	--	--	--	
Turkmenistan[e]	--	--	--	--	--	--	--	--	--	
Ukraine	1,429	--	15	90 [e]	--	--	--	--	15 [e]	
United Kingdom	--	--	60	300 [e]	--	--	--	--	--	
Uzbekistan	--	--	--	NA	--	--	96	96	--	
Total, Europe and Central Eurasia	12,100	13,800	9,550	3,410	14,100	4,590	2,280	3,320	711	
Share of world total	12.7%	5.4%	20.4%	41.6%	7.6%	14.7%	13.6%	19.4%	23.5%	
United States	4,390	NA	2,070	--	--	--	1,170	962	39	
Share of world total	4.6%	NA	4.4%	--	--	--	7.0%	5.6%	1.3%	
World total	95,000	256,000	46,800	8,200	185,000	31,200	16,800	17,100	3,030	

See footnotes at end of table.

TABLE 4—Continued
EUROPE AND CENTRAL EURASIA: PRODUCTION OF SELECTED MINERAL COMMODITIES IN 2012[1,2]

(Thousand metric tons unless otherwise specified)

Country	Gold, mine output (kilograms)	Iron and steel			Metals—Continued				Manganese ore, mine output, metal content	Mercury, mine output, metal content (metric tons)
		Iron ore, mine output, metal content	Pig iron and direct-reduced iron	Steel, crude	Mine output, metal content	Lead				
						Metal, refined				
						Primary[3]	Secondary			
Albania	--	--	--	500 [e]	--	--	--	--	--
Armenia	2,896	--	--	--	--	--	--	--	--
Austria	--	686 [e]	5,751	7,421	--	--	25	--	--
Azerbaijan	1,563	114 [e]	--	268	--	--	--	--	--
Belarus	--	--	--	2,869	--	--	--	--	--
Belgium	--	--	4,072	7,386	--	--	88	--	--
Bosnia and Herzegovina	--	872 [e]	750	700	4 [e]	--	3 [e]	--	--
Bulgaria[e]	5,200	--	--	632 [5]	12	68	--	11	--
Croatia	--	--	--	109 [e]	--	--	--	--	--
Cyprus[e]	--	--	--	--	--	--	--	--	--
Czech Republic	--	--	3,936	5,072	--	--	30 [e]	--	--
Denmark-Greenland	2,800 [e]	--	--	--	--	--	--	--	--
Estonia	--	--	--	--	--	--	8	--	--
Finland	10,814	--	12 [e]	3,759	--	--	--	--	--
France	--	--	9,532	15,609	--	--	75 [e]	--	--
Georgia[e]	2,300	--	--	--	NA	--	--	110	--
Germany	--	47	27,608	42,661	--	134 [e]	290 [e]	--	--
Greece[e]	--	550	--	2,000	18	--	10	--	--
Hungary	--	--	1,229	1,543	--	--	--	13 [e]	--
Iceland	--	--	--	--	--	--	--	--	--
Ireland	--	--	--	--	47	--	18 [e]	--	--
Italy	--	--	9,424	27,257	--	--	138	--	--
Kazakhstan	39,903	14,326	2,707	2,610	38	88	--	390 [e]	--
Kosovo	--	--	--	--	5	--	--	--	--
Kyrgyzstan	10,333	--	5,909	6,867	--	--	--	--	75
Latvia	--	--	--	800 [e]	--	--	--	--	--
Lithuania	--	--	--	--	--	--	--	--	--
Luxembourg	--	--	--	2,232	--	--	--	--	--
Macedonia	--	--	--	216	34	--	--	--	--
Malta	--	--	--	--	--	--	--	--	--
Moldova	--	--	--	317	--	--	--	--	--
Montenegro	--	--	--	45	--	--	--	--	--
Netherlands	--	--	5,909	6,867	--	--	17 [e]	--	--
Norway	--	3,421	--	600 [e]	--	--	--	--	--
Poland	916	--	3,944	8,539	58 [e]	19 [e]	121 [e]	--	--
Portugal[e,p]	--	10	100	1,400	--	--	3	--	--

See footnotes at end of table.

TABLE 4—Continued
EUROPE AND CENTRAL EURASIA: PRODUCTION OF SELECTED MINERAL COMMODITIES IN 2012[1,2]

(Thousand metric tons unless otherwise specified)

| Country | Gold, mine output (kilograms) | Iron and steel | | Metals—Continued | Lead | | | Manganese ore, mine output, metal content | Mercury, mine output, metal content (metric tons) |
| | | Iron ore, mine output, metal content | Pig iron and direct-reduced iron | Steel, crude | Mine output, metal content | Metal, refined | | | |
						Primary[3]	Secondary		
Romania[e]	--	--	3,417[5]	3,811[5]	--	7[5]	3	--	--
Russia[e]	217,800[5]	61,400	55,700	70,400[5]	93	85	--	33	50
Serbia	900	--	312	345	2[e]	--	1[e]	--	--
Slovakia	546	--	3,519	4,403	--	--	--	--	--
Slovenia	--	--	--	632	--	--	10[e]	--	--
Spain[p]	3,600	--	3,570	15,600	--	--	125[e]	--	--
Sweden	6,015	17,186	5,253	4,326	64	62[e]	44[e]	--	--
Switzerland	--	--	--	1,400[e]	--	--	3	--	--
Tajikistan	2,401	--	--	--	1[e]	--	--	--	32[e]
Turkmenistan	--	--	--	--	--	--	--	--	--
Ukraine	--	45,100[e]	28,514	32,394	--	--	14[e]	396[e]	--
United Kingdom	--	--	7,183	9,579	(4)[e]	157[e]	155[e]	--	--
Uzbekistan	93,000[e]	--	--	736	--	--	--	--	--
Total, Europe and Central Eurasia	401,000	144,000	182,000	285,000	375	620	1,180	952	157
Share of world total	14.8%	10.3%	15.8%	18.3%	7.2%	12.4%	24.1%	6.1%	10.3%
United States	235,000	33,500	32,100	88,700	345	111	1,110	--	NA
Share of world total	8.7%	2.4%	2.8%	5.7%	6.6%	2.2%	22.7%	--	NA
World total	2,700,000	1,440,000	155,000	1,560,000	5,200	4,980	4,900	15,600	1,520

See footnotes at end of table.

TABLE 4—Continued
EUROPE AND CENTRAL EURASIA: PRODUCTION OF SELECTED MINERAL COMMODITIES IN 2012[1,2]

(Thousand metric tons unless otherwise specified)

Metals—Continued

Country	Nickel		Platinum-group metals, refined, primary and secondary (kilograms)		Silver, mine output, metal content (metric tons)	Tin (metric tons)		Titanium (metric tons)	
	Mine output, metal content	Refinery products, metal content	Palladium	Platinum		Mine output, metal content	Metal, primary[3]	Ilmenite, TiO$_2$ content	Metal sponge, metal content
Albania	3 [e]	--	--	--	--	--	--	--	--
Armenia	--	--	--	--	22	--	--	--	--
Austria	--	1 [e]	--	--	--	--	--	--	--
Azerbaijan	--	--	--	--	1	--	--	--	--
Belarus	--	--	--	--	--	--	--	--	--
Belgium[e]	--	--	--	--	--	--	--	--	--
Bosnia and Herzegovina	--	--	--	--	--	--	--	--	--
Bulgaria	--	--	--	--	55 [e]	--	--	--	--
Croatia	--	--	--	--	--	--	--	--	--
Cyprus[e]	--	--	--	--	--	--	--	--	--
Czech Republic	--	--	--	--	--	--	--	--	--
Denmark-Greenland[e]	--	--	--	--	--	--	--	--	--
Estonia	--	--	--	--	--	--	--	--	--
Finland	80 [e]	46	--	830 [e]	128	--	--	--	--
France	--	15	--	--	--	--	--	--	--
Georgia	--	--	--	--	1 [e]	--	--	--	--
Germany	--	--	26,114	16,670	--	--	--	--	--
Greece[e]	14	19 [5]	--	--	32	--	--	--	--
Hungary	--	--	--	--	--	--	--	--	--
Iceland	--	--	--	--	--	--	--	--	--
Ireland	--	--	--	--	6 [e]	--	--	--	--
Italy	(4)	--	--	--	963 [5]	--	--	15,000	21,000
Kazakhstan[e]	--	--	--	--	--	--	--	--	--
Kosovo	4	--	--	--	--	--	--	--	--
Kyrgyzstan	--	--	--	--	--	--	--	--	--
Latvia[e]	--	--	--	--	--	--	--	--	--
Lithuania	--	--	--	--	--	--	--	--	--
Luxembourg	--	19	--	--	--	--	--	--	--
Macedonia	--	--	--	--	--	--	--	--	--
Malta	--	--	--	--	--	--	--	--	--
Moldova	--	--	--	--	--	--	--	--	--
Montenegro	--	--	--	--	--	--	--	--	--
Netherlands[e]	--	--	--	--	--	--	--	--	--
Norway[e]	(4)	92 [5]	--	--	--	--	--	400	--
Poland[e]	--	--	15	25	1,149 [5]	--	--	--	--
Portugal[p]	--	--	--	--	27	42	--	--	--

See footnotes at end of table.

TABLE 4—Continued
EUROPE AND CENTRAL EURASIA: PRODUCTION OF SELECTED MINERAL COMMODITIES IN 2012[1,2]

(Thousand metric tons unless otherwise specified)

Country	Nickel		Platinum-group metals, refined, primary and secondary (kilograms)		Metals—Continued Silver, mine output, metal content (metric tons)	Tin (metric tons)		Titanium (metric tons)	
	Mine output, metal content	Refinery products, metal content	Palladium	Platinum		Mine output, metal content	Metal, primary[3]	Ilmenite, TiO$_2$ content	Metal sponge, metal content
Romania	--	--	--	--	--	--	--	--	--
Russia	255	258	82,400	30,200	1,679	100 [e]	500 [e]	--	42,000
Serbia	--	--	22	3	5	--	--	--	--
Slovakia	--	--	--	--	(4)	--	--	--	--
Slovenia	--	--	--	--	--	--	--	--	--
Spain	6	--	--	--	4 [e]	--	--	--	--
Sweden	--	--	--	--	309	--	--	--	--
Switzerland	--	--	--	--	--	--	--	--	--
Tajikistan	--	--	--	--	2	--	--	--	--
Turkmenistan	--	--	--	--	--	--	--	--	--
Ukraine	--	12 [e]	--	--	--	--	--	145,640	8,500 [e]
United Kingdom	--	34	--	--	--	--	--	--	--
Uzbekistan	--	--	--	--	60 [e]	--	--	--	--
Total, Europe and Central Eurasia	363	496	109,000	47,700	4,440	142	500	161,000	71,500
Share of world total	10.8%	36.3%	38.8%	32.1%	4.6%	0.1%	0.2%	4.2%	48.2%
United States	--	--	12,400	3,700	1,060	--	--	167,000	--
Share of world total	--	--	4.4%	2.5%	1.1%	--	--	4.4%	--
World total	3,360	1,370	280,000	149,000	97,200	273,000	305,000	3,830,000	148,000

See footnotes at end of table.

TABLE 4—Continued
EUROPE AND CENTRAL EURASIA: PRODUCTION OF SELECTED MINERAL COMMODITIES IN 2012[1,2]

(Thousand metric tons unless otherwise specified)

Country	Metals—Continued				Industrial minerals			
	Tungsten, mine output, metal content (metric tons)	Zinc (metric tons)		Ammonia, N content	Cement, hydraulic	Diamond, natural, gemstones and industrial (thousand carats)	Phosphate rock, P$_2$O$_5$ content	Potash, K$_2$O equivalent	Salt
		Mine output, metal content	Metal, primary and secondary						
Albania	--	--	--	--	2,000 [e]	--	--	--	NA
Armenia	--	10,700	--	--	438	--	--	--	38
Austria	625 [e]	--	--	400 [e]	4,455	--	--	--	1,064
Azerbaijan	--	--	--	815	1,966	--	--	--	5
Belarus	--	--	290,000	830	4,906	--	--	4,840	1,700 [e]
Belgium[e]	--	--	--	--	6,800	--	--	--	--
Bosnia and Herzegovina	--	7,000 [e]	--	--	846	--	--	--	862
Bulgaria	--	12,116	72,000	320 [e]	1,900 [e]	--	--	--	2,100
Croatia	--	--	--	350 [e]	1,244	--	--	--	18
Cyprus	--	--	--	--	1,080	--	--	--	--
Czech Republic	--	--	--	200 [e]	3,650 [e]	--	--	--	600 [e]
Denmark-Greenland	--	--	--	--	16 [e]	--	--	--	--
Estonia	--	--	--	--	450 [e]	--	--	--	--
Finland[e]	--	51,467 [5]	314,742 [5]	72	1,300	--	307	--	5,457
France	--	161,000	161,000	2,644	18,000	--	--	--	29
Georgia[e]	--	NA	--	150	870	--	--	--	--
Germany	--	41,824	169,000 [e]	2,823	32,432	--	--	3,767	14,445
Greece	--	--	--	130 [e]	11,000 [e]	--	--	--	192
Hungary	--	--	--	300 [e]	1,900 [e]	--	--	--	--
Iceland	--	--	--	--	146	--	--	--	NA
Ireland	--	337,500	--	--	2,200 [e]	--	--	--	3,098
Italy	--	--	97	2,365	26,200	--	350 [e]	--	464
Kazakhstan	--	369,700	319,900	--	7,800 [e]	--	--	--	--
Kosovo	--	3,800	--	--	600 [e]	--	--	--	--
Kyrgyzstan	--	--	--	--	900 [e]	--	--	--	1 [e]
Latvia	--	--	--	--	--	--	--	--	--
Lithuania	--	--	--	918	1,015	--	--	--	--
Luxembourg	--	--	--	--	1,217	--	--	--	--
Macedonia	--	28,000 [e]	--	--	683	--	--	--	--
Malta	--	--	--	--	--	--	--	--	6 [e]
Moldova	--	--	--	--	1,500 [e]	--	--	--	--
Montenegro	--	--	--	--	--	--	--	--	12 [e]
Netherlands[e]	--	--	257,000	NA	2,700	--	--	--	NA
Norway	--	--	152,647	320	1,700 [e]	--	--	--	--
Poland[e]	--	89,000	161,000	2,100	15,919 [5]	--	--	--	3,916 [5]
Portugal[p]	763	30,008	--	244 [e]	7,200 [e]	--	--	--	520

See footnotes at end of table.

TABLE 4—Continued
EUROPE AND CENTRAL EURASIA: PRODUCTION OF SELECTED MINERAL COMMODITIES IN 2012[1,2]

(Thousand metric tons unless otherwise specified)

Country	Metals—Continued					Industrial minerals			
	Tungsten, mine output, metal content (metric tons)	Zinc (metric tons)		Ammonia, N content	Cement, hydraulic	Diamond, natural, gemstones and industrial (thousand carats)	Phosphate rock, P_2O_5 content	Potash, K_2O equivalent	Salt
		Mine output, metal content	Metal, primary and secondary						
Romania	--	--	300	115 [e]	8,082	--	--	--	2,240 [e]
Serbia	--	NA	--	130 [e]	1,831	--	--	--	17
Slovakia	--	--	--	486	2,915	--	--	--	6
Slovenia	--	--	--	--	1,200 [e]	--	--	--	--
Spain	--	--	490,000 [e]	--	20,000 [p]	--	--	436 [e]	4,385
Sweden	--	188,300	--	--	3,000 [e]	--	--	--	--
Switzerland	--	--	--	30 [e]	4,467	--	--	--	528
Tajikistan	--	--	--	--	232	--	--	--	28
Turkmenistan [e]	--	--	--	280	1,900	--	--	--	220
Ukraine	--	--	--	4,160 [e]	9,801	--	--	--	6,189
United Kingdom [e]	--	--	--	1,100	8,500 [r]	--	--	770	6,000
Uzbekistan [e]	--	--	61,100	1,300	6,800	--	187	(4) [5]	--
Total, Europe and Central Eurasia	4,410	1,350,000	2,710,000	32,900	295,000	34,900	5,340	15,400	56,000
Share of world total	6.0%	10.2%	21.8%	24.4%	7.7%	26.8%	8.2%	43.7%	20.7%
United States	--	738,000	261,000	8,730	74,900	--	8,590	900	37,200
Share of world total	--	5.6%	2.1%	6.5%	2.0%	--	13.2%	2.6%	13.8%
World total	73,800	13,300,000	12,400,000	135,000	3,810,000	130,000	65,200	35,200	270,000

See footnotes at end of table.

TABLE 4—Continued
EUROPE AND CENTRAL EURASIA: PRODUCTION OF SELECTED MINERAL COMMODITIES IN 2012[1,2]

(Thousand metric tons unless otherwise specified)

Country	Mineral fuels and related materials					
	Coal			Natural gas, dry (million cubic meters)	Petroleum, crude (thousand 42-gallon barrels)	Uranium, U$_3$O$_8$ content (metric tons)
	Anthracite	Bituminous	Lignite			
Albania[e]	--	--	1	8	5,900	--
Armenia	--	--	--	--	--	--
Austria	--	--	--	1,729	5,896	--
Azerbaijan	--	--	--	17,242	320,667	--
Belarus	--	--	--	218	1,660	--
Belgium[e]	--	--	--	--	--	--
Bosnia and Herzegovina	--	2,314	12,312	--	--	--
Bulgaria[e]	--	--	32,000	396 [5]	170	--
Croatia	--	--	--	2,013	4,479	--
Cyprus	--	--	--	--	--	--
Czech Republic	--	10,796	43,710	204	1,020	262 [e]
Denmark-Greenland	--	--	--	9,000 [e]	75,701	--
Estonia	--	--	--	--	--	--
Finland	--	--	--	1,100 [e]	5,949	--
France	--	240	--	8	364	--
Georgia[e]	1,792	8,978	185,432	10,660	19,200	59 [e]
Germany	--	--	62,335	--	662	--
Greece	--	--	9,297	2,280	4,410 [e]	--
Hungary	--	--	--	--	--	--
Iceland	--	--	--	350 [e]	--	--
Ireland	--	--	--	8,400 [e]	36,865	--
Italy	--	107,911	7,748	40,129	576,200	24,648
Kazakhstan	--	--	8,028	--	--	--
Kosovo	--	133	1,051	19	657	2,150 [e]
Kyrgyzstan	--	--	--	--	--	--
Latvia	--	--	--	--	822 [e]	--
Lithuania	--	--	--	--	--	--
Luxembourg	--	--	7,310	--	--	--
Macedonia	--	--	--	--	--	--
Malta	--	--	--	--	--	--
Moldova	--	--	2,000 [e]	--	--	--
Montenegro	--	--	--	80,787	8,212	--
Netherlands	--	1,583	--	106,710 [6]	694,230	--
Norway	--	79,855	64,280	5,782	5,000	--
Poland	--	--	--	--	1,725	--
Portugal[p]	--	--	33,500	10,933 [5]	31,000	88
Romania[e]						

See footnotes at end of table.

TABLE 4—Continued
EUROPE AND CENTRAL EURASIA: PRODUCTION OF SELECTED MINERAL COMMODITIES IN 2012[1,2]

(Thousand metric tons unless otherwise specified)

Country	Mineral fuels and related materials					
	Coal		Lignite	Natural gas, dry (million cubic meters)	Petroleum, crude (thousand 42-gallon barrels)	Uranium, U_3O_8 content (metric tons)
	Anthracite	Bituminous				
Russia	11,400	276,500	78,100	655,000	3,615,000	3,348
Slovakia	--	--	2,292	110	105	--
Slovenia	--	--	4,321	2	2	--
Spain	4,073	2,252	--	44 [p]	323	--
Sweden	--	--	--	--	--	--
Switzerland	--	--	--	--	--	--
Tajikistan	--	412	--	11	87	--
Turkmenistan	--	--	--	69,000 [e]	79,915	--
Ukraine	20,763	64,600	200 [e]	19,318	24,110	1,132 [e]
United Kingdom	1,101	15,687	--	57,000 [e]	368,139	--
Uzbekistan	--	253	3,600	62,911	23,067	3,190
Total, Europe and Central Eurasia	39,100	572,000	596,000	1,160,000	5,920,000	34,900
Share of world total	5.4%	9.8%	49.9%	34.0%	20.7%	51.6%
United States	2,150	848,000	71,600	717,000	2,370,000	1,880
Share of world total	0.3%	14.6%	6.0%	21.0%	8.3%	2.8%
World total	731,000	5,820,000	1,190,000	3,420,000	28,600,000	67,600

[e]Estimated; estimated data, U.S. data, and world totals are rounded to no more than three significant digits; may not add to totals shown. [p]Preliminary. NA Not available. -- Zero or zero percent.

[1]Some of the individual entries in this table may differ from those that appear in individual country production tables elsewhere in this volume owing to the inclusion in this table received at a later date.

[2]Totals may not add due to independent rounding. Table includes data available as of May 7, 2014.

[3]Primary production also includes undifferentiated (primary and secondary) production for some countries listed.

[4]Less than ½ unit.

[5]Reported figure.

[6]Reported as total methane sales.

TABLE 5

EUROPE AND CENTRAL EURASIA: HISTORIC AND PROJECTED BAUXITE MINE PRODUCTION, 2005–2019[1]

(Thousand metric tons)

Country	2005	2010	2012	2015[e]	2017[e]	2019[e]
Albania	--	--	--	--	--	--
Bosnia and Herzegovina	1,032	844	800	900	900	900
Greece	2,441		1,816	1,900	2,000	2,000
Hungary	535	365	255	250	250	250
Italy	300	--	--	--	--	--
Kazakhstan	4,815	5,310	5,170	5,300	5,400	5,500
Montenegro	672	61	--	--	--	--
Romania	--	--	--	--	--	--
Russia	5,000	5,688	5,700	5,800	5,900	6,000
Total	14,800	12,300	13,700	14,200	14,500	14,700

[e]Estimated. -- Zero.

[1]Estimated data and totals are rounded to no more than three significant digits; may not add to totals shown.

TABLE 6

EUROPE AND CENTRAL EURASIA: HISTORIC AND PROJECTED PRIMARY AND SECONDARY ALUMINUM METAL PRODUCTION, 2005–2019[1]

(Thousand metric tons)

Country	2005	2010	2012	2015[e]	2017[e]	2019[e]
Austria	150	375	373	400	450	450
Azerbaijan	32	--	55	70	75	80
Bosnia and Herzegovina	131	118	126	160	160	160
Bulgaria	5	12	--	--	--	--
Czech Republic	30	40	50	50	50	50
Denmark-Greenland	20	25	25	25	25	25
Finland	34	21	20	20	20	20
France	664	540	533	530	540	540
Germany	1,366	1,014	1,045	1,000	1,000	1,000
Greece	163	137	165	140	140	140
Hungary	82	234	150	150	150	150
Iceland	273	826	803	800	1,400	1,400
Italy	1,314	1,414	1,112	1,300	1,400	1,400
Kazakhstan	--	226	249	250	255	260
Macedonia	4	--	--	--	--	--
Montenegro	117	82	90	90	90	90
Netherlands	391	300	300	300	300	300
Norway	1,376	1,060	1,070	1,100	1,100	1,100
Poland	66	16	11	11	11	11
Portugal	18	18	18	18	18	18
Romania	246	258	265	280	280	280
Russia	3,647	3,947	3,924	4,000	4,000	4,000
Serbia	(2)	2	--	--	--	--
Slovakia	162	163		160	160	160
Slovenia	139	58		58	58	58
Spain	637	651	651	650	650	650
Sweden	133	104	159	160	160	160
Switzerland	238	25	--	--	--	--
Tajikistan	380	349	273	320	350	350
Ukraine	244	155	105	100	100	100
United Kingdom	574	498	360	300	300	300
Uzbekistan	3	--	--	--	--	--
Total	12,600	12,700	11,900	12,400	12,600	12,700

[e]Estimated. -- Zero.

[1]Estimated data and totals are rounded to no more than three significant digits; may not add to totals shown.

[2]Less than 1/2 unit.

TABLE 7

EUROPE AND CENTRAL EURASIA: HISTORIC AND PROJECTED COBALT MINE PRODUCTION, 2005–2019[1]

(Co content in metric tons)

Country	2005	2010	2012	2015[e]	2017[e]	2019[e]
Finland	6,158	9,413	10,547	12,000	12,000	12,000
Kazakhstan	--	--	--	--	--	--
Russia	6,300	6,200	6,300	6,300	6,400	6,400
Total	12,500	15,600	16,800	18,300	18,400	18,400

[e]Estimated. -- Zero.

[1]Estimated data and totals are rounded to no more than three significant digits; may not add to totals shown.

TABLE 8

EUROPE AND CENTRAL EURASIA: HISTORIC AND PROJECTED COPPER MINE PRODUCTION, 2005–2019[1]

(Cu content in thousand metric tons)

Country	2005	2010	2012	2015[e]	2017[e]	2019[e]
Albania	2	2	5	5	5	5
Armenia	19	31	41	70	90	110
Azerbaijan	--	(2)	1	2	3	6
Bulgaria	112	105	108	100	100	100
Finland	16	15	NA	16	16	16
Georgia	10	7	7	7	8	8
Kazakhstan	402	381	419	440	450	460
Macedonia	22	8	10	10	10	10
Poland	512	481	479	480	480	480
Portugal	90	74	74	76	80	85
Romania	15	5	7	7	7	7
Russia	700	703	883	900	950	1,000
Serbia	27	28	41	50	60	60
Spain	5	46	75	80	85	90
Sweden	98	77	82	85	85	85
Uzbekistan	104	90	96	100	105	110
Total	2,130	2,050	2,330	2,430	2,530	2,600

[e]Estimated. -- Zero. NA Not available.

[1]Estimated data and totals are rounded to no more than three significant digits; may not add to totals shown.

[2]Less than 1/2 unit.

TABLE 9
EUROPE AND CENTRAL EURASIA: HISTORIC AND PROJECTED REFINED COPPER METAL PRODUCTION
(PRIMARY AND SECONDARY), 2005–2019[1]

(Thousand metric tons)

Country	2005	2010	2012	2015[e]	2017[e]	2019[e]
Armenia	10	8	10	10	11	11
Austria	72	114	114	110	110	110
Belgium	383	381	380	380	380	380
Bulgaria	61	215	226	230	230	230
Cyprus	--	3	4	4	4	4
Czech Republic	--	--	--	--	--	--
Finland	125	146	148	150	150	150
Germany	638	704	686	700	700	700
Hungary	--	--	--	--	--	--
Italy	32	2	8	7	7	7
Kazakhstan	418	323	367	390	400	410
Macedonia	--	--	2	2	2	2
Norway	39	3	36	36	35	34
Poland	560	547	566	570	570	570
Romania	21	4	--	--	--	--
Russia	958	830	875	890	900	900
Serbia	27	22	35	40	50	50
Slovakia	--	46	49	50	50	50
Spain	302	270	270	270	270	270
Sweden	222	191	219	220	230	240
Ukraine	25	20	15	15	15	15
Uzbekistan	104	90	96	98	99	100
Total	4,000	3,920	4,110	4,170	4,210	4,230

[e]Estimated. -- Zero.

[1]Estimated data and totals are rounded to no more than three significant digits; may not add to totals shown.

TABLE 10

EUROPE AND CENTRAL EURASIA: HISTORIC AND PROJECTED GOLD MINE PRODUCTION, 2005–2019[1]

(Kilograms)

Country	2005	2010	2012	2015[e]	2017[e]	2019[e]
Armenia	1,400	974	2,896	3,300	3,400	3,500
Azerbaijan	--	1,900	1,563	2,000	2,500	3,000
Bulgaria	3,868	3,300	5,200	6,000	8,000	9,000
Denmark-Greenland	1,000	1,600	2,800	3,000	3,000	4,000
Finland	3,747	7,628	10,814	11,000	11,000	11,000
France	1,500	1,500	--	--	--	--
Georgia	1,620	2,000	2,300	2,300	2,350	2,400
Kazakhstan	17,875	30,272	39,903	45,000	50,000	54,000
Kyrgyzstan	16,751	18,072	10,333	18,000	18,000	18,000
Macedonia	400	--	--	--	--	--
Poland	510	776	916	900	900	900
Romania	400	400	--	--	--	--
Russia	164,186	189,000	217,800	230,000	235,000	240,000
Serbia	650	700	1,000	1,000	1,000	1,000
Slovakia	109	534	546	550	600	600
Spain	3,400	3,500	3,600	3,700	3,800	3,800
Sweden	6,600	6,242	6,015	6,300	6,300	6,300
Tajikistan	1,927	2,049	2,401	2,700	2,900	3,100
Ukraine	180	--	--	--	--	--
Uzbekistan	84,210	90,000	93,000	100,000	105,000	110,000
Total	310,000	360,000	401,000	436,000	454,000	471,000

[e]Estimated. -- Zero.

[1]Estimated data and totals are rounded to no more than three significant digits; may not add to totals shown.

TABLE 11

EUROPE AND CENTRAL EURASIA: HISTORIC AND PROJECTED BENEFICIATED IRON ORE PRODUCTION (MINE OUTPUT), 2005–2019[1]

(Fe content in thousand metric tons)

Country	Average iron content	2005	2010	2012	2015[e]	2017[e]	2019[e]
Austria	32%	665	662	686	700	700	700
Azerbaijan	57%	4	33	114	130	140	150
Bosnia and Herzegovina	42%	702	588	872	900	900	900
Bulgaria	50%	--	--	--	--	--	--
Czech Republic	29%	--	--	--	--	--	--
Germany[2]	11%	38	41	47	45	45	45
Greece	38%	575	560	550	500	500	400
Kazakhstan	57%	11,100	13,800	14,326	14,500	14,600	14,700
Norway	62%	420	3,105	3,200	3,200	3,000	3,000
Portugal	36%	12	14	14	14	14	14
Romania	52%	69	--	--	--	--	--
Russia	59%	56,100	56,600	61,400	62,000	63,000	64,000
Slovakia	34%	182	--	--	--	--	--
Spain	38%	--	--	--	--	--	--
Sweden	60%	15,300	16,750	17,186	18,000	18,000	18,000
Ukraine	55%	37,700	43,000	45,100	46,000	46,000	46,000
United Kingdom	54%	(3)	--	--	--	--	--
Total	XX	123,000	135,000	143,000	146,000	147,000	148,000

[e]Estimated. XX Not applicable. -- Zero.

[1]Estimated data and totals are rounded to no more than three significant digits; may not add to totals shown.

[2]Iron ore is used domestically as an additive in cement and other construction materials but is of too low a grade to use in the steel industry.

[3]Less than 1/2 unit.

TABLE 12

EUROPE AND CENTRAL EURASIA: HISTORIC AND PROJECTED CRUDE STEEL PRODUCTION, 2005–2019[1]

(Thousand metric tons)

Country	2005	2010	2012	2015[e]	2017[e]	2019[e]
Albania	87	390	500	600	600	600
Austria	7,031	7,206	7,421	7,800	7,500	7,200
Azerbaijan	286	129	268	290	310	325
Belarus	2,076	2,672	2,869	3,000	3,050	3,100
Belgium	10,420	7,973	7,386	7,800	7,500	7,500
Bosnia and Herzegovina	283	591	700	700	700	700
Bulgaria	1,969	744	632	600	600	600
Croatia	74	95	109	90	90	90
Czech Republic	6,189	5,180	5,072	5,200	5,200	5,200
Finland	4,738	4,023	3,759	4,000	4,000	4,000
France	19,481	15,414	15,609	15,800	15,800	15,800
Germany	44,524	43,830	42,661	43,000	43,000	43,000
Greece	2,266	1,839	2,000	2,000	2,000	2,000
Hungary	2,005	1,678	1,543	900	900	900
Italy	29,061	25,750	27,257	30,000	30,000	32,000
Kazakhstan	4,477	3,338	2,610	3,000	3,200	3,400
Latvia	550	655	800	800	800	800
Luxembourg	2,194	2,563	2,232	2,500	2,500	2,500
Macedonia	326	292	216	200	200	200
Moldova	1,016	242	317	600	800	1,000
Montenegro	104	45	45	42	40	40
Netherlands	6,919	6,651	6,867	6,800	6,800	6,800
Norway	701	514	600	650	650	650
Poland	8,336	7,996	8,539	9,000	10,000	10,000
Portugal	1,400	1,350	1,400	1,400	1,400	1,400
Romania	6,280	3,724	3,417	4,000	4,000	4,000
Russia	66,186	66,800	70,400	72,000	73,000	74,000
Serbia	1,286	1,254	345	900	1,700	1,700
Slovakia	4,242	4,580	4,403	4,450	4,500	4,500
Slovenia	583	606	632	650	650	650
Spain	17,800	16,340	15,600	16,000	16,500	17,000
Sweden	5,692	4,844	4,326	4,800	4,800	4,800
Switzerland	1,158	1,330	1,400	1,400	1,500	1,600
Ukraine	38,541	33,559	32,394	33,000	33,500	34,000
United Kingdom	13,210	9,709	9,579	9,500	9,500	9,500
Uzbekistan	607	731	736	740	740	740
Total	312,000	285,000	285,000	294,000	298,000	302,000

[e]Estimated.

[1]Estimated data and totals are rounded to no more than three significant digits; may not add to totals shown.

TABLE 13

EUROPE AND CENTRAL EURASIA: HISTORIC AND PROJECTED NICKEL MINE PRODUCTION, 2005–2019[1]

(Ni content in metric tons)

Country	2005	2010	2012	2015[e]	2017[e]	2019[e]
Albania	NA	3	3	3	3	3
Finland	3,386	4,400	80,000	80,000	80,000	80,000
Greece	23,210	13,800	14,000	14,000	14,000	14,000
Kazakhstan	193	500	450	400	400	400
Kosovo	--	9,080	4,400	4,400	4,200	4,000
Norway	100	351	351	350	350	350
Russia	277,177	269,277	255,000	250,000	245,000	250,000
Spain	5,386	5,400	6,200	6,600	6,900	7,200
Ukraine	6,000	--	--	--	--	--
Total	315,000	303,000	360,000	356,000	351,000	356,000

[e]Estimated. NA Not available. -- Zero.

[1]Estimated data and totals are rounded to no more than three significant digits; may not add to totals shown.

TABLE 14

EUROPE AND CENTRAL EURASIA: HISTORIC AND PROJECTED PALLADIUM MINE PRODUCTION, 2005–2019[1]

(Kilograms)

Country[2]	2005	2010	2012	2015[e]	2017[e]	2019[e]
Poland	10	15	15	15	15	15
Russia	97,400	84,700	82,400	81,000	80,000	80,000
Serbia	19	22	22	20	20	20
Total	97,400	84,700	82,400	81,000	80,000	80,000

[e]Estimated.

[1]Estimated data and totals are rounded to no more than three significant digits; may not add to totals shown.

[2]Palladium production for Finland and Norway has not been estimated.

TABLE 15

EUROPE AND CENTRAL EURASIA: HISTORIC AND PROJECTED PLATINUM MINE PRODUCTION, 2005–2019[1]

(Kilograms)

Country	2005	2010	2012	2015[e]	2017[e]	2019[e]
Finland	678	718	830	850	850	850
Poland	20	25	25	25	25	25
Russia	29,000	25,700	30,200	32,000	34,000	36,000
Serbia	3	--	3	3	3	3
Total	29,700	26,400	31,100	32,900	34,900	36,900

[e]Estimated. -- Zero.

[1]Estimated data and totals are rounded to no more than three significant digits; may not add to totals shown.

TABLE 16

EUROPE AND CENTRAL EURASIA: HISTORIC AND PROJECTED OF TIN MINE PRODUCTION, 2005–2019[1]

(Sn content in metric tons)

Country	2005	2010	2012	2015[e]	2017[e]	2019[e]
Portugal	30	20	42	45	50	60
Russia	3,000	144	100	1,000	2,000	3,000
Total	3,030	164	142	1,050	2,100	3,100

[e]Estimated.

[1]Estimated data and totals are rounded to no more than three significant digits; may not add to totals shown.

TABLE 17

EUROPE AND CENTRAL EURASIA: HISTORIC AND PROJECTED TIN METAL PRODUCTION
(PRIMARY AND SECONDARY), 2005–2019[1]

(Metric tons)

Country	2005	2010	2012	2015[e]	2017[e]	2019[e]
France	1,500	1,500	--	--	--	--
Russia	5,500	1,381	700	--	1,000	3,000
Total	7,000	2,880	700	0	1,000	3,000

[e]Estimated. -- Zero.

[1]Estimated data and totals are rounded to no more than three significant digits; may not add to totals shown.

TABLE 18

EUROPE AND CENTRAL EURASIA: HISTORIC AND PROJECTED DIAMOND PRODUCTION, 2005–2019[1]

(Thousand carats)

Country	2005	2010	2012	2015[e]	2017[e]	2019[e]
Russia:						
Gem grade	23,000	17,800	19,900	21,000	21,500	22,000
Industrial grade	15,000	15,000	15,000	15,000	15,000	15,000
Regional total	38,000	32,800	34,900	36,000	36,500	37,000

[e]Estimated.

[1]Estimated data and totals are rounded to no more than three significant digits; may not add to totals shown.

TABLE 19

EUROPE AND CENTRAL EURASIA: HISTORIC AND PROJECTED LITHIUM PRODUCTION, 2005–2019[1]

(Li content in metric tons)

Country	2005	2010	2012	2015[e]	2017[e]	2019[e]
Portugal	26,185	40,110	20,700	41,000	44,000	44,000

[e]Estimated.

[1]Estimated data and totals are rounded to no more than three significant digits; may not add to totals shown.

TABLE 20

EUROPE AND CENTRAL EURASIA: HISTORIC AND PROJECTED SALABLE COAL PRODUCTION, 2005–2019[1,2]

(Thousand metric tons)

Country	2005	2010	2012	2015[e]	2017[e]	2019[e]
Albania	3	3	1	1	1	1
Austria	14	--	--	--	--	--
Bosnia and Herzegovina	9,144	10,976	12,312	12,000	12,000	12,000
Bulgaria	24,909	29,700	34,300	34,000	34,000	34,000
Czech Republic	61,903	55,124	54,506	55,000	55,000	55,000
Georgia	5	241	240	270	290	310
Germany	202,621	182,303	196,202	190,000	185,000	180,000
Greece	73,585	53,600	62	62	63	64
Hungary	9,580	9,114	9,297	9,500	9,500	9,500
Kazakhstan	86,586	106,568	115,659	140,000	160,000	180,000
Kosovo	6,391	7,958	8,028	8,500	8,500	8,500
Kyrgyzstan	396	575	1,184	1,500	1,800	2,100
Macedonia	6,949	6,583	7,310	8,000	9,000	9,000
Montenegro	1,297	1,938	2,000	2,000	2,000	2,000
Norway	300	1,685	1,583	1,600	1,600	1,500
Poland	159,039	133,238	144,135	144,000	144,000	144,000
Romania	34,201	30,000	33,500	34,000	35,000	35,000
Russia	282,881	321,600	366,000	380,000	390,000	400,000
Serbia	34,993	38,598	38,728	39,000	39,000	39,000
Slovakia	2,511	2,378	2,292	2,300	2,300	2,100
Slovenia	4,539	4,430	4,321	4,400	4,100	4,000
Spain	19,350	8,430	6,325	6,000	5,500	5,500
Tajikistan	99	200	412	600	750	900
Ukraine	74,559	75,200	85,700	91,000	95,000	100,000
United Kingdom	20,498	18,159	16,788	18,000	18,000	18,000
Uzbekistan	3,003	3,100	3,600	4,200	4,600	5,000
Total	1,120,000	1,100,000	1,140,000	1,190,000	1,220,000	1,250,000

[e]Estimated. -- Zero.

[1]Estimated data and totals are rounded to no more than three significant digits; may not add to totals shown.

[2]Includes anthracite, bituminous, and run-of-mine lignite.

The Mineral Industry of Albania

By Mark Brininstool

Albania's mineral deposits included chromite, copper, limestone, and petroleum, but it was not a significant producer of mineral commodities on a global or regional scale. In 2012, Albania's gross domestic product (GDP) increased by 1.3%. In 2011 (the most recent year for which data were available), industrial production made up about 11% of the GDP, and the mineral sector accounted for about 16% of the value of industrial production and about 2% of the GDP (Institute of Statistics of the Republic of Albania, 2013; International Monetary Fund, 2013, p. 153).

Production

Data on mineral production in Albania in 2012 were unavailable, and all production data for mineral commodities in 2012 were estimated. Copper production in 2011 more than doubled, most likely a result of a production capacity increase at the Munella Mine. The mine's owner, Ekin Maden Ticaret ve Sanayi A.S. (Ekin Maden) of Turkey, reported that capital investments at the Munella Mine increased the mine's production capacity to 300,000 metric tons per year (t/yr) of copper ore in the first quarter of 2011 from 120,000 t/yr in 2010. Cement output was estimated to have increased by about 11% because 2012 was the second full year of operation for Antea Cement Sh.A.'s cement plant after it was inaugurated in September 2010. Estimated cement production as a whole was still significantly lower than estimated cement production capacity. Crude oil production was estimated to have increased by 18% owing to increased production by Bankers Petroleum Ltd. and Stream Oil and Gas Ltd. In calendar year 2012, Bankers Petroleum produced 15,020 barrels per day (bbl/d) of crude petroleum, and in fiscal year 2012 (which ran from December 2011 through November 2012), Stream Oil and Gas produced an estimated 1,160 bbl/d of crude petroleum. The Government-owned firm Albpetrol Sh.a. may have had some unreported crude petroleum production, but it is believed that its active production facilities were operated by Bankers Petroleum or Stream Oil and Gas through production-sharing agreements (Antea Cement Sh.A., 2010; Stream Oil and Gas Ltd., 2012, p. 5; Bankers Petroleum Ltd., 2013, p. 3; Ekin Maden Ticaret ve Sanayi A.S., 2013).

Structure of the Mineral Industry

In table 2, the production capacity at Kurum International Sh.A.'s lime plant was increased to 72,000 t/yr from 36,000 t/yr. The increase was based on new information available on Kurum's Web site. It was not known if production increased along with the increase in capacity. Table 2 is a list of major mineral industry facilities (Kurum International Sh.A., 2013).

Outlook

Albania has the potential to increase its production of metals and mineral fuels as exploration projects continue and as foreign companies increase investments in Albania. Further investment will depend on economic and social stability as well as on the improvement of infrastructure.

References Cited

Antea Cement Sh.A., 2010, Antea Cement, one of the biggest industrial investments in Albania: Antea Cement Sh.A., September 13. (Accessed May 17, 2011, at http://www.anteacement.com/default.asp?entryID=45&siteID=1&pageid=15&tablepageid=3&langid=2.)

Bankers Petroleum Ltd., 2013, Management's discussion and analysis: Bankers Petroleum Ltd., March 13, 19 p. (Accessed June 27, 2013, at http://www.bankerspetroleum.com/en/components/investor/mda_ye_2012_-_3r8i9s3final.pdf.)

Ekin Maden Ticaret ve Sanayi A.S., 2013, Mining: Ekin Maden Ticaret ve Sanayi A.S. (Accessed July 18, 2013, at http://www.ekinmaden.com/.)

Institute of Statistics of the Republic of Albania, 2013, Gross domestic product by economic activities (1996–2011 at current prices): Institute of Statistics of the Republic of Albania. (Accessed June 27, 2013, at http://www.instat.gov.al/.)

International Monetary Fund, 2013, World economic outlook: International Monetary Fund, April, 184 p. (Accessed June 27, 2013, at http://www.imf.org/external/pubs/ft/weo/2013/01/pdf/text.pdf.)

Kurum International Sh.A., 2013, Kurum International Sh.A.: Kurum International Sh.A., undated. (Accessed May 7, 2013, at http://www.kurumholding.com.tr/default.asp?PG=EN02030201.)

Stream Oil and Gas Ltd., 2012, Management's discussion and analysis— Year ended November 30, 2012: Stream Oil and Gas Ltd., 19 p. (Accessed June 27, 2013, at streamoilandgas.com/_resources/financials/SKO_MDA_YE_2012_28Mar13.pdf.)

TABLE 1
ALBANIA: PRODUCTION OF MINERAL COMMODITIES[1]

(Metric tons unless otherwise specified)

Commodity[2]		2008	2009	2010	2011	2012 [e]
METALS						
Chromite, gross weight (18% to 42% Cr_2O_3)		225,373 [r]	283,558 [r]	328,322 [r]	330,938 [r]	330,000
Copper:						
Ore, gross weight		105,000	114,286	139,926 [r]	305,284 [r]	305,000
Cu content or ore (1.6% average)[e]		1,700 [r]	1,800 [r]	2,200 [r]	4,900 [r]	4,900
Iron and steel:						
Metal, ferroalloys, ferrochromium		11,916	7,556	23,233 [r]	28,694 [r]	29,000
Steel:						
Crude steel, secondary		380,000	440,000	390,000 [r,e]	463,620 [r]	500,000
Rolled steel		194,000	216,000	235,882 [r]	295,333 [r]	300,000
Nickel, ore:						
Iron-nickel and nickel-silicate ores		353,290	68,840 [r]	269,300 [r]	270,000 [r,e]	270,000
Ni content of ores[e]		4,000	1,000	3,000	3,000	3,000
INDUSTRIAL MINERALS						
Cement, hydraulic	thousand metric tons	918	1,110	1,300	1,800	2,000
Clay, kaolin	do.	649 [r]	796 [r]	795 [r]	974 [r]	1,000
Gypsum		87,261 [r]	71,276 [r]	77,400 [r]	80,000 [e]	80,000
Lime[e]		25,000	25,000	25,000	25,000	25,000
Limestone	cubic meters	3,837,529	3,271,617 [r]	2,363,445 [r]	2,400,000 [r,e]	2,400,000
Silica sand[e]	do.	12,077 [3]	12,000	12,000	12,000	12,000
MINERAL FUELS AND RELATED MATERIALS						
Coal, lignite		1,500	2,000 [r]	2,500 [r]	1,200 [r]	1,200
Gas, natural, gross production[e]	thousand cubic meters	6,200 [3]	7,900 [3]	8,000	8,000	8,000
Petroleum:						
Crude[e]	thousand 42-gallon barrels	4,000	4,200	5,000	5,000 [r]	5,900
Refinery products:						
Bitumen		92,000	80,000 [r,e]	32,600 [r]	44,000 [r]	44,000
Coke[e]		47,000 [3]	40,000 [r]	20,000 [r]	20,000 [r]	20,000
Fuels[e]	thousand 42-gallon barrels	1,400	1,200 [r]	600 [r]	600 [r]	600

[e]Estimated; estimated data are rounded to no more than three significant digits. [r]Revised. do. Ditto.
[1]Table includes data available through June 27, 2013.
[2]In addition to the commodities listed, a variety of industrial minerals and construction materials (common clay, dolomite, lime, olivinite, salt, sand and gravel, and stone) are believed to have been produced, but output is not reported quantitatively, and available information is inadequate to make reliable estimates of output. Also, a small amount of bauxite may have been produced.
[3]Reported figure.

TABLE 2
ALBANIA: STRUCTURE OF THE MINERAL INDUSTRY IN 2012

(Thousand metric tons unless otherwise specified)

Commodity		Major operating companies	Location of main facilities	Annual capacity[e]
Cement		Fushe Kruje Cement Factory, Sh.p.k. (Seament Holding)	Fushe Kruje, 35 kilometers north of Tirana	1,320
Do.		Elbasan Cement Factory (Seament Holding)	Elbasan, 32 kilometers southeast of Tirana	775
Do.		Antea Cement Sh.A. (Titan Cement S.A.)	Kruje, 35 kilometers north of Tirana	1,500
Do.		Colacem Albania Sh.p.k. (Colacem S.p.A.)	Balldre, northwestern Albania	NA[1]
Chromite		Albanian Chrome, Sh.p.k. (ACR) (DCM DECOmetal, 100%)	Mine in Bulquize, 40 kilometers northwest of Tirana	85
Do.		do.	Pogradec (including Katjiel and Pojske Mines)	NA
Do.		Numerous small producers	Mostly concentrated near Bulquize	NA
Copper:				
Mine		Beralb Sh.a. (Ekin Maden Ticaret ve Sanayi A.S.)	Munella Mine, 25 kilometers from Fushe-Arrez	300
Do.		do.	Lakrosh Mine, 14 kilometers from Fushe-Arrez	150
Concentrate		do.	Fushe-Arrez Flotation Plant	50
Ferrochromium		Albanian Chrome Sh.p.k. (ACR) (DCM DECOmetal, 100%)	Elbasan, 32 kilometers southeast of Tirana	36
Lime		Kurum International Sh.A.	Elbasan	72
Natural gas	million cubic meters	Albpetrol Sh.a. (Government owned)	Gasfields in southwestern Albania	NA
Do.	do.	Stream Oil and Gas Ltd.	Delvina gasfield in southern Albania	NA[2]
Petroleum:				
Crude	42-gallon barrels per day	do.	Ballsh-Hekal, Cakran-Mollaj, and Gorischt-Kocul oilfields in southwestern Albania	1,160[3]
Do.	do.	Bankers Petroleum Ltd.	Oilfields at Kucova and Patos Marinza, east of Fier in south-central Albania	15,000[3]
Refined	do.	Albanian Refining and Marketing Organization Sh.a. (ARMO)	Refineries at Ballsh and Fier	NA
Steel		Kurum International Sh.A.	Electric arc furnace plant at Elbasan	600

[e]Estimated. Do., do. Ditto. NA Not available.

[1]Construction of the Colacem Albania Sh.p.k. cement plant reportedly began in 2009, but it was not known if construction was completed or if production began at the plant.

[2]Stream Oil and Gas Ltd. did not produce any natural gas in 2012.

[3]Annual capacity estimated based on 2011 production.

THE MINERAL INDUSTRY OF ARMENIA

By Elena Safirova

Armenia ranked seventh in the world in mine output of molybdenum in 2012. Besides molybdenum, Armenia produced such other metals as copper, gold, silver, and zinc, and industrial minerals and products thereof, which included cement, diatomite, gypsum, limestone, and perlite. The country also produced aluminum foil from aluminum imported from Russia, and ferromolybdenum, molybdenum metal, and rhenium salts (ammonium perrhenate and potassium perrhenate) from local ores. It also had developed a diamond-cutting industry based on imported diamond. Armenia possesses resources of copper, gold, iron ore, lead, molybdenum, and zinc. It also has resources of construction materials, such as basalt, granite, limestone, marble, and tuff; semiprecious stones, such as agate, jasper, and obsidian; and other nonmetallic minerals, such as bentonite, diatomite, perlite, and zeolites (Arm3a.org, 2013; Polyak, 2013).

The country had almost no domestic fuel production; most domestically produced electricity was generated by one nuclear powerplant and several hydroelectric powerplants. Armenia imported uranium fuel for its nuclear powerplant and natural gas from Russia. Armenia also had been receiving natural gas from Iran through a direct pipeline (built in 2009) between the two countries and, since 2006, liquefied natural gas (LNG) that was transported in tanker trucks. In 2012, Armenia and Iran were in the process of building a new oil pipeline between the two counties that was expected to be completed in 2014 (Black Sea-Kavkaz.org, 2012; Mineral.ru, 2012a).

Minerals in the National Economy

In 2012, Armenia's real gross domestic product (GDP) increased by 7.2% compared with an increase of 4.7% in 2011. The nominal GDP in 2012 amounted to $10.1 billion.[1] The share of industrial production in the total GDP was 37.7%, and the share of the mineral industry in total industrial production was 17.0%. In 2012, industrial production increased by 8.8% in real terms compared with that of 2011. In 2011 (the latest year for which the data were available), mining of metallic ores dominated the mining and quarrying sector, accounting for 97.3% of the value of production in this sector (National Statistical Service of the Republic of Armenia, 2012; U.S. Central Intelligence Agency, 2013).

Government Policies and Programs

In November 2011, Armenia's Parliament adopted a new mining code and a package of accompanying bills that were to go into force in the beginning of 2012. The mining code regulates use of all mineral resources except oil and gas, radioactive materials, and fresh underground water. The new code contains major regulations concerning the use of mineral resources, exploration and mining, the procedures for obtaining mining and exploration licenses, and the environmental responsibility of companies. One of the accompanying bills passed by the Parliament outlines environmental fees and fees for the use of mineral resources; in particular, it replaces the fees paid for the use of mineral resources with royalty payments. Under the new law, metals mining is taxed based on the metal content of the produced concentrates as opposed to the gross volume of the extracted ore, which was used as a taxation base previously. The new code also establishes the criteria for issuing mining licenses. Under the new code, the mining project description submitted for obtaining a license must contain a plan for environmental monitoring, measures for social development of the local community, and a plan for projected closure of the mine and the demolition of the equipment when the resources are depleted. During mine exploitation, the mining companies are to make payments to a restoration fund, which is to be used to restore the land after the mine closure. The new code also modifies the royalty computation scheme. The new formula for computing royalties assumes a floor level of 4% and increases when profits constitute a higher percentage of total sales. The companies expected the total royalties to reach between 10% and 12% of the metal content value. In the first quarter of 2012, the total amount of royalties collected by the Government was 6.3 billion drams (about $16 million), which was 2.3 times higher than the amount collected in 2011 (Arka.am, 2011; News.am, 2011, 2012).

The Government expected that the new code would stimulate production by the mineral industry and result in Government tax revenues in 2013 that would be about 3.5 billion drams ($8.8 million) higher than in 2011. Armenian environmentalists, on the other hand, were against the new code because it does not prohibit mining in the forests and other lands of national significance, does not specify how the mining companies are to treat mine tailings, and changes the taxation mechanism so that the companies do not pay any taxes on produced tailings and mining waste (Manukyan, 2011).

Mineral Trade

In 2012, the country's exports, which were valued at $1.38 billion, were much lower than the country's imports of $4.26 billion. Mineral commodities contributed a significant share of the country's export revenue. The main export commodities were diamond, energy (electric power), foodstuffs, nonferrous metals, pig iron, unwrought copper, and other mineral products. Overall, exports of industrial minerals accounted for $403 million, or 29.2% of the country's export revenue; ferrous and nonferrous metals and articles made out of them accounted for $342 million (24.8%); and precious

---
[1]Where necessary, values have been converted from Armenian drams (AMD) to U.S. dollars (US$) at an annual average exchange rate of AMD0.002489=US$1.00 for 2012, AMD0.0026845=US$1.00 for 2011, and to European euros (€) at an average annual exchange rate of AMD0.0019363=1€ for 2012.

metals and precious stones accounted for $173 million (12.5%). The main export partners of Armenia were Russia (which accounted for 20.2% of Armenia's export revenue), Bulgaria (9.4%), Belgium (9.2%), Iran (7.9%), Germany (7.6%), the United States (6.3%), Canada (6.2%), the Netherlands (5.8), Switzerland (5.2%), and Georgia (5.1%). In 2012, Armenia's imports of mineral products included rough diamond, natural gas, and petroleum. The main trade partners for imports were Russia (which provided 24.9%, by value, of Armenia's imports), China (9.4%), Germany (6.2%), Iran (5.1%), Turkey (5.0%), Italy (4.0%), and the United States (3.6%) (National Statistical Service of the Republic of Armenia, 2012; U.S. Central Intelligence Agency, 2013).

Production

In 2012, Armenia produced 497% more bentonite than it produced in 2011. Production of molybdenum metal increased by 39%; that of molybdenum concentrate, by 35%; caustic soda, by 30%; zinc concentrate, by 26%; copper concentrate, by about 23%; and copper metal, by about 13.5%. At the same time, production of salt decreased by 21%; estimated production of rhenium, by 12.5%; and silver, by 12%. Data on other mineral production are in table 1.

Commodity Review

Metals

Aluminum.—The ARMENAL aluminum foil rolling mill was one of the leading production facilities in Armenia and the only producer of aluminum foil in the Caucasus and Central Asia regions. It was formed from the Kanaker aluminum plant in Yerevan in 2000 and became a part of United Company RUSAL's packaging division. In 2012, ARMENAL produced 26,264 metric tons (t) of aluminum foil, which was an increase of 3.8% compared with the 2011 production volume. In 2012, the plant fulfilled about 1,800 export orders, which was a 2.7% increase compared with the number of orders filled in 2011. During the year, worker productivity at ARMENAL increased by 5.5%. In 2012, the plant paid 1,641 million drams (about $4.1 million) in taxes, which was a 15% increase compared with that of 2011. The plant employed about 700 workers at an average monthly wage of 260,000 drams ($650), which was about 2.2 times higher than the average wage in Armenia. The plant had several small efficiency-improving projects underway and was planning to increase production to between 30,000 and 31,000 metric tons per year (t/yr) of aluminum foil in the near future (MetalDaily.ru, 2013a; United Company RUSAL, 2013).

Copper and Molybdenum.—The leading producer of copper and molybdenum concentrates in Armenia was the Zangezur copper-molybdenum complex (ZCMC) followed by the Agarak copper-molybdenum mining and processing complex (ACMC). The ZCMC was developing the Kajaran deposit, which had resources that were estimated to be 4.5 million metric tons (Mt) of copper and 722,000 t of molybdenum. The owners of ZCMC in 2012 were Cronimet Mining GmbH of Germany, which owned 60% of the shares; OAO Pure Iron Plant (15%); and Armenian Molybdenum Production LLC (AMP) and Zangezur Mining LLC (12.5% each). Since privatization in 2004, the company had invested about $300 million in production facilities, and another $150 million was spent on social projects. As of 2012, ZCMC was processing about 17 million metric tons per year (Mt/yr) of ore and was planning to increase its annual processing capacity to 25 Mt/yr in the next several years (Aniarmenia.ru, 2012; Regnum.ru, 2012f).

The ACMC included the Agarak Mine and a beneficiation plant. The ACMC was acquired by GeoProMining, Ltd. (GPM) of Russia in 2007 and had been undergoing expansion and modernization since then. In April 2012, ACMC announced that it had completed construction of the new beneficiation plant and acquired 19 new production vehicles. The new plant was equipped with new flotation machines that would increase the annual capacity by about 20% to 3.5 Mt/yr and would increase the company's ability to extract metals by between 3% and 5% (Mineral.ru, 2012b; Yerkramas.org, 2012; GeoProMining, Ltd., 2013).

In 2012, ZAO Teghut was preparing for production of copper and molybdenum at the Teghut deposit, which was scheduled to start in 2014. The Teghut copper-molybdenum deposit was the second largest copper-molybdenum deposit in Armenia; its resources were estimated to be 1.6 Mt of copper and 100,000 t of molybdenum. The company expected to mine between 8 and 12 Mt/yr of ore and to produce more than 70,000 t of copper concentrates with between 28% and 30% copper content and 1,200 t of molybdenum concentrates with 50% molybdenum content. The Vallex Group, of which ZAO Teghut was a part, was planning to invest a total of $320 million in the project, and, as of March 2012, the company had already invested $130 million. The Vallex Group also included ZAO Armenian Copper Programme (ACP), which managed the copper smelter in Alaverdi, and Base Metals Co., which operated the Drmbon gold-copper deposit in Nagorno-Karabakh. Environmental activists in Armenia and abroad were demanding that the Government annul the Teghut project because it posed a threat to the Teghut Forest. Although the company offered to plant trees in the areas adjacent to the forested area that would be cut, the ecologists responded that this measure would jeopardize biodiversity because the Teghut Forest is home to 55 species of mammals (21 of which are endangered), 86 species of birds, and 191 types of plants (9 of which are endangered). By yearend, it was still unclear whether or not the construction of the Teghut Mine would proceed as planned (table 2; MinerJob.ru, 2012c; Regnum.ru, 2012e, g).

In March 2012, the Vallex Group signed an agreement with the Government of Nagorno-Karabakh regarding the mining of another significant copper-molybdenum deposit, the Kashen deposit, which is located in the Martakert region, 15 kilometers (km) east of the already exploited Drmbon Mine. The Kashen deposit reportedly contained about 41 Mt of ore. About 1.6 Mt/yr of ore from the Kashen Mine was expected to be processed. The Vallex Group was planning to invest about $80 million in the project, and the mine was expected to open in 2015 (Regnum.ru, 2012c, d).

ACP, which was a part of the Vallex Group, was an Armenian producer of crude copper metal. ACP processed copper concentrates from ZAO Base Metals, which was located in Nagorno-Karabakh, and from the ZCMC. In 2012, ACP increased its crude copper metal production by 13.5% to 10,075 t. ACP's entire output was exported to Europe (Express.am, 2013a).

OAO Pure Iron Plant (PIP), which was located in Yerevan, was the major producer of ferromolybdenum and molybdenum metal in Armenia. The plant processed mostly molybdenum concentrate produced by the ACMC. Cronimet Mining owned 51% of the PIP, and the other 49% was owned by the residents of Armenia. In 2012, the plant reduced its ferromolybdenum production slightly by 0.3% to 2,997.7 t and increased its production of molybdenum metal by 38.8% to 675 t. In 2012, the company produced ferromolybdenum metal valued at 22.1 billion drams (about $55 million) and exported most of it, for 22.01 billion drams (about $54.8 million). The production of molybdenum metal was valued at 7.6 billion drams (about $18.9 million), and the company exported all of it. Cronimet Mining was the company involved in the exporting of molybdenum products produced at the PIP. The PIP employed about 600 people, and the average monthly wage rate was 220,000 drams (about $548). The plant was planning to increase its production of ferromolybdenum in 2013 to 3,200 t, and that of molybdenum metal, to 750 t (MetalDaily.ru, 2011; Express.am, 2013b; News.am, 2013).

AMP was another leading molybdenum producer in Armenia; the company was also located in Yerevan and processed copper concentrate produced at the ZCMC. In 2012, AMP produced 2,840 t of ferromolybdenum valued at 20.1 billion drams ($50 million). In addition to molybdenum, the company produced ammonium perrhenate with a rhenium content of 69.4%. AMP was planning to open a new hydrometallurgical plant in the summer of 2013 that would produce molybdenum metal. The company's investment in the new plant would exceed 1 billion drams (about $2.5 million). AMP employed about 500 workers at an average wage rate of 208,000 drams per month (about $518 per month) (Newsarmenia.ru, 2013; Versia.am, 2013).

Gold.—The Ararat Gold Recovery Co. (AGRC), which was also known as GPM Gold, continued to mine the Sotk (Zod) deposit. As of 2011, AGRC was a subsidiary of GPM. AGRC had a gold processing facility in the city of Ararat. In October, the company announced that it had invested $45 million in technical modernization of the Sotk Mine and completed the first stage of innovations. These included acquisition of 30 units of mining equipment and remodeling of buildings used for vehicle maintenance and machine oil storage. The next step in technical modernization would cost a total of $100 million and include application of the Albion Process™ technology and completion of a new beneficiation plant in Ararat City. The Albion Process™ was initially developed by Xstrata plc of the United Kingdom and was first used to treat refractory gold and silver concentrates in the Dominican Republic; Armenia would be the second place in the world where the technology is applied to treat refractory gold. The main benefit of the Albion Process™ is that it allows processing of a variety of ore types. The ores found in Sotk are of the sulfide type, and it is usually more expensive to extract gold from such ores.

GPM expected to increase annual gold production to 3.7 t in 2013 and 4.6 t in 2015 from 1.3 t in 2011 (MinerJob.ru, 2012b; GeoProMining, Ltd., 2013).

In 2012, ZAO Geoteam of Armenia, which represented Lydian International Ltd. of the United Kingdom, invested $52 million in exploration and preparation for development of the Amulsar gold deposit located in the south of Armenia. The company had already invested $30 million in exploration between 2006 and 2012. Geoteam expected that the company's total investment in exploration and development would reach $250 million and that another $200 million would be invested in mining. According to company data, the deposit contained about 2.4 million troy ounces (75 t) of gold and 9.6 million troy ounces (300 t) of silver. The company planned to conduct open pit mining and recovery using the heap-leach method; it was planning to build a processing plant and to start mining in 2015. The expected life of the mine was 16 years. It was expected that, in the mining stage, Geoteam would pay about $86 million in taxes, which would be used to support the local economy (Mineral.ru, 2012c; MinerJob.ru, 2012a).

Residents of the nearby city of Jemruk were concerned about possible harmful effects of the future Amulsar Mine, not only on the environment, but on the livelihoods of the people of the area as well. The city of Jemruk is a mineral water resort, and the proximity of the mine could damage the water quality and reduce the city's attractiveness to tourists. Geoteam responded to those environmental concerns by stating that the Amulsar deposit is located 13 km from the city, that the deposit does not contain significant amounts of either uranium or lead, and that the ore processing was unable to harm Sevan Lake, which is considered a natural treasure of Armenia. The mining and metallurgical complex was planned to be located more than 4 km from Spandaryan Lake, from which it was thought pollutants could leak into Sevan Lake. Moreover, the company conducted simulations showing that all particles that would be emitted during explosions used during the mining stage would settle within a 1-km radius of the mine, which would be far away from Jemruk city. As of 2012, the Ministry of Environmental Protection was continuing to consider the environmental safety aspects of the Amulsar project (Arminfo.am, 2012; Mineral.ru, 2012d: Regnum.ru, 2012b).

Another company mining gold in Armenia was ZAO Dundee Precious Metals Kapan (DPMK) (formerly known as ZAO Deno Gold Mining), which was a subsidiary of Dundee Precious Metals Inc. of Canada. DPMK was mining the Shahumian deposit located in the Kapan ore field. In 2012, the company continued work to assess the total resources of the deposit. In addition to gold, the ore contained copper, silver, and zinc. The company was producing two concentrates—copper-silver-gold and zinc-silver-gold. In 2012, DPMK reduced gold production by 19%, to 679 kilograms (kg) from 836 kg in 2011. Silver production decreased by 13.5% to 13,968 kg; copper, by 18% to 1,357 t; and zinc, by 22% to 6,996 t. In 2012, the company's sales decreased by 26% to $54 million and the gross profits plummeted to $3.4 million from $25.7 million in 2011. In 2008, independent consulting company Coffey International Ltd. of Australia had produced a resource estimate based on exploration results from the Soviet period and limited drilling work conducted previously by Deno Gold Mining. According to Coffey's report, the probable resource with gold content between 0 and 2.0 grams per metric ton (g/t) was estimated to be 1,836 t, and the probable resource with gold content between 1.0 and 2.0 g/t was estimated to be 400 t. Gold content ranges were between 0.37 and 1.68 g/t; silver, between 6.5 and 20.87 g/t; copper, between 0.09% and 0.19%; and zinc, between 0.32% and 0.83%. To obtain a more precise estimate of the deposit's resources, DPMK decided to conduct drilling of the entire deposit. The company was planning to complete the new estimate of the deposit's resources by the end of 2012, but it was not known if this task was completed on schedule (MinerJob.ru, 2012a; Dundee Precious Metals Inc., 2013; Erndjakyan, 2013).

Iron and Steel.—A new metallurgical plant was under construction in the city of Charentsavan in the Kotaiks region, which is located 25 km from Yerevan, and it would be the first such plant in Armenia. The construction was being done by ASCE Group Armenia, and the equipment was expected to be supplied by Siemens VAI branches located in Austria, Italy, and Russia; the Russian branch was to provide automation of the two electric arc furnaces already owned by the ASCE Group. The new plant was expected to produce steel-reinforcing bar (rebar) from metal scrap collected in Armenia. The rolling mill of the plant would have the capacity to produce 125,000 t/yr of steel. The total cost of construction was estimated to be €4 million (about $5 million), and the cost of equipment would amount to another €20 million (about $25 million). ZAO Ameriabank of Armenia was financing $15 million of the total cost of the project, and, in November 2011, the Government allowed ASCE Group to delay payment of the value-added tax on imported equipment for up to 3 years. The total area of the plant was planned to be 100,000 square meters, and it was expected to employ about 300 workers. The project was scheduled to be completed in 2013 (Regnum.ru, 2012a; MetalDaily.ru, 2013b).

Mineral Fuels and Related Materials

Uranium.—In 2012, ZAO Armenian Russian Mining Co. was continuing exploration for uranium in Syunik Marz and Vayots Dzors Marz Provinces. The company, a joint enterprise of the Government of Armenia and the Government of Russia, was created in 2008, with the goal of exploration for and subsequent mining of uranium in Armenia. Initially, Russia invested $3 million that was expected to fund exploratory drilling in Syunik Marz. According to various preliminary estimates, uranium resources in Armenia were thought to be between 30,000 and 100,000 t. In 2010, residents of the Syunik region (led by environmental activists) held demonstrations against the uranium program in Armenia because of environmental concerns (Kavkaz-uzel.ru, 2010; Zarobyan, 2010).

Outlook

In the next few years, Armenia is likely to continue with its development of facilities used to process copper, gold, and molybdenum. Several new gold and iron ore mining projects were underway in 2012 and could be operational in the next 3 to 5 years. The mining code that went into effect in the beginning of 2012 is likely to make the mineral industry of Armenia more attractive to potential investors. At the same time, many residents of the country appear concerned about the potential harm mining projects could do to environmentally sensitive areas in the country. Over time, the resolution of tensions between mining companies and environmental activists could become more difficult. Success of mining and mineral processing projects in Armenia will depend on the ability of the Government to provide a solid legal basis for reconciling often contradictory goals of economic development and environmental protection.

References Cited

Aniarmenia.ru, 2012, V 2012 godu "Zangezurskiy medno-molibdenovyy kombinat" planiruet uvelichit' proizvodstvo mednogo kontsentrata na 25% [In 2012, Zangezur copper-molybdenum complex is planning to increase production of copper concentrate by 25%]: Aniarmenia.ru, March 14. (Accessed August 23, 2013, at http://aniarmenia.ru/novosti/a-w-news/economics/331-ec126?tmpl=component&print=1&layout=default&page=.)

Arka.am, 2011, Parlament Armenii prinyal kodeks "O Nedrakh", nesmotrya na protest ekologov [Armenia's Parliament adopted the Mining Code, despite the protests of ecologists]: Arka.am, November 28. (Accessed August 23, 2013, at http://www.arka.am/ru/news/politics/29383.)

Arm3a.org, 2013, Kratkaya informatsia ob Armenii [Brief information about Armenia]: Arm3a.org. (Accessed August 23, 2013, at http://www.arm3a.org/?lang=ru&go=about_armenia.)

Arminfo.am, 2012, Kompaniya "Geoteam" gotova I dal'she vypolnyat' obyazatel'stva po provedeniyu ekspertiz po Amulsarskomu zolotinisnomu mestorozhdeniyu [Geoteam company is ready to continue fulfilling its obligations related to examination of the Amulsar gold deposit]: Arminfo.am, August 10. (Accessed August 23, 2013, at http://www.arminfo.am/index.cfm?objectid=E388DA50-E2CF-11E1-A0DFF6327207157C.)

Black Sea-Kavkaz.org, 2012, Armyanskaya energetika: perspektivy I regional'naya rol' [Armenian Energy—perspectives and regional role]: Black Sea-Kavkaz.org, August. (Accessed August 23, 2013, at http://bs-kavkaz.org/2012/08/armyanskaya-energetika-perspektivy-i-regionalnaya-rol/.)

Dundee Precious Metals Inc., 2013, Operations—Producing mines—Kapan: Dundee Precious Metals Inc. (Accessed August 23, 2013, at http://www.dundeeprecious.com/English/operations/producing-mines/kapan/default.aspx.)

Erndjakyan, Artur, 2013, "Dandi Prishs Metals Kapan" (byvshaya Deno Gold Mining) sokratila proizvodstvo zolota v 2012 godu na 18,8% [In 2012, Dundee Precious Metals Kapan (formerly Deno Gold Mining) reduced gold production by 18.8%]: Arminfo.am, July 11. (Accessed August 23, 2013, at http://www.arminfo.am/index.cfm?objectid=EB4CC960-EA1F-11E2-A9410EB7C0D21663.)

Express.am, 2013a, Metallurgicheskaya kompaniya ACP uvelichila proizvodstvo chernovoy medi [ACP metal producer increased production of primary copper]: Express.am, July 23. (Accessed August 23, 2013, at http://www.express.am/news/view/metallurgicheskaj%D0%B0-kompanij%D0%B0-asp-uvelichila-proizvodstvo-chernovoi-medi.html.)

Express.am, 2013b, Zavod "Chistoe Zhelezo" uvelichil proizvodstvo molibdena [Pure Iron Plant increased its molybdenum production]: Express.am, January 30. (Accessed August 23, 2013, at http://www.express.am/news/view/zavod-chistoe-zhelezo-uvelichil-proizvodstvo-molibdena.html.)

GeoProMining, Ltd., 2013, Armenia: GeoProMining, Ltd. (Accessed August 23, 2013, at http://www.geopromining.com/en/our-business/geography/armenia/.)

Kavkaz-uzel.ru, 2010, Zhiteli Syunikskoy oblasti Armenii mitinguyut protiv razrabotki uranovykh rudnikov [Residents of Syunik Oblast' of Armenia protest against development of uranium mines]: Kavkaz-uzel.ru, November 10. (Accessed August 23, 2013, at http://www.kavkaz-uzel.ru/articles/176864.)

Manukyan, Karina, 2011, Akop Sanasaryan: Kodeks "O Nedrakh" vnov' stolknul interesy chinovnikov I zelenykh aktivistov [Akop Sanasaryan—Mining Code again invokes contradictions between Government officials and green activists]: Arminfo.am, November 22. (Accessed August 23, 2013, at http://arminfo.info/russian/interview/article/22-11-2011/09-50-00.)

MetalDaily.ru, 2011, V 2011 godu "Erevanskiy zavod chistogo zheleza" proizvel 3 tysyachi ton ferromolibdena [In 2011, Yerevan Pure Iron Plant produced 3 thousand tons of ferromolybdenum]: MetalDaily.ru, December 22. (Accessed August 23, 2013, at http://www.metaldaily.ru/news/news61321.html.)

MetalDaily.ru, 2013a, Armyanskaya "dochka" "Rusala" v 2012 godu uvelichila proizvodstvo fol'gi na 3,8% [In 2012, Armenian subsidiary of RUSAL increased production by 3.8%]: MetalDaily.ru, February 11. (Accessed August 23, 2013, at http://www.metaldaily.ru/news/news70642.html.)

MetalDaily.ru, 2013b, Pervyy v Armenii metallurgicheskiy zavod budet ukomplektovan oborudovaniem Siemens [The first Armenian steel mill will be equipped by Siemens]: MetalDaily.ru, July 13. (Accessed August 23, 2013, at http://www.metaldaily.ru/news/news65977.html.)

Mineral.ru, 2012a, Armeniya i Iran stroyat nefteprovod [Armenia and Iran are building an oil pipeline]: Mineral.ru, April 14. (Accessed August 23, 2013, at http://www.mineral.ru/News/48285.html.)

Mineral.ru, 2012b, GeoProMining podgotovila tekhnicheskuyu bazu dlya vnedreniya tekhnologii "Al'bion" [GeoProMining prepared technical basis for using the Albion technology]: Mineral.ru, October 17. (Accessed August 23, 2013, at http://www.mineral.ru/News/50502.html.)

Mineral.ru, 2012c, Lydian International narastila resursnuyu bazu proekta Amulsar [Lydian International increased resource base of the Amulsar project]: Mineral.ru, March 10. (Accessed August 23, 2013, at http://www.mineral.ru/News/52132.html.)

Mineral.ru, 2012d, Osvoeniyu zolotonosnogo mestorozhdeniya Amulsar (Armeniya) meshayut protest ekologov [Ecologist's protests impede development of the Amulsar gold deposit]: Mineral.ru, February 28. (Accessed August 23, 2013, at http://www.mineral.ru/News/47781.html.)

MinerJob.ru, 2012a, Dundee Precious Metals zavershit predvaritel'nuyu otsenku otkrytoi razrabotki Kapanskogo zoloto-polimetallicheskogo mestorozhdeniya v 3-em kvartale [Dundee Precious Metals will complete its preliminary estimate of the open pit development for Kapan gold-polymetallic deposit in the 3d quarter]: MinerJob, January 16. (Accessed August 23, 2013, at http://www.minerjob.ru/viewnew.php?id=22107.)

MinerJob.ru, 2012b, Kompaniya GeoProMining zavershila ocherednoy etap modernizatsii aktivov v Armenii [GeoProMining company completed the next step for modernization of its assets in Armenia]: MinerJob.ru, October 17. (Accessed August 23, 2013, at http://www.minerjob.ru/viewnew.php?id=22107.)

MinerJob.ru, 2012c, V Armenii v 2014 godu nachnutsya razrabotki Tegutskogo mestorozhdeniya [In Armenia, development of Teghut deposit will start in 2014]: MinerJob, March 30. (Accessed August 23, 2013, at http://www.minerjob.ru/viewnew.php?id=21068.)

National Statistical Service of the Republic of Armenia, 2012, Statistical yearbook of Armenia: National Statistical Service of the Republic of Armenia. (Accessed August 23, 2013, at http://www.armstat.am/en/?nid=45&year=2012.)

News.am, 2011, Parlament prinyal v tretyem i okonchatel'nom chtenii kodeks "O Nedrakh" [The Parliament of Armenia adopted the third and final version of the Mining Code]: News.am, November 28. (Accessed August 23, 2013, at http://news.am/rus/news/83595.html.)

News.am, 2012, Novyi nalog na gornorudnuyu promyshlennost' dal pervye rezul'taty – rost postupleniy v 2,3 raza [The new mining tax brought the first results—Revenue increased 2.3 times]: News.am, May 1. (Accessed August 23, 2013, at http://news.am/rus/news/103510.html.)

News.am, 2013, Yerevanskiy zavod chistogo zheleza planiruet proizvodstvo sverkhchistogo molibdena [Yerevan Pure Iron plant plans to produce superpure molybdenum]: News.am, February 8. (Accessed August 23, 2013, at http://news.am/rus/news/139430.html.)

Newsarmenia.ru, 2013, Novyi gidrometallurgicheskiy tsekh po proizvodstvu chistogo molibdena budet otkryt letom v Armenii [In Armenia, a new hydrometallurgical plant producing pure molybdenum will open in summer]: Newsarmenia.ru, April 8. (Accessed August 23, 2013, at http://www.newsarmenia.ru/economy/20110408/42430121.html.)

Polyak, D.E., 2013, Molybdenum: U.S. Geological Survey Mineral Commodity Summaries 2013, p. 106–107.

Regnum.ru, 2012a, "Ameriabank" profinansiruet stroitel'stvo staleprokatnogo proizvodstva v armyanskom Charentsavane [Ameriabank will finance construction of rolled steel production in Armenia Charenzavan]: Regnum.ru, April 5. (Accessed August 23, 2013, at http://www.regnum.ru/news/1517756.html.)

Regnum.ru, 2012b, "Geoteam" schitaet opaseniya ekologov v svyazi s razrabotkoy Amulsarskogo mestorozhdeniya absurdnymi [Geoteam thinks that the concerns of ecologists related to Amulsar deposit development are absurd]: Regnum.ru, November 30. (Accessed August 23, 2013, at http://www.regnum.ru/news/1472889.html.)

Regnum.ru, 2012c, Ekspluatatsiya Kashenskogo rudnika v Karabakhe: plyusy I minusy [Exploitation of Kashen Mine in Karabakh—Pros and cons]: Regnum.ru, July 28. (Accessed August 23, 2013, at http://www.regnum.ru/news/1556179.html.)

Regnum.ru, 2012d, Kompaniya "Valleks" investiruet v razrabotku Kashenskogo mestorozhdeniya v Karabakhe $80 mln [Vallex invests $80 million in development of Kashen Mine in Karabakh]: Regnum.ru, June 26. (Accessed August 23, 2013, at http://www.regnum.ru/news/1545517.html.)

Regnum.ru, 2012e, Konstitutsionnyy sud v Armenii rassmotrel vopros Tekhutskogo mestorozhdeniya [Constitutional court of Armenia considered the case of Teghut deposit]: Regnum.ru, August 23. (Accessed August 23, 2013, at http://www.regnum.ru/news/1564267.html.)

Regnum.ru, 2012f, Modernizatsiya Agarakskogo kombinata oboshlas' GeoProMining v $30 mln – interv'yu (Armeniya) [Reconstruction of the Agarak complex cost GeoProMining $30 million—Interview (Armenia)]: Regnum.ru, April 18. (Accessed August 23, 2013, at http://www.regnum.ru/news/1521998.html.)

Regnum.ru, 2012g, WB and EBRD instsenirovali obsuzhdeniya po samomu krupnomu mestorozhdeniyu zolota v Zakavkaz'e [WB and EBRD pretended to discuss the largest gold deposit in Zakavkaz'e]: Regnum.ru, April 19. (Accessed August 23, 2013, at http://www.regnum.ru/news/1522485.html.)

United Company RUSAL, 2013, ARMENAL: United Company RUSAL. (Accessed August 23, 2013, at http://www.rusal.ru/en/about/9.aspx.)

U.S. Central Intelligence Agency, 2013, Armenia, in The world factbook: U.S. Central Intelligence Agency, July 10. (Accessed August 23, 2013, at https://www.cia.gov/library/publications/the-world-factbook/geos/am.html.)

Versia.am, 2013, Premier-ministr Armenii posetil erevanskuyu kompaniyu "Armenian molibden prodakshn" [Armenian Prime Minister visited Armenian molybdenum production company in Yerevan]: Versia.am, March 27. (Accessed August 23, 2013, at http://www.versia.am/economic-16.)

Yerkramas.org, 2012, Agarakskiy medno-molibdenovyy kombinat zavershil stroitel'stvo novogo tsekha [Agarak copper-molybdenum complex completed construction of a new plant]: Yerkramas.org, April 13. (Accessed August 23, 2013, at http://www.yerkramas.org/2012/04/13/agarakskij-medno-molibdenovyj-kombinat-zavershil-stroitelstvo-novogo-cexa/.)

Zarobyan, Nikita, 2010, Dobycha urana v Armenii – tsena riska vysoka [Uranium production in Armenia—Risk comes at a high price]: Yerkramas.org, March 22. (Accessed August 23, 2013, at http://www.yerkramas.org/2010/03/22/dobycha-urana-v-armenii-cena-riska-vysoka.)

TABLE 1
ARMENIA: PRODUCTION OF MINERAL COMMODITIES[1]

(Metric tons unless otherwise specified)

Commodity		2008	2009	2010	2011	2012
METALS						
Aluminum, foil		22,694	21,456	24,617	25,314 r	26,264
Copper:						
Concentrate, Cu content		18,800	23,233	31,062	33,597	41,220
Blister, smelter, primary		6,480	6,858	7,644	8,876	10,075
Ferromolybdenum		5,323	5,144	5,126	5,525	5,836
Gold, mine output, Au content	kilograms	1,359	944	974	2,736 r	2,896
Molybdenum:						
Concentrate, Mo content		4,472	4,365	4,335	4,817	6,500
Metal		520	500	469	486	675
Rhenium e	kilograms	400	400	400	400	350
Silver	do.	40,434	52,876	68,428	25,227 r	22,200
Zinc, concentrate, Zn content		4,200	3,800	7,808	8,475	10,700
INDUSTRIAL MINERALS						
Barite e		600	500	550	600	600
Caustic soda		4,476	1,138	960	63	82
Cement	thousand metric tons	770	467	488	422	438
Clays:						
Bentonite		40,000	38,000	1,397	835	4,987
Bentonite, powder		1,100	1,000	1,100	1,100	1,200
Diamond, cut	carats	100,945	49,573	65,000	65,000	67,000
Diatomite e		200	180	220	220	220
Gypsum		45,900	40,100	38,700	34,000	32,000
Limestone	thousand metric tons	18,000	15,000	18,000	18,000	17,500
Perlite e		35,000	35,000	95 r, 2	229 r, 2	181 [2]
Salt		37,300	29,400	29,400	35,600	38,000

e Estimated; estimated data are rounded to no more than three significant digits. r Revised. do. Ditto.

[1] Table includes data available through August 15, 2013.

[2] Reported figure.

TABLE 2
ARMENIA: STRUCTURE OF THE MINERAL INDUSTRY IN 2012[1]

(Metric tons unless otherwise specified)

Commodity		Major operating companies, main facilities, or deposits	Location or deposit name	Annual capacity[e]
Aluminum, rolled and foil		ARMENAL (formerly Kanaker aluminum plant) (United Company RUSAL)	Kanaker	25,000
Cement		Araratcement Factory CJSC	Ararat region	NA
Copper:				
Mine output, Cu content	million metric tons	Agarak copper-molybdenum mining and processing complex (ACMC) [GeoProMining, Ltd. (GPM)]	Agarak	3.5
Do.		Dundee Precious Metals Kapan (Dundee Precious Metals Inc.)	Kapan	NA
Do.	million metric tons	Zangezur copper-molybdenum complex (ZCMC) [Cronimet Mining GmbH, 60%; OAO Yerevan Pure Iron Plant, 15%; Armenian Molybdenum Production LLC (AMP), 12.5%; Zangezur Mining LLC, 12.5%]	Kajaran	17
Blister		ZAO Armenian Copper Programme (ACP) (Vallex Group)	Alaverdi	15,000
Diamond, cut stones		Aghavni diamond-cutting works[2]	Nor Geghi	NA
Do.		Amma group diamond-cutting works[2]	Artashat	NA
Do.		Andranik-Dashk diamond-cutting works	Nor Hachyn	NA
Do.		Arevakn diamond producing plant	do.	NA
Do.		Diamond Company of Armenia (DCA)	Yerevan	NA
Do.		Diamond Tech	Talin	NA
Do.		Lori diamond-cutting works	Nor Hachyn	NA
Do.		Lusampor[2]	Melik'gyugh	NA
Do.		Punji diamond-cutting works[2]	Yerevan	NA
Do.		Sapphire diamond-cutting works	Nor Hachyn	NA
Do.	thousand carats	Shoghakan gem-cutting plant	do.	120
Gold	kilograms	Ararat Gold Recovery Co. (AGRC) [GeoProMining, Ltd. (GPM)]	Sotk, Zod	2,000
Do.		Megradzor deposit	Meghradzor	NA
Do.		Lichkvazkoye, Shaumyanskiy Rayon, Sotkskoye, and Terterasarskoye deposits	NA	NA
Do.		ZAO Dundee Precious Metals Kapan (DPMK) (Dundee Precious Metals Inc.)	Shahumian deposit	NA
Iron ore		Hrazdan deposit	Hrazdan region	NA
Molybdenum:				
Mine output, Mo content		Agarak copper-molybdenum mining and processing complex (ACMC) [GeoProMining, Ltd. (GPM)]	Agarak	2,000
Do.		Zangezur copper-molybdenum complex (ZCMC) [Cronimet Mining GmbH, 60%; OAO Yerevan Pure Iron Plant, 15%; Armenian Molybdenum Production LLC (AMP), 12.5%; Zangezur Mining LLC, 12.5%]	Kajaran	20,400
Metal, ferromolybdenum		Armenian Molybdenum Production LLC (AMP) (Cronimet Mining GmbH, 51%, and Armenian residents, 49%)	NA	3,600
Do.		OAO Yerevan Pure Iron Plant	Yerevan	NA
Perlite	thousand metric tons	Aragats perlite mining-beneficiation complex	Aragats deposit	1,110
Rhenium		Agarak copper-molybdenum mining and processing complex (ACMC) [GeoProMining, Ltd. (GPM)]	Agarak	NA
Do.		Zangezur copper-molybdenum complex (ZCMC) [Cronimet Mining GmbH, 60%; OAO Yerevan Pure Iron Plant, 15%; Armenian Molybdenum Production LLC (AMP), 12.5%; Zangezur Mining LLC, 12.5%]	Kajaran	NA
Zinc, mine output, Zn content		Dundee Precious Metals Kapan (Dundee Precious Metals Inc.)	Kapan	NA

[e]Estimated; estimated data are rounded to no more than three significant digits. Do., do. Ditto. NA Not available.

[1]Many location names have changed since the breakup of the Soviet Union. Many enterprises, however, are still named or commonly referred to based on the former location name, which accounts for discrepancies in the names of enterprises and that of locations.

[2]Current existence of enterprise cannot be confirmed.

THE MINERAL INDUSTRY OF AUSTRIA

By Steven T. Anderson

During 2012, the Erzberg open pit iron ore mine at Eisenerz in the State of Styria and the underground tungsten mine at Mittersill in the State of Salzburg continued to be the only metal mines in operation in Austria. This was not the case with mining of industrial minerals, however, and the country still produced dolomite, gypsum, kaolin, lime, limestone, magnesite, salt, silica (quartz) sand, talc, and other industrial mineral products. Excluding production (if any) in the United States[1] in 2012, Austria was estimated to have been the fourth-ranked producer of magnesite in the world and the sixth-ranked producer of tungsten and to have accounted for 3.6% and about 1%, respectively, of the world's production. The country was also estimated to have accounted for approximately 1% of the world's production of natural gypsum in 2012 (including production by the United States). Both natural gas and crude petroleum were produced in Austria, although the latest year in which the country produced coal was 2006. Austria had a significant mineral (including metals) processing sector, as well as a significant capacity to recover metals from secondary sources (tables 1, 2; Crangle, 2013; Kramer, 2013; Shedd, 2013).

Minerals in the National Economy

According to preliminary data for 2012, the approximate value of the country's marketed mineral industry production was $34.5 billion[2] [about 8.7% of the gross domestic product (GDP)] compared with a revised value of $37.6 billion (9% of the GDP) in 2011. Of the total in 2012, the value of marketed production by the natural gas and petroleum sector (including production of petroleum refinery products) was $15.5 billion (about $16 billion in 2011); the ferrous metals sector (including manufacturing of iron and steel and possibly including production of ferroalloys), about $9.2 billion (about $10.4 billion in 2011); the nonferrous metals sector, about $5 billion ($5.9 billion in 2011); the building materials, ceramics, and stone sector, $4.1 billion (a revised figure of $4.6 billion in 2011); and the mining and quarrying sector, about $0.6 billion ($0.7 billion in 2011). According to an index of the real value of production with a base year of 2010, the real value of production by the nonferrous metals sector decreased by 8.9% in 2012 compared with that of 2011; that of the building materials, ceramics, and stone sector, by 5.7%; the mining and quarrying sector, by 5%; the natural gas and petroleum sector, by 4.8%; and t the ferrous metals sector, by 1.9% (International Monetary Fund, 2013; Österreichisches Institut für Wirtschaftsforschung, 2013a, c; Wirtschaftskammer Österreich, 2013, p. 50).

In 2012, there were 41,313 employees in the mineral industry, and they accounted for about 1.2% of the total number of employees in the country compared with 40,220 and about 1.2%, respectively, in 2011. During 2012, average employment in the ferrous metals sector was about 14,200 workers (14,100 in 2011); in the building materials, ceramics, and stone sector, about 13,500 (13,750); the nonferrous metals sector, about 5,940 (5,760); the natural gas and petroleum sector, about 2,260 (2,340); and the mining and quarrying sector, about 1,750 (1,750) (Österreichisches Institut für Wirtschaftsforschung, 2013b; Wirtschaftskammer Österreich, 2013, p. 33–36, 47, 51).

In the absence of a detailed mineral trade balance, more aggregated data were used to indicate that the value of Austria's exports of raw materials (including nonfuel minerals, but also including nonmineral raw materials, like wood) decreased to about $4.9 billion in 2012 from about $5.5 billion in 2011, and that of the country's imports of raw materials decreased to $8.1 billion from about $9.3 billion. Also, the value of exports of fuels and energy (including mineral fuels and also including electricity) increased to $5.8 billion in 2012 from $5.7 billion in 2011, and that of imports of fuels and energy increased to about $22.2 billion from a revised figure of $21.8 billion. Thus, Austria's trade balance for energy, fuels, and raw materials (including most of the mineral trade balance as a subset) was –$19.6 billion in 2012 compared with a revised figure of about –$19.9 billion in 2011. Detailed information about whether petroleum refinery products are included in the above trade balance was not available, and other processed mineral products (such as pig iron and steel) are not included (International Monetary Fund, 2013; Wirtschaftskammer Österreich, 2013, p. 60–61).

Because processed metals and industrial mineral products accounted for a greater proportion of the total value of output of the country's mineral industry than did mineral raw materials, it is also useful to look at the trade data that are available for nonfuel mineral-based manufactured products specifically. Austria's exports of manufactured ferrous metals (including iron and steel and possibly including ferroalloys) decreased in value to about $9.2 billion in 2012 from a revised figure of about $10.2 billion in 2011; nonferrous metals (including such products as aluminum and tungsten carbide, metal, and oxide powders), to about $4.6 billion from $5.2 billion; and nonmetallic mineral products (estimated to include such intermediate products as cement), to $2.6 billion from about $2.8 billion. Imports of ferrous metals decreased in value to $4.9 billion from about $5.8 billion in 2011; nonferrous metals, to $4.6 billion from $6.1 billion; and industrial mineral products, to $2.3 billion from about $2.4 billion. Thus, Austria's trade balance for processed mineral products increased to about $4.6 billion in 2012 from $3.9 billion in 2011; the trade surplus in this sector of the mineral industry was not enough to overcome the country's overwhelming trade deficit in energy,

---

[1]U.S. data were withheld to avoid disclosing company proprietary data.
[2]Where necessary, values have been converted from euro area euros (€) to U.S. dollars (US$) at an annual average exchange rate of €0.7187=US$1.00 for 2011 and €0.7775=US$1.00 for 2012. All values are nominal, at current prices, unless otherwise stated.

fuels, and raw materials, however (International Monetary Fund, 2013; Wirtschaftskammer Österreich, 2013, p. 60–61).

Government Policies and Programs

The basis of Austria's mining law is the Mineralrohstoffgesetz (MinroG) (Federal Law BGBl. I no. 38/1999), or "Mineral Resources Law," which came into effect on January 1, 1999, replacing the country's previous mining law (BGBl. 259/1975) that had been in effect since April 11, 1975. As of the end of 2012, the MinroG had been amended by Federal Laws BGBl. I no. 21/2002, BGBl. I no. 112/2003, BGBl. I no. 85/2005, BGBl. I no. 84/2006, BGBl. I no. 113/2006, BGBl. I no. 115/2009, BGBl. I no. 65/2010, BGBl. I no. 111/2010, and BGBl. I no. 144/2011; and by the publication BGBl. I no. 83/2003. The MinroG applies to the exploration for, production of, and processing of minerals in the country; the use of workings of unused mines; and the exploration for, locating of, and evaluation of the suitability of such geologic structures as caverns for holding or storing substances, such as liquid and gaseous mineral fuels. Three environmental laws that were directly applicable to mining and other mineral production and processing operations in the country were the Remediation Act of 1989 (BGBl. no. 299/1989), as last amended in 2011 by BGBl. I no. 15/2011; the Environmental Information Act of 1993 (BGBl. no. 495/1993), as last amended in 2012 by BGBl. I no. 50/2012; and the Environmental Impact Assessment Act of 2000 (BGBl. no. 697/1993), as last amended in 2012 by BGBl. I no. 77/2012 (Bundeskanzleramt Österreich, 2010; Schmelz and Rajal, 2012; Bundesministerium für Wirtschaft, Familie und Jugend, 2013, p. 26–33; undated; Rohöl-Aufsuchungs Aktiengesellschaft, undated).

According to the MinroG, Austrian mineral resources are divided into three main categories, as follows:
- Bergfreie—For resources in this category, the holder of the mining license has ownership of those minerals in the deposit for which the holder has a license to mine. The mineral raw materials in this category that are currently being produced in Austria are metallic ores, such as iron ore and tungsten (scheelite); oil shale; and many industrial minerals, including clays (such as bentonite and kaolin), diabase, graphite, gypsum, limestone and marble that contain at least 95% calcium carbonate, magnesite, talc, and silica sand that contains at least 80% SiO_2.
- Bundeseigene—The resources in this category are state owned, no matter who is awarded a license to extract and produce them. Mineral fuels, such as oil and natural gas, and related materials, such as uranium, are included under this classification. Also, all salt, whether contained in brines, in solution, or in rock salt, is owned by the state.
- Grundeigene—The resources in this category are owned by the owner of the land. They include the stone, sand, and gravel not included in the first category and feldspar. The owner of the land must still obtain a license before producing any of these mineral commodities (Bundesministerium für Wirtschaft, Familie und Jugend, 2013, p. 21–26).

In fall 2012, the Austrian Government founded an "Austrian Raw Materials Alliance" to increase the dialogue among the Government, industry, scientists, and stakeholders with an interest in ensuring a secure supply of raw materials for the country in the long-term future. The alliance proposed that the first point of emphasis to reduce imports and to increase a secure supply of raw materials should be on increasing the volume of recovery of critical mineral materials from secondary sources (recycling) in Austria (Bundesministerium für Wirtschaft, Familie und Jugend, 2013, p. 6–7).

Production

Data on Austria's mineral production are in table 1. In 2012, the production of tungsten in concentrate (W content) was estimated to have decreased by 11.5% compared with that of 2011. Wolfram Bergbau und Hütten AG (a subsidiary of Sandvik AB of Sweden) had been mining slightly lower tungsten ore grades at the Mittersill Mine since 2009 in order to increase the mine's productive lifespan and to optimize the recovery of tungsten from the ore. According to preliminary data, the company decreased production of tungsten ore (gross weight) in 2012 by 11% compared with that of 2011 through a 2-month halt of production in response to a decrease in demand, primarily for tungsten carbide powder (table 1; Bundesministerium für Wirtschaft, Familie und Jugend, 2013, p. 12, 15–18, 38–39).

In 2012, production of crude graphite decreased by about 76% compared with that of 2011 as production by Grafitbergbau Kaisersberg GmbH continued to fluctuate with demand in the small market served by the company; production of kaolin decreased by about 24% during the same timeframe in response to a decrease in demand from the small market served (and an apparent lack of opportunity to serve a broader market). Production of rock salt increased by about 31%, but this contributed only a small [53-metric-ton (t)] increase to the change in the total annual tonnage of salt production in Austria. Almost all of the salt produced in the country was in the form of salt brines, and salt brine production decreased by 16% compared with that of 2011 mostly in response to decreased demand for de-icing compounds as a consequence of the mild winter weather conditions in Central Europe and Western Europe at the beginning of 2012. Production of crude magnesite decreased by about 10% compared with that of 2011 owing to slight decreases in production at most mines in Austria following record-high production in the country in 2011. The decrease in production of magnesite was mostly in response to a decrease in demand for refractories from the steel manufacturing sector, which was one of the leading demand sectors for magnesite produced in Austria (table 1; Bundesministerium für Wirtschaft, Familie und Jugend, 2013, p. 12, 15–18, 40, 43–46, 50–51; K+S Aktiengesellschaft, 2013, p. 88, 92–94, 119).

In 2012, the value of marketed production in the construction materials, ceramics, and related industrial minerals sector of Austria increased by 0.22% from that of 2011. Within this broad sector, the value of marketed production of minerals and materials used primarily in industrial applications (including abrasives, ceramics, and refractories) increased by 1.35% from that of 2011, and the value of marketed production of minerals and materials primarily used in construction (including cement, sand and gravel, and natural stone) decreased by 0.06% from that of 2011. During this same timeframe, the changes in the data for the volumes of production in table 1 are consistent with the data for the value of marketed production of these industrial minerals and materials. The two most export-intensive subsectors were abrasives and refractories, and they benefited from an increase in export demand (mostly from Germany). The subsectors most affected by imports (mostly from lower cost countries) were ceramics and natural stone (table 1; Bundesministerium für Wirtschaft, Familie und Jugend, 2013, p. 11–18, 47–50, 68–74; Fachverband der Stein- und Keramischen Industrie Österreich, 2013, p. 7–9, 19–22, 25–26; Österreichisches Institut für Wirtschaftsforschung, 2013d).

Production of oil shale increased by 309% compared with that of 2011 owing to the reopening of an underground oil shale mine in the State of Tirole. The oil extracted from this shale was used in the production of cosmetics and pharmaceutical products; it was not used as a fuel or further refined to produce fuels. The country's production of natural gas also showed a notable increase (8.7%) compared with that of 2011 owing to an estimated 50.3% increase in natural gas production by Rohöl-Aufsuchungs Aktiengesellschaft (table 1; Bundesministerium für Wirtschaft, Familie und Jugend, 2013, p. 15–18, 40–42, 51–61).

Structure of the Mineral Industry

Table 2 is a list of major mineral industry facilities. Many mineral producers and processors (including most of the producers of industrial minerals in Austria) are not listed in table 2 owing to the lack of available information concerning the production capacities of the many small- and medium-scale ("Mittelstand") family-owned companies that produce minerals in the country. In 2012, there were reportedly 1,124 mining and quarrying operations and 3 operations that produced natural gas and (or) crude petroleum compared with 1,215 and 3, respectively, in 2011. Of the mining and quarrying operations in 2012, 1,121 produced industrial minerals (1,213 in 2011), including 1,106 open pit mines or quarries (1,198 in 2011), 11 underground (nonsalt) industrial mineral mines (11 in 2011), and 4 underground salt mines (5 in 2011); 2 mines produced iron ore and micaceous iron oxide (2 in 2011); and 1 mine produced nonferrous metals (tungsten) in both 2011 and 2012. Almost all the mineral companies operating in Austria were privately owned, but the Government owned 100% of the currently nonproducing coal company Graz-Koflacher Eisenbahn und Bergbaugesellschaft and 31.5% of the oil and gas company OMV Austria Exploration & Production GmbH (table 2; Bundesministerium für Wirtschaft, Familie und Jugend, 2012, p. 8; 2013, p. 17).

Commodity Review

Metals

Aluminum. By 2015, AMAG Austria Metall AG planned to complete construction of a new hot-rolling aluminum plant and expand the production capacity at its existing aluminum production facilities in Ranshofen. The company expected these investments to result in an increase of its Austrian production capacity to slightly more than 225,000 metric tons per year (t/yr) of aluminum metal compared with 150,000 t/yr in 2012 (AMAG Austria Metall AG, 2012, p. 8–9, 17, 26–28, 46, 52; Gesamtverband der Aluminiumindustrie e.V., 2012).

Tungsten.—During the 2-month stoppage of production in 2012, Wolfram Bergbau und Hütten made adjustments and preparations to increase tungsten production capacity at the Mittersill Mine, including widening the main ramp to the underground portion of the mine, installing a larger crusher, and increasing the processing capacity at the presorting stage. Accordingly, Wolfram Bergbau und Hütten could have the capacity to produce and process about 450,000 t/yr of tungsten ore (gross weight) in the future, but information concerning actual production or a production plan for 2013 (and beyond) was not available (Bundesministerium für Wirtschaft, Familie und Jugend, 2013, p. 38–39; Wolfram Berbau und Hütten AG, undated).

Industrial Minerals

Magnesium Compounds.—In 2012, sales of refractory materials to the worldwide steel manufacturing sector (including to the Austrian steel producer voestalpine AG) accounted for about 61% of RHI AG's global revenues compared with a revised figure of 63% in 2011. The company's production of raw materials accounted for 13% of total revenues in 2012 compared with 12% in 2011; about 79% of the revenues for the company's raw materials division in 2012 were intragroup revenues for supplying the company's own demand for raw materials compared with 82% in 2011. In 2012, RHI decreased its total tonnage of refractory products sold by 5% from that of 2011, and lower demand from the steel manufacturing sector in the European Union was a main reason for this decrease. The company's revenues from sales of refractory products in Austria decreased by 5%, but detailed data on the tonnages produced or sold by RHI just in Austria were not available. By the end of 2012, RHI had completed expansions of mines in Austria and Turkey to help the company's own production account for 80% of the magnesia raw materials (mostly magnesite) the company uses to produce refractory products. The goal of this 80%-self-sufficiency strategy was to allow RHI to be as independent as possible from the international markets for these raw materials, and the company reported that about 70% of the current commercially exploitable deposits of the magnesia raw materials it used are located in China, North Korea, and Russia (Bundesministerium für Wirtschaft, Familie und Jugend, 2013, p. 12, 44–46; RHI AG, 2013, p. 1–3, 7, 30–34, 42–43, 54, 72, 115).

Mineral Fuels and Related Materials

Natural Gas.—In 2012, Austria's natural gas consumption decreased to about 8.6 billion cubic meters from about 9 billion cubic meters in 2011 owing mostly to a 19% decrease in natural gas consumption by thermal power stations in the country; domestic production of natural gas increased to 1.7 billion cubic meters from 1.6 billion cubic meters in 2011. Consequently, the country's dependence on imports of natural gas could have decreased by about 0.5 billion cubic meters in 2012, but net imports of natural gas actually decreased by 1.24 billion cubic meters owing to the use of natural gas from storage (Bundesministerium für Wirtschaft, Familie und Jugend, 2013, p. 12–18, 52–62, 79; OMV Aktiengesellschaft, 2013, p. 37–47).

Petroleum and Petroleum Refinery Products.—In 2012, the volume of the country's imports of crude petroleum increased to about 52 million barrels (Mbbl) from 51 Mbbl in 2011. Data on exports of crude petroleum were not available. Additionally, Austria imported about 6.05 million metric tons (Mt) of petroleum refinery products and exported 2.4 Mt compared with about 6.11 Mt and about 2.24 Mt, respectively, in 2011 (Bundesministerium für Wirtschaft, Familie und Jugend, 2013, p. 12–18, 52–62, 79; Fachverband der Mineralölindustrie Österreichs, 2013, p. 2, 11, 21; OMV Aktiengesellschaft, 2013, p. 37–43, 48–51).

Outlook

Based upon data available through October 2013, Austria could increase production of crude steel to about 7.86 Mt in 2013, or by about 6% compared with that of 2012. This could increase demand for refractory products in Austria, but steel production in Germany (an important demand sector and export destination for Austrian refractory products) could decrease slightly or remain about the same as in 2012. Consequently, production of crude magnesite in Austria could increase to about 820,000 t (by about 5%) in 2013 compared with 779,000 t in 2011. If Wolfram Bergbau und Hütten produces at about its estimated production capacity following the 2012 improvements at the Mittersill Mine (450,000 t/yr gross weight of tungsten ore) and continues to mine about the same grade of tungsten as in 2011 and 2012, then Austria could produce about 750 t/yr of tungsten in concentrates (W content) in 2013 and beyond. The country could return to producing about 1,100 t/yr of tungsten in concentrates (W content) if the company is able to mine tungsten grades similar to the average tungsten grade it mined at Mittersill from 2008 through 2010. In addition, if Wolf Minerals Ltd.'s redevelopment of the Hemerdon Mine in the United Kingdom is completed by the end of 2013 (as planned), then Wolfram Bergbau und Hütten could increase its production of value-added tungsten metal products owing to an offtake agreement it has with Wolf Minerals to acquire some future production of tungsten in concentrates from the Hemerdon Mine (table 1; Swanepoel, 2012; Bundesministerium für Wirtschaft, Familie und Jugend, 2013, p. 12, 38–39, 44–46; Wolfram Berbau und Hütten AG, undated; World Steel Association, undated).

The construction sector study group, Euroconstruct forecast that the value of production of the commercial building (industrial) construction sector in Austria could increase by 0.9% in 2013 and by 1.1% in 2014; housing (residential) construction sector, 1% in 2013 and 0.7% in 2014; construction activity in the civil engineering sector could decrease by 0.6% in 2013 and increase by 1.2% in 2014. The Government planned to subsidize thermal refurbishment of commercial buildings through 2014. In 2013 and beyond, Government programs are expected to have a more noticeable effect on increasing demand for industrial minerals used in construction, and on Austria's production of these minerals and mineral products, possibly including cement, clays (such as brick clay and clays used in manufacturing ceramics for buildings and households), diabase, gypsum, construction sand and gravel, and stone (Bundesministerium für Wirtschaft, Familie und Jugend, 2013, p. 11–17; Fachverband der Stein- und Keramischen Industrie Österreich, 2013, p. 7–9, 19–22).

References Cited

AMAG Austria Metall AG, 2012, Annual report 2011: Ranshofen, Austria, AMAG Austria Metall AG, May 3, 147 p.

Bundeskanzleramt Österreich, 2010, BGBl. I Nr. 65/2010—Änderung des emissionsschutzgesetzes für kesselanlagen und des Mineralrohstoffgesetzes: Vienna, Austria, Bundesgesetzblatt für die Republik Österreich, Teil I, August 18, 3 p.

Bundesministerium für Wirtschaft, Familie und Jugend, 2012, Österreichisches montan-handbuch 2012: Vienna, Austria, Bundesministerium für Wirtschaft, Familie und Jugend, November 7, 291 p.

Bundesministerium für Wirtschaft, Familie und Jugend, 2013, Österreichisches montan-handbuch 2013: Vienna, Austria, Bundesministerium für Wirtschaft, Familie und Jugend, July 11, 298 p.

Bundesministerium für Wirtschaft, Familie und Jugend, [undated], Rechtsgrundlagen Bergbau—Mineralrohstoffgesetz: Vienna, Austria, Bundesministerium für Wirtschaft, Familie und Jugend. (Accessed October 28, 2010, at http://www.bmwfj.gv.at/EnergieUndBergbau/RechtsgrundlagenBergbau/Seiten/default.aspx.)

Crangle, R.D., Jr., 2013, Gypsum: U.S. Geological Survey Mineral Commodity Summaries 2013, p. 70–71.

Fachverband der Mineralölindustrie Österreichs, 2013, Mineralölbericht 2012: Vienna, Austria, Fachverband der Mineralölindustrie Österreichs, September, 55 p.

Fachverband der Stein- und Keramischen Industrie Österreich, 2013, Geschäftsbericht 2012/13: Vienna, Austria, Fachverband der Stein- und Keramischen Industrie Österreich, April 22, 31 p.

Gesamtverband der Aluminiumindustrie e.V. 2012, AMAG invests 220 million at Ranshofen site: Duesseldorf, Germany, Gesamtverband der Aluminiumindustrie e.V. news release, April 2, 1 p.

International Monetary Fund, 2013, Austria, in World economic outlook database: International Monetary Fund, October. (Accessed December 4, 2013, via http://www.imf.org/external/pubs/ft/weo/2013/02/weodata/index.aspx.)

Kramer, D.A., 2013, Magnesium compounds: U.S. Geological Survey Mineral Commodity Summaries 2013, p. 96–97.

K+S Aktiengesellschaft, 2013, Financial report 2012: Kassel, Germany, K+S Aktiengesellschaft, March 14, 228 p.

OMV Aktiengesellschaft, 2013, Annual report 2012: Vienna, Austria, OMV Aktiengesellschaft, March 20, 160 p.

Österreichisches Institut für Wirtschaftsforschung, 2013a, Übersicht 5.1.2—Produktionsindex in der industrie nach fachverbänden—Veränderung gegen das vorjahr in prozent: Vienna, Austria, Österreichisches Institut für Wirtschaftsforschung. (Accessed December 4, 2013, at http://www.wifo.ac.at/cgi-bin/tabellen/transtb2.cgi?2+2+industrie_FVB.print++++++0++0.)

Österreichisches Institut für Wirtschaftsforschung, 2013b, Übersicht 5.1.3—Beschäftigte in der industrie nach fachverbänden: Vienna, Austria, Österreichisches Institut für Wirtschaftsforschung. (Accessed December 4, 2013, at http://www.wifo.ac.at/cgi-bin/tabellen/transtb2.cgi?2+3+industrie_FVB.print++++++0++0.)

Österreichisches Institut für Wirtschaftsforschung, 2013c, Übersicht 5.1.9—Produktionswert in der industrie nach fachverbänden—Wert der abgesetzten produktion: Vienna, Austria, Österreichisches Institut für Wirtschaftsforschung. (Accessed December 4, 2013, at http://www.wifo.ac.at/cgi-bin/tabellen/transtb2.cgi?2+9+industrie_FVB.print++++++0++0.)

Österreichisches Institut für Wirtschaftsforschung, 2013d, Übersicht 7.2—Bauwirtschaft ÖNACE 2008—Veränderung gegen das vorjahr in prozent: Vienna, Austria, Österreichisches Institut für Wirtschaftsforschung. (Accessed December 4, 2013, at http://www.wifo.ac.at/cgi-bin/tabellen/transtb2.cgi?2+2+interbau++++++0++0.)

RHI AG, 2013, Annual report 2012: Vienna, Austria, RHI AG, April 4, 136 p.

Rohöl-Aufsuchungs Aktiengesellschaft, [undated], History: Rohöl-Aufsuchungs Aktiengesellschaft. (Accessed October 29, 2011, at http://www.rag-austria.at/en/company/portrait/history.html.)

Schmelz, Christian, and Rajal, Bernd, 2012, Austria—Getting the deal through—Environment 2012: London, United Kingdom, Law Business Research Ltd., November 19, p. 3–11. (Accessed November 20, 2012, at http://schoenherr.eu/news-publications/pdfs/Getting the Deal Through Environment 2012 Austria.pdf.)

Shedd, K.B., 2013, Tungsten: U.S. Geological Survey Mineral Commodity Summaries 2013, p. 176–177.

Swanepoel, Esmarie, 2012, Wolf signs offtake deal for Hemerdon: Mining Weekly, Creamer Media (pty) Ltd., April 20. (Accessed November 2, 2012, at http://www.miningweekly.com/article/wolf-signs-offtake-agreement-for-hemerdon-2012-04-20.)

Wirtschaftskammer Österreich, 2013, Statistical yearbook 2013: Vienna, Austria, Wirtschaftskammer Österreich, May, 96 p.

Wolfram Berbau und Hütten AG, [undated], Refining plant St. Martin: St. Martin, Austria, Wolfram Berbau und Hütten AG. (Accessed February 11, 2011, at http://www.wolfram.at/wolfram_at/wEnglisch/unternehmen/index3223655460navidW2618000000000.html?navid=1.)

World Steel Association, [undated], Crude steel production: Brussels, Belgium, World Steel Association. (Accessed December 16, 2013, at http://www.worldsteel.org/statistics/crude-steel-production.html.)

TABLE 1
AUSTRIA: PRODUCTION OF MINERAL COMMODITIES[1]

(Thousand metric tons unless otherwise specified)

Commodity		2008	2009	2010	2011	2012
METALS						
Aluminum, metal, secondary	metric tons	158,958	282,944	374,837	385,551 [r]	372,769
Copper, metal, secondary:						
Smelter	do.	94,200	90,800	92,200	92,200	95,000 [p]
Refined	do.	106,668 [r]	96,240 [r]	113,705 [r]	112,610 [r]	113,578
Iron and steel:						
Iron ore, including micaceous iron oxide:						
Gross weight		2,033	2,002	2,069	2,207 [r]	2,142
Fe content		650	641	662	706 [r]	686 [e]
Metal:						
Pig iron		5,795	4,353	5,621	5,815	5,751
Ferroalloys, electric arc furnace, unspecified		723	588	637	650 [e]	650 [e]
Crude steel		7,594	5,662	7,206	7,474	7,421
Semimanufactures, hot-rolled products		6,850	5,394	6,621	6,874 [r]	6,850 [e]
Lead, refined, secondary	metric tons	26,902	22,197	25,499	26,208 [r]	24,504
Nickel, including Ni content of ferroalloys[e]	do.	800	700	1,000 [r]	1,000 [r]	1,000
Tungsten ore and concentrate:						
Ore:						
Gross weight	do.	434,296	344,851	429,748	423,790	376,460 [p]
W content[e]	do.	1,270	1,010	1,110	975	898
Concentrate:						
Gross weight	do.	4,627	3,436	3,812	3,380	2,760
W content	do.	1,122	887	977	706 [r]	625 [e]
INDUSTRIAL MINERALS						
Aluminum oxide, fused[e]	metric tons	15,000	10,000	11,500	12,000	12,000
Cement:						
Clinker		3,996	3,428	3,097	3,176 [r]	3,206
Hydraulic		5,309	4,646	4,254	4,427	4,455
Clays:						
Kaolin, crude	metric tons	49,527	83,980	58,956	56,976 [r]	43,174
Unspecified, possibly including bentonite, brick clay, and illite		2,473	1,866	1,860	1,927 [r]	1,794
Diabase (of basaltic rocks)		2,410	2,098	1,762	2,083 [r]	1,881
Feldspar, byproduct of silica processing[e]	metric tons	27,000	27,000	27,000	27,000	27,000
Graphite, crude	do.	250	750	420	925 [r]	219
Gypsum and anhydrite, crude		1,087	911	872	815 [r]	792
Lime, including quicklime		909	725	774	801 [r]	800 [e]
Of which, marketed		612	507	492	528 [r]	525 [e]
Magnesite:						
Crude		837	545	757	868 [r]	779
Sintered or dead burned		290	230	264	293 [r]	270 [e]
Caustic calcined		50	21	52	57 [r]	55 [e]
Mica[e, 2]	metric tons	3,420	2,840	3,430 [r]	3,590 [r]	3,400
Nitrogen, N content of ammonia[e]		400	370	400	400	400
Salt (NaCl):						
Brines, gross	thousand cubic meters	2,912	3,460	3,608	3,809 [r]	3,193
Evaporated, mechanical heating process		867	1,035	1,072	1,150 [r]	952
Rock	metric tons	503	50	95	169 [r]	222
Mine output, NaCl content		874	1,038	1,083	1,270 [r]	1,064
Sand and gravel:						
Dolomite, loose rocks and gravel		3,151	2,790	2,620	2,870 [r]	2,646
Quartz (silica) sand		2,175	1,200	939	898 [r]	814
Sand and gravel, unspecified		27,718	25,722	24,128	25,046 [r]	23,701

See footnotes at end of table.

TABLE 1—Continued
AUSTRIA: PRODUCTION OF MINERAL COMMODITIES[1]

(Thousand metric tons unless otherwise specified)

Commodity		2008	2009	2010	2011	2012
INDUSTRIAL MINERALS—Continued						
Sodium compounds, manufactured, n.e.s.[e, 3]		250	275	285	300[r]	300
Stone:						
Amphibolite		1,808	1,780	1,670	1,318[r]	1,145
Basalt, not included in diabase		1,797	1,744	1,473	1,791[r]	1,363
Dolomite		4,409	3,967	3,915	3,711[r]	3,469
Gneiss		1,668	1,431	1,505	1,435[r]	1,503
Granite and granulite		3,315	3,078	2,340	3,034[r]	2,704
Limestone, including marble		23,758	22,074	21,190	21,571[r]	21,073
Marl		1,826	1,508	1,149	1,484[r]	1,073
Quartz, quartzite, and pegmatite		327	377	294	285[r]	315
Serpentinite		1,690	1,751	2,013	1,484[r]	1,311
Other, including conglomerate and sandstone		61	22	38	47[r]	31
Sulfur, byproduct of petroleum and natural gas	metric tons	8,016	12,007	9,873	9,669[r]	10,329
Talc and leucophyllite (white mica), crude	do.	154,577	111,388	138,367	132,018[r]	134,665
MINERAL FUELS AND RELATED MATERIALS						
Coke		1,410	1,281	1,388	1,356[r]	1,346
Natural gas:						
Marketable (net)	million cubic meters	1,544	1,580[r]	1,713	1,591[r]	1,729
Natural gas liquids[4]	thousand 42-gallon barrels	836	972	927	846[r]	830
Oil shale	metric tons	114	144	176	132[r]	540
Petroleum:						
Crude[5]	thousand 42-gallon barrels	6,066	6,371	6,167	5,900[r]	5,896
Refinery products:[4]						
Liquefied petroleum gas	do.	1,134	1,068	1,011	1,175[r]	826
Gasoline	do.	15,100[r]	14,800[r]	12,900[r]	13,800[r]	13,500
Kerosene and jet fuel	do.	3,750	2,480	3,780	4,880[r]	4,890
Diesel fuel	do.	23,200	23,600	20,400	25,100	27,100
Distillate fuel oil	do.	5,730[r]	5,870	4,960[r]	4,170[r]	6,050
Residual fuel oil	do.	6,600	5,540	5,070	4,920[r]	4,110
Lubricants and miscellaneous oils	do.	943	723	672	501	451
Bitumen, bituminous mixtures, and other residues	do.	2,690	2,550	1,770	2,280	2,220
Other (unspecified)	do.	3,030[r]	2,020[r]	1,510[r]	1,220[r]	2
Total	do.	62,200[r]	58,700[r]	52,100[r]	58,000[r]	59,100

[e]Estimated; estimated data are rounded to no more than three significant digits; may not add to totals shown. [p]Preliminary. [r]Revised. do. Ditto.

[1]Table includes data available through February 12, 2014.

[2]Estimated from reported exports minus imports of mica.

[3]Not elsewhere specified. Data could include production of soda ash and sodium sulfate.

[4]Figure converted to barrels from metric tons according to a converson factor and reflects the significant digits of the conversion factor.

Source: U.S. Energy Information Administration, 2008, International Energy Annual—Table C.1, General Conversion Factors: Washington, DC, U.S. Energy Information Administration. June–December. (Accessed March 7, 2010, at http://www.eia.doe.gov/emeu/iea/tablec1.html.)

[5]All figures were converted to barrels from metric tons according to a converson factor of 7.040 barrels of crude oil per metric ton.

Source: U.S. Energy Information Administration, [undated], International Energy Statistics—Austria: Washington, DC, U.S. Energy Information Administration. (Accessed March 7, 2010, at http://tonto.eia.doe.gov/cfapps/ipdbproject/IEDIndex3.cfm?tid=94&pid=57&aid=32.)

TABLE 2
AUSTRIA: STRUCTURE OF THE MINERAL INDUSTRY IN 2012

(Thousand metric tons unless otherwise specified)

Commodity	Major operating companies and major equity owners	Location of main facilities	Annual capacity
Alumina, fused	Treibacher Schleifmittel GmbH (Imerys S.A., 100%)	Plant at Villach, State of Carinthia	60
Aluminum	AMAG Austria Metall AG (B&C Industrieholding GmbH, 34.1%; Raiffeisenlandesbank Oberösterreich AG, 16.5%; AMAG Employees Private Foundation, 11.1%; Esola Beteiligungsverwaltungs GmbH, 5%; Oberbank Industrie- und Handelsbeteiligungsholding GmbH, 5%; Treibacher Industrieholding GmbH, 5%; AMAG Management, 0.4%; free floating shares, 27%)	Secondary ingot plant at Ranshofen, State of Upper Austria	150
Do.	Hammerer Aluminium Industries GmbH	Secondary extrusion plant at Ranshofen, State of Upper Austria	80
Do.	Hydro Aluminium Nenzing GmbH (Norsk Hydro ASA, 100%)	Secondary plant at Nenzing, State of Vorarlberg	59
Do.	Speedline Aluminium Giesserei GmbH (Swiss Alu Trading AG, 100%)	Secondary plant at Schlins, State of Vorarlberg	49
Do.	Aluminum Lend GmbH (Salzburger Aluminium AG, 100%)	Secondary ingot plant at Lend, State of Salzburg	40
Do.	NEUMAN Aluminium Austria GmbH (CAG Holding GmbH, 100%)	Secondary plant at Marktl, State of Styria	16
Do.	Bavaria Industriekapital AG	Secondary plant at Gleisdorf, State of Styria	NA
Do.	Georg Fischer Automotive AG	Secondary plant at Altenmarkt, State of Salzburg; Secondary plant at Herzogenburg, State of Lower Austria	NA
Do.	Nemak Linz GmbH (Tenedora Nemak S.A. de C.V., 100%)	Secondary plant at Linz, State of Upper Austria	NA
Do.	Almaxal Brüder Tschirk GmbH	Secondary plant at Neudörfl, State of Burgenland	NA
Do.	Almeta Metallumschmelzwerk GmbH	Secondary plant at Vienna; secondary plant at Sollenau, State of Lower Austria	NA
Calcium carbonate, ground	Omya GmbH (Omya AG, 100%)	Plant at Gummern, State of Carinthia	2,500
Do.	do.	Plants at Golling, State of Salzburg; Neu Pirka, State of Styria; and Ulmerfeld-Hausmening, State of Lower Austria	NA
Cement	Lafarge Perlmooser AG (Lafarge S.A., 70%, and Strabag SE, 30%)	Plant at Mannersdorf, State of Lower Austria; plant at Retznei, State of Styria; grinding plant at Kirchbichl, State of Tirole	2,200
Do.	Wietersdorfer & Peggauer Zementwerke GmbH	Plant at Peggau, State of Styria; Plant at Wietersdorf, State of Carinthia	1,100
Do.	Gmundner Zement Produktions- und Handels GmbH	Plant at Gmundnen, State of Upper Austria	800
Do.	Kirchdorfer Zementwerk Hofmann GmbH	Plant at Kirchdorf, State of Upper Austria	800
Do.	Zementwerk LEUBE GmbH	Plant at Gartenau, State of Salzburg	770
Do.	Wopfinger Baustoffindustrie GmbH	Plant at Wopfing, State of Lower Austria	300
Do.	Holcim (Wien) GmbH (Holcim Ltd., 100%)	Plant at Vienna	300
Do.	Holcim (Vorarlberg) GmbH (Holcim Ltd., 100%)	Lorüns grinding plant and cement plant at Bludenz, State of Vorarlberg	200
Chalk	Mühlendorfer Kreidefabrik Margit-Hoffman Ostenhof KG (Omya AG, 100%)	Plant at Müllendorf, State of Burgenland	NA
Clays, including brick clay	Wienerberger AG	Clay mines at Göllersdorf, State of Lower Austria; at Rotenturm and Stoob, State of Burgenland; and at Apfelberg and Weißkirchen, State of Styria	NA
Clays, kaolin, and silica sand	Österreichische Kaolin- und Montanindustrie AG	Mines at Weinzierl and Kriechbaum; processing plant at Aisthofen, State of Upper Austria	170
Coal	Graz-Koflacher Eisenbahn und Bergbaugesellschaft GmbH (Government, 100%)	Oberdorf Mine, Bärnbach, State of Styria (closed)	1,200

See footnotes at end of table.

TABLE 2—Continued
AUSTRIA: STRUCTURE OF THE MINERAL INDUSTRY IN 2012

(Thousand metric tons unless otherwise specified)

Commodity	Major operating companies and major equity owners	Location of main facilities	Annual capacity
Copper, refined, secondary	Montanwerke Brixlegg AG (A-Tec Industries AG, 100%)	Plant at Brixlegg, State of Tirole	110
Diabase, basalt	Diabaswerk Saalfelden GmbH (STRABAG SE, 100%)	Mine and plant at Saalfelden, State of Salzburg	NA
Do.	Klöcher Basaltwerke GmbH & Co KG (ASAMER Holding AG, 100%)	Mines and plants at Klöch and Oberhaag, State of Styria	NA
Feldspar	Quarzwerke Österreich GmbH (Quarzwerke GmbH, 100%)	Mine and plant at St. Georgen an der Gusen, State of Upper Austria	NA
Ferroalloys, FeV, FeMo, FeNi	Treibacher Industrie AG	Plant at Althofen, State of Carinthia	15 [e]
Graphite, natural	Graphitbergbau Mühldorf Mörth GmbH	Trandorf Mine at Weinberg and extended to Weinbergwald, State of Lower Austria; mine at Eichenwald, State of Styria	15
Do.	Grafitbergbau Kaisersberg GmbH	Kaisersberg Mine, State of Tirole	3
Gypsum and anhydrite, natural	Moldan Baustoffe GmbH & Co. KG (Salzburger Sand- & Kieswerke GmbH, 100%)	Abtenau and Moosegg Mines, near Kuchl bei Hallein, State of Salzburg	300
Do.	Saint-Gobain Rigips Austria GmbH (Compagnie de Saint-Gobain, 100%)	Mine at Grundlsee and main plant at Bad Aussee, State of Styria; Mine and plant at Puchberg, State of Lower Austria	250
Do.	Knauf GmbH	Hinterstein Mine, Spital am Pyhrn, State of Upper Austria; Mines at Dörfelstein and Tragöß-Oberort, and plant at Weißenbach bei Liezen, State of Styria	160
Do.	Gipswerk Schretter & Cie. GmbH	Mine at Weißenbach am Lech and plant at Vils, State of Tirole	NA
Iron ore	VA Erzberg GmbH (voestalpine AG, 100%)	Erzberg Mine at Eisenerz, State of Styria	3,000
Iron oxide, micaceous	Kärntner Montanindustrie GmbH	Mine near Waldenstein, State of Carinthia	NA
Lead	Bleiberg Bergwerks-Union AG (Metall Gesellschaft, 74%)	Smelter at Brixlegg, State of Tirole	55
Lime	voestalpine Stahl AG (voestalpine AG, 100%)	Limestone mine near Kremsmauer Mountain, and plant at Steyrling, State of Upper Austria	360
Do.	Zementwerk LEUBE GmbH	Plant at Gartenau, State of Salzburg	150 [e]
Magnesite, crude	Veitsch-Radex GmbH & Co. (RHI AG, 100%)	Mine and plant at Breitenau, State of Styria; Mine at Eichberg, State of Lower Austria; Am Bürgl Mine, area near Weissenstein, State of Tirole; mine and processing plant at Millstätter Alpe, State of Carinthia	800
Do.	Styromagnesit Steirische Magnesitindustrie GmbH	Angerer, Kaintaleck and Wieser Mines, and plant near Oberdorf an der Laming, State of Styria; Wald Mine in the Schoberpass, State of Styria	75
Do.	CEMEX Austria AG (CEMEX S.A.B. de C.V., 100%)	Mine and plant at Veitsch, State of Styria	NA
Do.	PRONAT Steinbruch Preg GmbH (Schotter- und Betonwerk Karl Schwarzl Betriebsgesellschaft m.b.H., 100%)	Magnesite and dunite (olivine rock) mine at Gulsen, and plant at Preg, State of Styria	NA
Natural gas million cubic meters	OMV Austria Exploration & Production GmbH [OMV Aktiengesellschaft (Free floating shares, 48.5%; Government, 31.5%; International Petroleum Investment Co. 20%), 100%]	Main fields in the Vienna basin, State of Lower Austria, and some fields in the State of Upper Austria	1,500
Do. do.	Rohöl-Aufsuchungs Aktiengesellschaft (EVN AG, 50.025%; E.ON Ruhrgas E&P GmbH, 29.975%; Steirische Gas-Wärme GmbH, 10%; Salzburg AG, 10%)	Main fields in the State of Upper Austria, and some fields in the State of Lower Austria and the State of Salzburg	550 [e]
Nitrogen, N content of ammonia	Borealis Agrolinz Melamine GmbH (Borealis AG, 100%)	Plant at Linz, State of Upper Austria	498
Oil shale	Tiroler Steinölwerke Albrecht GmbH & Co. KG	Mine in the Bächental, near Pertisau am Achensee, State of Tirole	NA

See footnotes at end of table.

TABLE 2—Continued
AUSTRIA: STRUCTURE OF THE MINERAL INDUSTRY IN 2012

(Thousand metric tons unless otherwise specified)

Commodity		Major operating companies and major equity owners	Location of main facilities	Annual capacity
Petroleum, crude	thousand 42-gallon barrels	OMV Austria Exploration & Production GmbH [OMV Aktiengesellschaft (Free floating shares, 48.5%; Government, 31.5%; International Petroleum Investment Co., 20%), 100%]	Main fields in the Vienna basin, State of Lower Austria, and some fields in the State of Upper Austria	5,500 [e]
Do.	do.	Rohöl-Aufsuchungs Aktiengesellschaft (EVN AG, 50.025%; E.ON Ruhrgas E&P GmbH, 29.975%; Steirische Gas-Wärme GmbH, 10%; Salzburg AG, 10%)	Main fields in the State of Upper Austria, and some fields in the State of Lower Austria and the State of Salzburg	750 [e]
Rare-earth chemicals and oxides		Treibacher Industrie AG	Plant at Althofen, State of Carinthia	NA
Salt, NaCl content		Salinen Austria AG	Mines at Bad Ischl and Hallstatt, and evaporite saltworks at the Ebensee, State of Upper Austria; mine at Hallein-Dürrnberg, State of Salzburg; mine at Hall in Tirol, State of Tirole; mine at Altaussee, State of Styria	1,100
Silica sand		Krempelbauer-Quarzsandwerk St. Georgen Hentschläger & Co. KG	Burger and Knoll-Wizany Mines at Luftenberg, Krempelbauer and Poscher Mines at St. Georgen, and Treffling Mine at Aigen-Engerwitzdorf, State of Upper Austria	NA
Do.		Quarzwerke Österreich GmbH (Quarzwerke GmbH, 100%)	Mine and plant at Melk, State of Lower Austria; mine and plant at St. Georgen an der Gusen, State of Upper Austria	NA
Do.		Quarzsande GmbH (Zementwerk LEUBE GmbH, 100%)	Mine and plant at Eferding, mine at Bruck-Waasen, and mine at Wolfsegg, State of Upper Austria	NA
Steel, crude		voestalpine Stahl GmbH (voestalpine AG, 100%)	Plant at Linz, State of Upper Austria	6,000
Do.		voestalpine Stahl Donawitz GmbH Co & KG (voestalpine AG, 100%)	Plant at Donawitz (near Leoben), State of Styria	1,500
Do.		Breitenfeld Edelstahl AG	Plant at Mitterdorf im Mürztal, State of Styria	300
Do.		Böhler Edelstahl GmbH & Co KG (voestalpine AG, 100%)	Plant at Kapfenberg, State of Styria	150 [e]
Talc and leucophyllite (white mica)		Naintsch Mineralwerke GmbH (Imerys S.A., 100%)	Talc mines at Lassing and Rabenwald, and plant at Oberfeistritz, State of Styria; talc and mica mine at Kleinfeistritz, and a plant at Weisskirchen, State of Styria	200 [e]
Do.		Aspanger Bergbau und Mineralwerke GmbH & Co. KG (Wietersdorfer & Peggauer Zementwerke GmbH, 100%)	Leucophyllite mine and mica processing plant at Aspangberg-Zöbern, State of Lower Austria	NA
Tungsten:				
Ore (scheelite), gross weight		Wolfram Bergbau und Hütten AG (Sandvik AB, 100%)	Mine at Mittersill and processing plant at Bergla, in the Felbertauerntal, State of Salzburg	425 [e]
Concentrate, W content	metric tons	do.	do.	1,800 [e]
Carbide, powders	do.	do.	Primary and secondary chemical treatment and sintering plant at St. Martin, in the Sulmtal, State of Styria	3,000 [e]
Carbide and metal, powders	do.	Treibacher Industrie AG	Plant at Althofen, State of Carinthia	NA
Metal, powders	do.	Plansee SE (Plansee Holding AG, 100%)	Plants at Liezen, State of Styria, and at Reutte, State of Tirole	NA
Do.	do.	Wolfram Bergbau und Hütten AG (Sandvik AB, 100%)	Primary and secondary chemical treatment and sintering plant at St. Martin, in the Sulmtal, State of Styria	3,600 [e]
Oxides	do.	do.	do.	NA

[e]Estimated; estimated data are rounded to no more than three significant digits. Do., do. Ditto. NA Not available.

The Mineral Industry of Azerbaijan

By Elena Safirova

Azerbaijan produced a wide range of metals and industrial minerals, including alumina, aluminum, iron ore, and steel. Its major importance as a world mineral producer, however, was based on its crude oil industry and, more recently, its natural gas industry. The country had been a significant crude oil producer for more than a century, but the focus since independence in 1991 was on developing offshore resources in the Caspian Sea. Oilfield and gasfield development was concentrated in two projects—the Azeri-Chirag-Guneshli (ACG) offshore oilfield complex and the Shah-Deniz offshore gasfield (U.S. Energy Information Administration, 2013).

Minerals in the National Economy

In 2012, the real gross domestic product (GDP) of Azerbaijan decreased by 2.3%. The nominal GDP amounted to $68.73 billion, and industrial production contributed 63.8% to the total GDP. Mining and quarrying accounted for 78.8.% of the country's industrial output whereas the rest of the industrial output was produced by manufacturing (15.3%) and electricity, heating, and water production and distribution (5.9%). In 2012, industrial production decreased by 2.3% and production by the mining and quarrying sector decreased by 4.2% compared with that of 2011 primarily because of decreased production of oil and natural gas condensate (State Statistical Committee of the Republic of Azerbaijan, 2012, 2013; U.S. Central Intelligence Agency, 2013).

In 2012, Azerbaijan exported about $34.2 billion worth of goods and services. Of that amount, $29.5 billion (86%) came from crude oil exports; other significant export commodities were natural gas (4.7%), diesel fuel (2.8%), and kerosene (0.7%). The main export partners of Azerbaijan were Italy (which received 23.2% of Azerbaijan's total exports), India (7.9%), France and Indonesia (7.4% each), Israel (7.0%), the United States (6.7%), Germany and Russia (4.0% each), and Greece (3.5%). Azerbaijan's total imports in 2012 were valued at about $9.7 billion, and the main imported commodities included chemicals, foodstuffs, machinery and equipment, metals, and petroleum products. The country's major import partners during the year were Turkey (which provided 15.8% of Azerbaijan's imports), Russia (14.3%), Germany (8.1%), the United States (7.4%), China (6.5%), Ukraine (5.6%), the United Kingdom (5.1%), and Kazakhstan (3.5%). With net exports of $24.5 billion, Azerbaijan was able to continue investing in infrastructure, stabilizing the economy, and reducing poverty in the country (State Statistical Committee of the Republic of Azerbaijan, 2013; U.S. Central Intelligence Agency, 2013).

Production

In 2012, production of lime for construction in Azerbaijan increased by 2,100% because of the construction of a new modern plant. Estimated primary and secondary aluminum production and alumina output increased by 175% and 142%, respectively, owing to the start of production at a new aluminum smelter. Output of sand for construction increased by 66%; that of gypsum, by 49%; cement, by 38%; crude steel, by 14%; bentonite, by 9.1%; and natural gas, by 5.4%. Production of caustic soda was reduced by more than 99% and essentially stopped. Production of marketable salt decreased by 72%; silver, by 46%; steel pipes, by 37%; metal content of copper ore, by 18%; and gold, by 12%. Other data on mineral production are in table 1.

Commodity Review

Metals

Aluminum.—In the middle of 2011, the OJSC Azerbaijan Aluminum (Azeral) plant in Sumqayit restarted production of primary aluminum. Prior to that, both the Azeral aluminum plant in Sumqayit and the Ganja alumina plant had been idle for several years. During 2012, Azerbaijan exported an estimated 50,000 metric tons (t) of primary aluminum and received $94.9 million in revenue (Rustambekov, 2013).

In January, Azeral's new aluminum plant in Ganja started operations. The new plant has an annual smelter capacity of 50,000 metric tons per year (t/yr) of primary aluminum and cost $230 million to build. Early in 2013, the company was expected to complete the second stage of the project, which would increase the plant's annual production capacity to 100,000 t/yr. The construction of the new plant started in 2008 but was delayed because of the global economic crisis (Kavkasia.net, 2012; Mamedov, 2012).

Gold, Silver, and Copper.—In 2009, Anglo Asian Mining PLC (Anglo Asian) of the United Kingdom began gold production at the Gedabek gold, silver, and copper mine, which is located about 55 kilometers (km) from the city of Ganja. In 2012, the company reported producing 1,562.8 kg of gold, 625.8 kg of silver, and 502 t of copper. Anglo Asian was controlled by R.V. Investment Group Services (51% interest), and the Government of Azerbaijan [through the Ministry of Ecology and Natural Resources (MENR)] owned a 49% interest. The original production-sharing agreement between Anglo Asian and the Government, which was signed in 1997, included development of six deposits in southwestern Azerbaijan—the Gedabek, the Gosha Bulag, the Gyzyl Bulag, the Ordubad, the Soyutlu, and the Vezhnali fields. The Ordubad, the Soyutlu, and the Vezhnali fields are located in the breakaway region of Nagorno-Karabakh where conflicts with ethnic Armenians took place from 1988 to 1994. According to the contract, Anglo Asian was planning to mine a total of 400 t of gold, 2,500 t of silver, and 1,500 t of copper. According to the agreement, the Government was to receive its share of profits in gold, which would help the country build up its gold reserves (Anglo Asian Mining PLC, 2013; Interfax.az, 2013a; News.mail.ru, 2013).

In April and then in June, the resource estimates for the Gedabek deposit were updated. According to the latest estimates, the Gedabek deposit contained 1.139 grams per metric ton (g/t) gold and 9.456 g/t silver. The total contained metal in the resources of Gedabek were estimated to be 23.1 t of gold and 39.7 t of silver. In 2012, the company started to build an agitation-leaching plant to complement an existing heap-leach processing operation for the purpose of increasing the gold recovery and processing rate at Gedabek. Regular mining operations at another deposit, the Gosha Bulag deposit, were expected to begin at the end of 2013. Annual gold production at Gosha Bulag was expected to reach between 500 and 650 kg (News.day.az, 2012a).

In November, another gold producer, Azerbaijan International Mineral Resources Operating Co. Ltd (AIMROC), began its operations in Azerbaijan. The company was mining the Chovdar polymetallic deposit which, in addition to gold, contained copper, lead, and silver. Gold resources of the deposit were estimated to be 40 t, and the company projected a mine life of between 8 and 10 years. AIMROC was a consortium of Fargate Mining Corp. of Panama, Globex International LLP of the United Kingdom, Londex Resources S.A. of Panama, Mitsui Mineral Development Engineering Co. Ltd. of Japan, and Willy & Meyris S.A. of Panama. In December 2006, the Government and AIMROC signed an agreement under which AIMROC would develop mining projects at the Chovdar, the Dagkesman, the Garadag, the Geydar, and the Kokhnemyadan ore fields and Kurekch ore basin; the Government owned a 30% interest in these projects (Mineral.ru, 2012e).

Industrial Minerals

Lime.—In 2012, the overall production of construction materials in Azerbaijan increased by 17%. Production of lime used in construction increased by 22-fold to 49,000 t. In December 2011, a new construction materials plant that was operated by AAC Company opened 50 kilometers (km) southwest of Baku. Two major products produced by the plant were aerated concrete and construction lime. Aerated concrete is a light, seismically resistant, thermoproof and soundproof, and environment-friendly material that was gaining popularity for a variety of construction projects. The plant had a combined annual lime production capacity of 65,000 t/yr and was expected to produce up to 180,000 cubic meters per year of aerated concrete. Raw materials used in production are mostly local quartz sand and limestone. In 2012, the plant employed 130 workers (1news.az, 2011; Trend.az, 2011; News.day.az, 2012b; Interfax.az, 2013b).

Mineral Fuels

Natural Gas.—In 2012, Azerbaijan produced and sold as a commodity 17,242 million cubic meters of natural gas, which was an increase of 5.4% compared with the level of output in 2011. As of January 2013, according to the Oil and Gas Journal, Azerbaijan's gas reserves were approximately 990 billion cubic meters. Almost all Azerbaijani gas was produced in two offshore fields—the ACG complex and the Shah-Deniz field (U.S. Energy Information Administration, 2013).

The Shah-Deniz natural gas and condensate field started producing at the end of 2006. The field is located on the deepwater shelf of the Caspian Sea, where the water depth reaches 500 meters. The total resources of Shah-Deniz are estimated to be 1.2 trillion cubic meters of natural gas and 240 billion of gas condensate. The field was being developed by a consortium of companies led by BP p.l.c. of the United Kingdom. According to the initial agreement signed in 1996, BP (the project operator) and Statoil ASA of Norway each had a 25.5% interest. State Oil Corp. of the Republic of Azerbaijan (GNKAR), Naftiran Intertrade Co. (NICO), Total S.A. of France, and OAO Lukoil of Russia each had a 10% share, and Türkiye Petrolleri Anonim Ortaklığı (TPAO) had a 9% share. Shah-Deniz was expected to reach full production capacity in 2017 and to start supplying European customers with natural gas sometime in 2019. In March, Azerbaijan and Turkey signed an intergovernmental agreement regarding the construction of the Trans Anatolian Natural Gas Pipeline (TANAP), which would transport the Shah-Deniz gas to Europe through Turkey. TANAP was planned to have an annual capacity of 16 billion cubic meters and was expected to cost $5 billion to build. It was expected that GNKAR and Botas of Turkey would form a TANAP consortium, and that later they might invite other participants to join the consortium. According to the preliminary plans, construction of the TANAP would begin in 2014 and be completed by 2018 (Mineral.ru, 2012a).

Azerbaijan was planning to increase its natural gas production rapidly in the next decade; it was expecting to produce 20 billion cubic meters per year by 2015 and to double production by 2025. In addition to ACG and Shah-Deniz, other promising natural gas fields in the country included the Absheron, the Babek, and the Umid fields. Babek reportedly had estimated resources of 400 billion cubic meters, followed by Absheron and Umid, which had 340 billion and 200 billion cubic meters, respectively. According to GNKAR, in 2012, a total of about 50 companies from 20 countries were involved in natural gas extraction from 40 different deposits in Azerbaijan (Lenta.ru, 2012; Mineral.ru, 2012b).

Petroleum.—In 2012, the production of crude oil in Azerbaijan decreased to 43.0 Mt, or by 5.8% compared with that of 2011. The major source of crude oil in the country was the ACG field, which had been in operation for 15 years. The ACG field is located about 100 km east of Baku in the Caspian Sea and covers 430 square kilometers. The ACG field had an estimated 5 billion barrels of reserves; it produced mostly Azeri Light, which is a medium-light and sweet crude that is valued for its middle-distillate yield. During the 15-year period, the total investment in ACG reached $24.8 billion. As of 2012, the ACG complex had 86 drilling wells, of which 56 were used for oil extraction and the others were water and gas pressure wells. The ACG petroleum project was developed by a consortium of companies lead by BP, which had a 35.78% share in the project. Other participants included State Oil Company of Azerbaijan Republic (SOCAR) (11.65%), Chevron Corp. of the United States (11.27%), INPEX Corp. of Japan (10.96%), Statoil of Norway (8.56%), Exxon Mobil Corp. of the United States (8.01%), TPAO of Turkey (6.75%), Itochu Corp. of Japan (4.3%), and Hess Corp. of the United States (2.72%) (Mineral.ru, 2012c, d).

Azeri crude oil was refined domestically at two refineries—the Azerneftyag refinery and the Heydar Aliyev refinery. The total (combined) refining capacity of both refineries was about 400,000 barrels per day (about 20 Mt/yr). Modernization of both refineries was projected to cost between $600 millon and $700 million. In May, Azerbaijan announced that the country had made the decision to switch to Euro 3 emission standards in its gasoline production starting in 2013; this upgrade in production standards would cost the country $1 billion. The restrictions on auto imports to support the new gasoline standards began in 2012 (Rustambekov and Rzaev, 2012).

By 2018, Azerbaijan was planning to build a new oil refinery in Sangachaly that would have the capacity to produce 15 Mt/yr. Another oil refinery that would refine Azerbaijani oil was under construction by Azerbaijan in Ceyhan, Turkey; the planned capacity of the Ceyhan refinery was 10 Mt/yr. In 2012, only a small fraction of the crude oil was being refined at local refineries. Most of the crude oil was exported by way of pipelines. Azerbaijan had three export pipelines—the Baku-Tbilisi-Ceyhan (BTC), the Baku-Novorossiysk, and the Baku-Supsa—and about 80% of the petroleum was exported through the BTC pipeline (U.S. Energy Information Administration, 2013).

Outlook

Azerbaijan's strong economic growth from 2003 through 2008 was fueled by increased crude oil exports and opened new opportunities for the country. In the past few years, the country has made serious attempts to diversify its economy. It has started developing new polymetallic deposits containing gold, silver, and copper; it is reviving its steel and aluminum production capacities; and it has made great strides in continuing to develop its resources of natural gas. The country is also reinvesting the proceeds from exporting hydrocarbons in other economic sectors, such as in construction. Azerbaijan is investing resources in building petroleum processing and petroleum transporting facilities, both domestically and abroad, to provide export opportunities for Azerbaijani oil as well as to expand the national petroleum industry beyond extraction of crude oil.

In the next few years, it is likely that oil production will not increase as fast as it did in the previous 10 years but it will nonetheless have moderate and controlled growth rates. Natural gas production, on the other hand, has the potential to double in the next 10 years. Gold and copper mining is likely to increase when the Gedabek and the Gosha Bulag Mines reach their production capacities (U.S. Central Intelligence Agency, 2013).

References Cited

1news.az, 2011, OOO "AAC" provelo uspeshnye ispytaniya proizvodstva gazobetona [OOO AAC successfully conducted testing of aerated concrete production]: 1news.az, July 19. (Accessed November 1, 2013, at http://1news.az/economy/20110719035056597.html.)

Anglo Asian Mining PLC, 2013, The cash generative gold producer—Annual report and accounts 2012: Anglo Asian Mining PLC, 52 p. (Accessed November 1, 2013, at http://www.angloasianmining.com/media/pdf/Anglo Asian Mining plc_Annual Report 2012.pdf.)

Interfax.az, 2013a, Dobycha zolota v Azerbaidzhane za 2012 god sokratilas' na 12%, serebra na 49% [In 2012, production of gold in Azerbaijan decreased by 12%, silver, by 49%]: Interfax.az, January 18. (Accessed November 1, 2013, at http://interfax.az/view/563584/ru.)

Interfax.az, 2013b, Proizvodstvo stroimaterialov v Azerbaidzhane v 2012 godu uvelichilos' na 17% [In 2012, production of construction materials in Azerbaijan increased by 17%]: Interfax.az, January 25. (Accessed November 1, 2013, at http://interfax.az/view/564315/ru.)

Kavkasia.net, 2012, V Azerbaidzhane otkrylsya kompleks Gyandzhinskogo alyuminievogo zavoda [The Ganja aluminum production complex opened in Azerbaijan]: Kavkasia.net, January 21. (Accessed November 1, 2013, at http://kavkasia.net/Azerbaijan/2012/1327205326.php.)

Lenta.ru, 2012, Azerbaidzhan reshil udvoit' dobychu gaza k 2025 godu [Azerbaijan decided to double its natural gas production by 2025]: Lenta.ru, April 10. (Accessed November 1, 2013, at http://lenta.ru/news/2012/04/10/gas.)

Mamedov, Enver, 2012, Rastsvet Tsvetmeta [Blossoms of non-ferrous metallurgy]: RegionPlus.az. (Accessed November 1, 2013, at http://www.regionplus.az/ru/articles/view/1537.)

Mineral.ru, 2012a, Baku i Ankara namereny podpisat' na sleduyushey needle soglashenie po transanatoliyskomu gazoprovodu [Baku and Ankara intend to sign an agreement on Transanatolian pipeline next week]: Mineral.ru, March 14. (Accessed November 1, 2013, at http://www.mineral.ru/News/47928.html.)

Mineral.ru, 2012b, Dobycha gaza v Azerbaidhzane v 2015 godu sostavit 20 mlrd kub.m [In 2015, Azerbaijan will produce 20 billion cubic meters of gas]: Mineral.ru, June 7. (Accessed November 1, 2013, at http://www.mineral.ru/News/48875.html.)

Mineral.ru, 2012c, Na 'Azeri-Chirag-Gyuneshli' 15 let tomu nazad byla dobyta pervaya neft' [Azeri-Chirag-Guneshli produced its first oil 15 years ago]: Mineral.ru, November 8. (Accessed November 1, 2013, at http://www.mineral.ru/News/50700.html.)

Mineral.ru, 2012d, Na mestorozhdeniyah 'Azeri-Chirag-Gyuneshli' ekspluatiruetsya 86 skvazhin [Azeri-Chirag-Guneshli deposits exploit 86 drill holes]: Mineral.ru, August 27. (Accessed November 1, 2013, at http://www.mineral.ru/News/49841.html.)

Mineral.ru, 2012e, Nachal dobychu vtoroi proizvoditel' zolota v Azerbaidzhane [Second gold producer started mining in Azerbaijan]: Mineral.ru, November 24. (Accessed November 1, 2013, at http://www.mineral.ru/News/50904.html.)

News.day.az, 2012a, Azerbaidzhan budet bol'she dobyvat' zolota, serebra I medi [Azerbaijan will produce more gold, silver, and copper]: News.day.az, November 1. (Accessed November 1, 2013, at http://news.day.az/economy/363902.html.)

News.day.az, 2012b, V Azerbaidzhane poyavilsya novyi stroitel'nyi brend [Azerbaijan has a new construction brand]: News.day.az, January 11. (Accessed November 1, 2013, at http://news.day.az/economy/310352.html.)

News.mail.ru, 2013, V Azerbaidzhane ustanovili record dobychi zolota [Azerbaijan set a record in gold production]: News.mail.ru, September. (Accessed November 1, 2013, at http://news.mail.ru/inworld/azerbaijan/economics/14748967.)

Rustambekov, Bakhram, 2013, Azerbaidzhan sokratil eksport alyuminia [Azerbaijan reduced its aluminum export]: 1news.az, September 24. (Accessed November 1, 2013, at http://1news.az/economy/20130924042623752.html.)

Rustambekov, Bakhram, and Rzaev, Timur, 2012, Obnarodovana data nachala proizvodstva v Azerbaidzhane benzina standarta 'Evro-3' [The start date for EURO-3 gasoline production in Azerbaijan is announced]: 1news.az, October 23. (Accessed November 1, 2013, at http://www.1news.az/economy/oil_n_gas/20121023021031139.html.)

State Statistical Committee of the Republic of Azerbaijan, 2012, Industry of Azerbaijan: State Statistical Committee of the Republic of Azerbaijan. (Accessed November 1, 2013, at http://www.stat.gov.az/menu/6/statistical_yearbooks/source/stat-yearbook_2013.zip.)

State Statistical Committee of the Republic of Azerbaijan, 2013, Statistical yearbook of Azrbaijan: State Statistical Committee of the Republic of Azerbaijan. (Accessed November 1, 2013, at http://www.stat.gov.az/menu/6/statistical_yearbooks/source/industry_2013.zip.)

Trend.az, 2011, Gazobeton obespechit ekologicheskuyu bezopasnost' stroitel'nykh ob'ektov Azerbaidzhana [Aerated concrete will provide ecological security for construction projects in Azerbaijan]: Trend.az, August 3. (Accessed November 1, 2013, at http://www.trend.az/capital/business/1913663.html.)

U.S. Central Intelligence Agency, 2013, Azerbaijan, *in* The world factbook: U.S. Central Intelligence Agency, June 8. (Accessed November 1, 2013, at https://www.cia.gov/library/publications/the-world-factbook/geos/aj.html.)

U.S. Energy Information Administration, 2013, Azerbaijan: U.S. Energy Information Administration Country Analysis Brief, September 10. (Accessed November 1, 2013, at http://www.eia.gov/countries/cab.cfm?fips=AJ&trk=p2.)

TABLE 1

AZERBAIJAN: PRODUCTION OF MINERAL COMMODITIES[1]

(Metric tons unless otherwise specified)

Commodity		2008	2009	2010	2011	2012
METALS						
Alumina		164,879	9,600	--	6,200	15,000 e
Aluminum, primary and secondary		61,607	--	--	20,000 e	55,000 e
Copper ore, metal content		--	--	184	611 r	502
Gold	kilograms	--	333	1,900	1,775	1,563
Iron ore, marketable:						
Gross weight		28,100	--	57,800	214,300	215,000 e
Fe content e		14,900	--	32,900	113,600	114,000
Silver	kilograms	--	--	1,500	1,217	653
Steel:						
Crude		74,800	78,874	128,600	234,000 r	267,700
Pipes		28,196	6,918	36,545	98,500	61,800
INDUSTRIAL MINERALS						
Bentonite		40,700	10,581	18,073	55,000 r	60,000
Bromine e		3,500	3,400 r	3,500	3,500	3,500
Caustic soda		20,635	7,041	6,220	9,800	86
Cement		1,594,900	1,286,300	1,278,800	1,425,000	1,966,000
Gypsum		38,375	45,630	49,200	100,800	150,500
Iodine e	kilograms	300,000	300,000	300,000	350,000	350,000
Lime, construction		1,318	684	802	2,229	49,000
Limestone		1,363,978	1,228,775	1,173,863	1,200,000 e	1,100,000 e
Salt, marketable		7,527	5,466	4,449	18,848	5,345
Sand, construction		1,247,200	877,200	1,178,000	1,335,200	2,211,200
Sulfuric acid		39,400	12,400	10,100	15,500	15,100
MINERAL FUELS AND RELATED MATERIALS						
Natural gas	million cubic meters	16,337	16,325	16,673	16,361 [2]	17,242
Petroleum:						
Crude:						
In gravimetric units		44,720,275	50,416,000	50,838,000	45,626,000	42,982,000
In volumetric units e	42-gallon barrels	325,000,000	351,000,000	352,000,000	331,610,000 [3]	320,667,000 [3]
Refinery products:						
In gravimetric units		6,885,300	6,032,790 r	6,169,600 r	5,150,000 e	4,800,000
In volumetric units	42-gallon barrels	58,807,000	51,520,000 r	52,688,000 r	43,981,000 r	40,992,000

eEstimated; estimated data are rounded to no more than three significant digits. rRevised. -- Zero.

[1]Table includes data available through October 30, 2013.

[2]Only natural gas sold as a commodity.

[3]Reported figure.

TABLE 2
AZERBAIJAN: STRUCTURE OF THE MINERAL INDUSTRY IN 2012[1]

(Metric tons unless otherwise specified)

Commodity	Major operating companies and major equity owners	Locations or deposit names	Annual capacity[e]
Alumina	Ganja refinery	Ganja	450,000
Aluminum	OJSC Azerbaijan Aluminum [Azeraluminum (Azeral)] (Det. AL Aluminum)	Sumqayit	60,000
Do.	OJSC Azerbaijan Aluminum [Azeraluminum (Azeral)]	Ganja smelter	50,000
Alunite ore	Zaglik alunite mining directorate	Zaylik, Dashcasan region	600,000
Cement	NA	Plants in Karadagly and the Tavuzcay region	2,000,000 [2]
Clays, bentonite	NA	Dash-Salakhlinskoye deposit	100,000
Copper ore	Karadagskiy complex	Samkir region	30,000
Gold kilograms	Anglo Asian Mining PLC [R.V. Investment Group Services, 51% and Government, 49%]	Gedabek	2,000
Do.	Azerbaijan International Mineral Resources Operating Co. Ltd. (AIMROC)	Chovdar deposit, near Ganja	NA
Iodine and bromine	NA	Plants in Baku, Karadagly, and Neftcala	NA
Iron ore, marketable	Dashkasan mining directorate	Daskasan region	50,000
Lime	AAC Co.	Plant in Baku region	65,000
Natural gas, processing	NA	Plant in Karadagly region	NA
Petroleum and natural gas:			
Crude petroleum and gas condensate	Azerbaijan International Operating Co. (AIOC), in conjunction with BP p.l.c., Chevron Corp., State Oil Company of Azerbaijan Republic (SOCAR), Total S.A., Inpex Corp., Statoil ASA, Exxon Mobil Corp., Türkiye Petrolleri A.O. (TPAO), Itochu Corp., Devon Energy Corp., and Delta Hess (joint venture of Delta Oil and Hess Corp.)	Azeri-Chirag-Guneshli (ACG) offshore oilfields in the Caspian Sea	55,000,000
Natural gas billion cubic meters	International consortium consisting of BP p.l.c., Statoil ASA, OAO Lukoil, Oil Industries' Engineering and Construction (OIEC), State Oil Company of Azerbaijan Republic (SOCAR), Total S.A., and Türkiye Petrolleri A.O. (TPAO)	Shah-Deniz gas condensate field	17.5
Refined petroleum	NA	Azerneftyag refinery in Baku	12,000,000 [3]
Do.	NA	Heydar Aliyev Baku refinery	8,000,000 [3]
Rock salt	NA	Hehram and Pusyan deposits, Naxcivan region	2,500,000
Steel:			
Crude	Baku Steel Works	Baku	400,000
Pipe, tubes	Azerboru JSC	Sumqayit	400,000
Ingots	Baku Steel Casting	Baku	NA

[e]Estimated. Do. Ditto. NA Not available.
[1]Many location names have changed since the breakup of the Soviet Union. Many enterprises, however, are still named or commonly referred to based on the former location name, which accounts for discrepancies in the names of enterprises and that of locations.
[2]Capacity estimates are totals for all enterprises that produce cement.
[3]Capacity for crude petroleum distillation.

THE MINERAL INDUSTRY OF BELARUS

By Elena Safirova

Belarus' mineral production enterprises included a potash mining company, three metallurgical steel plants, a nitrogen production enterprise, and two crude oil refineries. Belarus was the third-ranked country among the world's potash producers following Canada and Russia (Jasinski, 2013). The country's only mineral production enterprise that played a major role in world markets was its potash mining firm OAO Belaruskali. Although Belarus does not have significant sources of fuel minerals on its territory, it had a number of energy infrastructure establishments (oil pipelines, gas pipelines, and two large oil refineries) that positioned the country as an important player in the export of oil and gas to Europe from Russia.

In 2012, Belarus continued to depend on Russia for its domestic energy needs and for the supply of crude oil to operate its two refineries. The Government continued to work with Russia, and also to establish relationships with other countries that could serve as alternative energy sources for Belarus, and to modernize its domestic industry to diversify the domestic supply of and demand for energy products. Together, Belarus, Kazakhstan, and Russia formed a Customs Union, which means that the members do not impose protectionist customs duties on each other. As a result, Belarus was able to pay reduced prices for natural gas compared with the prices paid by nonmember countries. Belarus moved toward establishing closer working relationships with other countries (such as Azerbaijan, Turkmenistan, and Venezuela) that could supply oil and natural gas should Belarus need them. The country was also trying to diversify its domestic energy supply. In 2012, Belarus was in the early stages of building a new 2,400-megawatt (MW)-capacity nuclear powerplant in the Grodno region that was expected to become operational by 2018. The country was also ramping up domestic exploration for oil and shale gas. Because the new production lines of Belarusian cement plants work on coal, this would alleviate domestic demand for natural gas (Interfax.by, 2012i; 2013b).

Minerals in the National Economy

In 2012, the country's real gross domestic product (GDP) increased by 1.5%, and the nominal GDP amounted to $63.3 billion.[1] The industrial production of Belarus contributed 31.8% to the Republic's GDP, out of which the mineral sector accounted for 1.4%. The total value of industrial production increased by 5.7%. In 2012, the value of mineral industry output decreased by 2.0% compared with that of 2011; the combined value of metallurgical production and products made out of metal increased by 4.5%, and the value of nonmetal mineral products decreased by 3.1% compared with that of 2011

---

[1] Where necessary, values have been converted from Belarusian rubles (BYR) to U.S. dollars (US$) at an annual average exchange rate of BYR8,335.86=US$1.00 for 2012 and from euro area euros (€) to U.S. dollars (US$) at an annual average exchange rate of €0.809=US$1.00 for 2012.

(National Bank of the Republic of Belarus, 2013; National Statistical Committee of the Republic of Belarus, 2013a, b).

The total value of foreign direct investment (FDI) in the Belarusian economy in 2012 was $10.4 billion, which was a 21.8% decrease compared with FDI in 2011. The mineral sector received only 1.0% of the total FDI. Russia provided 46.6% of all FDI in the Belarusian economy and was the main source of foreign investment in that year (National Statistical Committee of the Republic of Belarus, 2013b).

In 2012, Belarus exported $51.9 billion worth of goods and services and imported $49.0 billion. The main export category was mineral products, which accounted for 36.2% of the total export revenue. Other important export categories were chemicals (21.7%), equipment and machinery (17.9%), agricultural products and food (10.6%), and metals (5.5%), among others. The major export partner of Belarus was Russia, which received 35.4% of all exports, by value. It was followed by the Netherlands (16.5%), Ukraine (12.1%), Latvia (7.1%), Germany (3.8%), Lithuania (2.6%), and Poland (2.1%). The main import category was mineral products, which accounted for 39.4% of the total value of imports. It was followed by equipment and machinery (22.9%), chemicals (12.4%), metals (10.1%), and agricultural products and food (7.8%). The major import partner of Belarus was Russia, which supplied 59.3% of Belarus's goods and services imports, by value. Other significant import partners were Germany (5.9%), China (5.1%), Ukraine (5.0%), Poland (2.9%), and Italy (2.1%) (National Statistical Committee of the Republic of Belarus, 2013a, b).

Production

In 2012, Belarus increased production of all steel products except steel cord—output of steel pipes increased by 8.0% to 226,900 metric tons (t), and that of crude steel, by 3.24% to 2.87 million metric tons (Mt). Other minerals for which production increased included cement, the output of which increased by 6.6% to 4.9 Mt; refined petroleum, by 5.8% to 21.7 Mt; and sulfuric acid, by 4.0% to 936,000 t. At the same time, production of peat for horticultural use decreased by 36.2% to 269,000 t; potash, by 8.8% to 4.8 Mt in K_2O equivalent; steel cord, by 6.6% to 87,900 t; and peat for fuel use, by 5.1% to 2.7 Mt. Other production data are in table 1.

Structure of the Mineral Industry

Most of the mineral industry enterprises were consolidated under the State Concern for Oil and Chemistry, known as Belneftekhim. Belneftekhim included Belaruskali, which was one of the leading potash producers in the world; OAO Grodno Azot, which specialized in the production of ammonia, nitrogenous fertilizers, and sulfuric acid; two oil refineries (OAO Naftan and OAO Mozyr NPZ), which together had a total annual throughput capacity of 22 Mt; and almost 50 other organizations. Belarus had adopted an industry privatization plan and created a list of enterprises that could be privatized. The list included only smaller production facilities, however, and excluded all of the country's enterprises of national significance in terms of contribution to Belarus's GDP (Romanchuk, 2011).

In 2012, Belarus continued discussions with potential foreign investors concerning the sale of one of the flagship enterprises, such as Belaruskali or one of the country's petroleum refineries. The economic situation in Belarus, however, was much better than in the years since, and the focus of the discussions was on a potential sale of a minority share (between 10% and 25%) of Belaruskali. The Government stated that the true market value of Belaruskali was between $30 billion and $40 billion and said that it was not willing to sell a part of the company for a price corresponding to a lower total value. Negotiations and discussions with companies from Middle East countries, China, Europe, and India were reported, but as of yearend, no agreements had been reached (Interfax.by, 2012b–d, f).

Commodity Review

Metals

Iron and Steel.—The OAO Byelorussian Steel Works (BMZ) was the predominant producer of iron and steel in Belarus. In 2012, BMZ produced 2.7 Mt of steel, 2.2 Mt of rolled steel, 124,600 t of steel pipes, and 87,900 t of steel cord. Production of steel increased by 3.2% compared with that in 2011; rolled steel, by 1.2%; and steel pipes, by 6.3%. Production of steel cord, however, decreased by 6.6%. BMZ was a minimill that was originally built in 1984 and, as of 2012, employed 15,000 workers. In 2011, it was transformed from a Government enterprise into a corporation, but a 100% ownership was retained by the Government (OAO Byelorussian Steel Works, 2013).

In August, together with 13 other companies, BMZ became a part of a newly formed holding company named Belarusian Metallurgical Co. (BMK). BMZ was expected to serve as a managing company of BMK. Other companies that joined BMK were OJSC Rechitsa Metizny plant, JSC Mogilev metallurgical works, Minsk bearing plant, OJSC Belvtorchermet, OJSC Belcvetmet, OJSC NII Bearing, and OAO BelNIILIT. The main reason for the creation of the company was to achieve cost reduction through close cooperation among the enterprises in obtaining loans, as well as in logistics, marketing, personnel management, and research and development. BMK would have three divisions—metallurgical production, raw materials provision, and research and development (Interfax.by, 2012h).

In mid-2012, BMZ started construction of a new rolling mill. BMZ expected that, after the completion of the new mill, the company would be able to increase its rolled-steel output and start making new products, such as rolled steel designed for rebar production. The first stage of the project would result in a new mill with a production capacity of 700,000 metric tons per year (t/yr) of rolled steel; in the second stage, the production capacity would be increased to 1 million metric tons per year (Mt/yr). The project was being financed by the Eurasian Bank for Development (EABR) and OAO Belarusbank, which would provide $174.3 million and €176.5 million ($218.2 million), respectively. BMZ was expected to be able to export about 75% of the output from the new mill. The mill was planned to come online in December 2014 and to provide 200 new jobs at BMZ (Interfax.by, 2013a).

In December, Minsk Motor Works (MMZ), which was the only producer of diesel engines in Belarus, started construction of a new plant that would produce cast iron. The total cost of the project was expected to be $175 million, including $47.6 million for the cost of construction. After completion of the first stage of the project in 2014, the capacity of the new plant was expected to be 18,000 t/yr. By that time, the new plant would employ about 370 people. At the second stage, MMZ was planning to purchase and install additional equipment, including an automated line for the production of ingots. By the time that the second stage is completed in 2017, the total plant employment was expected to be 930 workers. MMZ was planning to use cast iron in its production of diesel engines. The total design capacity of the new plant was planned to be 50,000 t/yr of cast iron (Interfax.by, 2012e).

Industrial Minerals

Cement.—In 2012, Belarusian cement plants produced 4.91 Mt of cement, which was a 6.6% increase compared with the 2011 production level. Belarus had three main cement producers—OAO Belarusian Cement Plant (BCZ), which was also known as the Kostyukovichi cement plant; OAO Krasnoselskstroymaterialy, which was also known as the Volkowysk cement plant, and OAO Krichevtsementnoshifer (Ernst & Young, 2013).

In July 2007, the Belarusian Government made a decision to increase the production capacity of its cement plants by more than 5 Mt/yr. In particular, the Government planned to build a new production line with production capacity of 1.8 Mt/yr at each of the three plants. As a result, the total capacity of all Belarusian cement plants was expected to double. The new lines were designed to use dry cement production technology that had not been used before in Belarus. In addition, a decision was made to switch the existing production lines from natural gas to coal as the primary energy source, which would reduce production costs and, ultimately, the cost of cement. China International Trust and Investment Corp. (CITIC) obtained a contract to build all three plants. Although the construction of all three plants was delayed, reportedly because of quality and technological issues with the equipment being installed, the first line (at Krasnoselsk) came online in April and the second one (at BCZ) became operational in July. The new line in Krichev was expected to be commissioned in June 2013 (Interfax.by, 2012j).

In January, the Government announced that it had decided to compensate all three cement plants for a portion of interest paid on loans that were taken out for construction of the new plants. For the loans taken out in Belarusian rubles, compensation would amount to 50% of the current refinancing rate offered by the National Bank of Belarus. For the loans taken out in foreign currency, the Government offered compensation in the amount of 50% of the actual rate at which the loans were taken.

The funds paid as compensation were to be used for investment projects undertaken by the cement plants (Interfax.by, 2012g).

By 2014, when all enterprises were projected to produce cement at levels close to capacity, total production of cement in Belarus was expected to reach about 9.2 Mt. Domestic demand for cement in the country was somewhere between 4 and 5 Mt/yr, and Belarus would export the rest of the output, primarily to Russia. The cement plants were planning to create a cement holding company that would focus its activity on marketing and logistics involved in exporting about 5 Mt of the plants' output (TUT.by, 2012).

Potash.—OAO Belaruskali was one of the world's leading producers of potash fertilizers and had a 15% share of the world market. In 2012, the production of potash in Belarus decreased to about 4.84 Mt of potash in K_2O equivalent, or by 8.8%. Historically, potash was the leading export product from Belarus. In 2012, however, the export of potash fertilizers decreased by 21.9%, and the revenue from potash exports decreased by 17.9% to $2.7 billion. Only 55.7% of the potash produced was exported compared with 88.5% in 2011. Belaruskali was planning to increase production in 2013 by between 10% and 13% from the 2012 production level (Mineral.ru, 2013; OAO Belaruskali, 2013).

Belaruskali's Starobin potash deposit contains magnesium salt, rock salt, and sylvinite. Commercial levels of potash occur at depths of 400 to 1,200 meters (m) and deeper. The thickness of individual beds of potash varies from 4 to 20 m. In July, Belaruskali started operations at the Beryozovskiy section of the deposit, which increased the total annual capacity to 10.3 Mt/yr from 8.8 Mt/yr in 2011. The company was planning to further increase the capacity of the Beryozovskiy section further in 2014. Belaruskali's total capacity to produce potash fertilizers was expected to reach 11 Mt/yr by 2015 (Mineral.ru, 2012a, b; OAO Belaruskali, 2013).

In 2011, the Government held a tender for development of the Petrikovskoye potash salts deposit, which is located in the Gomel region and had proven resources of 236 Mt in K_2O equivalent. Belaruskali was named the winner of the tender but was required to find an investor to form a public company, with the Government's share set at a minimum of 25%. During 2012, the position of Belaruskali with respect to selecting the investor had changed. In July, Belaruskali was about to start accepting applications from prospective co-investors and was finishing up the detailed description of the investment project. In October, however, the decision was made that Belaruskali would be able to develop the deposit using a combination of its own funds and loans. In October, the Government issued a decree that amended the results of the 2011 tender and relieved Belaruskali from the obligation to bring in a foreign investor to develop the Petrikovskoye deposit. Preliminary construction work was expected to begin in 2013; mining of the deposit was expected to start in December 2019. The mine was expected to reach its full planned capacity by December 2021 (Interfax.by, 2012a).

During 2012, Belaruskali and its Russian counterpart, OAO Uralkali, were discussing the creation of a joint trading company, Soyuzkali. Before 2012, ZAO Byelorusskaya Kaliynaya Kompaniya (BKK) was the primary trader of Belarusian and Russian potash fertilizers on the international market, and it held a 43% share of world potash exports. Soyuzkali would be registered in Switzerland, and Uralkali and Belaruskali would each have a 50% share in the new company. Both parties saw potential benefits from rebranding and restyling of the BKK, primarily because the new company would have access to less expensive loans. Also, it was expected that the new trading company would take on additional functions of trading nitrogen and phosphorus fertilizers. Soyuzkali was expected to be registered in the beginning of 2013 and to start trading by the middle of the year (Mineral.ru, 2012c).

Outlook

Belarus is expected to continue to be a major supplier of potash to world markets and to develop additional mines in the Starobin and Petrikovskoye deposits. In 2013, it is also likely to become a regional exporter of cement. Although Belarus did not sell any of its flagship state enterprises in 2012, Belarus still may decide to sell some of them, such as Belaruskali, BMZ, Grodno Azot, and the Mozyr NPZ and Naftan refineries, in the future, depending on the country's financial situation. If some of those facilities are privatized, the direction of enterprise development may be affected. After the sale of its Beltransgaz pipeline to OAO Gazprom in 2011, the Belarusian Government has secured for the country advantageous natural gas prices that will likely provide some stability to the country's overall economy for the next few years. The future direction of Belarus' energy sector is likely to depend on political relations with Russia and on the country's ability to develop and maintain a reliable business network with countries outside of the Commonwealth of Independent States community.

References Cited

Ernst & Young, 2013, Obzor tsementnoy otrasli stran Tamozhennogo soyuza [An overview of the cement sector in Customs Union member countries]: Ernst & Young. (Accessed September 5, 2013, at http://www.ey.com/Publication/vwLUAssets/EY-Cement-industry-in-Customs-Union-countries-Rus/$FILE/EY-Cement-industry-in-Customs-Union-countries-Rus.pdf.)

Interfax.by, 2012a, "Belarus'kaliy" mozhet ne privlekat' vneshnie investitsii dlya razrabotki Petrikovskogo mestorozhdeniya khlorkaliya -- ukaz [Belaruskali is allowed not to use foreign investment for development of Petrikovskoye potash deposit -- decree]: Interfax.by, October 22. (Accessed September 5, 2013, at http://www.interfax.by/news/belarus/118852.)

Interfax.by, 2012b, Belarus obsuzhdaet prodazhu Indii 10–15% aktsiy "Belaruskaliya"—V.Semashko [Belarus is discussing selling 10%–15% of Belaruskali shares to India—V.Semashko]: Interfax.by, November 16. (Accessed September 5, 2013, at http://www.interfax.by/news/belarus/120408.)

Interfax.by, 2012c, Indiya po-prezhnemu interesuetsya priobreteniem "Belarus'kaliya" – posol Indii [India is still interested in buying Belaruskali—India's ambassador]: Interfax.by, January 24. (Accessed September 5, 2013, at http://www.interfax.by/news/belarus/105463.)

Interfax.by, 2012d, Lukashenko: Tsena "Belaruskaliya" – $30–$32 mlrd, kontrol'nyi paket prodavat'sya ne budet [Lukashenko—The price of Belaruskali is $30–$32 billion, control of the company is not for sale]: Interfax.by, July 17. (Accessed September 5, 2013, at http://www.interfax.by/news/belarus/113973.)

Interfax.by, 2012e, MMZ pristupil k stroitel'stvu zavoda po vypusku vysokotochnogo chugunnogo lit'ya stoimost'yu 110.6 mln evro [MMZ started construction of a €110.6 million high-precision cast iron plant]: Interfax.by, December 6. (Accessed September 5, 2013, at http://www.interfax.by/news/belarus/121528.)

Interfax.by, 2012f, Pravitel'stvo prodolzhaet peregovory o prodazhe miniritarnoy doli "Belarus'kaliya" – V. Semashko [The Government continues negotiations about selling a minority share in Belaruskali—V. Semashko]: Interfax.by, March 19. (Accessed September 5, 2013, at http://www.interfax.by/news/belarus/107943.)

Interfax.by, 2012g, Pravitel'stvo vozmestit tsementnym zavodam chast' protsentov po kreditam [The Government will compensate the cement plants for a portion of interest paid on loans]: Interfax.by, January 18. (Accessed September 5, 2013, at http://www.interfax.by/news/belarus/105203.)

Interfax.by, 2012h, Sozdan holding "Belorusskaya Metallurgicheskaya Kompaniya" [Belarusian Metallurgical Co. holding is created]: Interfax.by, August 30. (Accessed September 5, 2013, at http://www.interfax.by/news/belarus/116164.)

Interfax.by, 2012i, V Belarusi mozhno dobyvat' neft', slantsevyi i poputnyi gaz, no na zalezhi priridnogo gaza nadezhdy net–uchenye [In Belarus one can produce oil shale and petroleum gas, but there is no hope to find natural gas—Scientists]: Interfax.by, April 6. (Accessed September 5, 2013, at http://www.interfax.by/news/belarus/114494.)

Interfax.by, 2012j, V Belarusi prodleny sroki vvoda novykh liniy na trekh tsementnykh zavodakh [In Belarus, opening dates for three new cement lines are postponed]: Interfax.by, January 26. (Accessed September 5, 2013, at http://www.interfax.by/news/belarus/105575.)

Interfax.by, 2013a, BMZ privlekaet ot EABRa i Belarusbanka 284 mln evro na stroitel'stvo hovogo prokatnogo stana [Belarus is taking €284 million from EABR and Belarusbank for construction of the new rolling mill]: Interfax.by, February 21. (Accessed September 5, 2013, at http://www.interfax.by/news/belarus/125659.)

Interfax.by, 2013b, Glava Rosatoma posetit stroitel'stvo belorusskoy AES [The head of Rosatom will visit the construction site of Belarusian nuclear powerplant]: Interfax.by, January 30. (Accessed September 5, 2013, at http://www.interfax.by/news/belarus/124506.)

Jasinski, S.M., 2013, Potash: U.S. Geological Survey Mineral Commodity Summaries 2013, p. 122–123.

Mineral.ru, 2012a, "Belarus'kaliy" uvelichil dobychu na 2 mln t rudy v god [Belaruskali increased production by 2 million tons a year]: Mineral.ru, July 6. (Accessed September 5, 2013, at http://www.mineral.ru/News/49218.html.)

Mineral.ru, 2012b, "Belarus'kaliy" vvel v ekspluatatsiyu pervuyu ochered' Beryozovskogo rudnika [Belaruskali put into operation the first line of Beryozovsky Mine]: Mineral.ru, July 10. (Accessed September 5, 2013, at http://www.mineral.ru/News/49263.html.)

Mineral.ru, 2012c, Potentsial'nye vygody sozdaniya 'Soyuzkaliya' [Potential benefits of Soyuzkali creation]: Mineral.ru, August 29. (Accessed September 5, 2013, at http://www.mineral.ru/News/49876.html.)

Mineral.ru, 2013, "Belarus'kaliy v 2012 g. snizil valyutnuyu vyruchku na 17,9% -- do 2,7 mlrd dol. [In 2012, Belaruskali reduced its export revenue by 17.9% to $2.7 billion]: Mineral.ru, February 25. (Accessed September 5, 2013, at http://www.mineral.ru/News/51983.html.)

National Bank of the Republic of Belarus, 2013, Official average exchange rate of the Belarusian ruble versus foreign currencies for 2012: National Bank of the Republic of Belarus. (Accessed September 5, 2013, at http://www.nbrb.by/engl/statistics/Rates/AvgRate/?yr=2012.)

National Statistical Committee of the Republic of Belarus [Belstat], 2013a, Belarus v tsifrakh [Belarus in figures]: Minsk, Belarus, National Statistical Committee of the Republic of Belarus. (Accessed September 5, 2013, at http://belstat.gov.by/homep/ru/publications/belarus_in_figures/2013/belarus_in_figures_2013.rar.)

National Statistical Committee of the Republic of Belarus [Belstat], 2013b, Statisticheskoe obozrenie Belarusi 2012 [2012 Belarus statistical review]: Minsk, Belarus, National Statistical Committee of the Republic of Belarus. (Accessed September 5, 2013, at http://belstat.gov.by/homep/ru/publications/review/January-December- 2012.php.)

OAO Belaruskali, 2013, Home page: OAO Belaruskali (Accessed September 5, 2013, at http://www.kali.by/russian/bel_main.html.)

OAO Byelorussian Steel Works, 2013, Production: OAO Byelorussian Steel Works. (Accessed September 5, 2013, at www.belsteel.com/about/production.php.)

Romanchuk, Yaroslav, 2011, Prodavaemoe i neprodavaemoe belorusskoi ekonomiki [Sellables and unsellables of the Belarusian economy]: Nauchno-Issledovatel'skiy Tsentr Mizesa [von Mises Research Center], December 26. (Accessed September 5, 2013, at http://liberty-belarus.info/Privatizatsiya/Strasti-po-privatizatsii2011.html.)

TUT.by, 2012, Belarus planiruet eksportirovat' bole 5 mln ton tsementa v 2013 gody [In 2013, Belarus is planning to export more than 5 million tons of cement]: TUT.by, April 30. (Accessed September 5, 2013, at http://news.tut.by/economics/286766.html.)

TABLE 1
BELARUS: PRODUCTION OF MINERAL COMMODITIES[1]

(Thousand metric tons unless otherwise specified)

Commodity[2]		2008	2009	2010	2011	2012
METALS						
Steel:						
Crude		2,660	2,449	2,672	2,779	2,869
Rolled		2,387	2,299	2,458	2,457	2,596
Pipes	metric tons	145,000	107,400	183,200	210,100	226,900
Cord	do.	96,500	68,900	92,900	94,100	87,900
INDUSTRIAL MINERALS						
Cement		4,219	4,350	4,531	4,604	4,906
Lime		900	787	805	741	719
Nitrogen, N content of ammonia	metric tons	743,400	828,600	835,900	803,900	815,200
Potash, K_2O equivalent		4,968	2,485	5,223	5,306	4,840
Salt[3]	metric tons	1,476,000	1,695,100	1,700,000	1,700,000 e	1,700,000 e
Sulfuric acid		857	833	891	900 e	936
MINERAL FUELS AND RELATED MATERIALS						
Natural gas	million cubic meters	203	205	213	222	218
Peat:						
Horticultural use		395	272	241	422 r	269
Fuel use		2,361	2,216	2,352	2,823 r	2,679
Total		2,756	2,488	2,593	3,245 r	2,948
Petroleum:						
Crude		1,740	1,720	1,700	1,682	1,660
Refined		21,305	21,634	16,455	20,474	21,668

eEstimated; estimated data are rounded to no more than three significant digits; may not add to totals shown. rRevised. do. Ditto.
[1]Table includes data available through September 5, 2013.
[2]In addition to the commodities listed, Belarus had also produced dolomite and synthetic diamond, but available information is inadequate to make reliable estimates of output.
[3]Includes byproduct salt from potash production.

TABLE 2
BELARUS: STRUCTURE OF THE MINERAL INDUSTRY IN 2012

(Metric tons)

Commodity	Major operating companies and major equity owners	Location of main facilities	Annual capacity[e]
Cement	OAO Krasnoselskstroymaterialy	Hrodzyenskaya Voblasts'	2,700,000
Do.	OAO Krichevzementnoshifer	Mahylyowskaya Voblasts'	1,800,000
Do.	OAO Belarusian Cement Plant (BCZ)	do.	2,900,000
Diamond	Gomel Production Association Kristall	Homyel'skaya Voblasts'	NA
Nitrogen	OAO Grodno Azot (Belneftekhim)	Hrodzyenskaya Voblasts'	950,000 [1]
Peat, fuel use	Production at 31 enterprises that produce mainly briquets	All regions of the country	5,000,000 [2]
Petroleum:			
Crude	NGDU Rechitsaneft (Belneftekhim)	Rechitskoye, Ostashkovichskoye, Vishanskoye, Tishkovskoye, and Yuzhno-Ostashkovichskoye deposits, southeastern part of the country	2,000,000
Refined	OAO Mozyr NPZ (Government, 42.7%, and Slavneft, 42.5%)	Homyel'skaya Voblasts'	10,000,000 [3]
Do.	OAO Naftan (Novopolotsk NPZ)	Vitsyebskaya Voblasts'	12,000,000 [3]
Potash	OAO Belaruskali (Belneftekhim)	Starobin deposit, Minskaya Voblasts'	6,300,000 [4]
Steel:			
Crude	OAO Byelorussian Steel Works (BMZ) (Government, 100%)	Zhlobin, Homyel'skaya Voblasts'	2,700,000
Pipe	do.	do.	125,000
Rolled	do.	do.	2,300,000
Do.	OAO Mogilev Metallurgical Works [Byelorussian Steel Works (BMZ)]	Mahylyowskaya Voblasts'	120,000

[e]Estimated; estimated data are rounded to no more than three significant digits. Do., do. Ditto. NA Not available.
[1]N content of ammonia.
[2]Total peat for fuel use.
[3]Crude throughput.
[4]K_2O equivalent.

The Mineral Industries of Belgium And Luxembourg

By Alberto Alexander Perez

BELGIUM

Belgium was not a significant mineral producer, but it was a significant mineral processor and metals manufacturer. In 2011 (the latest year for which data were available), Belgium produced about 5.7% of the total world production of zinc. Belgium also produced 4.3% of the crude steel output in the European Union (EU) and 2.6% of the EU's cement production. It was also a significant cobalt producer, although available information was not sufficient to determine what percentage of EU or world output Belgium produced (Tolcin, 2013, p. 84.11; van Oss, 2013, p. 16.30–16.33; World Steel Association, 2013, p. 1).

In 2012, Belgium's gross domestic product (GDP) was $484.7 billion, which was a 0.2% decrease in real GDP compared with that of the previous year. The largest share of Belgium's GDP in 2012 was accounted for by services (77%) followed by industry (22.3%) and agriculture (0.7%) (U.S. Central Intelligence Agency, 2013).

Belgium's economy depended considerably on trade in goods and services, both for domestic consumption and for reexport. Belgium traded mostly with its EU partners; 70% of all Belgian exports went to EU members, and 67.6% of its imports came from EU members. Its main trading partners in 2012 were, in order of value, Germany, which accounted for 18% of Belgium's total exports and 14.2% of its imports; France, 16.1% of exports and 10.6% of imports; the Netherlands, 13% of exports and 20.9% of imports; the United Kingdom, 7.3% of exports and 5.5% of imports; the United States, 5.3% of exports and 6.1% of imports; Italy, 4.4% of exports; and Ireland, 4.4% of imports (European Commission, 2014e, f).

In 2012, Belgium exported $17.4 billion and imported $29.4 billion worth of goods and services to and from the United States, respectively. The traded goods were principally chemicals, machinery, miscellaneous manufactured goods, petroleum and coal products, and transportation equipment (U.S. Census Bureau, 2014a, b).

With respect to mineral commodity trade among EU countries and non-EU countries, Belgium received 2.9% of the EU's total imports of mineral fuels, lubricants, and related materials and supplied 9.2% of the EU's exports of these materials. Belgium also received 7.6% of the EU's imports of raw materials and supplied 5.6% of the EU's raw materials exports (European Commission, 2014a–d).

Belgium is a participant in the Benelux customs union, along with Luxembourg and the Netherlands. The Benelux customs union is an economic union aimed at reinforcing cross-border economic and legislative cooperation among the three countries (Benelux Parlement, 2014).

Minerals in the National Economy

Trading of diamond and the processing of metals were primary mineral industries in Belgium. The country had no economically exploitable reserves of coal or metallic ores in 2012.

Belgium imported substantial quantities of raw materials, and the metal processing industries, particularly steelmaking, were significant to the Belgian economy. Belgium was the 21st-ranked steel producer in the world and the 8th-ranked producer in the EU in 2012, measured by production tonnage. The country produced 7.4 million metric tons (Mt) of steel in 2012 compared with 8 Mt in 2011 (World Steel Association, 2013, p. 8–9).

Umicore S.A. was one of Europe's leading metal recyclers and processors; it had major facilities in Hoboken, Belgium, and was headquartered in Brussels. Nyrstar N.V. was a leading producer of zinc, by volume, in the world; it was headquartered in Balen, Belgium (Nyrstar N.V., 2013a, p. 52; Umicore S.A., 2013).

According to the Antwerp World Diamond Centre, Antwerp was the center of the world's open rough diamond market. Antwerp hosts 1,850 diamond companies and 4,500 diamond dealers, and about 10,000 people worked in the industry in the city (Antwerp World Diamond Centre, 2013a).

Production

In 2012, Belgium's production of cobalt increased by 32%, whereas production of pig iron decreased by 14% and zinc, by 11%. Only industrial minerals were mined. The refining of copper, minor metals (cadmium, cobalt, germanium, selenium, tellurium, and tin, among others), and zinc and the production of steel were the leading mineral processing industries in Belgium (table 1).

Structure of the Mineral Industry

Most facilities were privately owned either by Belgian companies or other EU companies. Among the most significant companies operating were Umicore, which had a catalysis division, an energy materials division, a performance materials division, and a recycling division in Belgium (as well as about another 76 industrial centers and 20 research centers throughout the world); and Nyrstar, which operated the Balen zinc smelter and the Overpelt plant and zinc alloy facility. Nyrstar also owned smelters in Auby, France; Budel, Netherlands; Clarksville, Tennessee; and Hobart and Port Pirie, Australia. The principal mining and mineral processing facilities in Belgium, with their locations and capacities, are listed in table 2 (Nyrstar N.V., 2013b; Umicore S.A., 2013).

Commodity Review

Metals

Iron and Steel.—In June 2012, ArcelorMittal reported that it had implemented a system for waste flue gas recovery at its plant in Ghent for the purpose of reducing emissions and lowering energy costs, as the gas recovered was used within the mill or sent to a local electricity generator. ArcelorMittal Gent estimated that by implementing the waste recovery system, the plant would reduce its energy consumption by 3% and reduce its CO_2 emissions by an equivalent of 170,000 metric tons per year (ArcelorMittal, 2012).

Zinc.—Nyrstar reported that production at the Balen smelter returned to normal in the second half of 2012 following a decrease in production during the first half of the year. In the first half of the year, production at the smelter was affected by a Belgian national industrial action and by an unplanned shutdown in the first quarter of 2012. In the second half of 2012, however, zinc metal production at the plant increased by 8% compared with production in the first half of 2011. Even so, zinc production for the full year 2012 ended 11% lower than that of the previous year (Nyrstar N.V., 2013a, p. 66).

Industrial Minerals

Gemstones.—The amount of Belgium's exports of polished diamond decreased by 18.91% compared with that of the previous year, and the value of these exports decreased by 17.63% to $997.9 million. The average price per carat of exported diamond was $1,866 in December 2012 (Antwerp World Diamond Centre, 2013b).

Mineral Fuels

Natural Gas and Petroleum.—The Antwerp Terminal and Processing Co. (ATPC), which was owned by Vitol Tank Terminals B.V. (VTTI) through its subsidiary Eurotank Belgium B.V. [part of the Vitol Group (Vitol)], announced plans to expand its capacity after achieving positive results for the first 15 months following its acquisition by VTTI. This expansion would increase the company's storage capacity by 500,000 cubic meters; no date was given as to when the capacity expansion would be completed (Vitol Tank Terminals B.V., 2011).

Outlook

Belgium's role as a leading mineral processor and major diamond trader is expected to continue, although its steel production is likely to decrease as a result of decreased demand. Belgium is also expected to remain significant in international and intra-European cargo handling of mineral products through its major ports (Antwerp, Ghent, Ostend, and Zeebrugge).

References Cited

Antwerp World Diamond Centre, 2013a, History: Antwerp World Diamond Centre. (Accessed July 14, 2013, at http://www.awdc.be/en/history.)

Antwerp World Diamond Centre, 2013b, Rough import and export figures show increase compared to December 2011: Antwerp World Diamond Centre. (Accessed November 30, 2012, at http://www.awdc.be/en/news?type=6.)

ArcelorMittal, 2012, Gent and Bremen cut 270,000 tonnes of emissions with gas recovery projects: ArcelorMittal news release, June. (Accessed July 22, 2013, at http://corporate.arcelormittal.com/news-and-media/news/2012/jun/20-06-2012.)

Benelux Parlement, 2014, What is Benelux: Brussels, Benelux Parlement. (Accessed August 5, 2014, at http://www.benelux-parlement.eu/en/benelux/benelux_intro.asp.)

European Commission, 2014a, Extra-EU28 trade of mineral fuels, lubricants and related materials (SITC 3), by member state—Share of exports by member state (%): European Commission. (Accessed August 5, 2014, via http://epp.eurostat.ec.europa.eu/tgm/refreshTableAction.do?tab=table&plugin=1&pcode=tet00056&language=en.)

European Commission, 2014b, Extra-EU28 trade of mineral fuels, lubricants and related materials (SITC 3), by member state—Share of imports by member state (%): European Commission. (Accessed August 5, 2014, via http://epp.eurostat.ec.europa.eu/tgm/refreshTableAction.do?tab=table&plugin=1&pcode=tet00056&language=en.)

European Commission, 2014c, Extra-EU28 trade of raw materials (SITC 2+4), by member state—Share of exports by member state (%): European Commission. (Accessed August 5, 2014, via http://epp.eurostat.ec.europa.eu/tgm/refreshTableAction.do?tab=table&plugin=1&pcode=tet00064&language=en.)

European Commission, 2014d, Extra-EU28 trade of raw materials (SITC 2+4), by member state—Share of imports by member state (%): European Commission. (Accessed August 5, 2014, via http://epp.eurostat.ec.europa.eu/tgm/refreshTableAction.do?tab=table&plugin=1&pcode=tet00064&language=en.)

European Commission, 2014e, Share of trade with the EU28—Share of exports to EU in total exports (%): European Commission. (Accessed August, 5, 2014, via http://epp.eurostat.ec.europa.eu/tgm/table.do?tab=table&plugin=1&language=en&pcode=tet00036.)

European Commission, 2014f, Share of trade with the EU28—Share of imports from EU in total imports (%): European Commission. (Accessed August, 5, 2014, via http://epp.eurostat.ec.europa.eu/tgm/table.do?tab=table&plugin=1&language=en&pcode=tet00036.)

Nyrstar N.V., 2013a, Annual report 2012: Nyrstar N.V., 210 p. (Accessed July 24, 2013, at http://www.nyrstar.com/_layouts/download.aspx?SourceUrl=/investors/en/Nyr_Documents/English/Nyrstar_Annual Report 2012_EN.pdf.)

Nyrstar N.V., 2013b, Smelting—Operational review: Nyrstar N.V. (Accessed July 14, 2013, http://www.nyrstar.com/operations/pages/smelting.aspx.

Tolcin, A.C., 2013, Zinc, *in* Metals and minerals: U.S. Geological Survey Minerals Yearbook 2011, v. I, p. 84.1–84.12. (Revised June 27, 2014.) (Accessed October 1, 2014, at http://minerals.usgs.gov/minerals/pubs/commodity/zinc/index.html#myb.)

Umicore S.A., 2013, Fact sheet—Umicore S.A.: Umicore S.A. (Accessed July 24, 2013, at http://tools.euroland.com/factsheet/b-unim/factsheethtml.asp.)

U.S. Census Bureau, 2014a, U.S. exports to Belgium by 5-digit end-use code: U.S. Census Bureau. (Accessed August 5, 2014, at https://www.census.gov/foreign-trade/statistics/product/enduse/exports/c4231.html.)

U.S. Census Bureau, 2014b, U.S. imports to Belgium by 5-digit end-use code: U.S. Census Bureau. (Accessed August 5, 2014, at https://www.census.gov/foreign-trade/statistics/product/enduse/imports/c4231.html.)

U.S. Central Intelligence Agency, 2013, Belgium, *in* The world factbook: U.S. Central Intelligence Agency. (Accessed July 24, 2013, at https://www.cia.gov/library/publications/the-world-factbook/geos/be.html.)

van Oss, H.G., 2013, Cement, *in* Metals and minerals: U.S. Geological Survey Minerals Yearbook 2011, v. I, p. 16.1–16.33. (Revised June 2013.) (Accessed October 2, 2014, at http://minerals.usgs.gov/minerals/pubs/commodity/cement/index.html#myb.)

Vitol Tank Terminals B.V., 2011, ATPC—The first 15 months: VTTI News. (Accessed November 30, 2012, at http://www.vtti.com/news_01.php?id=56.)

World Steel Association, 2013, World steel in figures 2013: Brussels, Belgium, World Steel Association, 30 p. (Accessed July 24, 2013, at http://www.worldsteel.org/dms/internetDocumentList/bookshop/WSIF_2013_spreads/document/WSIF_2013_spreads.pdf.)

LUXEMBOURG

In 2012, the iron and steel industry was Luxembourg's most economically important mineral industry, and steel was the country's main export commodity. Because it is a member of the Belgium Luxembourg Economic Union (BLEU), trade statistics for Luxembourg are inextricably linked with those of Belgium and, therefore, cannot be listed individually.

Production

Mining in Luxembourg consisted of small industrial mineral operations that produced mineral commodities only for domestic consumption. These minerals included dolomite, limestone, sand and gravel, and slate. Information on these operations was not readily available. Production data are in table 1.

Structure of the Mineral Industry

The principal mineral facilities in Luxembourg with their locations and capacities are listed in table 2. Most facilities were privately owned.

Commodity Review

Metals

Iron and Steel.—ArcelorMittal, which was headquartered in Luxembourg, was the world's leading steel manufacturer. It was more than two times larger, in terms of production quantity, than its nearest rival, Hebei Group of China (World Steel Association, 2013, p. 8).

ArcelorMittal Belval & Differdange S.A. agreed to revamp the Belval electric arc furnace (EAF) to increase the plant's production capacity and lower maintenance costs in Esch-sur-Alzette. This furnace, which was commissioned in 1997, would be refitted with a new lower shell, a new tilting frame, and a renewed upper shell. The revamping of the EAF would require the plant to be shut down for 2 weeks. Operations were expected to restart in March 2013, and the revamped furnace was expected to be commissioned in April 2013 (ArcelorMittal, 2012).

Outlook

Luxembourg is expected to continue to be a producer and exporter of steel. The country's industrial mineral production will likely continue to be limited to domestic consumption.

References Cited

ArcelorMittal, 2012, ArcelorMittal invests €6m in Belval: ArcelorMittal news release, October 23. (Accessed July 23, 2013, at http://corporate.arcelormittal.com/news-and-media/news/2012/oct/19-10-2012a.)

World Steel Association, 2013, World steel in figures: Luxembourg, World Steel Association, 28 p.

TABLE 1
BELGIUM AND LUXEMBOURG: PRODUCTION OF MINERAL COMMODITIES[1]

(Metric tons unless otherwise specified)

Country and commodity		2008	2009	2010	2011	2012[e]
BELGIUM[2]						
Metals:						
Cobalt, primary[3]		3,020	2,150[e]	2,600[e]	3,187	4,200[4]
Copper:						
Smelter, secondary		115,900	117,400	118,600	147,000	147,000
Refined, primary and secondary		395,800	373,700	381,000	380,000	380,000
Iron and steel:						
Pig iron	thousand metric tons	7,125	3,087	4,688	4,725	4,072[4]
Steel:						
Crude	do.	10,676	5,635	7,973	8,026	7,386[4]
Hot-rolled products	do.	11,792	7,172	9,649	10,012	9,800
Lead, refined, secondary		80,966	109,000	105,000[e]	88,129	87,958[4]
Zinc:						
Slab:						
Primary		239,000[e]	14,000	260,000	282,000	250,000[4]
Secondary, possibly remelted zinc[e]		40,000	40,000	40,000	40,000	40,000
Total		279,000[e]	54,000	300,000	322,000	290,000
Powder[e]		20,000	20,000	20,000	20,000	20,000
Industrial minerals:						
Barite		--[r]	--[r]	--[r]	--[r]	--[r]
Cement	thousand metric tons	6,969	6,113	5,990	6,844	6,800
Nitrogen, N content of ammonia	do.	830	830	830	830	830
Sulfur:[e]						
Byproducts:						
Elemental		225,000	225,000	225,000	225,000	225,000
Other forms		175,000	175,000	175,000	175,000	175,000
Total		400,000	400,000	400,000	400,000	400,000
Mineral fuels and related materials:						
Carbon black		--[r]	--[r]	--[r]	--[r]	--[r]
Coke, all types	thousand metric tons	2,545[r]	1,735[r]	2,133[r]	2,120[r]	2,100
Gas, manufactured	thousand cubic meters	463,000	463,000	463,000	463,000	463,000
Petroleum refinery products:						
Liquefied petroleum gas	thousand 42-gallon barrels	5,946	5,289	6,205	6,200[e]	6,200
Naphtha and white spirit	do.	14,300[e]	14,300[e]	14,300[e]	NA	NA
Gasoline	do.	34,257	32,338	30,186	30,100[e]	30,100
Kerosene	do.	14,758	16,294	15,950	15,900[e]	15,900
Kerosene, other	do.	283	466	511	510[e]	510
Distillate fuel oil	do.	96,425	88,289	93,075	93,000[e]	93,000
Refinery gas	do.	3,800[e]	3,800[e]	3,800[e]	NA	NA
Residual fuel oil	do.	43,701	34,432	35,150	35,100[e]	35,100
Bitumen	do.	8,600[e]	8,600[e]	8,600[e]	NA	NA
Total	do.	222,070	203,800	207,800	180,000	180,000
LUXEMBOURG						
Metals, steel:						
Crude	thousand metric tons	2,582	2,215	2,563	2,521	2,232[4]
Hot-rolled products	do.	2,837	2,910	1,941	2,220	2,000
Industrial minerals:						
Cement, hydraulic		1,091,000	1,000,000[e]	1,078,000	1,319,000	1,217,000[4]
Phosphates, Thomas slag:[e]						
Gross weight		475,000	475,000	475,000	475,000	475,000
P_2O_5 content		70,000	70,000	70,000	70,000	70,000

[e]Estimated; estimated data are rounded to no more than three significant digits; may not add to totals shown. [r]Revised. NA Not available. do. Ditto. -- Zero.

[1]Table includes data available through April 11, 2014.

[2]In addition to the commodities listed, Belgium produced a number of other metals, alloys, and industrial minerals, such as secondary aluminum, bismuth metal, kaolin, lime and dead-burned dolomite, quicklime, selenium, sodium sulfate, sulfuric acid, secondary tin metal, and worked and natural stone, for which only aggregate output figures were available.

[3]Production reported by N.V. Umicore S.A. includes production from China and South Africa.

[4]Reported figure.

TABLE 2
BELGIUM AND LUXEMBOURG: STRUCTURE OF THE MINERAL INDUSTRIES IN 2012

(Thousand metric tons unless otherwise specified)

Country and commodity		Major operating companies and major equity owners	Location of main facilities	Annual capacity
BELGIUM				
Cadmium, metal	metric tons	Umicore S.A./N.V.	Hoboken	1,800
Cement		Major companies, of which:	Plants, of which;	8,400
Do.		Cimenteries CBR SA (Heidelberg Cement Group)	Major plants at Lixhe, Mons/Obourg, Harmignies, and Ghent	(3,200)
Do.		Ciments d'Obourg SA (Holcim Group)	Plant at Obourg	(2,800) [1]
Do.		Compagnie des Ciment Belge (Ciments Francais S.A.)	Plant at Gaurain-Ramecroix	(2,400)
Cobalt	metric tons	Umicore S.A./N.V.	Refinery at Olen	500
Copper, secondary		Metallo-Chimique NV (Metallum Group)	Smelter at Beerse	80
Dolomite		SA Dolomeuse (Group Lhoist)	Quarry at Marche les Dames	500
Do.		do.	Plant at Marche les Dames	750
Do.		SA de Marche-les-Dames (Group Lhoist)	Quarries at Nameche	3,000
Do.		do.	Plant at Nameche	3,000
Do.		SA Dolomies de Merlemont (Group Lhoist)	Quarry at Philippeville	100
Lead, metal		Umicore S.A./N.V.	Smelter at Antwerp-Hoboken	90
Do.		do.	Refinery at Antwerp-Hoboken	125
Limestone		Carmeuse S.A. (privately owned)	Mines and plant at Engis	1,850
Do.		do.	Mines and plant at Frasnes	450
Do.		do.	Mines and plant at Maizeret	850
Do.		do.	Mines and plant at Moha	800
Do.		SA Transcar (Royal Volker Stevin)	Mines and plant at Maizeret	850
Petroleum, refined	42-gallon barrels per day	Total S.A.	Refinery at Antwerp	268,000
Do.	do.	ExxonMobil Petroleum & Chemical B.V.B.A. (Exxon Mobil Corp., 100%)	do.	239,000
Do.	do.	Antwerp Processing Co. (Vitol Group)	do.	125,000
Do.	do.	Belgian Refining Corp. (Guvnor Group)	do.	107,500
Do.	do.	PRA NV (Vitol Group)	do.	22,300
Salt		Zoutman NV	Plant at Roeselare	200
Sand, silica		SRC-Sibelco SA	Mines and plants at Lommel, Mol, and Maasmechelen	500
Steel:				
Crude		Various companies:	Of which:	
Do.		ArcelorMittal Liege (ArcelorMittal)	Plant at Liege	3,000
Do.		ArcelorMittal Gent (ArcelorMittal)	Plant at Ghent	3,000
Do.		NLMK La Louviere S.A. (NLMK Group)	Plant at La Louviere	900
Manufactured		Various companies:	Of which:	
Do.		NMLK Clabecq S.A. (NLMK Group)	Rolling mill at Clabecq	750
Do.		Industeel Belgium S.A. (ArcelorMittal)	Rolling mill at Charleroi	600
Do.		ArcelorMittal Genk (ArcelorMittal)	Galvanizing Plant at Genk-Zuid	360
Do.		Tubemeuse Industries S.A.	Tube mill at Flemalle	50
Tin		Metallo-Chimique NV (Metallum Group)	Smelter at Beerse	12
Zinc, metal		Nyrstar N.V.	Smelter and refinery at Balen/Overpelt	252
LUXEMBOURG				
Cement		Cimalux S.A. (Dyckerhoff AG)	Grinding plant at Esch-sur-Alzette	850
Do.		do.	Clinker plant at Rumelange	1,000
Steel		ArcelorMittal Belval and Differdange S.A. (ArcelorMittal)	Plants at Differdange, Esch-Belval, and Esch-Schifflange	5,320

Do., do. Ditto.

[1]Includes the capacity of the company SA Ciments de Haccourt.

THE MINERAL INDUSTRY OF BOSNIA AND HERZEGOVINA

By Yadira Soto-Viruet

Bosnia and Herzegovina's mineral industry was dominated by the mine output of bauxite, iron, and zinc. Mineral fuels produced in the country included brown coal, coke, and lignite. Other mineral commodities produced included barite, crushed stone, limestone, salt, and sand and gravel.

In 2012, Bosnia and Herzegovina's real gross domestic product (GDP) decreased by 1.10% compared with that of 2011. Mining and quarrying made up about 2.3% of the country's total GDP. In 2012, mining and quarrying made up about 12% of the total value of Bosnia and Herzegovina's imports and about 2% of the total value of exports. Imports of crude petroleum and natural gas were valued at $932 million and made up about 78% of the total value of mining and quarrying imports. Exports of metal ores were valued at $54 million and made up about 53% of the total value of mining and quarrying exports (Agency for Statistics of Bosnia and Herzegovina, 2013a, p. 2, 3; 2013b, p. 59).

Production

In 2012, barite production increased by 115% to 28 metric tons (t) from 13 t in 2011; production of ecaussine and other calcareous stones increased, by 66% to 234,120 t from 141,245 t; crude dolomite, by 46% to 127,774 t; gravel, by 23% to 1.1 million metric tons (Mt); sodium bicarbonate, by 22% to 58,620 t; and bauxite, by 13% to 800,316 t. Production of slate decreased by 88% to 30 t from 252 t; marble and travertine, by 62% to 692 t from 1,836 t; construction sand, by 54% to 499,916 t from 1.1 Mt (revised); crude kaolin, by 36% to 149,495 t; alumina, by 23% to 202,416 t; coke, by 22% to 696,231 t; and lime, by 19% to 397,802 t. Data on mineral production are in table 1.

Structure of the Mineral Industry

Table 2 is a list of the major mineral industry facilities.

Commodity Review

Bauxite and Alumina and Aluminum.—In 2012, high gas prices and low alumina and aluminum prices were a challenge for Alumina Factory Birac a.d. (Birac). The company was the only producer of alumina in the country. On August 1, Birac's operations were temporarily suspended owing to the company's outstanding debts with BH Gas d.o.o., which was the country's main gas distributor. On August 2, Birac resumed production after agreeing to pay part of its debt to BH Gas. The plant, which was owned by Ukio Bankas Investment Group of Lithuania, had an annual production capacity of 600,000 t of alumina (Sito-Sucic and Zuvela, 2012; SteelGuru, 2012).

Aluminij d.d. Mostar (Mostar) was Bosnia and Herzegovina's only aluminum producer and its leading exporter. On July 26, the company announced that it was shutting down 12.5% of its capacity, citing high power prices and low aluminum prices. In 2011, Mostar signed a contract with Hertwich Engineering of Austria for the purchase and installation of a new casting furnace and a new casting line of small (between 8 and 10 kilograms) alloy ingots at Mostar's casthouse at a cost of about $8 million. Mostar began operation of the new casting line in February 2013. Installation of a new 50-t-capacity melting furnace was also underway and was expected to be completed by April 2013. The company expected to increase its primary aluminum production capacity by up to 30,000 metric tons per year (Aluminij d.d. Mostar, 2011, 2012, 2013).

Iron and Steel.—In July, ArcelorMittal S.A. of Luxembourg through its subsidiary ArcelorMittal Zenica announced plans to implement two environmental projects at the company's blast furnace and coke plant at Zenica. The blast furnace project would include the installation of a de-dusting system at a cost of about $8 million. The new system would reduce hard particle and pollutant emissions by about 95% and was expected to be completed by yearend 2013. The coke plant project would include the installation of a new charging machine, at a cost of about $4 million, which would reduce solid particle emissions of coke from about 140 grams per metric ton (g/t) to about 5 g/t. The ArcelorMittal Zenica plant had a production capacity of about 1 million metric tons per year of steel (ArcelorMittal, 2012).

Outlook

Bosnia and Herzegovina forecasted an increase in the GDP of 0.5% in 2013 and 2.0% in 2014 (International Monetary Fund, 2013). The country's mineral industry will most likely continue to be a relatively minor producer of mineral commodities. Metals are expected to remain valuable export commodities for the country. In the short run, high gas and power prices will continue to present challenges to the alumina and aluminum sectors.

References Cited

Agency for Statistics of Bosnia and Herzegovina, 2013a, Gross domestic products of Bosnia and Herzegovina 2012—Production approach—First results: Agency for Statistics of Bosnia and Herzegovina, no. 3, July 25, 6 p. (Accessed October 28, 2013, at http://www.bhas.ba/saopstenja/2013/GDP_P_2012_001_bos.pdf.)

Agency for Statistics of Bosnia and Herzegovina, 2013b, International trade in goods of BiH: Agency for Statistics of Bosnia and Herzegovina, August, 82 p. (Accessed October, 28, 2013, at http://www.bhas.ba/tematskibilteni/robna eng.pdf.)

Aluminij d.d. Mostar, 2011, Aluminij engaged in six million euro worth modernization of the casthouse: Aluminij d.d. Mostar press release, December 14. (Accessed November 8, 2013, at http://www.aluminij.ba/en/news/383-aluminij-engaged-six-million-euro-worth-modernization-casthouse.)

Aluminij d.d. Mostar, 2012, Aluminij d.d. Mostar starts curtailment of 12.5% of the production capacity: Aluminij d.d. Mostar press release, July 26. (Accessed November 8, 2013, at http://www.aluminij.ba/en/news/478-aluminij-dd-mostar-starts-curtailment-125-production-capacity.)

Aluminij d.d. Mostar, 2013, Aluminij successfully started a new casting line of small ingots: Aluminij d.d. Mostar press release, February 8. (Accessed November 8, 2013, http://www.aluminij.ba/en/news/561-aluminij-successfully-started-new-casting-line-small-alloy-ingots.)

ArcelorMittal, 2012, Two new environmental projects launched at ArcelorMittal Zenica: Luxembourg, ArcelorMittal press release, July 4. (Accessed November 6, 2013, at http://corporate.arcelormittal.com/news-and-media/news/2012/jul/04-06-2012.)

International Monetary Fund, 2013, World economic outlook: International Monetary Fund, October, 249 p. (Accessed November 13, 2013, at http://www.imf.org/external/pubs/ft/weo/2013/02/pdf/text.pdf.)

Sito-Sucic and Zuvela, 2012, Bosnia alumina plant halts production, risks closure: Thomson Reuters, August 1. (Accessed November 13, 2013, http://www.reuters.com/article/2012/08/01/bosnia-birac-idUSL6E8J1K8W20120801.)

SteelGuru, 2012, Bosnia alumina plant resumes output: SteelGuru, August 5. (Accessed November 13, 2013, http://www.steelguru.com/metals_news/Bosnia_alumina_plant_resumes_output/277061.html.)

TABLE 1
BOSNIA AND HERZEGOVINA: PRODUCTION OF MINERAL COMMODITIES[1]

(Metric tons unless otherwise specified)

Commodity[2]	2008	2009	2010	2011	2012
METALS					
Alumina	294,455	191,792	269,414	261,874	202,416
Aluminum:					
Primary	123,000	96,000	118,000	130,875	126,000 [e]
Unwrought aluminum, including alloys	155,903	130,042	150,488	163,954 [r]	159,660
Bauxite	1,018,333	555,820	844,027	707,712	800,316
Iron and steel:					
Ore and concentrate:					
Gross weight	1,481,730	1,614,890	1,401,000	1,891,000	2,075,732
Fe content[e]	622,000	678,000	588,000	794,000	872,000
Metal:					
Crude steel	608,000	519,000	590,757	648,560	700,341
Ferroalloys, ferrosilicon[e]	640	470	870	1,800	--
Pig iron	243,000	482,469	620,935	684,734	749,539
Lead:					
Ores and concentrate, gross weight	6,029	3,781	5,811	6,648	7,000 [e]
Pb content[e]	3,300	2,100	3,200	3,700	3,700
Metal, smelter, secondary[e]	46,000 [r]	35,000 [r]	4,500 [r]	3,400 [r]	3,300
Silicon, metal[e]	12,400	11,000	17,300	17,500	15,900
Zinc:					
Ores and concentrate, gross weight	8,595	6,228	10,025	12,477	12,500 [e]
Zn content[e]	4,700	3,400	5,500	6,900	7,000
INDUSTRIAL MINERALS					
Barite	54	30	57	13	28
Cement	1,406,373	1,073,762	948,513	893,017	845,657
Clays:					
Bentonite	30,504	16,042	314	--	--
Kaolin, crude	259,325	148,384	41,808	232,147	149,495
Dolomite, crude	134,991	79,104	199,757	87,635	127,774
Graphite	272,084	133,819	45,079	-- [e]	-- [e]
Gypsum and anhydrite	150,039	74,302	64,570	71,870	73,665
Lime	215,787	280,939	339,429	488,577	397,802
Salt, all sources	555,122	556,089	662,631	833,734 [r]	862,017
Sand and gravel:					
Gravel	1,475,433	938,253	979,472	913,129	1,126,176
Sand, construction	175,527	156,128	572,452 [r]	1,095,486 [r]	499,916
Silica sand	702,018	524,752	227,721 [r]	118,978 [r]	121,491
Sodium compounds, sodium bicarbonate	19,441	21,944	35,986	47,847	58,620
Stone:					
Dimension:					
Marble and travertine	5,785	6,358	2,674	1,836	692
Ecaussine and other calcareous stone	181,493	64,186	66,133	141,245	234,120
Granite	23,764	18,755	--	--	--
Porphyry, basalt and other building stone	188,410	95,498	137,372	113,285	97,452
Slate	2,181	3,561	525	252	30
Crushed	4,370,598	3,577,927	3,776,726	4,369,575 [r]	3,711,065
Limestone, crushed and powdered	2,403,270	1,829,989	1,916,642 [r]	1,850,140 [r]	1,834,677

See footnotes at end of table.

TABLE 1—Continued
BOSNIA AND HERZEGOVINA: PRODUCTION OF MINERAL COMMODITIES[1]

(Metric tons unless otherwise specified)

Commodity[2]		2008	2009	2010	2011	2012
MINERAL FUELS AND RELATED MATERIALS						
Brown coal and lignite	thousand metric tons	11,244	11,515	10,976	12,738 [r]	12,312
Coke		576,785	609,377	919,962	886,911	696,231
Petroleum refinery products[3]	42-gallon barrels	853,000	8,240,000	8,920,000	9,880,000	8,590,000

[e]Estimated; estimated data are rounded to no more than three significant digits. [r]Revised. -- Zero.

[1]Table includes data available through November 4, 2013.

[2]In addition to commodities listed, calcined gypsum, common clay, crude ceramic clay, magnesite, manganese ore, soda ash, and steel semimanufactures may have been produced, but available information is inadequate to make reliable estimates of output.

[3]Data were converted to barrels from metric tons and were reported as follows: 2008—106,568; 2009—1,029,585; 2010—1,114,669; 2011—1,235,519; 2012—1,073,292.

TABLE 2
BOSNIA AND HERZEGOVINA: STRUCTURE OF THE MINERAL INDUSTRY IN 2012

(Thousand metric tons unless otherwise specified)

Commodity	Major operating companies and major equity owners	Location of main facilities	Annual capacity
Alumina	Alumina Factory Birac a.d. (Ukio Bankas Investment Group)	Plant at Zvornik	600.
Aluminum	Aluminij d.d. Mostar	Smelter at Mostar	120.[e]
Bauxite	A.D. Boksit Milici	Mine at Milici, west of Srebrenica	1,500.[e]
Cement	Tvornica Cementa Kakanj d.d. (HeidelbergCement AG)	Plant at Kakanj	400.
Do.	Fabrika Cementa Lukavac d.d.	Plant in Lukavac	800 cement, 600 clinker.
Coal:			
Brown	RMU Banovici	Opencast mines at Cubric, Grivice, and Turija, and underground mines Omazici and Separacija at Banovici	NA.
Do.	Zenica Group	Stara Jama, Raspotocje, and Stranjani Mines at Zenica	NA.
Do.	Durdevik Group	Potocari and Visca II opencast mines and Durdevik underground mine south of Zivinice	NA.
Do.	Kakanj Group	Vrtliste opencast mine at Kakanj	NA.
Do.	do.	Haljinic underground mine about 5 kilometers southeast of Kakanj	NA.
Do.	Breza Group	Sretno and Kamenice underground mines 20 kilometers northwest of Sarajevo	NA.
Do.	Abid Lolic Group	Grahovcici underground mine 10 kilometers west of Zenica	NA.
Do.	Tusnica Mine	Drage opencast mine at Livno	NA.
Do.	Rudnik i Termoelektrana Ugljevik (ZP Elektrokrajina a.d.)	Opencast mine at Ugljevik	NA.
Lignite	Kreka Group	Opencast mine at Dubrave	NA.
Do.	do.	Opencast mine at Sikulje	NA.
Do.	do.	Underground mines at Mramor, about 5 kilometers northeast of Lukavac, and Bukinje, located between Tuzla and Lukavac	NA.
Do.	EFT Rudnik i Termoelektrana Stanari d.o.o. (EFT Group)	Stanari opencast mine located 20 kilometers west of Doboj	1,100.[e]

See footnotes at end of table.

TABLE 2—Continued
BOSNIA AND HERZEGOVINA: STRUCTURE OF THE MINERAL INDUSTRY IN 2012

(Thousand metric tons unless otherwise specified)

Commodity		Major operating companies and major equity owners	Location of main facilities	Annual capacity
Coal:—Continued				
Lignite—Continued		Rudnik I Termoelektrana Gacko (ZP Elektrokrajina a.d.)	Opencast mine at Gacko	NA.
Do.		Gracanica Group	Dimnjace opencast mine at Gornji Vakuf-Uskoplje	NA.
Do.		Tusnica Mine	Opencast mine at Livno	NA.
Coke		Global Ispat Koksa Industrija d.o.o. Lukavac (Global Steel Holdings and Coke and Chemical Conglomerate)	Lukavac	700
Do.		ArcelorMittal Zenica (ArcelorMittal S.A.)	Plant at Zenica	NA.
Ferroalloys, ferrosilicon		Steelmin Ltd., 80%	Plant at Jajce	NA.[1]
Do.		B.S.I. d.o.o. (Metalleghe S.p.a.)	do.	18.
Iron ore		ArcelorMittal Prijedor (ArcelorMittal S.A., 51%) and Rudnici Zeljezne Rude "Ljubija" a.d.	Jezero and Buvac open pit mines at Ljubija	1,500.^{e}
Lead-zinc ore		NA	Mine and mill at Srebrenica	NA.[2]
Manganese ore		Rudnik Mangana Buzim (LM IMPEX)	Mine and concentrator at Buzim	NA.[2]
Petroleum, refined	42-gallon barrels per day	Rafinerija nafte Brod a.d. (OAO "NefthegazlnKor," 77.99%)	Oil refinery at Bosanski Brod	24,000.^{e}
Do.		Rafineriji ulja Modriča a.d. Modriča (OAO "NefthegazlnKor," 77%)	Oil refinery at Modrica	NA.
Pig iron		ArcelorMittal Zenica (ArcelorMittal S.A.)	Blast furnace at Zenica	NA.
Salt		Rudnik Soli Tuzla d.d.	Tuzla	NA.
Steel, crude		ArcelorMittal Zenica (ArcelorMittal S.A.)	Plant at Zenica	1,000.
Steel, crude, secondary		Jelsingrad Livar Steel Foundry a.d.	Banja Luka	NA.

^{e}Estimated. Do., do. Ditto. NA Not available.

[1] The ferralloys and ferrosilicon plant at Jajce was sold to Steelmin Ltd. of the United Kingdom in October 2011. Work to redevelop the plant was underway, and the plant was expected to be put into operation by 2013.

[2] Available information is not sufficient to determine if the operation is still active.

THE MINERAL INDUSTRY OF BULGARIA

By Yadira Soto-Viruet

Bulgaria's mineral industry included mine output of metal ores, mineral fuels (mainly coal), and a variety of industrial minerals. Industrial minerals produced in the country included fluorspar, gypsum, salt, and sand and gravel. Additionally, the metallurgical sector smelted and refined copper, lead, silver, steel, and zinc.

Minerals in the National Economy

In 2012, Bulgaria's real gross domestic product (GDP) increased by 0.8% compared with that of 2011. Bulgaria's industrial sector accounted for about 30.4% of the GDP. In 2011 (the latest year for which data were available), the gross value added of mining and quarrying activities was about $1.0 billion[1] (reported as BGN1.5 billion) and accounted for about 8% of the value added of industry and for 2% of the GDP. In 2012, about 24,600 people were employed in the mining and quarrying industry. In 2011 (the latest year for which data were available), 407 mining and quarrying enterprises were registered in Bulgaria. Of these 407 enterprises, 276 had no more than 9 employees, 86 had 10 to 49 employees, 28 had 50 to 249 employees, and 17 had more than 250 employees (International Monetary Fund, 2013, p. 48; National Statistical Institute of the Republic of Bulgaria, 2013a–e, p. 151; U.S. Central Intelligence Agency, 2013).

Mineral Trade

In 2012, the total value of Bulgaria's exports was about $26.7 billion (reported as BGN40.6 billion) compared with about $28.3 billion in 2011 (reported as BGN39.6 billion). The total value of Bulgaria's imports was about $32.7 billion (reported as BGN49.8 billion) compared with $32.7 billion (reported as BGN45.8 billion) in 2011. The country's major export trade partners were, in order of value, Germany (which received 10.2% of Bulgaria's exports), Italy (8.5%), Romania (8.1%), Greece (7.2%), Belgium (3.7%) and Spain (2.6%). Its major import trade partners were, in order of value, Germany (which supplied 11.1% of Bulgaria's imports), Italy (6.6%), Romania (6.5%), Greece (6.1%), and Spain (4.4%). Mineral fuel, lubricant, and related materials accounted for 17% of the total value of exports and 25% of the total value of imports (National Statistical Institute of the Republic of Bulgaria, 2013b).

Bulgaria's exports to the United States were valued at about $509 million in 2012 compared with about $416 million in 2011. Of this amount, petroleum products accounted about $62 million, and cement, lime, sand, and stone accounted for about $1 million. Imports from the United States were valued at about $248 million in 2012 compared with about $258 million in 2011; these included nearly $20 million in petroleum products, $2 million in excavating machinery, and $1 million in iron and steel mill products (U.S. Census Bureau, 2013a, b).

Production

In 2012, bentonite production increased by 44% to 78,000 metric tons (t) from 54,000 t (revised) in 2011; estimated vermiculite production, by 24% to 18,600 t from 15,000 t; estimated silver production, by 22% to 55,000 kilograms (kg) from 45,000 kg (revised); estimated gold production, by 18% to 5,200 kg from 4,400 kg (revised); zinc mine output, by 10% to 12,116 t from 10,977 t (revised); and the estimated output of petroleum refinery products, by 10% to 49.4 million barrels (Mbbl) from an estimated 45 Mbbl. Manganese production (in terms of content and gross weight) decreased by 75% to 10,600 t from 41,800 t (revised) and to 37,900 t from 149,400 t (revised), respectively; crude steel production decreased by 24% to 632,000 t from 834,000 t; zinc metal production, by 18% to 72,000 t from 88,000 t (revised); and natural gas production, by 11% to 396 million cubic meters from 443 million cubic meters (revised). Data on mineral production are in table 1.

Structure of the Mineral Industry

Table 2 is a list of major mineral industry facilities.

Commodity Review

Metals

Copper.—Aurubis AG of Germany, through its subsidiary Aurubis Bulgaria AD, owned the country's only copper smelting and refining facility. Aurubis Bulgaria, which is located in the town of Pirdop, produced about 226,000 t of refined copper and about 1.1 million metric tons (Mt) of sulfuric acid in 2012. The company investment project Aurubis Bulgaria 2014 was underway and was expected to be implemented by 2014. The project would include the installation of environmental protection equipment, such as equipment to reduce fugitive air emissions and facilities to treat wastewater (Aurubis AG, 2012, p. 30).

Copper production from the Chelopech underground mine increased by 29% to about 21,600 t (reported as 47.7 million pounds) from about 16,800 t (reported as 37 million pounds) in 2011. Dundee Precious Metals Inc. of Canada owned and operated the mine through its 100%-owned subsidiary Chelopech Mining EAD. The copper concentrate produced at the Chelopech Mine, which was located about 70 kilometers (km) east of the capital city of Sofia, was exported to be processed at Dundee's smelter in Tsumeb, Namibia. The Chelopech Mine was expected to produce between 19,500 and 20,900 t in 2013 (reported as between 43 and 46 million pounds) (Dundee Precious Metals Inc., 2012; 2013b, p. 4, 6, 15).

Gold.—In 2012, gold production from the Chelopech Mine increased by 28% to 3,752 kg (reported as 120,631 troy ounces)

---

[1]Where necessary, values have been converted from Bulgarian levs (BGN) to U.S. dollars (US$) at an annual average exchange rate of BGN1.40=US$1.00 for 2011 and BGN1.52=US$1.00 for 2012.

from 2,920 kg (reported as 93,881 troy ounces) in 2011. The increase was attributed to the completion of the mine and mill expansion that would allow Dundee to mine and process up to 2 million metric tons per year (Mt/yr) of ore. The gold concentrate produced at Chelopech was also exported to be processed at the Tsumeb smelter in Namibia. As of December 31, measured and indicated mineral resources at Chelopech were estimated to be 29.1 Mt at average grades of 4.0 grams per metric ton (g/t) gold, 9.4 g/t silver, and 1.3% copper, and inferred resources were estimated to be 9.3 Mt at average grades of 2.9 g/t gold, 10.6 g/t silver, and 0.9% copper. Gold production for 2013 was expected to be between 3,900 and 4,400 kg (reported as between 125,000 and 143,000 troy ounces]. The company also expected silver production from Chelopech to be between 5,660 and 6,070 kg (reported as between 182,000 and 195,000 troy ounces) (Dundee Precious Metals Inc., 2013a; 2013b, p. 6, 15; 2013c).

Dundee continued with its plan to develop the Chelopech pyrite recovery project, which would have the potential to recover about 40% to 50% of the contained gold in the mined Chelopech ore that was not being recovered in the current circuit. A preliminary economic assessment completed in the second quarter of 2012 reported that the project had the potential to produce about 400,000 metric tons per year (t/yr) of pyrite concentrate containing between about 2,300 and 2,800 kg of gold (reported as between 75,000 and 90,000 troy ounces); about 4,000 and 5,900 kg of silver (reported as between 130,000 and 190,000 troy ounces); and 2,000 and 2,700 t of copper (reported as between 4.5 million and 6.0 million pounds). At yearend 2012, Dundee signed an agreement with Xiangguang Copper Co. of China for the sale of 200,000 t/yr of pyrite concentrate between 2014 and 2016. Xiangguang Copper also agreed to purchase 3,000 metric tons per month (t/mo) of copper concentrate between March and July 2013 and 2,000 t/mo from July to December 2013. Production of pyrite concentrate was expected to begin by the fourth quarter of 2013 (Dundee Precious Metals Inc., 2013b, p. 6, 15; 2013c).

Dundee also continued with its plan to develop the Krumovgrad gold project in which it owned a 100% interest. The project was located about 3 km south of the town of Krumovgrad in southeastern Bulgaria. In January, a definitive feasibility study was completed by Balkan Mineral and Mining EAD, which was a subsidiary of Dundee. Measured and indicated mineral resources were estimated to be 7.99 Mt at average grades of 3.50 g/t gold and 2.00 g/t silver and inferred resources were estimated to be 0.40 Mt at average grades of 1.20 g/t gold and 1.00 g/t silver. The study reported that Krumovgrad had the potential to be developed as an open pit and to produce an average estimate of about 2,300 kilograms per year of gold (reported as 74,000 troy ounces per year). The company expected to begin production by 2015 (Balkan Mineral and Mining EAD, 2012, p. 10, 15–17; Dundee Precious Metals Inc., 2013b, p. 10; 2013d).

Lead and Zinc.—In March, the operations of the companies Gorubso AD and Lead and Zinc Complex Plc. (LZC), which were owned by Intertrust Holdings AD, were suspended owing to Intertrust Holding's financial debts and labor disputes over unpaid wages. In April, Intertrust Holdings listed both companies for sale, and Varba-Batanitsi AD acquired a more than 90% interest in Gorubso. In mid-April, Gorubso Madan Mine restarted operations, citing that all required safety measures had been taken care of and that production could be restarted. In December, Varba-Batanitsi, which was owned by KCM 2000 Group and Minstroy Holding AD, announced its plans to invest about $3 million in Gorubso Madan by 2013 (Energy Ecology Economy, 2012; Novinite.com, 2012b, c; International Lead and Zinc Study Group, 2013, p. 15; KCM 2000 Group, 2013a, p. 15; 2013b).

In August, after the LZC failed to make payment on its loans from the Bulgarian First Investment Bank (FIB), which was one of the three leading lenders to the LZC, the FIB obtained a court order and requested a tender process for the sale of a portion of the LZC to an outside investor. In September, Sofia-based Harmony 2012 Ltd. acquired a 50% interest in LZC, at a cost of about $6 million. No details as to when the complex would resume operations were available (Novinite.com, 2012a, d).

Industrial Minerals

Cement.—In Bulgaria cement was produced by four companies—Devnya Cement AD, Holcim Bulgaria AD, Vulcan Cement S.A., and Zlatna Panega Cement AD. These companies had a combined cement production capacity of 5.7 Mt/yr. The Devnya cement plant, which was owned by Italcementi Group of Italy, had an annual production capacity of about 2 Mt of cement. In March, Italcementi announced that the company had awarded the contract for the construction of the new cement line in Varna West Port to CBMI Construction Co. Ltd. of China [a subsidiary of China National Material Group Corp. Ltd. (Sinoma)]. The new production line was expected to be completed by 2015 and to produce 1.5 Mt/yr of cement. Construction of the new project began in April (Italcementi Group, 2012, 2013).

Mineral Fuels

Coal.—State-owned Bulgarian Energy Holding EAD, through its subsidiary Mini Maritsa Iztok Mines EAD, held 100% interest in the Troyanovo-1, the Troyanovo-3, and the Troyanovo-North Mines. In 2011 (the latest year for which data were available), Bulgarian Energy Holding's Maritsa Iztok Mines produced 33.0 Mt of lignite coal, about 90% which was used for thermal power generation in Bulgaria. The company had an annual capacity of about 35 Mt of coal and its property covered an area of about 240 square kilometers. As of January, proved and probable reserves at Maritsa were estimated to be about 968 Mt and 600 Mt, respectively. Other producers of lignite coal included the Beli Breg, the Chukurovo, and the Stanyantsi Mines (Mini Maritsa Iztok Mines EAD, 2011, p. 3, 9–10; 2013; Bulgaria Ministry of Economy, Energy, and Tourism, 2012, p. 7).

Natural Gas.—In October, Petroceltic International Plc of Ireland announced the completion of the merger with Melrose Resources plc of United Kingdom, which included the acquisition of the Kaliakra and the Kavarna gasfields located in the Galata exploration block in the western Black Sea. Natural gas production from the Kaliakra and the Kavarna fields was

396 million cubic meters (reported as 14 billion cubic feet) in 2012. The decrease in production was attributed to a decrease in production from the Kaliakra field in the second quarter of 2012. In 2011, proved and probable reserves at the Kaliakra and Kavarna fields were estimated by Melrose to be 934 million cubic meters (reported as 33 billion cubic feet) and 765 million cubic meters (reported as 27 billion cubic feet), respectively. In November, Petroceltic announced its plans to complete the Kaliakra-1 well, which had been temporarily suspended, by mid-2013. Petroceltic also announced its plans to drill the Kamchia-1 exploration well, which had targeted prospective resources of 765 million cubic meters (reported 27 billion cubic feet), by the first quarter of 2013. By yearend, the Government awarded Petroceltic an extension of its Galata exploration permit until July 2013 (Melrose Resources Plc., 2012, p. 12; Petroceltic International Plc, 2012; 2013, p. 23–25).

Petroleum.—LUKOIL Oil Co. of Russia, through its subsidiary LUKOIL Neftochim Bourgas AD, owned and operated the Burgas refinery, which had an annual capacity of about 215,000 barrels per day. In January, LUKOIL announced its plans to construct a heavy residue hydrocracking complex at Burgas. The first stage of the project included the construction of the 2.5-Mt/yr vacuum residue hydrocracker plant at a cost of about $1.5 billion. The company awarded the contract for the construction of the new plant to Technip Co. of Italy. Under the terms of the contract, Technip was to provide the engineering, procurement (of equipment and material), and construction services. The company expected to increase production of Euro-5 diesel by 1.2 Mt and to cease production of high-sulfur oil. The first stage of the project was expected to be completed by January 2015 (LUKOIL Neftochim Bourgas AD, 2012; Technip Co., 2012; Lukoil Oil Co., 2013, p. 8).

Outlook

Bulgaria forecasted a GDP growth rate of 1.2% for 2013 (International Monetary Fund, 2013, p. 48). Increased demand for and production of Bulgaria's mineral commodities are likely to depend mainly on the domestic and European economic outlook and are likely to remain modest in terms of world production. The construction plans at the Burgas refinery and the Devnya Cement plant, the expansion plans at Gorubso, and plans to develop Chelopech's pyrite recovery project are expected to strengthen the industry in the short run.

References Cited

Aurubis AG, 2012, Environmental report 2012: Aurubis AG, 60 p.

Balkan Mineral and Mining EAD, 2012, Krumovgrad gold project: Balkan Mineral and Mining EAD, 219 p. (Accessed September 5, 2013, at http://www.dundeeprecious.com/files/technical_reports/Krumovgrad 43-101 Tech Report_v001_o73pc6.pdf.)

Bulgaria Ministry of Economy, Energy, and Tourism, 2012, Bulletin on the state and development of the energy sector in the Republic of Bulgaria: Republic of Bulgaria Ministry of Economy, Energy, and Tourism, 20 p. (Accessed September 10, 2013, at http://www.mi.government.bg/files/useruploads/files/budget/bulletin_energy_2012_eng.pdf.)

Dundee Precious Metals Inc., 2012, Annual report 2011: Toronto, Ontario, Canada, Dundee Precious Metals Inc., 119 p. (Accessed November 21, 2013, at http://www.dundeeprecious.com/English/operations/producing-mines/Chelopech/recent-developments/default.aspx.)

Dundee Precious Metals Inc., 2013a, 2012 fourth quarter and annual results and 2013 guidance: Toronto, Ontario, Canada, Dundee Precious Metals Inc. press release, February 15, 6 p. (Accessed November 21, 2013, at http://www.dundeeprecious.com/files/PressReleases/News Release - Q4 2012 - news wire final_v001_h29131.pdf.)

Dundee Precious Metals Inc., 2013b, Annual report 2012: Toronto, Ontario, Canada, Dundee Precious Metals Inc., 130 p. (Accessed November 20, 2013, at http://www.dundeeprecious.com/files/annual_report/122195 DUNDEE AR Complete_final_v001_h3aq6j.pdf.)

Dundee Precious Metals Inc., 2013c, Chelopech gold/copper mine: Dundee Precious Metals Inc. (Accessed September 5, 2013, at http://www.dundeeprecious.com/English/operations/producing-mines/Chelopech/default.aspx.)

Dundee Precious Metals Inc., 2013d, Krumovgrad: Dundee Precious Metals Inc. (Accessed September 5, 2013, at http://www.dundeeprecious.com/English/operations/development-projects/krumovgrad/default.aspx.)

Energy Ecology Economy, 2012, Labour Minister urges termination of Gorubso Madan concession: Energy Ecology Economy, March 7. (Accessed September 11, 2013, at http://3e-news.net/show/21707_labour minister urges termination of gorubso madan concession_en/.)

International Lead and Zinc Study Group, 2013, New mine and smelter projects: Lisbon, Portugal, International Lead and Zinc Group, 70 p.

International Monetary Fund, 2013, World Economic Outlook: Washington, DC, International Monetary Fund, April, 184 p. (Accessed August 27, 2013, at http://www.imf.org/external/pubs/ft/weo/2013/01/pdf/text.pdf.)

Italcementi Group, 2012, Devnya Cement begins the realization of one of the largest investment products in Bulgaria during the last 20 years: Devnya, Bulgaria, Italcementi Group press release, March 14, 1 p. (Accessed September 5, 2013, at http://www.italcementigroup.com/NR/rdonlyres/98D6D02E-33E7-440E-B399-7AE5553236E8/0/Comunicato_devnya_UK.pdf.)

Italcementi Group, 2013, Bulgaria: Italcementi Group. (September 5, 2013, at http://www.italcementigroup.com/ENG/Italcementi+Group/A+global+presence/Bulgaria/Country+profile.htm.)

KCM 2000 Group, 2013a, Annual report 2012: KCM 2000 Group, 60 p. (Accessed September 9, 2013, at http://kcm2000.bg/files/p/27/kcm_annual_2012_en.pdf.)

KCM 2000 Group, 2013b, Varba-Batanitsi AD, KCM 2000 Group. (Accessed September 9, 2013, at http://www.kcm2000.bg/companies/verba-batantsi_ad/.)

LUKOIL Neftochim Bourgas AD, 2012, LUKOIL invested 1.5 billion USD in LUKOIL Neftochim Bourgas, LUKOIL Neftochim Bourgas AD press release, January 25. (Accessed September 10, 2013, at http://www.neftochim.bg/en/press-center/press-releases/lukoil-invested-1,-5-billion-usd-in-lukoil-neftohim-burgas.html.)

LUKOIL Oil Co., 2013, Annual report 2012: Lukoil Oil Co., 108 p. (Accessed September 10, 2013, at http://www.lukoil.com/materials/doc/Annual_Report_2012/Lukoil_GO_2012_eng.pdf.)

Melrose Resources Plc., 2012, Annual report and accounts 2011: Edinburgh, Scotland, Melrose Resources Plc., 78 p. (Accessed September 6, 2013, at http://www.petroceltic.ie/~/media/Files/P/Petroceltic-V2/Annual Reports/pdf/melrose-2011-ar.pdf.)

Mini Maritsa Iztok Mines EAD, 2011, Annual report 2011: Radnevo, Bulgaria, Mini Maritsa Iztok Mines EAD, 36 p. (Accessed September 12, 2013, at http://www.marica-iztok.com/files/profile/file_2_en.pdf.)

Mini Maritsa Iztok Mines EAD, 2013, Coal: Mini Maritsa Iztok Mines EAD. (Accessed September 12, 2013, at http://www.marica-iztok.com/bg/coals.php.)

National Statistical Institute of the Republic of Bulgaria, 2013a, Employees under labour contract by economic activity groupings and sector in 2012: National Statistical Institute of the Republic of Bulgaria. (Accessed August 27, 2013, at http://www.nsi.bg/otrasalen.php?otr=51&a1=2020&a2=2021&a3=2022#cont.)

National Statistical Institute of the Republic of Bulgaria, 2013b, Foreign trade of Bulgaria for 2012—Final data: National Statistical Institute of the Republic of Bulgaria, September 10, 14 p. (Accessed November 15, 2013, at http://www.nsi.bg/EPDOCS/fTrade2012_en_8S51V2G.pdf.)

National Statistical Institute of the Republic of Bulgaria, 2013c, Gross domestic product for the fourth quarter of 2012 and preliminary data for 2012: National Statistical Institute of the Republic of Bulgaria, March 6, 8 p. (Accessed November 18, 2013, at http://www.nsi.bg/EPDOCS/GDP2012q4_en_DBPNMIT.pdf.)

National Statistical Institute of the Republic of Bulgaria, 2013d, Number of non-financial enterprises by size in terms of employed and economic activity groupings: National Statistical Institute of the Republic of Bulgaria. (Accessed August 27, 2013, at http://www.nsi.bg/otrasalen.php?otr=71&a1=2474&a2=2475&a3=2476#cont.)

National Statistical Institute of the Republic of Bulgaria, 2013e, Statistical yearbook 2012: National Statistical Institute of the Republic of Bulgaria, September, 681 p. (Accessed November 18, 2013, at http://statlib.nsi.bg:8181/FullT/FulltOpen/SG_2011_2012_2013.pdf.)

Novinite.com, 2012a, 50% of troubled Bulgaria lead-zinc behemoth up for sale: Novinite.com, August 11. (Accessed September 16, 2013, at http://www.novinite.com/view_news.php?id=142216.)

Novinite.com, 2012b, Bulgaria's Gorubso Madan mines back to work: Novinite.com, April 17. (Accessed September 11, 2013, at http://www.novinite.com/view_news.php?id=138570.)

Novinite.com, 2012c, Bulgaria's Gorubso Madan mines to receive investments of BGN 4M in 2013: Novinite.com, April 17. (Accessed September 11, 2013, at http://www.novinite.com/view_news.php?id=145778.)

Novinite.com, 2012d, Bulgaria's largest non-ferrous plant sold in minutes: Novinite.com, September 14. (Accessed September 16, 2013, at http://www.novinite.com/view_news.php?id=143224.)

Petroceltic International Plc, 2012, Operations update—10 wells planned over next 12 months: Dublin, Ireland, Petroceltic International Plc press release, November 16, 5 p. (Accessed September 6, 2013, at http://www.petroceltic.com/~/media/Files/P/Petroceltic-V2/pdf/pr-2012/operations-update-12-11-06.pdf.)

Petroceltic International Plc, 2013, Annual report and accounts 2012: Dublin, Ireland, Petroceltic International Plc, 105 p. (Accessed September 9, 2013, at http://www.petroceltic.annualreport12.com/AR_2012.pdf.)

Technip Co., 2012, Technip awarded a major refining contract in Bulgaria: Paris, France, Technip Co. press release, January 25. (Accessed September 10, 2013, at http://www.technip.com/en/press/technip-awarded-major-refining-contract-bulgaria.)

U.S. Census Bureau, 2013a, U.S. exports to Serbia from 2003 to 2012 by 5-digit end-use code: U.S. Census Bureau. (Accessed September 12, 2013, at http://www.census.gov/foreign-trade/statistics/product/enduse/exports/c4870.html.)

U.S. Census Bureau, 2013b, U.S. imports from Serbia from 2003 to 2012 by 5-digit end-use code: U.S. Census Bureau. (Accessed September 12, 2013, at http://www.census.gov/foreign-trade/statistics/product/enduse/imports/c4870.html.)

U.S. Central Intelligence Agency, 2013, Bulgaria, *in* The world factbook: U.S. Central Intelligence Agency, August 22. (Accessed September 12, 2013, at https://www.cia.gov/library/publications/the-world-factbook/geos/bu.html.)

TABLE 1
BULGARIA: PRODUCTION OF MINERAL COMMODITIES[1]

(Metric tons unless otherwise specified)

Commodity[2]		2008	2009	2010	2011	2012
METALS						
Aluminum, metal, secondary		12,607	4,137	12,076	-- [r]	--
Bismuth, metal[e, 3]	metric tons	6	--	3	--	1
Cadmium, metal, smelter[e]		460	420	420	420	420
Copper:						
Ore, gross weight	thousand metric tons	27,191 [r]	26,936	27,581 [r]	28,214 [r]	28,300 [e]
Concentrate, Cu content[e]	do.	105	105	105	105	108
Metal, primary and secondary:						
Smelter		281,200 [r]	300,800 [r]	268,700 [r]	338,300 [r]	310,500
Refined, electrolytically		126,700	196,900	215,100	226,100 [r]	226,100
Gold, in concentrate[e]	kilograms	4,160 [4]	4,482 [4]	3,300 [r]	4,400 [r]	5,200
Iron and steel, metal:						
Pig iron for steelmaking	thousand metric tons	441	--	--	--	--
Ferroalloys[e]	do.	6 [4]	3 [4]	--	--	--
Steel, crude	do.	1,330	726	744 [r]	834	632
Semimanufactures[e]	do.	1,287 [4]	709 [4]	900 [r]	1,100 [r]	1,100
Lead:[e]						
Mine output, Pb content		15,000	15,000 [r]	12,000	12,000	12,000
Metal, refined, primary and secondary		91,000 [r]	83,000 [r]	81,000	71,000	68,000
Manganese ore:[5]						
Gross weight		64,600	28,500	131,600	149,400 [r]	37,900
Mn content		18,100	8,000	36,900	41,800 [r]	10,600
Silver, metal[e]	kilograms	50,000 [r]	43,000 [r]	42,000 [r]	45,000 [r]	55,000
Zinc:						
Mine output, Zn content		10,600 [r]	9,339 [r]	9,904 [r]	10,977 [r]	12,116
Metal, refined, primary and secondary		102,000 [r]	92,000 [r]	88,000 [r]	88,000 [r]	72,000

See footnotes at end of table.

TABLE 1—Continued
BULGARIA: PRODUCTION OF MINERAL COMMODITIES[1]

(Metric tons unless otherwise specified)

Commodity[2]		2008	2009	2010	2011	2012
INDUSTRIAL MINERALS						
Barite ore, run-of-mine[e]		40,000	14,300	350	120	--
Cement, hydraulic	thousand metric tons	4,903	2,662	1,966	1,882 [r]	1,900 [e]
Clays:						
Bentonite	do.	178	108	100 [e]	54 [r]	78
Kaolin, raw[e]	do.	1,530 [4]	939 [4]	900	900	900
Feldspar[e]	do.	90	80	80	80	80
Fluorspar[e]		--	--	--	31,800 [r]	32,000
Gypsum and anhydrite, crude	thousand metric tons	210	128	110 [r]	115 [r]	114
Lime, industrial	do.	1,422	950	1,309	1,495 [r]	1,500 [e]
Limestone[e]	do.	6,000 [r]	3,000 [r]	5,000 [r]	5,000 [r]	5,000
Nitrogen, N content of ammonia[e]	do.	350	320	320	320	320
Perlite	do.	7	15	-- [r]	-- [r]	4
Salt, all types	do.	2,100	1,300	1,900 [r]	2,200 [r]	2,100
Sand and gravel	do.	12,032 [r]	7,817 [r]	7,653 [r]	6,776 [r]	7,000 [e]
Silica, quartz sand[e]	do.	734 [4]	657 [4]	660	660	660
Sulfuric acid[e]		1,010,000	1,000,000	1,000,000	1,000,000	1,100,000
Vermiculite[e]		--	--	3,000	15,000	18,600
MINERAL FUELS AND RELATED MATERIALS						
Coal, marketable:[e]						
Bituminous	thousand metric tons	19 [4]	23 [4]	26	14 [r]	14
Brown	do.	2,643 [4]	2,244 [4]	2,200	2,300 [r]	2,300
Lignite	do.	26,008 [4]	25,015 [4]	27,500 [r]	34,500 [r]	32,000
Total	do.	28,670 [4]	27,282 [4]	29,700 [r]	36,800 [r]	34,300
Coke	do.	337	--	--	--	--
Natural gas, marketed	million cubic meters	218	17	74	443 [r]	396
Petroleum:[6]						
Crude	thousand 42-gallon barrels	170	175	170	160 [r]	170
Refinery products	do.	54,500	50,000	43,000 [r]	45,000	49,400

[e]Estimated; estimated data are rounded to no more than three significant digits; may not add to totals shown. [r]Revised. do. Ditto. -- Zero.

[1]Table includes data available through December 16, 2013.

[2]In addition to the mineral commodities listed, a variety of metals and industrial minerals, including calcinate sodium carbonate, refractory clays, sulfur, tin, and zeolites may have been produced, but available information is inadequate to make reliable estimates of output.

[3]Bismuth production was estimated based on net exports. Bismuth was produced as a byproduct of lead production.

[4]Reported figure.

[5]Reported by the International Manganese Institute.

[6]Figures were converted to barrels from production reported in thousand metric tons, as follows: crude production: 2008—23; 2009—24; 2010—23; 2011—22 (revised); and 2012—23. Refinery products: 2008—6,812; 2009—6,255; 2010—5,417; 2011—5,615; and 2012—6,171.

TABLE 2
BULGARIA: STRUCTURE OF THE MINERAL INDUSTRY IN 2012

(Thousand metric tons unless otherwise specified)

Commodity	Major operating companies and major equity owners	Location of main facilities	Annual capacity
Bentonite, mine output	Bentonite AD (S&B Industrial Minerals AD)	Kardjali	NA.
Cadmium	KCM A.D. (KCM 2000 Group)	Plovdiv	NA.
Do.	Lead and Zinc Complex Plc. (LZC) (Harmony 2012 Ltd., 50%)	Kardzhali	NA.[1]
Cement	Devnya Cement AD (Italcementi Group)	Devnya	2,000.
Do.	Vulkan Cement S.A. (Italcementi Group)	Dimitrovgrad	500.
Do.	Holcim Bulgaria AD (Holcim Ltd., 100%)	Beli Izvor	1,700.
Do.	Zlatna Panega Cement AD (Titan Group)	Zlatna Panega	1,500.
Coal:			
Bituminous	Balkan 2000 Mines EAD	Southeast of Tvarditsa, Sliven District	NA.
Brown	Otkrit Vagledobiv Mines EAD	Pernik coal basin, southwest of Sofia	NA.
Do.	Vagledobiv Bobov Dol EOOD	Bobov Dol coalfield	NA.
Do.	Other small producers	Cherno More Mine in the Black Sea coalfield and Vitren Mine in Katrishte deposit	NA.
Lignite	Mini Maritsa Iztok Mines EAD (state-owned Bulgarian Energy Holding EAD)	East Maritsa coal basin near Radnevo	3,500.
Do.	Other small producers	Beli Breg, Chukurovo, and Stanyantsi Mines	2,000.[e]
Copper:			
Concentrate, Cu content	Assarel-Medet JSC	Panagurishte, Pazardzhik District	50.
Do.	Ellatzite-Med AD (Geotechmin Co.)	Mine 8 kilometers south of Etropole, and concentrator near Mirkovo village	45.
Do.	Chelopech Mining EAD (Dundee Precious Metals Inc., 100%)	Chelopech	20.
Do.	Bradtze	Malko Turnovo	2.
Do.	Burgaskii Mines Ltd.	Zidorovo Mine at Burgas, near the Black Sea	1.
Metal:			
Smelter	Aurubis Bulgaria AD (Aurubis AG, 99.8%)	Pirdop	275.
Refinery	do.	do.	230.[e]
Fluorspar	Chiprovtzi Mine (Solvay S.A.)	Chiprovtzi, Montana Province	50,000.
Gold, in concentrate kilograms	Chelopech Mining EAD (Dundee Precious Metals Inc., 100%)	Chelopech	4,000.
Do. do.	Ellatzite-Med AD (Geotechmin Co.)	Mine 8 kilometers south of Etropole and concentrator near Mirkovo village	NA.
Do. do.	KCM A.D. (KCM 2000 Group)	Plovdiv	NA.
Kaolin, mine output	do.	Senovo, Rousse District	NA.
Lead-zinc:			
Concentrate, Pb-Zn content	Gorubso AD (KCM 2000 Group and and Ministroy Holding A.D.)	Kardjali	59 lead, 47 zinc.
Do.	Rudmetal JSC	Dimov Dol Mine, near Rudozem	3 lead, 2 zinc.
Metal:			
Pb, refined	KCM A.D. (KCM 2000 Group)	Plovdiv	65.
Do.	Lead and Zinc Complex Plc. (LZC) (Harmony 2012 Ltd., 50%)	Kardjhali	33.[1]
Zn, smelter	KCM A.D. (KCM 2000 Group)	Plovdiv	80.
Do.	Lead and Zinc Complex Plc. (LZC) (Harmony 2012 Ltd., 50%)	Kardjhali	28.[1]
Manganese ore	Obrochishte Mine (Euromangan AD)	Tsarkva village, 10 kilometers west of Balchik	NA.

See footnotes at end of table.

TABLE 2—Continued
BULGARIA: STRUCTURE OF THE MINERAL INDUSTRY IN 2012

(Thousand metric tons unless otherwise specified)

Commodity		Major operating companies	Location of main facilities	Annual capacity
Natural gas	million cubic meters	Melrose Resources Bulgaria EOOD (Petroceltic International Plc)	Kaliakra and Kavarna fields, in the Black Sea off the coast of Varna	400.
Do.	do.	Oil and Gas Exploration and Production Plc.	Bhutan, Bulgarevo, Dolni Dubnik, Durankulak, Marionov Geran, Selanovtzi, Staroseltzi fields	NA.
Perlite, mine output		S&B Industrial Minerals AD	Kardjali	NA.
Petroleum:				
Crude		Oil and Gas Exploration and Production Plc.	Bardarski Geran, Dolni Dubnik, Dolni Lukovit, Gorni Dubnik, Tjulenovo, Selanovtzi, Staroseltzi and other oilfields	NA.
Refined	42-gallon barrels per day	LUKOIL Neftochim Bourgas AD (LUKOIL Oil Co.)	Refinery at Burgas	215,000.
Silver:				
In concentrate	kilograms	Chelopech Mining EAD (Dundee Precious Metals Inc., 100%)	Chelopech	NA.
Metal	do.	KCM A.D. (KCM 2000 Group)	Plovdiv	100,000.^{e}
Steel, crude		Stomana Industry S.A. (Sidenor S.A., 100%)	Pernik	1,400.
Vermiculite, crude		Wolff and Muller Minerals Bulgaria OOD	Near Sofia	20.
Zeolite, mine output		S&B Industrial Minerals AD	Kardjali	NA.

^{e}Estimated. Do., do. Ditto. NA Not available.
^{1}Suspended.

THE MINERAL INDUSTRY OF CROATIA

By Harold R. Newman

Petroleum extraction and refining were the major economic activities of Croatia's mineral industry. Mineral resources included bauxite, clays, coal, gypsum, mica, natural asphalt, petroleum, and salt. The country remained reliant on mineral commodity imports for its industrial needs. Most of the output of industrial minerals was consumed by the domestic market.

The Energy and Mining Directorate (EMD) is responsible for the administrative and professional activities related to the energy and mining sectors, including drafting laws and regulations for these sectors. The EMD also coordinates energy and mining policy, including the country's energy development strategy, and provides advice for the implementation of energy policy (Energy and Mining Directorate, 2012a).

The Mining Act, which established Croatia's mining framework, was passed by the Government on June 19, 2009. In 2011, the Government passed amendments to the Mining Act to fill the gaps in the original mining framework. The amendments introduced, among other changes, the following four articles: Article 17, which provides for compensation to owners of property damaged as a result of the exploration and (or) exploitation by persons or legal entities that do not hold the required permits (These persons or legal entities will be responsible for the compensation.); Article 40, which obliges the holder of a license to submit a report on mineral reserves; Article 51, which requires bidders for public tenders for the exploration and (or) exploitation of mineral reserves to submit a statement that the bidder is not involved in any proceedings for unlawful exploration and (or) exploitation; and Article 136a, which states that a holder of the rights to mineral reserves will lose the rights automatically if exploration or exploitation is not commenced within the timeframe set out in the license. The amendments were in line with the overall tone of the Mining Act, which is designed to give control over mineral reserves back to the state (International Law Office, 2012).

The Energy Department carries out activities pertaining to the energy balance of the country (including analyzing the energy flows) and to the building of energy facilities. The Department proposes activities relating to energy efficiency. The Department also participates in the establishment of bilateral and multilateral agreements in the energy sector. The Energy Law of 2001, as amended in 2004 and 2007, contains measures aimed at ensuring a secure and reliable energy supply and efficient power generation and use, and addresses other key issues relevant to the energy sector. The Government's Energy Development Strategy, which was adopted in 2009 and covers the period 2009 to 2020, includes a commitment to include 20% of renewable energy in the country's total energy consumption by 2020 (Energy and Mining Directorate, 2012b, p. 8).

The Mining Department carries out administrative and oversight activities related to the exploration for and extraction of mineral raw materials, except clay, construction stone, and sand and gravel. The Mining Department also issues licenses for the exploration for and extraction of mineral raw materials, issues building permits for mining facilities and plants, and grants approvals for mining concessions and the extraction of mineral raw materials (Energy and Mining Directorate, 2012c).

The Hrvatski Geološki Institut (Croatian Geological Survey [CGS]), which is part of the EMD, is responsible for the analysis, collection, distribution, evaluation, and storage of geologic information. The CGS is the leading public research institute in Croatia in the field of geosciences and geological engineering. CGS collects geologic data for such purposes as exploration for mineral resources; environmental protection, including the protection of the fresh water supply; and urban planning (Croatian Geological Survey, 2012).

Minerals in the National Economy

In 2012, mineral production was not significant to Croatia's national economy. Croatia had a trade deficit with respect to mining and quarrying materials. Croatia's main trading partner was the European Union (EU), and its main export partners were Italy, which received 15% of Croatia's total exports; Bosnia (13%); Germany (11%); Slovenia (9%); and Austria (7%). Croatia's main import partners were Italy (17%), Germany (13%), Russia (8%), China (7%), and Slovenia (6%). The financial crisis in the EU that began in 2008 and continued in 2012 lowered external demand for Croatia's industrial exports. Economic recovery in the EU was expected to be slow and to continue to dampen private consumption. The Government was continuing with the process of entering the EU after a decade of carrying out the laws and reforms needed to bring Croatia's laws and standards in line with those of the EU. Croatia's entry into the EU was scheduled to take place on July 1, 2013 (U.S. Central Intelligence Agency, 2013).

Mineral Trade

U.S. trade in goods with Croatia was valued at $310 million in exports and $444 million in imports. U.S. exports to Croatia included, in order of value, metallurgical-grade coal, $80 million; other coal and fuels, $31 million; and petroleum products, $22 million. U.S. imports from Croatia included, in order of value, other petroleum products valued at $8 million; iron and steel products, $1 million; and iron and steel manufactures, $822,000 (U.S. Census Bureau, 2012a, b).

Production

In 2012, Croatia's mineral production was limited and continued at more or less the same levels as in 2011, with the exception of cement, gypsum, salt, sand and gravel, and silica sand, for which production decreased. Production of steel also deceased owing to the expansion and modernization work in progress at the CMC Sisak d.o.o. plant. There were no other significant changes in mineral production reported. Croatia no longer mined metallic ores, and metal production was based on domestic and foreign secondary raw materials. Industrial mineral production was sufficient to meet most of the country's domestic requirements. Mineral fuels, including natural gas and petroleum, were produced, although not in sufficient quantities to satisfy domestic demand (table 1).

Structure of the Mineral Industry

Table 2 is a list of the major mineral industry facilities.

Commodity Review

Metals

Aluminum.—TLM-TVP d.d. had a long history of producing and processing aluminum products. The plant's production process begins at the foundry cast house where the liquid metal is cast into aluminum blocks and then transferred to the hot-rolling mill. In the hot-rolling mill, the blocks are reduced to a thickness of from 3.5 millimeters (mm) to 15 mm. The strips are then transferred to the cold-rolling mill where they are rolled to sheets with final specifications of 0.5 mm to 4 mm in thickness. These products are used in construction and metal structures (TLM-TVP d.d., 2012).

Iron and Steel.—Commercial Metals Co, (CMC) of the United States announced that it had concluded the sale of its Croatian subsidiary, CMC Sisak d.d., to the Danieli Group of Italy for about $30.4 million. The transaction included the land, the steel works, and the rolling mills, but not the plant for the cold processing of steel. The CMC Sisak facility is located about 50 kilometers east of Zagreb. Danieli was one of the world's leading suppliers of steel products (Daily.tportal.hr, 2012).

In 2012, Adria Celik d.o.o. announced that it had started testing a steel production unit at the Zeljezara Split d.d. steel works. Adria Celik stated that the testing was a demanding process as the plant had been idle for about 4 years. Testing was to begin with the cold-testing phase, in which no steel is melted, followed by the hot-testing phase, during which time certain segments, such as the electric arc and cupola furnaces and the continuous casting and dedusting systems, would be tested. After the equipment and machinery have been tested and the necessary permits obtained, the steel mill was expected to begin regular production in 2013 (SeeNews, 2013).

Industrial Minerals

As of 2007 (the latest year for which these data were available), the mineral resource sites in Croatia included, in numerical order of the number of sites, crushed stone aggregates, 253 sites; dimension stone, 103 sites; sand and gravel, 82 sites; clay, 49 sites; bauxite, 15 sites; and gypsum, 9 sites. No information regarding the exploration and (or) exploitation status of these sites was available in 2012 (Slobodan and others, 2007).

Cement.—Holcim (Hrvatska) d.o.o., which was a subsidiary of Holcim Ltd. of Switzerland, announced that it would invest €1 million ($1.4 million[1]) for the reconstruction of a clinker cooler at Koromacno to increase thermal energy efficiency and decrease maintenance costs. In addition, Holcim aimed to advance its use of alternate fuels to more than 50% of its energy intake. In 2012, Holcim used coal, which was supplemented by emulsions, old tires, sawdust, solid recovered fuels, and waste oil as fuel for energy generation (Aggregate Research, 2012).

Mineral Fuels

Natural Gas and Petroleum.—INA-Industrija nafte d.d. (INA) was a leading Croatian mineral fuels company and was involved in natural gas and petroleum exploration and production, petroleum refining, and petroleum products distribution. INA conducted exploration and production both offshore in the Adriatic Sea and onshore in the Pannonian basin. INA also operated in Angola and Egypt (INA-Industrija nafte d.d., 2012a).

In 2012, INA and its partners operated 19 platforms in the Adriatic Sea. The Marcia and the North Adriatic fields hosted 17 platforms, 16 of which were petroleum production platforms and 1 of which was the Ivana K processing platform. The other two were natural gas production platforms in the Irena and the Izabela fields. Edina d.d., which was a joint venture between Edison Gas SpA of Italy and INA, was the operating company for both these fields. Five natural gas wells in these fields had been completed and put online (INA-Industrija nafte d.d., 2012b).

In 2012, INA owned and operated two petroleum refineries located at Rijeka and Sisak for the production of petroleum and refined products, as well as networks for the distribution of petroleum and refined products. The refinery at Rijeka was located near the port and had access to deep-draft ships and the Jadranski Naftovod plc (JANAF) petroleum pipeline system. The Rijeka refinery processed between 2 and 3 million metric tons per year (Mt/yr) of petroleum and other petroleum products for domestic and foreign markets. The Sisak refinery processed between 2 and 2.2 Mt/yr of petroleum and other petroleum products, which were sold on both the domestic and the export markets. Sisak processed domestic petroleum as well as petroleum imported through the Russian pipelines Druzba 1 and Druzba 2. Petroleum was also supplied by the JANAF pipeline (INA-Industrija nafte d.d., 2012c).

Outlook

Croatia is expected to remain a modest producer of mineral commodities, although increases in the production of industrial minerals could take place if the existing infrastructure is modernized. Mineral fuels are expected to remain the most important outputs of Croatia's mineral sector, although the country will most likely remain heavily dependent on imports of mineral fuels.

References Cited

Aggregate Research, 2012, Holcim Croatia posts operating loss amid challenging market conditions: Aggregate Research. (Accessed February 26, 2013, at http://www.aggregateresearch.com/print.aspx?ID=27523.)

Croatian Geological Survey, 2012, Department of Geology: Croatian Geological Survey. (Accessed September 21, 2013, at http://www.hgi-cgs.hr/eng/.)

---

[1]Where necessary, values have been converted from euro area euros (€) to U.S. dollars (US$) at a rate of €0.74=US$1.00.

Daily.tportal.hr, 2012, CMC sells Sisak steel mill to Danieli Group: Daily.tportal.hr. (Accessed September 22, 2013, at http://daily.tportal.hr/197739/CMC-sells-Sisak-steel-mill-to-Danieli-Group.html.)

Energy and Mining Directorate [Croatia], 2012a, Director General: Energy and Mining Directorate. (Accessed August 25, 2012, at http://www.kenny2.bnet.hr/mingorp.hr/defaulteng.aspx?id=27.)

Energy and Mining Directorate [Croatia], 2012b, Energy Department: Energy and Mining Directorate. (Accessed September 21, 2013, at http://www.kenny2.bnet.hr/mingorp.hr/defaultteng.aspx?id=114.)

Energy and Mining Directorate [Croatia], 2012c, Mining Department: Energy and Mining Directorate. (Accessed September 21, 2013, at http://kenny2.bnet.hr/mingorp.hr/defaultteng.aspx?id=115.)

INA-Industrija nafte d.d., 2012a, Offshore platforms and production: INA-Industrija nafte d.d. (Accessed September 24, 2013, at http://www.ina.print.aspx?id=2934.)

INA-Industrija nafte d.d., 2012b, Oil and gas exploration and production: INA-Industrija nafte d.d. (Accessed September 24, 2013, at http://www.ina.hr/default.aspx?id=2934.)

INA-Industrija nafte d.d., 2012c, Refining and marketing: INA-Industrija nafte d.d. (Accessed September 24, 2013, at http://www.ina.hr/default.aspx?id=294.)

International Law Office, 2012, Mining Act changes help to return control to the state: International Law Office. (Accessed December 2, 2012, at http://www.internationallawoffice.com/newsletters/detail.aspx?g=c4f5ef91-4305-4463.)

SeeNews, 2013, Croatia's Adria Celik testing steel production unit: SeeNews. (Accessed August 28, 2013, at http://www.seenews.com/news/croatia-s-adria-celik-testing-steel-production-unit-327471.)

Slobodan, Miko, Dragan, Vidic, Kruk, Boris, and Krasic, Dragan, 2007, Mineral resource management in Croatia: INFRA25708—Raw materials initiative—Thematic strategy on sustainable use of natural resources, Ljubljana, Slovenia, December 10–11, 2007, Presentation, 13 p. (Accessed August 30, 2012, at http://www.geo-zs.si/UserFiles/677/File/TAIEX/26_Slobodan Miko.pdf.)

TLM-TVP d.d., 2012, Welcome: TLM-TVP d.d. (Accessed September 22, 2013, at http://www.tlm-tvp.hr/onama_eng.asp.)

U.S. Census Bureau, 2012a, U.S. exports to Croatia by five-digit end-use code, 2002–2012: U.S. Census Bureau. (Accessed September 19, 2013, at http://www.census.gov/foreign-trade/statistics/product/enduse/exports/c4791.html.)

U.S. Census Bureau, 2012b, U.S. imports from Croatia by five-digit end-use code, 2002–2012: U.S. Census Bureau. (Accessed September 19, 2013, at http://www.census.gov/foreign-trade/statistics/product/enduse/imports/c4791.html.)

U.S. Central Intelligence Agency, 2013, Croatia—Economic overview, in The world factbook: U.S.Central Intelligence Agency. (Accessed September 21, 2013, at https://www.cia.gov/library/publications/the-world-factbook/geos/hr.html.)

TABLE 1

CROATIA: PRODUCTION OF MINERAL COMMODITIES[1]

(Metric tons unless otherwise specified)

Commodity[2]		2008	2009	2010	2011	2012[e]
METALS						
Aluminum:[e, 3]						
Alloys		31,582 [3]	30,000	30,000	32,000	31,000
Semimanufactures:						
Rolled		52,135 [3]	50,000	50,000	60,000	55,000
Extruded		6,287 [3]	6,000	6,000	8,000	8,000
Total		58,422 [3]	56,000	56,000	68,000	63,000
Steel:						
Crude, from electric furnaces		121,759	51,583	95,000	95,440 [r]	109,000
Semimanufactures, hot rolled[e]		121,000	51,000	103,000	100,000	50,000
INDUSTRIAL MINERALS						
Cement	thousand metric tons	3,637	2,823	5,078	2,570 [r]	1,244 [3]
Clays:						
Bentonite		19,759	NA	NA	NA	NA
Ceramic clay[e]		300,000	NA	NA	NA	NA
Gypsum and anhydrite, crude		329,649	221,888	248,675	231,008	182,557 [3]
Lime	thousand metric tons	541	350	330	271	300
Nitrogen, N content of ammonia[e]	do.	300	300	358 [3]	367 [3]	350
Pumice and related materials, volcanic tuff[e]	do.	15	15	15	15	20
Salt, all sources[e]		30,000	30,000	30,000	21,160 [r, 3]	18,342 [3]
Sand and gravel, excluding glass sand[e]	thousand metric tons	3,500 [r]	3,250 [r]	3,500 [r]	4,003 [r, 3]	3,682 [3]
Silica sand (quartz, quartzite, glass sand)		150,000 [e]	278,231	240,919	227,437	225,000
Stone:[e]						
Crushed and brown	thousand metric tons	18,000	17,652 [3]	13,270 [3]	13,033 [3]	14,000
Dimension stone		1,500,000	1,400,000	1,200,000	1,400,000	1,500,000
Sulfur, byproduct of petroleum		9,819	10,315	6,834	7,254	7,000

See footnotes at end of table.

TABLE 1—Continued
CROATIA: PRODUCTION OF MINERAL COMMODITIES[1]

(Metric tons unless otherwise specified)

Commodity[2]		2008	2009	2010	2011	2012[e]
MINERAL FUELS AND RELATED MATERIALS						
Carbon black		16,903	3,976	--	--	--
Natural gas, gross production	million cubic meters	2,729	2,705	2,727	2,471	2,013 [3]
Petroleum						
Crude, gross weight, includes condensate	thousand 42-gallon barrels	6,200	5,760	5,340	4,983 [r, 3]	4,479 [3]
Refinery products:	do.					
Distillate fuel oil		10,403	11,096	9,746	10,000 [e]	12,000
Residual fuel oil		7,519	7,592	6,242	6,500 [e]	7,000
Jet fuel		767	733	741	750 [e]	750
Liquefied petroleum gases		2,957	3,504	4,344	4,400 [e]	4,500
Motor gasoline		8,541	10,293	9,344	9,500 [e]	9,800
Other products		4,344	5,658	5,256	5,400 [e]	5,600
Total		34,531 [r]	38,876 [r]	35,673 [r]	36,600 [r, e]	39,700

[e]Estimated; estimated data are rounded to no more than three significant digits. [r]Revised. do. Ditto. NA Not available. -- Zero.
[1]Table includes data available through August 31, 2013.
[2]In addition to commodities listed, common clay and other industrial minerals may have been produced, but available information was inadequate to make reliable estimates of output.
[3]Reported figure.

TABLE 2
CROATIA: STRUCTURE OF THE MINERAL INDUSTRY IN 2012

(Thousand metric tons unless otherwise specified)

Commodity		Major operating companies	Location of main facilities	Annual capacity
Aluminum, semimanufactures		TLM-TVP d.d.	Sibenik	50 [e]
Do.		Top-Tvornica Olovni i Aluminijskih	Savska	NA
Carbon black		Petrokemija d.d.	Kutina	NA
Cement		Cemex Hrvatska d.d. (CEMEX S.A.B. de C.V., 100%)	Plants at Kastel Sucurac, Solin, and Solin Majdan	2,400
Do.		Holcim (Hrvatska) d.o.o. (Holcim Ltd., 100%)	Plant at Koromacno	1,000
Do.		Istra Cement International d.d. (Part of CALUCEM Group)	Plant at Pula	NA
Do.		Tvornica Cementa Umag d.o.o.	Cement plant at Umag	350
Do.	million cubic meters	Nasicecement d.d. (Nexe Grupa d.d.)	Nasice	1,000
Natural gas		INA-Industrija nafte d.d. (INA)	Natural gasfields at Molve, offshore platforms in the Adriatic Sea, and other locations	3,000
Petroleum:				
Crude	thousand 42-gallon barrels per day	do.	Oilfields at Kalinovac, Sandrovac, Struzec, Zutica, and other locations	20 [e]
Refined	do.	do.	Refinery at Rijeka (Urinj)	3,500
Do.	do.	do.	Refinery at Sisak	2,500
Salt		Solana Pag d.d.	Pag Island (marine salt)	NA
Steel, crude		CMC Sisak d.o.o. (Commercial Metals International AG, 100%)	Plant at Sisak	80
Do.		Zeljezara Split d.d. (Adria Celik d.o.o., 50%, and Techcom GmbH, 50%)	Plant at Split (closed)	185 [1]

[e]Estimated. Do., do. Ditto. NA Not available.
[1]Zeljezara Split d.d. stopped production in 2009, and production remained suspended throughout 2012. Production was expected to resume in 2013.

THE MINERAL INDUSTRY OF CYPRUS

By Harold R. Newman

Cyprus is the third largest island in the Mediterranean Sea after the islands of Sicily and Sardinia, and it is located in the northeastern corner of the Mediterranean Sea. The mineral resources of Cyprus[1] included asbestos, clays, copper, gypsum, lime, marble, sand, stone, and umber, which is an iron and manganese oxide. In 2012, the mining sector was small and had only a limited effect on the national economy.

The Geological Survey Department was placed under the auspices of the Ministry of Agriculture, Natural Resources, and Environment in the late 1960s and, in 2012, was responsible for the oversight of the country's mineral exploration programs and for evaluating the country's mineral resources. The Geological Survey Department served as the technical advisor to the Government for all matters related to the country's geology and also undertook research studies on behalf of the Government (Geological Survey Department, 2012a).

The Mines Service of the Ministry of Agriculture, Natural Resources and Environment administered mineral operations under the Mines and Quarries (Regulation) Law, chapter 270, 1959; the Mines and Quarries Regulations, 1958–1979; and the Cyprus Standard and Control of Quality Law 1975 (Ministry of Agriculture, Natural Resources and Environment, 2012).

The gross output of the mining and quarrying sectors in 2011 (the latest year for which data were available) was valued at €79.5 million ($105.3 million[2]). The output of mining and quarrying value-added products decreased to €46.4 million ($61.4 million). The number of people employed in the mining and quarrying sector was 534 (Statistical Service of the Republic of Cyprus, 2012).

International trade was important to the economy of Cyprus. The country's lack of energy resources, heavy industry facilities, and the raw materials required for the production of capital goods necessitated the importation of such items. The European Union (EU) and Cyprus's neighbors in the Middle East absorbed the majority of Cyprus's exports and supplied the majority of Cyprus's imports (U.S. Central Intelligence Agency, 2013).

U.S. exports to Cyprus in 2012 totaled $167 million and included, in order of value, $6.4 million in fuel oil, $1.3 million in finished metal shapes, $1.2 million in iron and steel products, and $246,000 in petroleum products (U.S. Census Bureau, 2012a). U.S. imports from Cyprus in 2012 totaled $29 million. These included, in order of value, $886,000 in sulfur and nonmetallic minerals, $321,000 in bauxite and aluminum, and $61,000 in cement, sand, and stone (U.S. Census Bureau, 2012b).

Production

The Troodos ophiolite complex dominated the central part of the island and constituted the geologic core of Cyprus. Directly associated with the Troodos ophiolite are asbestos, chromite, and massive sulfide mineral deposits. These deposits became exposed as a result of the uplift of the Troodos ophiolite which exposed the ore bodies to the surface, especially the massive sulfide deposit. Production of copper has taken place from these deposits since antiquity (Geological Survey Department, 2012b).

Bentonite, cement, sand and gravel, and stone were the major mineral commodities produced by the mineral industry of Cyprus. Other mineral commodities produced included common clay for brick and cement manufacture, gypsum, ocher, and umber. In 2012, the production of building stone, marble, and refined copper increased whereas that of cement, gypsum, lime, and umber decreased compared with that of 2011 (table 1).

Structure of the Mineral Industry

Table 2 is a list of the major mineral industry facilities, their locations, and their annual capacities. All facilities were privately owned.

Commodity Review

Metals

Copper.—Copper exploration in Cyprus was centered on the Troodos ophiolite complex. All of Cyprus's copper deposits that had been identified as of 2012 were volcanic-hosted massive sulfide (VMS) deposits that were under shallow cover rocks. EMED Mining Public Ltd. (EMED) was a Cyprus-based mineral development and exploration company that was focused on conducting exploration to justify the commencement of a full feasibility study for the Klirou copper-zinc property, which EMED considered to be a high-priority project. The project is located about 20 kilometers (km) southwest of Nicosia. The Klirou deposit extends from the surface to about 200 meters (m) below the surface and was potentially minable by open pit methods (EMED Mining Public Ltd., 2013).

The only copper mining/processing activity that was operating in 2012 was the Hellenic Copper Mines Ltd.'s operation at Skouriotissa. The Skouriotissa Mine's place in history was that it was thought to be the world's longest producing copper mine, with production dating back to about 4,000 years ago. The majority of the copper produced came from the processing of waste material from previous mining operations by the leaching and solvent extraction-electrowinning method to obtain metal from the low-grade copper waste material. The Skouriotissa facility produced 99.99%-pure copper metal cathodes (Mines Service, 2012).

Gold.—Northern Lion Gold Corp. of Canada was continuing to explore for gold in the Troodos complex and announced the

---

[1]Unless specifically stated, all data in this chapter are for the Republic of Cyprus in southern Cyprus because data related to areas of northern Cyprus administered by Turkish Cypriots were sparse or unavailable. The two areas have been separated since 1974.

[2]Where necessary, values have been converted from euro area euros (€) to U.S. dollars (US$) at a rate of €0.71=US$1.00.

approval of four additional mineral permits, which increased Northern Lion's tenure holdings by about 80 square kilometers. Two of the permits, Kato Lefkara and Pano Lefkara, were 100% owned by Northern Lion whereas the other two, Filousa and Perapedi, were held jointly with SES Sweden AB of Sweden. Initial exploration work on the Kato Lefkara and Pano Lefkara permits revealed surface gold mineralization. A followup core drilling program was planned to evaluate the new permits (Marketwire, 2012).

Industrial Minerals

The sedimentary rocks of Cyprus contain industrial minerals, including aggregates, bentonite, building stone, chalk, clay, gypsum, and limestone. In 2012, quarrying was conducted to obtain materials mainly for domestic use; however, information regarding the ownership, location, and output of most of these facilities was not readily available.

Cement.—Vassiliko Cement Works Public Ltd. was a significant producer and distributor of cement and clinker. The company produced several types of cement—ordinary Portland cement, low-alkali sulfate-resistant Portland cement, Portland composite cement, and white low-alkali limestone cement. Vassiliko also had a presence in the ready-mix concrete market and the quarrying sector through its subsidiaries (Vassiliko Cement Works Public Co. Ltd., 2012).

Vassiliko Cement announced that work had commenced on a new 6,000-metric-ton-per-day clinker line. Equipment included a circular blending bed, vertical mills for raw materials, and a kiln system with a five-stage double-string preheater, a calciner bypass, a rotary kiln, and a waste gas treatment plant (International Cement Review, 2012).

Clay and Shale.—Peletico Penta Ltd.'s operation consisted of mines at Pentacomo and Troulli and a bentonite processing plant at Pentacomo. Peletico Penta's plant's production process was redesigned for the production of bentonite bars through a dry compaction process. The activation of the bentonite was achieved without the use of soda and with a parallel decrease in energy demand and use of natural resources. Particles of the byproducts (collected dust) of the mining process were used as raw materials, which resulted in increased productivity and decreased production costs (Eco-Innovation Observatory, 2012).

Mineral Fuels

The energy policy of Cyprus was in line with the energy policy of the EU as a whole. Energy constituted one of the economically significant sectors of the country. The country's energy supply was highly dependent on imported fuels. The main goals of the country's energy policy were to safeguard competition in the marketplace, secure a reliable supply of energy, fulfill the energy demands of the country, and impose the least possible burden on the economy and the environment. The energy policy included promotion of natural gas, petroleum products, and renewable energy (Ministry of Energy, Commerce, Industry and Tourism, 2012).

Natural Gas and Petroleum.—The Government announced that 29 companies had made 15 bids for exploratory drilling in a second round of licensing to exploit offshore natural gas and petroleum deposits. Bids were submitted from 10 joint ventures and 5 companies from 15 countries. A final decision, which would involve 12 of 13 blocks, was expected in 2013.

Although Cyprus produced no natural gas or petroleum in 2012, Nobel Energy Inc. of the United States announced that it had discovered a natural gas field in Block 12 that had an estimated reserve of up to 226 billion cubic meters with an estimated value of $129 billion. Block 12 is located offshore the southeast coast of Cyprus, The discovery was the first of its kind in Cyprus and, if the field is developed, it could possibly supply all the domestic requirements for natural gas for the country (Rigzone, 2012).

Outlook

Domestic infrastructure construction is expected to continue to be the main source of domestic demand for cement, gypsum, sand and gravel, and stone and to be a significant contributor to the national economy. International demand is expected to continue to support Cypriot exports of bentonite, gypsum, ocher, and umber. Natural gas and petroleum exploration is expected to continue.

References Cited

Eco-Innovation Observatory, 2012, Dry compaction method for the creation of bentonite: Eco-Innovation Observatory. (Accessed August 17, 2012, at http://www.eco-innovation.eu/index.php?option=com_content&view= article&id=328%3Adry-compaction-method-for-the-creation-of-bentonite.)

EMED Mining Public Ltd., 2013, Projects: EMED Mining Public Ltd. (Accessed August 14, 2013, at http://www.emed-mining.com/projects/ cyprus.)

Geological Survey Department [Cyprus], 2012a, Mission: Geological Survey Department. (Accessed August 13, 2013, at http://www.moa.gov.cy/moa/gsd/ gsd.nsf//dmlMission_en/dmlMission_en?OpenDocument.)

Geological Survey Department [Cyprus], 2012b, Troodos: Geological Survey Department. (Accessed December 20, 2013, at http://www.moa.gov.cy/moa/ gsd/gsd.nsf/dmlTroodos_en/dmlTroodos_en?OpenDocument.)

International Cement Review, 2012, Cyprus: International Cement Review, January, p. 25.

Marketwire, 2012, Northern Lion samples 41 meters averaging 3.47 g/t gold in recently granted mineral permits, Republic of Cyprus: Marketwire. (Accessed August 14, 2013, at http://services.metalseconomics.com/MineSearch/News/ News.aspx?type=html&newsID=1008344775.)

Mines Service, 2012, Copper and gold: Ministry of Agriculture, Natural Resources and Environment. (Accessed December 19, 2013, at http://www.moa.gov.cv/ moa/Mines/MinesSrv.nsf/All/7E7B31B32C892C22574EA00.)

Ministry of Agriculture, Natural Resources and Environment, 2012, Mines Service: Ministry of Agriculture, Natural Resources and Environment. (Accessed August 13, 2013, at http://www.moa.gov.cy/moa/minesSrv.nsf/ dmlmines_en/dmlegmines_en?OpenDocument.)

Ministry of Energy, Commerce, Industry and Tourism, 2012, Energy Service: Ministry of Energy, Commerce, Industry and Tourism. (Accessed August 14, 2013, at http://www.mcit.gov.cy/mcit.nsf/dmlenergyservice_en/ dmlenergyservice_en?OpenDocument/.)

Rigzone, 2012, Cyprus secures 15 bids for gas exploration: Rigzone. (Accessed May 14, 2012, at http://www.rigzone.com/news/article_pf.asp?a_id=117812.)

Statistical Service of the Republic of Cyprus, 2012, Industrial statistics, 2012: Statistical Service of the Republic of Cyprus. (Accessed August 13, 2013, at http:/www.cyprus.gov.cy/mof/cystat/statistics.nsf/ Industry_A05_11-en-260912.)

U.S. Census Bureau, 2012a, U.S. exports to Cyprus by 5-digit end-use code: U.S. Census Bureau. (Accessed August 13, 2013, at http://www.census.gov/ foreign-trade/statistics/product/enduse/exports/c4910.html.)

U.S. Census Bureau, 2012b, U.S. imports to Cyprus by 5-digit end-use code: U.S. Census Bureau. (Accessed August 13, 2013, at http://www.census.gov/ foreign-trade/statistics/product/enduse/imports/c4910.html.)

U.S. Central Intelligence Agency, 2013, Economy—Overview, *in* The world factbook: U.S. Central Intelligence Agency. (Accessed September 18, 2013, at http://www.cia.gov/library/publications/the-world-factbook/geos/cy.html.)

Vassiliko Cement Works Public Co. Ltd., 2012, Our activities: Vassiliko Cement Works Public Co. Ltd. (Accessed August 14, 2013, at http://www.vassiliko.com/gb/ouractivities.)

TABLE 1
CYPRUS: PRODUCTION OF MINERAL COMMODITIES[1]

(Thousand metric tons unless otherwise specified)

Commodity[2]		2008	2009[e]	2010	2011[e]	2012
Cement, hydraulic		1,914	1,481 [3]	1,328	1,207 [3]	1,080
Clays:						
Bentonite	metric tons	155,125	152,722 [3]	162,169	160,625 [3]	160,180
Other:						
For brick and tile manufacture		490	480	210	160	120
For cement manufacture		635	400 [3]	445	405 [3]	375
Total		1,125	880	655	565	495
Copper, refined	metric tons	2,986	2,380	2,595	3,660	4,328
Gypsum, crude[4]	do.	405,500	317,000	333,300	335,000	327,800
Lime, hydrated	do.	14,285	12,000 [3]	9,951	9,824 [3]	4,551
Sand and stone:						
Limestone, crushed (Havara)		766	800	800	165	64
Marble, granules and chippings	metric tons	550	400	1,195	1,900	3,920
Marl, for cement production		2,595	2,600	1,805	1,640	1,500
Sand and gravel[5]		14,174	11,468 [3]	12,981	11,826 [3]	7,308
Stone, building[6]		71	70	97	84 [r,3]	89
Umber and ocher, for cement	metric tons	44,710	43,360 [3]	52,039	61,553 [3]	37,957

[e]Estimated; estimated data are rounded to no more than three significant digits; may not add to totals shown. [r]Revised. do. Ditto.
[1]Table includes data available through August 31, 2013.
[2]In addition to the commodities listed, small quantities of the mineral pigments ocher and terra verte are mined intermittently. Mineral production data from areas of northern Cyprus that are administered by Turkish Cypriots, and the production of fertilizers, perlite, and secondary metals from scrap are not included in this table because available information is inadequate to make reliable estimates of output.
[3]Reported figure.
[4]About 4,000 metric tons per year of gypsum was calcined.
[5]Includes crushed aggregate.
[6]Includes crude, semifinished, and worked stone.

TABLE 2
CYPRUS: STRUCTURE OF THE MINERAL INDUSTRY IN 2012

(Metric tons unless otherwise specified)

Commodity	Major operating companies and major equity owners	Location of main facilities	Annual capacity
Aluminum, semimanufactures:	Muskita Aluminum Industries Ltd.	Plant at Limassol	22,000
Cement	Vassiliko Cement Works Public Ltd.	Plant at Vassiliko, 5 quarries in the area	1,200,000
Do.	Cyprus Cement Co. Ltd.	Plant at Moni	400,000
Clay, bentonite	Peletico Penta Ltd.	Mines at Pentakomo and Troulli, plant at Pentakomo	NA
Do.	Hellenic Mining Co.	Nicosia	NA
Do.	Oryktako Ltd.	Mine at Kato Moni and processing plant at Malounda	10,000
Copper, refined	Hellenic Copper Mines Ltd.	Skouriotissa	5,000
Gypsum	Peletico Ltd.	Quarry and processing plant at Aradipou, near Larnaka	NA
Do.	Zeiplast Chemical Industries Ltd.	Near Moni	NA
Perlite	Peletico Ltd.	Expanded perlite facility at Larnaka	NA
Do.	Zeiplast Chemical Industries Ltd.	Expanded perlite facility at Moni	NA
Steel, semimanufactures	B.M.S. Metal Pipes Industries Ltd.	Tube and pipe mill, Paphos	15,000

Do. Ditto. NA Not available.

THE MINERAL INDUSTRY OF THE CZECH REPUBLIC

By Steven T. Anderson

In 2012, the Czech Republic was estimated to have been the 4th-ranked producer of kaolin in the world, the 11th-ranked producer of feldspar, and the 27th-ranked producer of crude steel, by tonnage. The country was estimated to have accounted for approximately 11% of the world's production of kaolin; of feldspar, slightly greater than 2%; bentonite, slightly greater than 1%; and industrial (silica) sand, about 1%. Coal, coke, and steel were the mineral commodities that were most significant to the country's domestic and regional markets. The Czech Republic was a significant Central European producer of heavy industrial goods manufactured by the country's chemical, machine building, and toolmaking industries. The production of coal for thermal powerplants and the use of nuclear power were significant sources of electricity and helped the country maintain a lower level of dependence on imported natural gas for electricity production than other countries in Central and Eastern Europe. Other mineral commodities produced in the country included cement, common sand and gravel, dolomite, garnet, gypsum, natural gas, and uranium (table 1; Czech Geological Survey, 2012, p. 90–91; Dolley, 2013; Tanner, 2013; Virta, 2013; World Steel Association, 2013b, p. 9).

Minerals in the National Economy

Based on estimated data from the Czech Statistical Office, the value added by the mining and quarrying sector contributed about 1.04% (about $2 billion[1]) to the country's gross domestic product (GDP) in 2012 compared with 1.21% ($2.6 billion) in 2011. In 2005 prices, the real value added by the sector decreased by 3% in 2012 compared with that of 2011, after decreasing by 7% in 2011 compared with that of 2010. In 2012, the average number of employees in the mining and quarrying sector was estimated to be about 43,300 compared with an average of about 46,500 employees in 2011, and the sector accounted for about 1% of the total number of employees in the Czech Republic, on average, during both years (Czech Statistical Office, 2013b; International Monetary Fund, 2013).

If converted to current U.S. dollars, the estimated value of the mineral industry trade balance (including trade in metal scrap and intermediate manufactured mineral products, such as cement, crude steel, and manufactured gas) for the Czech Republic was –$12.5 billion in 2012 compared with –$12.9 billion in 2011. This 3% reduction in the mineral trade deficit took place despite an estimated decrease in the value of the country's mineral industry exports to about $15.4 billion from $16.6 billion in 2011 because of a greater decrease in the value of imports of mineral industry products to about $27.9 billion from $29.5 billion in 2011. In Czech koruna, the estimated value of this same mineral industry trade deficit actually expanded by about 7.2% because the estimated value of imports of mineral industry products increased by about 4.8%. The reason for this discrepancy is that the estimated annual average nominal exchange rate changed significantly (by about 10.7%) during this timeframe (as described in footnote 1). In 2012, the three leading mineral product categories with respect to the Czech Republic's import expenditure were petroleum, petroleum products, and related materials ($8.4 billion), iron and steel ($6.9 billion), and gas, natural and manufactured (about $5.1 billion); the three leading mineral industry exports by the country were iron and steel (about $5.5 billion), nonmetallic mineral manufactures (about $2.8 billion), and petroleum, petroleum products, and related materials (about $1.7 billion). In 2012, mineral industry products accounted for about 20% of the total value of Czech goods imports and about 10% of the total value of the country's exports of goods (Czech Statistical Office, 2013a; International Monetary Fund, 2013).

Government Policies and Programs

Three main laws are applicable to the mineral industry in the Czech Republic. Act No. 44/1988 on the Protection and Use of Mineral Resources (the Mining Act), as amended, defines the minerals that are owned by the Government, establishes the authority of certain Government agencies with respect to mining activity, and sets out other rules on the management of mineral resources in the Czech Republic. The Czech National Council Act No. 62/1988 on Geological Work (the Geological Act), as amended, establishes the rules for prospecting and exploration of most mineral deposits. Act No. 61/1988 on Mining Operations, Explosives and on the State Mining Administration, as amended, defines appropriate mining methods. The Ministry of the Environment enforces environmental laws in the mining sector and has the authority to revoke exploration and mining leases if environmental laws are violated (Czech Geological Survey, 2012, p. 27–34).

In 1991, the Czech Government passed Government Resolution No. 444/1991, which established geographic limits on the expansion of coal and uranium mining. It was estimated that about 750 million metric tons (Mt) of brown coal reserves as well as some uranium reserves were located in areas where mining is restricted. A national energy policy document known as the State Energy Concept (SEC), which is a document with a 30-year outlook, was approved by Government Decision no. 211 and became the State Energy Policy on March 10, 2004. The SEC had been reviewed twice by the Ministry of Industry and Trade (MIT) before it was approved. The SEC 2004 included the following priorities: decrease the intensity of primary energy supplied (for domestic use) relative to the GDP [measured by the ratio of total primary energy supply (TPES) to GDP], maintain the current level of TPES, and comply with binding European Union emission limits in 2010. In 2009, the MIT prepared an updated draft of the SEC, which included

---
[1]Where necessary, values have been converted from Czech koruna (CZK) to U.S. dollars (US$) at an annual average exchange rate of about CZK17.7=US$1.00 for 2011 and CZK19.6=US$1.00 for 2012. All values are nominal, at current prices, unless otherwise stated.

the following strategies: achieve a balanced energy mix, with preferential use of all domestic energy resources, and maintain excess production of electricity; improve energy efficiency and reduce energy intensity, particularly in the building sector; increase energy security and the ability of the country to respond to energy supply disruptions; and minimize the effects of energy use on the environment. The update of the SEC was still being debated by the Government at the end of 2012, and it had asked for a strategic impact assessment of the update to the SEC before final approval (Ministry of Industry and Trade, Czech Republic, 2004, p. 1–9; Czech Coal Group, 2010, p. 75; International Energy Agency, 2010, p. 22–23, 45; Czech Geological Survey, 2012, p. 74–80, 87–92; Weiler, 2012; OECD Nuclear Energy Agency, 2013).

Production

In 2012, production of most minerals and materials used intensively in construction decreased significantly from that of 2011, including production of cement (–10%); feldspar substitutes, including for glass manufacturing (–32%) and glass sand (–13%); common sand and gravel (–12%); crude steel (–9%); and various forms of stone (between –12% and –32%). The production of other minerals used intensively in construction increased significantly during this same timeframe, however, including gypsum (+27%) and dolomite (+19%). In 2011, many state-funded construction projects were started that had been delayed because of adverse weather conditions in 2010, and housing construction also recovered from a down year in 2010. In 2012, however, there appeared to be a limited number of state-funded construction projects, and investors stopped preparation of infrastructure construction projects as repair and reconstruction funds became more limited and the policy priorities of the Czech Transportation Ministry became less clear (table 1; Czech Geological Survey, 2012, p. 51; International Cement Review, 2012a, b)

In 2012, changes in production of other industrial minerals were also mixed. Production of bentonite increased by 38% compared with that of 2011, and production of foundry sand increased by slightly greater than 24%. Production of gemstone-bearing rock decreased by between 29% and about 37% from that of 2011, however, and production of silica minerals decreased by 29%. Information was not available regarding the main causes of these disparate changes in industrial mineral production. In 2012, mine output of uranium decreased by about 12% compared with that of 2011 as a result of a continuing Government program to phase out uneconomic mining of uranium in the Czech Republic (table 1; Czech Geological Survey, 2012, p. 45–54, 74–76, 101–103).

Structure of the Mineral Industry

Table 2 is a list of major mineral industry facilities. In 2012, the only state-owned mining company that remained in the Czech Republic was DIAMO s.p., and even DIAMO was not intended to be a mining company. Rather, DIAMO was intended to coordinate and administer the ongoing remediation activities to restore the properties of former state-owned mines that are no longer producing. In 2012, DIAMO still produced some mined uranium, in addition to its remediation activities (Czech Geological Survey, 2012, p. 74–76, 101–103).

Commodity Review

Metals

Iron and Steel.—In 2012, the Czech Republic had no economically exploitable iron ore deposits and imported all iron ore products used in primary steel production; the country imported about 5.9 Mt of iron ore and concentrate compared with about 7.4 Mt in 2011. In 2011, ArcelorMittal Ostrava a.s. had reportedly laid off about 700 employees, and Evraz Vitkovice Steel a.s., 200, in order to decrease production more efficiently in 2012. The Czech Steel Federation anticipated that demand for steel from the construction sector would not recover until sometime in 2013, at the earliest (Czech Geological Survey, 2012, p. 201; World Steel Association, 2013a, p. 100; Czech Steel Federation, undated).

Industrial Minerals

Cement.—In 2012, production of cement in the Czech Republic decreased by about 400,000 t from that of 2011; this was very similar to the decrease in domestic consumption of cement by 370,000 t during the same timeframe. Coal was the leading fuel used in cement production in the country and accounted for about 44% of total fuel consumption for cement production, followed by other solid fuels (22.6%), biomass (20.6%), and used tires (7.6%); the remainder was accounted for by the use of liquid fuels, including heavy fuel oil and natural gas (Czech Cement Association, 2013).

Mineral Fuels and Related Materials

Uranium.—DIAMO remained the only domestic producer of uranium, and it supplied CEZ a.s. (the owner of the Czech Republic's two nuclear powerplants) with about one-third of the uranium it required. All domestically produced uranium was sent to Russia for processing into fuel. All nuclear fuel for the Dukovany Nuclear Power Station was purchased from the Russian firm OAO TVEL, whereas the Temelin Nuclear Power Station obtained its fuel from Westinghouse Electric Company LLC of the United States. CEZ's nuclear powerplants accounted for about 35% of all electricity in the Czech Republic in 2012 (CEZ a.s., 2013, p. 102–104; Czech Geological Survey, 2012, p. 101–103).

Outlook

Economic activity in the Czech Republic is expected to increase gradually during 2013 as external conditions improve. Dependence on imports of natural gas and petroleum is likely to continue to affect the trade balance negatively, but production of coal is likely to remain stable and to provide a significant portion of fuel for electricity generation (International Monetary Fund, 2012a, p. 1, 5; 2012b, p. 199).

References Cited

CEZ a.s., 2013, Annual report 2012: Prague, Czech Republic, CEZ a.s, March 31, 318 p. (Accessed July 25, 2013, at http://www.cez.cz/edee/content/file/investors/2012-annual-report/VZ2012aj.pdf.)

Czech Cement Association, 2013, Data 2012: Prague, Czech Republic, Czech Cement Association, April 15, 2 p. (Accessed July 18, 2013, at http://www.svcement.cz/includes/dokumenty/pdf/data_2012.pdf.)

Czech Coal Group, 2010, Yearly report of the Czech Coal Group—Business activities and sustainable development in 2009: Prague, Czech Republic, Czech Coal Group, 127 p. (Accessed September 12, 2012, at http://www.czechcoal.cz/en/ur/zprava/ur2009en.pdf.)

Czech Geological Survey, 2012, Mineral commodity summaries of the Czech Republic 2012: Prague, Czech Republic, Ministry of the Environment of the Czech Republic, October, 236 p.

Czech Statistical Office, 2013a, External trade of the Czech Republic—December 2012: Czech Statistical Office. (Accessed July 22, 2013, at http://www.czso.cz/csu/2012edicniplan.nsf/engpubl/6001-12-eng_m12_2012.)

Czech Statistical Office, 2013b, GDP by the production approach: Czech Statistical Office. (Accessed July 19, 2013, at http://apl.czso.cz/pll/rocenka/rocenkavyber.makroek_prod_en.)

Czech Steel Federation, [undated], Situation in 2011: Czech Steel Federation. (Accessed July 18, 2013, at http://www.hz.cz/en/situation-in-2011.)

Dolley, T.P., 2013, Sand and gravel (industrial): U.S. Geological Survey Mineral Commodity Summaries 2013, p. 138–139.

International Cement Review, 2012a, Czech Republic—Cautious outlook: International Cement Review, April, p. 14.

International Cement Review, 2012b, Czech Republic—Uncertain outlook for Czech cement: International Cement Review, July, p. 8.

International Energy Agency, 2010, Energy policies of IEA countries—Czech Republic 2010 Review: Paris, France, International Energy Agency, 151 p.

International Monetary Fund, 2012a, Czech Republic—2012 Article IV consultation report supplements; public information notice on the executive board discussion; and statement by the Executive Director for Czech Republic: Washington, DC, International Monetary Fund, April, 57 p.

International Monetary Fund, 2012b, World economic outlook: Washington, DC, International Monetary Fund, April, 250 p.

International Monetary Fund, 2013, Czech Republic, *in* World economic outlook database: International Monetary Fund, April. (Accessed July 19, 2013, via http://www.imf.org/external/pubs/ft/weo/2013/01/weodata/index.aspx.)

Ministry of Industry and Trade, Czech Republic, 2004, State energy policy of the Czech Republic: Prague, Czech Republic, Ministry of Industry and Trade, March 10, 54 p.

OECD Nuclear Energy Agency, 2013, Country profile—Czech Republic: Organisation for Economic Co-operation and Development, December 10. (Accessed February 11, 2014 at https://www.oecd-nea.org/general/profiles/czech.html.)

Tanner, A.O., 2013, Feldspar: U.S. Geological Survey Mineral Commodity Summaries 2013, p. 54–55.

Virta, R.L., 2013, Clays: U.S. Geological Survey Mineral Commodity Summaries 2013, p. 44–45.

Weiler, Eva, 2012, Czechs move cautiously towards more nuclear: IDN-IndepthNews news report, November 9. (Accessed February 11, 2014, at http://www.indepthnews.info/index.php/global-issues/1253-czechs-move-cautiously-towards-more-nuclear.)

World Steel Association, 2013a, Steel statistical yearbook 2013: Brussels, Belgium, World Steel Association, November 15, 117 p. (Accessed January 24, 2014, at http://www.worldsteel.org/publications/bookshop?bookID=fc4c14b2-eafd-45d4-8b56-b35c6a47531c.)

World Steel Association, 2013b, World steel in figures 2013: Brussels, Belgium, World Steel Association, May 30, 30 p. (Accessed July 17, 2012, at http://www.worldsteel.org/dms/internetDocumentList/bookshop/WSIF_2013_spreads/document/WSIF_2013_spreads.pdf.)

TABLE 1
CZECH REPUBLIC: PRODUCTION OF MINERAL COMMODITIES[1]

(Thousand metric tons unless otherwise specified)

Commodity[2]		2008	2009	2010	2011	2012
METALS						
Aluminum, metal, secondary		47	27 r	40 r	50 r	50 e
Iron and steel, metal:						
Pig iron		4,737	3,483	3,987	4,137	3,936
Steel, crude		6,387	4,594	5,180	5,583	5,072
Semimanufactures, hot-rolled products		5,286 r	3,957	4,625 r	4,616 r	4,276
Lead, metal, secondary		36	29	30	32	30 e
INDUSTRIAL MINERALS						
Cement, hydraulic		4,805 r	3,851 r	3,559 r	4,053 r	3,650 e
Clays:						
Bentonite		235 r	177	183	160	221
Brick clays and related materials		2,756	2,215	1,836	1,943	1,851
Kaolin, raw		3,833	2,886	3,493	3,606	3,318
Other		574	377	429	499	484
Diatomite		31	--	32	46	43
Dolomite		449	337	385	369	440
Feldspar		488	431	388	407	445
Feldspar substitutes, including nepheline syenite		36	23	19	22	15
Gemstones, crude:						
Moldavite-bearing rock		177	104	103	117	74
Pyrope-bearing rock		24	26	23	17	12
Graphite		3	--	--	--	--
Gypsum and anhydrite, crude		35	13	5	11	14
Lime, hydrated and quicklime		1,150	946 r	1,032 r	1,057 r	1,000 e
Nitrogen, N content of ammonia		175 r	173 r	160 r	189 r	200 e
Sand and gravel:						
Common sand and gravel		27,306	23,974	19,240	21,424	18,785
Foundry sand		702	374	473	395	491
Glass sand		1,151	990	888	976	849
Silica minerals, including quartz and quartzite		18	16	14	24	17
Stone:						
Crushed		44,277	41,307	37,270	36,717	32,535
Dimension		723	710	823	648	504
Limestone and other calcareous stones		11,465	9,488 r	9,828	11,244	9,858
Sulfur, byproduct, all sources e		45	40	40	40	40
Sulfuric acid		215	253 r	195 r	258 r	200 e
MINERAL FUELS AND RELATED MATERIALS						
Coal:						
Bituminous		12,197	10,621	11,193	10,967	10,796
Brown and lignite		47,872	45,616	43,931	46,848	43,710
Total		60,069	56,237	55,124	57,815	54,506
Coke, from coke ovens		3,399	2,295 r	2,548 r	2,588 r	2,350 e
Fuel briquets from brown coal e		156 [3]	150	140 r	150	140
Gas:						
Manufactured, all types e	million cubic meters	1,442 [3]	1,000	1,500	1,500	1,500
Natural, marketed	do.	168	180	201	187	204
Petroleum:						
Crude[4]	thousand 42-gallon barrels	1,600	1,470 r	1,173 [3]	1,105 [3]	1,020
Refinery products e, [5]	do.	58,000	52,000	58,000	53,000	55,000 e
Uranium:						
Mine output, U content	metric tons	290	286	259	252	222
U_3O_8 content e	do.	342	337	305	297	262
Concentrate production, U content	do.	261	243	237	216	219

e Estimated; estimated data are rounded to no more than three significant digits. r Revised. do. Ditto. -- Zero.

[1] Table includes data available through February 11, 2014.

[2] In addition to the commodities listed, ferrovanadium, secondary copper, secondary gold recovered from scraps, precious metals, and zinc metal may have been produced, but available information is inadequate to make reliable estimates of output.

[3] Reported figure.

[4] Figures were converted to barrels from production reported in thousand metric tons, as follows: 2008—236; 2009—217 (revised) ; 2010—173; 2011—163; and 2012—150.

[5] Estimated based on throughput reported in million metric tons, as follows: 2008—8.25; 2009—7.38; 2010—8.70 (estimated); and 2011—7.57 (estimated).

TABLE 2
CZECH REPUBLIC: STRUCTURE OF THE MINERAL INDUSTRY IN 2012

(Thousand metric tons unless otherwise specified)

Commodity	Major operating companies and major equity owners	Location of main facilities	Annual capacity
Aluminum, secondary	Alcan Decin Extrusions s.r.o.	Decin, northern Bohemia	NA
Do.	Kovohute Holdings DT- Mnisek Division (majority owned by Demonta Trade SE)	Mnisek pod Brdy	NA
Bentonite	KERAMOST a.s.	Most	NA
Do.	Sedlecky Kaolin a.s.	Bozicany	NA
Cement	Cement Hranice a.s. (Dyckerhoff AG, 100%)	Hranice	1,100
Do.	Ceskomoravsky Cement a.s. (HeidelbergCement AG, 100%)	Mokra	1,400 [e]
Do.	do.	Radotin	800 [e]
Do.	Holcim (Cesko) a.s. (Holcim Ltd., 100%)	Prachovice	1,200
Do.	Lafarge Cement a.s. (Lafarge S.A., 70%, and STRABAG SE, 30%)	Cizkovicka	1,200
Clay	LB Minerals s.r.o.	Horni Briza	NA
Do.	KERAMOST a.s.	Most	NA
Do.	Ceske Lupkove Zavody a.s.	Nove Straseci (refractory clay)	NA
Do.	P-D Refractories CZ a.s.	Velke Opatovice (refractory clay)	NA
Do.	RAKO-LUPKY s.r.o.	Lubna u Rakovnika	NA
Do.	Kaolin Hlubany a.s. (WBB Minerals, 94%)	Podborany	NA
Coal:			
Bituminous	OKD a.s. (New World Resources N.V.)	4 mines near Ostrava and Kravina in eastern Czech Republic	13,000 [e]
Brown	Dul Kohinoor a.s. (Czech Coal Group)	Centrum Mine in Marianske Radcice	350 [e]
Do.	Litvinovska uhelna a.s. (Czech Coal Group)	CSA Mine near Most	5,000 [e]
Do.	Severoceske doly a.s. (CEZ Group a.s., 100%)	Nastup Tusimice Mine southwest of Chomutov and Bilina Mine in Bilina	23,000 [e]
Do.	Sokolovska uhelna a.s.	Jiri and Druzba Mines at Sokolov	10,000 [e]
Do.	Vrsanska uhelna a.s. (Czech Coal Group)	Vrsany Mine just west of Most (contains the Vrsany and the Sverma sites)	10,000 [e]
Lignite	Lignit Hodonin s.r.o.	Hodonin, south of Moravia	500
Coke	ArcelorMittal Ostrava a.s.	Ostrava	1,500
Do.	OKK Koksovny a.s. (New World Resources N.V.)	Jan Sverma coking plant near Ostrava	400
Do.	do.	Svoboda coking plant near Ostrava	600
Do.	Trinecke Zelezarny a.s. (Moravia Steel a.s., 69%)	Trinec	700
Feldspar	LB Minerals s.r.o.	Horni Briza	NA
Do.	KMK Granit a.s.	Krasno	NA
Do.	Druzstvo DRUMAPO	Nemcicky	NA
Do.	Ceske sterkopisky spol. s.r.o.	Prague	NA
Do.	AGRO Brno - Turany a.s.	Brno	NA
Feldspar substitutes (including nepheline phonolite and syenite)	KERAMOST a.s.	Most	NA
Ferrovanadium	Nikom a.s. (Evraz Vitkovice Steel a.s.)	Vitkovice-Ostrava	NA
Gold, metal, secondary	Kovohute Pribram Nastupickna a.s.	Pribram	NA
Graphite	Grafitove doly Stare Mesto s.r.o.	Stare Mesto	NA
Kaolin	KERAMOST a.s.	Most	NA
Do.	Sedlecky Kaolin a.s.	Bozicany	NA
Do.	LB Minerals s.r.o.	Horni Briza	NA
Do.	Kaolin Hlubany a.s.	Podborany	NA
Do.	KSB s.r.o.	Bozicany	NA
Lead, metal, secondary, refined	Kovohute Pribram Nastupickna a.s.	Pribram	30
Natural gas million cubic meters	Gasfield operators in Brno and Ostrava regions, including: Moravske Naftove doly a.s. Ceska Naftarska Spol s.r.o. Green Gas DPB a.s. UNIGEO a.s.	Eastern/Southeastern Czech Republic, of which: Hodonin do. Paskov Ostrava-Hrabova	200 [e, 1]

See footnotes at end of table.

TABLE 2—Continued
CZECH REPUBLIC: STRUCTURE OF THE MINERAL INDUSTRY IN 2012

(Thousand metric tons unless otherwise specified)

Commodity		Major operating companies and major equity owners	Location of main facilities	Annual capacity
Petroleum:				
Crude	thousand 42-gallon barrels	Oilfield operators around Hodonin, including: Moravske Naftove doly a.s. Ceska Naftarska Spol s.r.o. UNIGEO a.s.	Location: Hodonin do. Ostrava-Hrabova	2,100 [e, 1]
Refinery	thousand 42-gallon barrels per day	Paramo a.s. (Unipetrol a.s.)	Refineries at Kolin and Pardubice	20 [e]
Do.	do.	Ceska Rafinerska (Unipetrol a.s., 51.2%, Eni International B.V., 32.5%, Shell Overseas Investments B.V., 16.3%)	Refineries at Litvinov and Kralupy nad Vltavou	165 [e]
Pig iron		ArcelorMittal Ostrava a.s. (ArcelorMittal, 100%)	Kunice-Ostrava	3,000
Do.		Trinecke Zelezarny a.s. (Moravia Steel a.s., 69%)	Trinec	2,100
Sand, industrial (glass and foundry)		Provodinske pisky a.s.	Provodin	NA
Do.		Sklopisek Strelec a.s.	Mladejov	NA
Do.		LB Minerals s.r.o.	Horni Briza	NA
Do.		Kalcit s.r.o.	Brno	NA
Do.		SEDOS doprava a.s.	Drnovice	NA
Do.		PEDOP s.r.o.	Lipovec	NA
Do.		SETRA s.r.o.	Brno	NA
Steel, crude		ArcelorMittal Ostrava a.s. (ArcelorMittal, 100%)	Kunice-Ostrava	3,000
Do.		Evraz Vitkovice Steel a.s.	Vitkovice-Ostrava	950
Do.		Pilsen Steel s.r.o. (OAO OMZ)	Plzen	150 [e]
Do.		Poldi Hutte s.r.o. (Scholz Edelstahl A.G.)	Kladno	120
Do.		Trinecke Zelezarny a.s. (Moravia Steel a.s., 69%)	Trinec	2,440
Do.		Vitkovice Heavy Machinery a.s.	Vitkovice-Ostrava	200
Do.		Zelezarny Hradek a.s. (Z-Group Steel Holding)	Hradek	NA
Do.		Zelezarny Veseli, a.s. (Z-Group Steel Holding)	Veseli nad Moravou	NA
Do.		Zelezarny Chomutov s.p. (Z-Group Steel Holding)	Chomutov	NA
Do.		ZDB Group a.s.	Bohumin	40 [e]
Uranium, U content	metric tons	DIAMO s.p. (Government, 100%)	Rozna I Mine at Dolni Rozinka	500

[e]Estimated. Do., do. Ditto. NA Not available.
[1]Annual capacity listed is total for all deposits, mines, and companies that produce the commodity.

THE MINERAL INDUSTRIES OF DENMARK, THE FAROE ISLANDS, AND GREENLAND

By Harold R. Newman

DENMARK

The mining, quarrying, and mineral processing sectors have not traditionally been significant contributors to Denmark's economy. Denmark's mineral resources are limited and composed mainly of industrial minerals and mineral fuels; therefore, the country's industrialized market economy depends on imported raw materials and foreign trade. Denmark is a member of the European Union (EU) and is located convenient to European trade routes through the Baltic Sea, the North Sea, and the Skagerrak Strait (U.S. Department of State, 2012).

Private ownership, exploration, development, and production of minerals are allowed under Danish law. The permitting procedures for mineral production are developed and administered at the county level. Regulations concerning the mineral industry are comparable with those of other EU countries (Ministry of Foreign Affairs, 2012).

In 2012, the EU remained Denmark's most significant trading zone—EU countries accounted for 71% of Denmark's external trade. The United States was Denmark's second-ranked non-EU trading partner after Norway and accounted for about 0.5% of Denmark's external trade (Statistics Denmark, 2012, p. 2).

U.S. exports to Denmark in 2012 totaled $2.2 billion, and U.S. imports from Denmark totaled $6.7 billion. U.S. exports to Denmark included, in order of value, finished metal shapes ($33.3 million), nonferrous metals ($17.1 million), petroleum products ($15.6 million), drilling and oilfield equipment ($15.5 million), and iron and steel products ($10.3 million) (U.S. Census Bureau, 2012a). U.S. imports from Denmark in 2012 included, in order of value, finished metal shapes ($77.6 million), iron and steel products ($51.6 million), other petroleum products ($48.2 million), and bauxite and aluminum [$2.3 million] (U.S. Census Bureau, 2012b).

Production

Denmark lacked economically exploitable metallic mineral resources; however, it had reserves of nonmetallic materials, such as chalk, clays (including bentonite) and kaolin, lime, peat, salt, and stone (including dimension stone and limestone). Denmark was the world's only commercial producer of moler, which is a natural mixture of diatomite and smectite clay that is used in filtration systems and insulation bricks. Petroleum production was declining, as reserves were being depleted. In 2012, petroleum production decreased for the 5th year in a row, dropping to about 76 million barrels (Mbbl) in 2012 from a level of about 105 Mbbl in 2008. Data on mineral production are in table 1.

Structure of the Mineral Industry

The Danish mineral industry was mostly privately owned. Table 2 is a list of the country's major mineral industry facilities, their capacities, and their locations.

Commodity Review

Metals

Iron and Steel.—NLMK DanSteel A/S, which was a subsidiary of NLMK International B.V. of the Netherlands, was the only steel plate producer in Denmark. DanSteel continued with construction of its new steel rolling mill, which was expected to be completed in 2013. The new mill would be able to produce plates up to 4 meters (m) in width. DanSteel's plates were used on bridges, offshore and onshore wind turbines, and ships. The investment cost was reported to be about 600 million Danish kroners (DKK) [$105 million[1]] (NLMK DanSteel A/S, 2012).

Industrial Minerals

Cement.—Aalborg Portland A/S (a subsidiary of Cementir Holdings S.p.A. of Italy) was the main producer of grey and white cement in Denmark. Aalborg operated seven kilns at its plant in Rordal, which had a combined capacity of 2.7 million metric tons (Mt) of gray cement and 850,000 metric tons (t) of white cement. The gray cement was sold mainly on the Danish market whereas the white cement was marketed on the international market. Aalborg also produced ready-mix concrete (Cementir Holdings S.p.A., 2012).

Diatomite.—Damolin A/S produced moler (diatomite) from its quarries at Fur and Mores Islands. Damolin had the capacity to process 230,000 cubic meters per year from five rotary kilns. In 2012, 80% of the production was exported. Diatomite is the key industrial mineral used in such products as absorbents for oil, cat litter, and granulated products. Diatomite is used in about 65% of the world's filtration systems (Damolin A/S, 2012).

Salt.—Akzo Nobel A/S's production of salt at Mariager was based on vacuum salt technology. The main materials are raw brine produced by solution mining in multi-effect evaporation plants located at Hvornum. The salt produced is suitable for the electrolytic production of chlorine, caustic lye, and sodium chlorate. Akzo Nobel was issued a new 30-year license to extract salt. This license extended the one that was due to expire in 2013 (Akzo Nobel A/S, 2012).

---

[1]Where necessary, values have been converted from Danish kroners (DKK) to U.S. dollars (US$) at an average rate of DKK5.5 = US$1.00.

Mineral Fuels and Other Sources of Energy

Natural gas and petroleum were the most valuable mineral commodities produced domestically. Production from the country's 19 active fields, however, was in decline. The Danish Energy Agency (DEA) announced that it would launch a new exploration licensing round sometime in the near future. The previous licensing round was held in 2005 to 2006. Activity was continuing offshore Denmark, where some existing fields were being further developed, several discoveries were being evaluated, and a new field, the Hejre natural gas and petroleum field, was scheduled to start production in 2015 (McLoughlin, 2012).

The Government announced that it had joined the Danish Underground Consortium (DUC) by taking a 20% stake in the operation. The DUC operated 16 fields that had a daily combined production of 14 million cubic meters of natural gas and 180,000 barrels of oil. Maersk A.S. would continue as the operator of the DUC (Gustafsson, 2012).

The Government's revenue from natural gas and petroleum production in 2012 was valued at DKK25.2 billion ($4.8 billion). This was a decrease of about 15% from 2011 when Government revenue totaled DKK30.3 billion ($5.4 billion). This decrease in revenue was attributable mainly to the declining production from the Danish natural gas and petroleum field. Even with the declining production, Denmark was expected to remain a net exporter of petroleum through 2020 and of natural gas through 2025 (Danish Energy Agency, 2012).

Denmark had two refineries—one in Kalundborg and the other in Frederica—that had a total (combined) crude distillation capacity of 172,000 barrels per day (bbl/d). The Kalundborg refinery was owned by Statoil Refining Denmark A/S and was Denmark's leading refinery. Statoil Refining processed primarily Norwegian crude but could process condensates and other types of crude oil as well. The Frederica refinery processed mostly Danish North Sea crude oil that was supplied by pipeline from Danish offshore wells. The primary markets, besides Denmark, were northwestern Europe and Scandinavia (International Energy Agency, 2012, p. 7).

Geothermal Energy.—The Geological Survey of Denmark identified significant geothermal resources in porous sandstone layers beneath the surface in Denmark. These resources are related mainly to the Mesozoic succession of the Danish basin and the Fennoscandian Border Zone, which had been discovered as a result of the drilling of about 60 deep wells that had been drilled either for geothermal energy, hydrocarbons, or natural gas storage. Denmark's geothermal plant located at Thisted in northwestern Denmark produced heat from water that was heated to 44° Celsius (111° Fahrenheit) by geothermal processes and pumped from the Upper Triassic Gassum sandstone aquifer at about a 1-kilometer (km) depth. The wide distribution of such underground reservoirs could make it possible for many of the existing district heating networks to make use of the high-efficiency geothermal heat (Vangkilde-Pedersen, Ditlefsen, and Højberg, 2012, p. 39–40).

Outlook

Further exploration of natural gas and petroleum reserves will likely continue in an effort to offset the country's declining production and help continue Denmark's role as a net exporter of natural gas and petroleum. Denmark is expected to remain a net exporter of petroleum until yearend 2020 and of natural gas until yearend 2025. Continued research in new technology and the testing of new exploration methods are expected to play a major role in Denmark's future natural gas and petroleum production. The Government is also likely to continue to consider the introduction of a long-term target of becoming fully independent of fossil fuels by the year 2050. Such a policy would likely encourage greater energy efficiency and growth in renewable energy production.

References Cited

Akzo Nobel A/S, 2012, Danish salt: Akzo Nobel A/S. (Accessed September 19, 2013, at http://www.akzonobel.com/Mariager/historie/dansk_salt/.)

Cementir Holdings S.p.A., 2012, Cement: Cementir Holdings S.p.A. (Accessed August 5, 2013, at http://www.cementirholding.it/denmark.php.)

Damolin A/S, 2012, About Damolin: Damolin A/S. (Accessed October 22, 2013, at http://www.damolin.com/Default.aspx?ID=771.)

Danish Energy Agency, 2012, State revenue from North Sea oil exceeded DKK 25 billion in 2012: Danish Energy Agency. (Accessed August 5, 2013, at http://www.ens.dk/en/info/news-danish-energy-agency/state-revenue-north-sea-oil-exceeded-25DKK-in-2012.)

Gustafsson, Katarina, 2012, Denmark takes 20% of oilfield operating group DUC: Market Watch. (Accessed October 22, 2013, at http://www.marketwatch.com/denmark-takes-20-of-oilfield-operating-group-duc-2012-07-09.)

International Energy Agency, 2012, Oil and gas security—Emergency response of IEA countries: International Energy Agency, 15 p.

McLoughlin, Patrick, 2012, Denmark's oil, gas output unlikely to see future recovery: Platts, McGraw Hill Financial. (Accessed August 4, 2013, at http://www.platts.com/latest-news/oil/London?ANALYSIS-Denmarks-oil-gas-output-unlikely-to-8109798.)

Ministry of Foreign Affairs [Denmark], 2012, Denmark in brief: Ministry of Foreign Affairs. (Accessed July 29, 2013, at http://www.denmark.dk/en/menu/About-Denmark/Denmark-In-Brief/.)

NLMK DanSteel A/S, 2012, The sole Danish steel plate producer creates growth: NLMK DanSteel A/S. (Accessed September 20, 2012, at http://www.dansteel.dk/1/541/3-march-2011---new-rolling-mill.html.)

Statistics Denmark, 2012, International trade in goods—Key figures: Statistics Denmark. (Accessed July 29, 2013, at http://www.dst.dk/en/Statistik/emner/udenrigshandel/udenrigshandel-med-varer.aspx.)

U.S. Census Bureau, 2012a, U.S. exports to Denmark by 5-digit end-use code, 2003–2012: U.S. Census Bureau. (Accessed August 3, 2013, at http://census.gov/foreign-trade/statistics/product/enduse/exports/c4099.html.)

U.S. Census Bureau, 2012b, U.S. imports from Denmark by 5-digit end-use code, 2003–2012: U.S. Census Bureau. (Accessed August 3, 2013, at http://census.gov/foreign-trade/statistics/product/enduse/imports/c4099.html.)

U.S. Department of State, 2012, Denmark: U.S. Department of State Fact Sheet. (Accessed July 28, 2012, at http://www.state.gov/r/pa/ei/bgn/3167.htm.)

Vangkilde-Pedersen, Thomas, Ditlefsen, Claus, and Højberg, A.L., 2012, Shallow geothermal energy in Denmark: Geological Survey of Denmark and Greenland Bulletin 26, 88 p.

FAROE ISLANDS

The Faroe Islands, which is a self-governing overseas administrative division of Denmark, had no significant identified mineral resources, although a small amount of crushed stone was thought to be produced for domestic consumption. The Faroese economy depended mainly on fishing and salmon farming and was aided by an annual subsidy of about 6% of the gross domestic product from Denmark. The main involvement of the Faroe Islands in the international mineral industry was as a market for imported materials, principally cement, fertilizer materials, and fuels.

Statoil ASA operated the Brugdan II exploration well, which is located about 80 km from the Faroe Islands between the North Atlantic Ocean and the Norwegian Sea. In November, the Faroese authorities authorized the suspension of drilling activities until 2013 owing to the harsh winter weather. Statoil has been operating in the Faroe Islands since 2000. In 2012, Statoil owned and operated six licenses that composed significant Faroe Islands acreage (Rigzone.com, 2012).

Outlook

In 2012, foreign petroleum companies continued to be engaged in geophysical exploration and exploration well drilling. Future discoveries in the Faroese area could make the eventual production of petroleum possible.

Reference Cited

Rigzone.com, 2012, Faroese well suspended due to bad weather: Rigzone.com. (Accessed November 28, 2012, at http://www.rigzone.com/news/oil_gas/a/122369/Faroese_Well_Suspended_due_to_Bad_Weather.)

GREENLAND

On January 1, 2010, the Inatsisartut Act No. 7 of December 7, 2009, on mineral resources and related activities [Mineral Resource Act] came into force. The Mineral Resource Act replaced the Consolidation Act No. 368 of June 18, 1998, on mineral resources in Greenland. The Mineral Resource Act establishes the framework for future development and control of mineral resources. There was broad political agreement within the Inatsisartut (Parliament) to support the development of the mineral industry (Government of Greenland, 2012).

In 2012, a total of 23 new mineral licenses were granted. Twelve of these were exploration licenses. In addition to the exploration licenses, six prospecting licenses were granted. Hudson Resources Inc. of Canada renewed two exploration licenses at its Sarfartog rare-earth elements (REE) project and NunaMinerals A/S of Denmark renewed two exploration licenses for base metals and diamond. Also, Avanna Exploration Ltd. renewed one exploration license for diamond (Bureau of Minerals and Petroleum, 2013, p. 5–6.).

Commodity Review

Metals

Gold.—In 2012, Angel Mining (Gold) A/S, which was a wholly owned subsidiary of Angel Mining plc of the United Kingdom, had commissioned and was operating the Nalunaq gold plant. The company expected to produce at a rate of about 700 kilograms per year of gold. A processing plant had been built and was thought to be the first underground cyanide leaching plant. Angel planned to minimize transport and refining costs by shipping gold dore. Angel planned further exploration programs to confirm the length and depth of the Nalunaq deposit. Nalunaq was classified as a narrow-vein mesothermal deposit (Angel Mining plc, 2012a).

Iron and Steel.—Red Rock Resources plc of the United Kingdom was continuing with exploration on the Melville Bugt iron ore project. A Joint Ore Reserves Committee (JORC) estimate listed estimated reserves of 67 Mt of iron ore grading 31.4% iron and 51.2% silica oxide. Twelve additional exploration targets had been identified and thought to have a potential tonnage of between 158 Mt and 470 Mt of iron ore grading between 27% and 47% iron. Red Rock's 1,570-square-kilometer (km^2) license area is located in northwestern Greenland about 150 km south of Qaanaag (Red Rock Resources plc, 2012).

Lead and Zinc.—Angel Mining was continuing with its exploration and development program to reopen the Black Angel Mine. The mine operated between 1973 and 1990, during which time about 12 Mt of ore was extracted with average grades of 4% lead and 12% zinc. Angel Mining expected to begin commercial production by yearend 2013. Also, Angel Mining was planning to undertake extensive exploration of its 259-km^2 license area to define other lead-zinc deposits thought to be in the area (Angel Mining plc, 2012b).

Industrial Minerals

Diamond.—NunaMinerals identified a kimberlite float at its Qaamasoq license and established, through testing, that the chemistry of the mantle-derived material was favorable for the occurrence of diamond. High concentrations of float occurred in a number of sites, notably in the 250-by-550-m Ullu (Nest) area. Rocks were characterized as having visible garnet. NunaMinerals was planning to follow up on these results in 2012 (NunaMinerals A/S, 2012).

Gemstones.—True North Gems Inc.'s Fiskenaesset ruby-sapphire project, which is located on the coast of Greenland about 160 km south of Nuuk, consisted of eight claim blocks covering 823 km^2. True North Gems was continuing with its exploration program and contracted with Greenland Mining Services to engineer, build, and fund the complete mine support infrastructure, including all mine-related buildings and open pit pre-stripping. True North Gems would retain responsibility for the design and funding of the ore processing circuit, for rough gem sorting and grading, and for gem marketing. The company stated that mining was expected to begin as early as 2014 (True North Gems Inc., 2012).

Rare Earths.—In 2012, Greenland Minerals and Energy Ltd. (GMEL) of Australia was continuing with its investigation of the Kvanefjeld deposit, which the company reported to be a significant REE deposit that also contains uranium and zinc. The Kvanefjeld deposit was second in size to the large Bayan Obo REE deposit in China. The main focus of GMEL's investigation was to develop an effective method of beneficiating the multielement ores. In September 2012 (the latest date for which data were available), the estimated indicated plus inferred resources of rare-earth oxides at a 150-parts-per-million U_3O_8 cutoff grade was 6.55 Mt in the Kvanefjeld deposit, 2.67 Mt in the nearby Sørensen deposit, and 1.11 Mt in the Zone 3 deposit; the deposits also contain resources of uranium and zinc (Greenland Minerals and Energy Ltd., 2012).

Mineral Fuels and Related Materials

Natural Gas and Petroleum.—In the Arctic region, which includes Greenland, natural gas and petroleum companies faced high costs, high risks, and lengthy lead times for development. The Arctic resource base is composed largely of natural gas and natural gas liquids, which are significantly more expensive to transport across long distances than petroleum.

Uranium.—The Government amended its standard terms for mineral exploration licenses and ceased a decades-old ban on uranium exploration to allow for the inclusion of radioactive elements as exploitable minerals. This represented a significant shift in Danish foreign policy, following 30 years of opposition to nuclear power. This amendment allowed GMEL to proceed with development of the Kvanefjeld REE deposit in 2012 and enabled it to conduct prefeasibility studies that demonstrated the potential for the development of a large-scale multielement mining operation at the Kvanefjeld deposit (Proactive Investors, 2011).

Outlook

Greenland has abundant mineral and other natural resources. More areas for exploration are expected to open up if global warming trends continue, as new mineral deposits are likely to be discovered as a result of the retreating ice. Finding new sources of hydrocarbons will continue to be very important for Greenland as possible sources of revenue, and offshore exploration is expected to increase as interest increases in this area. The country's independent status and the Government's encouragement are expected to continue to accelerate the development of the mineral industry in Greenland.

References Cited

Angel Mining plc, 2012a, Nalunaq: Angel Mining plc. (Accessed August 5, 2013, at http://www.angelmining.com/?page_id=282.)

Angel Mining plc, 2012b, Project overview: Angel Mining plc. (Accessed August 5, 2013, at http://www.angelmining.com/?page_id=278.)

Bureau of Minerals and Petroleum [Greenland], 2013, Greenland Mineral Exploration Newsletter no. 43: Bureau of Minerals and Petroleum, February, 8 p.

Government of Greenland, 2012, Greenland Parliament Act no. 7 of December 7, 2009, on mineral resources and mineral resource activities (the Mineral Resources Act): Government of Greenland. (Accessed July 24, 2014, at http://www.govmin.gl/index.php/about-bmp/legal-foundation.)

Greenland Minerals and Energy Ltd., 2012, Kvanefjeld—REEs, uranium, zinc: Greenland Minerals and Energy Ltd. (Accessed October 6, 2013, at http://www.ggg.gl/projects/kvanefjeld-rees-uranium-zinc/.)

NunaMinerals A/S, 2012, NunaMinerals—Qaamasoq confirmed as a diamond-play: NunaMinerals A/S, February 29. (Accessed August 5, 2013, at http://www.cisionwire.com/nuna-minerals-a-s-g/r/nunaminerals--qaamasoq-confirmed-as-a-diamond-play.)

Proactive Investors, 2011, Greenland Minerals and Energy—Greenland Government uranium decision "momentous" for Kvanefeld [sic]: Greenland Minerals and Energy Ltd. (Accessed August 5, 2013, at http://www.proactiveinvestors.com.au/companies/news/22973/Greenland-minerals-and-energy-greenland-government-uranium-decision-momentous-for-kvanefeld-22973.html.)

Red Rock Resources plc, 2012, Greenland: Red Rock Resources plc. (Accessed August 5, 2013, at http://www.rrrplc.com/projects/greenland/.)

True North Gems Inc., 2012, True North Gems signs partnership agreement to construct and operate the Aappaluttog ruby project: True North Gems Inc., November 20. (Accessed August 5, 2013, at http://www.truenorthgems.mwnewsroom.com/press-releases/true-north-gems-signs-partnership-agreement-to-construct-and-operate-the-aappaluttoq-ruby-project.)

TABLE 1

DENMARK AND GREENLAND: ESTIMATED PRODUCTION OF MINERAL COMMODITIES[1, 2]

(Metric tons unless otherwise specified)

Country and commodity[3]		2008	2009	2010	2011	2012
DENMARK						
Aluminum metal, secondary		25,000	25,000	25,000	25,000	25,000
Cement, hydraulic		16,092 [4]	15,780 [4]	16,000	16,000	16,000
Chalk, calcium carbonate	thousand cubic meters	2,000	2,735 [4]	2,700	2,600	2,600
Clays:[4]						
Bentonite		22,458	24,040 [r]	23,832 [r]	38,300	36,000
Other		5,000	5,000	5,000	5,000	5,000
Moler, extracted		252	202	225	225	225
Gas:						
Manufactured	million cubic meters	1,500	1,500	1,500	1,800	1,600
Natural	do.	9,564 [4]	9,600	8,438 [4]	9,000	9,000
Natural gas plant liquids	thousand 42-gallon barrels	50,000	47,000	45,000	45,000	42,000
Petroleum:						
Crude[4]	do.	104,573	97,455	90,338	80,665	75,701
Refinery products:						
Liquefied petroleum gas	do.	1,314 [4]	1,606 [4]	1,752 [r, 4]	1,700	1,600
Gasoline	do.	16,352 [4]	17,666 [4]	15,330 [r, 4]	16,000	16,000
Naphtha	do.	50	50	52 [r, 4]	50	50
Jet fuel	do.	3,942 [4]	3,212 [4]	3,416 [r, 4]	3,500	3,600
Distillate fuel oil	do.	23,068 [4]	24,674 [4]	22,703 [r, 4]	23,000	24,000
Refinery gas	do.	1,800	1,800	2,409 [r, 4]	2,200	2,000
Residual fuel oil	do.	8,870 [4]	8,139 [4]	7,665 [r, 4]	8,000	8,200
Total	do.	55,400	57,100 [r]	53,327 [r]	56,000	56,000
Salt, all forms		496,593 [4]	511,063 [4]	601,046 [4]	600,000	600,000
Sand and gravel	thousand metric tons	59,937 [4]	46,932 [4]	46,932 [4]	50,000	50,000
Stone, crushed	do.	384 [4]	312	640 [4]	542 [r, 4]	500
Sulfur, recovered		3,467 [4]	3,200	3,246 [r, 4]	3,045 [r, 4]	3,400
GREENLAND						
Gold	kilograms	1,518 [4]	1,600	1,600	1,800	2,800
Silver	do.	--	--	--	--	242

[r]Revised. do. Ditto. -- Zero.

[1]Estimated data are rounded to no more than three significant digits; may not add to totals shown.

[2]Table includes data available through July 31, 2013.

[3]In addition to the commodities listed, kaolin and peat were thought to be produced, but available information was inadequate to make reliable estimates of output.

[4]Reported figure.

TABLE 2
DENMARK AND GREENLAND: STRUCTURE OF THE MINERAL INDUSTRIES IN 2012

(Thousand metric tons unless otherwise specified)

Country and commodity		Major operating companies and major equity owners	Location of main facilities	Annual capacity
DENMARK				
Cement:				
Gray		Aalborg Portland A/S (Cementir Holding S.p.A.)	Plant at Rordal	2,700
White		do.	do.	850
Chalk (calcium carbonate)		A/S Faxe Kalkbrud	Quarries at Stevns and Sigerslev	250
Diatomite (moler)	thousand cubic meters	Damolin A/S	Quarries on Mors and Fur Islands	230
Lime		A/S Faxe Kalkbrud (Aalborg Portland Holding A/S)	Plant at Stubberup, near Fakse, on Zealand Island	200
Natural gas	million cubic meters	Maersk Olie og Gas A/S	Roar and Tyra gasfields, Danish North Sea	2,550
Petroleum:				
Crude	barrels per day	Dansk Underground Consortium	16 fields in the Danish North Sea	NA
Do.	do.	Maersk A.S.	4 fields in the Danish North Sea	NA
Do.	do.	DONG Energy AS	3 fields in the Danish North Sea	NA
Do.	do.	Hess Corp.	1 field in the Danish North Sea	NA
Refined	do.	Statoil A/S	Kalundborg	102,000
Do.	do.	A/S Dansk Shell	Fredericia	70,000
Salt		Akzo Nobel A/S	Mine (brine) at Hvornum, plant at Mariager	1,000
Steel, semimanufactures		NLMK DanSteel A/S (NLMK International B.V., 100%)	Plant at Frederiksvaerk (under modification—closed until 2013)	250
GREENLAND				
Gold	kilograms	Angel Mining plc	Nalunaq Mine at Nanortalik	6,000
Silver	do.	do.	do.	2,400

Do., do. Ditto. NA Not available.

The Mineral Industry of Estonia

By Alberto Alexander Perez

In 2012, Estonia's gross domestic product (GDP) increased by 3.2% compared with that of 2011. The GDP composition by sector was as follows: services, 66.4%; industry, 29.7%; and agriculture, 3.9%. In 2011 (the latest year for which data were available), mining and quarrying activities accounted for only about 1.2% of the total GDP (Statistics Estonia, 2012, p. 30, 188; 2013, p. 310; U.S. Central Intelligence Agency, 2013).

Production

Estonia is not rich in natural resources, but the country had one of the world's few rare-earth metals processing plants located outside of China. Estonia was also one of the few countries in the world that produced oil shale. Oil shale production was important for Estonia's economy because 80% of the oil shale extracted was used for the production of electrical and heat energy, and more than 90% of all electricity produced in Estonia was produced by the use of oil shale (Estonian Ministry of the Environment, 2013; Statistics Estonia, 2012, p. 30, 188; 2013, p. 310).

Structure of the Mineral Industry

Estonia was not a significant producer or processor of mineral commodities in the world. The country produced mostly construction materials, secondary lead from battery recycling, oil shale, and peat for domestic consumption. AS Kunda Nordic Tsement, which was a subsidiary of HeidelbergCement Sweden AB of Sweden (a subsidiary of HeidelbergCement AG of Germany) and of CHR Europe Holding BV of the Netherlands, was the only cement plant in Estonia. AS Tootsi Turvas (Tootsi), which was owned by Vapo OY of Finland, was Estonia's leading peat milling and export enterprise, measured by volume of output. In April 2011, Molycorp Minerals LLC (a subsidiary of Molycorp, Inc. of the United States) purchased a 90.023% share of rare-earth minerals producer Silmet AS for about $89 million; the company was renamed AS Molycorp Silmet. Eesti Energia Tehnoloogiatoostus had two shale oil plants in Narva with a combined capacity of 3,400,000 barrels per year (bbl/yr), and VKG Oil AS (part of Viru Keemia Group AS) had a shale oil plant in Kohtla-Jarve with a capacity of 1,840,000 bbl/yr. In December 2012, AS Nitrofert restarted ammonia and urea production at its plant at Kohtla-Jarve that had been idle for several years (Molycorp, Inc., 2011; Silmet AS, 2012a–d).

Industrial Minerals

Cement.—Kunda Nordic Tsement's plant in Kunda had been producing cement and construction aggregates since 1992. In 1999, the company invested in the construction of a local powerplant that operated on natural gas. In 2011 (the latest year for which data were available), the company had recorded sales of 838,000 metric tons (t) of cement and clinker (AS Kunda Nordic, 2014).

Nitrogen (Ammonia).—In December, AS Nitrofert (a fertilizer company based in Kohtla-Jarve) reopened plant operations in Ida-Viruma after a closure that had lasted 4 years because of poor economic returns in the ammonia and urea market. The company would again use natural gas imported from Russia as its raw material (Estonian Public Broadcasting, 2012).

Peat.—Tootsi was the major producer of peat in Estonia. In 2011 (the latest year for which data were available), the company produced an estimated 800,000 t of peat (including bales, block, fuel mill, horticultural, and sod). The company produced its peat in Ellamaa, Lavassaare, Peningi, Puhatu, and Ulila (AS Tootsi Turvas, 2014).

Rare Earths.—Molycorp Silmet began production of rare-earth metals in 1970 and had the capacity to produce up to 3,000 metric tons per year (t/yr) of rare-earth products and 700 t/yr of rare-metal products. The company produced cerium, lanthanum, neodymium, praseodymium, and samarium-europium-gadolinium products as well as niobium and tantalum metal chips, ingots, metallic hydrides, and powders (Molycorp, Inc., 2011; Silmet AS, 2012a–d).

Mineral Fuels and Other Sources of Energy

Oil Shale.—The volume of oil shale production in Estonia did not change much in the past 2 years. Oil shale was consumed mostly in powerplants and as a raw material for shale oil. Nearly 80% of the production of shale oil was exported, mostly to the Netherlands (34%), Belgium (21%), and Denmark (14%) (Statistics Estonia, 2013, p. 310). Eesti Energia AS produced about 94% of Estonia's oil shale. A significant amount of carbon dioxide is produced when oil shale is used to produce heat and electricity; consequently, Estonian law restricts mining of oil shale to 20 million metric tons per year (Mt/yr). One of the objectives the National Development Plan for Oil Shale Use in Estonia for 2008–2015 is to reduce oil shale production to 15 Mt/yr by 2015, so future production for domestic consumption of oil shale is likely to be reduced because of environmental concerns (Eesti Energia AS, 2012, p. 17–18; Statistics Estonia, 2013, p. 318).

Outlook

Estonia's reliance on oil shale is likely to continue unless the country finds an alternative fuel to use to produce electricity. Also, although the National Development Plan for Oil Shale Use calls for a reduction in oil shale production, Estonia's output of oil shale could increase as international demand for the oil produced from it increases and Estonia increases its exports in response.

References Cited

AS Kunda Nordic, 2014, HeidelbergCement in Estonia—Kunda Nordic—Key figures: AS Kunda Nordic. (Accessed October 10, 2014, at http://www.heidelbergcement.com/ee/en/kunda/firmast/votmenaitajad.htm.)

AS Tootsi Turvas, 2014. Finance indicators: AS Tootsi Turvas. (Accessed October 10, 2014, at http://www.vapo.ee/en/kontakt/finance-indicators/.)

Eesti Energia AS, 2012, Annual report 2011: Eesti Energia AS, 134 p. (Accessed July 31, 2013, at https://www.energia.ee/-/doc/10187/pdf/concern/annual_report_2011_eng.pdf.)

Estonian Ministry of the Environment, 2013, Mineral resources: Estonian Ministry of the Environment. (Accessed April 21, 2014, at http://www.envir.ee/445160.)

Estonian Public Broadcasting, 2012, Fertilizer producer Nitrofert resumes operations: Estonian Public Broadcasting. (Accessed October 10, 2014, at http://news.err.ee/v/economy/82ddbdce-b15c-4bc0-aede-23eb3e0b3e61.)

Molycorp, Inc., 2011, Molycorp acquires controlling stake in AS Silmet, expands operations to Europe, doubles near-term rare earth oxide production capacity: Molycorp, Inc. press release, April 4. (Accessed July 31, 2013, at http://us1.campaign-archive1.com/?u=a9e8676e87fad805702b98564&id=f30210c38c&e=[UNIQID].)

Silmet AS, 2012a, History: Silmet AS. (Accessed July 31, 2013, at http://www.silmet.ee/default.aspx?m1=48&m2=52&id=28&lang=1.)

Silmet AS, 2012b, Overview: Silmet AS. (Accessed July 31, 2013, at http://www.silmet.ee/default.aspx?m1=48&m2=52&id=28&lang=1.)

Silmet AS, 2012c, Production: Silmet AS. (Accessed July 31, 2013, at http://www.silmet.ee/default.aspx?m1=45&lang=1.)

Silmet AS, 2012d, Rare earth metals: Silmet AS. (Accessed July 31, 2013, at http://www.silmet.ee/default.aspx?m1=45&m2=85&lang=1.)

Statistics Estonia, 2012, Statistical yearbook of Estonia 2012: Statistics Estonia, July, 440 p. (Accessed July 31, 2013, at http://www.stat.ee/publication-download-pdf?publication_id=29873.)

Statistics Estonia, 2013, Statistical yearbook of Estonia 2013: Statistics Estonia, July, 436 p. (Accessed July 31, 2013, at http://www.stat.ee/publication-download-pdf?publication_id=34208.)

U.S. Central Intelligence Agency, 2013, Estonia, in The world factbook: U.S. Central Intelligence Agency. (Accessed July 31, 2013, at https://www.cia.gov/library/publications/the-world-factbook/geos/en.html.)

TABLE 1
ESTONIA: PRODUCTION OF MINERAL COMMODITIES[1]

(Metric tons unless otherwise specified)

Commodity[2]		2008	2009	2010	2011	2012[e]
Cement:						
Clinker (sold production)		324,000	314,000	209,000	381,000	380,000
Portland, other		808,000	326,200	375,000	451,000 [r, 3]	450,000
Clays:						
For brick	cubic meters	138,106	70,000	70,000	70,000	70,000
For cement	do.	33,494	15,000	15,000	15,000	15,000
Coke, electrode		35,380	29,900 [r]	22,400	22,000	22,000
Crushed stone used for concrete aggregates, for roadstone, and for other construction uses		7,891,000	5,400,300	5,752,600	6,196,300	6,200,000
Dolomite:						
For building	cubic meters	329,634 [3]	389,000 [r, 3]	390,000 [r, 3]	390,000	390,000
For finishing	do.	1,300 [r, 3]	3,200 [r, 3]	4,700 [r, 3]	4,500 [r, e]	4,500
For industry (technological limestone)	do.	146,000 [r, 3]	87,000 [r, 3]	74,000 [r, 3]	75,000 [r, e]	75,000
Fuel oil		444,800	489,300	524,300 [r]	559,900 [r]	560,000
Gravel, pebbles, shingle and flint	cubic meters	717,000	1,563,300 [r]	1,252,000 [r]	1,251,680	1,250,000
Lead, metal, secondary		10,000	9,176	10,718	7,840	7,800
Lime		59,400	24,100	27,200	36,100 [r, e]	36,000
Limestone:						
For building	cubic meters	2,627,741 [3]	1,200,000 [3]	1,200,000	1,200,000	1,200,000
For cement	do.	458,661 [3]	200,000 [3]	200,000	200,000	200,000
For industry (technological limestone)	do.	120,398 [3]	80,000 [3]	80,000	80,000	80,000
Niobium, metal, chips		NA	NA	NA	NA	NA
Nitrogen, N content of ammonia		78,912	-- [r, e]	-- [r, e]	--	--
Oil shale	thousand metric tons	16,117	14,939	17,934	18,700 [r]	18,700
Peat, all uses[3, 4]		732,700	859,700	965,000	926,700 [r]	927,000
Of which:						
For fuel		213,400	328,000	360,800	360,000	360,000
Briquets		67,500	45,300	83,600 [r]	74,800 [r]	74,800
Rare-earth metals[e]		3,000	3,000	3,000	3,000	3,000
Sand and gravel	cubic meters	4,750,800	3,000,000	3,000,000 [e]	NA	NA
Silica sand (technological sand)	do.	--	--	36,000	NA	NA
Sulfuric acid	kilograms	NA	--	--	--	--
Tantalum, metal, chips		NA	46	36	36 [e]	36

[e]Estimated; estimated data are rounded to no more than three significant digits. [r]Revised. do. Ditto. NA Not available. -- Zero.

[1]Table includes data available through July 30, 2013.

[2]In addition to the commodities listed, Estonia produces sulfur, but available information is inadequate to make reliable estimates of output.

[3]Reported figure.

[4]It can be assumed that the portion of total peat production not used as fuel is used in agricultural applications, although this is not specified.

TABLE 2
ESTONIA: STRUCTURE OF THE MINERAL INDUSTRY IN 2012

(Thousand metric tons unless otherwise specified)

Commodity		Major operating companies and major equity owners	Location of main facility	Annual capacity
Ammonia		AS Nitrofert	Plant in Ida-Viruma	100
Cement		AS Kundra Nordic Tsement (HeidelbergCement Sweden AB 75%, and CRH Europe Holding BV, 25%)	Plant in Kunda	NA
Peat		AS Tootsi Turvas (Vapo OY, 100%)	Ellamaa, Lavassaare, Peningi, Puhatu, and Ulila	NA
Rare earths	metric tons	AS Molycorp Silmet (Molycorp Minerals LLC, 90.02%)	Plant in Sillamae	3,000
Shale oil	barrels per year	Eesti Energia Tehnoloogiatoostus	2 plants in Narva, of which:	3,400,000
Do.	do.	do.	Enefit140	(1,500,000)
Do.	do.	do.	Enefit280	(1,900,000)
Do	do.	VKG Oil AS (Viru Keemia Grupp AS, 100%)	Plant in Kohtla-Jarve	1,840,000

Do. do. Ditto. NA Not available.

The Mineral Industry of Finland

By Alberto Alexander Perez

In 2012, Finland's real gross domestic product (GDP) measured in terms of purchasing power parity was $200.7 billion, which was a decrease of 0.2% from that of 2011. The leading contributor to Finland's GDP in 2012 was its services sector; industry accounted for 27.1% of the country's GDP. The principal products that Finland's industrial sector produced in 2012 were electronics, metal and metal products, paper products, scientific instruments, ships, and wood pulp. Finland was a member of the European Union (EU). Its main export partners were Sweden (which received 11.1% of Finland's exports, in terms of value), Russia (9.9%), Germany (9.4%), the Netherlands (6.4%), the United States (6.1%), the United Kingdom (5.1%), and China (4.6%). Its main import partners were Russia (which supplied 17.8% of Finland's imports, in terms of value), Sweden (14.8%), Germany (13.9%), the Netherlands (8%), and China (4.4%) (U.S. Central Intelligence Agency, 2013).

Minerals in the National Economy

In 2012, Finland had 46 mines and quarries that were regulated under the Finnish Mining Act, and several feasibility projects were ongoing. Metallic minerals mining and the processing and refining of metals were the principal areas of the Finnish mineral industry that had grown and demonstrated a potential to contribute to the exports of Finland. Employment had increased owing to mining operations, particularly in eastern and northern Finland. About 3,500 people were employed in the mineral sector in Finland in 2011, which was the latest year for which data were available. This number was expected to increase to 5,000 in the near future in line with the expected growth in the mineral industry, particularly in the area of metallic minerals. The Government of Finland had promoted the mineral industry through such measures as the building of infrastructure and developing areas where mining takes place, constructing roads and railways, and providing funding for research. Finland was a regionally significant processor and refiner of chromite, copper, nickel, and zinc. The principal facilities for the processing of copper and nickel were located at Harjavalta, those for the processing of chromium were located at Kemi, and those for the processing of zinc were located at Kokkola. Finland's deposits of chromite, cobalt, copper, iron, lead, nickel, and zinc were the foundation for the country's metal industry. Finland was the leading talc producer in Europe and the sixth-ranked talc producer in the world (Invest in Finland, 2011; Ministry of Employment and the Economy, 2014a; United Nations, 2013; Virta, 2013).

Government Policies and Programs

The Government of Finland regulates its mineral industry through two main laws: the Finnish Mining Act, which regulates the exploitation of metallic and industrial minerals in Finland, including soapstone and marble, and the Land Extraction Act, which regulates only the extraction of gravel and sand and the quarrying of natural stone. The objective of the Finnish Mining Act (621/2011) is to enable exploration and mining activities and regulate them so that they are carried out in a socially, economically, and ecologically sustainable way. The Act ensures that environmental, civil rights and landowner concerns are taken into account in the decisionmaking process for the development and exploration of any mining projects. The Act also takes other Finnish law into account in its application, in particular, Finland's Constitution and legislation concerning the Sami regions in northern Finland. Mining operators are subject to a number of permits. Under the revised Mining Act, which became effective on July 1, 2011, the right to exploit a deposit is based on a mining permit, and the review of permits is more comprehensive than under the original Mining Act. The mining operator's termination and reclamation obligations are also more extensive, including the requirement to provide a security deposit for the purpose of fulfilling reclamation obligations. The Finnish Safety and Chemicals Agency (Tukes) is the organization that grants and supervises the permits that are required by the Mining Act. Finnish law also provides environmental protection guidelines and requires several types of environmental permits for the exploitation of the mineral resources of the country (Ministry of Employment and the Economy, 2011, 2014b).

Production

Finland produced mostly base metals, gold, and platinum-group metals, as well as industrial minerals. The production of mineral commodities continued to be significant in terms of volume and contribution to the country's economy. In 2012, production of copper concentrate increased by 118%; silver metal, by 84.9%; feldspar, by 64%, gold metal mine output, by 27.8%, and nickel content of mine output, by 26.6%. Data on mineral production are in table 1.

Structure of the Mineral Industry

The Finnish mineral industry consists of the following two types of companies: (a) small stone quarry and sand and gravel pit operators, and (b) a group of large companies that operate international metal and industrial mineral operations and mines in Finland and abroad (United Nations, 2013).

Outokumpu Oyj (Outokumpu) and Rautaruukki Oyj (Ruukki) were the two leading metals manufacturing companies in Finland. They specialized in manufacturing steel and stainless steel. Outokumpu also operated the Kemi chromite mine in Lapland and, in addition to steel, also produced cadmium and ferroalloys. Outokumpu was no longer reporting mercury production in Finland, although some production as byproduct was likely. Outokumpu also had operations in the United States, Germany, Mexico, Sweden, and the United Kingdom.

Mondo Minerals Oyj (Mondo) of the Netherlands and Nordkalk Corp. (Nordkalk) were two of the principal industrial mineral producers in Finland. Mondo (a subsidiary of Advent International Corp. of the United States) was the second-ranked producer of talc, by volume, in the world. Mondo had its main mine and processing facilities in Sotkamo and Vuonos (Mondo Minerals Oyj, 2014a).

Nordkalk was a leading international producer of limestone (crushed and ground), concentrated calcite, quicklime, and slaked lime as well as dolomite and wollastonite, which Nordkalk extracted as a byproduct of the mining of limestone. Nordkalk had operations in 30 locations in nine countries as well as mines in five countries. In Finland, Nordkalk owned mines in Lappeenranta, Pargas, and Parainen.

First Quantum Minerals Ltd. (First Quantum) of Canada owned the Pyhasalmi copper mine. Finland was one of the few countries in Europe where copper was still being mined. The Pyhasalmi Mine had previously been owned by Inmet Mining Corp. of Canada; however, the company was purchased in 2011 by First Quantum (First Quantum Minerals Ltd., 2014).

Finland's mining companies were mostly privately owned, although the Government held an equity interest in some of the major mineral producers. The mineral industry operated on a free-market basis. The country's major mineral facilities and their annual capacities are listed in table 2.

Commodity Review

Metals

Chromium.—Outokumpu operated the Kemi chromite mine in Lapland and used the chromium for its production of stainless steel at its plant in Tornio. Outokumpu reported that the Kemi Mine had ore reserves of 33 million metric tons. The Kemi Mine was the only chromite mine within the EU (Outokumpu Oyj, 2014, p. 2).

Cobalt.—OM Group Inc. of the United States (OMG) announced in 2012 that it would divest itself of its advanced materials business, which included its cobalt production business. According to OMG, during 2012 and through the date of the sale of this section of the company in January 2013, it would continue to manufacture inorganic products using unrefined cobalt and other materials for automotive systems, construction and mining, industrial end markets, and the mobile energy storage and renewable energy markets. The divesture of the advanced materials business was to include the sale of the cobalt refinery facility in Kokkola, Finland, to the joint venture Freeport Cobalt OY, which was majority owned by a subsidiary of Freeport-McMoRan Copper & Gold Inc. of the United States (OM Group Inc., 2013, p. 5–6).

Copper.—Boliden AB's copper complex in Finland consisted of two plants—the copper smelter in Harjavalta, which produced copper anodes, and the copper refinery at Pori, where copper anodes were refined into copper cathodes. The complex was known as Boliden Harjavalta. The Harjavalta smelter had the capacity to produce 210,000 metric tons per year (t/yr) of copper, which was cast into copper anodes. Sulfur was recovered as a byproduct. The copper anodes were then shipped to the Pori refinery where the anodes were refined into copper cathodes. The capacity of the refinery was 155,000 t/yr. The refinery also produced gold and silver as byproducts. In 2012, the complex processed 516,027 metric tons (t) of copper concentrates, 247,709 t of nickel concentrates, and 124,527 t of copper cathodes (Boliden AB, 2013).

Gold.—Agnico-Eagle Mines Ltd. of Canada owned the Kittila Mine in the Lapland region. In 2012, the mine had a new record production of about 5,474 kilograms (kg) of gold content with an 88.3% recovery rate. This increase in production was owing to ongoing exploration and discovery and an increase in the processing capacity that, in 2012, expanded the identified Kittila mineralization in the Rimpi and the Roura deposit areas. The company was evaluating a 25% throughput expansion that could be operational by 2015. Further expansions were envisioned, as the deposit appeared to be significantly richer and thicker beneath the Rimpi zone (Agnico-Eagle Mines Ltd., 2013, p. 10).

Dragon Mining Ltd. of Australia owned and operated the Vammala plant located in the Sastamala region in southern Finland within the Tampere schist belt. The Vammala plant had the capacity to process 300,000 t/yr of ore and had crushing, milling, and flotation facilities that processed ore from the Orivesi and the Jokisivu gold mines, which are located 80 kilometers (km) to the northeast and 40 km to the southwest of the plant, respectively. In 2012, the Orivesi Mine produced 149,232 t of ore at an average grade of 3.50 grams per metric ton (g/t) gold and the Jokisivu Mine produced 141,443 t of ore at an average grade of 2.67 g/t gold. In 2012, the Vammala plant produced 684 kg of gold at a recovery rate of 76.8% (Dragon Mining Ltd., 2013).

Nickel.—The two main producers of mined nickel in Finland were Talvivaara Mining Co. plc (Talvivaara), which owned a polymetallic mine at Sotkamo, and Belvedere Resources Ltd. of Canada (Belvedere), which owned a mine and other installations in Hitura. Talvivaara reported that it was expecting to produce between 25,000 and 30,000 t of nickel in 2012. Talvivaara's Sotkamo nickel project was the world's first bioheap-leach project for nickel. It was centered on two polymetallic deposits—the Kolmisoppi and the Kuusilampi deposits, which are located about 30 km southwest of Sotkamo in eastern Finland. The deposits constitute one of the largest known nickel sulfide resources in Europe (Talvivaara Mining Co. plc, 2012, p. 7).

Belvedere produced about 2,200 t of nickel from its Hitura Mine in 2012. The mine had restarted operations in July 2010 but was likely be put on care-and-maintenance status in 2013 because keeping the mine active was not economically feasible given the projected prices of nickel in 2013 and beyond. Belvedere continued with its current expansion of the mine, however, and further expansion was also projected (Belvedere Resources Ltd., 2013, p. 13).

Industrial Minerals

Limestone.—Nordkalk Corp. which was part of the Rettig Group of Germany, was a leading producer of limestone and limestone-based products in the world. Nordkalk's largest production site (in terms of volume of production) in Finland was located in Lappeenranta, where the company had a quarry, a grinding plant, two flotation plants, and a lime kiln. Nordkalk subsidiary Suomen Karbonaatti Oy, which was also located in Lappeenranta, produced carbonate fillers and coating pigments (Nordkalk Corp., 2013, 2014).

Talc.—Mondo was a significant producer of talc in the world. In 2012, the company produced an estimated 396,000 t of talc concentrate. Mondo indicated that the talc ore found in Finland is a mixture of magnesite and talc, so that a separating process has to be applied to the ore (Mondo Minerals Oyj, 2014b).

Wollastonite.—Nordkalk was the only European producer of wollastonite in 2012. Nordkalk produced all its grades of wollastonite at its facilities in Lappeenranta. The company launched a new generation of high-aspect-ratio wollastonite fillers. The new fillers were designed for thermoplastic and thermoset applications (Nordkalk Corp., 2013).

Outlook

Finland's production of nickel and zinc is likely to increase, as projects to increase production capacity are expected to reach the production stage in the near future. The increased market interest in rare-earth minerals has reignited interest in areas of Finland that had previously been producing these minerals but had stopped because of economic and technical feasibility issues. Copper and silver production is expected to continue to be a significant element of the Finnish mineral industry, particularly as facilities are expanded to include multimetallic production projects. Market prices will determine whether expansion of the Finnish mineral industry continues in the long run.

References Cited

Agnico-Eagle Mines Ltd., 2013, Annual report 2012: Toronto, Ontario, Canada, Agnico-Eagle Mines Ltd., 102 p.

Belvedere Resources Ltd., 2013, Annual report 2012: Vancouver, British Columbia, Canada, Belvedere Resources Ltd., 27 p. (Accessed November 17, 2013, at http://www.belvedere-resources.com/assets/FS-Dec-31-2012.pdf.)

Boliden AB, 2013, Boliden Harjavalta: Boliden AB. (Accessed November 15, 2013, at http://www.boliden.com/Operations/Smelters/Harjavalta/.)

Invest in Finland, 2011, Finland sets new mining production record of 70 million tons: Invest in Finland. (Accessed September 28, 2012, at http://www.investinfinland.fi/articles/news/mining/finland-sets-new-mining-production-record-of-70-million-tons/49-371.)

Dragon Mining Ltd., 2013, Annual report 2012: Dragon Mining Ltd., 102 p. (Accessed November 19, 2013, http://www.dragon-mining.com.au/sites/default/files/dragon_ar2012_final_web_version_0.pdf.)

First Quantum Minerals Ltd., 2014, Pyhasalmi: First Quantum Minerals Ltd. (Accessed October 13, 2014, at http://www.first-quantum.com/Our-Business/operating-mines/Pyhasalmi/default.aspx.)

Ministry of Employment and the Economy [Finland], 2011, New mining act to enter into force on 1 July: Ministry of Employment and the Economy. (Accessed January 17, 2012, at http://www.tem.fi/index.phtml?105047_m=103119&l=en&s=4760.)

Ministry of Employment and the Economy [Finland], 2014a, The mining industry: Ministry of Employment and the Economy. (Accessed October 10, 2014, at https://www.tem.fi/en/enterprises/the_mining_industry.)

Ministry of Employment and the Economy [Finland], 2014b, Legislation regulating mining: Ministry of Employment and the Economy. (Accessed October 10, 2014, at https://www.tem.fi/en/enterprises/the_mining_industry/legislation.)

Mondo Minerals Oyj, 2014a, Who are we—An overview: Mondo Minerals Oyj. (Accessed October, 12, 2014, at http://www.mondominerals.com/en/the-talc-company/an-overview/.)

Mondo Minerals Oyj, 2014b, Products: Mondo Minerals Oyj. (Accessed October, 12, 2014, at http://www.mondominerals.com/en/talc-products/.)

Nordkalk Corp., 2013, Welcome to the world of Nordkalk wollastonite—Nordkalk has launched a new generation of wollastonite products: Nordkalk Corp. (Accessed November 10, 2013, at http://www.wollastonite.fi/default.asp?viewID=896.)

Nordkalk Corp., 2014, Lappeenranta: Nordkalk Corp. (Accessed October 13, 2014, at http://www.nordkalk.com/default.asp?viewID=1026&companyID=53.)

OM Group Inc., 2013, Annual report 2012: Cleveland, Ohio, OM Group Inc., 107 p.

Outokumpu Oyj, 2014, Chrome makes the steel stainless–Kemi Mine today: Outokumpu Oyj, 15 p. (Accessed October 12, 2014, at http://www.outokumpu.com/SiteCollectionDocuments/Outokumpu-Site-Visit-Kemi-Mine-presentation-11092013.pdf.)

Talvivaara Mining Co. plc, 2012, Annual results reviewed for year ended 31 December 2011, Finland: Talvivaara Mining Co. plc., February 16, 27 p. (Accessed November 3, 2013, at http://hugin.info/136227/R/1586406/497264.pdf.)

United Nations, 2013, Mining—General: United Nations, 8 p. (Accessed November 8, 2013, at http://www.un.org/esa/dsd/dsd_aofw_ni/ni_pdfs/NationalReports/finland/Mining.pdf.)

U.S. Central Intelligence Agency, 2013, Finland, in The world factbook: U.S. Central Intelligence Agency. (Accessed November 14, 2013, at https://www.cia.gov/library/publications/the-world-factbook/geos/fi.html.)

Virta, R.L., 2013, Talc and pyrophyllite: U.S. Geological Survey Mineral Commodity Summaries 2013, p. 160–161.

TABLE 1
FINLAND: PRODUCTION OF MINERAL COMMODITIES[1]

(Thousand metric tons unless otherwise specified)

Commodity		2008	2009	2010	2011	2012[e]
METALS						
Aluminum, metal, secondary	metric tons	24,706	17,885	20,736[r]	19,531	19,530
Chromite:						
Cr_2O_3 content		614	247	598[r]	693[r]	425[2]
Of which:						
Foundry sand		5	5	NA[r]	NA[r]	NA[r]
Lump ore		85	80	NA[r]	NA[r]	NA[r]
Total		90	85	NA[r]	NA[r]	NA[r]
Cobalt, refined	metric tons	6,301	4,665	9,413	10,441	10,547[2]
Copper:						
Concentrate, gross weight	do.	47,077	49,730	51,222[r]	47,802[r]	104,393[2]
Mine output, Cu content	do.	13,000	14,600	14,700	16,000	NA
Metal:						
Smelter	do.	174,354	139,710	153,853[r]	156,017[r]	156,000
Refined	do.	137,953	105,549	146,344[r]	148,639[r]	148,000
Gold, metal, mine output	kilograms	4,148[r]	5,749	7,628[r]	8,461[r]	10,814[2]
Iron and steel, metal:						
Ferroalloys, ferrochromium		234	123	283[r]	231[r]	288[2]
Pig iron	metric tons	2,943[r]	2,042	10,033[r]	12,145[r]	12,000
Steel, crude		4,418	3,078	4,023	3,985	3,759[2]
Mercury	kilograms	33,120	6,210	9,315[r]	--[r]	--
Nickel:						
Mine output, Ni content	metric tons	4,303	4,400	29,448[r]	63,209[r]	80,000
Metal, electrolytic	do.	51,936[r]	40,800	41,317[r]	49,823[r]	46,275[2]
Platinum	kilograms	214	265	718[r]	836[r]	830
Selenium, metal	do.	58,069	66,028	66,094[r]	88,231[r]	92,769[2]
Silver, metal	do.	59,375	60,019	64,751[r]	69,344[r]	128,200[2]
Zinc:						
Mine output, Zn content	metric tons	51,900	56,415	55,562	64,115	51,467[2]
Metal	do.	297,722	295,049	307,144[r]	307,352	314,742[2]
INDUSTRIAL MINERALS						
Cement, hydraulic		1,633	1,052	1,215[r]	1,387[r]	1,300
Feldspar	metric tons	45,250	2,312	28,013[r]	26,292[r]	43,124[2]
Lime		482	410	463[r]	456[r]	450
Mica:						
Biotite		57	54[r]	38[r]	32[r]	27[2]
Concentrate	metric tons	10,706	7,855[r]	13,809[r]	12,896[r]	12,112[2]
Nitrogen, N content of ammonia	do.	73,868	68,379	78,380[r]	72,352[r]	72,000
Phosphate rock apatite concentrate:						
Gross weight		780[e]	660	817[r]	870[r]	870
P_2O_5 content		NA	234	289	307	300
Pyrite, gross weight		510	679	706[r]	939[r]	940
Sodium sulfate		22	NA	NA	4[r]	4
Stone, crushed:						
Limestone and dolomite:						
Dolomite		NA	NA	NA	81[e]	81
For cement manufacture		1,807	1,132[r]	1,495[r]	1,600[r]	1,600
For agriculture		647	687	646[r]	450[r]	450
For lime manufacture		317	191[r]	234[r]	220[r]	220
Fine powders		650	650	650	NA	NA
Metallurgical[e]		1	1	1	NA	NA
Total		3,422	2,661[r]	3,026[r]	2,350[r,e]	2,350
Quartz silica sand		3,160	2,241	267[r]	312[r]	310

See footnotes at end of table.

TABLE 1—Continued
FINLAND: PRODUCTION OF MINERAL COMMODITIES[1]

(Thousand metric tons unless otherwise specified)

Commodity[2]	2008	2009	2010	2011	2012[e]
INDUSTRIAL MINERALS—Continued					
Sulfur:					
S content of pyrite	226	154	150 [e]	338	330
Byproduct:[e]					
Metallurgy	331	274	275	280	280
Petroleum	117	127	125	133 [2]	130
Total	448	401	400	410	410
Sulfuric acid	956	851	949 [r]	887 [r]	890
Talc	528	375 [r]	419 [r]	429 [r]	396 [r, 2]
Wollastonite metric tons	15,600	9,200 [r]	12,100 [r]	11,500 [r]	11,500
MINERAL FUELS AND RELATED MATERIALS					
Peat:					
For fuel use	6,933	5,576	7,533 [r]	6,847 [r]	6,800
For agriculture and other uses	1,552	876	867 [r]	674 [r]	670
Petroleum refinery products thousand 42-gallon barrels	95,325	95,000	88,137 [r]	90,686 [r]	90,000

[e]Estimated; estimated data are rounded to no more than three significant digits; may not add to totals shown. [r]Revised. do. Ditto. NA Not available.
-- Zero.

[1]Table includes data available through November 18, 2013.

[2]Reported figure.

TABLE 2
FINLAND: STRUCTURE OF THE MINERAL INDUSTRY IN 2012

(Thousand metric tons unless otherwise specified)

Commodity		Major operating companies and major equity owners	Location of main facilities	Annual capacity
Ammonia		Kemira Oyj (Government, 98%)	Plant at Oulu	75
Apatite		Kemira Agro Oyj (Government, 98%)	Mine and plant at Siilinjarvi	8,000
Cadmium, metal		Outokumpu Oyj (Government, 40%, and private investors, 12.3%)	Smelter at Kokkola	1
Cement		Finncement Oy (Irish Cement Ltd., 100%)	Plants at Lappeenranta and Parainen	1,020
Chromite		Outokumpu Oyj (Government, 40%, and private investors, 12.3%)	Mine at Kemi	1,000
Cobalt		Norilsk Nickel Harjavalta (MMC Norilsk Nickel, 100%)	Plant at Kokkola	NA
Copper:				
Ore, Cu content		First Quantum Minerals Ltd.	Mine at Pyhasalmi	10
Metal		Boliden Harjavalta AB (Boliden AB, 100%)	Smelter at Harjavalta	210
Do.		do.	Refinery at Pori	155
Feldspar		SP Minerals Oyj (Partek Corp., 50.1%, and SCR-Silbeco SA, 49.9%)	Mine and plant at Kemio	50
Ferrochrome		Outokumpu Oyj (Government, 40%, and private investors, 12.3%)	Smelter at Tornio	250
Gold:				
Ore, Au content	metric tons	Agnico-Eagle Mines Ltd.	Mine at Kittila	5
Do.	do.	Dragon Mining Ltd.	Mines at Orivesi and Jokisivu and plant in the Sastamala region	4
Do.	do.	Lappland Goldminers AB.	Pahtavaara Mine near Sodankyla	2
Metal	do.	Boliden AB	Smelter at Pori	4
Limestone		Nordkalk Corp. (Rettig Group, 100%)	Mines at Lappeenranta, Pargas, and Parainen	1,500
Do.		Rauma-Repola Oyj	Mine at Tornio	300
Mercury	metric tons	Outokumpu Oyj (Government, 40%, and private investors, 12.3%)	Smelter at Kokkola	150
Mica		Kemira Oyj (Government, 98%)	Mine at Siilinjarvi	10
Nickel:				
Ore, Ni content		Belvedere Resources Ltd.	Mine at Hitura	30
Do.		Talvivaara Mining Co. plc.	Mine at Sotkamo	20
Metal		Norilsk Nickel Finland (MMC Norilsk Nickel, 100%)	Smelter at Harjavalta	32
Do.		do.	Refinery at Harjavalta	50
Petroleum products	thousand barrels per day	Neste Oil Oyj, 50%, and Government, 50%	Plants at Naantali and Porvoo	NA
Phosphate-apatite		Yara International ASA.	Mine at Siilinjarvi	1,000
Quartz and quartzite		SP Minerals Oyj (Partek Corp., 50.1%, and SCR-Silbeco SA, 49.9%)	Mines at Kemio and Nilsia	250
Selenium	metric tons	Boliden AB	Smelter at Pori	35
Silver	do.	do.	do.	30
Steel:				
Crude		Rautaruukki Oyj (Government, 39.7%)	Plants at Halikko, Hameenlinna, Kankaanpaa, and Raahe	2,100
Do.		Fundia AB (Norsk Jenverk AS of Norway, 50%, and Rautaruukki, 50%)	Plants at Aminnefors, Dalsbruk, and Koverhar	850
Do.		Ovako AB (Triton Adviser Ltd., 100%)	Plant at Imatra	600
Stainless		Outokumpu Oyj (Government, 40%, and private investors, 12.3%)	Plant at Tornio	550
Talc		Mondo Minerals Oyj (Advent International Corp., 100%)	Mines at Lahnaslampi, Lipsavaara, and Horsmanaho	500
Wollastonite		Nordkalk Corp. (Rettig Group, 100%)	Mine and plant at Lappeenranta	40
Zinc:				
Ore, Zn content		First Quantum Minerals Ltd.	Mine at Pyhasalmi	25
Metal		Boliden AB	Smelter at Kokkola	260

Do., do. Ditto. NA Not available.

THE MINERAL INDUSTRY OF FRANCE

By Alberto Alexander Perez

France's gross domestic product (GDP) was $2.579 trillion in 2012, which was about the same as the revised value for its GDP in 2011. France had the third-largest GDP in the European Union (EU) after Germany and the United Kingdom. The output value of France's entire industrial sector accounted for about 18.8% of the GDP in 2012. The country was a significant processor of raw mineral materials and a manufacturer of industrial and consumer durable goods. France's heavy industries, which, among other product categories, produced automotive and aviation products, chemicals, and machine tools for domestic consumption and export, relied mainly on imported metal ores and concentrates and on imported industrial minerals and mineral fuels (U.S. Central Intelligence Agency, 2014).

Minerals in the National Economy

During at least the past 20 years, France gradually transitioned from being a mineral producer and processor of mineral commodities to being mainly a processor. Most mining, and certainly mining of metals, had ceased in metropolitan France. Owing to the size and structure of France's economy, the upstream input of minerals was key to the continued maintenance and growth of the country's heavy industries.

Government Policies and Programs

The French mining code was last modified on March 1, 2011. Most of the changes were aimed at simplifying the acquisition of exploration licenses and licenses for the development of future projects (Legifrance.gouv.fr, 2013).

The Ministry of Ecology and Sustainable Development is responsible for overseeing and regulating such environmental issues as agricultural runoff; air pollution from industrial and vehicle emissions; forest damage from acid rain; and water pollution from mining, mineral processing, and urban waste.

The Bureau de Recherches Géologiques et Minières [Bureau of Mining and Geological Research] (BRGM), which was France's geological survey, is the French institution that performs and develops geologic and mineral research in France and abroad. Its headquarters are located in Orleans.

Production

In 2012, the mineral industry of France produced at about the same level of output as in 2011. Production of alumina decreased by about 8.5%, and that of hydraulic cement, by an estimated 7.4%. Primary aluminum production increased by 4.5%, and pig iron and crude steel production, by 1.7% and 1%, respectively (table 1).

Mineral Trade

Most of France's demand for fuel and nonfuel mineral raw materials was met by imports. The major commercial partners of France were all members of the EU and included Belgium, Germany, Italy, and Spain. The United States was the leading non-EU commercial partner of France. In 2010 (the latest year for which data were available), exports from France to other countries in the EU[1] included iron and steel valued at $12 billion; petroleum and petroleum products, $8 billion; manufactured metals, $7.09 billion;[2] nonferrous metals, $4.695 billion; and metalliferous ores and metal scrap, $4.335 billion. Imports by France of goods originating from other countries in the EU included iron and steel valued at $15 billion; manufactured metals, $12.801 billion; natural gas, $12.6 billion; petroleum and petroleum products, $11.7 billion; nonferrous metals, $8.382 billion, and metalliferous ores and metal scrap, $2.136 billion. In contrast, France's leading mineral industry imports from a non-EU country in terms of value were petroleum and petroleum products valued at $48 billion; natural gas, $5.6 billion; nonferrous metals, $3.438 billion; and manufactured metals, $2.388 billion (Eurostat, 2011a; 2011b, p. 110–112, 118–120, 126–128, 134–138).

In terms of energy imports, France's imports of oil equivalent in 2010 (the latest year for which data were available) totaled 133.6 million metric tons. France's main energy suppliers were the countries of the Commonwealth of Independent States, Norway, and several African countries. Only a small percentage of energy imports originated from Middle East countries (Eurostat, 2013).

Structure of the Mineral Industry

Although France continued to maintain state monopolies in a number of sectors of the economy, principally in the energy production and transport sectors, state ownership in the mineral sector was minimal. In 2012, the French Government maintained partial ownership of the country's electricity generation and natural gas production and distribution facilities, as well as ownership of rail and public transportation systems in most French cities. Table 2 provides data on the major enterprises that produced metals, industrial minerals, and mineral fuels in France in 2012.

---

[1]In the European Commission's official reports, exports from one member country to other countries within the European Union (EU) are referred to as "dispatches," and imports by a member country from other countries in the EU are referred as "arrivals."

[2]Where necessary, values have been converted from euro area euros (€) to U.S. dollars (US$) at an average exchange rate of €.755=US$1.00.

Commodity Review

Metals

Aluminum.—In 2012, France's output of primary aluminum increased by 4.5% (table 1). Rio Tinto Ltd. of Australia was the country's sole producer of primary aluminum. Rio Tinto also operated facilities for the production of aluminum semimanufactures. Rio Tinto sold its Gardanne specialty alumina plant to H.I.G. Capital Europe, which in turn formed a new company called Alteo Holdings to manage the plant (Rio Tinto Alcan, 2012).

Ferroalloys.—The Brazilian company Vale S.A. reported that it had sold its manganese ferroalloys operations in Europe for $160 million to subsidiaries of Glencore International Plc. of the United Kingdom. The facilities included in the sale were Vale Manganese France SAS (located in Dunkerque, France) and Vale Manganese Norway AS (located in Mo I Rana, Norway) (Vale S.A., 2012, p. 4).

Iron and Steel.—France's output of pig iron decreased by 1.7%. Crude steel production decreased by 1% (table 1). Crude steel apparent use decreased by 13.3% (World Steel Association, 2013, p. 77).

Industrial Minerals

France produced a broad variety of industrial minerals. In 2012, the Imerys Group, which was a major French producer of industrial minerals, mined and processed ball clays, carbonates, feldspar, and red clays domestically and from deposits in such countries as China, Germany, Spain, the United States, and Vietnam for domestic use and export (Imerys S.A., 2012, p. 6–7).

Cement.—In 2012, cement production decreased by an estimated 7.4% and cement consumption, by 6.7% compared with the levels of production and consumption, respectively, in 2011. The decreases were owing to a 17% decrease in the residential construction sector.

France's principal cement manufacturers were Lafarge S.A. and Société des Ciments Français, which was a subsidiary of Italcementi S.p.A of Italy. In addition to their cement-producing facilities in France, both companies had major capital assets abroad. The other significant producers of cement in France were the Vicat Group, which had five plants with a total cement production capacity of 6 million metric tons per year (Mt/yr), and Holcim Ciments S.A, which had six plants and a total cement production capacity of 4.2 Mt/yr (table 2; Cembureau, 2013, p. 10).

Mineral Fuels and Other Sources of Energy

In 2012, nuclear energy accounted for an estimated 94% of primary electricity production. The principal sectors that consumed energy in France were, in order of consumption, the residential sector (44.5%), the transportation sector (31.9%), and the manufacturing and steelmaking industries and agricultural sectors combined (23.7%) (Institut National de la Statistique et des Études Économiques, 2014a).

Renewable energy production increased by 18.9% in 2012. Within the renewable energy sector, production of photovoltaic cell solar energy increased by 88.2% and production of wind power energy increased by 22.04% (Institut National de la Statistique et des Études Économiques, 2014b).

Natural Gas and Petroleum.—In 2012, France's domestic production of crude petroleum decreased by 8.6% compared with the output in 2011. Domestic production of petroleum products decreased by about 1.1% in 2012 compared with production in 2011 (U.S. Energy Information Administration, 2014).

Nuclear Energy.—Group Areva, which was the French Government-owned nuclear technology company, was building the first nuclear reactors in Western Europe in 20 years. Areva's reactor, which is called a Third Generation, or EPR (Evolutionary Power Reactor, or European Pressurized Reactor, as it is known in Europe), had helped the company compete for new construction contracts for nuclear powerplants in France and abroad.

In December 2011, at the International Thermonuclear Experimental Reactor (ITER) complex in Cadarche in the Provence-Alpes-Côte d'Azur region, the last segment of the seismic isolation pit basemat was poured, and construction of the support structure for the the Tokamak complex was reportedly progressing according to schedule; the reactor was expected to be commissioned by 2019. The seven participants in the ITER project were the United States, China, the EU, India, Japan, the Republic of Korea, and Russia. The project seeks to demonstrate the feasibility of producing nuclear power using nuclear-fusion-generated energy rather than nuclear-fission-generated energy (International Thermonuclear Experimental Reactor, 2011).

Outlook

The French economy has been slow to recover from its recession, and this has affected its industry and employment. Because France is principally a processor of minerals, the domestic rate of consumption of national goods and services and the demand for its manufactured goods abroad directly affect the French mineral industry and its expectations for growth. France will likely continue to import much of its ores and minerals for its manufactured goods industry, although the French Government has indicated that is interested in restarting the mining of mineral commodities in metropolitan France. The share of renewable energy in France's total consumption of energy continues to grow as the Government is investing and promoting renewable energy usage. Despite this increase, and although there is public interest in decreasing the role of nuclear energy in the country, nuclear power will very likely remain the focus of the Government's energy generation strategy for the near future.

References Cited

Cembureau, 2013, Activity report 2012: Brussels, Belgium: Cembureau, 42 p.

Eurostat, 2011a, External and intra European Union trade—Monthly statistics—Issue no. 06/2011: Brussels, Belgium, European Commission, 511 p. (Accessed October 17, 2011, at http://epp.eurostat.ec.europa.eu/portal/page/portal/international_trade/documents/ExtraIntraMonthlyEUTrade_ENVol06-2011.pdf.)

Eurostat, 2011b, External and intra-EU trade—A statistical yearbook data 1958—2010: Brussels, Belgium, European Commission, 410 p. (Accessed December 17, 2011, at http://epp.eurostat.ec.europa.eu/cache/ITY_OFFPUB/KS-GI-11-001/EN/KS-GI-11-001-EN.PDF.)

Eurostat, 2013, Energy production and imports: Brussels, Belgium, European Commission. (Accessed January 21, 2013, at http://epp.eurostat.ec.europa.eu/statistics_explained/index.php/Energy_production_and_imports.)

Imerys S.A., 2012, 2011 in brief: Paris, France, Imerys, S.A. 16 p. (Accessed December 12, 2012, at http://www.imerys.com/Scopi/Group/ImerysCom/imeryscom.nsf/pagesref/SPIT-8UKATZ/$file/RAIM011_PLAQUETTE_GB.pdf.)

Institut National de la Statistique et des Études Économiques, 2014a, Bilan energetique: Institut National de la Statistique et des Études Économiques. (Accessed May 1, 2014, at http://www.insee.fr/fr/themes/document.asp?reg_id=0&ref_id=T14F191.)

Institut National de la Statistique et des Études Économiques, 2014b, Production d'energie primaire d'origine renouvelable en 2012: Institut National de la Statistique et des Études Économiques. (Accessed May 1, 2014, at http://www.insee.fr/fr/themes/tableau.asp?reg_id=0&ref_id=NATTEF01331.)

International Thermonuclear Experimental Reactor, 2011, Seismic pit basemat completed: International Thermonuclear Experimental Reactor, December, 1 p. (Accessed December 11, 2012, at http://www.iter.org/proj/itermilestones.)

Legifrance.gouv.fr, 2013, Code Minier: Paris, France, Secrétariat Général du Gouvernement. (Accessed January 23, 2013, at http://www.legifrance.gouv.fr/affichCode.do;jsessionid=F1CDF2189FB439A9EB27C19AFA16B7A2.tpdjo07v_3?cidTexte=LEGITEXT000006071785&dateTexte=20130131.)

Rio Tinto plc, 2012, Rio Tinto receives binding offer for its specialty aluminas business: London, United Kingdom, Rio Tinto plc media release, March 28. (Accessed November 19, 2014, at http://www.riotinto.com/media/media-releases-237_1023.aspx.)

U.S. Central Intelligence Agency, 2014, France, in The world factbook: U.S. Central Intelligence Agency. (Accessed May 28, 2014, at https://www.cia.gov/library/publications/the-world-factbook/geos/fr.html.)

U.S. Energy Information Administration, 2014, France energy profile: U.S. Energy Information Administration. (Accessed May 1, 2014, http://www.eia.gov/cfapps/ipdbproject/iedindex3.cfm?tid=5&pid=alltypes&aid=1&cid=FR,&syid=2009&eyid=2012&unit=TBPD.)

Vale S.A., 2012, Vale concludes sale of manganese ferroalloy operations in Europe: Rio de Janeiro, Brazil, Vale S.A. (Accessed January 5, 2014, at http://saladeimprensa.vale.com/en/releases/interna.asp?id=21991.)

World Steel Association, 2013, Steel statistical yearbook 2012: Brussels, Belgium, World Steel Association, 112 p.

TABLE 1
FRANCE: PRODUCTION OF MINERAL COMMODITIES[1]

(Metric tons unless otherwise specified)

Commodity[2]		2008	2009	2010	2011	2012[e]
METALS						
Aluminum:						
Bauxite, gross weight[e, 3]	thousand metric tons	160	160	--	--	69 [4]
Alumina, metallurgical, gross weight[e]	do.	592	348	481	470 [r]	430 [4]
Metal:						
Primary	do.	389	345	356	334	349 [4]
Secondary	do.	209	138	184	191	184 [4]
Antimony, metal, including regulus[e]		500	500	500	500	500
Cadmium metal[e]		50	50	50	--	--
Cobalt, metal:		311	368	302	354	350
Gold, mine output, Au content[e]	kilograms	1,500	1,500	1,500	--	--
Iron and steel:						
Metal:						
Pig iron	thousand metric tons	11,372	8,104	10,137	9,698	9,532 [4]
Ferroalloys, electric furnace:[e]						
Ferromanganese	do.	47	46	138	131	131
Ferrosilicon	do.	100	20	27	59	59
Silicomanganese	do.	60	54	62	63	63
Silicon metal	do.	118	80	112	128	128
Other	do.	60	60	60	60	60
Total	do.	385	260	400	440	440
Steel:						
Crude	do.	17,900	12,840	15,414	15,780	15,609 [4]
Hot-rolled	do.	14,746	11,382	13,581	13,715	13,529 [4]
Lead, refined:[e]						
Primary		--	--	--	-- [r]	--
Secondary		82,000	82,000	82,000	53,887 [4]	75,000
Total		82,000	82,000	82,000	53,887 [4]	75,000
Nickel, refinery products, Ni content[5]		13,700 [r]	13,900 [r]	14,400 [r]	13,700 [r]	14,500
Tin, secondary[e]		1,500	1,500	1,500	--	--
Zinc metal, including slab and secondary		118,900	161,000	163,000	164,000	161,000

See footnotes at end of table.

TABLE 1—Continued
FRANCE: PRODUCTION OF MINERAL COMMODITIES[1]

(Metric tons unless otherwise specified)

Commodity[2]		2008	2009	2010	2011	2012[e]
INDUSTRIAL MINERALS						
Abrasives, undifferentiated[e]		270	270	270	270	270
Cement, hydraulic	thousand metric tons	21,400	18,300	17,998	19,433	18,000
Clays:						
Kaolin and kaolinitic clay (marketable)	do.	624	519	315	315	315
Refractory clay, unspecified[e]	do.	15	15	15	15	15
Diamond, synthetic, industrial[e]	thousand carats	3,600	3,600	3,600	3,600	3,600
Diatomite[e]	thousand metric tons	75	75	75	75	75
Feldspar, crude[e]	do.	650	650	650	650	650
Gypsum and anhydrite, crude	do.	3,500[e]	3,351	3,440	4,231	3,685[4]
Kyanite, andalusite, related materials[e]	do.	65	65	65	65	65
Lime, quick and hydrated, dead-burned dolomite[e]	do.	4,000	4,000	4,000	4,000	4,000
Mica[e]		20,000	20,000	20,000	20,000	20,000
Nitrogen, N content of ammonia[e]	thousand metric tons	800	2,970[4]	3,517[4]	3,500	2,644[4]
Pigments, mineral, natural, iron oxide[e]		1,000	1,000	1,000	1,000	76,196[4]
Phosphates, Thomas slag[e]	thousand metric tons	50	50	50	50	50
Pumice and other natural abrasives[e]	do.	270	270	270	270	270
Salt, all sources[e]	do.	6,240	6,200	5,867[4]	5,430	5,457[4]
Sodium compounds:[e]						
Soda ash	do.	1,000	1,000	1,000	1,000	1,000
Sodium sulfate	do.	120	120	120	120	120
Stone, sand and gravel:						
Chalk	do.	580[e]	1,294	1,765	2,733	1,702[4]
Dolomite, crude	do.	980[e]	777	700	393[r]	423[4]
Granite, crude	do.	370[e]	403	426	482	233[4]
Limestone, agricultural and industrial	do.	11,700[e]	8,302	9,102	10,666	10,216[4]
Marble and travertine, crude[e]	do.	150	150	150	150	150
Sand and gravel:						
Industrial sands		5,200[e]	7,442	8,498	6,286	8,880[4]
Other sand, gravel, and aggregates		165,000[e]	263,530	249,512	277,521	251,015[4]
Sandstone	thousand metric tons.	95[e]	109	100	100[e]	100
Slate, crude[e]		8,700	8,700	8,700	8,700	8,700
Sulfur, all sources[e]		650	650	650	650	650
Talc, crude[e]	thousand metric tons	420	420	420	420	420
MINERAL FUELS AND RELATED MATERIALS						
Asphaltic material[e]		20,000	11,675[4]	11,600	11,600	11,600
Carbon black		200,000[e]	178,777	203,563	134,329	134,000
Coal, briquets[e]	thousand metric tons	100	100	100	100	100
Coke, metallurgical[e]	do.	4,500	4,500	4,500	4,500	4,500
Gas, natural, marketed	million cubic meters	1,472	1,444	1,245	1,132	1,100
Petroleum:						
Crude	thousand 42-gallon barrels	7,117	6,624	6,606	6,508	5,949[4]
Refinery products:						
Liquefied petroleum gas	do.	33,860	29,236	24,346	24,300	24,300
Gasoline, all kinds	do.	141,195	133,225	115,596	115,000	115,000
Kerosene and jet fuel	do.	44,462	39,274	35,113	35,100	35,000
Distillate fuel oil	do.	275,148	246,959	224,950	224,900	220,000
Residual fuel oil	do.	73,342	61,137	59,313	59,300	59,000
Other products	do.	124,347	107,748	106,617	106,600	106,000
Total	do.	692,354	617,579	565,900	565,200	559,000

[e]Estimated; estimated data rounded to no more than three significant digits; may not add to totals shown. [r]Revised. do. Ditto. -- Zero.

[1]Table includes data available through January 5, 2014.

[2]In addition to the commodities listed, France produces germanium from domestic ores, but actual output is not regularly reported.

[3]Reprocessed bauxite not for metallurgical use.

[4]Reported figure.

[5]Excludes secondary production from nickel-cadmium batteries.

TABLE 2
FRANCE: STRUCTURE OF THE MINERAL INDUSTRY IN 2012

(Thousand metric tons unless otherwise specified)

Commodity		Major operating companies and major equity owners	Location of main facilities	Annual capacity
Alumina, metallurgical		Alteo Holdings, 100%	Plant at Gardanne	700
Aluminum		Rio Tinto Ltd.	Aluminum smelters, of which:	
Do.		do.	Saint-Jean-de-Maurienne, Savoie	120
Do.		do.	Dunkerque, Calais du Nord	250
Andalusite		Denain-Anzin Minéraux Réfractaire Céramique	Glomel Mine, Brittany	75
Antimony, metal		Produits Chimiques de Lucette	Plant at Le Genest, Mayeene Province	15
Barite		Barytine de Chaillac	Mine and plant at Chaillac	150
Do.		Société Industrielle du Centre	Mine at Rossigno, Indre Province	100
Cement		Four companies, of which the largest are:	80 plants, including:	26,700
Do.		Lafarge S.A.	14 plants, the largest of which is at St. Pierre-la-Cour (1,160)	10,000
Do.		Société des Ciment Français	Nine plants, the largest of which is at Gargenville (1,100)	7,500
Do.		Vicat Group	Five plants	6,000
Do.		Holcim Ciments S.A.S	Nine plants	5,900
Clay, kaolin		Groupe Mineral Harwanne (GMH)	Kaolin d'Arvor Mine, Quessoy	300
Cobalt, metal	metric tons	Société Métallurgique le Nickel (SLN)	Plant at Sandouville, near Le Havre	600
Copper, metal		Compagnie Générale d'Électrolyse du Palais	Electrolytic plant at Palais-sur-Vienne	45
Diatomite		Ceca S.A.	Mines and plants at Riom-les-Montagnne and St. Bauzille	100
Feldspar		Denain-Anzin Mineraux S.A. (Imerys Group)	Mine and plant at St. Chely d'Apcher	55
Ferroalloys		Comilog Dunkerque (ERAMET S.A., 100%)	Dunkerque	70
Do.		FerroPem S.A. (Grupo Ferro Atlantica, 100%)	Six plants	290
Do.		Glencore Manganese France S.A. (Glencore International Plc., 100%)	Plant at Dunkerque	140
Gypsum		S.A. de Matériel de Construction	Mine at Taverny	1,500
Indium		Nyrstar S.A.	Plant at Auby	48
Iron and steel, steel:				
Crude		ArcelorMittal Group	Plant at Dunkerque	6,700
Rolling mill		do.	Plant at Fos-sur-Mer	4,200
Do.		do.	Plant at Florange[1]	3,200
Do.		do.	Plant at Gandrange, Neuves Maisons	8,400
Mica		Denain-Anzin Minéraux S.A. (Imerys Group)	Mine at Ploemeur, Brittany	160
Natural gas	million cubic meters	Total Group	Gasfield and plant at Lacq	20,000
Nickel, metal		Société Métallurgia le Nickel (SLN)	Plant at Sandouville	16
Nitrogen, N content of ammonia		GPN S.A	Plant at Grandpuits, Grand-Quevilly, and Ottmarsheim	390
Petroleum:				
Crude	42-gallon barrels per day	Total S.A.	Paris Basin oilfields	1,000
Refined	do.	do.	Refineries at Gonfreville and La Mede	446,000
Do.	do.	Petroplus S.A.	Refinery at Petite-Couronne	285,000
Do.	do.	Total S.A.	Refinery at Feyzin	120,000
Do.	do.	do.	Refinery at Donges	200,000
Do.	do.	do.	Refinery at Grandpuits	96,000
Do.	do.	Ineos Group Ltd.	Refineries at Lavera	175,000
Do.	do.	Esso S.A.	Refineries at Fos-sur-Mer	62,000
Do.	do.	do.	Refineries at Gravenchon	237,000
Do.	do.	Cie. Rhenane de Raffinage (CRR)[2]	Refinery at Reichstett	80,000
Salt		Compagnie des Salins du Midi et des Salines de l'Est (Salins Group)	Mines and plants at Aigues-Mortes, Dax, Salin-de-Giraud, and Varangeville	2,500
Sulfur		Total S.A.	Byproduct from natural gas, Lacq plant	3,000
Talc		Talc de Luzenac S.A. (Imerys. S.A., 100%)	Trimouns Mine near Ariege, Pyrenees	350
Zinc, metal		Nyrstar S.A.	Plant at Auby	172

Do. do., Ditto.

[1]The Florange blast furnace was idle for all of 2012.

[2]Production operations terminated; conversion to petroleum product distribution in 2012.

The Mineral Industry of Georgia

By Elena Safirova

Prior to the proclamation of Georgian independence in 1991, a range of mineral commodities were mined in Georgia, including arsenic, barite, bentonite, coal, copper, diatomite, lead, manganese, zeolites, zinc, and others. The country's metallurgical sector produced ferroalloys and steel. Production of many of these mineral commodities ceased or had been significantly reduced since 1991 because many supply-demand chains were lost after the disintegration of the Soviet Union.

After the Rose Revolution of 2003, the new Georgian Government set out to reorient the economy toward privatization and free markets. The Government enacted a sweeping tax reform by reducing the types of taxes from 26 in 2003 to 6 in 2008, significantly cut the number of Ministries and state agencies, and dramatically reduced corruption. According to The World Bank, Georgia was one of the world's fastest reforming economies, and, in 2011, it ranked as the world's ninth-easiest country in which to do business (World Bank, The, 2013). During the past decade, the Government was also focused on attracting foreign capital, primarily in the form of foreign direct investment (FDI). The mineral industry, however, did not benefit from the steady economic growth of the past 8 years as much as did other sectors, such as the financial services and manufacturing sectors.

In the fall of 2012, after the Georgian "Dream Party" won the parliamentary elections, work in the mining industry was frequently interrupted by worker strikes. On October 15, employees of Georgian Manganese Holding, LLC in the city of Chiatura started a labor strike; the workers' demands included a 10% wage increase and improvement in working conditions. Later, workers at other Georgian Manganese plants joined with the miners at Chiatura. At the height of the strike, work at the company's five mines and three processing plants was halted, a total of 3,700 workers were participating in the strike, and worker demands were raised to a 100% wage increase. At the time, the average wage of the Chiatura miners was between 250 and 300 laris per month (between $150 and $180),[1] which was about one-third less than the average wages in Georgia. The strike continued for several weeks and the workers demanded to form a parliamentary committee to evaluate the working conditions of miners. In November, 1,600 workers at the Dzidziguri and the Mindeli coal mines in the city of Tkibuli started a preventive labor strike because the administration of the Saqnakhshiri, Ltd. was planning to reduce wages and lay off some workers. The Tkibuli strike continued for more than a week. As a result of the strikes, some of the workers' demands were met, but tensions between the workers and the companies remained (Minerjob.ru, 2012a, c, d; NewsGeorgia.ru, 2012).

In December, workers of the petroleum terminal in the seaport Kulevi, which was owned by the State Oil Company of Azerbaijan Republic (SOCAR), also struck, demanding wage increases and improvement in working conditions. The strike lasted for 4 days, and the workers union and the port administration were able to find a compromise. Also in December, 300 workers at the Rustavi metallurgical plant began a strike and demanded a wage increase, improvements in sanitary conditions at the plant, and provision of social services for workers. The administration partially satisfied workers' demands but threatened the strike participants with layoffs. Although strikes also took place in other industries during the period of October through December, it appeared that mineral production had, on average, lower wages and worse working conditions than other sectors of the Georgian economy and therefore the mineral sector had relatively more strikes (Metaltorg.ru, 2012).

Minerals in the National Economy

In 2012, the nominal gross domestic product (GDP) of Georgia increased by 9.6% compared with that of 2011, to $15.8 billion. The country's real GDP increased by 6.1% in 2012 compared with that of 2011. The share of industrial production in the GDP in 2012 was 17.2%. Mining and quarrying accounted for 5.0% of the value of industrial production. In 2012, the real value of production in mining and quarrying increased by 2.0% whereas the real value of manufacturing production increased by 16.4%, indicating that the Georgian economy was growing following the economic reforms of the previous decade, but that the mining sector was lagging behind other sectors of the economy (National Statistics Office of Georgia, 2013c; U.S. Central Intelligence Agency, 2013).

In 2012, foreign direct investment (FDI) decreased by 23% to $865 million from $1,117 million in 2011, and a total of 66 countries invested in the Georgian economy. Japan was the leading investor in Georgia (provided 20% of the total FDI received by Georgia in 2012), followed by Azerbaijan (17%), the Netherlands (15%), the United Kingdom (8%), China (6%), and the Czech Republic (5%). The FDI in mining was $12.7 million, or 0.5% of the total FDI in the country; this was a 68.5% increase compared with the FDI in 2011 (Bizzone.info, 2013; National Statistics Office of Georgia, 2013b).

Mineral Trade

In 2012, Georgia ran a substantial trade deficit—the total value of its exports ($2.38 billion) was greatly exceeded by the total value of its imports ($7.84 billion). The country's major export trade partners were, in order of value, Azerbaijan (which received 26.4% of Georgia's exports), Armenia (11.0%), the United States (9.5%), Ukraine (7.0%), Turkey (6.0%), Canada (4.4%), Bulgaria (2.9%), Kazakhstan (2.6%), Belgium (2.5%), and Italy (2.2%). Its major import trade partners were, in order of value, Turkey (which supplied 17.8% of Georgia's

---

[1]Where necessary, values have been converted from Georgian laris (GEL) to U.S. dollars (US$) at an annual average exchange rate of GEL1.651=US$1.00 for 2012 and GEL1.687=US$1.00 for 2011.

imports), Azerbaijan (8.1%), Ukraine (7.6%), China (7.2%), Germany (6.9%), Russia (6.0%), Japan (4.0%), Bulgaria and Italy (3.5% each), and Romania (3.3%). Mineral commodities, especially metals, played a significant role in the country's exports. Ferroalloys accounted for 11.0% of the country's total export value; nitrogenous mineral or chemical fertilizers, 5.8%; unwrought gold, 5.0%; and copper ores and concentrates, 2.3%. Among the country's imports, the largest category was petroleum and petroleum oils, which made up 12.1% of the total. Petroleum gases contributed another 3.2% (National Statistics Office of Georgia, 2013a).

Ferroalloys were the most valuable mineral product exported from Georgia. In 2012, Georgia exported 227,700 metric tons (t) of ferroalloys and received for them $260.5 million. This was an increase of 14.2% in terms of export volume but only 2.2% in terms of export value compared with the volume and value of ferroalloys exports in 2011, respectively. Out of the 227,700 t of ferroalloys exported in 2012, 115,700 t (50.1%) was shipped to the United States; 19,400 t (8.5%), to Ukraine; 16,500 t (7.3%), to Egypt; 15,700 t (6.9%), to Turkey; and 15,500 t (6.8%), to Canada (National Statistics Office of Georgia, 2013a).

Petroleum and petroleum oils were the most valuable import category. In 2012, Georgia imported 915,700 t of petroleum and petroleum oils valued at $951 million, which was a 0.8% increase in terms of volume and a 4.4% increase in terms of cost compared with those in 2011. Of the 915,700 t of petroleum and petroleum oils imported, 341,000 t (37.3%) was shipped from Azerbaijan; 219,000 (24.0%), from Romania; 187,000 (20.5%), from Bulgaria; 67,000 (7.3%), from Russia; and 30,000 (3.3%), from Italy. Georgia also imported 1.4 million metric tons (Mt) of petroleum gases and other gaseous hydrocarbons valued at $252.7 million, which was an 11.2% increase in terms of volume and a 6.8% increase in terms of cost compared with those in 2011. The major supplier of petroleum gases to Georgia was Azerbaijan, which supplied 88.7%; Russia, which supplied 1.1% of Georgia's petroleum gas imports, was a distant second (National Statistics Office of Georgia, 2013a).

Government Policies and Programs

In November, the Ministry of Economy and Sustainable Development announced that the taxes on the use of mineral resources would be reviewed and could be revised as a result of the review. As of 2012, the taxes payable to the Government for mining natural resources were set at 0.9 lari ($0.55) per gram of gold, 136 lari ($82.4) per metric ton of copper, 0.012 lari ($0.007) per ton of manganese ore (assuming 1% grade), and 3 lari ($1.8) per cubic meter of mineral water. The Ministry stated that the tax rates for copper, gold, manganese, mineral water, and timber may be changed. At the same time, the Ministry underscored that Georgia was to adhere to its international obligations and treaties and to protect the rights of all businesses in the country (Kirtskhalia, 2012; MinerJob.ru, 2012b).

Production

Most of the data in table 1 were estimated because 2012 production data for most mineral commodities were not available. Production of mined copper increased by an estimated 17.5%; that of gold, by an estimated 15%; and silver, by an estimated 8.3%. Output of silicomanganese and mined zinc increased by an estimated 6.7% each. At the same time, production of manganese decreased by an estimated 5.2%, and that of coal, by an estimated 4%. Other production data are in table 1.

Commodity Review

Metals

Copper and Gold.—The Madneuli polymetallic deposit is situated in the Bolnisi region in southern Georgia about 80 kilometers (km) south of Tbilisi near the borders with Armenia and Azerbaijan. The Madneuli Mine was established in 1975 and had a long history as a precious metals producer in the region. The main ore types at Madneuli are barite-polymetallic ore, copper-barite ore, copper-zinc ore, gold-copper ore, and quartzite ore. The Madneuli complex included an open pit mine, a crushing facility, and a processing plant that used flotation to produce copper concentrate. The main pit had a total depth of 350 meters.

Since 2005, the Madneuli Mine was mostly (99.16%) owned by GeoProMining, Ltd. (GPM); the remaining shares were held by the former and current employees of the Madneuli Mine. GPM's mining license for the Madneuli Mine was valid through April 2014. In October 2011 and February 2012, however, the Ministry of Energy and Natural Resources conducted auctions for exploration licenses for deposits of barite, copper, and gold in the Bolnisi, the Dmanisi, the Marneuli, the Tetritskaro, and the Tsalk regions and for a mining license in the Kvemo Kartli region. The exploration and mining licenses were obtained by the Mining Investment Co. of Russia, which was a part of the Capital Group. The GPM did not obtain licenses for the new deposits and felt that its growth opportunities in Georgia were highly limited, so it decided to sell its assets in Georgia and to focus on operations in the other two countries where it had assets—Armenia and Russia. In June, it was announced that GPM had sold two of its Georgian holdings—JSC Madneuli and Quartzite Ltd.—to RMG Rich Metals Group for a total of $120 million (Apsny.ge, 2012a; Civil.ge, 2012). The new owner renamed the two companies JSC RMG Copper (formerly JSC Madneuli) and RMG Gold (formerly Quartzite Ltd.). According to the companies' Web sites, both RMG Copper and RMG Gold were in the process of reorganization at yearend 2012. RMG Rich Metals announced that it had invested $10 million in infrastructure and modernization efforts and stressed the importance of environmental protection and social programs at its enterprises. According to some sources, RMG Rich Metals is also a part of the Capital Group (Apsny.ge, 2012c; Gvimradze, 2012).

Manganese.—For more than a century, Georgia had mined manganese ore from the Chiatura deposit. A portion of the ore was used to produce manganese ferroalloys (ferromanganese and silicomanganese) at the Zestafoni ferroalloys plant, which was located 28 km from the Chiatura deposit. Chiatura Manganese included four mines and three open pit quarries; the enterprise's annual production capacity was about 400,000 t/yr (Felman Trading Inc., 2012).

Since 2006, both the Chiatura Manganese Mine and the Zestafoni plant had been a part of Georgian Manganese Holding, LLC. In October 2012, Georgian American Alloys, Inc. of the United States acquired 100% ownership interest in the Chiatura Manganese Mine, the Zestafoni ferroalloys plant, and the Vartsikhe hydroelectric facility, which powered the Chiatura Mine and the Zestafoni plant. Felman Trading, Inc. was expected to continue to serve as the primary distributor of silicomanganese produced by Georgian Manganese (Georgian American Alloys, Inc., 2013; JSC RMG Copper, 2013).

In February, the Georgian Government announced a repeat auction for the mining rights for manganese at a section of the Schkmer deposit located in the Racha region in northern Georgia. The auction was to take place in March. The starting price of a 20-year license was set at 1 million laris (about $600,000) with a deposit of 200,000 laris (about $120,000). The Government announced that the main criteria for selecting the winner of the tender were the offered price and a package of environmental measures the companies would undertake. The Schkmer deposit is the country's second largest manganese deposit after the Chiatura deposit. The resources of the section to be licensed were estimated to contain more than 1 Mt of manganese. The Government had previously tried to auction the Schkmer deposit in 2008 and 2011 but was unable to find a buyer (Apsny.ge, 2012b; Infogeo.ru, 2012).

Industrial Minerals

Nitrogen.—In September, construction of a new carbamide (urea) plant started in the village of Kulevi, which is located in western Georgia on the shore of the Black Sea. The plant would be built within the Poti Free Industrial Zone (Poti FIZ) and would provide jobs for the residents of Kulevi, Poti, and Supsa. The plant was expected to provide jobs for a total of 1,500 people, 300 of which would be physically located at the plant. The carbamide plant was planned to be built on a 24-hectare (59.3-acre) area. SOCAR Georgia Investment of Azerbaijan agreed to invest $100 million in the new construction; the total cost of the plant was expected to be $700 million. The plant was planned to have a capacity of 660,000 metric tons per year (t/yr) of carbamide and to start operations in 2016 (Bizzone.info, 2012a; Interfax.az, 2012).

Mineral Fuels

Oil and Natural Gas.—In 2012, foreign companies invested $68 million in exploration for petroleum and natural gas in Georgia and planned to invest at least an additional $150 million in 2013. In 2012, Georgia produced an estimated 50,000 t of petroleum and exported all of it, because the country did not have an operational oil refinery. The leading investors were Blake Oil and Gas Ltd. of the United Kingdom, which operated mostly in the Kakheti region in the eastern part of the country, and Jindal Petroleum Ltd. of India. According to Jindal Petroleum, each oil well would require between $5 million and $20 million in investment, and the company was planning to drill 10 wells (Apsny.ge, 2013).

Although Georgia did not produce natural gas on its own territory and produced only a limited volume of oil, the country spent the past decade trying to achieve the country's energy independence by diversifying its imports and building energy infrastructure. In 2006, the Baku-Tbilisi-Ceyhan oil pipeline opened; it was built to transport Caspian oil to the Turkish city of Ceyhan, which is located on the Mediterranean Sea. It was the first pipeline that would transport Azeri oil to the world market outside of the Russian pipeline network. The pipeline was 1,768 km in length, and 248 km of the pipeline was located in Georgia. BTC Pipeline Co., which operated the pipeline is owned by BP p.l.c. of the United Kingdom (30.1%), SOCAR (25%), Chevron Corp. of the United States (8.9%), Statoil ASA of Norway (8.71%), Türkiye Petrolleri Anonim Ortaklığı (TPAO) (6.53%), ENI S.p.A. of Italy and Total S.A. of France (5% each), Itochu Corp. of Japan (3.4%), ConocoPhillips Co. of the United States and Inpex Corp of Japan (2.5% each), and Hess Corp. of the United States (2.36%) (Jorbenadze, 2012; Rustambekov, 2012).

Another pipeline through Georgia, the Baku-Tbilisi-Erzurum gas pipeline, which opened in 2007, was to connect the Shah-Deniz oil and gas field in Azerbaijan to the Turkish city of Erzurum and eventually to transport natural gas from Azerbaijan to Europe using the Nabucco pipeline. The natural gas and crude oil pipelines were constructed in parallel, and some segments of the pipelines were put in place simultaneously, which helped to minimize the costs and environmental impact. The initial capacity of the pipeline was 12.5 billion cubic meters per year of natural gas. The BTE Pipeline was owned by BP and Statoil (25.5% each), Gazexport and the State Oil Corp. of the Republic of Azerbaijan (GNKAR), Naftiran Intertrade Co. (NICO) of Iran, Lukoil of Russia and Total (10% each), and TPAO (9%). In December 2012, Azerbaijan and Georgia started a project to expand the pipeline capacity to 25 billion cubic meters per year by 2017 and eventually to 45 billion cubic meters per year (NewsAzerbaijan.ru, 2012).

In June, representatives of SOCAR and Oil and Gas Corporation of Georgia signed an agreement to build a 29-km section of the future Kutaisi-Senaki gas pipeline in western Georgia. The section would connect the villages of Abasha and Senaki in western Georgia and was a part of the country's program on energy security. This section of the pipeline was expected to be completed within 9 months. The project was being financed by the U.S. Agency for International Development (Bizzone.info, 2012b).

Outlook

In the past decade, the Government of Georgia significantly improved the business climate in the country and attracted significant levels of FDI (World Bank, The, 2013). The mineral sector, however, was unable to fully take advantage of those changes. In the next 3 to 5 years, the mineral industry of Georgia is expected to have moderate but stable growth. Copper, ferroalloys, manganese, and steel are likely to remain the dominant mineral commodities in the short and medium terms. At the same time, a potential increase in domestic energy production could serve as a catalyst for faster future growth.

References Cited

Apsny.ge, 2012a, GeoProMaining prodal gruzinskie aktivy za 120 millionov dollarov [GeoProMining has sold its Georgian assets for $120 million]: Apsny.ge, June 19. (Accessed August 1, 2013, at http://www.apsny.ge/2012/eco/1340151201.php.)

Apsny.ge, 2012b, Gruziya povtorno popytaetsya nayti zhelayushikh zanyat'sya margantsem [Georgia will try again to find people who want to start a manganese business]: Apsny.ge, March 3. (Accessed August 1, 2013, at http://www.apsny.ge/2012/eco/1330819309.php.)

Apsny.ge, 2012c, Novyy vladelets "Madneuli" i "Kvartsita" prodolzit sotsial'nye I ekologicheskie programmy [The new owner of Madneuli and Quarzite will continue social and environmental programs]: Apsny.ge, July 27. (Accessed August 1, 2013, at http://www.apsny.ge/2012/eco/1343424126.php.)

Apsny.ge, 2013, V 2012 godu v razbedku gaza i nefti v Gruzii bylo investirovano 68 millionov dollarov [In 2012, $68 million was invested in oil and gas exploration in Georgia]: Apsny.ge, March 28. (Accessed August 1, 2013, at http://www.apsny.ge/2013/eco/1364495720.php.)

Bizzone.info, 2012a, SOCAR postroit gazoprovod v Gruzii [SOCAR will build a pipeline in Georgia]: Bizzone.info, May 19. (Accessed August 1, 2013, at http://bizzone.info/energy/2012/1337455438.php.)

Bizzone.info, 2012b, V Kulevi zalozheno stroitel'stvo karbamidnogo zavoda [Construction of a carbamide plant started in Kulevi]: Bizzone.info, September 18. (Accessed August 1, 2013, at http://bizzone.info/industry/2012/1348010356.php.)

Bizzone.info, 2013, V 2012 godu pryamye inistrannye investitsii v Gruzii osushestvleny iz 66 stran [In 2012, foreign direct investment in Georgia was made from 66 countries]: Bizzone.info, March 15. (Accessed August 1, 2013, at http://bizzone.info/stats/EFyFFFupul.php.)

Civil.ge, 2012, Litsenzia na dobychu poleznykh iskopaemykh prodana za 110,5 mln. lari [The license for mining resources is sold for 110.5 million laris]: Civil.ge, March 2. (Accessed August 1, 2013, at http://www.civil.ge/rus/article.php?id=23128.)

Felman Trading, Inc., 2012, Georgian Manganese: Felman Trading, Inc. (Accessed August 1, 2013, at http://felmantrading.com/en/producers/6/.)

Georgian American Alloys, Inc., 2013, Georgian American Alloys, Inc. acquires Georgian Manganese, LLC., Yahoo.com press release: April 22. (Accessed August 1, 2013, at http://finance.yahoo.com/news/georgian-american-alloys-inc-acquires-130000893.html.)

Gvimradze, Kote, 2012, Novyy vladelets [New owner]: AiF.ru, June 27. (Accessed August 1, 2013, at http://gazeta.aif.ru/online/tbilisi/510/14_02?print.)

Infogeo.ru, 2012, Gruziya ob'yavila povtornyi auktsion po Shkmerskomu mestorozhdeniyu margantsa [Georgia announced a repeat auction on Shkmer manganese deposit]: Infogeo.ru, February 29. (Accessed August 1, 2013, at http://www.infogeo.ru/metalls/news/?act=show&news=37957.)

Interfax.az, 2012, SOCAR vvedet k 2017 godu v stroy karbamidnyi zavod v Gruzii [By 2017, SOCAR will put a carbamide plant in Georgia in operation]: Interfax.az, September 19. (Accessed August 1, 2013, at http://interfax.az/view/552716.)

Jorbenadze, Irina, 2012, Gruziya: put' v obhod Rossii [Georgia—Path around Russia]: Rosbalt,ru, June 25. (Accessed August 1, 2013, at http://www.rosbalt.ru/exussr/2012/06/25/996372.html.)

JSC RMG Copper, 2013, Latest news: JSC RMG Copper. (Accessed August 1, 2013, at http://www.madneuli.ge/news-4-139.html.)

Kirtskhalia, N., 2012, Gruziya mozhet uvelichit' sbory za dobychu prirodnykh resursov [Georgia could increase taxes on mining mineral resources]: Trend.az, November 2. (Accessed August 1, 2013, at www.trend.az/print/2083510.html.)

Metaltorg.ru, 2012, Metallurgi Rustavi nachali zabastovku [Metallurgists at Rustavi started a strike]: Metaltorg.ru, December 26. (Accessed August 1, 2013, at http://www.metaltorg.ru/news/market_index.php?id=10079086&date=1356499260.)

MinerJob.ru, 2012a, Rabotniki krupneyshey v Gruzii ugol'noy Shakhty ob'yavili zabastovku [Workers at Georgia's largest coal mine started a strike]: MinerJob.ru, November 9. (Accessed August 1, 2013, at http://www.minerjob.ru/viewnew.php?id=22391.)

Minerjob.ru, 2012b, Sbor za pol'zovaniye pripodnymi resursami v Gruzii mozhet byt' peresmotren [The tax on use of mineral resources in Georgia may be changed]: Minerjob.ru, November 2. (Accessed August 1, 2013, at http://www.minerjob.ru/viewnew.php?id=22302.)

MinerJob.ru, 2012c, Schet bastuyushim v zapadnoy Gruzii poshel na tysyachi [People on strike in western Georgia are now counted by thousands]: MinerJob.ru, October 17. (Accessed August 1, 2013, at http://www.minerjob.ru/viewnew.php?id=22149.)

MinerJob.ru, 2012d, Tysyachi rabochikh i gornyakov bastuyut v Zapadnoy Gruzii [Thousands of workers and miners are on strike in western Georgia]: MinerJob.ru, October 19. (Accessed August 1, 2013, at http://www.minerjob.ru/viewnew.php?id=22128.)

National Statistics Office of Georgia [Geostat], 2013a, External trade of Georgia in 2012: National Statistics Office of Georgia, 139 p. (Accessed August 1, 2013, at http://geostat.ge/cms/site_images/ files/georgian/bop/Georgian External Trade 2012.pdf.)

National Statistics Office of Georgia [Geostat], 2013b, Foreign direct investments in 2012: National Statistics Office of Georgia. (Accessed December 10, 2012, at http://www.geostat.ge/index.php?action=page&p_id=140&lang=eng.)

National Statistics Office of Georgia [Geostat], 2013c, Gross domestic product of Georgia in 2012: National Statistics Office of Georgia. (Accessed August 1, 2013, at http://geostat.ge/index.php?action=wnews&lang=eng&npid=215.)

NewsAzerbaijan.ru, 2012, Baku i Tbilisi uvelichat moshnost' gazoprovoda Baku-Tbilisi-Erzerum [Baku and Tbilisi will increase capacity of Baku-Tbilisi-Erzerum pipeline]: NewsAzerbaijan, December 27. (Accessed August 1, 2013, at http://www.newsazerbaijan.ru/economic/20121227/298302239.html.)

NewsGeorgia.ru, 2012, V Tkibuli shakhtery dostigli konsensusa s administratsiey, no zabastobku ne prekratili [In Tkibuli, the miners reached a consensus with the administration but did not stop the strike]: NewsGeorgia.ru, November 15. (Accessed August 1, 2013, at http://www.newsgeorgia.ru/society/20121115/215339669.html.)

Rustambekov, Bakhram, 2012, Azerbaidzhan sokratil eksport nefti po truboprovodu Baku-Tbilisi-Dzheykhan [Azerbaijan reduced petroleum export by the Baku-Tbilisi-Ceyhan pipeline]: 1news.az, July 6. (Accessed August 1, 2013, at http://1news.az/economy/oil_n_gas/20120706093814125.html.)

U.S. Central Intelligence Agency, 2013, Georgia, in The world factbook: U.S. Central Intelligence Agency, May 3. (Accessed July 10, 2012, at https://www.cia.gov/library/publications/the-world-factbook/geos/gg.html.)

World Bank, The, 2013, Ease of doing business index: The World Bank. (Accessed August 1, 2013, at http://data.worldbank.org/indicator/IC.BUS.EASE.XQ.)

TABLE 1

GEORGIA: PRODUCTION OF MINERAL COMMODITIES[1]

(Metric tons unless otherwise specified)

Commodity[2]		2008	2009	2010	2011[e]	2012[e]
METALS						
Copper, mine output, Cu content of concentrate		11,000	9,800	6,700	6,300	7,400
Gold	kilograms	2,000	2,000	2,000	2,000	2,300
Iron and steel:						
Ferroalloys, electric furnace:						
Ferromanganese		11,342 [r]	1,838 [r]	824 [r]	195 [r, 3]	--
Silicomanganese		120,000	112,016	203,464	242,746 [r, 3]	257,421 [3]
Total		131,342 [r]	113,854 [r]	204,288 [r]	242,941 [r, 3]	257,421 [3]
Steel, rebar		NA	70,000	84,000	88,000	90,000
Manganese ore:[e]						
Gross weight		400,000	400,000	400,000	400,000	380,000
Mn content		116,000	116,000	116,000	116,000	110,000
Silver	kilograms	1,360	1,200	1,200	1,200	1,300
Zinc		NA [r]	NA [r]	NA [r]	NA [r]	NA
INDUSTRIAL MINERALS						
Cement[e]		450,000	870,368 [3]	856,880 [3]	860,000	870,000
Clays, bentonite[e]		5,000	5,000	5,000	4,800 [r]	4,900
Gypsum[e]		125	100	120	125 [r]	130
Nitrogen, N content of ammonia		150,000	150,000	150,000	145,000	150,000
Salt		30,000	30,000	30,000	28,000	29,000
MINERAL FUELS AND RELATED MATERIALS						
Coal, bituminous[e]		11,000	168,451 [3]	240,628 [3]	250,000	240,000
Natural gas	thousand cubic meters	7,910	12,200	7,900	7,900	8,000
Petroleum:						
Crude:[e]						
In gravimetric units		63,500	53,942 [3]	51,050 [3]	50,000	50,000
In volumetric units	42-gallon barrels	462,000	392,000	371,000	364,000 [r]	364,000
Refinery products:						
In gravimetric units		NA [r]	NA	NA	NA [r]	NA
In volumetric units	42-gallon barrels	NA [r]	NA	NA	NA [r]	NA

[e]Estimated data are rounded to no more than three significant digits; may not add to totals shown. [r]Revised. NA Not available. -- Zero.

[1]Table includes data available through May 6, 2013.

[2]In addition to the commodities listed, Georgia may have produced arsenic, barite, diatomite, iron ore, lead, perlite, and zeolites, but available information is inadequate to make reliable estimates of output.

[3]Reported figure.

TABLE 2
GEORGIA: STRUCTURE OF THE MINERAL INDUSTRY IN 2012

(Metric tons unless otherwise specified)

Commodity	Major operating companies and major equity owners[1]	Location or deposit names[1]	Annual capacity[e]
Arsenic:	Includes:	Location:	2,000 [2]
As content of ore	Racha mining and chemical plant	Lukhumi Mine, Ambrolauri region	
	Tsana mining and chemical plant	Tsana Mine, Lentekhi region	
Metal and compounds	Racha mining and chemical plant	Racha region	
	Tsana mining and chemical plant	Ts'ana region	
Barite	NA	Chordskoye deposit, Onis Raioni	70,000
Do.	RMG Copper (Rich Metals Group)	Madneuli Mine	NA
Barite-zinc ore	NA	Kvaisi Mine	NA
Bentonite	Includes:	Location:	200,000 [2]
	Askana LLC (Silver & Baryte Ores Mining Co., 97.7%)	Askanskoye Mine, Ozurget'i	
	NA	Gumbrskoye Mine, Gumbra region	
Cement	LLC Kartuli Cementi (LLC HeidelbergCement Caucasus Shared Services, 70%)	Kaspi and Rustavi	1,100,000
Do.	LLC SaqCementi (LLC HeidelbergCement Caucasus Shared Services, 75%)	Rustavi	500,000
Coal	Saqnakhshiri Ltd.	Akhaltsikhe, Tkibuli-Shaorskoye, and Tkvarchelskoye deposits in Akhalts'ikhis Raioni, Tkibuli, and Tqvarch'eli regions	300,000 [2]
Copper-gold ore	RMG Copper (Rich Metals Group)	Sakdrisi deposit, Bolnisi Region	12,000
Diatomite	NA	Kisatibskoye deposit, K'isat'ibi region	150,000
Ferroalloys:			
Ferromanganese	RMG Copper (Rich Metals Group)	Zestafoni ferroalloys plant, Zestap'onis Raioni	400,000
Silicomanganese	do.	do.	250,000
Manganese sinter	do.	do.	250,000
Gold	RMG Gold (Rich Metals Group)	Madneuli Mine	NA
Iron and steel, steel, rebar	Kutaisi metallurgical plant (OOO Eurasian Steel)	Kutaisi	100,000
Do.	Rustavi metallurgical plant (Georgian Steel Holding Group, 100%)	Rustavi	125,000
Do.	Geosteel (JSW Steel Ltd., 51%, and Georgian Steel Holding Group, 49%)	do.	180,000
Manganese ore	Chiaturamanganumi enterprise of Georgian Manganese Holding Limited LLC (Georgian American Alloys, Inc.)	Chiatura Mine	500,000
Nitrogen	JSC Azoti chemical plant	Rustavi	NA
Petroleum:			
Crude	Saknavtobi Oil and Gas Co. and most Georgian petroleum companies in joint ventures with Frontera Resources, Ioris Valley Oil & Gas Ltd., Ninotsminda Oil Co. Ltd., Georgian-British Oil Co. (GBOC), Anadarko Petroleum Corp., and GeoGeroil	About 60 wells that account for 98% of output in Mirzaani, Sup'sa, and Zemo T'elet'i regions	200,000 [2]
Do.	Canagro Ltd.	Sagarejo, eastern Georgia	NA

[e]Estimated; estimated data are rounded to no more than three significant digits. Do., do. Ditto. NA Not available.

[1]Many location names have changed since the breakup of the Soviet Union. Many enterprises, however, are still named or commonly referred to based on the former location name, which accounts for discrepancies in the names of enterprises and that of locations.

[2]Capacity estimate is the total for all enterprises that could produce that commodity.

THE MINERAL INDUSTRY OF GERMANY

By Steven T. Anderson

In 2012, Germany was a leading global exporter of industrial goods and services (including processed and fabricated mineral products). The country's mineral industry, however, depended heavily on imported mineral raw materials. Germany was the leading producer of lignite in the world, and essentially all the lignite consumed in the country was supplied by domestic production. Combustion of lignite accounted for 12.1% of total primary energy consumption in the country. Germany was dependent on imports of other mineral fuels for most of the remainder of its primary energy consumption, and combustion of petroleum and petroleum refinery products accounted for 33.1% of total primary energy consumption in Germany; that of natural gas, 21.6%; and that of anthracite and bituminous coal (hard coal), 12.2%. Renewable energy resources, such as wind power, accounted for 11.6% of total primary energy consumption; nuclear energy accounted for 7.9%; and other energy sources, including imported electricity, accounted for about 1.5%. Germany's metal processing sector relied on imports of metal ores and concentrates and reprocessing of metallic scrap and waste materials (both imported and produced domestically), because no metals were mined in sufficient amounts for metallurgical use in the country. Germany was also heavily reliant on imports of numerous industrial minerals and many refined metals (tables 1, 3, 4; AG Energiebilanzen e.V., 2013, p. 4; Bundesanstalt für Geowissenschaften und Rohstoffe, 2013, p. 12–32, 43–44, 109; Bundesministerium für Wirtschaft und Technologie, 2013, p. 5–12, 29–34).

In 2012, the country was estimated to have been the second-ranked producer of refined selenium in the world (and was estimated to have accounted for 32.5% of global production); the third-ranked producer of kaolin (13%), salt (6.6%), and refined lead (4%); the fifth-ranked producer of potash (9%) and bentonite (3.5%); the sixth-ranked producer of sulfur (5.3%); and the seventh-ranked producer of refined copper (3.4%) and crude steel (about 2.8%). Additionally, in 2011 (the latest year for which data were available), Germany either produced or was estimated to have produced between about 1% and probably not greater than 6% of the world's output of primary aluminum, barite, refined cadmium, cement, feldspar, natural gypsum, indium, crude iron, iron oxide pigments (natural), lime, magnesium compounds (as byproducts of potash mining), nitrogen (ammonia), silica (industrial sand and gravel), and zinc metal. It was estimated to have accounted for at least 5% of the world's total production capacity of alumina, fused aluminum oxide (abrasive), gallium (primary), graphite, magnesium metal (secondary), platinum metal (including secondary), rhenium metal (byproduct), strontium compounds, and titanium dioxide pigments (table 1; Bundesanstalt für Geowissenschaften und Rohstoffe, 2013, p. 13–16, 33–45, 125, 133, 136, 139, 142, 149, 153; International Copper Study Group, 2013, p. 15–17; International Lead and Zinc Study Group, 2013, p. 7–8, 44; U.S. Geological Survey, 2013, p. 6, 15, 17, 25, 35, 37, 39, 45, 55, 59, 71, 75, 79, 87, 93, 97, 113, 123, 135, 139, 143, 157, 159, 173; World Steel Association, 2013, p. 9).

The international competitiveness of the country's nonfuel mineral processing and fabrication sector relied primarily on such factors as a highly skilled labor force, research, development, and rapid assimilation of new technologies (including metal and other mineral materials recycling technologies), and on the development and maintenance of liberal trade relationships both within and outside the European Union (EU). Germany's position in the global mineral economy is predominantly that of a major consumer and processor of minerals, and this role continues to evolve as emerging economies grow and competition for mineral raw materials increases. In 2012, Germany was the world's third-ranked consumer of aluminum and copper, the fourth-ranked consumer of nickel and tin, and the fifth-ranked consumer of lead. Although the country remained one of the world's leading consumers of hard coal, crude petroleum, crude steel, and zinc, it was not among the top five consumers of these commodities (Bundesanstalt für Geowissenschaften und Rohstoffe, 2013, p. 6–26).

Minerals in the National Economy

In 2012, the total value of Germany's industrial output (including the value of output by the country's mineral industry, but not that of the construction sector) accounted for 23.3% ($793 billion[1]) of the gross domestic product (GDP) compared with 23.4% ($845 billion) in 2011. The value of marketed production by the country's metals processing sector (up to the foundry stage) accounted for about 3.4% ($117 billion) of the GDP compared with about 3.7% (about $135 billion) in 2011, and that of the minerals extraction sector (excluding coal) accounted for about 0.24% (about $8 billion) of the GDP compared with 0.24% ($8.7 billion) in 2011. Within the metals processing sector, the value of production of the nonferrous metals processing sector was $65 billion in 2012, and this was 8% lower than in 2011; slightly greater than $28 billion of the production by the nonferrous metals processing sector was marketed abroad (exported), and two-thirds of nonferrous exports went to other countries within the euro area (Bundesanstalt für Geowissenschaften und Rohstoffe, 2013, p. 11; International Monetary Fund, 2013; Statistisches Bundesamt, 2013a; 2013b, p. 8, 11).

One of Germany's competitive advantages is its vibrant industrial sector, which enables the country to maintain a highly skilled industrial workforce. In 2012, the number of employees in the country's metal processing sector increased to 242,571 from 238,826 in 2011, and in the nonfuel mining and quarrying

---

[1]Where necessary, values have been converted from euro area euros (€) to U.S. dollars (US$) at an annual average exchange rate of about €0.7187=US$1.00 for 2011 and €0.7775=US$1.00 for 2012. All values are nominal, at current prices, unless otherwise stated.

sector (including services), to 12,443 from 12,423 in 2011. In the coal mining sector the number of employees decreased to 32,449 from 37,730 in 2011; in the coking plant and petroleum processing sector, to 17,085 from 17,127 in 2011; and in the oil and natural gas extraction sector, to 3,019 from 3,076 in 2011 (Statistisches Bundesamt, 2013d).

According to the Bundesanstalt für Geowissenschaften und Rohstoffe, the 10 most valuable mineral raw materials produced in Germany in 2012 were, in decreasing order of value of production, lignite ($10.9 billion); natural gas (about $5.15 billion); crude petroleum (about $2.17 billion); construction sand and gravel ($1.8 billion); crushed stone, including chalk ($1.74 billion); potash ($1.5 billion); hard coal ($1.4 billion); rock salt and industrial brines, NaCl content ($745 million); dolomite, limestone, and marble (about $679 million); and kaolin ($656 million). Domestic production of mineral fuels appeared to have some economic benefit in addition to the simple value of output in that it helped to mitigate uncertainty in the domestic provision of electricity and distribution of imported mineral fuels (Bundesanstalt für Geowissenschaften und Rohstoffe, 2013, p. 13–16; International Monetary Fund, 2013).

According to Germany's Federal Statistics Office (DESTATiS), the country's estimated mineral trade balance in 2012 for all sectors of the mineral industry (including trade in intermediate mineral products, such as cement) was –$151 billion compared with a revised balance of –$158 billion in 2011 and –$118 billion in 2010. In 2012, Germany's mineral trade deficit decreased slightly compared with that of 2011 almost entirely because of a decrease in the total value of the country's mineral imports (to $234 billion from a revised value of about $241 billion), because there was only a slight decrease in the total value of its mineral exports (to about $82.7 billion from a revised value of $83 billion). The most costly mineral imports were crude petroleum and natural gas, and the most costly nonfuel mineral import was iron ore. The most valuable mineral export was petroleum refinery products, the most lucrative nonfuel mineral export was gold for commercial or industrial use, and the most valuable nonprecious metal export was iron and steel scrap. Germany's leading suppliers (by value) of mineral imports were Russia ($52 billion), Norway ($27 billion), and the Netherlands ($23 billion), mainly because these countries were significant sources of mineral fuels. Despite being the leading source of imported iron ore for Germany, Brazil ranked 10th as a supplier of mineral imports (overall), and the total value of Germany's mineral imports from Brazil in 2012 was about $4.5 billion (tables 3, 4; Bundesanstalt für Geowissenschaften und Rohstoffe, 2013, p. 17–21; International Monetary Fund, 2013; Statistisches Bundesamt, 2013c, p. 74–79).

Government Policies and Programs

Germany's main mining law is the Federal Mining Act (BGBl. IS. 1310), which was approved on August 13, 1980, and revised on December 9, 2006, through a slight revision to provisions of Article 11 (BGBl. IS. 2833). The country's production of some minerals (including gypsum and anhydrite, limestone and some other types of natural stone, peat, and some types of sand and gravel) was not directly regulated under the Federal Mining Act but was covered by a variety of other land-management and environmental regulations at both the Federal and the State levels. Also, the setup of the Federal Mines Inspectorate was not determined in the Federal Mining Act (although this inspectorate does enforce many of the regulations in the main mining law); the Federal Mines Inspectorate was established through Articles 83 and 84 of Germany's Constitution (Bundesministerium der Justiz, 2007, p. 1; Bundesanstalt für Geowissenschaften und Rohstoffe, 2013, p. 13–15; Bundesministerium für Wirtschaft und Technologie, 2013, p. 35–46).

The Environmental Impact Assessment Act (EIA Act) (BGBl. IS. 1757, 2797)—which was approved on June 25, 2005, and revised through slight changes to Article 2 (BGBl. IS. 3316) of the Act on December 21, 2006—is the environmental law that was most applicable to the mineral industry during 2012. This Act incorporates provisions of an older ordinance concerning the assessment of environmental impacts for mining projects (BGBl. IS. 1420), which was approved on July 13, 1990, and revised through slight changes to Article 8 (BGBl. IS. 2819) on December 9, 2006. The EIA Act also incorporates other older ordinances, such as one for the protection of groundwater against pollution caused by certain dangerous substances (BGBl. IS. 542), which was approved on March 18, 1997, and is still applicable to the use and disposal of many of the chemicals used in mining and mineral processing in Germany. The EIA Act requires environmental impact assessments for all domestic waste repositories created or used by the mineral industry. The Federal Mining Act actually stipulates how these repositories are to be constructed and operated (monitored) (Bundesministerium der Justiz, 2007, p. 30; Bundesanstalt für Geowissenschaften und Rohstoffe, 2013, p. 13–15; Bundesministerium für Wirtschaft und Technologie, 2013, p. 35–46).

Production

Data on mineral production are in table 1. Owing to fairly stable demand from the construction, engineering, and machinery sectors in 2012, production of most metals and industrial minerals in Germany remained about the same or decreased slightly compared with levels of production in 2011. An exception is the decrease of about 29% in production of rock salt and other salt brines, which was primarily owing to a supplier response in Germany to decreased demand for deicing compounds as a consequence of the mild winter weather conditions in Central and Western Europe at the beginning of 2012. The only increases of greater than 10% in the year-on-year production of other industrial minerals during this timeframe was a 70% increase in the production of evaporated salt and about a 14% increase in the production of marble and other calcareous dimension stone. Otherwise, annual production of manufactured phosphoric acid decreased by 37%; dolomite, 19%; fluorspar, about 17%; other clays, 14%; crude gravel (other than for construction), about 13%; and kaolin, 10%. Other than for salt, information was not available concerning the main causes of the significant changes in Germany's production of these individual industrial minerals (table 1; Bundesanstalt für Geowissenschaften und Rohstoffe, 2013, p. 42–45;

Bundesministerium für Wirtschaft und Technologie, 2013, p. 19–29; K+S Aktiengesellschaft, 2013, p. 88, 92–94, 119).

In 2012, increases of more than 10% in the annual production of metals compared with production in 2011 included a 47% increase in the production of direct-reduced iron; an estimated 17% increase in the production of tin alloys, and a 10% increase in the production of magnesium metal. Production of cobalt matte decreased by 26% and that of ferroalloys (other) decreased by 17% from 2011. Ferroalloys were used mostly in the production of specialty steels, and Germany's production of specialty steels decreased by 17.4% compared with that of 2011. Otherwise, information was not available concerning the main causes of the significant changes in Germany's production of these individual metals. The value of production of the entire nonferrous metals processing sector decreased by about 4% from that of 2011 owing to both an overall decrease in the sector's volume of production (in response to decreased demand, including from trading partners Italy and Spain) and a decrease in average prices during this timeframe (table 1; Bundesanstalt für Geowissenschaften und Rohstoffe, 2013, p. 11, 35–42).

With respect to mineral fuels in 2012, production of hard coal decreased by about 11% from that of 2011 owing to the continuing decrease in the hard coal mining subsidy, which led to the incremental closure of about one hard coal mine per year in the country during the past 4 years. The Government's program to eliminate the hard coal subsidy by the end of 2018 resulted in the closure of the Saar Mine on July 1, 2012. The 9% to 10% decrease in the production of natural gas was mostly as a result of decreasing reserves in the existing major gasfields in the country (table 1; Bundesanstalt für Geowissenschaften und Rohstoffe, 2013, p. 29–32; Bundesministerium für Wirtschaft und Technologie, 2013, p. 8–18; Landesamt für Bergbau, Energie und Geologie, 2013, p. 3, 9–10).

Structure of the Mineral Industry

Table 2 lists the major mineral industry facilities in Germany in 2012. Since the closure of the last metal mines in 1992, there has been no production of metallic ores with enough metal content for metallurgical use in Germany. The "iron ore" produced by Barbara Erzbergbau GmbH in 2012 had an estimated iron content of about 10.5% Fe (by volume), and was marketable only as an additive in construction materials (including as a cement additive). Many of the leading companies in the global metals processing sector owned and operated significant facilities in Germany, however. ThyssenKrupp AG (based in Duisburg, Germany) was the leading producer of crude steel in Germany and the 16th-ranked producer of crude steel in the world. Salzgitter AG (based in Salzgitter, Germany) was the second-ranked producer of crude steel in the country but was not among the top 40 producers of crude steel in the world. ArcelorMittal (based in Luxembourg) was the third-ranked producer of crude steel in Germany and the leading producer in the world. Aurubis was the leading producer of total refined copper in Germany and the EU, and Salzgitter held a 25% ownership interest in Aurubis. Aurubis was the second-ranked producer of copper cathodes in the world and the leading producer of secondary refined copper. Xstrata plc (based in Switzerland and registered in the United Kingdom) was the leading producer of zinc metal in Germany and the leading producer of mined zinc in the world. Norsk Hydro ASA of Norway was the second-ranked producer of aluminum in Germany and the fifth-ranked producer of primary aluminum in the world, and the company owned the largest single primary aluminum smelter in Germany (the Rheinwerk primary smelter at Neuss). Berzelius Metall GmbH (based in Stolberg, Germany) was the leading producer of primary lead in the country (table 2; Bundesanstalt für Geowissenschaften und Rohstoffe, 2013, p. 35–45; Stahlinstitut VDEh and Wirtschaftsvereinigung Stahl, 2013, World Steel Association, 2013, p. 8).

In October 2012, Advanced Metallurgical Group N.V. (AMG) of Belgium completed acquisition of 100% of Graphit Kropfmühl AG (the leading producer of graphite in Germany), including 100% ownership of RW Silicium GmbH (the only major producer of silicon metal in the country). In December 2012, AMG announced that it planned to achieve commercial production at Graphit Kropfmühl's reopened graphite mine (the only producer of natural graphite in Germany) by sometime in 2013 (table 2; Advanced Metallurgical Group N.V., 2013, p. 3–4, 12, 28–29).

Mineral Trade

Data on exports and imports of selected mineral commodities in 2012 are provided in tables 3 and 4, respectively. In 2012, the tonnage of Germany's imports of mineral fuels increased to 232 million metric tons (Mt) from about 227 Mt in 2011; metallic raw materials (including mineral ores and concentrates and other metallic raw materials, such as scrap metal), decreased to 62.4 Mt from 66.2 Mt in 2011; and industrial mineral raw materials decreased to 26 Mt from 30.7 Mt in 2011. Of the country's total volume of imports of mineral fuels in 2012, oil and natural gas imports accounted for 40% each, and the remaining 20% was almost entirely accounted for by imports of various types of coal, led by (in order of decreasing tonnage) steam coal, coking coal, and coke. Of the total volume of imports of metallic raw materials, imports of ores and concentrates accounted for 72%, and imports of iron ore accounted for about 88%. Scrap metal (all types) accounted for 14.4% of metallic raw materials, and iron and steel scrap accounted for the majority of scrap metal imports. Intermediate processed metals and refined metals accounted for slightly less than 10% of metallic raw materials. Imports of intermediate processed metals were led by imports of alumina and ferroalloys, and those of refined metals were led by imports of aluminum and refined copper. By far the leading industrial mineral imports (in decreasing order of tonnage) were gravel and stone pebbles, and these accounted for about 40% of the total volume of imports of industrial minerals (raw materials); rock salt, 8.8%; various types of sand, 8%; and limestone for cement manufacturing, 7.4% (tables 3, 4; Bundesanstalt für Geowissenschaften und Rohstoffe, 2013, p. 17–20).

By region, Germany sourced about 68% of the country's total imports of mineral raw materials from Europe (including Eastern Europe and Russia); South America, 11.5%; Africa, about 8.8%; North America, 5.3%; Asia, 3.9%; Australia-Oceania, 1.8%; and the rest of the world, less than 1%. Producers in Europe were the leading source of imports of mineral fuels and accounted for about 95% of Germany's imports of industrial minerals; those in South America accounted for the greatest share of the country's imports of metallic raw materials, especially of metal ores and concentrates (table 4; Bundesanstalt für Geowissenschaften und Rohstoffe, 2013, p. 19–21).

For its iron and steel sector in 2012, Germany imported 100% of the iron ore and concentrate that it used (about 39 Mt), and Brazil was the supplier of about 61% of these imports; it imported about 5.7 Mt of iron and steel scrap, which was supplied mainly by other European countries. For its specialty steel sector, Germany imported all the ores and concentrates that it used to produce some ferroalloys, recycled domestic and imported scrap containing ferroalloys, and directly imported ferroalloys; these ferroalloy imports included about 393,000 t of ferrochromium (of which South Africa was the leading supplier and accounted for about 54% of Germany's imports); about 248,000 t of ferrosilicon (of which the leading suppliers were Norway, 26%, and Iceland, about 16%); about 233,000 t of silicon metal (Norway, about 39%); 207,000 t of ferrosilicomanganese (Norway, about 27%, and India, 23%); 193,000 t of ferromanganese (Norway, about 29%, and South Africa, 28%); about 189,000 t of chromite ore and concentrate (South Africa, about 68%); and about 140,000 t of ferronickel (Ukraine, 63%) (table 4; Bundesanstalt für Geowissenschaften und Rohstoffe, 2013, p. 35–36).

For the country's nonferrous metals processing sector in 2012, some notable imports and sources included about 2.8 Mt of bauxite ore and concentrate (of which Guinea was the leading supplier, accounting for 77% of Germany's imports); 1.2 Mt of copper in ores and concentrates (Peru, 27%, and Chile, 21%); 652,000 t of titanium ore and concentrate (Norway, about 34%, and South Africa, about 23%); 325,000 t of zinc in ores and concentrates (Australia, about 44%); and about 228,000 t of lead in ores and concentrates (Sweden, about 26%). Some other important imports of metals (for alternative energy technologies, the automotive sector, and other industrial sectors) included about 43 t of palladium (of which Belgium and Russia were the leading suppliers, accounting for about 27% and 26%, respectively, of Germany's imports); 27 t of platinum (South Africa, about 37%); and about 4,160 t of rare-earth compounds, not including cerium compounds (the United States, about 53%) (table 4; Bundesanstalt für Geowissenschaften und Rohstoffe, 2013, p. 37–42).

In 2012, Germany exported about $50.5 billion of mineral raw materials, and the value of the country's exports of metallic raw materials accounted for 67.9% ($34.3 billion) of the total export value; energy raw materials (including electricity), 26.3% ($13.3 billion); and industrial minerals and other mineral raw materials that were not fuels and not metals, 5.8% (about $2.9 billion). Precious metals, including platinum-group metals (PGMs), accounted for 40% of the export value of metallic raw materials. In 2012, Germany's leading exports (by volume) of nonprecious processed metals and destinations in 2012 included 9.66 Mt of scrap iron and steel, of which about 19% was exported to the Netherlands and about 18% went to Italy; 2.34 Mt of ash and residue containing iron, of which 48% went to France; 969,000 t of aluminum scrap, of which 19% went to Italy and 15% went to Austria; 588,000 t of copper scrap, of which about 38% went to China; 455,000 t of aluminum hydroxide, of which 21% went to the Netherlands and about 18% went to the United States; about 369,000 t of aluminum oxide (alumina), of which 33% went to France; 277,000 t of refined copper (not alloyed), of which about 25% went to China; and 230,000 t of secondary aluminum (including alloys), of which about 22% to Austria and 19% went to France (table 3; Bundesanstalt für Geowissenschaften und Rohstoffe, 2013, p. 21; International Monetary Fund, 2013).

In order to protect private company information, the data on Germany's exports of potash are not available from the Government, so these data are not listed in table 3. In 2012, sales to other countries in Europe accounted for 35.7% of the total volume of sales by the potash and magnesium products unit of K+S Aktiengesellschaft; Asia, 21.6%; South America, 21.3%; Germany, 14.9%; Africa and Oceania, 4.1%; and North America, 2.4%. Although most of the production by this unit of K+S Aktiengesellschaft took place within Germany, the company also owned potash production and processing facilities in Canada, France, and possibly other countries. Thus, information was not available with regard to exactly how much of these sales actually represented Germany's exports (table 3; Bundesanstalt für Geowissenschaften und Rohstoffe, 2013, p. 21; K+S Aktiengesellschaft, 2013, p. 58–59).

Based on sales data for the first 9 months of 2012, Germany's mining-equipment manufacturing sector was on pace to increase sales of mining equipment by about 20% compared with that of 2011, despite receiving about one-half as many orders from China. The decrease in demand in China was offset by fourfold increases in demand for German mining equipment in Australia and Chile, and demand also increased significantly in other countries of Latin America (most notably in Colombia and Peru), Russia, and the United States. Because coal mining equipment was the leading type of mining equipment that Germany exported to China, the decrease in demand for mining equipment was estimated to have been mostly owing to decreased steel demand in the country, which negatively affected coking coal production in China. In 2012, China's increasing ability to produce its own mining equipment and the overall slowdown in the economic growth of the country also contributed to lower demand for Germany's exports of mining equipment compared with that of 2011 (Walker 2013, p. 4–5).

Commodity Review

Metals

Iron and Steel.—In 2012, global apparent steel use decreased by about 5.6% compared with that of 2011, including in Germany (by 8.8%) and in the EU as a whole (by about 5.5%); companies reported that demand for crude steel in Germany and the EU decreased steadily during the year. Production of crude steel in Germany decreased only by about 3.7% from that of 2011, however, and the country made up for the lower decrease in output (relative to the decrease in demand) with a decrease of 6.8% in imports of crude steel during this timeframe. In 2012, 43% of the total production of crude steel in the country was from secondary materials (recycled scrap) (tables 1, 4;

Bundesanstalt für Geowissenschaften und Rohstoffe, 2013, p. 35, 78; World Steel Association, 2013, p. 16).

Industrial Minerals

Graphite.—According to the Government, Graphit Kropfmühl had reopened the Kropfmühl Mine (Germany's only graphite mine) on June 21, 2012, after the company had decided to close the mine at the end of 2005 because of poor market conditions. In the interim, Germany continued to produce refined graphite from imports of crude graphite, but this production has not been included in table 1 to avoid double counting production of mined graphite in the countries that supply crude graphite to Germany. In 2012, Graphit Kropfmühl had reportedly produced some test samples of graphite from the mine before the company was acquired by AMG. This test production was not reported by the Government, possibly because it was too small to be marketable, or the production data could have been withheld to protect the single producer's private company information. At the end of 2012, AMG expected to ramp up production of graphite at the Kropfmühl Mine to some undisclosed level by the end of 2013 (table 1; Bundesanstalt für Geowissenschaften und Rohstoffe, 2013, p. 44; Advanced Metallurgical Group N.V. 2013, p. 28–29)

Mineral Fuels

Coal.—At the beginning of 2012, there was hard coal mining in the Ibbenbüren, the Ruhr, and the Saar coalfields, but it was uneconomical without a subsidy. The Government's program to eliminate the hard coal subsidy by the end of 2018 resulted in the closure of the Saar Mine on July 1, 2012, so there was no longer any production in Saarland by the end of the year. The West Mine (in the Ruhr coalfields) closed at the end of 2012, and the last of Germany's hard coal mines was expected to close in 2018. An economic consequence of decreasing production of hard coal domestically is that Germany will become more dependent on imported coke and coking coal from hard coal mines outside of the country, and this will subject sectors of the mineral industry, such as steel manufacturing, and other sectors of the economy to greater cost uncertainty. In 2012, Germany's imports of coke (from hard coal) and coking coal decreased by about 25% and 2%, respectively, at least partially owing to decreased demand by the steel manufacturing sector (Gesamtverbands Steinkohle e.V., 2011, p. 13–14, 16–32; Bundesanstalt für Geowissenschaften und Rohstoffe, 2013, p. 8, 32, 62, 67, 70–71, 118–120; Bundesministerium für Wirtschaft und Technologie, 2013, p. 11–12).

Reserves and Resources

At the end of 2012, Germany's reserves of lignite were estimated to be 40.4 billion metric tons (Gt) compared with about 40.5 Gt at yearend 2011; the country's proven and probable reserves of natural gas were estimated to have decreased to about 123 billion cubic meters compared with about 133 billion cubic meters in 2011; and its proven and probable reserves of crude petroleum were estimated to be 242 million barrels (Mbbl) (converted from a reported figure of about 32.5 Mt) compared with about 259 Mbbl (35.3 Mt) in 2011. At the end of 2012, K+S AG estimated company reserves of potash (K_2O content) in Germany to be about 132 Mt compared with 137 Mt at yearend 2011, its reserves of salt in Germany to be 115 Mt compared with about 118 Mt at yearend 2011, and reserves of kieserite (usable magnesium compounds contained in the company's potash deposits in Germany) to be about 109 Mt compared with 110 Mt at yearend 2011. Reliable information concerning additional reserves of industrial minerals in the country was not available (Bundesanstalt für Geowissenschaften und Rohstoffe, 2012, p. 30–37, 105, 108, 116; 2013, p. 28–33, 110, 113, 121; Bundesministerium für Wirtschaft und Technologie, 2013, p. 15–17, 151; K+S Aktiengesellschaft, 2012, p. 214–215; 2013, p. 216–217; Landesamt für Bergbau, Energie und Geologie, 2013, p. 41–43).

Assuming that the phaseout of the Government's subsidy of hard coal production will proceed according to the schedule as it stood in 2012, this policy will gradually increase the volumes of hard coal resources in Germany that are not economical to mine until the end of 2018, when hard coal production is expected to cease (owing to the end of the subsidy). In 2012, the Government had an estimate of how much hard coal would be produced in 2013 and through 2018 (conditional on the subsidies allowing it to be profitable to do so), and considered this estimated future production to represent a type of reserve figure for hard coal in Germany. Under this definition, the country's economically exploitable reserves of hard coal (in the presence of planned subsidies) were estimated to be about 36 Mt at the end of 2012 compared with about 48 Mt (of hard coal to be produced in 2012 and through 2018) at yearend 2011. The country's annual production of hard coal was expected to be the primary reason for annual decreases in these reserves until the end of 2018. In 2012, the country's hard coal resources remained at approximately 83 Gt (Bundesanstalt für Geowissenschaften und Rohstoffe, 2012, p. 111; 2013, p. 31, 115).

Outlook

To eliminate nuclear power gradually from Germany's energy mix by 2022 and still be on track to reduce greenhouse gas emissions by 80% in 2050 (compared with the level of emissions in 2010), about 38% of the electricity generated in Germany in 2030 is projected to come from renewable energy resources (compared with 20% in 2011, according to preliminary data); 23% from lignite (25% in 2011); about 20% from natural gas (about 14% in 2011); 15% from hard coal (about 19% in 2011); about 4% from heating oil, pumped storage, and other (5% in 2011); and 0% from nuclear power (about 17% in 2011). In energy equivalents, the direct implications of the realization of this scenario could be that Germany would consume about 50% more natural gas in the generation of electricity in 2030 than in 2011, about 7% less lignite, and about 19% less hard coal. In 2030, the country's entire hard coal demand would have to be satisfied with imports if the elimination of the hard coal subsidy results in zero production by 2030 (as expected), and approximately all the increase in natural gas consumption would also have to be satisfied through increased imports of natural gas. This projected 2030 energy mix would require approximately a 96% increase in electrical power generated from renewable energy resources compared with that of 2011. Indirect implications of increased consumption of renewable energy resources for the mineral industry could include increases in consumption of minerals used in wind turbines (including rare earths), in solar cells (including silicon and silver), and in other renewable energy technologies (Gesamtverbands Steinkohle e.V., 2011, p. 22–26; AG Energiebilanzen e.V., 2013, p. 25–27; Bundesministerium für Wirtschaft und Technologie, 2013, p. 4–5, 8, 16–18, 38, 49–51).

RWE Aktiengesellschaft (RWE) expected that the company's startup of its lignite-fired powerplant in Neurath (near Cologne) in 2012 would increase demand for (and the company's production of) lignite in Germany in 2013 and beyond, but RWE also decommissioned the last of its older lignite-fired powerplants at the end of 2012. Although this new powerplant at Neurath has a greater capacity to generate electricity from lignite than the decommissioned older one, the new powerplants are more efficient than the older ones. So, the net effect on the demand for lignite and the company's production in 2013 and beyond was uncertain. The Government's program to eliminate the hard coal subsidy by the end of 2018 left only three mines producing hard coal in Germany in 2013 (table 1; Gesamtverbands Steinkohle e.V., 2011, p. 14–18; Bundesanstalt für Geowissenschaften und Rohstoffe, 2012, p. 12–14, 32–35; RWE Aktiengesellschaft, 2013, p. 22, 33, 53, 62, 70, 120).

Future levels of production of fertilizer materials (such as potash) in Germany are expected to vary more with respect to fluctuations in demand outside of Europe than within Europe. Expected increases in the global population and in the level of prosperity in emerging market economies, including those of Latin America and Southeast Asia, are likely to increase food consumption and thus the intensity of land cultivation. Also, expected increases in meat consumption will likely drive the need for animal feed and therefore increase demand for almost all of Germany's fertilizer products even more than just an increase in the total level of food consumption. In 2013, however, K+S AG expects global demand for potash to increase from that in 2012 because there were lingering supply contract negotiations that constrained potash purchases in 2012. The company also forecast that most of the increase in potash sales volume would occur in the big contract sales markets of China and India, however, which make up a small percentage of the sales market for German production of potash, so Germany could produce approximately the same amount of potash in 2013 as in 2012 and possibly increase production by about 1% in 2014 (tables 1, 2; K+S Aktiengesellschaft, 2013, p. 132–145).

References Cited

Advanced Metallurgical Group N.V., 2013, Annual report 2012: Amsterdam, Netherlands, Advanced Metallurgical Group N.V., March 22, 158 p.

AG Energiebilanzen e.V., 2013, Energieverbrauch in Deutschland im Jahr 2012: Berlin, Germany, AG Energiebilanzen e.V., March, 40 p.

Bundesanstalt für Geowissenschaften und Rohstoffe, 2012, Rohstoffsituation 2011: Hannover, Germany, Bundesanstalt für Geowissenschaften und Rohstoffe, December, 153 p.

Bundesanstalt für Geowissenschaften und Rohstoffe, 2013, Rohstoffsituation 2012: Hannover, Germany, Bundesanstalt für Geowissenschaften und Rohstoffe, November, 155 p.

Bundesministerium der Justiz, 2007, Bundesberggesetz (BBergG): Berlin, Germany, Bundesministerium der Justiz, April 29, 95 p.

Bundesministerium für Wirtschaft und Technologie, 2013, Der Bergbau in der Bundesrepublik Deutschland 2012: Berlin, Germany, Bundesministerium für Wirtschaft und Technologie, November, 168 p.

Gesamtverbands Steinkohle e.V., 2011, Steinkohle jahresbericht 2011: Essen, Germany, Gesamtverbands Steinkohle e.V., October, 65 p.

International Copper Study Group, 2013, Copper Bulletin, v. 20, no. 3, March, 54 p.

International Lead and Zinc Study Group, 2013, World lead and zinc statistics: Lisbon, Portugal, International Lead and Zinc Study Group monthly bulletin, v. 53, no. 6, June, 76 p.

International Monetary Fund, 2013, Germany, in World economic outlook database: International Monetary Fund, October. (Accessed February 17, 2014, via http://www.imf.org/external/pubs/ft/weo/2013/02/weodata/index.aspx.)

K+S Aktiengesellschaft, 2012, Financial report 2011: Kassel, Germany, K+S Aktiengesellschaft, March 2, 226 p.

K+S Aktiengesellschaft, 2013, Financial report 2012: Kassel, Germany, K+S Aktiengesellschaft, March 14, 228 p.

Landesamt für Bergbau, Energie und Geologie, 2013, Erdöl und Erdgas in der Bundesrepublik Deutschland 2012: Hannover, Germany, Landesamt für Bergbau, Energie und Geologie, May, 62 p.

RWE Aktiengesellschaft, 2013, Annual report 2012: Essen, Germany, RWE Aktiengesellschaft, March 5, 234 p.

Stahlinstitut VDEh and Wirtschaftsvereinigung Stahl, 2013, Die größten stahlerzeuger in Deutschland 2012: Düsseldorf, Germany, Stahlinstitut VDEh and Wirtschaftsvereinigung Stahl, February, 1 p.

Statistisches Bundesamt, 2013a, Bruttowertschöpfung nach ausgewählten wirtschaftsbereichen—In jeweiligen preisen—Mrd. EUR: Wiesbaden, Germany, Statistisches Bundesamt. (Accessed August 14, 2013, at https://www.destatis.de/DE/ZahlenFakten/Indikatoren/Konjunkturindikatoren/VolkswirtschaftlicheGesamtrechnungen/vgr210.html.)

Statistisches Bundesamt, 2013b, Fachserie 4, reihe 3.1—Produzierendes gewerbe—Produktion des verarbeitenden gewerbes sowie des bergbaus und der gewinnung von steinen und erden 2012: Wiesbaden, Germany, Statistisches Bundesamt, April 16, 305 p.

Statistisches Bundesamt, 2013c, Fachserie 7, reihe 1—Außenhandel—Zusammenfassende übersichten für den außenhandel (vorläufige ergebnisse) 2012: Wiesbaden, Germany, Statistisches Bundesamt, June 3, 126 p.

Statistisches Bundesamt, 2013d, Monatsbericht im verarbeitenden gewerbe—Beschäftigte und umsatz der betriebe im verarbeitenden gewerbe—Deutschland, Jahre, Wirtschaftszweige (WZ2008 2-/3-/4-Steller): Wiesbaden, Germany, Statistisches Bundesamt. (Accessed August 14, 2013, via https://www-genesis.destatis.de/genesis/online/data.)

U.S. Geological Survey, 2013, Mineral commodity summaries 2013: U.S. Geological Survey, 198 p.

Walker, Simon, 2013, Best of Germany 2013—A supplement to Engineering & Mining Journal: Engineering & Mining Journal, July–August, 64 p.

World Steel Association, 2013, World steel in figures 2013: Brussels, Belgium, World Steel Association, May 30, 30 p.

TABLE 1
GERMANY: PRODUCTION OF MINERAL COMMODITIES[1]

(Thousand metric tons unless otherwise specified)

Commodity		2008	2009	2010	2011	2012
METALS						
Aluminum:						
Alumina		819	638	973	950 [e]	967
Aluminum hydroxide, Al_2O_3 equivalent		1,395	1,154	1,485	1,405	1,364
Metal:						
Primary		606	292	402	432	410
Secondary		721	561	611	634	635
Total[2]		1,327	853	1,014	1,067	1,045
Cadmium, metal, refinery[e]	metric tons	420	278	290	300	300
Cobalt, matte, including shavings and scrap	do.	913	654	829	671	497
Copper, metal:						
Smelter:						
Primary		295	286	379	346 [r]	340 [p]
Secondary		293	248	212	218 [r]	217 [p]
Total[2]		588	534	591	564 [r]	556 [p]
Refined:						
Primary		300	290	402	401 [r]	390 [p]
Secondary		389	379	302	308 [r]	296 [p]
Total[2]		690	669	704	709 [r]	686 [p]
Gallium, crude[e]	metric tons	25	20	30	30	30
Gold, metal, refined, including secondary	kilograms	NA	204,766 [3]	44,100 [e]	50,682	53,476
Indium, refined[e]	metric tons	10	10	10	10	10
Iron and steel:						
Ore, run of mine:[4]						
Gross weight		455	364	390	489 [r]	448
Fe content		48	38	41	51 [r]	47
Metal:						
Pig iron		29,111	20,104	28,560	27,943 [r]	27,048
Direct-reduced iron		520	380	450	380	560
Ferroalloys:						
Ferrochromium[e]	metric tons	26,960 [5]	13,667 [5]	18,300 [r]	18,500 [r]	17,800
Other	do.	5,000 [e]	6,336	9,200 [e]	9,985 [r]	8,248
Steel, crude		45,833	32,671	43,830	44,284 [r]	42,661
Semimanufactures		39,805	29,041	36,827	37,933	36,495
Lead, metal, refined:						
Primary		113	105	125	136 [r]	134 [e]
Secondary		302	286	279 [r]	293 [r]	290 [e]
Total		415	391	404 [r]	429 [r]	424 [e]
Magnesium, metal including castings		30	12	15	15	16
Platinum-group metals, metal, refined	metric tons	122	110 [e]	100 [e]	50	54
Selenium, contained metal[e]	do.	650	600	650	700 [r]	650
Silicon, metal	do.	29,092 [r]	27,620 [r]	30,105 [r]	30,134 [r]	29,000 [e]
Silver, metal, refined, including secondary	do.	1,783	1,616	1,768	1,886	1,753
Tin, alloys[e]	do.	6,114 [5]	5,003 [5]	7,000 [r]	6,000	7,000
Zinc, metal:						
Primary		211	134	144	142	139 [e]
Secondary		81	19	21	28 [r]	30 [e]
Total		292	153	165	170 [r]	169 [e]
INDUSTRIAL MINERALS						
Abrasives, manufactured:						
Corundum		95	49	83	90	83
Fused aluminum oxide, crude[e]		20	20	20	20	20
Silicon carbide[e]		20	20	20	20	20
Aluminum salt slag, Al_2O_3 equivalent[e]		200	150	200	200	200

See footnotes at end of table.

TABLE 1—Continued
GERMANY: PRODUCTION OF MINERAL COMMODITIES[1]

(Thousand metric tons unless otherwise specified)

Commodity		2008	2009	2010	2011	2012
INDUSTRIAL MINERALS—Continued						
Barite, marketable (contained $BaSO_4$)		79	46	56	55 [r]	52
Boron compounds, manufactured, including boric acid and oxide		204	130	163	157	149
Bromine compounds, including oxide[e]	metric tons	1,680	985	1,500	1,600 [r]	1,600
Cement:						
Clinker, intended for market		25,366	23,232	22,996	24,775	24,581
Hydraulic		33,581	30,441	29,915	33,540	32,432
Chalk, natural, including ground[e]		1,495 [5]	1,322 [5]	1,350	1,400	1,450
Clays, natural:						
Bentonite		414	326	363	375 [r]	366
Ceramic and refractory clays		4,229	3,711	3,978	4,027 [r]	4,399
Of which, fire clay and chamotte		267	250 [e]	246	253	270 [e]
Kaolin, marketable		3,622	4,514	4,560 [r]	4,899 [r]	4,399
Other, unspecified		182	193	198	200 [e]	172
Dolomite, neither burnt nor sintered		850 [e]	800 [e]	792	622	504
Feldspar, all uses[6]		3,616	3,698	5,203	5,483 [r]	5,321
Of which, feldspar for industrial uses[e]		161 [5]	201	203 [r]	218	205
Fluorspar, acid-grade		49	50	59	66 [r]	54
Gypsum and anhydrite:						
Natural		2,112	1,898	1,822	2,021	1,949
Byproduct of flue-gas desulfurization[e]		6,900	6,600	6,320 [r]	6,780 [r]	7,010
Lime, quicklime, dead-burned dolomite		7,313	5,945	6,856	7,113	6,672
Magnesium compounds, byproduct of potash mining		1,418	811	1,310	1,348 [r]	1,372
Mullite, synthetic[e]		15	15	15	15	15
Nitrogen, N content of ammonia		2,819	2,363	2,677	2,821 [r]	2,823
Peat, horticultural use	thousand cubic meters	7,629	8,364	7,759	7,911	8,205
Phosphoric acid, manufactured, P_2O_5 content		32	20	21	20	12
Pigments, iron oxide (including synthetic iron oxide)		251	209	234	223	204
Potash, K_2O content:						
Crude		4,046	2,208	3,630	3,827 [r]	3,767
Marketable		3,280	1,825	3,024	3,215 [r]	3,149
Salt, NaCl content, marketable:						
Evaporated salt, including marine salt		580	325	322	329 [r]	559
Industrial brines		9,084	9,798	8,752	8,066 [r]	7,506
Rock salt and other brines		6,169	8,816	10,602	9,048 [r]	6,381
Total[2]		15,833	18,939	19,676	17,432 [r]	14,445
Siliceous earth, marketable		52	43	49	53 [r]	50
Soda ash (Na_2CO_3), manufactured		2,715	2,291	2,539	2,668	2,627
Stone, sand and gravel:						
Stone, crude:						
Dimension, including partially worked		400 [r, e]	380	425	467	477
Of which, marble and other calcareous stone		250 [r, e]	247 [r]	287 [r]	314 [r]	356
Crushed, not including chalk		155,527 [r]	156,752 [r]	149,463	164,487	154,020
Dolomite and limestone, not for cement manufacture		21,300	19,000	18,000	18,400 [r]	17,600
Gravel, natural:						
Construction gravel		63,962	70,136	67,822	76,191	72,615
Crude, including flint and pebbles		12,631	10,442	9,693	11,043	9,639
Other gravel, including quartzite		11,911	NA	NA	NA	NA
Sand, natural:						
Construction sand		56,866	66,010	63,962	72,394	67,852
Silica sand, including glass sand and quartz sand		8,186	6,453	7,234	7,770	7,498
Other, including from granite and pegmatite		13,416	NA	NA	NA	NA
Total, sand and gravel		166,972	153,041	148,711	167,398	157,604
Strontium carbonate, manufactured[e]		80	100 [r]	120 [r]	130 [r]	120

See footnotes at end of table.

TABLE 1—Continued
GERMANY: PRODUCTION OF MINERAL COMMODITIES[1]

(Thousand metric tons unless otherwise specified)

Commodity		2008	2009	2010	2011	2012
INDUSTRIAL MINERALS—Continued						
Sulfur:						
Marketable		1,030	927	832	875 r	798
Byproduct:						
Metallurgy		2,458	2,137	2,266	2,394	2,373
Natural gas and petroleum		1,709	1,623	1,447	1,514	1,445
Total		4,167	3,760	3,713	3,908	3,818
MINERAL FUELS AND RELATED MATERIALS						
Carbon black		607	494	684	908	923
Coal:						
Anthracite and bituminous, marketable		17,171	13,766	12,900	12,059	10,770
Lignite		175,313	169,857	169,403	176,502	185,432
Coke:						
Of anthracite and bituminous coal		8,246	6,771	8,241	7,990 r	8,050
Of lignite		177	153	176	171	170
Fuel briquets of lignite		1,631	1,959	2,024	2,136	1,910
Gas:						
Manufactured:						
Blast furnace e	million cubic meters	9	6	9	9	9
Coke oven	do.	969	718	951	922 r, e	929 e
Total e	do.	978	724	960	931 r	938
Natural:						
Associated (byproduct of crude petroleum)	do.	100	90	81	80	78
Gross (non-associated)	do.	16,524 r	15,464	13,584	12,873	11,706
Marketable (dry or net)	do.	15,377	14,380	12,571	11,799 r	10,660
Petroleum:[7]						
Crude	thousand 42-gallon barrels	22,400	20,500	18,400	19,600	19,200
Refinery products:						
Liquefied petroleum gas	do.	36,390	33,490	33,180	32,860	33,010
Distillate fuel oil	do.	370,000	360,000	340,000	330,000	340,000
Residual fuel oil	do.	67,500	55,600	41,600	42,400	44,200
Gasoline, including aviation	do.	200,000	200,000	180,000	180,000	170,000
Kerosene and jet fuel	do.	36,500	35,200	37,400	38,100	40,000
Naphtha	do.	87,000	75,000	72,000	70,000	70,000
Refinery gas	do.	47,800	44,500	44,500	45,100	44,000
Bitumen, bituminous mixtures, and other residues	do.	33,900	34,300	32,800	34,600	33,000
Lubricants and miscellaneous oils	do.	17,000	16,000	18,000	17,000	17,000
Petroleum coke	do.	11,500	10,900	11,500	10,100	9,970
Mineral jelly, waxes, and paraffins	do.	1,300	800	900	900	1,000
Other	do.	8,290	6,040	8,630	6,590	8,330
Total e	do.	917,000	872,000	821,000	808,000	811,000
Uranium concentrate, U_3O_8 content		--	--	9	60 r	59

eEstimated; estimated data are rounded to no more than three significant digits; may not add to totals shown. PPreliminary. rRevised. do. Ditto.
NA Not available. -- Zero.
[1]Table includes data available through February 21, 2014.
[2]Data may not add to totals shown.
[3]Could include production in 2008.
[4]Iron ore is used domestically as an additive in cement and other construction materials but is of too low a grade to be used in the steel industry.
[5]Reported figure.
[6]All uses include use as gravel for road construction, and industrial uses include use in the manufacturing of ceramics.
[7]All figures through 2012 were converted to barrels from those reported in metric tons according to data from Mineralölwirtschaftsverband e.V., 2013, Jahresbericht—Mineralöl-Zahlen, 2012: Berlin, Germany, Mineralölwirtschaftsverband e.V., July, p. 48 and 79, and reflect the significant digits of the conversion factors.

TABLE 2
GERMANY: STRUCTURE OF THE MINERAL INDUSTRY IN 2012[1]

(Thousand metric tons unless otherwise specified)

Commodity	Major operating companies and major equity owners	Location of main facilities	Annual capacity
Abrasives (silicon carbide)	ESK-SiC GmbH	Plant at Grefrath, Cologne	36
Alumina	Almatis GmbH (Dubai International Capital LLC)	Plant at Ludwigshafen	NA
Do.	Nabaltec AG	Plant at Schwandorf	120
Do.	Aluminium Oxid Stade GmbH (DADCO Alumina & Chemicals Ltd., 100%)	Plant at Stade	1,050
Do.	Martinswerk GmbH (Albemarle Corp., 100%)	Plant at Bergheim	350
Do.	Rio Tinto Alcan (Rio Tinto plc, 100%)	Plant at Teutschenthal	17
Alumina, fused	Treibacher Schleifmittel GmbH (Imerys S.A., 100%)	Plant at Zschornewitz	NA
Aluminum	Hydro Aluminium Deutschland GmbH (Norsk Hydro ASA, 100%)	Rheinwerk primary smelter at Neuss	235
Do.	Metallhüttenwerke Bruch GmbH	Secondary foundry alloy plant at Dortmund; secondary cast alloy plants at Asperg and Bad Saeckingen	110
Do.	Aleris Recycling (German Works) GmbH (Aleris Corp., 100%)	Secondary smelters: Erftwerk at Grevenbroich, Innwerk at Toeging am Inn, and Neckarwerk at Deizisau	320
Do.	TRIMET Aluminium AG	Primary smelter at Essen-Borbeck	175 [e]
Do.	do.	Recycling plant and secondary smelter at Gelsenkirchen	160 [e]
Do.	do.	Recycling plant and secondary smelter at Harzgerode	40
Do.	Hamburger Aluminium-Werke GmbH (TRIMET Aluminium AG, 100%)	Primary smelter at Hamburg	133
Do.	Aluminiumwerk Voerde Aluminium GmbH (Klesch & Company Ltd., 100%)	Primary smelter at Voerde, North Rhine-Westphalia	130
Aluminum, hot-rolled products	Aluminium Norf GmbH [Novelis Inc. (Hindalco Industries Ltd., 100%), 50%, and Hydro Aluminium Deutschland GmbH, 50%]	Lippenwerk at Luenen (secondary) and rolling mill at Neuss	1,500
Aluminum salt slag	Alsa Technologies GmbH (Agor AG, 100%)	Plants at Hannover, Luenen, and Toeging	380
Do.	K+S Entsorgung GmbH (K+S Aktiengesellschaft, 100%)	REKAL plant at Sigmundshall	100
Arsenic, metal metric tons	PPM Pure Metals GmbH[2] (Recylex S.A., 100%)	Plant at Langelsheim	5
Do. do.	Reinstmetalle Osterwieck GmbH (PPM Pure Metals GmbH,[2] 100%)	Plant at Osterwieck	NA
Barite	Sachtleben Bergbau GmbH	Clara Mine in the Black Forest and plant at Wolfach, and Dreislar Mine at Medebach-Dreislar	87
Do.	Deutsche Baryt-Industrie Dr. Rudolf Alberti GmbH & Co. KG (Sachtleben Bergbau GmbH, 75%, and other private, 25%)	Wolkenhügel Mine[3] in the Harz Mountains and plant at Bad Lauterberg	50
Bentonite	Süd-Chemie AG (Clariant International Ltd., 100%)	Mining near Gammelsdorf, Bavaria, and plants at Duisburg, Heufeld, and Moosburg	500
Do.	S&B Industrial Minerals GmbH (S&B Industrial Minerals S.A., 100%)	Mining in region between Landshut and Mainburg, Bavaria	400
Do.	do.	Stollberg plant at Oberhausen	200 [e]
Do.	do.	Plant at Neuss	50
Do.	Kärlicher Ton- und Schamotte-Werke Mannheim & Co. KG (KTS)	Quarry at Muelheim-Kaerlich	50
Cadmium, metal:			
Primary (byproduct)	Metaleurop Zinkbetrieb GmbH & Co. KG (Xstrata plc, 100%)	Nordenham smelter, near Bremerhaven	160
Secondary	Accurec Recycling GmbH (I-met GmbH, 100%)	Battery recycling plant at Mülheim an der Ruhr	NA
Calcium carbonate, natural, ground	Alpha Calcit Fullstoff GmbH & Co. KG	Plant at Cologne	250
Do.	Omya GmbH (Omya AG, 100%)	Plants at Emden and Giengen-Burgberg	2,250
Do.	Omya Weil GmbH (Omya AG, 100%)	Plant at Weil am Rhein	NA
Do.	Eduard Merkle GmbH & Co. KG (Omya AG, 100%)	Plant at Blaubeuren-Altental	NA
Calcium carbonate, natural, including chalk	Vereinigte Kreidewerke Dammann KG (Omya AG, 100%)	Plants at Laegerdorf and Soehlde	500
Do.	Kreidewerk Rügen GmbH (Omya AG, 100%)	Quarries and plant at Sassnitz, on Ruegen Island	NA

See footnotes at end of table.

TABLE 2—Continued
GERMANY: STRUCTURE OF THE MINERAL INDUSTRY IN 2012[1]

(Thousand metric tons unless otherwise specified)

Commodity	Major operating companies and major equity owners	Location of main facilities	Annual capacity
Carbon black	Orion Engineered Carbons GmbH (Rhône Capital LLC, 50%, and Triton Advisors Ltd., 50%)	Kalscheuren plant at Cologne, and plant at Dortmund	NA
Cement	HeidelbergCement AG	Plant at Burglengenfeld; two plants at Ennigerloh; two plants at Geseke; plants at Koenigs Wusterhausen, Leimen, Paderborn, Mainz-Weisenau, and Schelklingen; the Lengfurt plant at Triefenstein; plant at Wetzlar	12,700
Do.	Dyckerhoff AG (Buzzi Unicem SpA, 88.37%, and other private, 11.63%)	Plants at Deuna, Geseke, Goellheim, Lengerich, Neuss, Neuwied, and the Amöneburg plant at Wiesbaden	7,200
Do.	SCHWENK Zement KG	Plants at Allmendingen, Bernburg, Heidenheim-Mergelstetten, and Karlstadt	6,900
Do.	CEMEX Deutschland AG (CEMEX S.A. de C.V., 100%)	Two plants at Beckum; plants at Dortmund, Duisburg, Eisenhuettenstadt, and Ruedersdorf	5,300
Do.	Holcim (Deutschland) AG (Holcim Ltd., 88.9%, and other private, 11.1%)	HANSA plant at Bremen, plants at Laegerdorf and Rostock, and the Höver plant at Sehnde	3,600
Do.	Lafarge Zement GmbH (Lafarge S.A., 100%)	Plants at Kall-Soetenich, Karsdorf, and Walzbachtal	3,400
Do.	Holcim (Baden-Württemberg) AG (Holcim Ltd., 100%)	Plant at Dotternhausen	1,600
Do.	TEUTONIA Zementwerk AG (HeidelbergCement AG, 94.2%, and other private, 5.8%)	Plant at Hannover	900
Do.	Märker Zement GmbH	Plants at Harburg and Lauffen	NA
Clays, including ball, ceramic, kaolinitic, and refractory clays	Sibelco Deutschland GmbH (S.C.R.- Sibelco NV, 100%)	25 quarries and 8 plants, including 2 at Ransbach and the Kannenbäckerland plant in Hoehr-Grenzhausen, Westerwald region; also including quarries and plants of Kaolin- und Tonwerke Seilitz-Loethain, Saxony region	2,000
Do.	Stephan Schmidt KG	Tonbergbau Grube Anton open pit mine, Dornburg-Langendernbach, Müllenbach and Thewald Mines, Hoehr-Grenzhausen; Wiesa-Thonberg and Cunnersdorf quarries, Kamenz-Wiesa, Westerwald	1,600
Do.	Marx Bergbau GmbH & Co. KG (Stephan Schmidt KG, 100%)	Lämmersbach and Meudt Mines, Ruppach-Goldhausen quarry, Dornburg-Langendernbach, Westerwald	350
Do.	Goerg & Schneider GmbH & Co. KG	Quarry and main plant at Boden, others at Mogendorf, Goddert, Siershahn, Wirges/Staudt, and Kettenbach/Taunus, Westerwald region; others in Saxony and Eifel regions	NA
Do.	Mittelhessische Tonbergbau GmbH (Goerg & Schneider GmbH & Co. KG, 50%, and Stephan Schmidt KG, 50%)	Quarry and plant in the Giessen/Lahn region	100
Do.	Rohstoffgesellschaft GmbH Ponholz	Mine and chamotte plant at Maxhuette-Haidoff, and Aufofweiher Mine, Bavaria	150
Do.	Adolf Gottfried Tonwerke GmbH	Quarries and plant near Grossheirath, Coburg, Bavaria	100
Do.	Erbsloh Lohrheim GmbH (Erbsloh family, 100%)	Mine at Lohrheim, Rheinland-Pfalz	30
Coal, anthracite and bituminous	Deutsche Steinkohle AG (RAG Aktiengesellschaft, 100%)	Augusta Victoria/Blumenthal, Prosper-Haniel, and West Mines, Ruhr region, North Rhine-Westphalia	11,000 [e]
Do.	do.	Saar Mine, Saar Basin, Saarland	1,500 [e]
Do.	do.	Ibbenbüren Mine, Steinfurt District, North Rhine-Westphalia	2,100
Coke	ThyssenKrupp Steel AG	Schwelgern plant at Duisburg	2,100
Do.	ArcelorMittal Bremen GmbH (ArcelorMittal, 99.88%, and other private, 0.12%)	Coking plant at the Prosper-Haniel Mine	2,000 [e]
Do.	Hüttenwerke Krupp Mannesmann GmbH (ThyssenKrupp Steel AG, 50%; Salzgitter AG, 30%; Vallourec & Mannesmann Tubes SA, 20%)	Plant at Duisberg-Huckingen steel complex	1,100

See footnotes at end of table

TABLE 2—Continued
GERMANY: STRUCTURE OF THE MINERAL INDUSTRY IN 2012[1]

(Thousand metric tons unless otherwise specified)

Commodity	Major operating companies and major equity owners	Location of main facilities	Annual capacity
Copper, refined	Aurubis AG (Salzgitter AG, 25%; institutional investors, 45%; other private investors, 30%)	Primary smelter and refinery and secondary plant at Hamburg	500 [e]
Do.	Hüttenwerke Kayser AG (Aurubis AG, 100%)	Secondary plant and refinery at Luenen	210 [e]
Dolomite	Rheinkalk Hagen-Halden GmbH & Co KG (Lhoist NV, 100%)	Steinbruch-Donnerkuhle quarry and Hönnetal plant at Menden, and plant at Hagen-Halden	7,500
Dolomite and lime	Geomin Erzgebirgische Kalkwerke GmbH	Underground mines at Hermsdorf and Lengenfeld	NA
Feldspar	Saarfeldspatwerke H. Huppert GmbH & Co. KG	Mine at Oberthal, Gudesweiler, Saarland	60
Do.	Gottfried Feldspat GmbH	Minc at Freihung-Thansuss, Weiden, Bavaria	15
Ferrochrome	Elektrowerk Weisweiler GmbH (Kermas Ltd., 100%)	Plant at Eschweiler-Weisweiler, near Aachen	30
Fluorspar	Sachtleben Bergbau GmbH	Clara Mine in the Black Forest and plant at Wolfach	55 [e]
Gallium metric tons	Ingal Stade GmbH (5N Plus Inc., 50%, and Neo Performance Materials Ltd., 50%)	Ingal plant at Stade	35
Do. do.	PPM Pure Metals GmbH[2] (Recylex S.A., 100%)	Plant at Langelsheim	NA
Gold, metal	Aurubis AG (Salzgitter AG, 25%; institutional investors, 45%; other private investors, 30%)	Primary smelter and refinery and secondary plant at Hamburg	NA
Do. metric tons	Hüttenwerke Kayser AG (Aurubis AG, 100%)	Secondary plant and refinery at Luenen	40 [e]
Do.	Heraeus Precious Metals GmbH & Co. KG	Primary smelter and refinery and secondary plant at Hanau	NA
Do.	Umicore AG & Co. KG (Umicore S.A., 100%)	Plant at Hanau	NA
Do.	Allgemeine Gold- und Silberscheideanstalt AG (Umicore S.A., 91.21%, and other, 8.79%)	Plant at Pforzheim	NA
Graphite, manufactured	GK Graphit Kropfmühl GmbH (Advanced Metallurgical Group N.V., 100%)	Plant at Kropfmuehl, Passau	20
Do.	do.	Plants at Bad Godesberg and Wedel, Holstein	8
Gypsum	VG-ORTH GmbH & Co. KG	Mine and plant at Stadtoldendorf, and plants at Osterode, Spremberg, and Witzenhausen	150
Do.	Gyproc GmbH (Etex Group S.A., 80%, and Lafarge S.A., 20%)	Mines and plant in Lower Saxony	110
Do.	Knauf Gips KG	Mines and plant at Iphofen	NA
Iron, blast furnace	ThyssenKrupp Steel AG	Two blast furnace plants at Hamborn and Schwelgern	12,000
Iron, direct reduced	ArcelorMittal Hamburg GmbH (ArcelorMittal, 100%)	Plant at Hamburg	600 [e]
Iron oxide, pigments	Lanxess AG	Plant at Krefeld-Uerdingen	300
Kaolin, feldspar, and quartz	Amberger Kaolinwerke GmbH—Eduard Kick GmbH & Co. KG (Quarzwerke GmbH, 100%)	Mines at Caminau, Hirschau, Kemmlitz, and Schnaittenbach, Bavaria	350
Do.	Gebrüder Dorfner GmbH & Co Kaolin- und Kristallquartzsand Werk KG	Mine near Hirschau, Bavaria	NA
Lead, metal	Weser Metall GmbH (Recylex S.A., 100%)	Primary and secondary smelter and refinery at Nordenham	145
Do.	Berzelius Metall GmbH [Eco-Bat Technologies Ltd. (Quexco Inc., 100%), 100%]	Secondary smelters at Braubach am Rhein and Freiberg/Sachsen	200
Do.	do.	Primary smelter at Stolberg	160 [e]
Do.	Johnson Controls Recycling GmbH (Johnson Controls Inc., 100%)	Battery recycling plant and secondary smelter at Krautscheid	120
Do.	Muldenhütten Recycling- und Umwelttechnik GmbH	Secondary smelter at Freiburg, Saxony	55
Do.	Aurubis AG	Refinery at Hamburg	50
Lead, oxide, Pb content	Weser Metall GmbH (Recylex S.A., 100%)	Primary and secondary smelter and refinery at Nordenham	20
Lignite	RWE Power AG (RWE Aktiengesellschaft, 100%)	Open pit mines in Rhenish mining area: Bergheim, Garzweiler, Inden, and Hambach	105,000
Do.	Vattenfall Europe Mining AG	Jänschwalde-Cottbus-Nord, Nochten, and Welzow-Süd Mines, Lausatian mining area	60,000
Do.	Mitteldeutsche Braunkohlengesellschaft AG	Profen and Vereinigtes Schleenhain Mines	25,000
Limestone	Harz-Kalk GmbH	Quarry at Ruebeland	2,000 [e]
Do.	Kalkwerk Bad Kösen GmbH	Quarry at Bad Koesen	2,000 [e]
Do.	Fels-Werke GmbH	Quarry at Kaltes Tal	2,000 [e]

See footnotes at end of table.

TABLE 2—Continued
GERMANY: STRUCTURE OF THE MINERAL INDUSTRY IN 2012[1]

(Thousand metric tons unless otherwise specified)

Commodity		Major operating companies and major equity owners	Location of main facilities	Annual capacity
Limestone—Continued		Schäfer Kalk GmbH & Co KG	Plants at Hahnstaetten, Stceden, Stromberg, and Grevenbrueck	3,000
Do.		Rheinkalk GmbH & Co KG (Lhoist NV, 100%)	Flandersbach quarry and plant at Wuelfrath, and lime plant at Menden-Hoennetal	7,500
Magnesium, metal, secondary		Norsk Hydro Magnesiumgesellschaft GmbH (Norsk Hydro ASA, 100%)	Plant at Bottrop	26
Do.		Aleris Recycling (German Works) GmbH (Aleris International Inc., 100%)	Plant at Toeging am Inn	15
Mullite, fused		Treibacher Schleifmittel Zschornewitz GmbH (Imerys S.A., 100%)	Plant at Zschornewitz	31
Mullite, sintered		Nabaltec AG	Plant at Schwandorf	10
Natural gas	million cubic meters	Mobil Erdgas-Erdöl GmbH (Exxon Mobil Corp., 100%), including any fields owned or operated by BEB Erdgas und Erdöl GmbH (Exxon Mobil Corp., 50%, and Royal Dutch Shell plc, 50%)	Goldenstedt, Hemmelte, Klosterseelte, Söhlingen, and other fields in Lower Saxony	14,000 [e]
Do.	do.	RWE-Dea AG (RWE Power AG, 100%)	Bötersen, Hemsbünde, Völkersen, and smaller fields in Lower Saxony; Inzenham-West Field, Bavaria	3,000 [e]
Do.	do.	Gaz de France Produktion Exploration Deutschland GmbH (Gaz de France S.A., 100%)	Salzwedel Field, Saxony-Anhalt; Schneeren and smaller fields in Lower Saxony	1,500 [e]
Do.	do.	Wintershall Holding AG (BASF AG, 100%)	A6/B4 Blocks offshore Schleswig Holstein; smaller fields in Lower Saxony	1,200 [e]
Do.	do.	EEG-Erdgas Erdöl GmbH (GDF Suez S.A., 100%)	Muehlhausen and other fields in Thüringen	50 [e]
Petroleum:				
Crude	thousand 42-gallon barrels	Wintershall Holding AG (BASF AG, 100%), 50%, and RWE-Dea AG (RWE Power AG, 100%), 50%	Mittelplate-Dieksand field in tidal flats of the North Sea offshore Schleswig-Holstein	15,500
Do.	do.	Wintershall Holding AG (BASF AG, 100%)	A6/B4 Blocks offshore Schleswig Holstein; Aitingen field, Bavaria; Emlichheim field, Lower Saxony; and smaller fields in Lower Saxony and Rheinland-Pfalz	2,000 [e]
Do.	do.	Gaz de France Produktion Exploration Deutschland GmbH (GDF Suez S.A., 100%)	Bramberge, Ruehlertwist, Scheerhorn, and Ringe fields in Lower Saxony; smaller fields in the States of Bavaria, Hamburg, Lower Saxony, and Mecklenburg-Western Pomerania	3,500 [e]
Do.	do.	Mobil Erdgas-Erdöl GmbH (Exxon Mobil Corp., 100%)	Barenburg, Ruehme, and Lueben fields, Lower Saxony; smaller fields in the States of Lower Saxony and Rheinland-Pfalz	1,800 [e]
Do.	do.	BEB Erdgas und Erdöl GmbH (Exxon Mobil Corp., 50%, and Royal Dutch Shell plc, 50%)	Georgsdorf, Meppen, and Ruehlermoor fields, west of the Ems River (Emsland), Lower Saxony	3,000 [e]
Refined	do.	Deutsche Shell AG	Refineries at Godorf, Hamburg, and Grasbrook	256,000 [e]
Do.	do.	Raffinerie Heide GmbH (Klesch & Co. SA, 100%)	Refinery near Heide, State of Schleswig Holstein	35,000 [e]
Do.	do.	Esso Deutschland GmbH (ExxonMobil Central Europe Holding GmbH, 100%)	Refineries at Karlsruhe and Ingolstadt	245,000 [e]
Do.	do.	Ruhr Oel GmbH (Petróleos de Venezuela S.A., 50%, and BP Gelsenkirchen GmbH, 50%)	Refinery at Gelsenkirchen	215,500 [e]
Do.	do.	BAYERNOIL Raffineriegesellschaft mbH (OMV AG, 45%; Ruhr Oel GmbH, 25%; AGIP Deutschland GmbH, 20%; Deutsche BP AG, 10%)	Refinery at Neustadt-Donau	145,000 [e]
Platinum-group metals, refined		Aurubis AG (Salzgitter AG, 25%; institutional investors, 45%; other private investors, 30%)	Primary smelter and refinery and secondary plant at Hamburg	NA
Do.		Heraeus Precious Metals GmbH & Co. KG	Primary smelter and refinery and secondary plant at Hanau	NA
Do.		Umicore AG & Co. KG (Umicore S.A., 100%)	Plant at Hanau	NA
Do.		Allgemeine Gold- und Silberscheideanstalt AG (Umicore S.A., 91.21%, and other, 8.79%)	Plant at Pforzheim	NA
Potash, K$_2$O content		K+S Kali GmbH (K+S Aktiengesellschaft, 100%)	Mines at Hattorf, Neuhof-Ellers, Niedersachen-Riedel, Sigmundshall, Unterbreizbach, Wintershall, and Zielitz	6,000

See footnotes at end of table.

TABLE 2—Continued
GERMANY: STRUCTURE OF THE MINERAL INDUSTRY IN 2012[1]

(Thousand metric tons unless otherwise specified)

Commodity		Major operating companies and major equity owners	Location of main facilities	Annual capacity
Salt (evaporated and rock)		esco - european salt company GmbH & Co. KG [K+S Salz GmbH (K+S Aktiengesellschaft, 100%)]	Bernburg Mine and evaporated salt works; Borth Mine and evaporated salt works near Wesel; Braunschweig-Lüneburg Mine near Helmstedt	5,300 [e]
Do.		Wacker Chemie AG	Stetten rock salt mine near Haigerloch	500
Do.		Südsalz GmbH (Südwestdeutsche Salzwerke AG, 90%, and Vereinigte Schweizerische Rheinsalinen AG, 10%)	Rock salt mine at Berchtesgaden and evaporated salt works at Bad Reichenhall, Bavaria; and mine at Heilbronn and evaporated salt works at Bad Friedrichshall-Kochendorf, Heilbronn district, State of Baden-Württemberg	5,000
Do.		Saline Luisenhall GmbH	Evaporated salt works at Göttingen	NA
Selenium, metal	metric tons	Retorte GmbH (Aurubis AG, 100%)	Plant at Röthenbach	2,500
Silica sand (industrial sand)		Quarzwerke GmbH	Mines and plants at Frechen, Gambach, Haltern, Hohenbocka, and Weferlingen	4,500 [e]
Do.		Amberger Kaolinwerke GmbH—Eduard Kick GmbH & Co. KG (Quarzwerke GmbH, 100%)	Mines and plants at Hirschau and Schnaittenbach	850
Siliceous earth, silica		Hoffmann Mineral and Co. KG	Mine and plant near Neuburg	55
Silicon, metal	metric tons	RW Silicium GmbH (Advanced Metallurgical Group N.V., 100%)	Four electric arc furnaces in plant at Pocking	32,000
Silver, metal		Aurubis AG (Salzgitter AG, 25%; institutional investors, 45%; other private investors, 30%)	Primary smelter and refinery and secondary plant at Hamburg	NA
Do.	metric tons	Hüttenwerke Kayser AG (Aurubis AG, 100%)	Secondary plant and refinery at Luenen	1,300 [e]
Do.	do.	Berzelius Metall GmbH [Eco-Bat Technologies Ltd. (Quexco Inc., 100%), 100%]	Secondary (lead) smelters at Braubach am Rhein and Freiberg/Sachsen; primary (lead) smelter at Stolberg	400 [e]
Do.		Heraeus Precious Metals GmbH & Co. KG	Primary smelter and refinery and secondary plant at Hanau	NA
Do.		Umicore AG & Co. KG (Umicore S.A., 100%)	Plant at Hanau	NA
Do.		Allgemeine Gold- und Silberscheideanstalt AG (Umicore S.A., 91.21%, and other, 8.79%)	Plant at Pforzheim	NA
Soda ash		Solvay S.A.	Plant at Rheinberg	NA
Steel, crude		ThyssenKrupp Steel AG (ThyssenKrupp AG, 100%)	Bruckhausen and Beeckerwerth plants, near Duisburg	12,000
Do.		Salzgitter AG	Plants at Peine and Salzgitter	6,400 [e]
Do.		Hüttenwerke Krupp Mannesmann GmbH (ThyssenKrupp Steel AG, 50%; Salzgitter AG, 30%; Vallourec & Mannesmann Tubes SA, 20%)	Plant at Duisberg-Huckingen	5,600
Do.		ArcelorMittal Bremen GmbH (ArcelorMittal, 99.88%, and other private, 0.12%)	Plant at Bremen	4,000
Do.		Saarstahl AG (Struktur-Holding-Stahl GmbH & Co KG, 74.9%, and Dillinger Hüttenwerke AG, 25.1%)	Plants at Burbach, Neunkirchen, and Voelklingen	3,000
Do.		AG der Dillinger Hüttenwerke (Saarstahl AG, 33.75%; ArcelorMittal, 30.08%; Struktur-Holding-Stahl GmbH & Co KG, 26.17%; Dillinger Hütte und Saarstahl mbH, 10%; other, 4.72%)	Plant at Dillingen	2,800
Do.		ArcelorMittal Eisenhüttenstadt GmbH (ArcelorMittal, 100%)	Plant at Eisenhuettenstadt	2,400
Do.		Badische Stahlwerke GmbH	Plant at Kehl	2,300 [e]
Do.		Brandenburger Elektrostahlwerk GmbH (RIVA FIRE S.p.A, 100%)	Plant at Brandenburg	1,700 [e]
Do.		ThyssenKrupp Nirosta (ThyssenKrupp Steel AG, 100%)	Plants at Bochum and Krefeld	1,600 [e]
Do.		ArcelorMittal Ruhrort GmbH (ArcelorMittal, 100%)	Plant at Duisburg	1,500 [e]
Do.		Georgsmarienhütte GmbH	Plants at Bous, Georgsmarienhuette, and Groeditz	1,300 [e]

See footnotes at end of table.

TABLE 2—Continued
GERMANY: STRUCTURE OF THE MINERAL INDUSTRY IN 2012[1]

(Thousand metric tons unless otherwise specified)

Commodity	Major operating companies and major equity owners	Location of main facilities	Annual capacity
Steel, crude—Continued	Stahlwerk Thüringen GmbH (Alfonso Gallardo S.A., 100%)	Plant at Unterwellenborn	1,100
Do.	Deutsche Edelstahlwerke GmbH	Plants at Siegen and Witten	1,100 [e]
Do.	Lech-Stahlwerke GmbH (Max Aicher GmbH & Co. KG, 100%)	Plant at Herbertshofen	1,100 [e]
Do.	ArcelorMittal Hamburg GmbH (ArcelorMittal, 100%)	Plant at Hamburg	1,100 [e]
Do.	Hennigsdorfer Elektrostahlwerk GmbH (RIVA FIRE S.p.A, 100%)	Plant at Hennigsdorf	1,000 [e]
Do.	Elbe-Stahlwerke Feralpi GmbH (Feralpi Siderurgica S.p.A., 100%)	Plant at Riesa	950 [e]
Strontium carbonate	Solvay & CPC Barium Strontium GmbH & Co. KG (Solvay S.A., 75%, and Chemical Products Corp., 25%)	Plant at Bad Hoenningen, near Hannover	95
Sulfur	Norddeutsche Erdgas-Aufbereitungs GmbH NEAG [BEB Erdgas und Erdöl GmbH (ExxonMobil Production Deutschland GmbH, 50%, and Royal Dutch Shell plc, 50%), 100%]	Natural gas desulfurization plants at Grossenkneten and Voigtei (near Nienburg-Weser), Lower Saxony	600
Sulfuric acid	Aurubis AG (Salzgitter AG, 25%; institutional investors, 45%; other private investors, 30%)	Acid plant, part of primary copper production facilities at Hamburg	2,500 [e]
Do.	BASF SE	Plant at Ludwigshafen	NA
Do.	Berzelius Metall GmbH [Eco-Bat Technologies Ltd. (Quexco Inc., 100%), 100%]	Plant near primary lead smelter at Stolberg	NA
Do.	Evonik Degussa GmbH (Evonik Industries AG, 100%)	Plant at Worms	NA
Do.	Lanxess AG	Plant at Leverkusen	NA
Do.	Weser Metall GmbH (Recylex S.A., 100%)	Acid plant near primary lead smelter and refinery at Nordenham	55
Do.	Metaleurop Zinkbetrieb GmbH & Co. KG (Xstrata plc, 100%)	Acid plant near primary zinc smelter and refinery at Nordenham	NA
Tin alloys, tinplate	ThyssenKrupp Rasselstein GmbH	Plant at Andernach	NA
Zeolites	Hans G. Hauri Mineralstoffwerk GmbH	Mine and plant at Boetzingen, near Freiburg	NA
Zinc, metal	Metaleurop Zinkbetrieb GmbH & Co. KG (Xstrata plc, 100%)	Nordenham Smelter, near Bremerhaven	160
Do.	Ruhr-Zink GmbH (GEA Group AG, 100%)	Refinery at Datteln[4]	140
Zinc, oxides	Harz Metall GmbH (Recylex S.A., 100%)	Waëlz rotary kilns at Oker-Goslar	80 [e]
Do.	Norzinco GmbH (Recylex S.A., 100%)	Secondary plant at Harlingerode	20
Zinc, powder	do.	do.	5

[e]Estimated; estimated data are rounded to no more than three significant digits. Do., do. Ditto. NA Not available.
[1]Table includes data available through March 7, 2014.
[2]In addition to producing arsenic as a byproduct of chemical manufacturing and gallium as a byproduct of aluminum production, PPM Pure Metals GmbH produces small quantities of germanium as a byproduct of processing imported ores and concentrates and small quantities of indium and tellurium as byproducts of zinc metal production by PPM's parent company, Recylex S.A.
[3]Closed in 2007.
[4]Closed at the end of 2008, and approximately 40% of total production of zinc metal at this refinery was from secondary materials.

TABLE 3
GERMANY: EXPORTS OF SELECTED MINERAL COMMODITIES IN 2012[1]

(Metric tons unless otherwise specified)

Commodity		Total	Destinations[e] United States	Other (principal)[2]
METALS				
Aluminum:				
Bauxite, ore and concentrate	thousand metric tons	35	--	Belgium 11; Netherlands 5.
Oxide		369,337	--	France 122,000.
Hydroxide		455,157	80,600	Netherlands 96,900.
Ash and residues containing aluminum		16,242	--	Austria 5,880; France 5,250; Spain 2,000.
Metal:				
Primary, not alloyed		42,580	--	France 17,200; Hungary 5,830; Luxembourg 5,580.
Primary, alloys, all forms	thousand metric tons	132	--	Austria 45; United Kingdom 24; Poland 15.
Secondary, including alloys		230,036	--	Austria 49,500; France 43,900.
Scrap		968,539	--	Italy 188,000; Austria 148,000; Netherlands 108,000.
Antimony:				
Metal, including alloys, all forms		103	--	France 39; Spain 24; Slovakia 12.
Ore and concentrate		5	--	Brazil, 100%.
Oxides		688	--	Belgium 126; Romania 82; Switzerland 80.
Arsenic, metal, including alloys, all forms		128	3	Belgium 100.
Bismuth, metal, crude, including scrap		92	--	Switzerland 50; United Kingdom 15.
Cadmium, metal, crude, powder, including scrap		379	--	Sweden 219; China 149.
Chromium:				
Chromate		815	--	Belgium 524; Austria 227.
Ore and concentrate		60,354	--	Russia 36,000.
Metal:				
Crude, including powder		1,384	--	Austria 227; France 177.
Scrap		4,737		Italy 4,610.
Cobalt:				
Ore and concentrate		8	--	Vietnam, 100%.
Oxides and hydroxides		84	--	France 16; Turkey 13; Spain 11.
Metal, including alloys, all forms		431	--	China 82; Italy 62.
Scrap		460	82	Luxembourg 112; France 111; United Kingdom 71.
Copper:				
Ore and concentrate	thousand metric tons	57	--	Sweden 55.
Ash and residue containing copper		16,807	--	Belgium 10,900; Canada 3,500.
Matte and speiss, including cement copper		2	--	Russia, 100%.
Metal:				
Unrefined		2,014	--	Belgium 1,450; Austria 441.
Refined, not alloyed		276,943	--	China 68,400; Belgium 50,100.
Alloys, all forms		14,579	--	Switzerland 2,320; China 1,970.
Scrap		587,935	--	China 226,000; Netherlands 118,000; Belgium 60,000.
Gallium, indium, and thallium, metal, including scrap		48	25	Switzerland 10; United Kingdom 10.
Germanium, metal, all forms		3	--	Russia 3.
Gold:				
Metal, including alloys, all forms	kilograms	197,894	--	Switzerland 121,000; United Kingdom 32,700; Unspecified 22,800.
Powder	do.	258	209	--
Waste, sweepings and scrap		956	--	Japan 929.
Iron and steel:				
Ore and concentrate	thousand metric tons	43	--	Denmark 20; Switzerland 15.
Ash and residue containing iron	do.	2,340	--	France 1,120; Netherlands 496; Luxembourg 255.
Metal:				
Pig iron, cast iron, related materials		171,304	--	France 31,900; Turkey 18,500; Poland 17,100.
Scrap	thousand metric tons	9,658	--	Netherlands 1,790; Italy 1,770; Luxembourg 1,270.
Sponge iron, powder		35,370	--	Austria 5,770; Sweden 4,100; Italy 4,070.
Ferroalloys:				
Ferrochromium		32,760	6,450	France 8,450; Austria 4,130; United Kingdom 3,640.
Ferromanganese		11,810	--	Poland 2,520; Austria 1,880; Czech Republic 1,830.
Ferromolybdenum		2,815	--	Sweden 926; Czech Republic 495.

See footnotes at end of table.

TABLE 3—Continued
GERMANY: EXPORTS OF SELECTED MINERAL COMMODITIES IN 2012[1]

(Metric tons unless otherwise specified)

Commodity		Total	Destinations[e]	
			United States	Other (principal)[2]
METALS—Continued				
Iron and steel—Continued:				
Metal—Continued:				
Ferroalloys—Continued:				
Ferronickel		391	--	Netherlands 202; Belgium 103; Italy 62.
Ferroniobium		228	--	France 35.
Ferrosilicomagnesium		2,062	--	Italy 483; Czech Republic 297; Brazil 254.
Ferrosilicomanganese		12,006	--	Poland 3,230; Czech Republic 2,520; France 2,440.
Ferrosilicon		67,352	--	Austria 13,500; France 10,800; Italy 8,150.
Ferrotitanium		4,687	--	Italy 708; Belgium 520.
Ferrotungsten		417	--	Austria 95; Italy 90; China 63.
Ferrovanadium		226	--	China 74; Netherlands 31; Venezuela 27.
Other ferroalloys		33,947	--	Italy 3,800.
Steel, crude		8,726	--	Belgium 4,520.
Lead:				
Ore and concentrate		60,356	--	China 57,400.
Ash, residues and slimes containing lead		20,761	--	Belgium 20,700.
Lead containing antimony		9,823	--	Czech Republic 4,750; Austria 3,770.
Metal:				
Refined		146,370	--	Italy 32,600; Austria 26,200; Spain 25,500.
Unrefined		22,396	--	Czech Republic 15,400; Belgium 5,530.
Scrap		7,982	--	Netherlands 2,930; India 1,580; Belgium 1,470.
Magnesium, metal, including alloys:				
Scrap		14,863	--	Austria 2,630; Czech Republic 2,470; Belgium 2,230.
Unwrought		12,935	464	United Kingdom 2,290; Romania 1,900; Italy 1,600.
Manganese:				
Ore and concentrate		4,388	--	France 1,460; Denmark 755; Belgium 645.
Manganite, manganate		287	--	Japan 80; Taiwan 79; Belgium 49.
Oxides		939	--	Poland 426; Netherlands 124.
Metal:				
Crude		12,978	--	Netherlands 3,300; Belgium 3,140; France 1,800.
Scrap		574	--	Republic of Korea 348; Netherlands 76; France 72.
Mercury		103	13	Netherlands 40; Spain 16.
Molybdenum:				
Ore and concentrate		3,738	--	South Africa 1,050; Vietnam 916; Hong Kong 564.
Metal, scrap		1,205	--	France 677; Austria 229; United Kingdom 171.
Nickel:				
Ore and concentrate		778	--	China 198; Belgium 181; Poland 181.
Matte, speiss, related materials		19,412	--	Canada 18,600.
Oxides and hydroxides		39	--	France 10; Austria 8; Sweden 5.
Ash and residue containing nickel		22	--	Sweden, 100%.
Metal, including alloys:				
Alloys, all forms		5,036	1,060	Austria 1,920; United Kingdom 730; Slovenia 589.
Unalloyed		4,807	--	Austria 2,020; Poland 817.
Scrap		7,727	1,280	Sweden 1,550; United Kingdom 1,050; Spain 920.
Niobium (columbium), ash and residue containing niobium and tantalum		22	--	Kazakhstan 13; United Kingdom 8.
Platinum-group metals:				
Metal, including alloys, all forms:				
Platinum	kilograms	15,738	4,030	Switzerland 4,800.
Palladium	do.	35,069	4,490	China 7,470; Belgium 6,170; Brazil 4,560.
Rhodium	do.	3,794	960	China 884; Hong Kong 524.
Iridium, osmium and ruthenium	do.	17,048	--	Singapore 11,400; Belgium 2,010.
Waste, sweepings and scrap		6,048	3,370	United Kingdom 1,400; Belgium 1,120.

See footnotes at end of table.

TABLE 3—Continued
GERMANY: EXPORTS OF SELECTED MINERAL COMMODITIES IN 2012[1]

(Metric tons unless otherwise specified)

Commodity		Total	United States	Destinations[e] Other (principal)[2]
METALS—Continued				
Rare-earth metals, including alloys:				
Cerium compounds		165	--	Austria 30; Italy 24; Republic of Korea 17.
Other compounds, all forms		534	--	Austria 401.
Metal		10	--	Saudi Arabia 3; Czech Republic 2; Turkey, 2.
Selenium, metal		343	--	Phillipines 44; Brazil 39; Mexico 37.
Silicon, metal		49,693	--	China 21,000.
Silver:				
Metal		1,963	--	United Kingdom 1,070; Unspecified 279.
Powder	kilograms	42,098	--	France 13,800; Greece 8,710.
Tin:				
Ash and residue containing tin		394	--	Belgium 334; Netherlands 60.
Metal:				
Alloys, all forms		1,352	--	Republic of Korea 379; Italy 201; Belgium 153.
Crude		1,223	--	Austria 278; Czech Republic 275; France 130.
Waste and scrap		1,141	--	Belgium 646; Netherlands 299; Poland 152.
Titanium:				
Ore and concentrate		13,792	--	Brazil 9,530; Mexico 3,250.
Metal:				
Powder		2,508	--	France 614; Italy 369.
Waste and scrap		6,663	2,290	United Kingdom 1,950; Ukraine 1,320.
Tungsten:				
Ore and concentrate		524	268	Vietnam 179; China 74.
Metal, waste and scrap		3,395	492	Austria 920; Sweden 384; Finland 367.
Vanadium, metal, including scrap		560	162	United Kingdom 122; Russia 80; Japan 57.
Zinc:				
Ore and concentrate		39,533	--	Belgium 18,100; France 16,400.
Ash and residue containing zinc		61,903	--	Belgium 35,500; Netherlands 18,100.
Matte and related materials		4,917	--	Belgium 2,300; Austria 1,410; Italy 649.
Oxide and peroxide		36,292	--	France 6,750; China 4,060.
Metal:				
Alloys, all forms		29,775	--	Austria 22,200.
Unalloyed		59,913	--	Austria 10,100; Poland 9,230; France 8,750.
Powder and dust		13,302	5,590	--
Scrap		73,200	--	China 30,100; Netherlands 14,000; Italy 9,370.
Zirconium:				
Metal, including alloys		146	24	France 40; Switzerland 32; Belgium 15.
Oxides, including germanium oxides		296	--	Austria 57; United Kingdom 44; Czech Republic 41.
Scrap		31	6	Spain 11; United Kingdom 11; Belgium 3.
INDUSTRIAL MINERALS				
Abrasives, natural:				
Corundum, emery, garnet, and so forth		9,725	--	Sweden 3,470; Norway 1,640; Switzerland 1,370.
Pumice		190,653	--	Netherlands 125,000; Luxembourg 35,500.
Asbestos, crude		10	--	Switzerland, 100%.
Borates, natural, including calcined		47	--	Turkey 16; United Kingdom 13; Czech Republic 12.
Cement	thousand metric tons	6,944	--	Netherlands 2,190; Belgium 917; France 910.
Chalk, natural		224,648	--	Netherlands 54,100; Poland 45,400; United Kingdom 35,900.
Clays, crude:				
Bentonite		82,504	--	Netherlands 22,400; Poland 12,300; Austria 9,570.
Ceramic and fire clays		7,093		Italy 3,530; Switzerland 2,260; Ukraine 745.
Chamotte or Dina's Earth		76,538	--	Italy 20,000; Czech Republic 11,900; Netherlands 10,000.
Kaolin		361,916	--	Austria 104,000; Italy 81,100; Poland 65,500.
Other, unspecified	thousand metric tons	2,288	--	Italy 824; Netherlands 693; Belgium 336.

See footnotes at end of table.

TABLE 3—Continued
GERMANY: EXPORTS OF SELECTED MINERAL COMMODITIES IN 2012[1]

(Metric tons unless otherwise specified)

Commodity		Total	Destinations[e]	
			United States	Other (principal)[2]
INDUSTRIAL MINERALS—Continued				
Diamond, natural:				
Gem, not set or strung	carats	137,301	21,300	Hong Kong 35,000.
Industrial stones	do.	4,921	--	Switzerland 3,200; United Kingdom 876.
Dust and powder	kilograms	3,019	--	Italy 779; Belgium 555.
Diatomite and other infusorial earth		22,190	--	China 3,400.
Feldspar		70,936	--	Italy 15,700; France 11,600; Czech Republic 7,800.
Fluorspar:				
Acid-grade		22,636	--	Czech Republic 5,610; Poland 3,490; France 3,350.
Metallurgical-grade		15,393	--	Czech Republic 5,530; France 3,160; Poland 2,340.
Graphite, natural		12,721	--	Czech Republic 3,170; Austria 1,980.
Gypsum and anhydrite, natural	thousand metric tons	1,946	--	Netherlands 306; Belgium 280; Sweden 204.
Kyanite and related materials:				
Andalusite, kyanite, sillimanite		6,689	--	Czech Republic 1,150; Hungary 963; Slovakia 843.
Mullite		10,598	1,930	Hungary 1,350; Italy 1,310; Poland 1,200.
Lime, hydrated		731,403	--	Netherlands 182; France 84,800; Belgium 76,100.
Lithium carbonate		2,259	--	Turkey 953; France 298.
Magnesium compounds:				
Magnesite, natural, including burned		73,535	--	France 15,700; Austria 13,000; Poland 11,300.
Epsomite		797,248	--	Malaysia 193,000; Indonesia 140,000; France 114,000.
Mica, natural, including splittings and waste		4,475	--	Poland 993; Brazil 792; Italy 492.
Peat, natural	thousand metric tons	2,078	--	Netherlands 1,020.
Phosphates:				
Crude		1,534	--	Poland 994; Netherlands 176.
Milled		208	--	Austria 103; Kazakhstan 98.
Phosphoric acid, all forms, P_2O_5 equivalent		9,178	--	Netherlands 4,820.
Precious and semiprecious stones, natural (other than diamond):				
Gem, not set or strung		824	--	Thailand 313; Hong Kong 180.
Dust and powder	kilograms	90	--	Poland 33; France 31; Kazakhstan 20.
Pyrite, unroasted		342	--	United Arab Emirates 90; Turkey 84; Poland 58.
Salt and brine	thousand metric tons	2,207	--	Belgium 313; Czech Republic 300; Italy 230.
Stone, sand and gravel:				
Basalt, lava rocks, and so forth		110,302	--	Netherlands 103,000.
Crushed rock, macadam		136,068	--	Switzerland 85,500; France 37,700.
Dimension stone:				
Dolomite and limestone		502,261	--	Luxembourg 228,000; Belgium 73,300; Poland 62,800.
Granite		60,015	--	Switzerland 49,500.
Marble, travertine, and so forth		214,499	--	China 139,000; Switzerland 45,500.
Limestone for cement	thousand metric tons	239	--	Luxembourg 140; Netherlands 44.
Quartz and quartzite		516,312	--	Netherlands 422,000.
Sand, natural	thousand metric tons	9,737	--	Netherlands 6,150; Belgium 1,990.
Sandstone		1,959	--	Netherlands 488; Switzerland 488; Austria 478.
Schist and shale		18,376	--	Belgium 7,350; Netherlands 6,540; Denmark 2,830.
Other natural stone, unspecified	thousand metric tons	16,912	--	Netherlands 9,860; Switzerland 2,030.
Sulfur		525,622	--	Belgium 120,000; Morocco 86,700; Israel 70,400.
Talc, steatite and soapstone, natural		5,275	--	Slovenia 2,080.
Vermiculite and perlite, natural		2,554	--	Poland 516; Austria 332; Czech Republic 324.

See footnotes at end of table.

TABLE 3—Continued
GERMANY: EXPORTS OF SELECTED MINERAL COMMODITIES IN 2012[1]

(Metric tons unless otherwise specified)

Commodity		Total	Destinations[e]	
			United States	Other (principal)[2]
MINERAL FUELS AND RELATED MATERIALS				
Asphalt and bitumen, natural		836	--	Turkey 403; Switzerland 287.
Coal:				
Anthracite and bituminous:				
Anthracite	thousand metric tons	167	--	Unspecified, 149.
Coke	do.	218	--	Belgium 110; Netherlands 40; Unpecified 31.
Semicoke, coking coal	do.	6	--	Poland, 5; Switzerland 1.
Other, including briquets	do.	103	--	Unspecified 31; Switzerland 30; Belgium 13.
Lignite	do.	1,584	--	Czech Republic 456; Belgium 382; Poland 190.
Coke of lignite		41,089	--	Austria 14,300; Czech Republic 8,630; Netherlands 6,160.
Gas, natural, gaseous	petajoules	1,188	--	Unspecified, 100%.
Petroleum, crude	do.	139	--	Bahamas 104; France 33.
Uranium, natural:				
Compounds, U content		98	--	Netherlands, 100%.
Crude, U content		55	--	Czech Republic 50.
Enriched, fissile isotopes	kilograms	24,584	8,600	France 6,420; United Kngdom 5,750.

[e]Estimated; estimated tonnages are rounded to no more than three significant digits; may not add to totals shown. do. Ditto. -- Less than 10%.

[1]Source: Bundesanstalt für Geowissenschaften und Rohstoffe, 2013, Table 2—Rohstoffsituation, 2012: Hannover, Germany, November.

[2]Destination country was estimated to have accounted for at least 10% of Germany's total exports of the mineral commodity.

TABLE 4
GERMANY: IMPORTS OF SELECTED MINERAL COMMODITIES IN 2012[1]

(Metric tons unless otherwise specified)

Commodity		Total	United States	Sources[e] Other (principal)[2]
METALS				
Aluminum:				
Bauxite, ore and concentrate	thousand metric tons	2,777	--	Guinea 2,140.
Oxide		476,257	--	Ireland 141,000; Netherlands 85,300; Jamaica 61,900.
Hydroxide		196,301	--	Spain 58,700; Ireland 57,100; France 38,500.
Ash and residue containing aluminum		184,491	--	France 30,800; Netherlands 20,300; Switzerland 20,300.
Metal:				
Primary, not alloyed		732,599	--	Netherlands 253,000; Iceland 123,000; Russia 110,000.
Primary, alloys, all forms	thousand metric tons	1,134	--	Netherlands 311; Norway 245; United Arab Emirates 11.
Secondary, including alloys		616,815	--	United Kingdom 178,000; Austria 85,700; Italy 76,500.
Scrap		575,926	--	Netherlands 123,000; Austria 74,300; France 59,900.
Antimony:				
Metal, including alloys, all forms		371	--	China 253; Belgium 71.
Ore and concentrate		6	--	Italy, 100%.
Oxides		6,013	--	China 1,950; France 1,950; Belgium 1,560.
Arsenic, metal, including alloys, all forms		43	--	China 40.
Bismuth, metal, crude, including scrap		968	--	Belgium 891.
Cadmium, metal, crude, powder, including scrap		69	1	Croatia 42; United Kingdom 18.
Chromium:				
Chromate		27,981	--	South Africa 15,700; Kazakhstan 6,070; Russia 4,030.
Ore and concentrate		188,783	--	South Africa 128,000; Turkey 51,900.
Metal:				
Crude, including powder		4,623	--	Russia 2,030; France 1,470; United Kingdom 582.
Scrap		1,679	--	Poland 635; Denmark 185; Sweden 171.
Cobalt:				
Ore and concentrate		128	--	Austria 44; Canada 32; United Kingdom 20.
Oxides and hydroxides		878	--	Finland 672; Belgium 176.
Metal, including alloys, all forms		2,513	392	Belgium 445; United Kingdom 430; Canada 304.
Scrap		1,416	--	United Kingdom 340; Poland 251; Austria 150.
Copper:				
Ore and concentrate	thousand metric tons	1,215	--	Peru 329; Chile 255; Argentina 186.
Matte and speiss, including cement copper		2,415	--	Ukraine 778; Morocco 604; Poland 536.
Ash and residue containing copper		61,283	15,400	Belgium 10,600; Italy 7,910.
Metal:				
Unrefined		45,708	--	Bulgaria 24,200.
Refined, not alloyed		702,576	--	Russia 190,000; Poland 145,000; Chile 109,000.
Alloys, all forms		37,491	--	United Kingdom 7,390; Spain 4,610; Belgium 4,050.
Scrap		655,895	--	Netherlands 74,800.
Gallium, indium, and thallium, metal, including scrap		63	7	United Kingdom 33; Slovakia 8.
Germanium, metal, all forms		7	1	China 3; Russia 2.
Gold:				
Metal, including alloys, all forms	kilograms	95,923	--	Unspecified 37,400; Switzerland 36,700.
Powder	do.	4,647	--	Switzerland 4,630.
Waste and sweepings		2,568	--	United Kingdom 560; Turkey 377.
Iron and steel:				
Ore and concentrate	thousand metric tons	38,908	--	Brazil 23,700; Sweden 5,060; Canada 4,400.
Ash and residue containing iron	do.	731	--	Austria 346; France 125; Belgium 97.
Pyrite, roasted		28,207	--	Finland 14,700; Russia 13,500.
Metal:				
Pig iron, cast iron, related materials		614,096	--	Russia 290,000; Brazil 87,800; South Africa 74,900.
Scrap	thousand metric tons	5,738	--	Netherlands 1,080; Poland 1,070; Czech Republic 1,000.
Sponge iron, powder		210,328	--	Netherlands 61,600; Trinidad & Tobago 53,600; Sweden 28,600.
Ferroalloys:				
Ferrochromium		392,515	--	South Africa 213,000; Unspecified 136,000.
Ferromanganese		192,818	--	Norway 55,300; South Africa 54,000; Spain 29,100.

See footnotes at end of table.

TABLE 4—Continued
GERMANY: IMPORTS OF SELECTED MINERAL COMMODITIES IN 2012[1]

(Metric tons unless otherwise specified)

Commodity	Total	Sources[e]	
		United States	Other (principal)[2]
METALS—Continued			
Iron and steel—Continued:			
Metal—Continued:			
Ferroalloys—Continued:			
Ferromolybdenum	16,934	--	Belgium 6,030; United Kingdom 3,030; Armenia 2,830.
Ferronickel	139,658	--	Ukraine 88,000; Venezuela 19,000; United Kingdom 14,900.
Ferroniobium	6,297	--	Brazil 4,530; Netherlands 919; Canada 655.
Ferrosilicochromium	8,573	--	Belgium 6,880; Unspecified 1,650.
Ferrosilicomagnesium	4,528	--	Slovenia 1,200; China 924; France 643.
Ferrosilicomanganese	206,967	--	Norway 56,500; India 48,000; France 21,300.
Ferrosilicon	248,310	--	Norway 64,600; Iceland 40,700; France 28,300.
Ferrotitanium	9,912	--	United Kingdom 3,100; Russia 2,200; Netherlands 1,890.
Ferrotungsten	1,010	--	China 502; Vietnam 245; Netherlands 167.
Ferrovanadium	5,096	--	Austria 2,900; South Africa 1,400.
Other ferroalloys	71,160	--	France 33,200; United Kingdom 10,900.
Steel, crude	27,853	--	Ukraine 11,900; Austria 4,620.
Lead:			
Ore and concentrate	228,444	--	Sweden 60,300; Australia 48,700; Ireland 47,500.
Ash, residues and slimes containing lead	173,826	--	France 135,000.
Lead containing antimony	23,760	--	Russia 9,930; Czech Republic 3,540; Sweden 3,140.
Metal:			
Refined	86,320	--	Belgium 28,700; United Kingdom 23,400; Poland 10,000.
Unrefined	38,496	--	United Kingdom 18,400; Belgium 6,850; Poland 6,390.
Scrap	28,058	--	Netherlands 8,280; Lithuania 5,810; France 4,410.
Magnesium, metal, including alloys:			
Scrap	20,493	--	China 11,500; Austria 3,320.
Unwrought	34,692	--	China 16,000; Czech Republic 4,580; Austria 4,200.
Manganese:			
Ore and concentrate	18,532	--	Netherlands 5,280; Brazil 4,370; Morocco 2,390.
Manganite, manganate	889	--	Spain 323; Austria 180; Netherlands 127.
Oxides	19,519	--	Greece 9,330; Spain 2,600.
Metal:			
Crude	34,380	--	China 22,600; Ukraine 3,640; Netherlands 3,510.
Scrap	381	--	Austria 228; Czech Republic 130.
Mercury	53	--	Peru 23; Portugal 9.
Molybdenum:			
Ore and concentrate	7,158	--	Netherlands 1,810; Belgium 1,320; United Kingdom 1,320.
Oxides and hydroxides, powder	2,477	--	Chile 1,490; Netherlands 379.
Molybdate compounds	319	128	Poland 68; France 35.
Metal:			
Crude	120	--	China 50; Russia 30.
Scrap	2,639	--	China 924; Austria 765; Armenia 668.
Nickel:			
Ore and concentrate	8,185	--	Netherlands 7,070.
Ash and residue containing nickel	9,680	--	Netherlands 3,160; France 1,510.
Matte, speiss, related materials	1,869	791	Netherlands 424; Japan 320; Brazil 316.
Oxides and hydroxides	609	--	Czech Republic 426.
Metal:			
Alloys, all forms	23,191	--	Indonesia 11,900; Russia 2,880.
Unalloyed	73,360	--	Russia 37,000; United Kingdom 11,600; Norway 8,510.
Scrap	14,191	--	Netherlands 5,410.
Niobium (columbium):			
Metal, powder containing niobium and rhenium	800	--	Brazil 764.
Ash and residue containing niobium and tantalum	12,305	48	Malyasia 10,300; Brazil 1,880.

See footnotes at end of table.

TABLE 4—Continued
GERMANY: IMPORTS OF SELECTED MINERAL COMMODITIES IN 2012[1]

(Metric tons unless otherwise specified)

Commodity		Total	United States	Sources[e] Other (principal)[2]
METALS—Continued				
Platinum-group metals:				
Metal, including alloys, all forms:				
Platinum	kilograms	27,259	3,710	South Africa 10,000; United Kingdom 4,310; Belgium 4,010.
Palladium	do.	42,705	4,700	Belgium 11,400; Russia 11,300; Switzerland 4,610.
Rhodium	do.	5,064	--	Belgium 1,940; South Africa 1,390; Russia 668.
Iridium, osmium, and ruthenium	do.	13,269	--	Belgium 7,580; South Africa 2,310; Japan 2,230.
Waste and scrap		8,021	--	France 1,120.
Rare earths:				
Cerium compounds		553	--	China 208; France 155; Estonia 76.
Other compounds, all forms		4,155	2,190	China 843; France 499.
Metal		290	--	China 264.
Selenium, metal		249	--	Sweden 61; Canada 51; Belgium 41.
Silicon, metal		232,639	--	Norway 91,200; France 36,500; Brazil 24,900.
Silver:				
Ore and concentrate		5,280	--	Peru 2,590; Mexico 919; Argentina 797.
Metal		1,262	--	Unspecified 862.
Powder	kilograms	201,655	98,000	Morocco 40,300; France 22,200.
Tantalum, metal:				
Powder		67	26	Kazakhstan 26.
Waste and scrap		109	30	Cyprus 20; Austria 16; United Kingdom 14.
Tin:				
Ash and residue containing tin		91	--	Belgium 45; Austria 26; Sweden 15.
Metal:				
Alloys, all forms		196	--	United Kingdom 93; Poland 45.
Crude		18,865	--	Indonesia 5,870; Peru 4,700; Belgium 3,910.
Waste and scrap		461	--	Netherlands 101; Czech Republic 71; Switzerland 60.
Titanium:				
Ore and concentrate		652,027	--	Norway 224,000; South Africa 149,000; Canada 123,000.
Oxide		17,359	--	France 4,010; Belgium 3,780; Netherlands 1,740.
Metal:				
Powder		6,937	--	Japan 1,240; Ukraine 1,210; Kazakhstan 1,080.
Waste and scrap		3,809	--	Italy 712; Switzerland 396; Austria 392.
Tungsten:				
Ore and concentrate		381	--	Bolivia 350.
Carbide		2,478	--	Austria 1,160; Unspecified 295; Canada 292.
Oxides and hydroxides		606	--	China 581.
Wolframate compounds		1,211	150	China 590; Russia 180.
Metal:				
Crude		104	--	China 31; Netherlands 30; United Kingdom 23.
Powder		1,816	--	Austria 982; Canada 340.
Waste and scrap		4,869	--	Czech Republic 506.
Vanadium, metal, including scrap		61	19	China 27; France 7.
Zinc:				
Ore and concentrate		324,954	51,700	Australia 142,000.
Matte and related materials		10,539	--	Netherlands 2,730; Austria 2,010; Belgium 1,740.
Oxide and peroxide		30,520	--	Austria 7,630; Peru 5,980; Netherlands 5,650.
Ash and residue containing zinc		27,108	--	Switzerland 10,700; France 3,770; Austria 2,950.
Metal:				
Alloys, all forms		64,642	--	Belgium 29,300; Luxembourg 9,050; Norway 7,950.
Unalloyed		365,788	--	Finland 135,000; Spain 99,100.
Powder and dust		6,717	--	Belgium 4,410.
Waste and scrap		25,536	--	Netherlands 9,270; Denmark 6,690; France 4,670.

See footnotes at end of table.

TABLE 4—Continued
GERMANY: IMPORTS OF SELECTED MINERAL COMMODITIES IN 2012[1]

(Metric tons unless otherwise specified)

Commodity		Total	United States	Sources[e] Other (principal)[2]
METALS—Continued				
Zirconium:				
Metal		105	43	France 35; China 21.
Oxides, includng germanium oxides		2,452	741	France 508; China 383; United Kingdom 378.
Waste and scrap		8	1	Belgium 2; France 2; Switzerland 2.
Zinkate and vanadate compounds		1,000	--	Austria 541; China 341.
INDUSTRIAL MINERALS				
Abrasives, natural:				
Corundum, emery, garnet, and so forth		13,296	--	India 8,700; Netherlands 2,410.
Pumice		14,654	--	Iceland 14,200.
Asbestos, crude		554	--	Brazil, 100%.
Barium compounds:				
Barite (barium sulfate)		237,120	--	China 182,000; Netherlands 27,000.
Witherite (barium carbonate)		134	--	Poland 119; Canada 15.
Borates, natural, including calcined		4,557	--	Unspecified 2,600; Belgium 1,840.
Cement	thousand metric tons	1,324	--	France 434; Czech Republic 291; Luxembourg 159.
Chalk, natural		176,673	--	France 103,000; Denmark 28,600; Belgium 27,000.
Clays, crude:				
Bentonite		450,851	--	Netherlands 134,000; Czech Republic 107,000; Italy 53,200.
Ceramic and fire clays		22,331	--	Czech Republic 12,800; Poland 4,580.
Chamotte or Dina's Earth		70,842	--	Luxembourg 20,800; Czech Republic 18,500; Netherlands 14,200.
Kaolin		625,827	115,000	Belgium 153,000; Czech Republic 145,000; United Kingdom 77,600.
Other, unspecified	thousand metric tons	71	--	Czech Republic 24; United Kingdom 15; Netherlands 8.
Diamond, natural:				
Gem, not set or strung	carats	311,770	--	India 116,000; Belgium 110,000; Israel 39,000.
Industrial stones	do.	158,572	--	India 43,400; Belgium 32,200; China 30,300.
Dust and powder	kilograms	18,883	2,040	China 5,610; Ireland 3,610; Republic of Korea 2,960.
Diatomite and other infusorial earth		42,122	12,400	Denmark 19,700.
Feldspar		109,158	--	Turkey 32,900; France 26,700; Czech Republic 15,500.
Fluorspar:				
Acid-grade		223,910	--	South Africa 79,900; China 51,100; Namibia 46,300.
Metallurgical-grade		56,958	--	United Kingdom 42,700; China 8,940.
Graphite, natural		43,349	--	China 20,400; Unspecified 8,710; Brazil 6,850.
Gypsum and anhydrite, natural	thousand metric tons	120	--	Austria 56; Belgium 24; France 22.
Iron oxide pigments		38,804	--	China 11,600; Belgium 9,930.
Kyanite and related materials:				
Andalusite, kyanite, sillimanite		54,045	--	South Africa 27,200; France 16,600.
Mullite		51,710	31,500	China 13,300.
Lime, hydrated		473,859	--	France 338,000; Czech Republic 53,100.
Lithium carbonate		6,064	843	Chile 5,050.
Magnesium compounds:				
Magnesite, natural, including burned		501,487	--	China 186,000; Netherlands 76,700.
Epsomite		258	--	Netherlands 215; Czech Republic 32.
Mica, natural, including splittings and waste		33,051	--	China 10,300; India 9,090; France 6,180.
Peat, natural	thousand metric tons	811	--	Lithuania 280; Latvia 208; Netherlands 132.
Phosphates:				
Crude		131,008	--	Israel 114,000.
Milled		2,347	--	France 674; Denmark 662; Belgium 636.
Phosphoric acid, all forms, P_2O_5 equivalent		163,907	--	Unspecified 79,500; Netherlands 26,700; Belgium 26,200.
Potash and potassium fertilizers, K_2O equivalent		2,056	--	United Kingdom 1,120; Netherlands 847.
Precious and semiprecious stones, natural (other than diamond):				
Gem, not set or strung		1,074	--	Brazil 481; South Africa 116.
Dust and powder	kilograms	963	--	China 560; Republic of Korea 196.

See footnotes at end of table.

TABLE 4—Continued
GERMANY: IMPORTS OF SELECTED MINERAL COMMODITIES IN 2012[1]

(Metric tons unless otherwise specified)

Commodity		Total	United States	Sources[e] Other (principal)[2]
INDUSTRIAL MINERALS—Continued				
Pyrite, unroasted		74,726	--	Finland 69,700.
Salt and brine	thousand metric tons	2,278	--	Netherlands 1,940.
Stone, sand and gravel:				
Basalt, lava rocks, and so forth		69,775	--	Norway 39,100; Italy 11,800; Netherlands 7,750.
Crushed rock, macadam		44,480	--	Netherlands 30,900; Switzerland, 13,600.
Dimension stone:				
Dolomite and limestone		780,883	--	Estonia 295,000; Belgium 260,000; United Kingdom 131,000.
Granite		153,254	--	Netherlands 33,400; Austria 22,400; Poland 18,700.
Marble, travertine, and so forth		60,376	--	Austria 24,000; Netherlands 10,100; Turkey 8,210.
Limestone for cement	thousand metric tons	1,909	--	Austria 592; Belgium 584; Poland 464.
Quartz and quartzite		188,401	--	Austria 69,100; Russia 61,800.
Sand, natural	thousand metric tons	2,089	--	France 1,300; Netherlands 313.
Sandstone		17,290	--	India 7,140; Poland 3,230; China 2,130.
Other natural stone, unspecified	thousand metric tons	10,096	--	Norway 3,670; United Kingdom 1,850; France 1,360.
Stone, sand and gravel—Continued:				
Schist and shale		41,738	--	France 28,600; Italy 5,130.
Sulfur		34,102	--	Norway 9,310; Poland 8,010; Belgium 4,470.
Talc, steatite and soapstone, natural		293,175	--	Netherlands 74,200; France 63,300; Austria 53,100.
Vermiculite and perlite, natural		105,059	--	Greece 89,900.
MINERAL FUELS AND RELATED MATERIALS				
Asphalt and bitumen, natural		7,974	2,030	Trinidad and Tobago 4,480.
Coal:				
Bituminous:				
Anthracite	thousand metric tons	1,399	2,790	Russia 867; Colombia 155.
Coke	do.	3,162	--	Poland 1,570; Belgium 386.
Semicoke, coking coal	do.	9,486	2,790	Australia 4,210; Canada 1,530.
Other, including briquets	do.	32,641	6,990	Russia 9,630; Colombia 8,880.
Lignite	do.	26	--	Czech Republic 26.
Coke of lignite		5,804	--	Italy 2,900; Austria 1,430.
Gas, natural, gaseous	petajoules	4,676	--	Unspecified, 100%.
Petroleum, crude	thousand metric tons	92,278	--	Russia 34,400.
Uranium, natural:				
Compounds, U content		3,921		France 2,430; United Kingdom 961.
Crude, U content		213	--	France 210.
Enriched, fissile isotopes	kilograms	13,655	--	Netherlands 4,040; United Kingdom 3,740; France 2,720.

[e]Estimated; estimated tonnages are rounded to no more than three significant digits; may not add to totals shown. do. Ditto. -- Less than 10%.

[1]Source: Bundesanstalt für Geowissenschaften und Rohstoffe, 2013, Table 2—Rohstoffsituation, 2012: Hannover, Germany, November.

[2]Source country was estimated to have accounted for at least 10% of Germany's total imports of the mineral commodity.

THE MINERAL INDUSTRY OF GREECE

By Harold R. Newman

Mining has been a part of Greek civilization since about 1,000 B.C. In 2012, some of Greece's important mineral resources included bauxite, bentonite, copper, gold, gypsum, and perlite. The mineral industry was composed of the sectors that mine and process metallic and nonmetallic minerals and mineral fuels. The output of the Greek mineral industry declined in both value and tonnage owing to the global economic downturn that started in 2009 and was continuing in 2012. Greece also had a high budget deficit and the European Union's (EU's) second highest external debt burden after Ireland (Industry Review, 2012).

In 2012, the mineral industry was regulated by the Mining Code, Legislative Decree 210/1973, as amended by a number of laws and ordinances on technical and procedural issues, such as law No. 669/1977 on the exploitation of ornamental rocks and industrial minerals; law No. 428/84, as amended by law No. 2115/93 on the exploitation of aggregates; and the Regulation on Mining and Quarrying Activities, which included the Health and Safety Regulation on Mining and Quarrying. Greece's mineral industry was also subject to the EU's Environmental Impact Directive and the EU Mining Waste Directive 2006/21/EC (United Nations, 2012).

Geologic studies in Greece were done primarily by the Institute of Geology and Mining Exploration Management (IGMEM), which was formerly the Institute of Geology and Mining Exploration. The IGMEM conducts applied geochemical analysis, basic geologic research, the exploration and evaluation of ore deposits and industrial minerals, hydrogeologic surveys, and water quality control. Most exploration activity was focused on northern Greece, which was thought to contain a significant amount of exploitable minerals (Institute of Geology and Mineral Exploration Management, 2012).

The challenging financial situation in Greece continued. In 2011 (the latest year for which data were available), the Government's public debt increased to about 165% of the GDP, which resulted in higher borrowing costs and loss of market access. Growth in 2012 was minus 6.9% of the GDP and the unemployment rate was an estimated 24%. Governments within the euro area provided the Greek Government with emergency short- and medium-term loans worth $147 billion so the country could make debt payments to its creditors (U.S. Central Intelligence Agency, 2012; U.S. Department of State, 2013).

Minerals in the National Economy

The mineral industry produced and (or) processed metals, industrial minerals, and mineral fuels. Greece was a supplier of several industrial minerals—most notably, bentonite, magnesite, and perlite. Most of the Greek companies that dealt in metal fabrication, mining of minerals, and refined metal production or processing were well established and had a strong export orientation. Although mineral industry activities have traditionally been an important segment of Greek industry, the significance of the mineral industry to the Greek economy has gradually decreased in the past 20 years.

The export and import of minerals commodities, however, continued to be relatively important to the Greek economy in 2012. U.S. exports to Greece included petroleum products valued at $59 million, copper valued at $42 million, iron and steel mill products valued at $1.5 million, and nonferrous metals valued at $617,000. U.S. imports from Greece included bauxite and aluminum valued at $54.5 million; cement, lime, and stone valued at $41 million; other petroleum products valued at $24.4 million; and fuel oil valued at $24.3 million (U.S. Census Bureau, 2012a, b).

Production

Greece was a global supplier of several industrial minerals, and production of these mineral commodities was closely tied to the export market (table 1). In 2012, Greece was the world's second-ranked producer of perlite after the United States and the world's sixth-ranked producer of pumice. It also was estimated to have produced about 9% of the world's bentonite and 1% of the world's bauxite. Greece was the only country that produced huntite, which is a carbonate mineral used as a fire retardant for polymers. Bauxite, which is the raw material needed for aluminum production, and lignite, which is used as a fuel in powerplants, were the country's two most abundant minerals in terms of reserves. In terms of the value of production, bauxite was the most important of Greece's mineral commodities (Bolen, 2013; Bray, 2013; Crangle, 2013; Virta, 2013).

Structure of the Mineral Industry

The major mineral commodities and the companies that produced them in 2012 are listed in table 2. Nearly all companies were privately owned; Government ownership was limited mainly to the mineral fuels sector.

Commodity Review

Metals

Bauxite and Alumina and Aluminum.—All Greece's major bauxite deposits are located in central Greece within the Parnassos-Ghiona geotectonic zone and on Evvoia Island. The three producers of bauxite in Greece were Delphi-Distomon S.A., Hellenic Mining Enterprises S.A., and S&B Industrial Minerals S.A. (S&B). The leading bauxite producer was S&B, which had an output capacity of 2 million metric tons year (Mt/yr) exclusively from underground sites located in the areas of Amfissa and Distomon. Delphi-Distomon and Hellenic Mining Enterprises supplied bauxite to the nonmetallurgical markets (table 2).

Gold.—Glory Resources Ltd. of Australia announced in 2012 that it had identified mineral resources at its Sapes gold project with a Joint Ore Reserves Committee (JORC) estimate of measured and indicated resources of 2.6 million metric tons (Mt) grading 9.8 grams per metric ton (g/t) gold. A JORC-compliant mineral resource estimate was to be completed following a proposed 6,000-meter (m) drilling program (Glory Resources Ltd., 2012).

Lead, Silver, and Zinc.—Eldorado Gold Corp. of Canada's Stratoni Mine was an underground operation located on the Chalkidiki Peninsula. It operated at a mining rate of about 18,000 metric tons per month of ore and produced lead, silver, and zinc concentrates. The Stratoni mineralization was classified as lead-silver-zinc carbonate replacement type mineralization, with galena, pyrite, and sphalerite as the main ore minerals. Resources at the Stratoni Mine were contained within the Mavres Petres ore body and had estimated proven and possible reserves of 1.8 Mt grading 8.5% zinc, 6.3% lead, and 177 g/t silver. The mine produced a lead-silver concentrate and a zinc concentrate by a conventional underground drift-and-fill method. The concentrates were shipped by sea to European facilities using either the Stratoni or the Thessaloniki port facilities (Eldorado Gold Corp., 2012).

Nickel.—Nickel laterite mineral resources were estimated to be about 250 Mt and were spread across three areas: central Evvoia, Neon Kokkinon, and northern Greece, in the area of Kastoria. Larco G.M.M. S.A., which was a leading producer of nickel in Europe and the only European user of domestic nickel ores, mined sedimentary-type nickel laterite by open pit and underground methods at its Agios Ioannis and Evia Mines near Larimna, and smelted the material at its plant in Larimna (Larco G.M.M. S.A., 2012).

Industrial Minerals

Cement.—Titan Cement Co. S.A. was a significant cement producer in Greece, and its four plants produced a combined total of about 6 Mt/yr. Domestic and export cement sales continued to fall in response to the decrease of construction activity and the slowdown in southeastern European markets. This resulted in a net €24.5 million ($32.6 million[1]) loss for 2012. Titan stated that it had no plans to move its headquarters from Greece (Global Cement, 2012).

Clay and Shale.—Greece was the world's second-ranked producer of bentonite after the United States. Extraction was conducted primarily on Milos Island by S&B. Bentonite deposits had also been found on the islands of Chios and Lesvos. S&B ranked first in the production of bentonite in Europe and was the leading exporter of bentonite in the world (Industrial Minerals, 2012).

Magnesium Compounds.—Grecian Magnesite S.A. (GM) operated three rotary kilns for crushed calcined magnesite (CCM) or dead-burned magnesite (DBM) with a combined capacity of 420 metric tons per day (t/d), a 50-t/d shaft kiln for CCM, and a 100-t/d double inclined shaft kiln. GM produced more than 50 different grades of CCM, DBM, and basic monolithic refractories for a wide range of applications (Grecian Magnesite S.A., 2012).

Mineral Fuels and Other Sources of Energy

Coal.—In 2012, the predominant fuel used in electricity generation in Greece was lignite, and Public Power Corp. S.A. (PPC) was Greece's major producer of lignite. PPC's leading lignite mines in Megalopolis and Ptolemais provided lignite for power generation. PPC's lignite-powered powerplants and hydroelectric powerplants had a combined capacity of 12.5 gigawatts and represented about 68% of the country's total installed capacity (Public Power Corp., 2012).

Petroleum.—Petra Petroleum Inc. of Canada announced that it had submitted an application in the Greece Open Door invitation process for hydrocarbon exploration rights. Petra Petroleum joined with Energean Oil & Gas S.A. to bid for the Ioannina Contract Area, which was a 4,187-square-kilometer area located onshore in northwestern Greece. The region was on trend with large discoveries and producing fields in Albania and is geologically similar to the producing regions of Italy. Energean Oil was the only natural gas and petroleum exploration and production company in Greece. It operated three petroleum fields—Epsilon, Prinos, and Prinos North in the Prinos Development Area (Petra Petroleum Inc., 2012).

Wind Energy.—According to Invest in Greece SA, the Government's Investment Promotion Agency, Greece's potential for wind power was among the largest in Europe. In 2012, about 1,400 megawatts (MW) was installed and operating, and the Government's target was for 7,500 MW to be installed by 2020. Wind energy was among the top energy priorities of the Government, which had indicated that it was committed to the increased usage of renewable energy (Invest in Greece SA, 2012).

Outlook

The economic outlook for 2012 is not expected to improve greatly, even though the Government is expected to continue its efforts to reform the economy and address the debt issue. Greece is expected to remain a major supplier of bentonite, perlite, and pumice in the international market. The industrial minerals sector will likely continue to be a small but important part of the country's revenue earnings. Development of mineral resource projects in the northern part of Greece is likely to continue along with mineral fuel exploration efforts offshore Greece.

References Cited

Bolen, W.P., 2013, Perlite: U.S. Geological Survey Mineral Commodity Summaries 2013, p. 116–117.
Bray, E.L., 2013, Bauxite and alumina: U.S. Geological Survey Mineral Commodity Summaries 2013, p. 26–27.
Crangle, R.D, Jr., 2013, Pumice and pumicite: U.S. Geological Survey Mineral Commodity Summaries 2013, p. 124–125.
Eldorado Gold Corp., 2012, Stratoni: Eldorado Gold Corp. (Accessed July 9, 2013, at http://www.eldoradogold.com/s/Stratoni.asp.)
Global Cement, 2012, Titan posts Euro24.5m loss in 2012: Global Cement. (Accessed July 28, 2013, at http://www.globalcement.com/news/itemlist/tag/Titan.)
Glory Resources Ltd., 2012, Glory targets big resource jump: Glory Resources Ltd. (Accessed September 2, 2013, at http://www.gloryresources.com.au/wp-content/uploads/2012/08/Glory-Resources-Glory-targets-big-Sapes-resources-jump-280812.pdf.)

---

[1]Where necessary, values have been converted from euro area euros (€) to U.S. dollars (US$) at an average rate of €1.00=US$1.32.

Grecian Magnesite S.A., 2012, Profile: Grecian Magnesite S.A. (Accessed July 9, 2013, at http://www.grecianmagnesite.com/index.php?page=&category_id=28.)

Industrial Minerals, 2012, Greece seeks mineral lifeboat: Industrial Minerals, no. 532, January, p. 40.

Industry Review, 2012, Mining in Greece to 2015: Industry Review. (Accessed July 7, 2013, at http://industryreviewstore.blogspot.com/2012/02/greek-mining-industry-market.html.)

Institute of Geology and Mineral Exploration Management, 2012, About Institute of Geology and Mineral Exploration Management (IGMEM): Institute of Geology and Mineral Exploration Management. (Accessed July 7, 2013, at http://www.linkedin.com/company/institute-of-geology-and-mineral-exploration-igme.)

Invest in Greece SA, 2012, Wind: Invest in Greece SA. (Accessed July 9, 2013, at http://www.investingreece.gov.gr/?pid=36&sectorID=48&la=1.)

Larco G.M.M. S.A., 2012, Uses of nickel: Larco G.M.M. S.A. (Accessed July 9, 2013, at http://www.larco.gr/nickel.php.)

Petra Petroleum Inc., 2012, Petra Petroleum Inc. announces application for exploration license in Greece: Petra Petroleum Inc. (Accessed July 9, 2013, at http://www.petrapetroleum.com/pdf/PTLNRJul11.pdf.)

Public Power Corp., 2012, PPC today: Public Power Corp. (Accessed July 9, 2013, at http://www.dei.gr/Default.aspx?t=1001&nt=18&lang=2.)

United Nations, 2012, Policy and regulations—Main features of national mining codes or mineral industry code: United Nations, p. 18–19. (Accessed July 7, 2013, at http://www.un.org/esa/dsd/dsd_aofw-ni/ni_pdfs/NationalReports/greece/Greece-CSD18-19_chapter_II-mining.pdf.)

U.S. Census Bureau, 2012a, U.S. exports to Greece by 5-digit-end-use code 2003–2012: U.S. Census Bureau. (Accessed July 28, 2013, at http://www.census.gov/foreign-trade/statistics/product/enduse/exports/c4840.html.)

U.S. Census Bureau, 2012b, U.S. imports to Greece by 5-digit-end-use code 2003–2012: U.S. Census Bureau. (Accessed July 28, 2013, at http://www.census.gov/foreign-trade/statistics/product/enduse/imports/c4840.html.)

U.S. Central Intelligence Agency, 2012, Greece, in The world factbook: U.S. Central Intelligence Agency. (Accessed July 7, 2013, at http://www.cia.gov/library/publications/the-world-factbook/geos/countrytemplate_gr.html.)

U.S. Department of State, 2013, Investment climate statement-Greece—Bureau of Economic and Business Affairs: U.S. Department of State. (Accessed September 2, 2013, at http://www.state.gov/e/eb/rls/othr/ics/2013/204649.htm.)

Virta, R.L., 2013, Clays: U.S. Geological Survey Mineral Commodity Summaries 2013, p. 44–45.

TABLE 1

GREECE: PRODUCTION OF MINERAL COMMODITIES[1]

(Metric tons unless otherwise specified)

Commodity[2]		2008	2009	2010	2011	2012[e]
METALS						
Aluminum:						
Bauxite		2,176,300	1,935,000	1,902,000	2,300,000 [r, e]	1,816,000 [3]
Alumina, Al_2O_3		771,769	718,797	725,000	809,700 [r]	784,400 [3]
Metal, primary		162,339	134,737	136,765 [r]	167,490 [r]	165,046 [3]
Chromite, ore, crude[e]		1,400	1,400	1,400	1,200	1,200
Iron and steel:						
Iron ore and concentrate, nickeliferous, Fe content[e]		570,000	560,000	560,000	550,000	550,000
Metal:						
Ferroalloys, ferronickel, gross weight		83,200	41,300	42,000	40,000	40,000
Steel, crude	thousand metric tons	2,477	2,082	1,839	1,993	2,000
Lead:[e]						
Mine output, Pb content		23,314 [3]	17,027 [3]	12,200	16,592 [3]	18,062 [3]
Metal, secondary	thousand metric tons	11	10	10	10	10
Nickel:						
Ore, Ni content of nickeliferous iron ore	thousand metric tons	16,640	10,203	13,837	14,100	14,000
Metal, Ni content of ferronickel		18,600	8,269	13,960	18,527 [r]	18,632 [3]
Silver, mine output, Ag content[e]	kilograms	35,500	30,177 [3]	29,000	30,000	32,000
Zinc, mine output, Zn content by analysis		20,300 [e]	18,126	19,967	39,127 [r]	41,824 [3]
INDUSTRIAL MINERALS						
Cement, hydraulic[e]	thousand metric tons	11,361 [3]	11,160 [3]	11,000	11,000	11,000
Clays:						
Bentonite, crude, includes attapulgite and sepiolite		1,389,800	926,186	1,381,643 [r]	1,188,442 [r]	1,235,105 [3]
Kaolin, crude		4,360	--	--	--	--
Feldspar		62,000	55,737	23,050	10,200 [r]	12,000
Gypsum and anhydrite		865,000	730,000	700,000	587,000 [r]	746,000 [3]
Magnesite:						
Ore, crude		361,165	380,834	396,000	541,813 [r]	351,266 [3]
Huntite, crude[e]		19,600 [3]	10,652 [3]	12,000	23,800 [r]	24,200 [3]

See footnotes at end of table.

TABLE 1—Continued
GREECE: PRODUCTION OF MINERAL COMMODITIES[1]

(Metric tons unless otherwise specified)

Commodity[2]		2008	2009	2010	2011	2012[e]
INDUSTRIAL MINERALS—Continued						
Nitrogen, N content of ammonia[e, 4]		130,000	130,000	130,000	130,000	130,000
Perlite:						
Crude		1,100,000 [e]	862,935	760,000 [e]	842,870 [r]	876,396 [3]
Screened		500,000 [e]	398,451	400,000 [e]	507,235 [r]	450,000 [3]
Pozzolan, Santorin earth[e]		1,059,000 [3]	830,000 [3]	850,000	350,000 [r]	285,000 [3]
Pumice		828,000	381,000	400,000 [e]	468,960 [r]	385,917 [3]
Salt, all types		220,000	189,000	190,000 [e]	174,500 [r]	191,970 [3]
Silica		64,521	37,905	40,000 [e]	1,671 [r]	--
Stone, marble[e]	cubic meters	347,526 [3]	255,516 [3]	250,000	285,000 [r]	320,000 [3]
Sulfur, S content of mixed sulfide ore		264,299	225,054	230,000 [e]	214,943 [r]	227,197 [3]
MINERAL FUELS AND RELATED MATERIALS						
Coal, lignite	thousand metric tons	64,521	61,800	53,600	58,400	62,335 [3]
Petroleum:						
Crude	thousand 42-gallon barrels	478	628	636	676	662 [3]
Refinery products:[e]						
Liquefied petroleum gas	do.	7,665 [3]	7,519 [3]	8,030 [r, 3]	8,000	8,000
Gasoline	do.	35,077 [3]	34,419 [3]	36,865 [r, 3]	32,000	32,000
Naphtha	do.	8,400	8,400	8,400	8,400	8,400
Jet fuel	do.	14,600	12,410 [3]	12,775 [r, 3]	14,000	14,000
Kerosene	do.	157 [3]	438 [3]	146 [r, 3]	200	200
Distillate fuel oil	do.	48,910 [3]	46,691 [3]	51,468 [r, 3]	47,000	47,000
Refinery gas	do.	4,800	4,800	4,800	4,800	4,800
Residual fuel oil	do.	39,055 [3]	38,435 [3]	37,814 [r, 3]	42,000	42,000
Bitumen	do.	4,000	3,960 [3]	4,000	4,000	4,000
Other	do.	1,800	1,876 [3]	1,800	1,800	1,800
Total	do.	164,000 [r]	159,000 [r]	166,000 [r]	162,000 [r]	162,000

[e]Estimated; estimated data are rounded to no more than three significant digits; may not add to totals shown. [r]Revised. do. Ditto. -- Zero.

[1]Table includes data available through October 30, 2013.

[2]In addition to the commodities listed, dolomite, lignite briquets, manganese, and other crude construction materials are produced, but available information is inadequate to make reliable estimates of output levels.

[3]Reported figure.

[4]Estimate based upon installed capacity (see table 2).

TABLE 2
GREECE: STRUCTURE OF THE MINERAL INDUSTRY IN 2012

(Thousand metric tons unless otherwise specified)

Commodity	Major operating companies and major equity owners	Location of main facilities	Annual capacity
Alumina, Al_2O_3	Aluminium S.A (Mytilineos Holdings S.A., 53%)	Agios Nikolaos, Boeotia area	1,100
Aluminum	do.	do.	180
Barite, $BaSO_4$	S&B Industrial Minerals, S.A. (Eliopoulos-Kyriakopoulos Group)	Milos Island (closed)	1
Bauxite	do.	Mines at Amfissa and Distomon, plants at Phocis and Itea	2,000
Do.	Delphi-Distomon S.A. (Mytilineos Holdings S.A.)	Mines at Amfissa and Distomon	800
Do.	Hellenic Mining Enterprises. S.A.	Mines at Aga Marina, Lamia	500
Bentonite:			
Crude	Mediterranean Bentonite Co. S.A. (Industria Chimica Mineraria S.p.A.)	Surface mines on Milos Island	20
Do.	Mykobar Mining Co. S.A. (Silver & Baryte Ores Mining Co. S.A.)	Mines at Adamas, Milos Island	300
Do.	do.	Plants at Adamas, Milos Island	200
Do.	S&B Industrial Minerals, S.A. (Eliopoulos-Kyriakopoulos Group)	Mines at Adamas, Milos Island	600
Processed	do.	Plant at Voudia Bay, Milos Island	400
Cement	Halkis Cement Co. S.A. (Lafarge Group, 89%)	Micro-Vathi plant, west-central (closed)	3,000
Do.	Halyps Cement S.A. (Ciments Français Group)	Paralia Aspropyrgos plant, Athens	800
Do.	Heracles General Cement S.A. (Lafarge Group)	Plants at Halkis Evia, Milaki Evia, and Volos	9,600
Do.	Titan Cement Co. S.A.	Elefsis plant, Athens area	400
Do.	do.	Kamari plant, Boeotia	2,600
Do.	do.	Patras plant, northern Peloponnesus	1,900
Do.	do.	Salonica plant, Salonica	1,650
Chromite	Financial-Mining-Industrial and Shipping Corp. (FIMISCO)	Tsingeli Mine, Volos	25
Ferroalloys, ferronickel, Ni content	Larco G.M.M. S.A.	Larimna metallurgical plant	25
Gold, Au in concentrate kilograms	Hellas Gold S.A. (European Goldfields Ltd.)	Kassandra Mines (Olympias and Stratoni), northeastern Chalkidiki	5,000
Gypsum	Lava Mining and Quarrying Co. S.A.	Altsi, Crete Island	500
Do.	Titan Cement Co. S.A.	do.	280
Hunite/Hydromagnesite	Microfine S.A.	Mines in Kozani basin	100
Lead, mine, Pb in concentrate	Hellas Gold S.A. (European Goldfields Ltd.)	Kassandra Mines (Olympias and Stratoni), northeastern Chalkidiki	30
Lignite	Public Power Corp. (PPC) (Government)	Aliveri Mine, Euboea Island	420
Do.	do.	Megalopolis Mine, central Peloponnesus	7,000
Do.	do.	Ptolemais Mine, near Kozani	28,000
Magnesite, concentrate	Grecian Magnesite S.A. (GM)	Mine and plant at Gerakini and Kalives, Chalkidiki, northern Greece	200
Manganese, battery-grade MnO_2	Eleusis Bauxite Mines Mining, Industrial and Shipping S.A. [National Bank of Greece (OAE)]	Nevrokopi, Drama	10
Marble, slab and tile cubic meters	Aghia Marina Marble Ltd.	Various areas of northern Greece	NA
Do. do.	Michelakis Marble S.A.	Kavala	NA
Do. do.	Gourlis Group	Quarries at Levadia, Neurokopi, and Tiseo	NA
Natural gas million cubic meters per day	Public Petroleum Corp. (PPC) (Government)	Prinos offshore gasfield and oilfield, east of Thasos Island	125
Do. do.	Energean Oil and Gas S.A.	South Kavala gasfield, east of Thasos Island	NA
Nickel, ore	Larco G.M.M. S.A.	Agios Ioannis Mine, near Larimna	700
Do.	do.	Evia Mine, near Larimna	1,500
Nitrogen, N content of ammonia	Phosphoric Fertilizers S.A.	Nea Karvall	150

See footnotes at end of table.

TABLE 2—Continued
GREECE: STRUCTURE OF THE MINERAL INDUSTRY IN 2012

(Thousand metric tons unless otherwise specified)

Commodity		Major operating companies and major equity owners	Location of main facilities	Annual capacity
Perlite		S&B Industrial Minerals, S.A. (Eliopoulos-Kyriakopoulos Group)	Mines on Milos Islands; plant at Pireaus	650
Do.		Otavi Minen Hellas S.A. (Otavi Minen AG)	Milos Island	150
Petroleum, crude	42-gallon barrels per day	Energean Oil and Gas S.A.	Prinos offshore oilfield, east of Thassos Island	NA
Petroleum, refined	do.	Hellenic Aspropyrgos Refinery S.A.	Aspropyrgos	95,000
Do.	do.	Motor Oil (Hellas) Corinth Refineries S.A.	Aghii Theodori, Corinth	170,000
Do.	do.	Petrola Hellas S.A.	Eleusis	100,000
Do.	do.	Thessaloniki Refining Co. A.E.	Thessaloniki	76,000
Pozzolan (Santorin earth)		Lava Mining and Quarrying Co. (Heracles General Cement Co.)	Xylokeratia, Milos Island	600
Do.		Titan Cement Co. S.A.	do.	300
Pumice		Lava Mining and Quarrying Co. (Heracles General Cement Co.)	Yali Island	1,000
Quartz (microcrystalline)		do.	Adamas, Milos Island	150
Steel, crude		Halyvourgia Thessalias S.A. (Manessis Bros. and Voyatzis S.A., 65%, and National Investment Bank for Industrial Development, 35%)	Steelworks at Volos	1,500
Do.		Sidenor Steel Products Manufacturing S.A.	Steelworks at Thessaloniki and Almyros	2,800
Do.		Halyvourgiki, Inc.	Steelworks at Eleusis	1,200
Do.		Hellenic Steel Co.	Steelworks at Thessaloniki	1,000
Do.		Corinth Pipeworks S.A (CPW)	Steelworks at Thisvi	700
Zinc, mine, Zn in concentrate		Hellas Gold S.A. (European Goldfields Ltd.)	Kassandra Mines (Olympias and Stratoni), northeastern Chalkidiki	30

Do., do. Ditto. NA Not available.

THE MINERAL INDUSTRY OF HUNGARY

By Steven T. Anderson

Since the fall of communism in Hungary at the end of 1989, many large state-owned industrial companies were either closed or transformed into smaller privatized companies. In particular, many large uneconomic coal mines were closed in the country. Mining operations for crude construction materials (including aggregates, crushed rock, dimension stone, gravel, and sand) and other industrial minerals continued production, however. In 2012, Hungary was estimated to be the fifth-ranked producer of perlite (mostly for use in construction) in the world and to have accounted for about 4% of global production. With respect to metallic minerals, bauxite was still mined in the country; alumina was produced from the bauxite, as was gallium (as a byproduct of alumina refining), and Hungary could have accounted for about 2% of the world's production of gallium. Manganese ore was also produced in 2012, but the country accounted for no more than 1% of global mine production. Hungary continued to produce mineral fuels and related materials, but imports still accounted for about two-thirds of the country's total energy consumption (table 1; Bolen, 2013; Corathers, 2013; Hungarian Central Statistical Office, 2013b; Jaskula, 2013; U.S. Central Intelligence Agency, 2014; Encyclopedia of the Nations, undated).

Minerals in the National Economy

In 2012, the value[1] of production by the mining and quarrying sector accounted for 0.35% ($439 million) of the gross domestic product (GDP) compared with about 0.3% (about $417 million) of the GDP in 2011; the value of output by the coke and petroleum refinery products manufacturing sector accounted for 6.4% ($8 billion) of the GDP in 2012 compared with about the same percentage (about $9 billion) in 2011. In 2012, the value added to the GDP by the entire industrial sector accounted for 23% of the GDP, but this had mostly to do with manufacturing assembly activities for export. The country was not a major consumer of nonfuel minerals. The value added to the GDP by the construction sector accounted for about 3.8% of the GDP, and this sector appeared to use at least some domestically produced construction materials (Hungarian Central Statistical Office, 2013c, d; U.S. Department of Commerce, 2013, p. 35).

In 2012, Hungary's trade balance for crude materials (including nonfuel minerals) was $1.1 billion compared with $0.8 billion in 2011; and that for mineral fuels, related materials, and energy (including electricity) was –$8.1 billion compared with about –$8.5 billion in 2011. In 2012, the value of the country's imports of mineral fuels and related materials (including electricity) accounted for 13% of the total value of all imports compared with about 12% in 2011.

Hungarian Oil and Gas Co. plc. (MOL) was a participant in the Nabucco natural gas pipeline development project, which could provide Hungary with additional sources (possibly including Azerbaijan) and routes (through Turkey, Bulgaria, and Romania) for importing natural gas. Currently, Hungary obtains more than 80% of its imports of natural gas from Russia. MOL was also in discussions to become a participant in the South Stream pipeline project, however, which could provide an additional route for importing natural gas (possibly directly across the bed of the Black Sea to Bulgaria, then through Bulgaria and Serbia), but the source of the natural gas would be Russia (RT.com, 2012, 2013; Schneeweiss and Shiryaevskaya, 2012; Wiesmann, 2012; Hungarian Central Statistical Office, 2013a; Konstantinova, 2013; Molnár, 2013, p. 35, 40–41, 76; U.S. Central Intelligence Agency, 2014).

Government Policies and Programs

In 2012, the main mining law was Act No. 48, which came into effect in 1993, but it has been amended many times (including by Ministerial Decree no. 81/2012, which was promulgated in 2012 and provided additional regulation of the country's mining concession tendering procedure). The Mining Law and related amendments, decrees, and codes apply to all mineral commodities, including mineral fuels and related materials. The Mining Law defines the Government's legal basis for estimating reserves, for providing the geologic and technical information needed to outline concession tender conditions; for determining environmental risks associated with mining; for temporarily stopping mine production; and for regulating exploration, mine operation, and mineral processing, as well as overseeing mine closures and mine site remediation. According to the Mining Law, all mineral raw materials and geothermal energy are state-owned as long as they remain in their natural place of occurrence, but they become the property of the extractor upon extraction (and utilization). The Mining Law also sets up a schedule of new royalty rates, including 12% on the value of production of oil, natural gas, and carbon dioxide; 5%, nonmetallic hard minerals (other than hard mineral fuels, such as coal); and 2%, other hard minerals and geothermal energy. Through the end of 2012, information concerning whether or not there had been any changes to these royalty rates was not available (Katona and Fodor, 1998; United Nations Department of Social and Economic Affairs, Division for Sustainable Development, 2009; Hungarian Office for Mining and Geology, undated).

On February 14, 2012, the Government published a new National Energy Strategy. Five main areas were identified for the Government to focus on in order to try to achieve the objectives set forth in the strategy:
- increasing energy savings and energy efficiency,
- increasing the share of renewable energies,

---

[1]Where necessary, values have been converted from Hungarian forints (HUF) to U.S. dollars (US$) at an annual average exchange rate of about HUF200.67=US$1.00 for 2011 and HUF225.02=US$1.00 for 2012. All values are nominal, at current prices, unless otherwise stated.

- integrating the Central European grid network and constructing required cross-border capacities,
- maintaining existing nuclear capacities, and
- utilizing domestic coal and lignite resources in an ecofriendly manner for power generation

(Hungarian Ministry of National Development, 2012; Molnár, 2013).

Production

In 2012, production of aluminum in Hungary decreased by 19% compared with that of 2011 in response to decreasing prices in Europe, which Alcoa Inc. of the United States reported were significantly affected by uncertainty in commodity markets as a result of difficulties in managing the sovereign debt of some countries in the region. Also, aluminum demand in most sectors appeared to be lower in 2012 than in 2011 in Europe. The country's production of crude steel decreased by about 12% from that in 2011 in response to a further decrease in demand for steel in Europe, as well as owing to lower cost efficiencies at some of Hungary's steel plants relative to other European producers. Production of manganese (gross weight) in Hungary also decreased by 12% (or by about 14% in terms of Mn content), probably in response to decreased demand from the steel sector during this timeframe. In 2010 and 2011 (the latest years for which this information was available), the global steel sector accounted for approximately 90% of the total consumption of manganese in the world (table 1; Gesamtverband der Aluminiumindustrie e.V., 2012; ISD Dunaferr Co. Ltd., 2012; PR Newswire, 2012; Roskill Information Services Ltd., 2012; Alcoa Inc., 2013, p. 52–54, 62, 76, 90, 129; Gulyas, 2013).

In 2012, total production of clays, not including bentonite, was estimated to have decreased by 29% compared with that of 2011, but detailed production information by type of clay was not available for 2012. Consequently, the 34% decrease in production of unspecified clays listed in table 1 could be at least partially attributable to decreases in the production of kaolin or refractory clays as well. Information was also not available concerning the main reasons for the greater than 10% decreases in production of dolomite, gravel, peat, quartzite, and sandstone (table 1).

Structure of the Mineral Industry

Table 2 is a list of major mineral industry facilities. In September 2012, the Government categorized Magyar Aluminium Ltd. (MAL) as a company of strategic importance; at the end of February 2013, it also announced that a state-owned company (Nemzeti Reorganizacios Nonprofit LLC) would take over the management and operation of MAL after a Hungarian court ordered that MAL be liquidated to pay off company creditors and others who had sued for damages resulting from the collapse of a tailings dam on October 4, 2010, and MAL's reduced production following the accident. Some analysts expected that these developments could have been precursors to an attempt to fully nationalize MAL, but definitive information regarding possible nationalization plans or a timeline for any possible nationalization of MAL was not available. On September 30, 2012, the Government held approximately a 25% ownership interest in MOL. Information concerning any other government ownership of companies in the mineral industry was not available (table 2; Associated Free Press, 2012; Dékány, 2013b; Hungarian Oil and Gas Co. plc, 2013, p. 12; Szabó and Vitéz, 2013; Velkei, 2013).

Commodity Review

Metals

Aluminum, Bauxite and Alumina, and Gallium.—Alcoa projected that there could be increased cost uncertainty in 2013 and beyond for such energy-intensive industries as primary aluminum production owing to increasing regulation of greenhouse gas (GHG) emissions in Europe. The company expected increased GHG regulations following scientific reports that attributed a significant proportion of the current trend in global warming to be caused by human activities. Electricity is an essential input into the production of alumina (and gallium, as a byproduct of alumina refining) and primary aluminum, and energy expenditures are an extremely important part of the production costs for these commodities, especially in Europe. On February 5, 2013, MAL announced plans to close the Halimba bauxite mine in northwestern Hungary, but the company planned to continue to process ore and concentrates produced at its nearby Bakony Mine, as well as to import ore and concentrates. Production of bauxite at the Halimba Mine had been decreasing, especially since the tailings dam failure in October 2010, and MAL reported that it was no longer possible to operate the mine profitably at the current lower scale of production. Decreased production at the Halimba Mine was likely to have been the main cause of the 8% decrease in bauxite production in Hungary in 2012 compared with that in 2011. According to a supply contract that was renewed in 2011, MAL could have imported about 150,000 t of bauxite ore in 2012 from Rudnici Boksita Jajce (a bauxite mining company located in Montenegro), in which MAL reportedly owned a majority stake. Through 2012, about 80% of the ore and concentrates that MAL processed to produce alumina (and gallium) every year had been mined domestically (table 1; SeeNews, 2011; Alcoa Inc., 2013, p. 36, 42, 52–54, 117; AME Group, 2013; Budapest Times, The, 2013; Mining Journal, 2013; Worldal.com, 2013; Magyar Aluminium Ltd., undated).

Industrial Minerals

Cement.—In 2012, production of cement was estimated to have decreased by 5% compared with that of 2011 possibly because increasing production at the NOSTRA cement plant, which was inaugurated in September 2011, may not have been enough to compensate for the decrease in national cement production resulting from the closure of the Hejöcsaba cement plant during the last quarter of 2011. In 2012, Holcim Ltd. announced that it planned to close its plant at Labatlan sometime in 2013. In 2012, the value added to the GDP by the construction sector decreased by 6% compared with that of 2011, but this could have been mostly owing to the decrease in the prices of output by the sector, on average. According to a volume index of production by the construction sector, the physical volume of production by the sector decreased by only about 2.6% during this timeframe. Nonetheless, domestic demand for construction materials probably decreased in 2012 compared with that of 2011, and STRABAG SE reported that a return to growth in the sector in 2013 and beyond could require Government stimulus funding and approval of funds from the European Union (EU). In 2013, factors that could increase demand for construction materials in the civil engineering sector were expected to be metro construction in Budapest, construction projects to provide new information technology and water management services in the country, and investments in the energy sector (table 1; Hungarian Cement Association, 2011; Zadravecz, 2011; International Cement Review, 2012; Holcim Ltd., 2013, p. 74–75, 207; Hungarian Central Statistical Office, 2013c, e; Maheshwari, 2013; Mining Journal, 2013; STRABAG SE, 2013, p. 4, 14, 20, 89, 98).

Mineral Fuels and Related Materials

Coal, Natural Gas, and Uranium.—Since at least about 2008, production of coal at the Markushegy Mine has decreased steadily, and the mine had been found not to be profitable by the end of 2012. In January 2013, the European Commission decided to authorize public funding to aid in the process of closing the Markushegy Mine by the end of 2014 and help mitigate any social or environmental effects from the mine closure. Facing decreasing coal reserves and substantial uncertainty concerning access to natural gas pipelines that were being proposed in 2012, the Government announced that it would enter into a joint venture with Wildhorse Energy Ltd. of Australia to develop a uranium mining project in the Mecsek hills, which could include reopening an old uranium mine near Pecs. If successful, this project could help feed the planned service-life extension project of MVM Paks Nuclear Power Plant Ltd. In addition, the Government had reportedly included measures in its new National Energy Strategy to encourage development of more modern coal mining based on pure coal technology and of new oil and natural gas resources in Hungary (Eddy, 2012; Hungarian Ministry of National Development, 2012; European Commission, 2013; Makan, 2013; Mining Journal, 2013; Molnár, 2013; BudapestTelegraph.com, 2014; MVM Paks Nuclear Power Plant Ltd., undated).

Outlook

If the Government does nationalize MAL (and keeps it operating), or liquidates it by selling it to a new owner who can obtain the same environmental permissions that MAL has to keep it operating, then Hungary could continue to produce alumina, bauxite, and gallium at about the same levels as in 2012. Under no scenario—nationalization and 100% state ownership of MAL, new and at least partially private ownership of MAL, or the same ownership of MAL as in 2012—was a complete return to 2008 levels of production projected. The construction sector was expected to increase output slightly in 2013, but not by more than 1% above that of 2012, and it could possibly increase by between 3% and 4% more in 2014 and again in 2015, especially if the Government budgets more stimulus funds for the sector. Production of cement and other construction materials could increase by about those same percentages through 2015. Also, construction of the Hungarian section of the South Stream pipeline, including a new natural gas storage facility, could begin in 2015, which would significantly increase demand for construction materials in the country above what was projected in 2012 for the period 2015 through 2017, when the pipeline was projected to begin supplying natural gas to Hungary (Hungarian Cement Association, 2011; Associated Free Press, 2012; BudapestTelegraph.com, 2013; Dékány, 2013a; Maheshwari, 2013; Mining Journal, 2013; Szabó and Vitéz, 2013).

References Cited

Alcoa Inc., 2013, Annual report on Form 10–K for the fiscal year ended December 31, 2012: New York, New York, Alcoa Inc., February 19, 172 p.

AME Group, 2013, MAL Magyar Aluminium Termelo es Kereskedelmi Zrt.: AME Group Key Events & Announcements, September 17. (Accessed January 29, 2014, at http://www.ame.com.au/Website/Content/News.html?commoditycode=Ala.)

Associated Free Press, 2012, Plant managers on trial for Hungarian toxic mud spill: PhysOrg.com, September 25. (Accessed November 19, 2012, at http://phys.org/news/2012-09-trial-hungarian-toxic-mud.html.)

Bolen, W.P., 2013, Perlite: U.S. Geological Survey Mineral Commodity Summaries 2013, p. 116–117.

BudapestTelegraph.com, 2013, South Stream to start supplying gas to Hungary in 2017: Budapest [Hungary] Telegraph, December 13. (Accessed January 29, 2014, at http://www.budapesttelegraph.com/news/579/south_stream_to_start_supplying_gas_to_hungary_in_2017.)

BudapestTelegraph.com, 2014, Lázár—Expansion of Paks plant can cut price of electricity: Budapest [Hungary] Telegraph, January 17. (Accessed January 29, 2014, at http://www.budapesttelegraph.com/news/608/lazar:_expansion_of_paks_plant_can_cut_price_of_electricity.)

Budapest Times, The, 2013, MAL to be liquidated: The Budapest [Hungary] Times, March 7. (Accessed January 26, 2014, at http://budapesttimes.hu/2013/03/07/mal-to-be-liquidated/.)

Corathers, L.A., 2013, Manganese: U.S. Geological Survey Mineral Commodity Summaries 2013, p. 100–101.

Dékány, Lóránt, 2013a, Gazprom looks for Hungarian storage facility: Budapest [Hungary] Telegraph, November 13. (Accessed January 29, 2014, at http://www.budapesttelegraph.com/news/533/gazprom_looks_for_hungarian_storage_facility.)

Dékány, Lóránt, 2013b, Hungarian government determined to protect MOL—Prime minister says: Budapest [Hungary] Telegraph, October 16. (Accessed January 29, 2014, at http://www.budapesttelegraph.com/news/505/hungarian_government_determined_to_protect_mol_-_prime_minister_says_.)

Eddy, Kester, 2012, Hungary prepares uranium mine: The Financial Times [London, United Kingdom], June 27. (Accessed November 7, 2012, at http://blogs.ft.com/beyond-brics/2012/06/27/hungary-prepares-uranium-mine/#axzz2rqSEeRAa.)

Encyclopedia of the Nations, [undated], Hungary—Mining: Flossmoor, Illinois, Advameg, Inc. (Accessed November 7, 2012, at http://www.nationsencyclopedia.com/Europe/Hungary-MINING.html.)

European Commission, 2013, State aid—Commission approves aid for closure of coal mine in Hungary: Brussels, Belgium, European Commission press release, January 23. (Accessed January 29, 2014, at http://europa.eu/rapid/press-release_IP-13-35_en.htm.)

Gesamtverband der Aluminiumindustrie e.V., 2012, Aluminium business situation in autumn 2012—German aluminium industry produces less: Düsseldorf, Germany, Gesamtverband der Aluminiumindustrie e.V. news release, November 20. (Accessed January 28, 2014, at http://www.aluinfo.de/index.php/gda-news-en/items/aluminium-business-situation-in-autumn-2012---german-aluminium-industry-produces-less.html.)

Gulyas, Veronika, 2013, Hungary steel maker Dunaferr to lay off 1,500 people: The Wall Street Journal, August 13. (Accessed January 27, 2014, at http://online.wsj.com/article/BT-CO-20130813-703730.html.)

Holcim Ltd., 2013, Annual report 2012; Jona, Switzerland, Holcim Ltd., February 25, 220 p.

Hungarian Cement Association, 2011, Hungary—Building on traditions: International Cement Review, December, p. 22–28.

Hungarian Central Statistical Office, 2013a, Commodity pattern of external trade in HUF—2001–2012: Budapest, Hungary, Hungarian Central Statistical Office, September 2. (Accessed January 25, 2014, at http://www.ksh.hu/docs/eng/xstadat/xstadat_annual/i_qkt006.html.)

Hungarian Central Statistical Office, 2013b, Energy balance—1990–2012: Budapest, Hungary, Hungarian Central Statistical Office, July 3. (Accessed January 25, 2014, at http://www.ksh.hu/docs/eng/xstadat/xstadat_annual/i_qe001.html.)

Hungarian Central Statistical Office, 2013c, Value and distribution of gross value added by industries—NACE Rev. 2—1995–2012: Budapest, Hungary, Hungarian Central Statistical Office, September 30. (Accessed January 25, 2014, at http://www.ksh.hu/docs/eng/xstadat/xstadat_annual/i_qpt002c.html.)

Hungarian Central Statistical Office, 2013d, Value of industrial production by subsections—NACE Rev. 2—2001–2012: Budapest, Hungary, Hungarian Central Statistical Office, August 14. (Accessed January 25, 2014, at http://www.ksh.hu/docs/eng/xstadat/xstadat_annual/i_oia006a.html.)

Hungarian Central Statistical Office, 2013e, Volume indices of construction activities of the national economy by contractors—1990–2012: Budapest, Hungary, Hungarian Central Statistical Office, December 16. (Accessed January 25, 2014, at http://www.ksh.hu/docs/eng/xstadat/xstadat_annual/i_oe002.html.)

Hungarian Ministry of National Development, 2012, National Energy Strategy 2030 published: Budapest, Hungary, Hungarian Ministry of National Development, February 15. (Accessed January 26, 2014, at http://www.kormany.hu/en/ministry-of-national-development/news/national-energy-strategy-2030-published.)

Hungarian Office for Mining and Geology, [undated], Law in English: Budapest, Hungary, Hungarian Office for Mining and Geology. (Accessed November 7, 2012, at http://www.mbfh.hu/home/html/index.asp?msid=1&sid=0&hkl=435&lng=1.)

Hungarian Oil and Gas Co. plc., 2013, 2012 Annual report: Budapest, Hungary, Hungarian Oil and Gas Co. plc, April 25, 103 p.

International Cement Review, 2012, Hungarian closure: International Cement Review, November, p. 14.

ISD Dunaferr Co. Ltd., 2012, Profitable Dunaferr is the aim: Dunaújváros, Hungary, ISD Dunaferr Co. Ltd. press release, June 4. (Accessed January 28, 2014, at http://www.dunaferr.hu/en/mediacentre/press-release/326-profitable-dunaferr-is-the-aim.)

Jaskula, B.W., 2013, Gallium: U.S. Geological Survey Mineral Commodity Summaries 2013, p. 58–59.

Katona, Gabor, and Fodor, Bela, 1998, Introduction of the mining royalty system in Hungary: Nonrenewable resources, v. 7, no. 1, p. 3–5.

Konstantinova, Elizabeth, 2013, Nabucco says has 'A lot to negotiate' over Shah Deniz accord: New York, New York, Bloomberg L.P., January 10. (Accessed January 14, 2013, at http://www.bloomberg.com/news/2013-01-10/nabucco-shah-deniz-have-a-lot-to-negotiate-dolezal-says-1-.html.)

Magyar Aluminium Ltd., [undated], History: Ajka, Hungary, Magyar Aluminium Ltd. (Accessed January 28, 2014, at http://english.mal.hu/engine.aspx?page=bemutatkozunk.)

Maheshwari, Shushmul, 2013, CEE cement industry—Heading for a new dawn: worldcement.com, Palladian Publications Ltd., January 10. (Accessed January 24, 2014, at http://www.worldcement.com/news/BRICs/articles/Cement_industry_Central_Eastern_Europe_Russia_1.aspx.)

Makan, Ajay, 2013, Azerbaijan gas decision to disappoint Brussels: The Financial Times [London, United Kingdom], September 19. (Accessed September 20, 2013, at http://www.ft.com/intl/cms/s/0/0f9a3c44-212a-11e3-a92a-00144feab7de.html.)

Mining Journal, 2013, Focus—Central Europe—Rising to the challenge: Mining Journal, May 17, p. 16–25

Molnár, Kata, 2013, Security of gas supply and dependency on imports: London, United Kingdom, International Law Office, Globe Business Publishing Ltd., January 21. (Accessed February 1, 2013, at http://www.internationallawoffice.com/newsletters/detail.aspx?g=22d1f290-544f-4a08-a5ac-32d37c593b34&redir=1.)

MVM Paks Nuclear Power Plant Ltd., [undated], About us: Paks, Hungary, MVM Paks Nuclear Power Plant Ltd. (Accessed January 29, 2014, at http://paksnuclearpowerplant.com/about-us.)

PR Newswire, 2012, Outlook positive for manganese producers: New York, New York, PR Newswire Association LLC, February 1. (Accessed November 16, 2012, at http://www.prnewswire.com/news-releases/outlook-positive-for-manganese-producers-138474619.html.)

Roskill Information Services Ltd., 2012, Manganese—Global industry markets and outlook 2012—12th edition: London, United Kingdom, Roskill Information Services Ltd., January 31. (Accessed January 28, 2014, at http://www.roskill.com/reports/steel-alloys/manganese.)

RT.com, 2012, Gazprom and partners kick off construction of South Stream pipeline: Moscow, Russia, Autonomous Nonprofit Organization "TV-Novosti", December 13. (Accessed January 14, 2013, at http://rt.com/business/russia-south-stream-launch-506/.)

RT.com, 2013, Russia launches South Stream gas pipeline in Serbia: Moscow, Russia, Autonomous Nonprofit Organization "TV-Novosti", November 25. (Accessed January 26, 2014, at http://rt.com/business/serbia-gazprom-pipeline-launch-238/.)

Schneeweiss, Zoe, and Shiryaevskaya, Anna, 2012, Nabucco faces 'terminal blow' as Hungary woos Russia link: New York, New York, Bloomberg L.P., April 24. (Accessed November 19, 2012, at http://www.bloomberg.com/news/2012-04-24/nabucco-faces-terminal-blow-as-hungary-woos-russia-link.html.)

SeeNews, 2011, Montenegrin bauxite mining co Rudnici Boksita renews export deal with Hungary's MAL: Sofia, Bulgaria, All Data Processing Ltd. (Accessed January 28, 2014, at http://wire.seenews.com/news/montenegrin-bauxite-mining-co-rudnici-boksita-renews-export-deal-with-hungarys-mal-190657.)

STRABAG SE, 2013, Annual report 2012: Villach, Austria, STRABAG SE, April 30, 180 p.

Szabó, Gábor, and Vitéz, I.F., 2013, It'll take time and ingenuity to nationalize the Hungarian Aluminum Company: Budapest [Hungary] Telegraph, March 16. (Accessed January 26, 2014, at http://www.budapesttelegraph.com/news/292/it'll_take_time_and_ingenuity_to_nationalize_the_hungarian_aluminum_company.)

United Nations Department of Social and Economic Affairs, Division for Sustainable Development, 2009, Hungary—Mining: New York, New York, United Nations, July 22, 8 p. (Accessed November 7, 2012, at http://www.un.org/esa/dsd/dsd_aofw_ni/ni_pdfs/NationalReports/hungary/Mining.pdf.)

U.S. Central Intelligence Agency, 2014, Hungary, in The world factbook: U.S. Central Intelligence Agency, January 7. (Accessed January 25, 2014, at https://www.cia.gov/library/publications/the-world-factbook/geos/hu.html.)

U.S. Department of Commerce, 2013, Doing business in Hungary—Country commercial guide for U.S. companies 2013: U.S. and Foreign Commercial Service and U.S. Department of State, May 31, 99 p.

Velkei, Tamás, 2013, Ajka alumina plant accident—Three years on the courts are still out: Budapest [Hungary] Telegraph, October 5. (Accessed January 29, 2014, at http://www.budapesttelegraph.com/news/491/ajka_alumina_plant_accident_-_three_years_on_the_courts_are_still_out.)

Wiesmann, Gerrit, 2012, RWE set to quit Nabucco gas pipeline: The Financial Times [London, United Kingdom], December 2. (Accessed December 3, 2012, at http://www.ft.com/intl/cms/s/0/278a8582-3c94-11e2-a6b2-00144feabdc0.html.)

Worldal.com, 2013, Troubled alumina maker MAL cuts off 250 jobs: Zhengzhou City, China, Henan Zhonglv Information Co. Ltd., September 18. (Accessed January 29, 2014, at http://www.worldal.com/news/eu/2013-09-18/137948739144007.shtml.)

Zadravecz, Zsófia, 2011, Europe's youngest cement plant: International Cement Review, December, p. 30–35.

TABLE 1
HUNGARY: PRODUCTION OF MINERAL COMMODITIES[1]

(Thousand metric tons unless otherwise specified)

Commodity[2]		2008	2009	2010	2011	2012
METALS						
Alumina, gross weight, calcined basis		299	185	214	165 r, e	150 e
Aluminum, unwrought, including secondary[e]		280 r	184 [3]	234 [3]	185 r	150
Bauxite, gross weight		511	317 r	365 r	278	255
Gallium[e]	kilograms	5,100	3,400	4,000 r	5,000 r	4,600
Iron and steel, metal:						
Pig iron		1,289	1,050	1,325	1,317 r	1,229
Steel:						
Crude		2,097 r	1,403 r	1,678 r	1,746 r	1,543
Semimanufactures		2,196	1,452	1,594	1,765 r	1,928
Manganese ore, run-of-mine:						
Gross weight		50	43	55	58	51
Mn content		13 r	13 r, e	15	16 r	13
INDUSTRIAL MINERALS						
Cement, hydraulic[e]		3,544 [3]	2,800 r	2,100 r	2,000 r	1,900
Clays:						
Bentonite, raw		7	5	17 r, e	17	17
Chamotte, refractory clays[e]		200	209 [3]	82 [3]	86	80
Kaolin, beneficiated		584	266	239	248	250 e
Other, unspecified		4,900 e	1,851	1,271	1,780 r, e	1,170 e
Diatomite[e]	metric tons	1,500	1,000	1,300	1,309 [3]	1,287 [3]
Gypsum and anhydrite	do.	15,940 r	19,766	20,000 e	0 r	0
Lime, calcined[e]		250	210 [3]	260 [3]	250	230
Nitrogen, N content of ammonia[e]		300	300	300	300	300
Peat, agricultural use[4]		90	85 r, e	54 r	82	60
Perlite		67 e	82	71	71 r	72
Quartzite[e]	metric tons	1,300	1,000	700 r	540 r, 3	313 [3]
Sand and gravel:						
Gravel		25,000 e	23,496	19,157	18,350 r	16,160
Sand:						
Common		5,400	12,095	5,902	4,943 r	4,600 e
Foundry		100	111	137	144 r	150 e
Glass (silica)		220	85	271	287 r	300 e
Stone:						
Dimension, all types[e]		5,700	5,500	5,500	2,919 r	2,826 [3]
Dolomite		6,200 e	4,393	2,503	668 r	549
Limestone		3,200 e	5,082	2,874	4,580 r	4,435
Marl		700 e	506	25	57 r	61
Sandstone[e]		15,000	10,000	15,000	15,150 [3]	9,970 [3]
Sulfur, byproduct, elemental, all sources[e]		65	60	60	60	60
Sulfuric acid[e]		80	75	75	75	75
MINERAL FUELS AND RELATED MATERIALS						
Coal:						
Brown		1,373	973	911	758	859
Lignite		8,041	8,027	8,203	8,801	8,438
Total		9,414	9,000	9,114	9,559	9,297
Coke, metallurgical[e]		999	746	1,100 r	820 r	870
Gas, natural, net (marketable)	million cubic meters	2,703	2,748	2,600	2,667 r	2,280

See footnotes at end of table.

TABLE 1—Continued
HUNGARY: PRODUCTION OF MINERAL COMMODITIES[1]

(Thousand metric tons unless otherwise specified)

Commodity[2]		2008	2009	2010	2011	2012
MINERAL FUELS AND RELATED MATERIALS—Continued						
Petroleum:						
Crude[5]	thousand 42-gallon barrels	5,580 [r]	5,550 [r]	5,025 [r]	4,470	4,410
Refinery:[e]						
Motor fuel (including aviation fuel)	do.	13,600 [r]	11,200 [r]	10,700 [r]	11,200 [r]	12,000
Distillate fuels	do.	9,820	8,070	8,760	9,200	9,870
Kerosene	do.	2,270	1,870	1,760	1,850	1,980
Gas oils	do.	31,100 [r]	25,600 [r]	28,800 [r]	30,200 [r]	32,400
Other fuel oils	do.	870 [r]	715 [r]	367 [r]	385 [r]	414
Lubricating oils	do.	1,510	1,240	1,190	1,250	1,340
Liquefied propane and butane	do.	3,300	2,710	2,500	2,630	2,820
Petroleum jelly, paraffin wax, and other waxes	do.	384	316	328	344	370
Petroleum coke, bitumen, and residues	do.	6,580	5,410	5,120	5,380	5,770
Total	do.	69,400 [r]	57,100 [r]	59,500 [r]	62,400 [r]	67,000

[e]Estimated; estimated data are rounded to no more than three significant digits; may not add to totals shown. [r]Revised. do. Ditto.

[1]Table includes data available through February 18, 2014.

[2]In addition to the commodities listed, talc, urea, and a variety of other industrial minerals and construction materials may have been produced, but available information is inadequate to make reliable estimates of output.

[3]Reported figure.

[4]Data before 2011 may include production of alginite and (or) paludal materials (including paludal mud).

[5]Figures were converted to thousand 42-gallon barrels from production reported in thousand metric tons.

TABLE 2
HUNGARY: STRUCTURE OF THE MINERAL INDUSTRY IN 2012

(Thousand metric tons unless otherwise specified)

Commodity	Major operating companies and major equity holders	Location of main facilities	Annual capacity
Alumina	Magyar Aluminium Ltd. (MAL)	Ajka Timfoldgyar plant, about 120 kilometers southwest of Budapest, near Lake Balaton	400
Alumina, fused	Motim Electrocorundum Ltd.	Plant at Mosanmagyarovar	50
Aluminum	Alcoa-Köfém Kft (Alcoa Inc., 100%)	Székesfehérvár ingot plant	NA
Bauxite	Magyar Aluminium Ltd. (MAL)	Bakony and Halimba Mines, 5 kilometers south of Ajka, northwest Hungary	NA
Bentonite	Bentonit Hungaria Kft (S&B Industrial Minerals S.A., 100%)	Mines and plant at Egyhazaskeszo	NA
Cement	Duna-Drava Cement Kft. (HeidelbergCement AG, 50%, and Schwenk Zement KG, 50%)	Plants at Beremend, 30 kilometers south of Pecs, and Vac, 35 kilometers north of Budapest	2,500
Do.	Holcim Hungaria Zrt. (Holcim Ltd.)	Plant at Labatlan	500
Do.	Lafarge Cement CE Holding GmbH (Lafarge S.A., 70%, and STRABAG SE, 30%)	NOSTRA plant at Kiralyegyhaza, southwestern Hungary	1,000
Clays	Agyag-Asvany Kft.	Two opencast mines at Felsopeteny	NA
Coal:			
Brown coal	Vertes Power Plant Ltd. (Magyar Villamos Muvek Zrt., 96.59%)	Markushegy Mine at Oroszlany, 55 kilometers west of Budapest	1,400 [e]
Lignite	Mátrai Erömü Zrt. (MÁTRA) (RWE AG, 50.9%; Magyar Villamos Muvek Zrt., 25.5%; EnBW AG, 21.7%)	Thorez opencast mine at Visonta, 80 kilometers northeast of Budapest	4,700 [e]
Do.	do.	Opencast mine at Bukkabrany, 130 kilometers northeast of Budapest	4,000 [e]
Coke	ISD Kokszolo Ltd. (ISD Dunaferr Co. Ltd.)	Dunaujvaros, 60 kilometers south of Budapest	1,000
Iron, pig iron	ISD Dunaferr Co. Ltd. (Industrial Union of Donbass)	do.	1,400
Manganese	Mangán Mining and Processing Ltd.	Úrkút manganese ore mines, 120 kilometers southwest of Budapest	NA
Natural gas	Hungarian Oil and Gas Co. plc. (MOL)	Oil and gas fields in southern and southwestern Hungary	NA
Perlite	Perlit 92 Kft	Palhaza, northeastern Hungary; opencast mine and processing plant	NA
Petroleum:			
Crude 42-gallon barrels per day	Hungarian Oil and Gas Co. plc. (MOL)	Oil and gas fields in southern and southwestern Hungary	14,800 [e]
Refined	Duna Refinery [Hungarian Oil and Gas Co. plc. (MOL), 100%]	Szazhalombatta, 25 kilometers southwest of Budapest	8,100
Pig iron	ISD Dunaferr Co. Ltd. (Industrial Union of Donbass)	Dunaujvaros, 60 kilometers south of Budapest	1,300
Silica	Uveg-Asvany Banyaszati Ipari Kft.	Mine and plant at Fehevaresugo	NA
Steel, crude:			
Primary	ISD Dunaferr Co. Ltd. (Industrial Union of Donbass)	Dunaujvaros, 60 kilometers south of Budapest	1,600
Secondary	OAM OZD Steelworks Ltd.	120 kilometers northeast of Budapest	360
Do.	Dam 2004 Acel-es Hengermu Kereskedemi es Szolgaltato Ltd.[1]	Diosgyor, 145 kilometers northeast of Budapest	550

[e]Estimated. Do., do. Ditto. NA Not available.

[1]Stopped production in December 2008.

THE MINERAL INDUSTRY OF ICELAND

By Harold R. Newman

Iceland is the third-largest island in Europe after Great Britain and Greenland and is one of the most active volcanic regions on Earth; it has more than 100 volcanoes, of which more than 25 have erupted in recent history. Iceland consists mainly of basaltic rock of Quaternary and Tertiary ages. It lies astride the Mid-Atlantic Ridge, which is a part of an undersea mountain system that was formed by magma pushing up through the gap formed along the spreading boundary of the European and North America tectonic plates. Because of its active volcanism, Iceland is the largest island on the Mid-Atlantic Ridge (Iceland on the Web, 2012; Kious and Tilling, 2012).

The Government had applied for membership in the European Union (EU) in 2009 and was continuing its efforts to join the EU. Nations applying for EU membership must negotiate 35 policy chapters with the 27-nation EU, and Iceland had completed the negotiation of 18 chapters by yearend 2012. Many of the Government policies were already in line with EU policies owing to Iceland's membership in the European Economic Area. More difficult negotiations were expected in the future when the EU and the Government open talks on fishing, which was a major source of revenue for the country. Also, there was disagreement concerning the country's whaling tradition (EUbusiness Ltd., 2012).

Iceland had only a few proven mineral resources, which included base metals and industrial minerals. The country was dependent on imports to meet domestic demand for most mineral commodities. The country accounted for about 2% of global production of primary aluminum and about 1% of global production of ferrosilicon (Bray, 2013; Corathers, 2013).

Most of Iceland's production of aluminum and ferrosilicon was exported. Because of the country's geographic proximity to the EU and membership in the European Free Trade Association, most of Iceland's trade was with Europe (U.S. Central Intelligence Agency, 2013).

Iceland was an open economy, with goods and services accounting for 70.5% of the gross domestic product in 2012. U.S. exports to Iceland included petroleum products valued at $18.9 million; metallurgical-grade coal valued at $9.9 million; and iron and steel products valued at $707,000 (U.S. Census Bureau, 2012a). U.S. imports from Iceland included finished metal shapes, except steel, valued at $15.2 million; steelmaking and ferroalloying materials valued at $8.2 million; industrial chemicals valued at $1.3 million; and miscellaneous nonferrous minerals valued at $267,000 (U.S. Census Bureau, 2012b).

Production

Metallic minerals in Iceland were not available in sufficient quantities to make mining feasible with existing technology. The country's aluminum and ferrosilicon industries relied on imported materials and inexpensive geothermal and hydroelectric energy. Aluminum was Iceland's leading mineral commodity followed by ferrosilicon. In 2012, production of both aluminum and ferrosilicon increased. The country's domestic production of industrial minerals included cement, crushed stone, pumice, salt, sand and gravel, and scoria (table 1).

Structure of the Mineral Industry

The only significant change in the structure of Iceland's mineral industry in 2012 was the full purchase of the Elkem Iceland silicon facility by a Chinese company, China National Blue Star. The ownership of the shares in the other mineral enterprises remained about the same as in 2011 (table 2).

Commodity Review

Metals

Aluminum.—In 2012, aluminum smelting, which is very power intensive, was the most economically important industry in the country. Three plants were in operation: Alcoa Inc. of the United States's Fjaröaál smelter at Reydarfjordur; Century Aluminum Co. of the United States's smelter at Grundartangi; and Rio Tinto Alcan of Canada's Reykjavik [ISAL] smelter at Straumsvik. Aluminum production accounted for about one-seventh of the country's economic output (Info Iceland, 2012).

Alcoa announced that the construction of a new potlining facility at Fjaröaál had been completed at a cost of $35 million. Construction was initiated in 2010, and the facility took about 18 months to complete. Hatch Engineering Co. of Canada was the construction company. Bids were invited for the operation of the potlining facility. Alcoa's smelter at Fjaröaál was proving to be an economic asset for eastern Iceland (Alcoa Inc., 2012).

Rio Tinto Alcan was increasing its production capabilities and updating equipment in its casthouse in Straumsvik to increase the smelter's production capacity to 230,000 metric tons per year (t/yr) from 190,000 t/yr, along with increasing the efficiency of the purification equipment. The output also would be changed from rolling slabs to extrusion billets, which would require modifications to the vesting equipment and installation of three continuous homogenizing furnaces (HRV Engineering, 2012).

Silicon.—Elkem A.S. of Norway announced the sale of its subsidiary Elkem Iceland to China National Blue Star for $2 billion. ChemChina Corp. of China held an 80% share in China National Blue Star and the Blackstone Group of the United States held a 20% share. The purchase included the acquisition of Elkem Carbon, Elkem Foundry Products, Elkem Silicon Materials, and Elkem Solar (Iceland Review, 2012).

Industrial Minerals

Pumice.—Jardefnaidnadur ehf (JEI), which was known for its research of aggregates in southern Iceland, also mined pumice in the Mount Hekla region about 100 kilometers from Reykjavik. The Hekla pumice is a lightweight porous stone of volcanic nature. The stone was mined from an open pit operation, and the raw material was transported to the harbor area for processing. The processing consists of grinding and screening of the pumice to various grain sizes and washing it to remove any sludge particles. In 2012, JEI's main export markets were Belgium, Denmark, the Netherlands, and the United States (Jardefnaidnadur ehf, 2012).

Mineral Fuels and Other Sources of Energy

Geothermal and Hydroelectric Energy.—Iceland was at the forefront in the use of renewable energy resources, and it had one of the largest potential sources of renewable energy in the world. Iceland has a significant amount of large-scale power potential from geothermal sources and hydropower generation. The country's economically viable electric-power-generating potential was estimated to be 50,000 gigawatthours per year (GWh/yr), of which only 8,500 GWh/yr was produced in 2012 (Iceland Trade Directory, 2012).

Petroleum.—The Government announced the awarding of two exploration and production licenses in its offshore Dreki Area, northeast of the country. The Dreki Area is a potentially hydrocarbon-rich region that lies within Iceland's Exclusive Economic Zone (EEZ). Faroe Petroleum plc and Valiant Petroleum plc of the United Kingdom were the two companies to be awarded licenses in the EEZ. The Ministry of Petroleum and Energy of Norway could participate in the licenses up to a share of 25%. The offshore license area is on a ridge that forms part of the Jan Mayen microcontinent between the conjugate margins of both the East Greenland and the Norwegian continental shelves where several significant natural gas and petroleum fields have been developed (Rigzone, 2012).

Outlook

Aluminum and ferroalloy production are expected to continue to dominate the mineral sector of Iceland. Iceland's future development will continue to depend on its utilization of its abundant hydroelectric and geothermal power. No base-metal mining is expected to take place in the near future. Petroleum exploration is expected to continue offshore Iceland. The Ministry of Industry and Commerce is expected to continue to seek ways to improve the competitiveness of the Icelandic mineral industry by increasing its variety and productivity.

References Cited

Alcoa Inc., 2012, Alcoa Fjardaal potlining facility now in operation—An investment of 36 million dollars: Alcoa Inc. (Accessed December 17, 2013, at http://www.alcoa.com/iceland/en/news/whats_new/2012/2012_06_potlining.asp.)

Bray, E.L., 2013, Aluminum: U.S. Geological Survey Mineral Commodity Summaries 2013, p. 16–17.

Corathers, L.A., 2013, Silicon: U.S. Geological Survey Mineral Commodity Summaries 2013, p. 144–145.

EUbusiness Ltd., 2012, Iceland takes quick steps towards EU membership: EUbusiness Ltd. (Accessed July 2, 2012, at http://www.eubusiness.com/news-eu/iceland-fish.hcv.)

HRV Engineering, 2012, Rio Tinto Alcan-IPU project: HRV Engineering. (Accessed December 26, 2013, at http://www.hrvengineering.com/Projects/RIOTintoalcanIPUProject/.)

Iceland on the Web, 2012, Iceland nature—Geology of Iceland: Iceland on the Web. (Accessed December 16, 2013, at http://www.iceland.vefur.is/iceland_nature/geology_of_iceland/index.htm.)

Iceland Review, 2012, Iceland ferrosilicon plant in Chinese ownership: Iceland Review. (Accessed December 18, 2013, at http://www.icelandreview.com?icelandreview/daily_news/?cat_id=16567&ew_0_a_id+372483.)

Iceland Trade Directory, 2012, Energy in Iceland: Iceland Trade Directory. (Accessed December 26, 2013, at http://www.icelandexport.is/english/industry_sectors_in_iceland/energy_in_iceland/.)

Info Iceland, 2012, Aluminum: Info Iceland. (Accessed December 26, 2013, at http://www.infoiceland.is/aluminium.html.)

Jardefnaidnadur ehf, 2012, About JEI: Jardefnaidnadur ehf. (Accessed December 26, 2013, at http://www.jei.is/index.html.)

Kious, W.J., and Tilling, R.L., 2012, This dynamic Earth—The story of plate tectonics (online edition, ver. 1.17): U.S. Geological Survey. (Accessed February 18, 2014, at http://pubs.usgs.gov/gip/dynamic/dynamic.html.)

Rigzone, 2012, Faro, Valiant, Petoro awarded Icelandic blocks: Rigzone. (Accessed December 6, 2013, at http:www.rigzone.com/news/article_pf.asp?a_id=122536.)

U.S. Census Bureau, 2012a, U.S. exports to Iceland by 5-digit end-use code: U.S. Census Bureau. (Accessed December 17, 2013, at http://www.census.gov/foreign-trade/statistics/product/enduse/exports/c4000.html.)

U.S. Census Bureau, 2012b, U.S. imports from Iceland by 5-digit end-use code: U.S. Census Bureau. (Accessed December 17, 2013, at http://www.census.gov/foreign-trade/statistics/product/enduse/imports/c4000.html.)

U.S. Central Intelligence Agency, 2013, Iceland, in The world factbook: U.S. Central Intelligence Agency. (Accessed December 25, 2013, at https://www.cia.gov/library/publications/the-world-factbook/geos/ic.html.)

TABLE 1
ICELAND: PRODUCTION OF MINERAL COMMODITIES[1]

(Metric tons)

Commodity[2]	2008	2009	2010	2011	2012
Aluminum, metal, primary[3]	761,204	817,963	825,803	780,853	802,827
Cement, hydraulic[e, 4]	138,000	138,000	140,000	142,000	146,000 [5]
Ferrosilicon	107,882	112,983	114,231	120,076	131,818

[1]Table includes data available through December 31, 2013.
[2]In addition to the commodities listed, other materials were thought to be produced, including pumice, salt, sand and gravel, scoria, and stone, crushed; however, information is inadequate to make reliable estimates of output.
[3]Ingot and rolling billet production.
[4]Sales.
[5]Reported number.

TABLE 2
ICELAND: STRUCTURE OF THE MINERAL INDUSTRY IN 2012

(Thousand metric tons)

Commodity	Major operating companies and major equity owners	Location of main facilities	Annual capacity
Aluminum	Alcoa Inc.	Fjaröaál smelter at Reydarfjordur	347
Do.	Reykjavik [ISAL] (Rio Tinto Alcan, 100%)	Straumsvik	230
Do.	Century Aluminum Co.	Grundartangi	260
Cement	Sementsverksmidja Rikisins (Government, 100%)	Akranes	115
Ferrosilicon	Elkem Iceland (Elkem A/S)	Plant at Grundartangi	100
Fertilizer	Aburdarverksmidja Rikisins (Government, 100%)	Gufunes	60
Pumice	Jardefnaidnadur ehf	Mount Hekla	210
Do.	Pumice Products Ltd. (BM Valla Ltd., 100%)	do.	32
Salt	Icelandic Salt Co. (Akzo Nobel NV, 58%)	Plant at Svartsengi	5

Do., do. Ditto.

THE MINERAL INDUSTRY OF IRELAND

By Alberto Alexander Perez

In 2012, the gross domestic product (GDP) of Ireland was $210.4 billion, which was an increase of about 0.9% compared with that of 2011. The banking and housing crisis that had affected the country since 2009 continued to depress the economy and particularly affected Ireland's industrial and services sectors, including the mineral industry. The general unemployment rate rose to 14.7%. In 2012, Ireland's total exports decreased by 6% to $119 billion[1] from $126 billion in 2011. Ireland accounted for 0.5% of all imports of mineral fuels, lubricants and related materials to the European Union (EU) and 0.3% of all such exports from the EU to non-EU countries. Ireland also received 0.4% of the imports of raw materials to the EU and provided 0.9% of the exports of raw materials from the EU. Ireland traded mostly with its EU partners, as 58.9% of all Irish exports went to EU members and 66.9% of its imports came from EU members.

In 2012, the United States imported $33.4 billion and exported $7.4 billion worth of goods and services from and to Ireland, respectively, principally chemicals, computers and electronic products, miscellaneous manufactured goods, and nonelectrical machinery. Ireland's main export partners, in terms of their share of the total value of all exports, were the United States (18%), the United Kingdom (17.4%), Belgium (15.6%), Germany (8.4%), Switzerland (5.8%), and France (5%). Ireland was part of the European Monetary Union, and, consequently, its official currency was the euro (€) (U.S. Central Intelligence Agency, 2013; European Commission, 2014a–f; U.S. Census Bureau, 2014).

Minerals in the National Economy

According to Ireland's Department of Communications, Energy and Natural Resources, mining in Ireland is highly cyclical and responsive to international commodity price movements. The mineral industry employed directly an estimated 1,300 people and produced an output valued at about $556 million in 2012. There were five operating mines in Ireland. They were, in order of the value of output, the Tara lead and zinc mine in Navan, Co. Meath, which was owned by Boliden Tara Mines Ltd. (a subsidiary of Boliden AB of Sweden) and was the leading zinc mine, by production tonnage, in the EU; the Lisheen lead and zinc mine, which was owned by Vedanta Resources plc of the United Kingdom; the Drummond and Knocknacran gypsum mines in Co. Monaghan, which were owned by Irish Gypsum Ltd. (a subsidiary of Saint Gobain Group of France); and the Galmoy lead and zinc mine in Co. Kilkenny, which was owned by Galmoy Mines Ltd. (a subsidiary of Lundin Mining Corp. of Canada) (Department of Communications, Energy and Natural Resources, 2012, 2013; U.S. Central Intelligence Agency, 2013).

Production

In 2012, Ireland's mine production of lead and zinc decreased by 5% and 1%, respectively. Alumina production remained at the same level as in 2011. Ireland was the principal zinc producer in the EU and a significant producer of lead. Ireland was ranked 10th in the world (by tonnage of output) in the mine production of zinc and 13th in the mine production of lead (table 1; Department of Communications, Energy and Natural Resources, 2013, p. 1; Guberman, 2013; Tolcin, 2013).

Structure of the Mineral Industry

Although Ireland was a significant producer of lead and zinc in the EU, the mining and the mineral-processing industry contributed only a small percentage of the country's GDP. Companies were mostly privately owned. The Exploration and Mining Division of the Department of Communications, Energy and Natural Resources is responsible for both the technical management of the state mineral licensing and leasing system and for promoting the mineral industry. The major mineral industry facilities and their capacities are listed in table 2 (Department of Communications, Energy and Natural Resources, 2013, p. 1).

CRH plc is an Irish conglomerate that was formed by the union of Cement Ltd. and Roadstone Ltd. to form Cement Roadstone Holdings plc. The name was changed to CRH plc in 1987. In 2012, CRH owned several Irish construction materials companies that produced principally cement, construction aggregates, lime, and limestone. CRH had previously owned Premiere Periclase plc (a seawater-magnesia-producing plant) but sold the operation to RHI AG of Austria in the second half of 2011 (CRH plc, 2012, p. 45; 2014).

Commodity Review

Metals

Bauxite and Alumina.—In 2012, United Company RUSAL (RUSAL) of Russia reported production of 1,927,000 metric tons (t) at its alumina refinery located on Aughinish Island on the south side of the Shannon estuary near Limerick City. The refinery was operated by Aughinish Alumina plc and was producing at 99% of capacity. RUSAL also reported that a 3-year contract had been signed with Norsk Hydro ASA of Norway to buy bauxite from Mineracao Rio do Norte S.A. (MRN) (a subsidiary of Vale S.A. of Brazil). RUSAL estimated that, with this supply, the Aughinish alumina refinery would have enough sweetening bauxite to achieve its production targets through 2015. Sweetening bauxite is a bauxite mineral of a higher quality than that used during refining; it is added during the alumina-making process to boost alumina production (United Company RUSAL, 2013, p. 15, 32).

---

[1]Where necessary, values have been converted from euro area euros (€) to U.S. dollars (US$) at an average exchange rate of €.7778=US$1.00.

Lead and Zinc.—Boliden Tara Mine's operation in Navan, Co. Meath, produced about 170,000 t of zinc and about 35,000 t of lead in 2012. Since the mine began its operations in 1977, production had totaled 80.7 million metric tons (Mt) at an average grade of 8.2% zinc and 1.9% lead. The mine's Joint Ore Reserves Committee (JORC)-classified ore reserves (proven and probable) were 14 Mt grading 7.2% zinc and 1.7% lead. The mine employed 718 people in 2012 (Boliden AB, 2013, p. 89, 117; Department of Communications, Energy and Natural Resources, 2013, p. 1).

Galmoy Mines ceased underground mining operations at its mine in Galmoy in October 2012. The total mine production of lead and zinc for 2012 amounted to 142,000 t at grades of 14% zinc and 2.4% lead. The ore was processed at the Lisheen Mine operations located in Co. Kilkenny (Lundin Mining Corp., 2013; Department of Communications, Energy and Natural Resources, 2013, p. 1).

In 2012, Vedanta Resources reported that the Lisheen Mine had produced 1.4 Mt of ore grading 11% zinc and 1.95% lead. The mine produced 321,000 t of zinc concentrates grading 53.3% zinc and 41,000 t of lead concentrates grading 60.3% lead. Since mining began at Lisheen in 1999, a total of about 19.72 Mt of ore grading an average of 11.8% zinc and 2% lead had been mined (Department of Communications, Energy and Natural Resources, 2013, p. 1).

Industrial Minerals

CRH reported that it owned 2 cement installations, a lime quarry, and 102 other types of quarries or pits in Ireland from which it extracted aggregates for construction, including sand and gravel. All the company's other extraction operations were located abroad, although it had processing facilities in Ireland to which it imported raw materials and at which it manufactured construction materials and components (CRH plc, 2013, p. 20).

Gypsum.—Irish Gypsum extracted gypsum from two sites in southern Co. Monaghan. The company reported that it had extracted less than 300,000 t of gypsum in 2012. The continued collapse of the construction market in Ireland appeared to be a leading reason for the continued decrease in gypsum production. The gypsum was crushed and blended locally and then supplied to Irish Gypsum's plaster mill and plaster board plant in Kingscourt (Department of Communications, Energy and Natural Resources, 2013, p. 1).

Outlook

Ireland will likely remain a major producer of zinc ore and a significant producer of alumina and lead ore in the EU. Exploration activity for gold, lead, and zinc is also expected to continue.

Oil and gas exploration and further development of the energy and renewable energy sectors will likely become a priority in Ireland in the near future. Consequently, interest in exploring for petroleum and natural gas offshore Ireland is likely to increase.

References Cited

Boliden AB, 2013, Annual report 2012; Stockholm, Sweden, Boliden AB, 132 p.
CRH plc, 2012, Annual report on Form 20–F: CRH plc. (Accessed August 4, 2014, at http://www.crh.com/docs/2012-annual-report/20-f.pdf?sfvrsn=4.)
CRH plc, 2013, Annual report on Form 20–F: CRH plc. (Accessed August 4, 2014, at http://www.crh.com/docs/annual-report-2013/2013-20-f.pdf?sfvrsn=4.)
CRH plc, 2014, Overview: CRH plc. (Accessed August 4, 2014, at http://www.crh.com/our-group/group-profile/history.)
Department of Communications, Energy and Natural Resources, 2012, Minister brief—Natural resources: Dublin, Ireland, Department of Communications, Energy and Natural Resources. (Accessed August 15, 2013, at http://www.dcenr.gov.ie/Corporate+Units/Minister/Ministers+Brief/Natural+Resources.htm.)
Department of Communications, Energy and Natural Resources, 2013, Ireland—Exploration and mining news: Dublin, Ireland, Department of Communications, Energy and Natural Resources, May 1, 6 p. (Accessed August 15, 2013, at http://www.dcenr.gov.ie/NR/rdonlyres/5695C011-D2A6-407A-A104- DE9A4909187A/0/IndustryNews_1stMay2013.pdf.)
European Commission, 2014a, Extra-EU28 trade of mineral fuels, lubricants and related materials (Sitc 3), by member state—Share of exports by member state (%): European Commission. (Accessed August 5, 2014, at http://epp.eurostat.ec.europa.eu/tgm/refreshTableAction.do?tab=table&plugin=1&pcode=tet00056&language=en.)
European Commission, 2014b, Extra-EU28 trade of mineral fuels, lubricants and related materials (Sitc 3), by member state—Share of imports by member state (%): European Union. (Accessed August 5, 2014, at http://epp.eurostat.ec.europa.eu/tgm/refreshTableAction.do?tab=table&plugin=1&pcode=tet00056&language=en.)
European Commission, 2014c, Extra-EU28 trade of raw materials (SITC 2+4), by member state—Share of exports by member state: European Commission. (Accessed August 5, 2014, at http://epp.eurostat.ec.europa.eu/tgm/refreshTableAction.do?tab=table&plugin=1&pcode=tet00064&language=en.)
European Commission, 2014d, Extra-EU28 trade of raw materials (SITC 2+4), by member state—Share of imports by member state: European Commission. (Accessed August 5, 2014, at http://epp.eurostat.ec.europa.eu/tgm/refreshTableAction.do?tab=table&plugin=1&pcode=tet00064&language=en.)
European Commission, 2014e, Share of trade with the EU28, exports: European commission. (Accessed August, 5, 2014, at http://epp.eurostat.ec.europa.eu/tgm/printTable.do?tab=table&plugin=1&language=en&pcode=tet00036&printPreview=true#.)
European Commission, 2014f, Share of trade with the EU28, imports: European commission. (Accessed August, 5, 2014, at http://epp.eurostat.ec.europa.eu/tgm/printTable.do?tab=table&plugin=1&language=en&pcode=tet00036&printPreview=true#.)
Guberman, D.E., 2013, Lead: U.S. Geological Survey Mineral Commodity Summaries 2013, p. 90–91.
Lundin Mining Corp., 2013, Production statistics: Lundin Mining Corp. (Accessed August 15, 2013, at http://www.lundinmining.com/s/ProductionStats.asp.)
Tolcin, A.C., 2013, Zinc: U.S. Geological Survey Mineral Commodity Summaries 2013, p. 188–189.
United Company RUSAL, 2013, Annual report 2012: Moscow, Russia, United Company RUSAL, 236 p.
U.S. Census Bureau, 2014, Trade in good with Ireland 2012: U.S. Census Bureau. (Accessed August 5, 2014, at https://www.census.gov/foreign-trade/balance/c4190.html.)
U.S. Central Intelligence Agency, 2013, Ireland, in The world factbook: U.S. Central Intelligence Agency. (Accessed August 15, 2013, at https://www.cia.gov/library/publications/the-world-factbook/geos/ei.html.)

TABLE 1
IRELAND: PRODUCTION OF MINERAL COMMODITIES[1]

(Thousand metric tons unless otherwise specified)

Commodity[2]		2008	2009	2010	2011	2012[e]
METALS						
Alumina		1,890	1,240	1,864	1,926	1,927 [3]
Lead:						
Mine output, Pb content	metric tons	50,200	43,000	39,100	50,700	47,000 [3]
Metal, refined, secondary[e]	do.	20,000	19,000	19,000	18,000	18,000
Silver, mine output, Ag content	kilograms	8,462	8,000	3,818	6,109	6,000
Zinc, mine output, Zn content	metric tons	398,200	357,000	342,500	344,000	337,500 [3]
INDUSTRIAL MINERALS						
Cement, hydraulic[e]		3,900	2,600	2,290	2,200	2,200
Gypsum[e]		600	400	300	300	300 [3]
Lime[e]		300	300	300	300	300
Sand and gravel[e, 4]		25,000	20,000	10,000	7,000	7,000
Stone and other quarry products:[e]						
Limestone	million metric tons	1	1	1	1	1
Other[5]	metric tons	60,000	45,000	40,000	25,000	25,000
MINERAL FUELS AND RELATED MATERIALS						
Gas, natural, marketed	million cubic meters	502	413	402	356	350
Peat:[e, 6]						
For horticultural use		500	500	500	500	500
For fuel use, milled peat[7]		3,800	3,800	4,991 [3]	3,707 [3]	3,700
Total		4,300	4,300	5,500	4,200	4,200
Briquets		300	300	238	192 [3]	190
Petroleum refinery products:[8]						
Liquefied petroleum gas	thousand 42-gallon barrels	387	380	482	566	560
Naphtha[e]	do.	900	900	237 [3]	144 [3]	140
Gasoline, motor	do.	4,797	3,687	3,996	4,360	4,300
Distillate fuel oil	do.	8,726	7,702	9,616	9,481	9,400
Residual fuel oil	do.	7,469	6,023	8,154	8,991	9,000
Refinery fuel and losses	do.	872	694	980	752	750
Total[e]	do.	23,200	19,400	23,465	24,294	24,100

[e]Estimated; estimated data are rounded to no more than three significant digits; may not add to totals shown. do. Ditto.
[1]Table includes data available through August 15, 2013.
[2]In addition to the commodities listed, Ireland also produces lime, limestone, seawater magnesia, and significant quantities of synthetic diamond. Output, however, is not quantitatively reported on a regular basis, and general information is inadequate to make reliable estimates of output.
[3]Reported figure.
[4]Excludes output by local authorities and road contractors.
[5]Includes clays for cement production, fire clay, granite, marble, rock sand, silica rock, and slate.
[6]Includes production by farmers and by the Bord Na Mona (Government Peat Board).
[7]Includes milled peat used for briquet production.
[8]From imported crude oil.

TABLE 2
IRELAND: STRUCTURE OF THE MINERAL INDUSTRY IN 2012

(Thousand metric tons unless otherwise specified)

Commodity		Major operating companies and major equity owners	Location of main facility	Annual capacity
Aggregates		Roadstone Wood Ltd. (CRH plc, 100%)	Tallagh, South Dublin	10,600
Alumina		Aughinish Alumina plc (United Company RUSAL, 100%)	Aughinish Island, Co. Limerick	1,990
Cement		Irish Cement Ltd. (CRH plc, 100%)	Plants in Limerick and Co. Louth	3,500
Diamond, industrial		Element Six Ltd. (De Beers Group, 100%)	Shannon, Co. Clare	NA
Do.		Sandvik Hyperion AB (Sandvik AB, 100%)	Dublin	NA
Gypsum		Irish Gypsum Ltd. (Saint Gobain S.A. 100%)	Mines in Knocknacran and Drummond, Co. Monaghan	NA
Lead-zinc, concentrate		Vedanta Lisheen Mining Ltd. (Vedanta Resources plc, 100%)	Lisheen Mine, Co. Kilkenny	187
Do.		Boliden Tara Mines Ltd. (Boliden AB, 100%)	Tara Mine, Navan, Co. Meath	215
Do.		Galmoy Mines Ltd.[1] (Lundin Mining Corp., 100%)	Galmoy Mine, Co. Kilkenny	135
Lime		Clogrennane Lime Ltd. (CRH plc, 100%)	Plants in Co. Carlow and Co. Claire	500
Limestone		Roadstone Wood Ltd. (CRH plc, 100%)	Tallagh, South Dublin	NA
Magnesia		Premier Periclase Ltd. (RHI Group, 100%)	Drogheda, Co. Louth	73
Natural gas	million cubic meters	Star Energy Group plc. (Petroliam Nasional Berhad (PETRONAS) 100%)	Kinsale Head field, Celtic Sea	2,100
Peat		Bord Na Mona [Government Peat Board]	Production mainly in the Midlands	4,200
Petroleum, refined	42-gallon barrels per day	ConocoPhillips Whitegate Refinery Ltd.	Whitegate, near Cork	71,000

Do. Ditto. NA Not available.

[1]Mine closed at end of 2012.

THE MINERAL INDUSTRY OF ITALY

By Alberto Alexander Perez

In 2012, Italy's gross domestic product (GDP) amounted to an estimated $1.838 trillion, which was a decrease of 2.4% compared with that of 2011. In terms of its GDP, the country ranked fourth in the European Union (EU) after Germany, the United Kingdom, and France. The Italian industrial sector was dependent on imported nonfuel and fuel mineral inputs. In 2012, Italy produced a significant amount of industrial mineral commodities—in particular, its Carrara and Sienese marbles and its clays. Italy was the 14th-ranked manufacturer of cement in the world, as measured by tonnage of production. Italy, similar to France, Germany, the United Kingdom, and other highly industrialized countries, was undergoing a shift towards a more services-based economy. In 1970, its industry's share of the GDP was 30% whereas that of services was 51.9%. In 2012, industry accounted for only 24.2% of the GDP, and services accounted for 73.8% (Cembureau, 2013, p. 7; U.S. Central Intelligence Agency, 2014).

Minerals in the National Economy

Italy's mineral industry produced such metals as copper, iron and steel, lead, and zinc, all of which were important materials for the country's manufacturing sector. All raw materials (ores and concentrates) used to produce these and other metals were imported or were obtained from secondary scrap recovery. Italy was a significant world producer of a variety of industrial minerals, including cement, clays, feldspar, lime, marble, pumice, and sand and gravel (table 1).

Italy was highly dependent on imported mineral fuels. Eni S.p.A. (Eni), which was 30% owned by the Italian Government, was the country's leading petroleum and natural gas company (table 2).

Government Policies and Programs

The Government played a significant role in the economy through regulation of ownership of large financial and industrial companies; privatizations and regulatory reform in accordance with EU directives, however, had reduced that role in recent years. Italy's basic mining legislation is mining law No. 1443 of July 29, 1927, which gives subsoil ownership of minerals to the state. The reimbursement of the state by mining concessionaires is regulated by law No. 752 of June 10, 1982. Quarrying operations are regulated by law No. 44 of September 1982.

All petroleum and gas upstream activities are supervised by the Ufficio Nazionale Minerario per gli Idrocarburi e la Geotermia [National Office for Mining, Hydrocarbons, and Geothermal Resources] (UNMIG), which operates within the Ministero dello Sviluppo Economico [Ministry of Economic Development]. After various organizational changes, the seven offices that compose UNMIG now all operate within the Directorate General for Energy and Mineral Resources.

By presidential decree, in 2007, the Committee for Hydrocarbons and Mineral Resources (CIRM) was set up to carry out some of the UNMIG's duties. The committee performs technical advisory tasks related to mining, hydrocarbons, and mineral royalties (Ministero dello Sviluppo Economico, 2010).

Environmental issues in Italy were focused on three main problem areas—air pollution from industrial emissions, such as sulfur dioxide; water pollution of coastal and inland rivers from industrial and agricultural effluents; and such natural hazards as avalanches, landslides, land subsidence in Venice, and volcanic eruptions (U.S. Central Intelligence Agency, 2014).

Production

In 2012, pig iron production decreased by 4.2% compared with that of 2011 to 9.42 million metric tons (Mt), and crude steel production decreased by 5.1% to 27.3 Mt. In the industrial minerals sector, cement production decreased by 20.9%. Crude petroleum production remained about the same at 36.9 million barrels (table 1).

Structure of the Mineral Industry

The Italian Government has ultimate control of Italy's mineral industry. Mineral resources, by law, are the property of the Italian nation, but private and mixed public and private entities were the principal owners of Italy's mineral industry. Government (public) ownership continued mainly in the mineral fuels sector (table 2).

Mineral Trade

In 2010 (the latest year for which detailed data were available), exports of nonferrous metals[1] from Italy to other countries within the EU totaled $5.4 billion[2] and those of metalliferous ores and metal scraps totaled $922 million. Imports of goods to Italy from other countries in the EU included manufactured metals valued at $6.2 billion, nonferrous metals valued at $8.3 billion, and metalliferous ores and metal scrap valued at $3.6 billion. In contrast, Italy's largest mineral industry imports, in terms of value, from a non-EU country (not including mineral fuels) were iron and steel (valued at $8.6 billion) and nonferrous metals ($5.7 billion). The leading exports from Italy to non-EU countries were iron and steel and manufactured metals that were valued at $5.3 billion and $6.6 billion, respectively (Eurostat, 2012, p. 110–112, 118–120, 126–128, 134–136).

---

[1]The classifications stated are from the United Nations Statistics Division's Standard International trade classification, Revision 4 (Series M, no. 34, Rev. 4, March 2006).

[2]Where necessary, values have been converted from euro area euros (€) to U.S. dollars (US$) at an average exchange rate of €0.755=US$1.00.

The most significant component of Italy's mineral trade in 2010 (the latest year for which data were available) was the net import of energy; Italy imported 149,536,000 metric tons (t) of oil equivalent, which was a decrease of 4.8% compared with the level of energy imports in 2009. Of this amount, Italy imported its crude oil, mainly from (in order of tonnage) the Commonwealth of Independent States (principally from Russia, Azerbaijan, and Kazakhstan) as well as from Libya, Iraq, Iran, Saudi Arabia, Norway, and various other sources (Eurostat, 2010, 2013).

Commodity Review

Metals

Aluminum and Bauxite and Alumina.—Italy did not produce alumina in 2012. United Company RUSAL (RUSAL) of Russia continued its suspension of operations at its Eurallumina facility in Italy owing to the high cost of running the facilities (particularly the cost of electricity), and no production was reported from the plant in 2012 (table 1; United Company RUSAL, 2012, p. 5).

On January 9, 2012, Alcoa Inc. (Alcoa) of the United States announced the permanent closing of the Porto Vesme alumina plant. This closure was expected to have a major effect on Italy's production of alumina, as the Fusina aluminum smelter had been in curtailment since 2010. Italy had been found in violation of EU law by subsidizing electricity to the Sardinian plants and was instructed to recover the amount of the subsidy from the companies involved, mainly Eurallumina S.p.A. The country's chief producers of alumina and primary aluminum were Eurallumina and Alcoa, respectively (table 2; Alcoa Inc., 2012, 2013).

Copper.—The main copper producers in Italy were KME Group S.p.A., which had operations at Fornaci di Barga and at Scrivia under its subsidiary KME Italy S.p.A., and Simar S.p.A (Simar) (a member of the Cordofin Group S.p.A.), which had a refinery at Porto Marghera, near Venice. Simar produced mainly copper-zinc-titanium alloys (KME Group S.p.A., 2010, p. 4; Simar S.p.A., 2010). For its copper production, Italy imported small amounts of copper concentrate and relied mainly on imports of copper metal and on scrap recovery.

Iron and Steel.—Lucchini S.p.A. (Lucchini), a subsidiary of OAO Severstal of Russia, stated that it was temporarily shutting down its Piombino blast furnace in December in response to market conditions. The Piombino plant, which was Lucchini's main production site, had the capacity to produce 2.5 million metric tons per year of steel. The shutdown was likely to last about 40 days. Piombino had already shut down twice in 2012, in May and August; those closures each lasted 2 to 3 weeks (Thomson Reuters, 2012).

In July, the ILVA steel plant in Taranto, which was owned by the Riva Group, was shut down following a Government probe into whether dioxin and other chemicals from the plant had spread and affected the population in the Taranto region. Taranto prosecutors ordered the seizure of the production sections responsible for the making of coke and the storage of minerals. By the end of the year, ILVA was still engaged in a protracted court battle over the full reopening of the plant (Di Giorgio, 2012).

Lead and Zinc.—Glencore International plc of Switzerland remained the country's principal processor (smelter and refinery) of lead and zinc (table 2). The company operated a plant at Porto Vesme, Sardinia.

Industrial Minerals

In 2012, Italy remained a leading European producer and a significant world producer of such industrial minerals as feldspar (24.7% of world output), pumice (18%), gypsum (3%), bentonite (2%), and lime (2%) (Crangle, 2013a, b; Miller, 2013; Tanner, 2013; Virta, 2013).

Cement.—Italian cement production decreased by 20.9% owing to a generalized and continued decrease in construction activity, which was driven by the contraction of the economy in 2012. Reductions of 10.6% and 9.1% in the civil engineering works sector and nonresidential construction sector, respectively, were the most significant, although a reduction of 6.3% in the residential construction sector was quite significant as well (Cembureau, 2013, p. 11).

Outlook

Italy's manufacturing industries remain those most affected by the decrease in domestic demand. The construction sector continues to be affected by low investment. Italy is one of the largest EU members in terms of population and the size of its industrial sector; however, the economic recession continues to affect its mineral and industrial production sectors. Italy will likely continue to rely on imported and recycled primary mineral materials for its industrial sector. The country is also likely to continue to rely on major imports of mineral fuels despite potential increases in domestic mineral fuels production from new deposits coming onstream in the near term. Economic reforms and austerity programs may continue to curtail increases in domestic consumption.

References Cited

Alcoa Inc., 2012, Alcoa programma un piano di riduzione delle capacità produttive o fermate degli smelter in Italia e Spagna, parte del piano di ristrutturazione del suo business primario globale: Alcoa Inc., January 9. (Accessed December 10, 2012, at http://alcoaportovesme.info/Comunicato-Stampa.)

Alcoa Inc., 2013, Alcoa to close Fusina smelter in Italy: Alcoa Inc., June 28. (Accessed April 30, 2014, at http://www.alcoa.com/global/en/news/news_detail.asp?page=20130628000134en&newsYear=2013.)

Cembureau, 2013, Activity report 2011: Brussels, Belgium, Cembureau, 46 p.

Crangle, R.D., Jr., 2013a, Gypsum: U.S. Geological Survey Mineral Commodity Summaries 2013, p. 70–71.

Crangle, R.D., Jr., 2013b, Pumice and pumicite: U.S. Geological Survey Mineral Commodity Summaries 2013, p. 124–125.

Di Giorgio, Massimiliano, 2012, Italy partly closes ILVA steel plant after pollution probe: Thomson Reuters, July 26. (Accessed April 30, 2014, at http://www.reuters.com/article/2012/07/26/steel-italy-idUSL6E8IQEBD20120726.)

Eurostat, 2010, Statistical aspects of the oil economy in 2009: Brussels, Belgium, European Commission, 9 p. (Accessed December 12, 2012, at http://epp.eurostat.ec.europa.eu/cache/ITY_OFFPUB/KS-QA-10-031/EN/KS-QA-10-031-EN.PDF.)

Eurostat, 2012, External and intra European Union trade—Monthly statistics—Issue number 12/2011: Brussels, Belgium, European Commission, 511 p. (Accessed January 21, 2013, at http://epp.eurostat.ec.europa.eu/portal/page/portal/international_trade/documents/ExtraIntraMonthlyEUTrade_ENVol12-20111.pdf.)

Eurostat, 2013, Energy production and imports: Brussels, Belgium, European Commission. (Accessed January 21, 2013, at http://epp.eurostat.ec.europa.eu/statistics_explained/index.php/Energy_production_and_imports.)

KME Group S.p.A., 2010, Annual report 2009—Group profile: Florence, Italy, KME Group S.p.A., 222 p.

Miller, M.M., 2013, Lime: U.S. Geological Survey Mineral Commodity Summaries 2013, p. 92–93.

Ministero dello Sviluppo Economico, 2010, Decreto del Presidente della Repubblica 14 Maggio 2007, n. 78: Ministero dello Sviluppo Economico. (Accessed November 5, 2010, at http://unmig.sviluppoeconomico.gov.it/unmig/norme/78dpr07.htm.)

Simar S.p.A., 2010, Company profile: Simar S.p.A. (Accessed January 5, 2014, at http://www.simarzincorame.com/company.htm.)

Tanner, A.O., 2013, Feldspar: U.S. Geological Survey Mineral Commodity Summaries 2013, p. 54–55.

Thomson Reuters, 2012, Italy's Lucchini to halt steel furnace temporarily: Thomson Reuters, November 23. (Accessed April 30, 2014, at http://www.reuters.com/article/2012/11/23/italy-steel-idUSL5E8MN9P220121123.)

United Company RUSAL, 2012, UC RUSAL announces full year production results for 2011: Moscow, Russia, United Company RUSAL press release, February 13, 6 p. (Accessed January 5, 2014, at http://www.rusal.ru/upload/uf/1bd/13 02 2012 RUSAL_Production results_2011.pdf.)

U.S. Central Intelligence Agency, 2014, Italy, in The world factbook: U.S. Central Intelligence Agency. (Accessed January 5, 2014, at https://www.cia.gov/library/publications/the-world-factbook/geos/it.html/.)

Virta, R.L., 2013, Clays: U.S. Geological Survey Mineral Commodity Summaries 2013, p. 44–45.

TABLE 1
ITALY: PRODUCTION OF MINERAL COMMODITIES[1]

(Metric tons unless otherwise specified)

Commodity		2008	2009	2010	2011	2012
METALS						
Aluminum:						
Alumina, calcined basis		1,327,566	752,873	--	--	--
Metal:						
Primary		188,400	170,600	168,000	141,000 r	110,000
Secondary[2]		1,282,917	826,977	1,246,236	1,049,101	1,002,814
Total		1,471,317	997,577	1,414,236	1,190,101 r	1,112,814
Bismuth, metal[e]		5	5	5	5	5
Copper, metal, refined, all kinds[e]		24,200	6,500	1,800 r	7,600 r	7,700
Iron and steel, metal:						
Pig iron	thousand metric tons	10,377	5,692	8,555	9,838	9,424
Ferroalloys, electric furnace:						
Ferromanganese		8,500	5,500	17,000	18,000	18,000
Ferrosilicon[e]		10,000	10,000	10,000	10,000	10,000
Silicomanganese		25,500	17,000	22,900	24,600	24,600
Other[e]		10,000	10,000	10,000	10,000	10,000
Total		54,000	42,500	59,900	62,600	62,600
Steel, crude	thousand metric tons	30,600	19,848	25,750	28,735	27,257
Lead:						
Mine output, Pb content[e]		800	800	800	800	--
Metal, refined:						
Primary		42,400 r	17,000 r	--	--	--
Secondary		157,500 r	132,000 r	150,000 r	149,500 r	138,400
Total		199,900 r	149,000 r	150,000 r	149,500 r	138,400
Zinc, metal, primary		107,100	103,400 r	104,700 r	110,200 r	97,200
INDUSTRIAL MINERALS						
Barite[e]		5,000	3,500	3,500	3,500	3,500
Bromine[e]		300	300	300	300	--
Cement, hydraulic	thousand metric tons	43,000	36,317	34,408	33,120	26,200
Clays, crude:						
Common clay	do.	4,472	6,324	5,900	4,750	3,777
Bentonite	do.	281	146	111	102	144
Refractory, excluding kaolinitic earth	do.	942	844	844 e	844 e	878
Ball clay	do.	591	1,070	612	638	746
Fuller's earth[e]	do.	3	3	3	3	3
Kaolin[e]	do.	5	5	6	8	8
Diatomite[e]		25,000	25,000	25,000	25,000	25,000
Feldspar[e]	thousand metric tons	4,700	4,700	4,700	4,700	4,700
Gypsum	do.	5,450 e	5,101	4,441	5,939	2,563

See footnotes at end of table.

TABLE 1—Continued
ITALY: PRODUCTION OF MINERAL COMMODITIES[1]

(Metric tons unless otherwise specified)

Commodity		2008	2009	2010	2011	2012
INDUSTRIAL MINERALS—Continued						
Lime, hydrated, hydraulic, and quicklime[e]	thousand metric tons	6,000	6,000	5,800	5,800	5,800
Magnesia[e]	do.	100	100	100	100	100
Nitrogen, N content of ammonia[e]	do.	460	460	460	460	2,365 [3]
Perlite[e]		60,000	60,000	60,000	60,000	60,000
Pigments, mineral, iron oxides, natural		500 [e]	105	117	112	118
Pumice and related materials:[e]						
Pumice	thousand metric tons	30	30	30	30	30
Pozzolan	do.	4,000	4,000	4,000	4,000	4,000
Salt	do.	2,200 [e]	3,471	4,006	2,912	3,098
Sand and gravel	do.	210,000 [e]	164,218	150,996	164,844	123,049
Silica sand	do.	14,000 [e]	19,759	17,656	16,369	13,946
Sodium compounds, n.e.s.:[e, 4]						
Soda ash	thousand metric tons	500	500	500	500	500
Sodium sulfate	do.	125	125	125	125	125
Stone:						
Calcareous:						
Alabaster[e]	do.	7,000	7,000	7,000	7,000	7,000
Chalk[e]	do.	228	228	200	200	200
Dolomite	do.	1,700 [e]	1,601	1,572	1,139	1,069
Limestone for lime and cement	do.	32,900	41,090	38,440	37,269	28,524
Marble and travertine, crude	do.	4,600 [e]	4,604	5,500	3,516	3,348
Crushed and broken[5]	do.	60,000	84,718	80,336	76,793	74,169
Granite	do.	1,480	1,009	1,614	1,585	1,734
Sandstone	do.	397	451	497	440	281
Slate	do.	288	52	55	61	44
Sulfur:[e]						
From metallurgy	do.	90	90	90	90	90
From hydrocarbons	do.	650	650	650	650	650
Talc and related materials		112,000	112,000	110,000	110,000	110,000
MINERAL FUELS AND RELATED MATERIALS						
Asphalt and bituminous rock, natural	thousand metric tons	1,800 [e]	1,030	1,454	2,169	2,100 [e]
Coke, metallurgical[e]	do.	4,000	4,000	4,000	4,000	4,000
Gas, natural	million cubic meters	9,260	8,127	8,296	8,438	8,400 [e]
Natural gas liquids[e]	thousand 42-gallon barrels	350	350	350	350	350
Petroleum:						
Crude	do.	36,300	30,215	35,040	36,201	36,865
Refinery products	do.	719,499	665,541	688,646	688,600	688,000

[e]Estimated; estimated data are rounded to no more than three significant digits; may not add to totals shown. [r]Revised. do. Ditto. -- Zero.

[1]Table includes data available through January 5, 2014.

[2]Unwrought aluminum alloys in secondary form.

[3]Reported figure.

[4]Not elsewhere specified.

[5]Output of limestone and serpentine for dimension stone is included with "Stone: Crushed and broken."

TABLE 2
ITALY: STRUCTURE OF THE MINERAL INDUSTRY IN 2012

(Thousand metric tons unless otherwise specified)

Commodity	Major operating companies and major equity owners	Location of main facilities	Annual capacity
Aluminum	Alcoa Inc.	Smelter at Fusina	44
Barite	Bariosarda S.p.A. (Ente Mineraria Sarda)	Barega and Mont 'Ega Mines on Sardinia	100
Do.	Edem S.p.A. (Government)	Mines at Val di Castello, Lucca	20
Do.	Edemsarda S.p.A. (Soc. Imprese Industriali)	Mines at Su Benatzu, Sto. Stefano, and Peppixeddu, Sardinia	20
Do.	Societá Mineraria Baritina S.p.A	Mines at Marigolek, Monte Elto, and Primaluna, near Milan	20
Bauxite	Sardabauxiti S.p.A. (Cogein S.p.A., 40%; Comtec S.p.A., 40%; Icofin Co., 20%)	Mine at Olmedo, Sardinia	350
Bentonite	Industria Chimica Carlo Laviosa S.p.A	Mines and plant on Sardinia and a plant near Pisa	250
Cement	11 companies, of which the largest are:		
Do.	Italcementi Fabbriche Riunite Cemento S.p.A.	18 plants, of which the largest are Calusco, Monselice, and Collefero	15,000
Do.	Buzzi Unicem Group	11 plants, of which Guidonia, Lugagnano, Morano, Piacenza, S'Arcangelo di Romagna, and Settimello are the largest	9,000
Do.	Cementerie del Tirreno S.p.A.	6 plants at Arquasta Scivia, Livorno, Maddaloni, Napoli, Spoleto, and Taranto	5,300
Copper:			
Refined	Simar S.p.A. (Cordifin S.p.A.)	Refinery at Porto Marghera	60
Refined, secondary	KME Italy S.p.A. (KME Group S.p.A.)	Refinery at Fornaci di Barga and Scrivia	24
Do.	Sitindustrie S.p.A.	Refinery at Pieve Vergonte	22
Feldspar	At least 5 companies, of which the largest are:	Locations:	1,500
Do.	Maffei S.p.A.	Surface mines at Pinzolo and Campiglia	(200)
Do.	do.	Underground mine at Vipiteno	(300)
Do.	Miniera di Fragne S.p.A.	Surface mine at Alagna Valsesia	(60)
Do.	Sabbie Silicee Fossanova S.p.A.	Surface mine at Fossanova	(30)
Gypsum	Fassa S.r.l.	Plant at Moncalvo, Asti	90
Lead, metal	Glencore International plc	Refinery at San Gavino, Sardinia	100
Do.	do.	Kivcet smelter and Imperial smelter at Porto Vesme, Sardinia	80
Iron	Altiforni e Ferriere di Servola S.p.A. (Lucchini S.p.A.)	Pig iron and coke plant at Trieste	500
Lignite	Ente Nazional per l'Energia Electrica	Surface mine at Santa Barbara (closed)	1,000
Lime	Unicale S.p.A.	Plants in Lombardy region	500
Magnesium, metal	Societa Italiana Magnesio S.p.A.	Plant at Bolzano	8
Marble	A number of companies, of which the largest include:	Locations:	2,000
Do.	Mineraria Marittima Srl	Quarries in the Carrara and Massa areas	(500)
Do.	Industria dei Marmi Vicentini S.p.A.	do.	(300)
Do.	Figaia S.p.A.	do.	(100)
Nitrogen, N content of ammonia	Hydro Agri S.p.A.	Plant at Ferrara	410
Petroleum:			
Crude	Eni S.p.A.	Oilfields: offshore Sicily, in the Adriatic Sea, and onshore in Po River Valley	90
Refined thousand 42-gallon barrels per day	Various companies	About 14 refineries	2,000
Potash, ore	Industria Sali Otassici e Affini per Aziono S.p.A.	Underground mines at Corvillo, Pasquasia, and Racalmuto, Sicily.	1,300
Do.	Sta. Italiana Sali Alcalini S.p.A. (Italkali)	Underground mines at Casteltermini and Pasquasia, Sicily	700

See footnotes at end of table.

TABLE 2—Continued
ITALY: STRUCTURE OF THE MINERAL INDUSTRY IN 2012

(Thousand metric tons unless otherwise specified)

Commodity	Major operating companies and major equity owners	Location of main facilities	Annual capacity
Pumice	Pumex S.p.A.	Quarries, Lipari Island, north of Sicily	600
Do.	Sta. Siciliana per l'Industria ed il Commercio della Pomice di Lipari S.p.A. (Italpomice S.p.A.)	do.	200
Pyrite	Nuova Solmine S.p.A.	Underground mines at Campiano and Niccioleta	900
Salt, rock	Sta Italiana Sali Alcalini S.p.A. (Italkali)	Underground mines at Petralia, Racalmuto, and Realmonte, Sicily	4,000
Do.	Solvay S.p.A.	Underground mines at Buriano, Pontteginori, and Querceto, Tuscany	2,000
Steel	ILVA S.p.A. (Riva Group)	5 steel plants, the largest of which is in Taranto (1,500)	4,000
Do.	Riva Acciaio S.p.A. (Riva Group)	7 steel plants	7,000
Do.	Luccini S.p.A. (OAO Sverstal)	Steel plant in Piombino	2,500
Do.	Acciaierie e Ferriere Vicentine Beltrame S.p.A. (AFV-Beltrame S.p.A.)	Steel plant at Vicenza	1,000
Talc	Luzenac Val Chisone S.p.A.	Mines at Pinerolo, near Turin, and an open pit mine in Orani, Sardinia	120
Do.	IMI FABI S.p.A.	Mine in Orani, Sardinia	20
Zinc, metal	Glencore International plc	Plant at Porto Vesme, Sardinia	120
Do.	Pertulosa Sud S.p.A.	Plant at Crotone, Calabria	100

Do., do. Ditto.

The Mineral Industry of Kazakhstan

By Elena Safirova

Kazakhstan produced a diverse range of mineral commodities and was the world's leading producer of uranium (37% of world output); the second-ranked producer of chromite (16% of world output); the fourth-ranked producer of titanium sponge (11% of world output) and magnesium metal (3% of world output); and the fifth-ranked producer of rhenium (6% of world output). The country was also a significant producer of bauxite, cadmium, copper, gallium, and zinc. The mineral industry accounted for a significant share of the country's gross domestic product (GDP) and foreign trade revenue; petroleum and natural gas were the leading commodities in terms of production value. Kazakhstan's Government promoted the development of the mineral industry and owned interests in a number of significant mineral-commodity-producing companies (Bedinger, 2013; Bray, 2013; Edelstein, 2013; Jaskula, 2013; Kramer, 2013; Papp, 2013; Polyak, 2013; Tolcin, 2013a, b; U.S. Energy Information Administration, 2013).

Minerals in the National Economy

In 2012, Kazakhstan's real GDP increased by 5.1% compared with that of 2011, and the nominal 2011 GDP was valued at $203.5 billion.[1] Total industrial production was valued at $113 billion, of which $68.7 billion (60.8% of the value of industrial production) was from mineral extraction (which included $57.9 billion from the extraction of crude petroleum, $3.3 billion from the mining of nonferrous metal ores, $1.6 billion from the mining of iron ores, $1.3 billion from the extraction of coal and lignite, and $548 million from the extraction of natural gas). Metallurgy contributed $13.1 billion to industrial output, of which ferrous metallurgy contributed $4.8 billion. As of January 1, 2013, Kazakhstan had a total of 2,240 enterprises engaged in mining and mine development; 189 of these enterprises were joint firms with foreign partners and 209 were fully owned by foreigners (Agency of Statistics of the Republic of Kazakhstan, 2013a, b).

Mining and metallurgy attracted significant amounts of fixed capital investment and foreign direct investment in Kazakhstan. Total fixed capital investment in industrial production was about $19.2 billion, of which $10.7 billion (55.7%) was investment in mining and $1.8 billion (9.4%) was investment in metallurgy. Investment in crude petroleum and natural gas production made up about 67% of total fixed capital investment in the mineral industry. In 2011 (the latest year for which data were available), gross foreign direct investment totaled about $23.5 billion, of which investment in mining and mine development accounted for $5.4 billion (investment in crude petroleum and natural gas made up about $3.5 billion of the total mining investment), and investment in metallurgy and in production of finished products made out of metals accounted for $2.4 billion (Agency of Statistics of the Republic of Kazakhstan, 2013a).

Government Policies and Programs

In August, the Government of Kazakhstan adopted a new document called "A Concept for the Development of the Geological Industry of the Republic of Kazakhstan through 2030." The document describes the current state of the exploration and mining industries and outlines goals and principles for their future development. According to the document, Kazakhstan's proven reserves of copper and polymetallic ores have been in decline and would be able to supply the country's industry for only another 10 to 15 years. Since 2000, proven reserves—including increases in reserves brought about through new exploration—of copper decreased by 2.4 million metric tons (Mt) (or by 5.8%); zinc, by 7.2 Mt (19.8%); lead, by 1.1 Mt (6.5%); and bauxite, by 54.5 Mt (15%). In other words, in the past several decades there was an imbalance between additions to reserves from exploration and development and removals from reserves by mining, and this trend would need to be reversed to avoid significant depletion of reserves (Ministry of Industry and New Technologies, 2012; TengriNews.kz, 2012).

The "Concept" document recommends forming a Government structure designed to assist in effective exploration for and the rational use and timely replenishment of mineral reserves. The document assumes that the first 2 years (2013 and 2014) would be spent on preparing to reorganize the mineral industry; in particular, preparing new technical and legal documents aimed at creating a new legal and technical foundation for reorganization of the exploration industry. In addition, the Government would create a new research institute, conduct marketing research, and train specialists. Then, during the period of 2015 to 2020, massive national and regional exploration work would be conducted, and during the period of 2020 to 2030, the main tasks would be prospecting and development. The document notes that an important component of revitalization of the exploration industry will be attracting leading foreign companies to Kazakhstan's mineral exploration industry and training young geologists in the country (MinerJob.ru, 2012b; Ministry of Industry and New Technologies, 2012).

In December, the President of Kazakhstan gave an order to cancel the Government moratorium on the issuance of new mining licenses. The moratorium was introduced in 2008 and was motivated by the Government's understanding of the need for the country to adopt a new mining code and to cut down on corruption surrounding the process of issuing licenses. Since 2008, new licenses had been available only to firms that had formed joint ventures with Tau-Ken Samruk, which was the vertically integrated Government-owned company working in the spheres of mining and metallurgy.

---

[1]Where necessary, values have been converted from Kazakhstani tenge (KZT) to U.S. dollars (US$) at an annual average exchange rate of KZT149.11=US$1.00.

The Ministry of Industry and New Technologies (MINT) was also discussing the possibility of offering companies mining licenses without going through the standard competition procedures in exchange for companies agreeing to make large investments in the mining projects. The MINT also noted that, despite its rich mineral resources, Kazakhstan attracts less than 1% of world investment in metallic deposits. The Ministry expected that legislation supporting the expedited procedures for issuing mining licenses could be written and adopted in 2013 (Mineral.ru, 2012h; MinerJob.ru, 2012e, f).

Production

Output of mineral commodities generally remained close to the levels of output in 2011. Production of mined silver and silver metal increased by 48% each; barite concentrate, gold metal, and salt, by 27% each; silicon metal, by an estimated 25%; ferrosilicochromium and estimated marketable barite, by 15% each. On the other hand, ferrosilicon production decreased by 70%. Production of crude steel decreased by 29%, and that of rolled steel, by 23%. Refined lead production decreased by 21%; production of gallium, by 16%; and that of pig iron, by 14% (table 1).

Structure of the Mineral Industry

The four most significant producers of nonfuel mineral commodities in Kazakhstan were Eurasian Natural Resources Corp. plc (ENRC) of the United Kingdom (aluminum, ferroalloys, and iron ore), Kazakhmys plc of the United Kingdom (copper and zinc), the state-owned company Kazatomprom JSC (uranium and rare metals), and Kazzinc JSC (lead and zinc, and byproducts, such as minor metals and gold). ENRC and Kazakhmys were both listed on the London Stock Exchange, and Kazzinc was majority owned by Glencore International plc of Switzerland, which was also listed on the London Stock Exchange. In 2012, Glencore increased its share of Kazzinc to 69.61% from 50.3% (Xstrata plc, 2013).

Although ENRC and Kazakhmys had headquarters in the United Kingdom, both companies were originally Kazakhstani companies, and a combination of Kazakhstani nationals and the Government of Kazakhstan still owned a majority of the shares of both companies. The core assets of all four companies were obtained in the early to mid-1990s when Kazakhstan's mining and metals production facilities were privatized. Each company controlled a majority of Kazakhstan's output of at least one mineral commodity, and Kazatomprom controlled all production of uranium in Kazakhstan; private companies were able to participate in the uranium industry only through partnerships with Kazatomprom (Eurasian Natural Resources Corp. plc, 2013, p. 65, 94; Kazakhmys plc, 2013, p. 121).

Commodity Review

Metals

Chromium.—Kazakhstan produced about 16% of the world's output of chromite and was the world's second-ranked producer behind South Africa. Of the two producers of chromite in Kazakhstan, TNK Kazchrome, which was a division of ENRC, was by far the leading producer, producing 3.59 million metric tons per year (Mt/yr) of marketable chromite output (a 5% increase compared with that of 2011) and 1.13 Mt of ferrochromium (a 1% increase compared with that of 2011). ENRC reported Australasian Joint Ore Reserves Committee (JORC)-compliant proven reserves of 59.0 Mt of ore grading 41.6% Cr_2O_3 and probable reserves of 153.4 Mt of ore grading 40.9% Cr_2O_3 (Eurasian Natural Resources Corp. plc, 2013; Papp, 2013).

The second-ranked producer of chromite in Kazakhstan was Oriel Resources Ltd., which was a subsidiary of OAO Mechel of Russia. The complex was designed eventually to reach production of between 600,000 and 700,000 metric tons per year (t/yr) of chromite concentrate. Chromite ore produced by Oriel was sent to Mechel's Tikhvin ferroalloys plant, which was located in Tikhvin, Leningrad Oblast, Russia. Mechel estimated that Oriel had reserves of about 16.8 Mt of ore with an average grade of 42.2% Cr_2O_3 (Kazakhstan Today, 2013).

Copper.—Kazakhmys was the dominant producer of copper ore and metals in Kazakhstan. In 2012, the company produced 306,100 metric tons (t) of copper in concentrate, 294,000 t of refined copper cathodes, and 23,900 t of wire rods. The average crude ore copper grade was 0.95%, compared with ore grades of 1.01% in 2011 and 1.18% in 2008. The decrease in the overall copper content of concentrate production in Kazakhstan was mainly the result of the declining ore grades at Kazakhmys's mines. Other copper producers in Kazakhstan included Aktyubins Copper Co. of Russia, JSC Polymetal of Russia, and Kazzinc (Mineral.ru, 2012e; Kazakhmys plc, 2013).

Kazakhmys planned to increase its production of copper in concentrates to more than 500,000 t/yr by 2018 with the development of new mines, including the Aktogai and the Bozshakol projects. Bozshakol was in the mine construction stage, and production was planned to begin in 2015, with expected output of about 100,000 t/yr of copper contained in concentrate from 2015 to 2030, and 60,000 t/yr from 2031 to 2056. Development of Bozshakol was expected to cost $1.8 billion. Mine construction at the Aktogai site was expected to begin in the first half of 2013, and open pit mining was expected to commence in 2015. The life expectancy of the Aktogai Mine was projected to be 50 years. During first 10 years, the mine was expected to produce 104,000 t/yr of copper cathode equivalent, on average, and during all 50 years, the average output was projected to be 72,000 t/yr. The total cost of the Aktogai project was estimated to be $2 billion (Mineral.ru, 2012f; MinerJob.ru, 2012c, d).

In June, Kazakhmys began an $80 million renovation project at its Nikolaevskaya plant. The renovation was expected to increase the plant capacity to 2.2 Mt/yr of ore from 1.8 Mt/yr as well as to increase the extraction efficiency of copper, lead, and zinc from ore into concentrates. In addition, after the renovation, the plant would be able to extract precious metals, which, before the renovation, were left within the copper, lead, and zinc concentrates. The renovation was expected to be completed in the beginning of 2013 (Mineral.ru, 2012n; MinerJob.ru, 2012h).

In April, Central Asia Metals plc of the United Kingdom opened a hydrometallurgical plant in the village of Kounrad, which is located near the city of Balkhash in the Karaganda region. The plant was built to recover copper from the waste dumps created by the operations of the Kounrad open pit copper mine that was in operation between 1936 and 2005. The recovery of copper was achieved by in situ leaching followed by solvent extraction-electrowinning (SX–EW). The plant's design capacity was 10,000 t/yr of copper cathodes, and the company expected to reach this production level in 2013. In 2012, Central Asia Metals produced 6,586 t of copper cathodes at the plant. The company reported that it had spent $42 million on the plant construction instead of the planned $47 million and had completed construction within 18 months, as planned. Central Asia Metals was planning to evaluate a potential expansion of its operations in the summer of 2013 (Mineral.ru, 2012a; MinerJob.ru, 2012k; 2013; Central Asia Metals, Plc., 2014).

In August, Orsu Metals Corp. of the United Kingdom, obtained a license for development of the Karchiga polymetallic deposit. Orsu Metals was planning to begin construction of a processing plant that would take between 12 and 15 months to complete. The estimated capital costs of the project were $147 million, and the company was planning to begin mining in the first quarter of 2014. The Karchiga deposit's resources were estimated to be 10 Mt of sulfide ore containing 166,700 t of copper and 1.4 t of gold. The new processing plant would have a capacity to process 750,000 t/yr of ore and to produce 11,800 t/yr of copper concentrate and 2,800 t/yr of copper cathodes (MinerJob.ru, 2012g).

In December, KGHM Polska Miedz S.A. of Poland (a subsidiary of KGHM International Ltd.) applied for an exploration license for a large area in the Karaganda region. The company expected the area to contain significant resources of copper, gold, silver, and other metals. KGHM was ranked second among the copper producers in Europe and ninth among the world producers. The company was planning to increase its refined copper production worldwide to 700,000 t in 2018 from 527,000 t in 2012. KGHM also owned other mineral deposits in Chile, Canada, and the United States (Mineral.ru, 2012p; MinerJob.ru, 2012i).

Gold.—In 2012, Kazakhstan produced 39,903 kilograms (kg) of unprocessed and semiprocessed gold, which was an 8.3% increase compared with the output in 2011. Kazzinc was the leading gold producer in the country and produced 17,400 kg, which was a 30.8% increase compared with the company's output in 2011. Kazakhmys—ranked a distant second—produced 4,012 kg, which was an 8.5% increase in output compared with that of 2011. Some of the other producers of mined gold included AK Altynalmas, GMK Kazakhaltyn, JSC Polymetal of Russia, Nord Gold N.V. (which was a gold producing subsidiary of OAO Severstal of Russia), Polyus Gold International Ltd. of Russia, and TOO Yubileynoye (Mineral.ru, 2012b; 2013a; IA Novosti—Kazakhstan, 2013; Murtazin, 2014).

In 2010, the President of Kazakhstan announced that the country would increase mined gold output to 70 t/yr by 2015. Reaching 70 t/yr of gold production by 2015 compared with about 40 t in 2012 would be a significant increase in production, and a few important projects were in development. Kazakhmys planned to begin production in 2015 at its Bozshakol copper project, which was estimated to have a contained gold resource of about 160 t and could significantly increase Kazakhmys's gold production (Kazakhmys plc., 2013).

As of 2012, Kazakhstan had two gold refineries. One of them was a part of the Ust-Kamenogorsk metallurgical complex and a division of Kazzinc. It had the capacity to produce 8 t/yr of gold and 300 t/yr of silver and to produce ingots of Good Delivery standard. Good Delivery standard is a set of rules issued by the London Bullion Market association; the Good Delivery standard rules include minimum fineness (995 parts per thousand), required dimensions of the bullion, and content and appearance of special marks imprinted on the bullion. Another refinery was located in central Kazakhstan and belonged to Kazakhmys. It had the capacity to refine 10 t/yr of gold and 650 t/yr of silver that met the requirements of Kazakhstan's national standard, which were lower than the Good Delivery standard. In June, Kazakhstan began construction of a third gold refinery in Astana. The planned capacity of the new refinery was 25 t/yr of gold and 50 t/yr of silver, and the total cost of the project was estimated to be $30 million. The construction of the new refinery was to be completed by the end of 2013 (Mineral.ru, 2012t).

Lead and Zinc.—Kazakhstan was the world's eighth-ranked zinc producer and was also a modest world producer of lead. Kazzinc was the leading zinc producer in Kazakhstan; the company was formed in 1997 by a merger of three leading nonferrous metals producers in eastern Kazakhstan—the Ust-Kamenogorsk lead and zinc plant, the Leninogorsk polymetal plant, and the Zyryanovskiy lead plant. As of 2012, Kazzinc employed more than 22,000 workers and had annual revenue of $2.8 billion. In 2012, Kazzinc produced 301,300 t of zinc metal, which constituted only a slight increase compared with the 300,800 t of output produced in 2011. In August, Kazzinc completed the reconstruction of its lead production line. The reconstruction was expected to lead to higher productivity and a reduction in energy costs and would allow for use of secondary lead in production (Mineral.ru, 2012m; 2013b; Tolcin, 2013b).

During 2012, Glencore increased its holding in Kazzinc to 69.61% from 50.7% by purchasing an 18.91% share from the Verny Capital Group in October. In the beginning of 2012, Glencore had been planning to increase its holding in Kazzinc to 93%, but that deal fell through. As of yearend, the Verny Capital Group continued to own about 30% of Kazzinc (Mineral.ru, 2012c, d).

In addition to Kazzinc, other producers of zinc in Kazakhstan included Kazakhmys, TOO Nova Zinc, and TOO ShalkiyaZinc Ltd. In 2012, Kazakhmys produced 151,600 t of zinc in concentrate, and TOO Nova Zinc produced 34,250 t of zinc in concentrate and 3,590 t of lead in concentrate. ShalkiyaZinc halted its mine production in 2008 because of the reduction in the world zinc prices and had not yet resumed production. Meanwhile, ShalkiyaZinc was continuing with construction of a new processing plant in Kentau with an annual capacity of 4 Mt/yr of ore throughput; ShalkiyaZinc expected to resume production at the mine when the market conditions improve. In 2011, SAT & Co. Holding purchased an 84.3% share of ShalkiyaZinc for $50 million, and in February 2012, SAT & Co.

increased its shares to 98.56%. Assets of ShalkiyaZinc, which is located in southern Kazakhstan, include the Shalkiya underground mine, the Talap lead and zinc deposit, and a beneficiation plant located 165 kilometers from the Shalkiya Mine (Mineral.ru, 2012i; Chelyabinsk Zinc Plant, 2014).

Nickel.—SAT & Co. was planning to build a nickel processing plant in eastern Kazakhstan by 2015. The new plant was expected to employ a unique technology that would enable processing of nickel oxide ores with low nickel content. The cost of construction was estimated to be between $200 million and $250 million, and the company was to use both loans and its own funds. In addition, the Government was to invest in related infrastructure, training, and social development. The construction was to start in the fourth quarter of 2012 (Mineral.ru, 2012q).

In 2009, SAT & Co. acquired a 100% share in TOO Kaznickel, which had exploration and mining licenses for the Gornostayevskoye cobalt and nickel deposit. According to the 1999 estimates, probable reserves of the deposit were 20.4 Mt of ore containing 173,000 t of nickel and 12,000 t of cobalt. The processing plant's full capacity was expected to be 1.3 Mt/yr of ore. The first stage of the project was to be completed in mid-2015. By that time, the plant would be processing 500,000 t/yr of ore to produce 2,500 t/yr of nickel matte. By 2017, the plant was expected to reach full capacity and to increase nickel matte production to 4,000 t/yr. The plant's output would be exported to China and Russia for further processing (Mineral.ru, 2012q).

Titanium.—The AO Ust-Kamenogorsk titanium-magnesium plant (UKTMK) was the only titanium producer in Kazakhstan and the only fully integrated titanium producer in the world. The plant's production cycle included all production stages; that is, from mining to the output of the final products. UKTMK's main outputs included titanium sponge, primary magnesium in ingots, and titanium ingots and alloys. As of 2012, all output of UKTMK was exported (Listopad, Ivashenko, and Chervonyi, 2010; Kazenergy.com, 2013).

To increase the rate of vertical integration of its production processes, UKTMK participated in two joint ventures with foreign partners. One of those ventures was UKAD, which was a joint venture between UKTMP and ERAMET S.A. of France for the production of semifinished titanium products, such as titanium rods and forged pieces. UKAD's new $70 million plant, which was equipped with a 4,500-t forging press, officially opened in Saint Georges de Mons, France, in September 2011. The plant was expected to supply titanium rods, sheets, and forged pieces to the aerospace, medical, nuclear, and oil and gas industries (Kazenergy.com, 2013).

UKTMK formed another joint venture, SP TOO POSUK Titanium, with POSCO of the Republic of Korea to build a plant to produce titanium slabs using an electron beam furnace in East Kazakhstan Province. Construction of the new $70 million plant began in October 2011 and was expected to be completed in 2014. The plant's designed production capacity was 6,000 t/yr of titanium slabs. UKTMP was to supply the titanium sponge raw materials to be used at the plant, and POSCO would process the slabs into plates at its production facilities in the Republic of Korea (Kazenergy.com, 2013).

In December, TOO Tenir-Logistics announced that it was planning to start production of titanium dioxide in Kazakhstan. The plant would be located in close proximity to the Tymlay titanium and magnesium deposit in Jambyl region. The plant's designed capacity would be 10,000 t/yr of pigment titanium dioxide and 10,000 t/yr of silicon dioxide, and the total cost of construction was projected to be $28.8 million. At yearend, Tenir-Logistics was looking for investors, and the project timeline was not known (Mineral.ru, 2012s; MinerJob.ru, 2012j).

Industrial Minerals

Rare Earths.— In November, Summit Atom Rare Earth Co. (SARECO), which was a joint venture between AO NAK Kazatomprom and Sumitomo Corp. of Japan, began production of rare-earth concentrates at its newly constructed plant. The plant cost $30 million to build and was the first plant in Kazakhstan to focus on production of rare earths. The facility was located in Stepnogorsk and was expected to have an initial production capacity of 1,500 t/yr of total rare-earth oxides (REOs). In 2013, the company was planning to produce about 1,000 t of REOs and to reach 1,500 t of REO capacity in 2014. About 40% of the rare-earth elements are heavy rare-earth elements, mainly dysprosium (Mineral.ru, 2012l).

In the beginning, the SARECO plant expected to use uranium tailings accumulated in Kazakhstan as a raw material for its rare-earth production. In the future, however, it was planning to switch to a dedicated rare-earth deposit as an input source, but, as of 2012, no specific deposit had been selected. If a reliable raw material source can be found, SARECO was planning to expand the production capacity of the plant to 3,000 t/yr of REO by 2015 and to between 5,000 t/yr and 6,000 t/yr by 2017. In addition to production of total REOs, the company was planning to add a rare-earth separating line and a line for manufacturing magnets from rare-earth metals. SARECO was planning to export its output of REO concentrates to Japan, where they would be used for production of electric and hybrid automobiles (MinerJob.ru, 2012a).

Mineral Fuels and Related Materials

Uranium.—In 2012, Kazakhstan produced 20,900 t of uranium and remained the world's leading producer of uranium, accounting for about 37% of world uranium mine output. Kazakhstan has no nuclear powerplants, and all mined uranium was exported. AO NAK Kazatomprom, including through its shares in joint ventures, produced 11,900 t of uranium, which corresponded to 20% of world uranium production; it exported 9,260 t of uranium. According to the leadership of Kazatomprom, Kazakhstan has vast resources and could be able to increase its uranium production to 30,000 t/yr within the next 3 years. To achieve this goal, Kazakhstan would both expand its current production capacity and intensify the in situ leaching processes employed at almost all the country's uranium operations (Mineral.ru, 2012j, o).

Kazatomprom also continued forming joint ventures with foreign partners. In December, Kazatomprom announced that Ulba Conversion—a joint venture of Kazatomprom (51%)

and Cameco Corp. of Canada (49%)—would build a uranium processing plant in Ust-Kamenogorsk. The plant would include both refinement and enrichment production stages. The plant would have a capacity to produce 12,000 t/yr of uranium hexafluoride (UF_6). The new plant was scheduled be constructed by 2016 and to reach its planned capacity by 2018. In another joint project, in December, OAO TVEL of Russia and Kazatomprom signed key documents on a uranium enrichment joint venture. The project involved the creation of a new SP ZAO Center for Uranium Enrichment (CUE) in Russia. The center would be located in the city of Novouralsk in Sverdlovskaya Oblast and would be located within the OAO Urals Electrochemical Complex. The first shipments of uranium by the joint venture were expected to begin in the second half of 2013 (Mineral.ru, 2012g, k, r).

Outlook

Interest in Kazakhstan's mineral industry will likely continue to increase along with an increase in the number of projects aimed at exploiting the country's significant mineral resources. Projects involving copper, gold, rare metals, rare-earth metals, and uranium could be of particular interest. The number of exploration projects underway in Kazakhstan indicate the potential for future increases in production of mineral commodities in the country, but any future development will depend on a variety of factors, including mineral commodity prices and the development of Government policies and programs to encourage the growth of the industry. In particular, if the Government is successful in implementing its new program on geologic exploration, the country is likely to become a stronger leader in mineral production.

References Cited

Agency of Statistics of the Republic of Kazakhstan, 2013a, Kazakhstan v 2012 godu—Statisticheskij ezhegodnik Kazakhstana [Kazakhstan in 2012—Statistical yearbook of Kazakhstan]: Astana, Kazakhstan, Agency of Statistics of the Republic of Kazakhstan, December, 503 p.

Agency of Statistics of the Republic of Kazakhstan, 2013b, Promyshlennost' Kazakhstana i ego regionov, 2008–2012 [Industry of Kazakhstan and its regions, 2008–2012]: Astana, Kazakhstan, Agency of Statistics of the Republic of Kazakhstan, 217 p.

Bedinger, G.M., 2013, Titanium and titanium dioxide: U.S. Geological Survey Mineral Commodity Summaries 2013, p. 172–173.

Bray, E.L., 2013, Bauxite and alumina: U.S. Geological Survey Mineral Commodity Summaries 2013, p. 26–27.

Central Asia Metals Plc., 2014, Kounrad: Central Asia Metals Plc. (Accessed May 5, 2014, at http://www.centralasiametals.com/projects/kounrad.)

Chelyabinsk Zinc Plant, 2014, Akzhalsk deposit: Chelyabinsk Zinc Plant. (Accessed May 5, 2014, at http://www.zinc.ru/resources/akzhalsk.)

Edelstein, D.L., 2013, Copper: U.S. Geological Survey Mineral Commodity Summaries 2013, p. 48–49.

Eurasian Natural Resources Corp. plc, 2013, Annual report and accounts 2012: London, United Kingdom, Eurasian Natural Resources Corp. plc, 160 p.

IA Novosti—Kazakhstan, 2013, Kazakhstan v 2012 godu uvelichil proizvodstvo zolota na 8,6% -- do 40 tonn [In 2012, Kazakhstan increased gold production by 8.6% to 40 t]: IA Novosti—Kazakhstan, January 16. (Accessed May 5, 2014, at http://newskaz.ru/economy/20130116/4623509.html.)

Jaskula, B.W., 2013, Gallium: U.S. Geological Survey Mineral Commodity Summaries 2013, p. 58–59.

Kazakhmys plc, 2013, Annual report and accounts 2012: Kazakhmys plc, April 10, 212 p.

Kazakhstan Today, 2013, Na mirovom rynke khroma za poslredniye dva mesyatsa tseny povysilis' (obzor) [In the past two months, the world prices of chromium increased (a review)]: Kazakhstan Today, April 8. (Accessed May 5, 2014, at http://www.kt.kz/rus/economy/na_mirovom_rinke_hroma_za_poslednie_dva_mesjaca_cece_povisilisj_obzor__1153570698.html.)

Kazenergy.com, 2013, Flagman titano-magnievoy industrii Kazakhstana [The leader of Kazakhstan's titanium and magnium industry]: Kazenergy.com. (Accessed May 5, 2014, at http://www.kazenergy.com/2-57-2013/9010-2013-04-29-09-56-31.html.)

Kramer, D.A., 2013, Magnesium metal: U.S. Geological Survey Mineral Commodity Summaries 2013, p. 98–99.

Listopad, D.A., Ivashenko, V.P., and Chervonyi, I.F., 2010, Mirovoy rynok titana I perspektiv ego razvitiya [The world market of titanium and perspectives on its development]: Teoria i praktika metallurgii [Theory and Practice of Metallurgy], v. 5–6, p. 16–21.

Mineral.ru, 2012a, Central Asia Metals gotovitsya k postavkam medi s zavoda v Kazakhstane [Central Asia Metals is preparing for copper deliveries from a plant in Kazakhstan]: Mineral.ru, May 20. (Accessed February 19, 2014, at http://www.mineral.ru/News/48653.html.)

Mineral.ru, 2012b, EZ OTM nadeetsya na affinazh 20 t zolota Kazakhstana [EZ OTM hopes to process 20,000 metric tons of gold from Kazakhstan]: Mineral.ru, August 28. (Accessed February 19, 2014, at http://www.mineral.ru/News/49851.html.)

Mineral.ru, 2012c, Glencore do kontsa goda uvelichit dolyu v Kazzinke tol'ko do 69,61% [By yearend, Glencore will increase its share in Kazzinc only to 69.61%]: Mineral.ru, September 27. (Accessed February 19, 2014, at http://www.mineral.ru/News/50243.html.)

Mineral.ru, 2012d, Glencore uvelichil dolyu v kazakhskom proizvoditele tsinka Kazzinc do 69,61% [Glencore increased its share in Kazakhstani zinc producer Kazzinc to 69.61%]: Mineral.ru, December 10. (Accessed February 19, 2014, at http://www.mineral.ru/News/50444.html.)

Mineral.ru, 2012e, Kazakhmys po itogam pervogo kvartala 2012 g. sokratil vypusk medi I tsinka [Kazakhmys reduced copper and zinc consumption in the first quarter of 2012]: Mineral.ru, April 27. (Accessed February 19, 2014, at http://www.mineral.ru/News/48444.html.)

Mineral.ru, 2012f, Kazakhmys proizvel v 2011 g. chut' bole 300 tys t. medi [In 2011, Kazakhmys produced a bit more than 300,000 metric tons of copper]: Mineral.ru, February 1. (Accessed February 19, 2014, at http://www.mineral.ru/News/47473.html.)

Mineral.ru, 2012g, Kazakhstan k 1 iyunya primet reshenie po tsentru obogasheniya urana v Novoural'ske [Kazakhstan will make a decision about the uranium enrichment center by June 1]: Mineral.ru, May 28. (Accessed February 19, 2014, at http://www.mineral.ru/News/48739.html.)

Mineral.ru, 2012h, Kazakhstan otmenyaet moratoriy na razrabotku nedr [Kazakhstan cancels the mining moratorium]: Mineral.ru, December 1. (Accessed February 19, 2014, at http://www.mineral.ru/News/51000.html.)

Mineral.ru, 2012i, Kazakhstanskiy holding SAT & Company dovel svoyu dolyu v ShalkiyaZinc do 98,56% [SAT & Co. Holding of Kazakhstan increased its share in ShalkiyaZinc to 98.56%]: Mineral.ru, February 1. (Accessed February 19, 2014, at http://www.mineral.ru/News/47468.html.)

Mineral.ru, 2012j, Kazakhstan sposoben za dva-tri goda narastit' dobychu urana do 30 tys. t. v god [Kazakhstan can increase uranium production to 30,000 metric tons per year in 2 to 3 years]: Mineral.ru, June 21. (Accessed February 19, 2014, at http://www.mineral.ru/News/49036.html.)

Mineral.ru, 2012k, Kazatomprom i Cameco do 2018 g. nachnut stroit' zavod po affinazhu urana v Ust-Kamenogorske [Kazakhstan and Cameco will start building a uranium refinery in Ust-Kamenogorsk before 2018]: Mineral.ru, December 21. (Accessed February 19, 2014, at http://www.mineral.ru/News/51273.html.)

Mineral.ru, 2012l, Kazatomprom i Sumitomo Corporation otkryli zavod po proizvodstvu redkozemel'nykh metallov [Kazatomprom and Sumitomo Corp. opened a plant producing rare-earth metals]: Mineral.ru, November 7. (Accessed February 19, 2014, at http://www.mineral.ru/News/50684.html.)

Mineral.ru, 2012m, Kazzinc rekonstruiroval proizvodstvo svintsa [Kazzinc has remodeled its zinc production]: Mineral.ru, August 28. (Accessed February 19, 2014, at http://www.mineral.ru/News/49845.html.)

Mineral.ru, 2012n, Korporatsiya Kazakhmys nachala modernizatsiyu Nikolaevskoy obogatitel'noy fabriki v BKO [Kazakhmys corporation started reconstruction of its Nikolaevskaya beneficiation plant in Eastern Kazakhstan]: Mineral.ru, June 27. (Accessed February 19, 2014, at http://www.mineral.ru/News/49103.html.)

Mineral.ru, 2012o, Po itogam 2012 g. ob'em dobychi urana v Kazakhstane sostavil 20,9 tys t. [In 2012, Kazakhstan produced 20,900 metric tons of uranium]: Mineral.ru, January 25. (Accessed February 19, 2014, at http://www.mineral.ru/News/51585.html.)

Mineral.ru, 2012p, Polska Miedz ishet mednoe syr'e v Kazakhstane [Polska Miedz is looking for copper ore in Kazakhstan]: Mineral.ru, November 12. (Accessed February 19, 2014, at http://www.mineral.ru/News/50844.html.)

Mineral.ru, 2012q, SAT & Company postroit zavod po proizvodstvu nickelya v Vostochnom Kazakhstane [SAT & Company will build a nickel plant in eastern Kazakhstan]: Mineral.ru, August 13. (Accessed February 19, 2014, at http://www.mineral.ru/News/49665.html.)

Mineral.ru, 2012r, TVEL i Kazatomprom podpisali klyuchevye dokumenty po sovmestnomu proektu po obogasheniyu urana [TVEL and Kazatomprom signed key documents for the joint uranium enrichment project]: Mineral.ru, November 26. (Accessed February 19, 2014, at http://www.mineral.ru/News/50916.html.)

Mineral.ru, 2012s, V Kazakhstane planiruyut nachat' proizvodstvo dioksida titana [Kazakhstan plans to start production of titanium dioxide]: Mineral.ru, December 18. (Accessed February 19, 2014, at http://www.mineral.ru/News/51212.html.)

Mineral.ru, 2012t, V Kazakhstane postroyat novyi affinazhnyi zavod [Kazakhstan will build a new gold refinery]: Mineral.ru, May 14. (Accessed February 19, 2014, at http://www.mineral.ru/News/48583.html.)

Mineral.ru, 2013a, Kazzinc v 2012 g. uvelichil proizvodstvo zolota ns 30,8% [In 2012, Kazzinc increased gold production by 30.8%]: Mineral.ru, February 14. (Accessed February 19, 2014, at http://www.mineral.ru/News/51842.html.)

Mineral.ru, 2013b, Kazzinc v 2012 g. uvelichil vyruchku na 26% [In 2012, Kazzinc increased its revenue by 26%]: Mineral.ru, March 6. (Accessed February 19, 2014, at http://www.mineral.ru/News/52098.html.)

MinerJob.ru, 2012a, AO "NAK Kazatomprom" i Sumitomo otkryli zavod po vypusku redkozemel'nykh metallov [AO NAK Kazatomprom and Sumitomo Corp. opened a rare-earth metals plant]: MinerJob.ru, November 8. (Accessed February 19, 2014, at http://www.minerjob.ru/viewnew.php?id=22368.)

MinerJob.ru, 2012b, Geologi kazakhstana poluchayut garantirovannuyu rabotu na dva desyatiletiya vpered [Kazakhstan's geologists got guaranteed jobs for the next two decades]: MinerJob.ru, September 14. (Accessed February 19, 2014, at http://www.minerjob.ru/viewnew.php?id=21846.)

MinerJob.ru, 2012c, Kazakhmys nachinaet stroitel'stvo gorno-obogatitel'nogo kombinata na mestorozhdenii Bozshakol' [Kazakhmys is starting construction of a mining and beneficiation plant at the Bozshakol deposit]: MinerJob.ru, June 3. (Accessed February 19, 2014, at http://www.minerjob.ru/viewnew.php?id=21391.)

MinerJob.ru, 2012d, Kazakhmys planiruet na Aktogae proizvodit' samuyu deshevuyu med' [Kazakhmys plans to produce the cheapest copper at Aktogai]: MinerJob.ru, December 7. (Accessed February 19, 2014, at http://www.minerjob.ru/viewnew.php?id=22851.)

MinerJob.ru, 2012e, Kazakhstan nameren privlekat' vedushie zarubezhnye kompanii dlya vndreniya novykh tekhnologiy v geologicheskuyu otrasl' [Kazakhstan intends to attract leading foreign companies for implementation of new technologies in the geologic industry]: MinerJob.ru, July 24. (Accessed February 19, 2014, at http://www.minerjob.ru/viewnew.php?id=21536.)

MinerJob.ru, 2012f, Kazakhstan reshil predostavlyat' mestorozhdeniya v obmen na investitsii [Kazakhstan decided to offer deposits in exchange for investments]: MinerJob.ru, July 5. (Accessed February 19, 2014, at http://testserver.minerjob.ru/viewnew.php?id=21396.)

MinerJob.ru, 2012g, Kazakhstan utverdil proekt britanskoy Orsu po osvoeniyu mednogo mestorozhdeniya Karchiga [Kazakhstan confirmed the project of British company Orsu for development of the Karchiga copper deposit]: MinerJob.ru, August 20. (Accessed February 19, 2014, at http://minerjob.ru/viewnew.php?id=21684.)

MinerJob.ru, 2012h, Korporatsiya Kazakhmys nachala modernizatsiyu Nikolaevskoy obogatitel'noy fabriki v VKO [Kazakhmys Corp. started modernization of the Nikolaevskaya beneficiation plant located in eastern Kazakhstan]: MinerJob.ru, June 25. (Accessed February 19, 2014, at http://www.minerjob.ru/viewnew.php?id=21360.)

MinerJob.ru, 2012i, Pol'skyuyu KGHM interesuyut resursy Kazakhstana [Polish KGHM is interested in Kazakhstan's resources]: MinerJob.ru, December 14. (Accessed February 19, 2014, at http://www.minerjob.ru/viewnew.php?id=22960.)

MinerJob.ru, 2012j, V Kazakhstane planiruyut nachat' sobstvennoe proizvodstvo dioksida titana [Kazakhstan plans to open its own production of titanium dioxide]: MinerJob.ru, December 13. (Accessed February 19, 2014, at http://www.minerjob.ru/viewnew.php?id=22929.)

MinerJob.ru, 2012k, V Kazakhstane zarabotal novyi mednyi zavod [A new copper plant started operations in Kazakhstan]: MinerJob.ru, May 4. (Accessed February 19, 2014, at http://www.minerjob.ru/viewnew.php?id=21163.)

MinerJob.ru, 2013, Central Asia Metals uspeshno proizvela med' v Kazakhstane [Central Asia Metals successfully produced copper in Kazakhstan]: MinerJob.ru, January 14. (Accessed February 19, 2014, at http://forum.minerjob.ru/viewnew.php?id=23159.)

Ministry of Industry and New Technologies, 2012, O Kontseptsii razvitiya geologicheskoy otrasli respubliki Kazakhstan do 2030 goda [About the concept for the development of the geologic industry of the Republic of Kazakhstan through 2030]: Ministry of Industry and New Technologies, August 13. (Accessed May 5, 2014, at http://www.mint.gov.kz/index.php?id=437&lang=ru.)

Murtazin, Azat, 2014, "Kazakhmys" nedoschitalsya zolota [Kazakhmys's gold fell short]: Kursiv.kz, February 5. (Accessed May 5, 2014, at http://www.kursiv.kz/news/details/kompanii/Kazahmys-nedoschitalsya-zolota/.)

Papp, J.F., 2013, Chromium: U.S. Geological Survey Mineral Commodity Summaries 2013, p. 42–43.

Polyak, D.E., 2013, Rhenium: U.S. Geological Survey Mineral Commodity Summaries 2013, p. 130–131.

TengriNews.kz, 2012, Razvedannykh zapasov medi i polimetallov v Kazakhstane ostalos' na 10 – 15 let [Proven reserves of copper and polymetals will last 10 to 15 years]: TengriNews.kz, July 24. (Accessed May 5, 2014, at http://tengrinews.kz/kazakhstan_news/razvedannyih-zapasov-medi-i-polimetallov-v-kazahstane-ostalos-na-10-15-let-217792.)

Tolcin, A.C., 2013a, Cadmium: U.S. Geological Survey Mineral Commodity Summaries 2013, p. 36–37.

Tolcin, A.C., 2013b, Zinc: U.S. Geological Survey Mineral Commodity Summaries 2013, p. 188–189.

U.S. Energy Information Administration, 2013, Kazakhstan: U.S. Energy Information Administration country analysis brief, October 28. (Accessed February 19, 2014, at http://www.eia.gov/countries/cab.cfm?fips=KZ.)

Xstrata plc, 2013, Annual report 2012: Xstrata plc, 152 p. (Accessed May 5, 2014, at http://www.glencorexstrata.com/assets/Uploads/2012-Annual-Report-FINAL.pdf.)

TABLE 1
KAZAKHSTAN: PRODUCTION OF MINERAL COMMODITIES[1]

(Metric tons unless otherwise specified)

Commodity[2]		2008	2009	2010	2011	2012
METALS						
Aluminum:						
Alumina		1,600,000	1,608,000	1,639,000	1,670,000	1,510,000
Bauxite, gross weight		5,160,100	5,130,000	5,310,200 r	5,495,200 r	5,170,200
Metal, primary		106,000	127,000	226,000	249,000	249,000
Antimony, Sb content of concentrate		890	597	785	800	865
Beryllium		NA	NA	NA	NA	2,526
Bismuth:[e]						
Mine output, Bi content		150	-- [3]	-- [3]	-- [3]	--
Metal, refined		--	90	150	150	150
Cadmium, metal[e]		1,100	1,300	1,400	1,300	1,200
Chromite, marketable ore		3,552,000	3,544,000	3,200,000 r	3,800,000 r	3,590,000
Copper:						
Mine output, Cu content of concentrate		421,700 r	406,100 r	380,600 r	405,300 r	419,200
Metal:						
Smelter, undifferentiated		392,575	332,854	318,637	302,975	302,183
Refined, primary		398,411	312,767	323,368	338,524 r	367,161
Gallium	kilograms	18,666	18,702	18,702	18,703	15,711
Gold:						
Mine output, Au content	do.	20,825	22,839	30,272 r	36,846 r	39,903
Metal, refined	do.	8,205	10,279	13,456 r	16,672 r	21,133
Iron and steel:						
Iron ore, marketable:						
Gross weight		21,486,300	22,281,300	24,229,100	24,812,800	25,209,800
Fe content[e]		12,200,000	12,700,000	13,800,000	14,100,000	14,326,000 [3]
Metal:						
Pig iron		3,106,000	2,996,000	2,984,000	3,141,100 r	2,707,000
Ferroalloys:						
Ferrochromium		1,220,315	1,173,286	1,311,302	1,289,917	1,305,343
Ferrosilicochromium		133,828	60,829	159,765	143,296	164,853
Ferrosilicon		54,964	33,100	4,813	1,683	494
Silicomanganese		179,939	200,374	224,627	232,039	251,530
Other		1,473	1,205	1,283	1,754	1,845
Total		1,590,519	1,468,794	1,701,790	1,668,689	1,724,065
Steel:						
Crude		4,243,582	3,324,300 r	3,338,000	3,699,300 r	2,610,000
Finished, rolled		2,826,202	2,990,167	2,899,800 r	3,107,900 r	2,402,300
Lead:						
Concentrate, Pb content		38,800	33,600 r	35,400 r	38,800 r	38,100
Refined, primary and secondary		105,766	80,994	103,110	111,249	88,099
Magnesium, metal, primary[e]		21,000	21,000	21,000	21,000	21,000
Manganese:						
Ore:						
Gross weight		2,485,000	2,457,400	3,044,700 r	2,963,000	2,675,000
Mn content[e]		600,000	520,000	610,000	590,000	595,000
Concentrate:						
Gross weight		1,117,200 r	982,400 r	1,094,400 r	1,096,300 r	1,070,500
Mn content[e]		400,000	360,000	390,000	390,000	390,000
Nickel, Ni content of laterite ore[e]		500	500	500	500	450
Niobium, metal		NA	NA	NA	NA	43
Rhenium[e]	kilograms	5,500	3,000	3,000 r	3,000 r	3,000
Silicon, metal[e]		--	--	1,500	8,000	10,000
Silver:						
Mine output, Ag content	kilograms	645,627	618,141	552,060 r	650,649 r	963,182
Refined	do.	628,763	613,544	548,990 r	646,685 r	958,495
Tantalum, metal		NA	NA	NA	NA	213

See footnotes at end of table.

TABLE 1—Continued
KAZAKHSTAN: PRODUCTION OF MINERAL COMMODITIES[1]

(Metric tons unless otherwise specified)

Commodity[2]		2008	2009	2010	2011	2012
METALS—Continued						
Titanium:						
Ilmenite and leucoxene[e]		25,000	25,000	25,000	25,000	25,000
Sponge		26,000	16,800	14,500	20,700	21,000 [e]
Zinc:						
Concentrate, Zn content		387,400 [r]	398,400 [r]	405,300 [r]	376,700 [r]	369,700
Smelter, primary and secondary		365,572	327,873	318,858	319,847 [r]	319,900
INDUSTRIAL MINERALS						
Asbestos, all grades		230,100	230,000	214,100	223,100	241,200
Barite:						
Ore and concentrate		492,200	306,000	358,000	466,200 [r]	590,100
Marketable[e]		170,000	170,000	200,000	200,000	230,000
Boron[e]	thousand metric tons	30	30	30	30	30
Cement		5,837,300	5,694,100	6,686,300	7,642,100	7,800,000 [e]
Fluorspar[e]		64,300 [r, 3]	65,000 [r]	65,000 [r]	65,000	65,000
Gypsum[e]		696,900 [3]	700,000	700,000	700,000	700,000
Lime		905,917	798,180	881,225	958,231	1,000,000 [e]
Phosphate rock, beneficiated:						
Gross weight		1,226,000	1,225,000	1,600,000 [e]	1,600,000 [e]	1,600,000 [e]
P_2O_5 content[e]		280,000	280,000	350,000	350,000	350,000
Rare-earth elements, rare-earth oxide content		--	--	--	--	50 [e]
Salt		438,047 [r]	222,942	276,131 [r]	364,222 [r]	463,960
Sulfur, byproduct:[e]						
Metallurgy		300,000	300,000	300,000	300,000	300,000
Natural gas and petroleum		1,732,600 [3]	2,200,000	2,400,000	2,400,000	2,400,000
Total		2,030,000	2,500,000	2,700,000	2,700,000	2,700,000
MINERAL FUELS AND RELATED MATERIALS						
Coal:						
Bituminous	thousand metric tons	106,296	91,042	99,285	103,015	107,911
Lignite	do.	4,777	5,084	7,283	8,368	7,748
Total	do.	111,073	96,126	106,568	111,383	115,659
Coke		2,687,700	2,552,000	2,526,800 [r]	2,663,300 [r]	2,569,300
Natural gas:						
Nonassociated gas	thousand cubic meters	18,708,000	18,132,000	17,595,000	19,305,000	20,308,800
Associated gas	do.	14,181,000	17,809,000	19,811,000	20,199,000	19,820,100
Total	do.	32,889,000	35,941,000	37,406,000	39,504,000	40,128,900
Petroleum:						
Crude oil and gas condensate[4]	42-gallon barrels	514,000,000	556,000,000	578,000,000	582,000,000	576,200,000
Refinery products[5]	do.	93,400,000	92,900,000	101,600,000	106,200,000	108,400,000
Uranium:						
U content		8,521	14,020	17,803	19,451	20,900
U_3O_8 content		10,049	16,534	20,995	22,939	24,648

[e]Estimated; estimated data are rounded to no more than three significant digits; may not add to totals shown. [r]Revised. do. Ditto. NA Not available. -- Zero.

[1]Table includes data available through January 31, 2014.

[2]In addition to the commodities listed, Kazakhstan may also have produced a number of other mineral products, including cesium, cobalt, germanium, indium, molybdenum, scandium, selenium, tellurium, and vanadium, but information is inadequate to estimate production.

[3]Reported figure.

[4]Figures were converted to barrels from metric tons, which were reported as follows: 2008—70,671,000; 2009—76,482,600; 2010—79,517,700; 2011—80,039,100; and 2012—79,224,500.

[5]Figures were converted to barrels from thousand metric tons, which were reported as follows: 2007—11,384; 2008—11,791; 2009—11,717; 2010—12,794; 2011—13,393; and 2012—13,668.

TABLE 2
KAZAKHSTAN: STRUCTURE OF THE MINERAL INDUSTRY IN 2012[1,2]

(Metric tons unless otherwise specified)

Commodity	Major operating companies, main facilities, or deposits	Location or deposit names	Annual capacity[e]
Alumina	Aluminium of Kazakhstan JSC [Eurasian Natural Resources Corp. plc (ENRC)]	Pavlodar	1,600,000
Aluminum, primary	Kazakhstan Aluminium Smelter JSC [Eurasian Natural Resources Corp. plc (ENRC)]	do.	250,000
Barite	Vostochnoye Rudoupravleniye LLP	Shyganak, Zhambyl Province	NA
Do.	Zhartas LLC	Zhambyl Province	25,000
Do.	Stroyservice LLC	Kentau District, South Kazakhstan Province	30,000
Do.	Zhairemsky GOK[3] JSC [Eurasian Natural Resources Corp. plc (ENRC)]	Ushkatyn III, Zhairem, and Zhumanai deposits near Zhairem	NA
Do.	JSC Yuzhpolimetall	Kentau District, South Kazakhstan Province	NA
Do.	Barite Oil Kentau LLC	Kentau District, South Kazakhstan Province	NA
Bauxite	Kazakhstan Aluminium Smelter JSC [Eurasian Natural Resources Corp. plc (ENRC)]	Torgai and Krasnooktyabrsk mining complexes, Kostanay Province	5,400,000
Beryllium, metal	Ulba Metallurgical Plant JSC (Kazatomprom JSC)	Oskemen (also known as Ust-Kamenogorsk)	NA
Bismuth, metal	Ust-Kamenogorsk metallurgical complex [Kazzinc JSC (Glencore International plc, 69.61%)]	do.	NA
Do.	Chimkent metallurgical plant (JSC Yuzhpolimetall)	Shymkent	NA
Cadmium	do.	do.	NA
Do.	Ust-Kamenogorsk metallurgical complex [Kazzinc JSC (Glencore International plc, 69.61%)]	Oskemen (also known as Ust-Kamenogorsk)	NA
Chromite, marketable ore containing about 50% Cr_2O_3 content	TNK Kazchrome [a subsidiary of Eurasian Natural Resources Corp. plc (ENRC)]	Khromtau, Aktobe Province	3,600,000
Do.	Oriel Resources Ltd. (OAO Mechel)	Voskhod GOK,[3] Khromtau, Aktobe Province	600,000
Copper:			
Mining, recoverable, Cu content	Kazakhmys plc:		
	Central Region:		
	Konyrat Mine	Karagandy Province	11,800 [4]
Do.	Sayak I and III Mines	do.	23,500
Do.	Shatyrkul Mine	Zhambyl Province	12,700
Do.	Abyz Mine	Karagandy Province	5,710
Do.	Nurkazgan Mine	do.	20,000
Do.	Akbastau Mine	East Kazakhstan Province	9,000
Do.	East Region:		
	Artemyevsky Mine	do.	25,000
Do.	Belousovsky Mine	do.	2,700
Do.	Irtyshsky Mine	do.	5,750
Do.	Nikolayevsky Mine	do.	25,700
Do.	Orlovsky Mine	do.	86,200
Do.	Yubileyno-Snegirikhinsky Mine	do.	14,200
Do.	Zhezkazgan Region:		
	Annensky Mine	Karagandy Province	25,000
Do.	East Mine	do.	35,000
Do.	North Mine	do.	28,000
Do.	South Mine	do.	30,000
Do.	Stepnoy Mine	do.	30,000
Do.	West Mine	do.	23,300
Do.	Zhomart Mine	do.	60,000

See footnotes at end of table.

TABLE 2—Continued
KAZAKHSTAN: STRUCTURE OF THE MINERAL INDUSTRY IN 2012[1,2]

(Metric tons unless otherwise specified)

Commodity	Major operating companies, main facilities, or deposits	Location or deposit names	Annual capacity[e]
Copper—Continued:			
Mining, recoverable, Cu content—Continued	Kazzinc JSC (Glencore International plc, 69.61%): Ridder complex:		
	Ridder-Sokolny Mine	East Kazakhstan Province	NA
Do.	Shubinsky Mine	do.	2,750
Do.	Tishinsky Mine	do.	15,000
Do.	Zyrianovsk complex:		
	Maleevsky Mine	15 kilometers north of Zyryanovsk	40,000
Do.	Grekhovsky Mine	NA	NA
Do.	Aktyubinsk Copper Co. TOO (CJSC Russian Copper Co.)	50th Anniversary of October Mine, at Koktau, Aktobe Province	NA
Do.	JSC Polymetal	Varvarinskoye deposit, Kostanay Province	NA
Concentrate, Cu content	Kazakhmys plc: Central Region:		
	Balkhash concentrator	Karagandy Province	40,000
Do.	Karagaily concentrators:		28,000
	Abyz	do.	
	Akbastau	do.	
	Kosmurun	do.	
Do.	Nurkazgan concentrator	do.	15,000
Do.	East Region:		
	Orlovsky concentrator	do.	70,000
Do.	Belousovsky concentrator	East Kazakhstan Province	13,000
Do.	Irtyshsky concentrator	do.	6,000
Do.	Nikolayevsky concentrator	do.	30,000
Do.	Zhezkazgan Region:		
	Satpayev concentrator	do.	30,000
Do.	Zhezkazgan No. 1 concentrator	do.	88,800
Do.	Zhezkazgan No. 2 concentrator	do.	95,000
Do.	Kazzinc JSC (Glencore International plc, 69.61%): Ridder complex: Ridder concentrator	Karagandy Province	10,000
Do.	Zyrianovsk complex: Zyrianovsk concentrator	do.	10,000
Do.	Aktyubinsk Copper Co. TOO (CJSC Russian Copper Co.)	50th Anniversary of October Mine, at Koktau, Aktobe Province	55,000
Do.	JSC Polymetal	Varvarinskoye deposit, Kostanay Province	NA
Metal	Kazakhmys plc mines or plants: Central Region:		
	Balkhash smelter	Karagandy Province	250,000
Do.	Balkhash refinery	do.	250,000
Do.	Zhezkazgan Region:		
	Zhezkazgan smelter	do.	250,000
Do.	Zhezkazgan refinery	do.	250,000
Do.	Ust-Kamenogorsk metallurgical complex [Kazzinc JSC (Glencore International plc, 69.61%)]	Oskemen (also known as Ust-Kamenogorsk)	70,000
Do.	Central Asia Metals plc	Karagandy Province	10,000
Ferroalloys:			
Ferrochrome:			
High-, medium-, and low-carbon FeCr containing 69% Cr	Aktobe plant {Kazchrome [Eurasian Natural Resources Corp. plc (ENRC)]}	Aktobe	450,000
High-carbon FeCr containing 69% Cr	Aksu plant {Kazchrome [Eurasian Natural Resources Corp. plc (ENRC)]}	Aksu	850,000
Ferrosilicon	do.	do.	NA
Ferrosilicochromium	do.	do.	NA
Silicomanganese	do.	do.	NA
Do.	Taraz Metallurgical Plant LLP (SAT & Co.)	Taraz, Zhambyl Province	NA
Do.	Temirtau Electrometallurgical Complex	Temirtau, Karagandy Province	NA

See footnotes at end of table.

TABLE 2—Continued
KAZAKHSTAN: STRUCTURE OF THE MINERAL INDUSTRY IN 2012[1, 2]

(Metric tons unless otherwise specified)

Commodity	Major operating companies, main facilities, or deposits	Location or deposit names	Annual capacity[e]
Gallium	Aluminium of Kazakhstan JSC [Eurasian Natural Resources Corp. plc (ENRC)]	Pavlodar	NA
Gold, mined	Kazzinc JSC (Glencore International plc, 69.61%)	Northern Kazakhstan	NA
Do.	Kazakhmys JSC	do.	NA
Do.	Polyus Gold International, Ltd.	do.	NA
Do.	JSC Polimetal	do.	NA
Do.	Nord Gold N.V.	Suzdal Mine	NA
Do.	GMK Kazakhaltyn	Northern Kazakhstan	NA
Do.	AK Altynalmas	Eastern Kazakhstan	NA
Do.	TOO Yubileynoye	Aktobe Province	NA
Indium	Kazzinc JSC (Glencore International plc, 69.61%)	NA	NA
Iron and steel:			
Pig iron	ArcelorMittal Temirtau	Temirtau, Karagandy Province	5,700,000
Steel, crude	do.	do.	6,000,000
Iron ore, marketable, gross weight	JSC Sokolov-Sarbai Mining Production Association [Eurasian Natural Resources Corp. plc (ENRC)]	4 open pit mines and 1 underground mine in Kostanay Province	20,000,000
Do.	TOO Orken (ArcelorMittal Temirtau)	Karagandy Province	5,000,000
Lead:			
Mining, recoverable Pb content of ore	Kazzinc JSC (Glencore International plc, 69.61%): Ridder complex: Shubinsky Mine	15 kilometers east of Ridder	630
Do.	Tishinsky Mine	15 kilometers southwest of Ridder	15,000
Do.	Zyrianovsk complex: Maleevsky Mine	15 kilometers north of Zyryanovsk	26,000
Do.	TOO ShalkiyaZinc Ltd.	Shalkiya Mine, 15 kilometers northeast of Zhanakorgan	NA
Concentrate, Pb content	Kazzinc JSC (Glencore International plc, 69.61%): Ridder concentrator	Ridder, East Kazakhstan Province	NA
Do.	Zyrianovsk concentrator	Zyryanovsk, East Kazkahstan Province	NA
Do.	TOO ShalkiyaZinc Ltd.	Kentau concentrating plant, South Kazakhstan Province	NA
Do.	TOO Nova Zinc (JSC Chelyabinsk Zinc Plant)	Akzhal	4,000
Metal	Chimkent metallurgical plant (JSC Yuzhpolimetall)	Shymkent	NA
Do.	Ust-Kamenogorsk metallurgical complex [Kazzinc JSC (Glencore International plc, 69.61%)]	Oskemen (also known as Ust-Kamenogorsk)	130,000
Magnesium, metal	Ust-Kamenogorsk titanium-magnesium plant	do.	NA
Manganese, crude ore	Facilities:	Locations:	NA[5]
	Kazmarganets {Kazchrome JSC [Eurasian Natural Resources Corp. plc (ENRC)]}	Tur and East Kamys Mines, Karagandy Province	
	Zhairemsky GOK[3] JSC [Eurasian Natural Resources Corp. plc (ENRC)]	Perstenevsky, Ushkatyn III, Zhomart and Zapadny Zhomart Mines near Zhairem	
	Atasurda mining and processing complex (TOO Orken)	Atasu	
	TOO Arman 100	170 kilometers east of Zhezkazgahan, Karagandy Province	
	Temirtau electrometallurgical complex	Temirtau, Karagandy Province	
Minor metals (indium, selenium, tellurium, thallium, and so forth)	Belogorskiy rare-metals plant	Asubulak, East Kazakhstan Province	NA[6]
Do.	Chimkent metallurgical plant (JSC Yuzhpolimetall)	Shymkent	NA[6]
Do.	Ust-Kamenogorsk metallurgical complex [Kazzinc JSC (Glencore International plc, 69.61%)]	Oskemen (also known as Ust-Kamenogorsk)	NA
Natural gas million cubic meters	Companies:	Locations:	NA[5]
	Tengizchevroil (Chevron Corp., 50%; KazMunaiGas JSC, 20%; ExxonMobil Kazakhstan Inc., 25%; LukArco B.V., 5%)	Tengiz and Korolev fields	
	Karachaganak Petroleum Operating B.V. (BG Group plc, 29.25%; ENI S.p.A., 29.25%; Chevron Corp., 18%; OAO Lukoil, 13.5%; KazMunaiGas JSC, 10%)	Karachaganak field	
	Additional production at smaller fields	NA	

See footnotes at end of table.

TABLE 2—Continued
KAZAKHSTAN: STRUCTURE OF THE MINERAL INDUSTRY IN 2012[1,2]

(Metric tons unless otherwise specified)

Commodity	Major operating companies, main facilities, or deposits	Location or deposit names	Annual capacity[e]
Niobium, metal	Ulba Metallurgical Plant (Kazatomprom JSC)	Oskemen (also known as Ust-Kamenogorsk)	NA
Petroleum:			
Crude	Tengizchevroil (Chevron Corp., 50%; KazMunaiGas JSC, 20%; ExxonMobil Kazakhstan Inc., 25%; LukArco B.V., 5%)	Tengiz and Korolev fields	NA [3]
	Karachaganak Petroleum Operating B.V. (BG Group plc., 29.25%; ENI S.p.A., 29.25%; Chevron Corp., 18%; OAO Lukoil, 13.5%; KazMunaiGas JSC, 10%)	Karachaganak field	
	CNPC AktobeMunaiGas (China National Petroleum Corp., 85.42%)	Aktobe Province	
	PetroKazakhstan Inc. (China National Petroleum Corp., 67%, and KazMunaiGas JSC, 33%)	South Turgai basin	
	Mangistaumunaigaz JSC	Mangistau Province	
	Ozenmunaigas (KazMunaiGas JSC)	do.	
	Embamunaigas (KazMunaiGas JSC)	Western Kazakhstan	
	JV Kazgermunai LLP (KazMunaiGas JSC)	Kyzylorda Province	
	JSC Karazhanbasmunai (CITIC Group and KazMunaiGas JSC)	Mangistau Province	
	North Buzachi oilfield	do.	
	Additional producers	NA	
Refined, crude oil throughput 42-gallon barrels per day	JSC Pavlodar Oil Chemistry Refinery (KazMunaiGas JSC, 58%)	Pavlodar	120,000
Do. do.	Atyrau Refinery (KazMunaiGas, 99.49%)	Atyrau	100,000
Do. do.	PetroKazakhstan Inc. (China National Petroleum Corp., 67%, and KazMunaiGas JSC, 33%)	Shmykent	110,000
Phosphate rock, beneficiated	Chulaktau mining and processing complex (Kazphosphate LLC)	Chulaktau, Zhambyl Province	NA
Do.	Karatau mining and processing complex (Kazphosphate LLC)	Zhanatas, Zhambyl Province	NA
Do.	Temir Service LLP (Sunkar Resources plc)	Chilisai deposit, northwestern Kazakhstan	400
Rare-earth metals, products	SARECO (AO NAK Kazatomprom and Sumitomo Corp.)	Stepnogorsk	1,500
Rhenium:			
Ammonium perrhenate (containing 69.2% Re)	Zhezkazganredmet (RedMet) (Government owned)	Zhezkazgan, Karagandy Province	NA
In tailings from copper ore processing	Balkhash copper mining-metallurgical complex (Kazakhmys plc)	Karagandy Province	NA
Silicon, metal	Silicium Kazakhstan LLP	Karaganda	12,500
Silver, refined	Facilities:	Locations:	1,000 [5]
	Chimkent metallurgical plant (JSC Yuzhpolimetall)	Shymkent	
	Ust-Kamenogorsk metallurgical complex [Kazzinc JSC (Glencore International plc, 50.7%)]	Oskemen (also known as Ust-Kamenogorsk)	
	Balkhash refinery (Kazakhmys plc)	Karagandy Province	
Tantalum, metal	Ulba Metallurgical Plant JSC (Kazatomprom JSC)	Oskemen (also known as Ust-Kamenogorsk)	NA
Titanium:			
Ore	Tioline LLP	Obuhovskoye deposit, just north of Kokshetau, Akmola Province	NA
Do.	Satpaevsk Titanium Mines Ltd. (Ust-Kamenogorsk titanium-magnesium plant, 49%)	Bektemir deposit, East Kazakhstan Province	NA
Do.	Shokash deposit	Aktobe Province	NA
Metal (sponge)	AO Ust-Kamenogorsk titanium-magnesium plant (UKTMK)	Oskemen (also known as Ust-Kamenogorsk)	35,000

See footnotes at end of table.

TABLE 2—Continued
KAZAKHSTAN: STRUCTURE OF THE MINERAL INDUSTRY IN 2012[1, 2]

(Metric tons unless otherwise specified)

Commodity	Major operating companies, main facilities, or deposits	Location or deposit names	Annual capacity[e]
Uranium, U content	Companies:	Locations:	19,500 [5]
	Akbastau JV (Kazatomprom JSC, 50%, and Uranium One Inc., 50%)	Blocks 1, 3, and 4 of the Budenovskoye deposit, Sozak Region, South Kazakhstan Province	
	Appak LLP (Kazatomprom JSC, 65%; Sumitomo Corp., 25%; Kansai Electric Power Co. Inc., 10%)	West Mynkuduk Mine of the Mynkuduk deposit, Sozak Region, South Kazakhstan Province	
	Baiken-U LLP (Kazatomprom JSC, 60%, and Japanese consortium, 40%)	Block No. 2 of the Kharassan deposit, Zhanakorgan Region, Kyzylorda Province	
	Betpak Dala JV (Uranium One Inc., 70%, and Kazatomprom JSC, 30%)	Akdala Mine and Site No. 4 (South Inkai) Mine of the Inkai deposit, Sozak Region, South Kazakhstan Province	
	Inkai JV (Cameco Corp., 60%, and Kazatomprom JSC, 40%)	Blocks 1, 2, and 3 of the Inkai deposit, Sozak Region, South Kazakhstan Province	
	Karatau LLP (Kazatomprom JSC, 50%, and Uranium One Inc., 50%)	Block No. 2 of the Budenovskoye deposit, Sozak Region, South Kazakhstan Province	
	Katco JV (Areva Group, 51%, and Kazatomprom JSC, 49%)	Tortkuduk Mine and Block No. 1 of the South Moinkum deposit, Sozak Region, South Kazakhstan Province	
	JSC Ken Dala.kz (Kazatomprom JSC, 100%)	Central Mynkuduk deposit, Sozak Region, South Kazakhstan Province	
	Kyzylkum LLP (Japanese consortium, 40%; Uranium One Inc., 30%; Kazatomprom JSC, 30%)	Block No. 1 of the Kharassan deposit, Zhanakorgan Region, Kyzylorda Province	
	Mining Company LLP (Kazatomprom JSC, 100%): Mining Group No. 6 LLP	North and South Karamurun Mines, Shieli and Zhanakorgan Regions, Kyzylorda Province	
	Stepnoye Mining Group LLP	Uvanas and East Mynkuduk Mines, Sozak Region, South Kazakhstan Province	
	Taukent Mining Chemical Plant LLP	Kanzhugan and South Moinkum Mines, Sozak Region, South Kazakhstan Province	
	Semizbai-U (Kazatomprom JSC and its subsidiary, Mining Company LLP, 51%, and China Guangdong Nuclear Power Group, 49%)	Irkol Mine in Kyzlorda Province and Semizbai Mine, on the border of North Kazakhstan and Akmola Province	
	Stepnogorsk Mining-Chemical Complex LLP (Kazatomprom JSC, 100%)	Shantobe Mine of the Vostok and Zvezdnoe deposits, 300 kilometers west of Stepnogorsk	
	JV Zarechnoye JSC (Kazatomprom JSC, 49.67%, and JSC Atomredmetzoloto, 49.67%)	Zarechnoye and South Zarechnoye deposits, Olrarski Region, South Kazakhstan Province	

See footnotes at end of table.

TABLE 2—Continued
KAZAKHSTAN: STRUCTURE OF THE MINERAL INDUSTRY IN 2012[1,2]

(Metric tons unless otherwise specified)

Commodity	Major operating companies, main facilities, or deposits	Location or deposit names	Annual capacity[e]
Zinc:			
Mine output, Zn content	Kazakhmys plc:		
	East Region complex:		
	Artemyevsky Mine	East Kazakhstan Province	90,000
Do.	Belousovsky Mine	do.	NA
Do.	Irtyshsky Mine	do.	18,000
Do.	Nikolaevsky Mine	do.	20,000
Do.	Orlovsky Mine	do.	78,200
Do.	Yubileyno-Snegirikhinsky Mine	do.	16,500
Do.	Central Region complex: Abyz Mine	Karagandy Province	13,500
Do.	Kazzinc JSC (Glencore International plc, 69.61%):		
	Ridder complex:		
	Ridder-Sokolny Mine	East Kazakhstan Province	NA
Do.	Shubinsky Mine	do.	4,000
Do.	Tishinsky Mine	do.	65,000
Do.	Shaimerden deposit	Kostanay Province	NA
Do.	Zyrianovsk complex: Maleevsky Mine	do.	135,000
Do.	TOO Nova Zinc (JSC Chelyabinsk Zinc Plant)	Akshatau, Karagandy Province	NA
Do.	TOO ShalkiyaZinc Ltd.	Kyzylorda Province	NA
Concentrate, Zn content	Kazakhmys plc:		
	East Region complex:		
	Artemyevsky concentrator	do.	55,000
Do.	Belousovsky concentrator	do.	5,800
Do.	Irtyshsky concentrator	do.	11,000
Do.	Nikolaevsky concentrator	do.	36,000
Do.	Orlovsky concentrator	do.	60,000
Do.	Karaganda Region complex: Karagaily concentrator	Karagandy Province	8,000
Do.	TOO Nova Zinc (JSC Chelyabinsk Zinc Plant)	Akshatau, Karagandy Province	35,000
Do.	TOO ShalkiyaZinc Ltd.	Kyzylorda Province	NA
Do.	Kazzinc JSC (Glencore International plc, 69.61%):		
	Ridder concentrator	do.	NA
Do.	Zyrianovsk concentrator	Zyryanovsk, East Kazakhstan Province	NA
Metal	Kazzinc JSC (Glencore International plc, 69.61%):		
	Ridder zinc refinery	East Kazakhstan Province	110,000
Do.	Ust-Kamenogorsk metallurgical complex	do.	190,000

[e]Estimated; estimated data are rounded to no more than three significant digits. Do., do., Ditto. NA Not available.
[1]Table includes data available through January 31, 2014.
[2]Many location names have changed since the breakup of the Soviet Union. Many enterprises, however, are still named or commonly referred to based on the former location name, which accounts for discrepancies in the names of enterprises and that of locations.
[3]GOK is the abbreviation for gorno-obogatitelnyi kombinat, which translates as "mining and beneficiation complex."
[4]Production at the Konyrat (formerly known as the Kounrad Mine) was stopped in 2005 owing to high production costs.
[5]Capacity estimates are totals for all enterprises that produce that commodity.
[6]It is unknown which, if any, rare metals are still being produced at this facility.

The Mineral Industry of Kosovo

By Mark Brininstool

Kosovo has deposits of aggregates and construction minerals, bauxite, chromium, lead and zinc, lignite, magnesite, nickel, and silver, but the production volume of most minerals was small by regional and world standards. Industry activity for almost all minerals stopped in the late 1990s during fighting between ethnic Albanian guerilla groups and Serbian security forces, but production of most mineral commodities was restarted and increased significantly during the past decade. On February 17, 2008, the Serbian Province of Kosovo declared its independence after having been under the administration of the United Nations' Interim Administration Mission in Kosovo since 1999 (Independent Commission for Mines and Minerals, 2005).

Minerals in the National Economy

The International Monetary Fund reported that Kosovo's real gross domestic product increased by 2.1% in 2012, and it increased by 5% in 2011. In 2011 (the latest year for which trade data were available), mineral commodities, including ferronickel, industrial minerals, lead, and zinc, were an important source of export revenue for Kosovo's economy and were the leading export and import goods. Exports of base metals and articles of base metals continued to be the leading export category in terms of value, and they totaled about €190.1 million ($136.9 million[1]), or about 61% of the total value of exports. The value of base-metal exports increased by about 3% compared with that of 2010. The value of exports of mineral products decreased by 0.5% to €38.6 million ($27.8 million) and made up about 12% of total exports. ("Mineral products" included industrial minerals, mineral fuels, and metal ores, but not processed metals.) Mineral products were the leading import category and were valued at about €538.1 million ($387.4 million) and made up 22% of the total value of imports (Central Bank of the Republic of Kosovo, 2013, p. 99–100; International Monetary Fund, 2013, p. 153).

Production

Production of zinc concentrate increased by 31%; silica sand, by 14%; and lead concentrate, by 11%. Marl output decreased by 51%; nickel ore, by 45%; sand and gravel, by 42%; clay by 23%; and limestone, by 21% (table 1). In 2012, Arsi Sh.p.k. of Albania began production of chromite near Llapceve. Production figures for lead and zinc metal were revised to zero for the series in table 1. Production of lead and zinc metal had been reported in 2009, but it is now known that there had been no production since 1999 (table 1; Arsi Sh.p.k., 2013).

Structure of the Mineral Industry

In 2011 and 2012, steps were taken to begin the privatization of the Trepca Complex, which had been an important regional producer of lead and zinc before production was halted during the conflict between ethnic Albanian guerilla groups and Serbian security forces in 1999. Trepca restarted production of lead and zinc ores in 2005, but in 2012, its output was still much lower than its pre-1999 output. The initial privatization efforts focused on selling seven of Trepca's nonmining facilities, but the privatization of these assets was put on hold in 2012 owing to complications from unsettled debts and opposition to the privatization by Serbia, which does not recognize Kosovo's independence (Privatisation Agency of Kosovo, 2011, p. 17; 2012, p. 15–16; SETimes.com, 2012).

In 2012, Arsi began production of chromite near Llapceve and had a number of exploration licenses in Brezovice, Crepule, Hasalar, Qafe e Prushit, Qerret, and Rahovec. The company was the only active chromite exploration company in Kosovo and claimed to own 80% of the known chromite deposits in Kosovo. Table 2 is a list of major mineral industry facilities (Arsi Sh.p.k., 2013).

Outlook

Kosovo is expected to remain a modest producer of mineral commodities but is likely to continue to increase production if the economy continues to develop and if infrastructure improves. Because Kosovo mainly exports its mineral products, the level of foreign demand will also be an important factor in the development of the mineral industry. The Trepca Complex could make Kosovo a regionally significant producer of lead and zinc, but major investments would be necessary to expand production.

References Cited

Arsi Sh.p.k., 2013, Arsi's chromium and base metal projects: Arsi Sh.p.k., February, 20 slides. (Accessed May 13, 2013, at http://www.arsi-group.com/wp-content/uploads/2013/04/Arsi-Presentation.pdf.)

Central Bank of the Republic of Kosovo, 2013, Monthly statistics bulletin—December 2012: Central Bank of the Republic of Kosovo, no. 137, January, 130 p. (Accessed May 29, 2013, at http://www.bqk-kos.org/repository/docs/2013/MSB no 137.pdf.)

Independent Commission for Mines and Minerals, 2005, Mineral deposits: Independent Commission for Mines and Minerals. (Accessed May 6, 2013, at http://www.kosovo-mining.org/kosovoweb/en/mining/minerals.html.)

International Monetary Fund, 2013, World economic outlook: International Monetary Fund, April, 184 p. (Accessed May 29, 2013, at http://www.imf.org/external/pubs/ft/weo/2013/01/pdf/text.pdf.)

Privatisation Agency of Kosovo, 2011, Annual report, January–December 2010: Privatisation Agency of Kosovo, 56 p. (Accessed May 16, 2013, at http://www.pak-ks.org/repository/docs/raporti_anglisht_FINAL.pdf.)

---

[1]Where necessary, values have been converted from euro area euros (€) to U.S. dollars (US$) at the rate of €0.72=US$1.00 in 2011.

Privatisation Agency of Kosovo, 2012, Annual report, January–December 2011: Privatisation Agency of Kosovo, 66 p. (Accessed May 16, 2013, at http://www.pak-ks.org/repository/docs/Annual_Report_2011_ENG_FINAL_(2).pdf.)

SETimes.com, 2012, Privatisation of Kosovo's Trepca mine delayed: SETimes.com, August 6. (Accessed May 16, 2013, at http://www.setimes.com/cocoon/setimes/xhtml/en_GB/features/setimes/features/2012/08/06/feature-01.)

TABLE 1
KOSOVO: PRODUCTION OF MINERAL COMMODITIES[1]

(Metric tons unless otherwise specified)

Commodity[2]		2008	2009	2010	2011	2012
METALS						
Chromite (Cr_2O_3, 26%)		--	--	--	--	2,000
Ferroalloys, ferronickel:[e]						
Gross weight		24,300	27,700	30,400	68,300	68,300
Ni content of ferronickel		5,600	6,360	7,000	15,700	15,700
Lead and zinc:						
Ore, gross weight		141,769 [r]	158,137	185,842	233,115 [r]	225,490
Lead content of ore		5,800 [r]	6,500 [r]	7,700 [r]	7,100 [r]	6,700
Zinc content of ore		5,000 [r]	5,600 [r]	6,400 [r]	6,600 [r]	6,600
Concentrate:						
Lead concentrate:						
Gross weight		--	4,285	8,232	7,471 [r]	8,298
Pb content		--	3,000	5,700	4,500 [r]	5,300
Zinc concentrate:						
Gross weight		--	5,332	8,678	7,409 [r]	9,695
Zn content		--	2,500	4,100	2,900 [r]	3,800
Metal:						
Pb, refined		--	-- [r]	-- [r]	-- [r]	--
Zn, refined		--	-- [r]	-- [r]	-- [r]	--
Nickel:						
Ore, wet		677,426 [r]	918,709	779,506	683,855	379,151
Ni content of ore		8,500 [r]	10,500	9,100 [r]	7,700	4,400
INDUSTRIAL MINERALS						
Cement[e]	thousand metric tons	590	600	600	600	600
Clay[e, 3]		85,000	200,000 [r]	93,580	158,000 [r]	122,000
Limestone	cubic meters	1,917,196	2,164,589	2,606,047	3,282,964 [r]	2,595,382
Marl	do.	322,007	291,829	302,630	259,616	128,161
Pumice and related materials, volcanic tuff[e]	do.	45,005	58,788	60,000	60,000	60,000
Sand and gravel, excluding glass sand		45,463 [r]	46,085 [r]	18,533 [r]	32,819	22,249
Silica sand (glass sand)	do.	27,325	20,181	25,178	19,711	22,490
MINERAL FUELS AND RELATED MATERIALS						
Lignite	thousand metric tons	7,885 [r]	7,839 [r]	7,958	8,212	8,028

[e]Estimated; estimated data are rounded to no more than three significant digits. [r]Revised. do. Ditto. -- Zero.

[1]Table includes data available through June 4, 2013.

[2]In addition to the commodities listed, other aggregates and construction materials were thought to have been produced, but available information is inadequate to make reliable estimates of output.

[3]Estimate based on reported production in cubic meters: 2008—35,382; 2009—83,294; 2010—38,925; 2011—65,930; and 2012—50,757.

TABLE 2
KOSOVO: STRUCTURE OF THE MINERAL INDUSTRY IN 2012

(Thousand metric tons)

Commodity	Major operating companies and major equity owners.	Location of main facilities	Annual capacity^e
Bauxite	Bauxite Mine Volljak (Government owned)	Grebnik Mine, about 50 kilometers east of Pristina and 17 kilometers north of Orahovac	NA
Cement	Sharrcem Sh.p.k. [Sharr Beteiligungs GmbH (Titan Group)]	Hani Elezit, on border with Macedonia	600
Chromite	Arsi Sh.p.k.	Llapceve, about 12 kilometers north of Rahovec	NA
Coal, lignite	Kosovo Energy Corp. J.S.C. (Government owned)	Bardh and Mirash Mines, just west of Pristina	NA
Lead and zinc:			
Ore	Trepca Complex (Government owned)	Stan Terg Mine	NA
Do.	do.	Crnac Mine, northern Kosovo	NA
Do.	do.	Artana Mine at Novo Brdo	NA
Do.	do.	Belo Brdo Mine, northern Kosovo	NA
Concentrate	do.	Concentrator at Tuneli i Pare near Mitrovica	NA
Do.	do.	Concentrator at Kizhnica, about 8 kilometers southeast of Pristina	NA
Do.	do.	Concentrator at Leposavic in northern Kosovo	NA
Metal:			
Lead	do.	Lead smelter at Zvecan	NA
Zinc	do.	Zinc smelter at Mitrovica	NA
Nickel:			
Ore	Ferronikeli (Cunico Resources NV)	Mines at Chikatovo (Dushkaja and Suke Mines) and Glavitca	NA
Metal[1]	do.	Ferronickel plant at Gllogovac, about 20 kilometers west of Pristina	12

^{e}Estimated; estimated data are rounded to no more than three significant digits. Do., do. Ditto. NA Not available.
[1]Nickel in ferronickel.

THE MINERAL INDUSTRY OF KYRGYZSTAN

By Elena Safirova

Kyrgyzstan was a major world producer of mercury and uranium in 2012, and gold was the primary mineral (in terms of value) mined in the country. Other minerals being mined included antimony, clay, coal, fluorspar, gypsum, limestone, natural gas, petroleum, sand and gravel, silica, and silver. Kyrgyzstan has deposits of other minerals that were not being mined. They include arsenic, bauxite, copper, iron ore, lead, rare-earth metals, sulfur, tin, tungsten, and zinc (Russian American Chamber of Commerce in the USA, 2007; Chunuev, 2013; Polyak, 2013; U.S. Central Intelligence Agency, 2013; Virta, 2013; Welcome.kg, 2013).

Minerals in the National Economy

In 2012, Kyrgyzstan's real gross domestic product (GDP) decreased by 0.9%; nominal GDP was $6.47 billion.[1] Industrial production decreased by 20% to $3.51 billion and contributed 44% to the total value of the GDP. In 2012, the country had 1,929 industrial enterprises that employed 158,000 workers. The value of Kyrgyzstan's exports decreased in 2012 by 15.5% to $1.89 billion, and the value of the county's imports increased by 26.1% to $5.37 billion. The ratio of exports to imports decreased to 35.2% from 52.6% in 2011 (National Statistical Committee of the Kyrgyz Republic, 2012, 2013; U.S. Central Intelligence Agency, 2013).

The main export commodities included cotton, electricity, garments, gold, machinery, meat, mercury, shoes, tobacco, uranium, and wool. Kyrgyzstan's leading export partners (by volume) were Uzbekistan, which received 28.8% of Kyrgyzstan's exports, Kazakhstan (22.0%), Russia (14.6%), China (7.0%), the United Arab Emirates (6.3%), and Afghanistan (5.7%). The main import commodities were chemicals, foodstuffs, oil and gas, and machinery and equipment. The country's primary import partners were China, which supplied 55.9% of Kyrgyzstan's imports, Russia (17.7%), and Kazakhstan (6.4%) (National Statistical Committee of the Kyrgyz Republic, 2013; U.S. Central Intelligence Agency, 2013).

Government Policies and Programs

In the past decade, the Government actively worked on attracting investments in its mineral industry. As of the beginning of 2012, Kyrgyzstan had 1,001 active licenses for exploration, development, and mining of mineral resources. Out of this total, 370 licenses were for mineral exploration and development and 631 were for mining. Only a few projects, however, resulted in mineral production. Corruption in the previous Government administrations, which had issued most of the licenses, led to the practice of giving licenses to mining companies that were not qualified to conduct efficient development and mining operations. Many residents in Kyrgyzstan expressed concern that, despite the high number of licenses, the country was not mining minerals on a large scale and therefore was not getting the full benefits from its natural resources. To address these concerns, the State Agency on Geology and Mineral Resources (Gosgeolagenstvo) revoked 550 licenses in 2010 and 292 licenses in 2011. In 2012, the Agency revoked 13 licenses (Chunuev, 2013).

In 2012, multiple disagreements arose between prospecting and mining companies and the local population. On many occasions, locals blocked roads leading to mining properties, seized vehicles and equipment used for mining, conducted demonstrations, and engaged in fights with miners. The most common complaints of the local population were related to environmental concerns, violations of license agreements and of safety regulations, and the lack of financial contribution to localities where the mining companies operated. In addition, several Government officials were accused of corruption. At yearend, eight criminal cases were under investigation. The charges involved stealing of large amounts of gold-containing concentrates; the organization of a black market for issuing licenses for prospecting, development, and mining; and unreported smuggling from the country of large amounts of mined minerals. One of the alleged criminals was the previous head of the Gosgeolagenstvo, who was accused of using his position for illegal licensing (Chunuev, 2013).

In July, the Gosgeolagenstvo announced an auction for 12 deposits. Of the 12 deposits, 6 contained alluvial gold, another 5 contained gold and other metals; and one contained coal. The sites included the Pereval'noye, the Terek, the Terekkhan, and the Togolok gold deposits. Gosgeolagenstvo developed a list of 16 criteria for selecting investors, including the speed of processing plant construction, the hiring of Kyrgyz nationals, the use of safe development and production technologies, and the size of investment. To adhere to the principles of transparency, the 12 deposits were to be auctioned on August 28, and the auction was to be translated on state television; prior to the auction date, 49 companies submitted applications for the auction. On the day of the auction, however, the auction was halted by a group of people who, introducing themselves as "people's activists," took over the television transmission, chanting "The Fatherland is not for sale." The auction was halted and was rescheduled for a later date. Because of the changes in the mining code and adoption of new procedures for scheduling and conducting auctions and tenders, the auction was expected to be held sometime in 2013; all applications submitted for the previous auction were annulled (Mineral.ru, 2012a; MinerJob.ru, 2012c).

In September, the President of Kyrgyzstan signed into law a new mining code that was adopted by the country's Parliament in June. The main principles of the new code are the protection of investments as a type of private property, noninvolvement of

---

[1] Where necessary, values have been converted from Kyrgyzstani soms (KGS) to U.S. dollars (US$) at the average annual exchange rate of KGS47.01=US$1.00 for 2012 and KGS46.14=US$1.00 for 2011.

national and local authorities in the management decisions of private enterprises; protection of the rights of mining companies to make mining decisions; and provision of exclusive rights for transition of licenses from exploration to mining. The new code specifies the "single window" principle for mineral rights; that is, licenses specify the plot of land and mineral rights as a part of a single package. The new code also describes how mining contributes to the socio-economic development of localities, outlines the "social packet" that should be included in the application for exploration and mining licenses; establishes renewal fees for continued holding of licenses, and specifies the responsibilities of the Government, mining companies, and the local authorities. The new code provides for environmental protection of both the localities where particular mining entities operate and of the entire country. Overall, the new mining code is intended to improve the investment climate in the country and to clearly specify a set of fees that mining companies are obligated to pay to the central and local governments for the use of natural resources (Knews.kg, 2012).

In addition to the mining code, relevant changes were made to the tax code. Starting on January 1, 2013, gold-mining companies are levied a tax on revenue, instead of a tax on profits; the tax rate is linked to the world gold price. For example, at the world gold price of $1,724.8 per troy ounce (the price as of November 13, 2012), the enterprises' revenues were to be taxed at the rate of 11%. The Mining and Metallurgy workers' Union of Kyrgyzstan was concerned that this taxation change was likely to hit hardest less profitable enterprises, such as OAO Kyrgyzaltyn, which was the country's second-ranked gold producer.

In November, the Parliament approved a bill that would impose customs duties on ore exports that contain precious metals. The bill was intended to encourage the construction of ore-processing plants in Kyrgyzstan. The bill assumed that the initial customs duty rate would be 5% in 2013 and would gradually be increased to 30% by 2017. The reason for the introduction of a customs duty was a sharp increase in ore exports in the recent years—in 2010, exports of ores containing precious metals totaled only 300 metric tons (t), whereas in 2011, 41,000 t of ores containing precious metals was exported (Mineral.ru, 2012c).

Production

In 2012, coal production increased by 41.1%, the estimated production of lime increased by 15.4%, and production of salt increased by 12.5%. Production of gold decreased by 44.6%; that of mercury, by 33.7%; and natural gas, by 30.5%. The estimated mine output of antimony decreased by 20%, and that of uranium, by 10%. These and other data on mineral production are in table 1.

Commodity Review

Metals

Gold.—As of November 2012, Kyrgyzstan had 60 known gold deposits with combined resources of 448 t of gold. Only a few of the deposits, however, were mined. The largest of the existing mines, the Kumtor gold mine, was located about 350 kilometers (km) southeast of Bishkek and about 60 km north of the border with China. The Kumtor Mine was operated by Centerra Gold Inc. of Canada. In 2012, Centerra Gold produced 315,000 troy ounces (9.8 t) of gold from the mine, which was a sharp (46%) reduction compared with its output in 2011. The reason for the reduction was an unexpected movement of ice from a glacier into the mine's open pit. In 2013, Centerra was planning to return to its previous production level of between 17.1 and 18.9 metric tons per year (t/yr). In 2012, output from the Kumtor Mine contributed 5.5% to the GDP of Kyrgyzstan and accounted for 18.9% of the country's total industrial production, and Centerra Gold paid $103.2 million in taxes and customs duties to the Government on the output from the mine. Mining operations at Kumtor were carried out by traditional open pit mining methods (Centerra Gold Inc., 2013a–c; Lazenby, 2013).

In November, Centerra announced the expansion of the open pit at Kumtor, which resulted in a significant increase in the resources of the Kumtor deposit and included a 58% increase in proven and probable reserves to a total of 9.7 million troy ounces (302 t) of contained reserves. Accordingly, the company extended the expected life of the Kumtor Mine by 5 years, including the term of open pit mining through 2023 and the term of ore processing at the gold extracting plant through 2026. The company was expecting to produce an average of 20 t/yr of gold from the mine in the next 10 years and to increase the volume processed at the gold extracting plant by 18% by 2016 (Lazenby, 2013).

The relations between Centerra, the local population, and the Government, however, were complicated. In February, Kumtor's workers were on strike to demand a wage increase, and the strike lasted 10 days. According to some accounts, it was the 10-day idle period at the mine that led to the excessive ice accumulation that ultimately drastically reduced the mine's annual output. At the end of May, residents of Issyk-Kul Oblast', where the Kumtor Mine is located, blocked the highway leading to the mine and held a protest demonstration with about 600 participants. The protest participants handed their demands to the company's administration. The list of demands contained 13 points, ranging from the company's contribution to social needs of the population, such as orphanages, kindergartens, sports halls, and medical and veterinary care, to environmental demands and demands to contribute to infrastructure projects in the region. The blockage was ultimately removed in early June when, after negotiations with Government representatives, the company agreed to meet some of the demands (Mineral.ru, 2012d; MinerJob.ru, 2012b; Centerra Gold Inc., 2013d).

In June, some members of the Parliament called for nationalization of the Kumtor Mine. Although no voting on the matter took place and although many other Parliament members responded that Kyrgyzstan could not afford to nationalize the mine, the shares of Centerra lost about one-third of their value on June 25. By August, international banks reportedly increased the interest rates charged to businesses working in Kyrgyzstan. During the fall of 2012, both Centerra and the Government hired commissions of independent experts to evaluate the environmental damage, if any, inflicted by the company because of its operation of the Kumtor Mine. In December, the leadership of Kumtor Operating Co. announced that the Government had submitted to them the list of damages amounting to $152 million, which included damages from mine construction, the location of waste material and tailings, emission of harmful materials, damages to land resources, and use of water resources. At yearend, the outstanding issues between the Government and Kumtor Operating Co. were still not resolved (MinerJob.ru, 2012a).

Industrial Minerals

Rare Earths.—The Kutessai I, II, and III deposits in Chuy Valley contain rare-earth elements. During the Soviet times, 15 of the elements were mined by open pit method at Kutessai II. In 2009, Kutessai Mining Co. (Kutessai), whose parent company was Stans Energy Corp. of Canada, had obtained a 3-year license for mining Kutessai II from the Gosgeolagenstvo. During the next 3 years, the company was unable to start production because of a combination of technical difficulties and administrative delays, but gave 5.8 million soms ($126,000) as a charitable donation to the local government of Ak-Tyuz. In 2012, the original license was extended (Dudka, 2012).

In June, the committee on economic sector development of the Parliament recommended to the Gosgeolagenstvo that the extended license be annulled on the grounds that the company did not fulfill its obligations as specified by the original and the extended licenses. Also, the committee stated that the licenses were given to the company in violation of the laws in place at the time the licenses were issued, including the mining code. The Gosgeolagenstvo did not annul the license but was conducting negotiations with the company about a transfer of about 20% of the shares of the company to the Government. In August, the Gosgeolagenstvo confirmed the company's compliance with the licensing agreement. Also, Stans Energy filed a lawsuit against the Parliament's committee, and in November, the Inter-District Court of Bishkek ruled in favor of the company. It was not clear, however, if other attempts to annul the licenses would be initiated in the future (Mineral.ru, 2012b; Stans Energy Corp., 2012a, b).

Mineral Fuels

Coal.—In 2012, Kyrgyzstan produced 1,184,000 t of coal, which was a 41% increase compared with production in 2011. In Kyrgyzstan, coal appeared to be a cheaper and more reliable energy source compared with natural gas, which had to be imported. According to the Ministry of Energy, the country had about 70 deposits with a total resource potential of 2,260 million metric tons (Mt). In addition to bituminous coal and lignite, the country had about 260 Mt in proven and probable reserves of coking coal that was not being mined in 2012 (Xinhuanet.com, 2013).

Oil and Natural Gas.—In 2012, Kyrgyzstan was in the process of constructing two new oil refineries. The first of the refineries was being built in the city of Kara-Balta in Chuyskaya Oblast. By the original plan of 2009, the refinery was supposed to have a production capacity of only 100,000 t/yr (about 677,000 barrels per year) of petroleum. In 2010, however, when Jund China Petroleum Co. became a partner in the project, the planned capacity increased to about 800,000 t/yr [5.4 million barrels per year (Mbbl)]. The petroleum for the plant was planned to be exported from Kazakhstan. The total investment in the project was expected to be $250 million, and the costs were expected to be recouped in 6 years. The refinery was planned to produce gasoline (40%), diesel fuel (40%), liquefied gas, and lubricants. The depth of refining (the percentage of petroleum processed into usable substances) was planned to reach 94%, which is very high, especially for Central Asia. The refinery was expected to start operations in August 2013 (CentralAsiaOnline.com, 2012; Vasilivetskiy, 2012).

Another refinery was planned to be built by the State Oil Company of Azerbaijan Republic (SOCAR). The preliminary cost projected for the project was $150 million with a planned capacity of 2 million metric tons per year (Mt/yr) (15.7 Mbbl). The new refinery was expected to start production in 2014 (CentralAsiaOnline.com, 2012).

Outlook

Kyrgyzstan is actively trying to attract foreign investment into its mineral industry and to start production at new deposits. In 2012, the country was trying to improve transparency in the process of issuing licenses for exploration, prospecting, development, and mining of its natural resources. Its adoption of a new mining code is intended to provide a balance between the interests of mining companies, local populations, and the Government.

In the next 5 to 10 years, Kyrgyzstan is expected to start full-scale production at some of its gold deposits that were in preliminary stages of development during the past decade. When this happens, the tensions around the largest producing gold mine in the country, Kumtor, could subside and pave the way to increased exploitation of the natural resources of the country.

References Cited

Centerra Gold Inc., 2013a, Basic data: Centerra Gold Inc. (Accessed July 25, 2013, at http://www.kumtor.kg/en/deposit/production-basic-data.)

Centerra Gold Inc., 2013b, Basic operating results of Kumtor Operating Company in 2012: Centerra Gold Inc. (Accessed July 25, 2013, at http://www.kumtor.kg/en/itogi_2012.)

Centerra Gold Inc., 2013c, Contribution to the economy: Centerra Gold Inc. (Accessed July 25, 2013, at http://www.kumtor.kg/en/media-relations/contribution-to-the-kyrgyz-economy.)

Centerra Gold Inc., 2013d, Kumtor sent a letter to Kyrgyz Prime Minister. (Accessed July 25, 2013, at http://www.kumtor.kg/en/the-company-kumtor-sent-a-letter-to-the-prime-minister-of-the-kyrgyz-republic-which-sets-out-the-answers-to-the-demands-of-a-group-of-people-blocked-the-road-to-the-mine-kumtor.)

CentralAsiaOnline.com, 2012, Kyrgyzstan postroit 2 NPZ [Kyrgyzstan will build 2 oil refineries]: CentralAsiaOnline, April 27. (Accessed July 25, 2013, at http://centralasiaonline.com/ru/articles/caii/features/main/2012/04/27/feature-01.)

Chunuev, I.K., 2013, Ob itogakh taboty Gosudarstvennogo agenstva po geologii i mineral'nym resursam [About the work results of the State Agency for Geology and Mineral Resources]: Gosudarstvennoe Agenstvo po geologii I mineral'nym resursam Kyrgyzskoy respubliki, February 14. (Accessed July 25, 2013, at http://www.geology.kg/statistics/mining.)

Dudka, Irina, 2012, V Kyrgyzstane vydana litsenziya na pol'zovanie mestorozhdeniem redkozemel'nykh metallov Kutessay II [Kyrgyzstan issued a license for use of Kutessay II deposit]: 24.kg, August 27. (Accessed July 25, 2013, at http://www.chemmarket.info/ru/news/view/19655/.)

Knews.kg, 2012, V Kyrgyzstane vstupaet v silu novyy zakon o nedrakh [A new mining code goes into effect in Kyrgyzstan]: Knews.kg, August 14. (Accessed July 25, 2013, at http://www.knews.kg/econom/20287_v_kyirgyizstane_vstupaet_v_silu_novyiy_zakon_o_nedrah/.)

Lazenby, Henry, 2013, Centerra Gold production 40% down y-o-y, expects recovery in 2013: Mining Weekly Online, January 15. (Accessed July 25, 2013, at http://www.miningweekly.com/article/centerra-gold-production-40-down-y-o-y-expects-recovery-in-2013-2013-01-15.)

Mineral.ru, 2012a, Auktsion po prodazhe zolotorudnykh mestorozhdeniy v Kyrgyzstane sorvala tolpa, vorvavshayasya v zal c krikami "Rodina ne prodaetsya"[A crowd chanting "Fatherland is not for sale" thwarted the auction of gold deposits in Kyrgyzstan]: Mineral.ru, August 28. (Accessed July 25, 2013, at http://www.mineral.ru/News/49855.html.)

Mineral.ru, 2012b, Kirgizskie deputaty predlagayut annulirovat' litsenziyu "Kutessay Mayning" na mestorozhdenie Kutessay-2 [Kyrgyz Members of Parliament suggest annulling the license issued to Kutessay Mining for the Kutessay-2 deposit]: Mineral.ru, June 27. (Accessed July 25, 2013, at http://www.mineral.ru/News/49106.html.)

Mineral.ru, 2012c, Kyrgyzstan ponuzhdaet investorov stroit' pererabatybayushie zavody [Kyrgyzstan forces investors to build processing plants]: Mineral.ru, December 11. (Accessed July 25, 2013, at http://www.mineral.ru/News/51134.html.)

Mineral.ru, 2012d, Svedeniya o perekrytii trassy na rudnik Kumtor zhitelyami cela Barskaun (Kirgiziya) protivorechivy [Information about blockage of the highway leading to the Kumtor Mine by the residents of the village of Barskaun in Kyrgyzstan is conflicting]: Mineral.ru, June 1. (Accessed July 25, 2013, at http://www.mineral.ru/News/48810.html.)

MinerJob.ru, 2012a, Kirgizskie parlamentarii vystupayut za natsionalizatsiyu mestorozhdeniya "Kumtor" [Kyrgyz Parliament Members call to nationalize the Kumtor Mine]: MinerJob.ru, June 21. (Accessed July 25, 2013, at http://www.minerjob.ru/viewnew.php?id=21351.)

MinerJob.ru, 2012b, Na Kyrgyzskom rudnike "Kumtor" mozhet snizit'sya dobycha zolota [Kyrgyz Kumtor Mine could decrease gold production]: Miner.Job.ru, March 28. (Accessed July 25, 2013, at http://www.minerjob.ru/viewnew.php?id=21055.)

MinerJob.ru, 2012c, V Kirgizii proidet pervyi auktsion po realizatsii mestorozhdeniy poleznykh iskopaemykh [The first auction of mineral deposits will take place in Kyrgyzstan]: MinerJob.ru, August 1. (Accessed July 25, 2013, at http://www.minerjob.ru/viewnew.php?id=21585.)

National Statistical Committee of the Kyrgyz Republic, 2012, Industry of Kyrgyz Republic 2012: Bishkek, Kyrgyzstan, National Statistical Committee of the Kyrgyz Republic, 316 p. (Accessed July 25, 2012, at http://stat.kg/images/stories/docs/tematika/prom/Prom 2007-2011.pdf.)

National Statistical Committee of the Kyrgyz Republic, 2013, Kyrgyzstan—Brief statistical handbook 2009–2012: Bishkek, Kyrgyzstan, National Statistical Committee of the Kyrgyz Republic, 40 p. (Accessed November 21, 2012, at http://stat.kg/images/stories/docs/tematika/svod/Handbook 2009-2012.pdf.)

Polyak, D.E., 2013, Molybdenum: U.S. Geological Survey Mineral Commodity Summaries 2013, p. 106–107.

Russian American Chamber of Commerce in the USA, 2007, Kyrgyzstan—Mining industry overview: Russian American Chamber of Commerce in the USA. (Accessed July 25, 2013, at http://www.russianamericanchamber.com/en/services/mining/industry.htm.)

Stans Energy Corp., 2012a, Stans receives favourable judgments in all Kyrgyz cases: Stans Energy Corp., November 15. (Accessed July 25, 2013, at http://www.stansenergy.com/press-releases/stans-energy-receives-favourable-judgments-in-all-kyrgyz-cases.)

Stans Energy Corp., 2012b, State geological agency confirms Stans' compliance with licencing agreement: Stans Energy Corp., August 7. (Accessed July 25, 2013, at http://www.stansenergy.com/press-releases/state-geological-agency-confirms-stans-compliance-with-licencing-agreement.)

U.S. Central Intelligence Agency, 2013, Kyrgyzstan, in The world factbook: U.S. Central Intelligence Agency, July 10. (Accessed July 25, 2013, at https://www.cia.gov/library/publications/the-world-factbook/geos/kg.html.)

Vasilivetskiy, A., 2012, Kitay zakanchivaet stroitel'stvo NPZ v Kyrgyzskoy Kara-Balte [China is completing construction of the oil refinery in Kyrgyz Kara-Balta]: CentralAsia.ru, July 30. (Accessed July 25, 2013, at http://www.centrasia.ru/newsA.php?st=1343629620.)

Virta, R.L., 2013, Mercury: U.S. Geological Survey Mineral Commodity Summaries 2013, p. 102–103.

Welcome.kg, 2013, Poleznye Iskopaemye [Minerals]: Welcome.kg. (Accessed July 25, 2013, at http://www.welcome.kg/ru/kyrgyzstan/nature/pl2.)

Xinhuanet.com, 2013, V 2012 godu rost dobychi uglya v Kyrgyzstane sostavil 32,1 prots [In 2012, coal production growth in Kyrgyzstan was 32.1%]: Xinhuanet.com. (Accessed July 25, 2013, at http://russian.news.cn/economic/2013-01/25/c_132127352.htm.)

TABLE 1
KYRGYZSTAN: PRODUCTION OF MINERAL COMMODITIES[1, 2]

(Metric tons unless otherwise specified)

Commodity		2008	2009	2010	2011	2012
METALS						
Antimony:						
Mine output, Sb content[e]		700	700	700	1,500	1,200
Metal and compounds		--[r]	918	842	892[r]	924
Gold, mine output, Au content	kilograms	18,144[r]	16,978[r]	18,072[r]	18,648[r]	10,333
Mercury, metal	do.	--[r]	140,000[r]	98,700[r]	112,700[r]	74,700
INDUSTRIAL MINERALS						
Cement, hydraulic		1,218,100	579,400	759,700	1,016,600[r]	900,000[e]
Gypsum		55	50	51	57	59
Lime		8,700	4,700	6,500	2,600	3,000[e]
Salt[e]		900[r]	900[r]	900[r]	800[r]	900
Sands[e]	cubic meters	836,200[3]	800,000	850,000	850,000	800,000
MINERAL FUELS AND RELATED MATERIALS						
Coal:						
Bituminous		55,338	68,800	65,000	94,000	132,600
Lignite		437,263	538,100	510,000	745,000	1,051,400
Total		492,601	606,900	575,000	839,000	1,184,000
Natural gas	thousand cubic meters	19,800[r]	15,400	22,800	26,600[r]	18,500
Petroleum, crude:						
In gravimetric units		69,300[r]	75,100[r]	70,700[r]	77,000[r]	77,100
In volumetric units[e]	42-gallon barrels	516,000	562,000	602,000[r]	656,000[r]	657,000
Uranium, processed:						
U content		1,097	2,574	2,000	2,000[e]	1,800[e]
U_3O_8 content		1,309	3,071	2,385	2,385	2,150[e]

[e]Estimated; estimated data are rounded to no more than three significant digits; may not add to totals shown. [r]Revised. do. Ditto. -- Zero.

[1]Table includes data available through July 25, 2013.

[2]In addition to the commodities listed, Kyrgyzstan is thought to produce a number of other mineral commodities, including clays, copper, fluorspar, kaolin, mined mercury, molybdenum, gravel, silver, tin, and tungsten, but available information is not adequate to make reliable estimates of production.

[3]Reported figure.

TABLE 2
KYRGYZSTAN: STRUCTURE OF THE MINERAL INDUSTRY IN 2012[1]

(Metric tons unless otherwise specified)

Commodity	Major operating companies, main facilities, or deposits	Location or deposit names	Annual capacity[e]
Antimony:			
Sb content of ore	Kadamzhay mining and metallurgical complex (OAO KyrgyzAltyn, 100%), which included the Kadamzhay Mine and the Terek-Sayskiy Mine	Batkenskaya Oblast'	2,400 [2]
	Khaydarkan mining and metallurgical complex	Khaydarkan region	
Metal and compounds	Kadamzhay metallurgical facility (ATF Invest, a subsidiary of ATF Bank of Kazakhstan, 70.4%)	Kadamzhayskiy Rayon	28,000
Cement	Kantskiy cement plant	Kant	1,500,000
Coal	Seven underground mines and five open pits among the following deposits: Almalyk, Dzhergalan, Kara-Kiche-Kok-Yangak, Kyzyl-Kiya, Sulyukta, and Tashkumyr	Southwestern, central, and northeastern parts of the country	2,200,000 [2]
Copper	Talas Copper Gold Co.	Talasskaya Oblast'	NA
Fluorspar, concentrate	Khaydarkan mining and metallurgical complex	Khaydarkan deposit	5,000
Gold:			
Au content of ore	Kumtor Gold Co. (Centerra Gold Inc., 100%)	Kumtor deposit	22
Do.	OAO KyrgyzAltyn (Government, 100%)	Makmal deposit	3
Do. kilograms	Solton-Sary Mine	Naryn	500
Do.	Talas Gold	Jerooy-Bashi, Pereval; Talasskaya Oblast'	NA
Do.	Taldy-Bulak Levoberezhny deposit (Summer Gold Co., 40%, and Zijin Mining Group, 60%)	NA	NA
Do.	Ishtamberdy deposit (Lingbao Gold Co. Ltd.)	Chatkal region	NA
Do.	Unkurtash gold deposit (Highland Gold Mining Ltd.)	NA	NA
Do.	Bozymchak gold deposit (OcOO Kazakhmys Gold Kyrgyzstan)	Dzhalal-Abadskaya Oblast'	NA
Refined	Kara-Balta refinery	Chuyskaya Oblast'	NA
Mercury:			
Hg content of ore	Khaydarkan mining and metallurgical complex	Khaydarkan and Novoye deposits	700 [2]
Metal	do.	Khaydarkan deposit	1,000
Molybdenum, for nonmetallurgical uses	Kara-Balta mining and metallurgical complex	Chuyskaya Oblast'	NA
Do.	Molibden Joint Stock Co.	do.	NA
Natural gas million cubic meters	Kyrgyzazmunayzat	Approximately 300 wells; Changyr-Tash, Chigirchik Pereval, Izbaskentskoye, Kara-Agach, Mayluu-Suu, Susahoye, and Togap-Beshkenskoye deposits	100 [2]
Petroleum	do.	do.	150,000
Do.	Kyrgyz Petroleum Co.	Dzhalal-Abadskaya Oblast'	NA
Silver	Karagoyskoye deposit	Oshskaya Oblast'	NA
Do.	Kumyshtag deposit	Talasskaya Oblast'	NA
Tin	Novosibirsk Integrated Tin Works	Atdzhaylau deposit	150
Do.	do.	Trudovoye deposit	350
Do.	Tyanshanolovo mining and beneficiation complex	Sary-Dzhas field	NA
Do.	Uchkoshkon deposit	do.	NA
Tungsten	Enil'chek JSC mining enterprise	Atdzhaylau deposit	90
Do.	do.	Trudovoye deposit	95,600
Uranium, processed	Kara-Balta mining complex (GK Renova)	Zarechnoye deposit, Chuyskaya Oblast'	3,600
Do.	Linia Prava (LPU) (Nimrodel Resources, 90%)	Batken Leases, Southern Fergana Valley, Batkenskaya Oblast'	NA

[e]Estimated; estimated data are rounded to no more than three significant digits. Do., do. Ditto. NA Not available.
[1]Many location names have changed since the breakup of the Soviet Union. Many enterprises, however, are still named or commonly referred to based on the former location name, which accounts for discrepancies in the names of enterprises and that of locations.
[2]Capacity estimates are totals for all enterprises that produce that commodity.

THE MINERAL INDUSTRY OF LATVIA

By Alberto Alexander Perez

In 2012, Latvia's gross domestic product (GDP) increased by 5.6% compared with that of 2011. The services sector accounted for 69.6% of the GDP; the industrial sector, 25.2%; and the agricultural sector, 5.1%. Latvia's industrial growth rate in 2012 was 7.8%, and its main trade partners, in order of the volume of trade, were Lithuania, Russia, Germany, and Estonia (U.S. Central Intelligence Agency, 2013).

Latvia produced mainly industrial minerals and was not a significant world producer of any mineral commodities; it had, however, the Baltic States' only steel mill, JSC Liepājas Metalurgs. According to the company, 98% of the steel mill's production was for export, and this material accounted for 10% of total Latvian exports. The company invested in the construction of an electric arc furnace, and the first new production started in August 2011. By December 2012, the new furnace reached the designed production level and had a production capacity of 850,000 metric tons (JSC Liepājas Metalurgs, 2013a, b).

SIA CEMEX was the sole producer of cement in Latvia. In the summer of 2010, SIA CEMEX finished construction of a new cement plant with a capacity of 1.6 million metric tons per year at Broceni. SIA CEMEX planned to export its production to Belarus, Estonia, Finland, Lithuania, and Russia (CEMEX S.A.B. de C.V., 2013).

References Cited

CEMEX S.A.B. de C.V., 2013, Latvia: CEMEX S.A.B. de C.V. (Accessed July 19, 2013, at http://www.cemex.lv/eng/cx/cx_cl_hi.asp.)

JSC Liepājas Metalurgs, 2013a, Company: JSC Liepājas Metalurgs. (Accessed July 19, 2013, at http://lm.metalurgs.lv/?a=0&b=1.)

JSC Liepājas Metalurgs, 2013b, History: JSC Liepājas Metalurgs. (Accessed April 11, 2014, at http://lm.metalurgs.lv/?a=0&b=1&c=6.)

U.S. Central Intelligence Agency, 2013, Latvia, in The world factbook: U.S. Central Intelligence Agency. (Accessed July 18, 2013, at https://www.cia.gov/library/publications/the-world-factbook/geos/lg.html.)

TABLE 1
LATVIA: PRODUCTION OF MINERAL COMMODITIES[1]

(Metric tons)

Commodity[2]	2008	2009	2010	2011	2012[e]
Cement, portland[e]	310,000	650,000	1,100,000	1,100,000	1,200,000
Crushed rock	507,591	500,000[e]	1,375,197[r]	2,050,976[r]	2,000,000
Dolomite, crude (excluding calcined, crushed dolomite aggregate)	2,305,065	929,070	930,000[e]	930,000[e]	930,000
Gravel, pebbles, shingle and flint of a kind used for concrete aggregates, for road metaling, or for railway and other ballast	6,011,735	5,195,972	4,736,785[r]	5,386,702[r]	5,390,000
Gypsum[e]	230,000	230,000	230,000	230,000	230,000
Limestone	NA	NA	NA	NA	NA
Peat	923,404	1,163,803	1,119,417	1,378,681[r]	1,378,000[3]
Sand and gravel	2,222,504	2,292,848	1,388,188[r]	2,337,916[r]	2,338,000[3]
Sand, construction	2,153,000	1,314,535	1,320,000[e]	1,281,995[r, 4]	1,280,000
Silica sand, industrial[e]	12,000	12,000	12,000	12,000	12,000
Steel, crude[e]	635,000	692,000	655,000	515,000	800,000

[e]Estimated; estimated data are rounded to no more than three significant digits. [r]Revised. NA Not available.
[1]Table includes data available through July 19, 2013.
[2]In addition to the commodities listed, Latvia produced gypsum, silica, and industrial sand, but output was not reported, and information was not available to make reliable estimates of output.
[3]Reported figure.
[4]Production based on volume sold.

TABLE 2
LATVIA: STRUCTURE OF THE MINERAL INDUSTRY IN 2012

(Thousand metric tons)

Commodity	Major operating companies and major equity owners	Location of main facility	Annual capacity
Cement	SIA CEMEX (CEMEX S.A.B. de C.V.)	Plant in Broceni	1,600
Steel	JSC Liepājas Metalurgs	Plant in Liepāja	850

The Mineral Industry of Lithuania

By Alberto Alexander Perez

In 2012, Lithuania's real gross domestic product (GDP) increased by 3.6% compared with that of 2011. Lithuania's GDP composition by sector was as follows: services, 68.4%; industry, 28.4%; and agriculture, 3.2%. Lithuania was not a significant world producer of any mineral commodities, but crude petroleum and petroleum products made up a significant portion of the country's foreign trade value. In 2012, the majority of Lithuania's exports, in terms of value, went to European countries; exports to Russia accounted for 18.9% of Lithuania's total exports, which made Russia the principal trade partner of Lithuania. In 2011 (the latest year for which data were available), petroleum and petroleum products accounted for 26% of the total value of Lithuania's imports and 24% of the total value of the country's exports (Statistics Lithuania, 2012, p. 288–289, 376–384; U.S. Central Intelligence Agency, 2013).

Production

AB Orlen Lietuva, which was located near Mazaikiai about 90 km west of the petroleum terminals in Butinge, Klaipeda, and Ventspils, was the only petroleum refinery operating in the Baltic States; the refinery exported most of its production. AB Akmenes Cementas, which was located in Naujoji Akmene, was the only cement producer in Lithuania. Data on mineral production are in table 1.

Commodity Review

Industrial Minerals

Cement.—In August, Akmenes Cementas, which was partly owned by CEMEX S.A.B. de C.V. of Mexico, posted an increase in revenue of 22% compared with revenue in the same month in 2011. Cement output was reported to have increased by 7.7% as of mid-year. Akmenes Cementas indicated that the company's level of exports was similar to that of 2011; however, its exports to Belarus seem to have increased somewhat. The company was operating at nearly full capacity and sought to follow its previous strategy of exporting as much of its output as possible. The company stated that Lithuania's cement consumption was likely to decrease by 5%; however, because of the expected increase in exports, the company was projecting that its annual sales would increase by at least 10%, and its revenues, by 15%. The company stated that the upgrade project at the main installation in Naujoji Akmene was 60% complete and that the project was expected to be completed by the middle of 2013 (Global Cement, 2012a, b).

Mineral Fuels

Petroleum.—In November, AB Orlen Lietuva (a subsidiary of the Orlen Group) began a new stage in the implementation of a system of automated processes that the company had begun in 2006. In this newest stage, the company was to implement an automatic furnace ignition system. (The previous stage saw the implementation of a wireless automated control system that was less expensive to operate than the previous cable system.) The company stated that its goal in automating its processes was to implement an advanced process control system that would enable the sending of tasks automatically to the whole chain of operations based on the required amount of product to be produced (AB Orlen Lietuva, 2012).

Outlook

Exports, including exports of mineral commodities, are expected to play an increasingly significant role in the economy of Lithuania, especially because domestic demand is expected to remain weak for at least the near future. The Government's plans to invest further in infrastructure and the industrial base are likely to support this move towards a more export-oriented economy.

References Cited

AB Orlen Lietuva, 2012, New leap in Lietuva's automation: AB Orlen Lietuva. (Accessed July 18, 2013, at http://www.orlenlietuva.lt/EN/PressCenter/News/Pages/New-Leap-in-ORLEN-Lietuvas-automation.aspx?pageNumber=2.)

Global Cement, 2012a, Akmene Cement takings rise by 22% in first half: Pro Global Media Ltd., August 21. (Accessed July 18, 2013, at http://www.globalcement.com/news/item/1102-akmene-cement-takings-rise-by-22-in-first-half.)

Global Cement, 2012b, Exports drive 10% sales growth at Akmene Cement: Pro Global Media Ltd., August 21. (Accessed July 18, 2013, at http://www.globalcement.com/news/item/1176-exports-drive-10-sales-growth-at-akmene-cement.)

Statistics Lithuania, 2012, Statistical yearbook of Lithuania 2012: Statistics Lithuania, 696 p. (Accessed October 28, 2012, at http://www.stat.gov.lt/uploads/metrastis/1_LSM_2012.pdf.)

U.S. Central Intelligence Agency, 2013, Lithuania, *in* The world factbook: U.S. Central Intelligence Agency. (Accessed July 17, 2013, at https://www.cia.gov/library/publications/the-world-factbook/geos/lh.html.)

TABLE 1
LITHUANIA: PRODUCTION OF MINERAL COMMODITIES[1]

(Metric tons)

Commodity	2008	2009	2010	2011	2012
Ammonia, nitrogen content	759,000	472,400	434,100	869,500	918,400
Cement	1,075,581	583,100	834,000 r	996,300	1,015,000
Common clays and shales for construction use	624,470	246,398	250,000	282,400	300,100
Crushed granite	810,000	327,000 r	307,000	374,000	374,000 e
Crushed stone used for concrete aggregates, for roadstone, and for other construction use	6,896,987	2,426,773	2,500,000 e	2,500,000	2,500,000 e
Dolomite, crude (excluding calcined, crushed dolomite aggregate)	4,752	2,000 r, e	2,000 e	2,000	2,000 e
Limestone	1,625,089	911,900 r	928,000	1,179,000	1,354,000
Peat:					
Horticultural use	521,000	542,500	326,800	384,700	386,200
Fuel use e	15,000	15,000	15,000	15,000	15,000
Petroleum:					
Crude	127,658	115,000	114,500	113,900	112,000 e
Refinery products	8,814,800	8,012,300	8,579,400	8,796,900	8,458,600
Sand and gravel:					
Construction sand	5,055,172	3,100,000	3,104,000	5,776,800	5,081,200
Gravel, pebbles, shingle and flint	4,414,239	1,930,000	2,551,000	2,372,000	2,370,000 e
Silica sand, industrial	38,300	41,200	67,300 r	53,400	53,900
Stones, granules, chippings, and powder of, excluding marble	15,538	6,222	6,200 e	6,200	6,200 e
Sulfur	73,870	69,722	73,500	76,700	73,000
Sulfuric acid	1,051,400 r	1,155,849	1,150,000 e	1,150,000	1,150,000 e

eEstimated; estimated data are rounded to no more than three significant digits. rRevised.

[1]Table includes data available through July 18, 2013.

TABLE 2
LITHUANIA: STRUCTURE OF THE MINERAL INDUSTRY IN 2012

(Thousand metric tons unless otherwise specified)

Commodity	Major operating companies and major equity owners	Location of main facility	Annual capacity
Cement	AB Akmenes Cementas (CEMEX S.A.B. de C.V., 33.95%)	Plant in Naujoji Akmene	1,500
Petroleum, refined	AB Orlen Lietuva (Orlen Group)	Plant in Mazaikiai	15,000

THE MINERAL INDUSTRY OF MACEDONIA

By Yadira Soto-Viruet

Macedonia produced a number of metals, including copper, ferroalloys, and steel, as well as mine output of lead and zinc. Other mineral commodities produced in the country included bentonite, feldspar, gypsum, lignite, lime, and sand and gravel. Petroleum was imported and processed at the country's sole domestic refinery.

Minerals in the National Economy

In 2012, Macedonia's real gross domestic product (GDP) decreased by 0.3% compared with that of 2011. In 2011 (the latest year for which data were available), manufacturing made up 13.5% of the GDP and mining and quarrying made up only 1.5%. In 2012, mineral fuels, lubricants, and related materials made up about 21% of the value of Macedonian imports and about 6% of the value of exports. Macedonia's State Statistical Office listed ferronickel, flat-rolled steel products, and petroleum products as three of the country's five most significant export items, but data concerning the value of the trade of these goods in 2012 were not available. Based on the total volume of international commodity trade, the leading trade partners of Macedonia were Germany (17.2%), Greece (9.4%), Serbia (7.4%), Bulgaria (6.6%), and Italy (6.4%) (International Monetary Fund, 2013, p. 153; Republic of Macedonia State Statistical Office, 2013a, p. 30; 2013b, p. 1–2, 4).

Production

In 2012, the estimated production of copper content of concentrate increased by 37% to 10,400 metric tons (t) from 7,600 t (revised) in 2011 and the gross weight of the copper concentrate production increased by 26% to 45,266 t from 35,976 t. Ferronickel production increased by 11% to 83,700 t from 75,200 t, and nickel production increased by 11% to 19,247 t from 17,292 t. The production of sand and gravel increased significantly, to 124,442 t from 2,443 t. Silicomanganese production decreased by 72% to 14,179 t from 50,756 t; that of petroleum refinery products, by 65% to an estimated 1.9 million barrels (Mbbl) from an estimated 5.4 Mbbl; bentonite clay, by 52% to 6,900 t; agglomerated dolomite, by 50% to 2,606 t; and crude talc, by 47% to 286 t. Data on mineral production are in table 1.

Structure of the Mineral Industry

Table 2 is a list of major mineral industry facilities.

Commodity Review

Metals

Gold.—Euromax Resources Ltd. of Canada held 100% interest in the Ilovitza gold-copper project, which is located about 18 kilometers (km) east of Strumica in southeastern Macedonia. In July, the Government granted the company an exploitation concession agreement for the Ilovitza project and approved the project's environmental impact study. In October, a preliminary economic assessment was completed by Tetra Tech Inc. of the United States. The assessment reported that Ilovitza had the potential to be developed as an open pit mine and to process 8 million metric tons per year (Mt/yr) of sulfide material during a mine life of 19 years. The project would include the construction of roads, site buildings, tailings facilities, and power and water supplies. In late 2012, the company completed an 11,800-meter (m) diamond drilling program and expected to have a mineral resource update in mid-2013. A prefeasibility study was scheduled to be completed in the second quarter of 2013 (Euromax Resources Ltd., 2012a–c; Tetra Tech Inc., 2012, p. 2, 8, 9).

Genesis Resources Ltd. of Australia held a 62% interest in the Plavica project, which is located about 65 km east of the capital of Skopje in northeastern Macedonia. The project included seven exploration concessions and covers an area of about 185 square kilometers (km^2). In May, the company announced inferred mineral resources based on the Australian Joint Ore Reserves Committee standards. The inferred mineral resources were estimated to be 55.46 million metric tons (Mt) at an average grade of 1.0 gram per metric ton (g/t) gold at a cutoff grade of 0.75 g/t, 22.63 Mt at an average grade of 29.7 g/t silver at a cutoff grade of 20 g/t, and 7.98 Mt at an average grade of 0.43% copper at a cutoff grade of 0.4%. The company hired Golder Associates Pty Ltd. of Australia to conduct a scoping study for the project, which would include mine planning, mineralogy and metallurgy studies, and an environmental and social impact study for the region. In August, the company hired Spektra Jeotek Sanayi ve Ticaret A.S. of Turkey to conduct 7,000 m of reverse-circulation drilling and 2,500 m of diamond core drilling. In December, Genesis announced its plans to conduct additional drilling of 46,000 m of reverse-circulation drilling and 23,000 m of diamond core drilling by 2013. The company planned to complete a feasibility study; however, no further details as to when this study would take place were available (Genesis Resources Ltd., 2012a, b; 2013, p. 2).

Mineral Fuels

Lignite.—Production of lignite came from the Brod-Gneotino (commissioned in July), the Oslomej-East, the Oslomej-West, and the Sudovol Mines, which were operated by state-owned AD ELEM. In 2011 (the latest year for which data were available), AD ELEM produced 8.1 Mt of lignite compared with 6.6 Mt in 2010 (AD ELEM, 2010, 2011).

The Bitola thermal powerplant generated about 70% of the total electricity produced in Macedonia. The plant is located 8 km east of the city of Bitola and had a capacity of about 675 megawatts (MW). The Bitola powerplant used coal reserves from the Suvodol Mine, which were sufficient to supply the plant until 2014. The company expected to extend the coal reserves in Suvodol owing to the commissioning of the Brod-Gneotino Mine and the development of the deep underlying seam in the Suvodol-Bitola project; both projects are located in the Suvodol area. AD ELEM envisioned extending the operation of the Bitola plant to 2030. The deep underlying seam project was expected to be completed by 2013. The company also planned to develop new projects in the country, including the Mariovo lignite deposit and thermal powerplant and the Zivojno coal deposit (AD ELEM, 2012a, p. 2, 4–9, 13).

AD ELEM planned to construct the Mariovo powerplant with a capacity of about 300 MW. A prefeasibility study was underway, and the plant was expected to be completed by 2016. The plant would use coal from the Mariovo lignite deposit, which is located 30 km south of Prilep and covers an area of about 14 km^2. The deposit had estimated reserves of about 96.7 Mt of coal. The company reported that the Mariovo deposit had the potential to be developed as an underground mine and to produce about 2 Mt/yr of coal during a mine life of 30 years. Development of the Mariovo Mine was expected to be completed by 2015 (AD ELEM, 2012a, p. 6, 18–19; 2012c).

Zivojno Bitola is located in the Pelagonia Region about 35 km southeast of Bitola and 20 km from the Suvodol Mine and its extension, the Brod-Gneotino Mine. The deposit, which covers an area of about 25 km^2, was estimated between the years of 1966 and 1984 to contain reserves of about 100 Mt of coal. Based on a feasibility study conducted in 2004, the production capacity at Zivojno was estimated to be about 3 Mt/yr of coal. AD ELEM reported that studies were underway at Zivojno, including the classification and recategorization of the coal reserves, as well as engineering, geologic, and hydrogeologic investigations. The project was expected to be completed by 2017 (AD ELEM, 2012a, p. 8–9; 2012b).

Outlook

The Government of Macedonia forecasted an increase in the GDP of 2.0% in 2013. The modernization and rehabilitation of the Bitola and the Oslomej thermal powerplants, the expansion plans at the Suvodol project, and plans to develop the Mariovo and the Zivojno deposits are expected to strengthen the mineral industry in the short run. In the longer run, new projects in the nonfuel mineral sector, such as the possible development of new gold-copper deposits, are likely to attract foreign investment in the mineral sector and to increase interest in nonfuel mineral prospecting (International Monetary Fund, 2013, p. 153).

References Cited

AD ELEM, 2010, Annual report 2010: Skopje, Macedonia, AD ELEM. (Accessed August 14, 2013, at http://www.elem.com.mk/2010/main.html.)

AD ELEM, 2011, Annual report 2011: Skopje, Macedonia, AD ELEM. (Accessed August 14, 2013, at http://www.elem.com.mk/2011/start.html.)

AD ELEM, 2012a, Investment plant, 2012–2017: Skopje, Macedonia, AD ELEM, March, 48 p. (Accessed August 15, 2013, at http://www.elem.com.mk/images/stories/INVESTMENT_PLAN_2012_2017.pdf.)

AD ELEM, 2012b, Project concept note—Coal mine Zivojno Bitola: Skopje, Macedonia, AD ELEM, 4 p. (Accessed August 14, 2013, at http://www.elem.com.mk/images/stories/objekti/3_ProjectConceptNote_Zivojno_ANG.pdf.)

AD ELEM, 2012c, TPP Mariovo and mine Mariovo: Skopje, Macedonia, AD ELEM, 6 p. (Accessed August 14, 2013, at http://www.elem.com.mk/images/stories/objekti/2_Mariovo_ANG.pdf.)

Euromax Resources Ltd., 2012a, Euromax announces approval of Ilovitza EIS and preliminary economic assessment: Vancouver, British Columbia, Canada, Euromax Resources Ltd. press release, October 24, 4 p. (Accessed August 15, 2013, at http://www.euromaxresources.com/media/23873/release_241012.pdf.)

Euromax Resources Ltd., 2012b, Exploitation concession granted for Ilovitza project: Vancouver, British Columbia, Canada, Euromax Resources Ltd. press release, July 24, 2 p. (Accessed August 13, 2013, at http://www.euromaxresources.com/media/23852/release_24july_final.pdf.)

Euromax Resources Ltd., 2012c, Management's discussion and analysis for the year ended December 31, 2012: Vancouver, British Columbia, Canada, Euromax Resources Ltd., December 31, 2 p. (Accessed August 13, 2013, at http://www.euromaxresources.com/media/43754/mda_ye_2012.pdf.)

Genesis Resources Ltd., 2012a, JORC inferred resources for the Plavica project: Melbourne, Victoria, Australia, Genesis Resources Ltd., May 30, 4 p. (Accessed August 15, 2013, at http://www.genesisresourcesltd.com.au/attachments/investor-information/announcement/2012/Apr-Jun/20120607-GES_Plavica_Resource_120530_final.pdf.)

Genesis Resources Ltd., 2012b, Strategic plans—Market update: Melbourne, Victoria, Australia, Genesis Resources Ltd., August 14, 2 p. (Accessed August 15, 2013, at http://www.genesisresourcesltd.com.au/attachments/investor-information/announcement/2012/Jul-Sep/0816-GES_ASX_Market Update_120814_final_updated.pdf.)

Genesis Resources Ltd., 2013, Quarterly activities report—December 2012: Melbourne, Victoria, Australia, Genesis Resources Ltd., January 31, 18 p. (Accessed August 15, 2013, at http://www.genesisresourcesltd.com.au/attachments/investor-information/announcement/2013/Jan-Mar/130207-GES_Quarterly Activites Report Dec 2012 FINAL WITH APP 5B.pdf.)

International Monetary Fund, 2013, World economic outlook: Washington, DC, International Monetary Fund, April 13, 184 p. (Accessed August 12, 2013, at http://www.imf.org/external/pubs/ft/weo/2013/01/pdf/text.pdf.)

Republic of Macedonia State Statistical Office, 2013a, Gross domestic product, 2011: Republic of Macedonia State Statistical Office, April 3, 79 p. (Accessed August 12, 2013, at http://www.stat.gov.mk/Publikacii/3.4.13.02.pdf.)

Republic of Macedonia State Statistical Office, 2013b, The international trade volume of the Republic of Macedonia, January-December 2012: Republic of Macedonia State Statistical Office, February 5, 4 p. (Accessed August 12, 2013, at http://www.stat.gov.mk/pdf/2013/7.1.13.02.pdf.)

Tetra Tech Inc., 2012, Preliminary economic assessment on the Ilovitza gold project, Macedonia: Swindon, United Kingdom, Tetra Tech Inc., December 5. (Accessed August 13, 2013, at http://www.euromaxresources.com/media/45779/1298530100-rep-r0002-05_ilovitza_pea_final.pdf.)

TABLE 1
MACEDONIA: PRODUCTION OF MINERAL COMMODITIES[1]

(Metric tons unless otherwise specified)

Commodity[2]		2008	2009	2010	2011	2012
METALS						
Copper, mine and concentrator output:						
Ore, gross weight	thousand metric tons	4,240	3,767	4,199	4,118	4,435
Concentrate:						
Gross weight		38,337	35,430	37,678	35,976	45,266
Cu content[e]		8,400	7,600	7,900	7,600 [r]	10,400
Metal, refined		--	--	--	-- [r]	2,300
Iron and steel:						
Ferroalloys:						
Ferromanganese		12,623	--	--	--	--
Ferronickel (23% Ni), gross weight[e]		65,300	52,200	62,700	75,200	83,700
Ferrosilicon		42,674	7,657	30,044	56,167	42,402
Silicomanganese		54,931	--	36,705	50,756	14,179
Total		175,528	59,857	129,449	182,123	140,281
Steel:						
Crude, secondary		252,461	276,215	292,126	386,000	216,000
Semimanufactures		252,946	270,397	291,886	385,816	216,934
Lead, mine output, concentrate, Pb content[e]		35,000	38,000	38,000 [r]	36,000 [r]	34,000
Nickel, Ni content of FeNi		15,026	12,000	14,413	17,292	19,247
Zinc, mine output, concentrate, Zn content[e]		29,000	29,000 [r]	29,000 [r]	28,000 [r]	28,000
INDUSTRIAL MINERALS						
Cement	thousand metric tons	916	909	820	981	683
Clays, bentonite		22,890	15,350	12,798	14,466	6,900
Dolomite:						
Agglomerated		4,179	3,814	4,748	5,249	2,606
Not frayed, not calcined		75,855	78,523	116,290	125,700	129,120
Sintered		23,933	21,607	24,989	28,251	23,062
Feldspar, crude		28,920	19,377	23,188	25,032	17,168
Gypsum, crude		242,400	154,550	143,118	162,984	157,844
Lime		--	2,713	2,700 [e]	2,700 [e]	2,700 [e]
Limestone flux		827,100	694,968	1,063,839 [r]	1,142,662 [r]	818,559
Marl		1,083,830	560,170	749,750	861,666	954,495
Pumice and related materials, volcanic tuff		103,476	113,064	113,323	57,356	52,911
Sand and gravel, excluding glass sand		124,000 [r]	49,009 [r]	64,789 [r]	2,443 [r]	124,442
Silica sands (quartz sands or industrial sands)		131,735 [r]	112,106 [r]	116,079 [r]	125,949 [r]	125,900 [e]
Stone, excluding quartz and quartzite:						
Crushed and broken		33,023	59,715	130,105	104,209	60,403
Dimension, crude		71,819	69,082	78,603	64,320	64,384
Talc, crude		977	682	1,292	547	286
MINERAL FUELS AND RELATED MATERIALS						
Lignite	thousand metric tons	7,746	7,454	6,583	7,902	7,310
Petroleum, refinery products[e, 3]	thousand 42-gallon barrels	8,300	7,700	6,600	5,400	1,900

[e]Estimated data are rounded to no more than three significant digits; may not add to totals shown. [r]Revised. -- Zero.

[1]Table includes data available through August 5, 2013.

[2]In addition to commodities listed, secondary aluminum in small amounts, common clay, diatomite, and gold contained in copper concentrate also are thought to have been produced, but available information is inadequate to make reliable estimates of output.

[3]Figures were converted to barrels from production in thousand metric tons, which was reported as the following: 2008—1,036; 2009—963; 2010—829; 2011—680; and 2012—243.

TABLE 2
MACEDONIA: STRUCTURE OF THE MINERAL INDUSTRY IN 2012

(Thousand metric tons unless otherwise specified)

Commodity		Major operating companies and major equity owners	Location of main facilities	Annual capacity
Cement		Usje Cementarnica AD (Titan S.A., 95%)	Plant at Skopje	1,000 [e]
Copper:				
Ore		Bucim Mine (Solway Investment Group Ltd.)	Mine and mill at Bucim, west of Radovis	4,500
Metal		do.	Solvent extraction and electrowinning plant at Bucim, west of Radovis	3,000
Ferroalloys:				
Ferrosilicon	metric tons	Jugohrom Ferroalloys DOO (Camelot Group)	Plant at Jegunovce	66
Silicomanganese	do.	Skopski Leguri DOOEL	Plant at Skopje	56
Ferromanganese	do.	do.	do.	66
Ferronickel, Ni content of ferronickel	do.	Feni Industries (Cunico Resources)	Ferronickel plant at Kavadarci	22 [e]
Do.	do.	Skopski Leguri DOOEL	Plant at Skopje	500
Gold, mine output, Au in copper concentrate		Bucim Mine (Solway Investment Group Ltd.)	Mine and mill at Bucim, west of Radovis	NA
Lead, metal		MHK Zletovo (Metrudhem DOOEL)	Imperial smelter and refinery at Veles	NA [1]
Lead-zinc, concentrate		Sasa Mine (Solway Investment Group Ltd.)	Mill at Sasa, north of Makedonska Kamenica	NA
Lead-zinc ore		do.	Mine at Sasa, north of Makedonska Kamenica	1,000 [e]
Do.		Zletovo Mine (Indo Minerals and Metals DOOEL)	Mine and mill near Probistip	NA
Do.		Toranica Mine (Indo Minerals and Metals DOOEL)	Mine near Dolga Livada	NA
Lignite		AD ELEM (state owned)	Mine at Suvodol	6,500
Do.		do.	Mine at Oslomej	1,000
Do.		do.	Star Rudnik Mine at Oslomej	310 [e]
Do.		do.	Brod-Gneotino Mine, south of Suvodol near Brod	2,000
Nickel, ore		Feni Industries (Cunico Resources)	Opencast mine at Rzanovo, 32 kilometers south of Kavadarci	NA
Petroleum, refined	42-gallon barrels	OKTA A.D. Skopje (EL.P.ET Balkanike, 81.51%)	Oil refinery at Skopje	20,000 [e]
Steel, crude, secondary		Makstil A.D. Skopje (Duferco Group, 62%)	Plant at Skopje	360
Zinc, metal		MHK Zletovo (Metrudhem DOOEL)	Imperial smelter and refinery at Veles	NA [1]

[e]Estimated. Do., do. Ditto. NA Not available.

[1]MHK Zletovo, Macedonia's only producer of lead and zinc metal, was idled in 2003 or 2004 after bankruptcy proceedings were initiated. Attempts have been made to restart the plant since it was closed, but so far none of these attempts has been successful.

THE MINERAL INDUSTRY OF MALTA

By Harold R. Newman

The sedimentary rocks that formed the islands of the Maltese archipelago are composed mainly of limestone. The rock sequence is divided into five geologic formations: Upper Coralline Limestone, Greensand, Blue Clay Limestone, Globigerina Limestone, and Lower Coralline formation. The older sediments of the Lower Coralline Limestone were deposited about 35 million years ago whereas the more recent layers of the Upper Coralline Limestone were deposited about 7 million years ago. Also, there were some small areas of Pleistocene surface deposits. Malta had very few mineral resources (industrial minerals, metallic minerals, or mineral fuels) that were of economic significance (University of Malta, 2013).

The Malta Resources Authority (MRA) has regulatory responsibility relating to the energy, mineral, and water resources of the Maltese islands. The MRA was established by the Maltese Parliament through the Malta Resources Authority Act of 2000. Some of the mineral-related matters for which the MRA has oversight responsibility are oil exploration, quarry operators, and energy and water utilities (Malta Resources Authority, 2013a).

Production

The main mineral commodities produced in Malta in 2012 were limestone and evaporated (solar) salt, which were used locally, mostly in construction and lime making (table 1).

Structure of the Mineral Industry

Several small stone quarries operated on the islands of Gozo and Malta. Available information regarding the ownership of these quarries, however, as well as the amount of production, capacity, and the locations of the quarries was inadequate to prepare a Structure of the Mineral Industry table for Malta.

Mineral Trade

Malta is strategically located in the center of the Mediterranean Sea and is about 6 nautical miles off the main Mediterranean sea-route between Gibraltar and the Suez Canal, which is one of the major shipping lanes in the world. Because of its location, Malta has become an important transshipment center for major shipping lines. In 2004, the Malta Freeport Terminals Ltd. was established to develop the Malta Freeport, which offers modern transshipment facilities, storage, and various assembly and processing operations, including an oil terminal with bunkering facilities. In 2012, Malta Freeport ranked 12th among the top European ports and was the third-ranked transshipment and logistics center in the Mediterranean region. More than 95% of the Malta Freeport's container traffic was transshipment business, including petroleum and refined petroleum products (Malta Freeport Terminals Ltd., 2012).

The mineral-related economy of the country depended almost completely on imports, the reexport of raw materials and fuels, and the storage of crude petroleum. The European Union (EU) countries that were Malta's principal trading partners included Germany (14%), France (10.5%), Greece (7.7%), Italy (7.4%), and the United Kingdom (6.4%) for exports and Italy (32%), France (8.4%), the United Kingdom (8%), and Germany (6.9%) for imports (U.S. Central Intelligence Agency, 2013).

In 2012, exports to Malta from the United States were valued at $381 million. These exports included fuel oil valued at $245 million, natural gas liquids valued at $23 million, and petroleum products valued at $6 million. U.S. imports from Malta in 2012 were valued at $257 million. These imports included other petroleum products valued at $10 million and advanced iron and steel manufactures valued at $46,000. Malta depended almost completely on imports for its supply of raw materials and fuels (U.S. Census Bureau, 2012a, b).

Commodity Review

Mineral Fuels

Petroleum.—The Oil Exploration Department (OED) of the MRA was set up to implement and administer the provisions of the Petroleum Production Act, Chapter 156, as amended; the Continental Shelf Act, Chapter 194, as amended; and the Petroleum (Production) Regulations, Subsidiary Legislation 156.01, as amended. The OED administers the country's petroleum exploration, which includes promoting exploration, analyzing data, keeping samples and data, monitoring contractual obligations, and maintaining surveillance of exploration activity on Malta's Continental Shelf. Hydrocarbons have been produced for several decades in the Libya and Tunisia offshore areas, which are adjacent to Malta's marine area and are within the same geologic province (Malta Resources Authority, 2012b).

Mediterranean Oil and Gas Ltd. of the United Kingdom announced that it planned to start petroleum exploration drilling efforts by yearend 2013 on its license in Malta's offshore Area 4 where it has a production-sharing agreement with the Government. Mediterranean Oil and Gas stated that its decision was based on new seismological data that it had acquired in 2011 and 2012. The new data were gathered using a 9-kilometer-long cable that made use of sonar to explore for petroleum prospects below sea level. This €8.3 million ($11.1 million[1]) exploration project was the largest yet conducted in Malta by Mediterranean Oil and Gas (Malta Today, 2012).

---

[1]Where necessary, values have been converted from euro area euros (€) to U.S. dollars (US$) at a rate of €0.75=US$1.00.

Mediterranean Oil and Gas reported that an independent review by ERC Equipoise Ltd. of the United Kingdom had estimated 26 million barrels of petroleum resource potential. Mediterranean Oil and Gas stated that it intended to initiate drilling exploration by yearend 2013. The sea is 450 meters deep in the area, and the exploration well is expected to reach a depth of 2.5 kilometers and take 60 days to drill (Xuereb, 2012).

Outlook

International trade activities, including the transshipment and reexport of goods, such as petroleum, refined products, and other minerals, will continue to be significant to Malta's economy. The country is expected to continue offshore exploration for petroleum. Industrial minerals will continue to be produced for domestic consumption.

References Cited

Malta Freeport Terminals Ltd., 2012, About us: Malta Freeport Terminals Ltd. (Accessed June 17, 2013, at http://www.maltafreeport.com.mt/freeport/content.aspx?id=107934.)

Malta Resources Authority, 2013a, Directorate for Minerals Resources Regulation: Malta Resources Authority. (Accessed April 16, 2013, at http://www.mra.org.mt.)

Malta Resources Authority, 2013b, Oil Exploration Department: Malta Resources Authority. (Accessed April 16, 2013, at http://www.mra.org.mt.)

Malta Today, 2012, Mediterranean Oil and Gas to spud Malta oil well in December 2013: Malta Today, September 18. (Accessed April 17, 2013, at http://www.maltatoday.com.mt/ en/newsdetails/news/national/mediterranean-oil-gas-to-spud-malta-oil-well-in-December-2013-20120917.)

University of Malta, 2013, The geology of the Maltese Islands: University of Malta. (Accessed June 17, 2013, at http://www.um.edu.mt/science/physcs/smru/generalinformation/geologyofmalta.)

U.S. Census Bureau, 2012a, U.S. exports to Malta by 5-digit end-use code: U.S. Census Bureau. (Accessed March 27, 2012, at http://www.census.gov/foreign rade/statistics/product/enduse/exports/c4730.html.)

U.S. Census Bureau, 2012b, U.S. imports from Malta by 5-digit end-use code: U.S. Census Bureau. (Accessed March 27, 2012, at http://www.census.gov/foreign-trade/statistics/product/enduse/imports/c4730.html.)

U.S. Central Intelligence Agency, 2013, Malta, *in* The world factbook: U.S. Central Intelligence Agency. (Accessed April 16, 2013, at http://www.cia.gov/library/publications/the-world-faxtbook/geos/mt.html.)

Xuereb, Matthew, 2012, Potential 260 million barrels of oil off Malta: Times of Malta. (Accessed May 23, 2013, at http://www.timesofmalta.com/articles/view/20121015/ local/Potential-260-million-barrels-of-of-oil-off-malta.)

TABLE 1

MALTA: ESTIMATED PRODUCTION OF MINERAL COMMODITIES[1, 2]

(Cubic meters)

Commodity[3]	2008	2009	2010	2011	2012
Limestone	1,200,000	1,200,000	1,200,000	1,200,000	1,200,000
Salt, solar	6,000	6,000	6,000	6,000	6,000

[1]Estimated data are rounded to no more than three significant digits.

[2]Table includes data available through March 31, 2013.

[3]In addition to the commodities listed, small amounts of cement, fertilizer, lime, and plaster are produced, but available information information is inadequate to make reliable estimates of output.

THE MINERAL INDUSTRY OF MOLDOVA

By Elena Safirova

Moldova had a small mineral industry of limited regional significance that was engaged primarily in the mining and production of industrial minerals and mineral products, including cement, clays, gypsum, limestone, and sand and gravel. The country was dependent on imports for all its coal, natural gas, and oil supplies, which came mainly from Russia and Ukraine. Moldova's main mineral resources were industrial minerals used to produce construction materials and as an input for the cement, chemical, food processing, and glass industries. Small deposits of iron ore, natural gas, and oil had been explored but were found not to be economic to develop (U.S. Central Intelligence Agency, 2013).

As of the end of 2012, Moldova's dependence on imports of Russian natural gas resulted in the country owing an estimated $4.3 billion to Russian gas supplier Gazprom; the debt was largely for unreimbursed gas consumption by the Transnistria region. During the year, Gazprom charged Moldova from $380 to $390 per thousand cubic meters of natural gas, or about the same price that it charged Western European countries. The debt was complicating the renewal of the agreement between the Government and Gazprom concerning continued shipments of Russian gas to Moldova in 2013 and beyond. At the same time, Moldova was working on developing alternatives to Russia's energy sources. In August, the Prime Ministers of Moldova and Romania signed an agreement about energy security between the two countries and announced that the Yassy-Ungeny gas pipeline was to be completed by the end of 2013. The cost of the pipeline was estimated to be €20 million ($24.7 million)[1] and was expected to be financed by the European Union (EU) as a part of the EU's Moldova-Romania-Ukraine Joint Operational Program for the period 2007 through 2013. Another project expected to increase mutual energy security between Moldova and Romania was the Bălți-Suceava high-voltage electric power line. Moldova was also participating in the Energy and Biomass Project, which was being financed by the United Nations and the EU. According to the project managers, the use of hay as an alternative energy source could help meet about 25% of Moldova's energy needs (InoSMI.ru, 2012; RBC.ru, 2012a, b; Regnum.ru, 2012a, b; Vedomosti.md, 2012).

Minerals in the National Economy

In 2012, the nominal gross domestic product (GDP) of Moldova was $7.25 billion; real GDP decreased by 0.8% compared with that of 2011. Industrial production contributed 14.0% to the GDP. In 2012, industrial production decreased by 3.1% compared with that of 2011; the output in mining and quarrying, however, increased by 0.8%, whereas the output of the energy and manufacturing sectors decreased by 4.4% and 3.0%, respectively. The output of the chemical sector increased by 2.6%, and the production of other nonmetal mineral products (such as cement, concrete, glass, and gypsum) increased by 2.0% (National Bureau of Statistics of the Republic of Moldova, 2012, 2013; NOI.md, 2013; U.S. Central Intelligence Agency, 2013).

In 2012, the Moldovan trade deficit increased by 2.6% to $3.05 billion; the value of exports decreased by 2.5% to $2.16 billion, and that of imports was practically unchanged, having increased by 0.5% to $5.21 billion. In 2012, Moldova was a net importer of mineral commodities; the total value of mineral exports amounted to $33.1 million, and the total value of mineral imports was $1,217.6 million. Moldova was a net importer of base metals and articles made from them; the total value of these exports was $73.3 million, and the total value of imports of base metals and articles made from them was $284.4 million. No information on the export of specific mineral commodities was available. The major export categories were foodstuffs, machinery, and textiles, and the main export trade partners were Russia (which received 30.3% of Moldova's exports), Romania (16.5%), Italy (9.4%), Ukraine (5.7%), the United Kingdom (3.9%), and Belarus (3.7%). Moldova imported chemicals, machinery and equipment, mineral products and mineral fuels, and textiles. The country's major import trade partners were Russia (which supplied 15.7% of Moldova's imports), Romania (11.9%), Ukraine (11.4%), China (8.0%), Germany and Turkey (7.4% each), and Italy (6.3%) (National Bureau of Statistics of the Republic of Moldova, 2013; U.S. Central Intelligence Agency, 2013).

Production

Data on mineral production are in table 1.

Structure of the Mineral Industry

Table 2 lists major mineral industry facilities.

Commodity Review

Metals

Iron and Steel.—In 2012, OAO Moldovan metallurgical plant (MMZ), which was located in the Transnistria region, continued to struggle in the aftermath of the global economic crisis. The production of crude steel slightly decreased by 1.2% compared with that of 2011 to 316,682 metric tons (t) and remained very much below the plant's capacity level of about 1.1 million metric tons (Mt). The production of rolled steel increased to 356,754 t, or by 16.4% from its level in 2011. The plant was built in 1985 and specialized in the production of rolled steel from metal scrap. As of 2012, the plant was owned by EIM Energy and Investment Management Corp. (45.6%), Rumney Trust Reg. (45%),

---

[1]Where necessary, values have been converted from Moldovan lei (MDL) to U.S. dollars (US$) at an annual average exchange rate of MDL12.11=US$1.00 and from euro area euros (€) to U.S. dollars at an annual average exchange rate of €0.809=US$1.00 for 2012.

Decagon Avionics Ltd. (8.23%), and plant employees (1.17%) (OJSC Moldova Steel Works, 2013; RPInform.com, 2013).

In 2012, the plant had several periods of down time because of the reduced demand within the EU and because of problems with obtaining a sufficient amount of metal scrap. To work at full capacity, MMZ needed about 1.3 Mt of scrap metal; only about 250,000 t was obtained in Moldova, however, and the other scrap metal was imported. Although scrap metal from Ukraine had the best combination of price and transportation cost, Ukraine's export quota on scrap metal, which had been in effect since 2006, prevented MMZ from importing its raw material from Ukraine (Muntyan, 2012).

In July, Transnistria's authorities and MMZ signed a memorandum outlining the Transnistria region authorities' support of the metallurgical industry. According to the memorandum, tax rates for MMZ were reduced to 0.75% from 3.25% through the end of 2012. A long-term problem that MMZ had encountered since the start of the global economic crisis was that the Transnistrian authorities imposed business taxes per unit of product sales rather than per unit of business profits. As a result, even the enterprises that work at a loss have to pay business taxes. The authorities provided a temporary tax break to MMZ to preserve jobs in the region and allow the plant to recover fully from the economic crisis. In return, MMZ agreed to continue production without interruption through the end of 2012 (MetalDaily.ru, 2012; Nikitina, 2012; Regnum.ru, 2012c).

Industrial Minerals

Cement.—Despite the decrease in the GDP, cement production in Moldova in 2012 was estimated to have increased by about 7% to 1.5 Mt. The two leading cement producers in Moldova were Lafarge Ciment Moldova SA (a part of Lafarge S.A. of France), which was located in the northern part of Moldova, and the ZAO Rybnitsa cement complex (RCK), which was located in the Transnistria region (tables 1 and 2).

The RCK produced cement and lime for use in construction. The plant had the capacity to produce 1 million metric tons per year (Mt/yr) of cement but had not been able to produce at capacity since 2008. As of February 2012, RCK was planning to produce 424,000 t of cement during the course of the year. In the first 3 months of the year, RCK upgraded its lime production line and in April it opened a new rotating furnace for the production of lime. As a result of these changes, the energy efficiency of the company's lime production was increased, and during the month of May alone, the company saved 34,000 cubic meters of natural gas. As of 2012, the major shareholder of the plant was the Russian holding company Metalloinvest (Point.md, 2012; R-novosti, 2012; Zvdinvest.ru, 2012).

Outlook

In the next few years, Moldova's economy will likely return to prerecession production levels. The country is likely to remain a minor producer of mineral products, however, and to continue to focus on production of industrial minerals, especially construction materials. In the next decade, development of reliable energy sources and diversification of the energy supply are likely to be the most important and most challenging goals for the country.

References Cited

InoSMI.ru, 2012, Kak raspilit' 20 millionov evro na stroitel'stve bespoleznogo gazoprovoda? [How to split €20 million from construction of a useless pipeline?]: InoSMI, August 8. (Accessed June 14, 2013, at http://www.inosmi.ru/sngbaltia/20120808/196258446.html.)

MetalDaily.ru, 2012, Moldavskomu MZ nalogi na dokhody snizheny do 0,75% [Moldavian MZ has its income taxes reduced to 0.75%]: MetalDaily.ru, July 23. (Accessed June 14, 2012, at http://www.metaldaily.ru/news/news66164.html.)

Muntyan, Pavel, 2012, Moldavskiy metallurgicheskiy zavod ne smog zapustit' proizvodstvo [Moldovan metallurgical plant failed to restart production]: KP.md, February 20. (Accessed June 14, 2013, at http://www.kp.md/online/news/1087189/.)

National Bureau of Statistics of the Republic of Moldova, 2012, Statistical yearbook of the Republic of Moldova: National Bureau of Statistics of the Republic of Moldova, 560 p. (Accessed June 24, 2012, at http://www.statistica.md/public/files/publicatii_electronice/Anuar_Statistic/2012/anuar_2012_rus.pdf.)

National Bureau of Statistics of the Republic of Moldova, 2013, Moldova v Tsifrakh [Moldova in figures]: National Bureau of Statistics of the Republic of Moldova, 90 p. (Accessed June 14, 2013, at http://www.statistica.md/public/files/publicatii_electronice/Moldova_in_cifre/2013/Moldova_in_cifre_2013_rom_rus.pdf.)

Nikitina, Varvara, 2012, V Pridnestrov'ye metallurgicheskaya otrasl' poluchit podderzhku gosudarstva [In Transnistria, metallurgical sector will receive support from the government]: NewRegion2.ru, July 14. (Accessed June 14, 2013, at http://www.nr2.ru/pmr/395106.html/print/.)

NOI.md, 2013, Promyshlennoe proizvodstvo v Moldove v 2012 godu sokratilos' na 3,1% [In 2012, industrial production in Moldova decreased by 3.1%]: NOI.md, February 19. (Accessed June 14, 2013, at http://www.noi.md/ru/news_id/18992.)

OJSC Moldova Steel Works, 2013, About us: OJSC Moldova Steel Works. (Accessed June 14, 2013, at http://www.aommz.com/pls/web/web.main.show.)

Point.md, 2012, Rybnitskiy tsementnyi zavod uvelichivaet proizvodstvo [Rybnitsa cement plant is increasing production]: Point.md, January 31. (Accessed June 14, 2013, at http://point.md/ru/novosti/obschestvo/ribnickij-cementnij-zavod-uvelichivaet-proizvodstvo.)

RBC.ru, 2012a, Gazprom dogovorilsya s Moldaviey o postavkah i tranzite gaza [Gazprom and Moldova reached an agreement on shipments and transit of gas]: RBC.ru, November 19. (Accessed June 14, 2012, at http://top.rbc.ru/economics/19/11/2012/825634.shtml.)

RBC.ru, 2012b, Moldaviya planiruet prodlit' s 'Gazpromom' deistvuyushie gazovye kontrakty [Moldova plans to renew existing gas contracts with Gazprom]: RBC.ru, August 2012. (Accessed June 14, 2012, at http://www.rbc.ru/rbcfreenews/20120308115559.shtml.)

Regnum.ru, 2012a, Moldaviya prosit u RF 30% skidku na gaz, a Rumyniyu – uskorit' ob'edineniye gazovykh system dvuh stran [Moldova asks Russia to give it a 30% gas discount and asks Romania to speed up gas system unification of the two countries]: Regnum.ru, September 13. (Accessed June 14, 2013, at http://www.regnum.ru/news/1571124.html.)

Regnum.ru, 2012b, SShA prizvali Moldaviyu k srochnoy diversificatsii energoresursov [USA called on Moldova to quickly diversify its energy resources]: Regnum.ru, July 31. (Accessed June 14, 2013, at http://www.regnum.ru/news/1556964.html.)

Regnum.ru, 2012c, Vlasti Pridnestrov'ya okazhut podderzhku metallurgicheskoy otrasli vo izbezhaniye ee kraha [Transnistria's government will provide support to the metallurgical sector to avoid its collapse]: Regnum.ru, July 20. (Accessed June 14, 2013, at http://www.regnum.ru/news/1553769.html.)

R-novosti, 2012, Proizvodstvo izvesti [Lime production]: R-novosti, June 4. (Accessed June 14, 2013, at http://r-novosti.idknet.com/2012/06/04/proizvodstvo-izvesti/.)

RPInform.com, 2013, V 2012 g. "Moldavskiy Metallurgicheskiy Zavod" vyplavil men'she stali, chem v 2011-m [In 2012, Moldovan metallurgical plant produced less steel than in 2011]: RPinform.com, January 22. (Accessed June 14, 2013, at http://rpinform.com/ru/news/20130122/13348.html.)

U.S. Central Intelligence Agency, 2013, Moldova, in The world factbook: U.S. Central Intelligence Agency, June 20. (Accessed June 14, 2012, at https://www.cia.gov/library/publications/the-world-factbook/geos/md.html.)

Vedomosti.md, 2012, V Moldove sozdayut solomoprom [Moldova is creating its "industry based on hay"]: Vedomosti.md, January 17. (Accessed June 14, 2013, at http://www.vedomosti.md/news/V_Moldove_Sozdayut_Solomoprom.)

Zvdinvest.ru, 2012, Pridnestrov'ye. Rybnitskiy tsementnyi zavod uvelichivaet proizvodstvo. [Transnistria—Rybnitsa cement plant is increasing production]: Zvdinvest.ru, February 1. (Accessed June 14, 2013, at http://www.zvdinvest.ru/news/365.)

TABLE 1
MOLDOVA: PRODUCTION OF MINERAL COMMODITIES[1]

(Metric tons)

Commodity[2]	2008	2009	2010	2011	2012
METALS					
Steel:					
Crude	885,000	425,900	241,500	320,600	316,682
Rolled	816,000	440,900	231,400	306,500	356,754
INDUSTRIAL MINERALS					
Cement[e]	1,800,000 [r]	930,000 [r]	1,100,000 [r]	1,400,000 [r]	1,500,000
Clays, unspecified[e]	165,000	150,000	160,000	140,000	150,000
Gypsum	333,300	94,400	99,800	100,540	115,100
Lime[e]	12,000 [r]	4,000 [r]	5,000 [r]	5,500 [r]	12,000
Limestone	300,000 [r]	226,900 [r]	196,900 [r]	295,500 [r]	264,500
Sand and gravel	2,707,000 [r]	1,830,000 [r]	2,146,000 [r]	2,547,000 [r]	2,966,000

[e]Estimated; estimated data are rounded to no more than three significant digits. [r]Revised.

[1]Table includes data available through June 14, 2013.

[2]In addition to commodities listed, Moldova is thought to produce granite, natural gas, peat, and petroleum, but available information is not adequate to estimate production.

TABLE 2
MOLDOVA: STRUCTURE OF THE MINERAL INDUSTRY IN 2012

(Metric tons unless otherwise specified)

Commodity		Major operating companies and major equity owners	Location of main facilities	Annual capacity
Cement		Lafarge Ciment Moldova SA	Rezina	1,400,000
Do.		Rybnitsa Cement Complex (Metalloinvest Holding)	Rybnitsa, Transnistria region	1,100,000
Granite	thousand cubic meters	NA	Kosoutskoye deposit	150
Gypsum		CMC-Knauf joint venture	Kirovskoye deposit	850,000
Oil and natural gas:				
Oil		Redeco Moldova Oil and Gas Co.	Valeni oilfield	100,000
Natural gas	thousand cubic meters	do.	Victorovca gasfield	5,000
Sand and gravel		NA	71 mined deposits	NA
Steel, crude		OAO Moldovan metallurgical plant [EIM Energy and Investment Management Corp. (45.6%), Rumney Trust Reg. (45%), Decagon Avionics Ltd. (8.23%), and plant employees (1.17%)]	Rybnitsa, Transnistria region	1,100,000

Do., do. Ditto. NA Not available.

THE MINERAL INDUSTRY OF MONTENEGRO

By Harold R. Newman

Montenegro's mineral industry included the mining and processing of industrial minerals and coal. Metal production included primary aluminum smelting and crude steel production. Industrial mineral production included lime, marble blocks, and salt. In the first 8 months of 2012, the mining and quarrying sector recorded an increase in industrial output of 3.4% and a decrease in the value of the minerals produced of 2% (Central Bank of Montenegro, 2012, p. 11).

In 2012, the United States exported goods valued at $12.4 billion to Montenegro, including excavating machinery valued at $304 million; finished metal shapes valued at $124 million; and iron and steel products valued at $4 million. The United States imported $3.7 million worth of goods from Montenegro, including nonmonetary gold valued at $550 million; finished metal shapes and advanced manufactures (except steel) valued at $250 million; iron and steel manufactures (advanced) valued at $44 million; and cement, lime, sand, and stone (combined) valued at $38 million (U.S. Census Bureau, 2012a, b).

The Geological Survey of Montenegro (GSM) is a Government organization that was founded in 1945. GSM is composed of the following four departments: the Department of Regional Geology, Mineral Resources, and Mineral Resources Concessions; the Department of Hydrology, Engineering Geology, and Water Concessions; the Department of Mining Works, Research and Drilling; and the Department of Legal and Financial Affairs and Human Resources. The GSM's main duties are to conduct mineral resource investigations and geologic mapping. Two main laws deal separately with geology and mining—the Law of Geological Research and the Law of Mining, respectively. According to these laws, the mineral resources represent the country's natural wealth and are owned by the Government (Geological Survey of Montenegro, 2011).

Production

In 2012, production of aluminum was estimated to have decreased whereas production of crude steel was estimated to have increased. In 2012, alumina was not produced and production of bauxite ceased. Production of stone, including broken, crushed, and ornamental stone, was estimated to have increased (table 1).

Structure of the Mineral Industry

Table 2 is a list of major mineral industry facilities.

Commodity Review

Metals

Aluminum and Alumina and Bauxite.—The Parliament requested that the Government consider repossessing Kombinat Aluminijuma Podgorica AD (KAP) from the En+ Group Ltd. to preempt a default on state-backed loans that totaled €22 million ($30 million[1]). KAP operated alumina and primary aluminum production plants, a cast house, and a secondary aluminum plant. The company cast molten aluminum into standard ingots and T-ingots and was the leading industrial plant in Montenegro. The Government was seeking ways to maintain the KAP operation, which produced 90,000 metric tons (t) of primary aluminum in 2012 (Savic, 2012a).

The Government suspended bauxite mining operations at Bauxite Mines Niksic owing to Bauxite Mines' inability to pay utility costs and employee salaries. The company's financial loss from January to September 2011 was €8.6 million ($11.8 million). At yearend, negotiations between the Government and Neksan Ltd. of Russia were underway for Neksan to acquire Bauxite Mines (Montenegro365.com, 2012).

Copper, Lead, and Zinc.—Balamara Resources Ltd. of Australia announced that it had received a 25-year exploration and mining lease from the Government for its 100%-owned Monty copper, lead, and zinc project located a few kilometers (km) from Mojkovac and about 100 km from Podgorica. Balamara had developed a Joint Ore Reserves Committee (JORC) inferred ore reserves estimate that included 9.2 million metric tons (Mt) grading 3.7% zinc, 1.2% lead, and 0.36% copper. The Monty project included three deposits—Brskovo, Visnjica, and Zuta Pria—that had already been defined. Balamara was planning to bring the Monty project into production in 2014 (Balamara Resources Ltd., 2012).

Iron and Steel.—Toscelik Profil Ve Sac Endustriai AS bought the former Zeljezara AD Niksic (ZEHK) steel plant at auction for €15.1 million ($20.7 million). and changed the name to Toscelik Niksic A.D. Toscelik Profill. Toscelik (a subsidiary of Tosyali Group of Turkey) bought the plant after the plant failed to sell in four previous auctions. The court-appointed receivership administrator set the price at €15 million, which was one-half the price asked at the previous auction in January 2012. The steelworks went into bankruptcy in 2011 when workers demanded overdue wages (Savic, 2012b).

Mineral Fuels

Coal.—The Government was expected to award a concession for the exploitation of coal from the Maoce Basin and construction of a coal-fired powerplant. The Maoce basin is located in the northeastern area of Montenegro about 15 km from Pljevlje. Coal reserves were estimated to be 110 Mt of exploitable coal. The powerplant would have an estimated capacity of 500 megawatts (U.S. Department of State, 2012).

In 2012, the Ministry of Economy announced the first public call for awards for a concession contract for the exploration and production of hydrocarbons offshore Montenegro. The aim

---

[1]Where necessary, values have been converted from euro area euros (€) to U.S. dollars (US$) at a rate of €0.74=US$1.00.

of awarding the contracts was to increase knowledge of the Montenegro offshore area, which had not been well-explored. The search for natural gas and petroleum would be conducted in several phases, including exploration, appraisal, development, and production. The decision on awarding the exploration contract would be made by the Government (U.S. Department of State, 2012).

Outlook

Montenegro's mineral production is expected to remain modest, and the structure of mineral industry is not expected to change significantly in the near future. The Government is expected to continue to encourage companies to investigate the area of the southern Adriatic where several offshore prospects may contain commercial deposits of natural gas and petroleum.

References Cited

Balamara Resources Ltd., 2012, Monty project: Balamara Resources Ltd. (Accessed December 2, 2013, at http://www.balamara.com.au/monty-montenegro/.)

Central Bank of Montenegro, 2012, Microeconomic environment: Central Bank of Montenegro, September, p. 11. (Accessed November 20, 2012, at http://www.cb-cg.org/slike_i_fajovi/fajovi/fajovi_publikacije/bitencbcg/2012/bilten_cbcg_0912.pdf.)

Geological Survey of Montenegro, 2011, Our organisation: Geological Survey of Montenegro presentation, October, 21 p. (Accessed November 15, 2013, at http://www.balkangeo.net/docs/radusinovic.pdf.)

Montenegro365.com, 2012, Montenegro Government has suspended the production at Niksic-based bauxite mines: Montenegro365.com. (Accessed December 2, 2013, at http://www.montenegro365.com/business-and-company-news/1391-montenegro-government-has-suspended-the-production-at-niksic-based-bauxite-mines.)

Savic, Misha, 2012a, Montenegro set to take over KAP, Deripaska's En + fights back: Bloomberg L.P., March 1. (Accessed December 4, 2013, at http://www.bloomberg.com/news/2012-03-01/montenegrin-lawmakers-back-plan-to-take-kap-from-deripaska-s-en-.html.)

Savic, Misha, 2012b, Turkey's Tosyali Holding buys Montenegrin steel mill in Niksic: Bloomberg L.P., April 30. (Accessed December 7, 2013, at http://www.bloomberg.com/news/2012-04-30/turkey-s-tosyali-holding-buys-montenegrin-steel-mill-in-niksic.html.)

U.S. Census Bureau, 2012a, U.S. exports to Montenegro by 5-digit end use code: U.S. Census Bureau. (Accessed December 7, 2013, at http://www.census.gov/foreign-trade/statistics/product/enduse/exports/c4804.html.)

U.S. Census Bureau, 2012b, U.S. imports from Montenegro by 5-digit end use code: U.S. Census Bureau. (Accessed December 7, 2013, at http://www.census.gov/foreign-trade/statistics/product/enduse/imports/c4804.html.)

U.S. Department of State, 2012, 2012 investment climate statement—Montenegro: U.S. Department of State. (Accessed December 12, 2013, at http://www.state.gov/e/eb/rls/othr/ics/2012/191202.htm.)

TABLE 1
MONTENEGRO: PRODUCTION OF MINERAL COMMODITIES[1]

(Metric tons unless otherwise specified)

Commodity[2]		2008	2009	2010	2011	2012[e]
METALS						
Alumina		220,426	58,528	--	--	--
Aluminum, metal, ingot, primary		107,457	63,960	82,043	92,838	90,000
Bauxite		671,811	45,779	61,205	158,614	--
Iron and steel, crude steel		201,623	90,404	95,000 [e]	42,271	45,000
INDUSTRIAL MINERALS						
Gravel	cubic meters	146,381	74,368	49,517	50,000 [e]	50,000
Lime		9,839	4,497	839	3,448	3,000
Salt (sea water evaporate)		25,200	17,000	11,200	10,000 [e]	12,000
Stone, excluding quartz, quartzite, and dimension stone:						
Ornamental (marble blocks)	cubic meters	50,084	40,780	41,000 [e]	32,804	34,000
Crushed and broken	do.	179,521	65,015	65,000 [e]	60,000 [e]	65,000
Other, stone products	do.	109,436	51,373	39,921	40,000 [e]	42,000
MINERAL FUELS AND RELATED MATERIALS						
Coal, lignite		1,740,076	957,164	1,937,847	1,972,671	2,000,000

[e]Estimated; estimated data are rounded to no more than three significant digits. do. Ditto. -- Zero.

[1]Table includes data available through October 31, 2013.

[2]In addition to the commodities listed, additional industrial minerals were also likely produced, but available information is inadequate to make reliable estimates of output.

TABLE 2
MONTENEGRO: STRUCTURE OF THE MINERAL INDUSTRY IN 2012

(Thousand metric tons)

Commodity	Major operating companies and major equity owners	Location of main facilities	Annual capacity
Alumina	Kombinat Aluminijuma Podgorica (KAP) (En+Group Ltd., 29.365%, and Government, 29.365%)	Podgorica	280
Aluminum, primary	do.	do.	120
Bauxite	Bauxite Mines Niksic (Central European Aluminum Co., 31.82%, and Government of Montenegro, 31.82%)	Kutsko Brdo (suspended operations in 2012)	700 [e]
Coal	Rudnik Uglja A.D. Pljevlja	Pljevlja	2,000 [e]
Steel, crude	Toscelik Niksic A.D. (Tosyali Group, 100%)	Niksic	NA

[e]Estimated. do. Ditto. NA Not available.

The Mineral Industry of the Netherlands

By Alberto Alexander Perez

In 2012, the Netherlands' gross domestic product (GDP) was $773.1 billion, which was a decrease of 0.6% compared with that of the previous year. The Netherlands was a significant producer of nitrogen and salt, accounting for about 1.3% and 1.7% of world production, respectively. The main emphasis of the Dutch mineral industry was on trade and processing. In 2012, the Netherlands was a significant regional producer of natural gas and petroleum for the European market and a major transshipment center for mineral products that entered and left continental Europe. In 2012, the Port of Rotterdam was the busiest port in Europe in terms of the value and the volume of the cargo handled at the port (Apodaca, 2013; Kostick, 2013; Port of Rotterdam Authority, 2013, p. 6, 14; U.S. Central Intelligence Agency, 2013).

Minerals in the National Economy

The Staatstoezicht op de Mijnen [State Supervision of Mines] (SodM) is the agency within the Ministerie van Economische Zaken [Ministry of Economic Affairs] that oversees the production of minerals in the Netherlands and the Netherlands Continental Shelf. The agency is responsible for drafting and enforcing mining laws, mine safety, and mineral production regulations.

The mineral sector was dominated by natural gas and petroleum production, of which about 40% was from offshore fields. Mining was limited to the extraction of limestone, peat, and sand and gravel by quarrying and solution mining of salt in the eastern and northern areas of the country. In the nonfuel mineral sector, the Netherlands was engaged principally in downstream activities, including the chemical and metallurgical industries, which used mainly imported ores and industrial minerals (table 1; Staatstoezicht op de Mijnen, 2009).

Production

In 2012, crude steel production decreased by about 1% and pig iron production decreased slightly compared with production in 2011. The principal mineral commodities that were produced in the Netherlands were primary aluminum, cadmium, cement, crude petroleum, iron and steel, natural gas, nitrogen, salt, and refined zinc. Rotterdam remained important as a shipping and storage center. In 2012, the throughput (imports and exports) of the following mineral commodities were the most important in terms of total volume: crude petroleum, 98.3 million metric tons (Mt); mineral oil products, 81.8 Mt; iron ore and scrap, 32.7 Mt; and coal, 25.3 Mt (table 1; Port of Rotterdam Authority, 2013, p. 3).

Structure of the Mineral Industry

Mineral industry facilities in the Netherlands were mostly privately owned, although the Government continued to be involved in the energy sector through the regulation and oversight of petroleum and natural gas operations. Table 2 is a list of the major mineral industry facilities.

Commodity Review

Metals

Aluminum.— Zeeland Aluminium Co. BV (ZALCO), which was located in Vlissingen, filed for bankruptcy in December 2011. The company sold its anode facility to Century Aluminum Co. of the United States, which created Century Aluminium Vlissingen BV with these assets. Century expected to have the anode plant producing in 2013 with a projected production of 75,000 metric tons per year (t/yr); in the future, the production capacity of the plant was expected to be expanded to 145,000 t/yr. The rest of ZALCO was purchased by UTB Holding BV which restarted production of billets and rolling slabs of aluminum alloy at its foundry; the foundry was expected to be fully operational by mid-2013. ZALCO reported that its smelter was decommissioned as it was no longer viable for production (Century Aluminum Co., 2013a, b; Zeeland Aluminium Co. B.V., 2013).

Iron and Steel.—Tata Steel Group (Tata), which was the owner of Tata Steel Europe Ltd., announced in its annual report for 2011 that the Ijmuiden steel plant had begun a 5-year improvement program that was focused on enhancing production capacity, improving reliability, and reducing cost. Tata completed a second trial of the ULCOS Hisarna pig iron project at the plant. ULCOS (which stands for ultra-low carbon dioxide steelmaking) was a consortium of 48 European companies and organizations that had developed a process to produce iron that reduces carbon dioxide emissions by eliminating the need to pelletize iron ore and to produce coke from coal. Ijmuiden Tata Steel Europe Ltd. was previously known as Corus Group, but the name was changed officially in September 2010 (Noordhollandsdagblad, 2012; Tata Steel Group, 2012, p. 25).

Mineral Fuels

Petroleum.—Zeeland Refinery NV planned to upgrade the distillate hydrocracker at its refinery in Vlissingen, and contracted with a unit of Foster Wheeler AG's global engineering and construction group to do the upgrade. When completed, the project will maximize the throughput of the distillate hydrocracker by debottlenecking its reaction and fractionation sections. The work was slated to be completed by 2014 (Oil and Gas Journal, 2012).

Outlook

Public and private investment in the development of natural gas fields and the distribution of natural gas are likely to increase. The Netherlands is a leading natural gas distribution center in Europe, and this will likely continue as the Netherlands is expected to continue to be an exporter of natural gas in the region.

The Port of Rotterdam is expected to continue to be a leading European port, particularly in terms of container traffic, and to play a significant role in European trade.

References Cited

Apodaca, L.E., 2013, Nitrogen (fixed)—Ammonia: U.S. Geological Survey Mineral Commodity Summaries 2013, p. 112–113.

Century Aluminum Co., 2013a, Form 10–K—2012: Securities and Exchange Commission, 14 p.

Century Aluminum Co., 2013b, Our carbon anode facilities; Century Aluminum Co. (Accessed July 31, 2013, at http://www.centuryaluminum.com/carbon.php.)

Kostick, D.S., 2013, Salt: U.S. Geological Survey Mineral Commodity Summaries 2013, p. 134–135.

Noordhollandsdagblad, 2012, Succesvolle proeven bij Hisarna project Tata Steel Ijmuiden: Het Noordhollands Dagblad, HDC Media B.V. (Accessed July 31, 2013, at http://www.noordhollandsdagblad.nl/stadstreek/kennemerland/article19703093.ece.)

Oil and Gas Journal, 2012, Dutch refinery lets contract for Zeeland refinery: Oil and Gas Journal, September 11. (Accessed July 31, 2013, at http://www.ogj.com/ARTICLES/2012/09/DUTCH-REFINER-LETS-CONTRACT-FOR-ZEELAND-REFINERY.html.)

Port of Rotterdam Authority, 2013, Port statistics 2010–2011–2012: Port of Rotterdam Authority, 19 p. (Accessed July 31, 2013, at http://www.portofrotterdam.com/en/Port/port-statistics/Documents/Port-statistics-2012.pdf.)

Staatstoezicht op de Mijnen, 2009, Mission, vision and strategy: The Hague, Netherlands, Staatstoezicht op de Mijnen. (Accessed December 21, 2009, at http://www.sodm.nl/English/Organisation/Mission_vision_and_strategy.)

Tata Steel Group, 2012, 105th annual report 2011–2012: Tata Steel Group. (Accessed December 3, 2012, at http://www.tatasteeleurope.com/file_source/Functions/Finance/Documents/annual-report-2011-12.pdf.)

U.S. Central Intelligence Agency, 2013, Netherlands, in The world factbook: U.S. Central Intelligence Agency. (Accessed July 28, 2013, at https://www.cia.gov/library/publications/the-world-factbook/geos/nl.html.)

Zeeland Aluminium Co. B.V., 2013, Gieterij Zalco weer in productie: Zeeland Aluminium Co. B.V. (Accessed July 31, 2013 at http://www.zalco.nl/nieuws/gieterij-zalco-weer-in-productie.)

TABLE 1
NETHERLANDS: PRODUCTION OF MINERAL COMMODITIES[1]

(Metric tons unless otherwise specified)

Commodity[2]		2008	2009	2010	2011	2012
METALS						
Aluminum, metal, primary		317,000	300,000	300,000	300,000	110,000
Cadmium, metal, primary		527	490	560	570	560
Iron and steel:						
Pig iron, including blast-furnace ferroalloys (if any)		5,998,000	4,601,000	5,799,000	5,943,000	5,909,000
Steel:						
Crude		6,880,000	5,194,000	6,651,000	6,937,000	6,867,000
Semimanufactures		6,800,000[e]	5,100,000	6,523,000	6,765,000	6,700,000[e]
Lead, metal, refined, secondary[e]		16,000	16,000	17,000	17,000	17,000
Zinc, metal, primary		239,500	224,000	254,000	261,000	257,000
INDUSTRIAL MINERALS						
Cement, hydraulic[e]	thousand metric tons	2,700	2,700	2,700	2,700	2,700
MINERAL FUELS AND RELATED MATERIALS						
Gas, dry natural:						
Gross	million cubic meters	83,846	78,919	88,668	80,731	80,787
Marketed	do.	83,733	78,891	88,660	80,731	80,787
Petroleum:						
Crude	thousand 42-gallon barrels	12,230	9,302	7,300	8,121	8,212
Refinery products:						
Liquefied petroleum gas	do.	15,189	16,018	16,534	16,500	16,500[e]
Gasoline, motor	do.	59,442	60,037	63,145	63,000	63,000[e]
Naphtha and white spirit[e]	do.	90,000	90,000	90,000	NA	NA
Kerosene and jet fuel	do.	50,868	46,484	51,794	51,700	51,700[e]
Refinery fuel and loss[e]	do.	30,000	30,000	30,000	NA	NA
Diesel oil	do.	153,492	153,556	159,031	159,000	159,000[e]
Residual fuel oil	do.	55,626	54,800	63,218	63,000	63,000[e]
Unspecified	do.	116,999	121,416	101,653	100,000	100,000[e]
Total	do.	571,616	572,311	575,375[r]	453,000	453,000[e]

[e]Estimated; data are rounded to no more than three significant digits; may not add to totals shown. [r]Revised. NA Not available. do. Ditto.

[1]Table includes data available through July 31, 2013.

[2]In addition to the commodities listed, the Netherlands produced magnesium compounds, nitrogen, salt, sodium compounds, and construction materials, such as limestone, peat, and sand and gravel, as well as sulfur, elemental byproduct of metallurgy and of petroleum and natural gas, but output was not reported, and information was not available to make reliable estimates of output.

TABLE 2
NETHERLANDS: STRUCTURE OF THE MINERAL INDUSTRY IN 2012

(Thousand metric tons unless otherwise specified)

Commodity		Major operating companies and major equity owners	Location of main facilities	Annual capacity
Aluminum:				
Primary		Aluminum Delfzijl BV (Basemet B.V., a division of Klesch and Co. Ltd., 100%)	Smelter at Delfzijl	165
Secondary		Alumax Recycling BV	Smelter at Kerkade	50
Do.		Zeeland Aluminium Co. BV (ZALCO) (UTB Holding B.V. 100%)	Plant at Flushing (Vlissingen)	230
Cadmium	metric tons	Nyrstar NV (Zinifex Ltd. and Umicore NV)[1]	Plant at Budel	650
Calcium carbonate, ground		Omya Netherlands BV	Plant at Moerdijk	500
Cement		Eerste Nederlandse Cement Industrie NV (HeidelbergCement Group, 100%)	Plants at IJmuiden, Maastricht, and Rotterdam	3,700
Do.		Cementfabriek IJmuiden BV	Three plants at IJmuiden	1,600
Do.		Cementfabriek Rozenburg BV	Two plants at Rozenburg	920
Limestone		Ankerpoort NV (Lhoist SA, 100%)	Mines at Maastricht and Winterswijk	600
Magnesia		Nedmag Industries Mining & Manufacturing BV	Plant at Veendam	130
Do.		MAF Magnesite BV	Plant at Schiedam	40
Natural gas	million cubic meters	Nederlandse Aardolie Maatschappij BV (NAM) (Exxon Mobil Corp., 50%, and Royal Dutch Shell plc., 50%)	Groningen, Leeuwarden, Assen, and other onshore gasfields and several offshore wells in the North Sea	225
Petroleum:				
Crude	42-gallon barrels per day	BP p.l.c., ConocoPhillips Co., and Chevron Corp.	766 wells (204 producing), including the following North Sea fields: Haven, Helder, Helm, Hoorn, Kotter, Logger, and Rijn	83,500
Do.	do.	Nederlandse Aardolie Maatschappij BV (NAM) (Exxon Mobil Corp., 50%, and Royal Dutch Shell plc, 50%)	Onshore fields: Berkel, DeLier, Ijselmonde, Meerkapelle, Pernis, Pinacke, Rotterdam, Schoonebeck, West, Werkendam, and Zoetemeer	20,500
Do.	do.	Veba Oil and Gas Netherlands BV	Hanze field, North Sea	31,500
Refinery		Several companies, of which the four major ones are:	Refineries, including:	1,230,500
Do.		Netherlands Refining Co. (BP p.l.c., 69%, and Chevron Corp., 31%)	Rotterdam	(446,000)
Do.		Shell Nederland Raffinaderij BV	Pernis	(374,000)
Do.		Esso Nederland BV	Rotterdam	(175,000)
Do.		Total Raffinaderij Nederland NV	Vlissingen	(150,000)
Salt		Akzo Nobel Salt BV (Akzo Nobel NV, 100%)	Mines, of which:	4,100
Do.		do.	Hengelo	(2,100)
Do.		do.	Delfzijl	(2,000)
Sand, silica		Sigrano Nederland NV (Sibelco Group)	Mines and plants at Heerlin and Maastricht	500
Do.		Lieben Minërals BV	Mines at South Limburg	150
Sodium:				
Carbonate, synthetic		Brunner Mond Group BV	Plant at Delfzijl	380
Sulfate, synthetic		do.	do.	600
Steel		Tata Steel Europe Ltd. (Tata Steel Group)	Plant at IJmuiden	7,000
Zinc		Nyrstar NV (Zinifex Ltd. and Umicore NV)[1]	Plant at Budel	260

Do., do. Ditto.

[1] Nyrstar NV is an independent publicly traded company formed from the combined zinc and lead smelting and alloying business of Zinifex Ltd. and the zinc smelting and alloying business of Umicore NV.

The Mineral Industry of Norway

By Harold R. Newman

Norway's diverse geologic terrain contains a broad spectrum of mineral resources for possible exploration and development, including metals, industrial minerals, and mineral fuels. Norway's mineral resources included coal, iron ore, natural gas, nickel, petroleum, sand and gravel, stone, and titanium. The mines and quarries were mostly of regional significance and were located mainly along the coast. The natural gas and petroleum fields were located mainly offshore in the Norwegian area of the North Sea (U.S. Central Intelligence Agency, 2012).

Minerals in the National Economy

The country's natural gas and petroleum industries have continued to contribute significantly to Norway's national economy. In 2012, the petroleum sector accounted for the largest portion of the country's exports and about 26% of Government revenue. In anticipation of the eventual decrease in natural gas and petroleum production, the Government was saving a significant amount of revenue from petroleum exports in a sovereign wealth fund (SWF) valued at more than $700 billion. Norway's SWF was the second largest of all countries' SWFs after that of Luxembourg (U.S. Department of State, 2013).

Mineral Trade

Even though Norway was not a member of the European Union (EU), it participated in the EU's internal economic market, the European Economic Area. Mineral trade was important to the economy and, in terms of export value, petroleum was Norway's most significant mineral commodity in 2012. Norway was the world's seventh-ranked petroleum exporter and the leading petroleum exporter in Western Europe. The country was the world's fourth-ranked natural gas producer and the world's second-ranked exporter of natural gas after Russia (U.S. Energy Information Administration, 2012).

The U.S. trade in goods with Norway in 2012 totaled $3,501 million in exports and $6,566 million in imports for a negative trade balance of $3,065 million. U.S. exports to Norway included petroleum products valued at $199 million; drilling and oilfield equipment, $172 million; coal and other fuels, 93 million; fuel oil, $43 million; and finished metal shapes, $36 million. Norway's total exports to the United States included petroleum products valued at $1.8 billion; crude petroleum, $806 million; fuel oil, $550 million; liquefied petroleum gases, $346 million; and nickel, $201 million (U.S. Census Bureau, 2012a, b).

Production

Norway produced various mineral commodities, including aluminum, cadmium, cobalt, copper, ferroalloys, nickel, steel, and zinc metals; it was a global supplier of aluminum, ferroalloys, and petroleum. Production of natural gas increased in 2012, whereas production of primary aluminum, cobalt, and petroleum production decreased (table 1). Aggregates, limestone, nepheline syenite, and sand and gravel were some of Norway's more economically important industrial mineral raw materials. The country's production of ilmenite accounted for about 6% of world production (Bedinger, 2013).

Structure of the Mineral Industry

The Norwegian mineral industry was composed of a mixture of Government and privately owned operations. Table 2 lists the major mineral companies that were operating in Norway in 2012 and their respective mine and (or) plant locations and capacities.

Commodity Review

Metals

Cobalt, Copper, Gold, and Platinum-Group Metals.—Nordic Mining ASA's exploration efforts in northern Norway led to the discovery of magmatic mineralization in the Lokkarfjord and the Reinfjord areas on the Øksfjord Peninsula that contained cobalt, copper, gold, and platinum-group element mineralization. The Øksfjord Peninsula is part of the Seiland Igneous Province (SIP). The SIP shares many characteristics with other geologic formations that host deposits of copper, nickel, and platinum-group elements, such as the Bushveld Complex in South Africa, the Stillwater Complex in Montana, and the Fennoscandian Suhanko and Penikat intrusions in Finland. The SIP had not been significantly explored for minerals with commercial value (Nordic Mining ASA, 2012c).

Store Norske Gull AS, which was a subsidiary of Store Norske AS, was established to explore for gold on the Norwegian archipelago of Svalbard in the Arctic region. Store Norske had an exploration license area that consisted of 457 claims in Finnmark County and 52 claims in Troms County on the mainland of Spitsberg. The company's prospecting strategy included exploration for gold, nickel, and platinum-group elements (Store Norske Gull AS, 2012).

Arctic Gold AB of Sweden's main exploration project in northern Norway was the Bidjovagge gold prospect, which contained about nine partially mined ore bodies. Historical activity at Bidjovagge had discovered a mineral resource of about 1.4 million metric tons (Mt) with reported grades of 3.4 grams per metric ton (g/t) gold and 1.1% copper. Arctic Gold was investigating the possibility of expanding the mineral resources and proving an ore reserve that could be extracted using conventional methods (Arctic Gold AB, 2012).

Iron Ore.—Northern Iron Ltd. of Australia acquired the Sydvaranger iron project in 2007 for the production of magnetite ore concentrate to supply the European market. The project consisted of four magnetite iron deposits with Joint Ore Reserves Committee (JORC)-compliant estimated resources at Bjørnevatn, Fisketind Øst, Kjellmann, and Tverrdalen. Northern

Iron had an additional 20 prospects with iron mineralization located across a 12-kilometer (km) strike length. The Bjørnevatn Mine was in operation in 2012 and had a capacity of 2.8 million metric tons per year (Mt/yr) with an estimated mine life of 25 years (Northern Iron Ltd., 2012).

Nickel.—First Point Minerals Corp. of Canada announced that it had acquired two properties—the Fera prospect and the Leka prospect—after its exploration efforts identified anomalous nickel occurrences. The Fera prospect is located about 300 km north of Oslo, covers 152 square kilometers (km^2), and hosts several variable-size ultramafic bodies. The largest ultramafic body measures 3 by 5 km in area and hosts disseminated awaruite, which is a naturally occurring nickel-iron alloy. The second property, the Leka prospect, is located 195 km north of Trondheim and is 39 km^2 in area. The company discontinued exploration at the Leka prospect after a number of awaruite occurrences were determined not to be of economic significance and reduced its holdings at the Fera prospect. As of yearend 2012, the company had retained a 100% interest in seven licenses over a 70 km^2 area. Also, a comprehensive mapping and sampling program that started in mid-year 2012 was continuing at yearend (First Point Minerals Corp., 2012).

Silver.—In 2012, Dalradian Resources Inc. of Canada was continuing with its exploration programs in the northern and the southern parts of Norway where Dalradian held exploration licenses for 1.7 million hectares across three greenstone belts, as well as an area that hosts a historic silver mining district. Dalradian was engaged in data acquisition and analysis for all its concessions, and field programs were underway on its Kautokeino and Kongsberg concession areas (Dalradian Resources Inc., 2012).

Titanium.—In 2012, Nordic Mining was continuing with its proposed rutile mine development project at Engebøfjellet. The 2.5-km-long rutile-bearing eclogite body was reported to contain a mineral resource of 154 Mt of eclogite at an average grade of 3.8% rutile. The rutile is disseminated in the eclogite. The Engebøfjellet eclogite deposit is practically free from the usual occurrence of the radioactive elements thorium and uranium. The mining operation would be developed in two stages; first, as an open pit operation for a period of 10 to 15 years, and next, as an underground operation with a mine life of about 35 years. Nordic Mining reported that it was planning to produce 100,000 t/yr of rutile concentrate and 100,000 t/yr of garnet concentrate following a 2015 startup of the mine. The decision to produce byproduct garnet was based on new Government regulations that allow for increased use of silica minerals in abrasive applications (Nordic Mining ASA, 2012a).

Industrial Minerals

Fluorspar.—Tertiary Minerals plc of the United Kingdom announced that it had updated its JORC-compliant estimated inferred mineral resource at the Lassedalen project to 4 Mt grading 25% fluorspar. Tertiary was proceeding with detailed evaluation and with metallurgical testing to determine if acid-grade fluorspar could be produced at the Lassedalen project. Fluorspar is an essential raw material in the aluminum, chemical, and steel industries. The EU had noted in 2010 that fluorspar was one of the 14 minerals that had a high supply risk (Ollett, 2012).

Silica.—Nordic Mining announced that it was planning to produce high-quality quartz at its Nesodden deposit near Kvinnherad in western Norway. Nordic Mining was continuing with detailed mapping of the deposit, which had estimated mineral resources of 2.7 Mt of crystalline hydrothermal quartz in a 12.6-km-long vein that reaches a depth of 150 meters (m). The quartz vein is situated in Proterozoic age rocks south of the Hardanger Fault Zone (HFZ). The HFZ is a 600-km-long Caledonian ductile shear zone. The quartz vein is about 600 m long and about 15 m wide with a depth of 150 m (Nordic Mining ASA, 2012b).

Mineral Fuels

The energy situation in Norway was characterized by an abundance of two forms of energy: hydroelectric power and mineral fuel resources. Hydroelectric power met most of the domestic demand for energy; mineral fuels were produced mainly for export. Norway had a highly developed natural gas and petroleum sector. Natural gas production had been steadily increasing, and petroleum production was on the decline (table 1). Norway was the leading petroleum producer and exporter in Western Europe. At yearend 2011, Norway had the largest petroleum reserves in Western Europe with estimated proven reserves of 5.3 billion barrels (Gbbl) (U.S. Energy Information Administration, 2012).

The Government awarded 60 new production licenses to 42 companies in the latest Awards in Predefined Areas (APA) licensing round. Companies were awarded 34 licenses in the North Sea, 22 licenses in the Norwegian Sea, and 4 licenses in the Barents Sea. The licenses are situated in mature areas on the Norwegian Continental Shelf, where 27 of the 42 companies were already operating. The overall APA area was expanded to 196,625 km^2 (Hovland, 2012).

Coal.—Store Norske Spitsbergen Grubekompani A/S (SNSG) continued to be Norway's sole coal producer. The company's two mines, the Gruve 7 Mine and the Svea Nord Mine, are located on the Arctic archipelago of Svalbard, which is situated about midway between mainland Norway and the North Pole. Norway continued to be a net exporter of coal, of which more than 50% was exported to Germany (ArcticEcon, 2012).

Natural Gas.—Norway had estimated proven reserves of 2.3 trillion cubic meters of natural gas as of January 2012. Norway's natural gas production had been increasing every year since 1994. The annual increases had been sustained by incorporating new fields in the Barents Sea and the Norwegian Sea. Norway's single largest natural gas field was the Troll-Oseberg field (U.S. Energy Information Administration, 2012).

Eni S.p.A. of Italy started production from the offshore Marulk natural gas field located about 80 km from the coast. The Marulk field was the first Norwegian field that Eni had directly operated. Marulk is a natural gas and condensate field in the Cretaceous Lange and Lysing formations; it had estimated reserves of 1.2 billion cubic meters. Development of the field was through two production wells completed subsea and tied back to the floating production, storage, and offloading vessel on the Nome oilfield, which is located about 25 km northeast of the two wells (Eni S.p.A., 2012a).

Petroleum.—Norway, which has the largest petroleum reserves in Western Europe, was reported to have 5.3 Gbbl of proven reserves as of January 2012. All the reserves were located offshore on the NCS, which is divided into three sections: the Barents Sea, the North Sea, and the Norwegian Sea. The bulk of production had taken place in the North Sea, and smaller amounts had been produced in the Barents Sea and the Norwegian Sea (U.S. Energy Information Administration, 2012).

Faroe Petroleum plc of the United Kingdom announced that it had been awarded new prospective exploration licenses under the 2012 APA license round on the Norwegian Continental Shelf. Faroe was awarded five licenses in the North Sea area where the company already held a number of licenses, including the 2011 Butch discovery well (MBendi Information Services (Pty) Ltd., 2013a).

Lundin Petroleum AB of Sweden announced that its wholly owned subsidiary Lundin Norway AS was awarded seven exploration licenses in the 2012 APA licensing round. The awarded licenses included four licenses in the North Sea, two licenses in the Norwegian Sea, and one license in the Barents Sea. Lundin Norway would be the operator of five of the licenses and Lundin Petroleum would be the operator of two of the licenses. Lundin Petroleum was an independent natural gas and petroleum exploration and production company with assets primarily in Europe (MBendi Information Services (Pty) Ltd., 2013b).

In 2012, StatoilHydro ASA announced that it would spend about $3 billion to finish completing 40 wells. StatoilHydro reported that its exploration activity level in 2012 was about the same as in 2011. Statoil completed 41 exploration wells in 2011, of which 22 were discoveries (Eni S.p.A., 2012b).

Wintershall Holdings GmbH of Germany was continuing its growth in northern Europe with the receipt of three new licenses in the North Sea. Wintershall already held 43 licenses, of which 24 licenses were as operator. As such, Wintershall was one of the leading licenses holders on the NCS. The company was proceeding with the Edvard Grieg, the Knaar, and the Maria development projects. Wintershall's Skarfjell discovery in April 2012 was estimated to have between 60 million and 160 million barrels of recoverable oil (Dupre, 2013).

Outlook

Norway's hydrocarbons sector is likely to continue to be a significant source of revenue to the Government as petroleum exploration continues in the Barents Sea, the Norwegian Sea, and new offshore exploration areas in the Arctic region. The Norwegian Petroleum Directorate is expected to continue with efforts to open up new offshore areas, particularly in the Arctic region. Norway is expected to continue to obtain nearly all its electricity from hydropower; however, other renewable resources, such as wind power, are being investigated. Industrial minerals are expected to continue to be important to the Nation's domestic economy.

References Cited

ArticEcon, 2012, Coal mining in Svalbard: Store Norske & Arktikugol—Norway, April 3. (Accessed June 25, 2013, at http://www.articecon.wordpress.com/2012/04/03/coal-mining-in-svalbard-store-norske-arktikugol-norway.)

Arctic Gold AB, 2012, About Arctic Gold: Arctic Gold AB. (Accessed June 22, 2013, at http://www.arcticgold.se/about.html.)

Bedinger, G.M., 2013, Titanium mineral concentrates: U.S. Geological Survey Mineral Commodity Summaries 2013, p. 174–175.

Dalradian Resources Inc., 2012, Dalradian provides update on Norwegian exploration program: Dalradian Resources Inc. (Accessed June 22, 2013, at http://www.dalradian.com/investor-centre/news-releases/news-releases-details/2012/Dalradian-Provides-Update-on-Norwegian-Exploration-Programs1130791/default.aspx.)

Dupre, Robin, 2013, Wintershall awarded three new licenses in North Sea: Rigzone.com. (Accessed September 8, 2013, at http://www.rigzone.com/news/oil_gas/a/123575 Wintershall_Awarded_Three_New_ Licenses_In_North_Sea.)

Eni S.p.A., 2012a, Eni starts up production at Marulk field: Rigzone.com. (Accessed September 8, 2013, at http://www.rigzone.com/news/oil_gas/a/116702.)

Eni S.p.A., 2012b, Statoil to spend $3B on 40 wells in 2012: Rigzone.com, April 2. (Accessed February 13, 2012, at http://www.rigzone.com/news/article_pf.asp?a_id=114961.)

First Point Minerals Corp., 2012, Norway–Fera property: First Point Minerals Corp. (Accessed September 8, 2013, at http://www.firstpointminerals.com/s/OtherNickel-Iron.asp?ReportID=599755.)

Hovland, K.M., 2012, Norway awards 60 new oil production licenses: Dow Jones & Company Inc. (Accessed May 29, 2012, at http://www.rigzone.com/news/article_pf.asp?a_id=114325.)

MBendi Information Services (Pty) Ltd., 2013a, Faroe Petroleum announces new exploration licenses awarded in Norway: MBendi Information Services (Pty) Ltd. (Accessed January 16, 2013, at http://www.mbendi.com/a_sndmsg/news_view.asp?l=130509&PG=35.)

MBendi Information Services (Pty) Ltd., 2013b, Lundin Petroleum awarded seven licenses in the Norwegian licensing round: MBendi Information Services (Pty) Ltd. (Accessed January 16, 2013, at http://www.mbendi.com/a_sndmsg/news_view.asp?=130510&PG=35.)

Nordic Mining ASA, 2012a, Engebø rutile: Nordic Mining ASA. (Accessed June 22, 2013, at http://www.nordicmining.com/engeboe-rutile-english/category317.html.)

Nordic Mining ASA, 2012b, Kvinnherad quartz: Nordic Mining ASA. (Accessed June 22, 2013, at http://www.nordicmining.com/nordic-mining-kvinnherad-quartz/category276.html.)

Nordic Mining ASA, 2012c, Øksfjord Peninsula: Nordic Mining ASA. (Accessed June 22, 2013, at http://www.nordicmining.com/new-nickel-copper-and-platinum-group-element-discoveries-in-north-norway/category139.html.)

Northern Iron Ltd., 2012, Projects: Northern Iron Ltd. (Accessed June 22, 2013, at http://www.northerniron.com.au/projects.)

Ollett, John, 2012, Tertiary Minerals updates Lassedalen fluorspar resource: Industrial Minerals, January 18. (Accessed February 15, 2012, at http://www.indmin.com/print.aspx?ArticleId=2963243.)

Store Norske Gull AS, 2012, Properties and claim areas: Store Norske Gull AS. (Accessed June 22, 2013, at http://www.snsk.no/properties-and-claim-areas.147127.en.html.)

U.S. Census Bureau, 2012a, U.S. exports to Norway by 5-digit end-use code, 2003–2012: U.S. Census Bureau. (Accessed May 26, 2012, at http://www.census.gov/foreign-trade/statistics/product/enduse/exports/c4039.html.)

U.S. Census Bureau, 2012b, U.S. imports from Norway by 5-digit end-use code, 2003–2012: U.S. Census Bureau. (Accessed May 26, 2012, at http://www.census.gov/foreign-trade/statistics/product/enduse/imports/c4039.html.)

U.S. Central Intelligence Agency, 2012, Norway in, The world factbook: U.S. Central Intelligence Agency. (Accessed May 22, 2012, at https://www.cia.gov/library/publications/the-world-factbook/geos/countrytemplate.html.)

U.S. Department of State, 2013, Norway—Factsheet: U.S. Department of State. (Accessed May 11, 2012, at http://www.state.gov/r/pa/ei/bgn/3421.htm.)

U.S Energy Information Administration, 2012, Norway—Country analysis brief overview: U.S Energy Information Administration. (Accessed May 16, 2013, at http://www.eia.gov/countries/cab.cfm?fips=NO.)

TABLE 1
NORWAY: PRODUCTION OF MINERAL COMMODITIES[1]

(Thousand metric tons unless otherwise specified)

Commodity		2008	2009	2010	2011	2012[e]
METALS						
Aluminum:						
Primary	metric tons	1,368,000	1,139,000 [r]	1,060,000	1,982,000 [r]	1,800,000
Secondary[e]	do.	350,000	350,000	300,000	300,000	250,000
Cadmium, metal	do.	178	249	300	309	300
Cobalt, metal, refined	do.	3,719	3,510	3,208	3,067	2,969 [2]
Copper, metal, refined, primary and secondary	do.	32,000	33,900	32,000 [e]	32,000 [e]	36,000
Iron and steel:						
Iron ore and concentrate, gross weight[3]		746	896	3,266	2,532 [r]	3,421 [2]
Metal:						
Ferroalloys:						
Ferromanganese		308 [r]	197 [r]	297 [r]	338 [r]	300
Ferrosilicomanganese		262 [r]	231 [r]	249	266 [r]	250
Ferrosilicon, 75% basis[e]		185 [2]	233 [2]	230	230	250
Silicon metal[e]		155	150	175	175	150
Steel, crude		560	579	514	620	600
Mercury[e]	metric tons	33 [2]	30	25	25	25
Nickel:						
Mine output, concentrate, Ni content	do.	377	369	351 [r]	339 [r]	351 [2]
Metal, primary	do.	88,741	88,577	92,100 [r]	92,427 [r]	91,687 [2]
Titanium:						
Ilmenite concentrate		915	671	864	870 [r]	831 [2]
TiO$_2$ content		403	289	371	400 [e]	400
Zinc, metal, primary	metric tons	145,469	137,622	147,775	153,200 [r]	153,000 [2]
INDUSTRIAL MINERALS						
Cement, hydraulic[e]		1,800	1,700	1,700	1,800	1,700
Clays		279	227	230 [e]	230 [e]	225
Feldspar		62	71	56	25 [r]	-- [2]
Graphite, flake	metric tons	4,100	4,562	6,270	7,789 [r]	6,992 [2]
Lime, hydrated, quicklime[e]		110	100	100	100	125
Mica, flake[e]	metric tons	2,000 [r]	1,000	--	--	--
Nepheline syenite		346	270 [e]	327	330	320 [2]
Nitrogen, N content of ammonia		350	300	300	300	300
Olivine sand		2,554	1,267	2,560	2,237 [2]	1,650 [2]
Sand and gravel		14,817	13,047	13,011	13,215	14,262 [2]
Stone, crushed[r, 3]		52,338	51,378	54,134	63,855	67,670
Sulfur, byproduct:[e]						
Metallurgical		95 [r]	90 [r]	80	90 [r]	90
Petroleum		20	20	20	19 [r]	20
Total		115 [r]	110 [r]	100	109 [r]	110
Talc, soapstone, steatite	metric tons	30,000 [r, 3]	23,350 [r, 3]	6,400	8,191	7,983 [2]
MINERAL FUELS AND RELATED MATERIALS						
Coal, all grades		3,429	2,437	1,685	1,640 [r]	1,583
Gas, natural, marketed[4]	million cubic meters	99,200	99,000	105,280	101,376 [r]	106,710 [2]
Peat, for agricultural use[e]	do.	497 [2]	480	490	500	500
Petroleum:						
Crude[5]	thousand 42-gallon barrels	901,550	854,830	777,450	732,555 [r]	694,230 [2]
Refinery products:						
Naphtha[e]	do.	10,000	10,000	10,000	10,000	10,000
Gasoline	do.	20,640 [r]	32,215 [r]	28,142 [r]	28,000	28,000
Kerosene	do.	5,420 [r]	4,752 [r]	3,942 [r]	4,000 [r]	5,000
Distillate fuel oil	do.	45,479 [r]	43,995 [r]	43,545 [r]	44,000 [r]	45,000
Residual fuel oil	do.	17,411 [r]	13,140 [r]	12,301 [r]	12,000	12,000
Other products	do.	1,800 [r]	1,700 [r]	1,750 [r]	1,800 [r]	1,800
Total[e]	do.	101,000 [r]	106,000 [r]	99,700 [r]	99,800 [r]	102,000

See footnotes at end of table.

TABLE 1—Continued
NORWAY: PRODUCTION OF MINERAL COMMODITIES[1]

eEstimated; estimated data are rounded to no more than three significant digits; may not add to totals shown. rRevised. do. Ditto. -- Zero.

[1]Table includes data available through June 30, 2013.
[2]Reported figure.
[3]Source: British Geological Survey.
[4]Reported as total methane sales.
[5]Excluding natural gas liquids.

TABLE 2
NORWAY: STRUCTURE OF THE MINERAL INDUSTRY IN 2012

(Thousand metric tons unless otherwise specified)

Commodity	Major operating companies and major equity owners	Location of main facilities	Annual capacity
Aluminum	Hydro Aluminium ANS (Norsk Hydro ASA, 70%)	Smelters at Ardal, Hoyanger, Karmoy, and Sunndal	600
Do.	do.	Rolling mill at Holmestrand	90
Do.	Alcoa Inc.	Smelters at Farsund and Mosjoen	250
Do.	Sor-Norge Aluminium A/S 50%, and Hydro Aluminium ANS, 49%)	Smelter at Odda	50
Cadmium metric tons	Norzink A/S (Outokumpu Oyj, 100%)	Smelter at Eitrheimsneset	5
Cement	Norcem A/S	Plants at Brevik and Kjopsvik	2,150
Coal	Store Norske Spitsbergen Grubekompani A/S	Mines at Longyearbyen and Svea	450
Cobalt	Nikkelverk A/S (Xstrata plc, 100%)	Smelter at Kristiansand	5
Copper:			
Ore, Cu content	Nikkel og Olivin A/S (Outokumpu Oyj, 100%)	Mine at Narvik	1
Metal	Nikkelverk A/S (Xstrata plc, 100%)	Smelter at Kristiansand	40
Dolomite	Franzefoss Bruk A/S	Mine at Ballagen	350
Do.	Norwegian Holding A/S	Mines at Hammerfall, Logavlen, and Kvitblikk	500
Feldspar	Franzefoss Bruk A/S	Mine at Lillesand	100
Ferroalloys	Elkem Salten (Elkem A/S, 100%)	Ferrosilicon plant at Straumen	90
Do.	Elkem Bjolvefossen (Elkem A/S, 100%)	Ferrosilicon plant at Alvik	60
Do.	Elkem Thamshavn (Elkem A/S, 100%)	Ferrosilicon plant at Orkanger	60
Do.	Finnfjord Smelteverk A/S, Rana Metal (FESIL ASA, 100%)	Ferrosilicon plant at Mo i Rana	110
Do.	A/S Hafslung Metal (FESIL ASA,100%)	Ferrosilicon plant at Sarpsborg	75
Do.	Ila og Lilleby Smelteverk (FESIL ASA, 100%)	Ferrosilicon plant at Finnsnes	20
Do.	Oye Smelteverk (Tinfos Jernverk A/S, 100%)	Silicomanganese plant at Kvinesdal	235
Graphite, flake	Skaland Graphite AS	Skalane Mine and plant at Skaland	12
Iron, metal	Ulstein Jernstoperi A/S	Hordvikneset	10
Iron ore	Rana Gruber A/S (Norsk Jernverk Holding A/S, 100%)	Mine at Mo i Rana	2,000
Do.	Arctic Bulk Minerals A/S	Mine and plant at Kirkenes	1,500
Do.	Northern Iron Ltd.	Mine at Bjørnevatn	2,800
Lime	Hylla Kalkverk (Nikolai Bruch A/S, 100%)	Verdal/Trondheim mine and plant	80
Do.	A/S Norsk Jernverk	Plant at Mo i Rana	48
Do.	Ardal og Sunndal Verk A/S	More og Romsdal plant at Surnadal	20
Do.	Breivik Kalkverk A/S	Alesund plant at Larsnes	20
Do.	Mjoendalen Kalkfabrik	Plant at Asen/Drammen	7
Limestone	Norcem A/S	Dalen, Bjorntvedt, and Kjopsvik Mines	1,600
Do.	Vardelskalk A/S (Franzefoss Burk A/S, 100%)	Sandvika Mine	800
Do.	Breivik Kalkverk A/S	Visnes and Glaerum Mines	500
Magnesium	Norsk Hydro ASA (Government, 51%)	Plants at Porsgrunn and Sauda	50
Manganese, alloys	Eramet SA	do.	500

See footnotes at end of table.

TABLE 2—Continued
NORWAY: STRUCTURE OF THE MINERAL INDUSTRY IN 2012

(Thousand metric tons unless otherwise specified)

Commodity		Major operating companies and major equity owners	Location of main facilities	Annual capacity
Natural gas	million cubic meters	Statoil ASA	Gama, Gullfaks, Sleipner Ost, and Statfjord fields	12,270
Do.	do.	ConocoPhillips Skandinavia A/S (operator)	Ekofisk field	9,900
Do.	do.	Elf Petroleum Norge A/S	Frigg, Heimdal, and Ost-Frigg fields	5,750
Do.	do.	Statoil ASA	Mikkel field	2,100
Do.	do.	Total S.A., 40%; Petoro S.A., 30%; Marathon Petroleum Norge AS, 20%; Norsk Hydro Produksjon A/S, 10%	Skirne field	1,550
Do.	do.	Esso Norge A/S	Odin field	1,000
Do.	do.	Amoco Norway A/S	Hod and Valhallfields	910
Nepheline syenite		North Cape Mineral A/S (Unimin Corp., 84%)	Mine at Stjernoy	350
Nickel:				
Ore, concentrate, Ni content		Nikkel og Olivin A/S (Outokumpu Oyj, 100%)	Mine at Narvik	5
Do.		Titania A/S (Kronos Norge A/S, 100%)	Mine at Tellnes	0.5
Metal		Nikkelverk A/S (Xstrata plc., 100%)	Smelter at Kristiansand	85
Olivine		Sibelco Nordic AS	Grubse and Raubergvik Mines and plant at Aheim	2,500
Do.		do.	Stranda Mine and plant	300
Do.		Franzefoss Bruk A/S	Lefdal Mine at Bryggja	500
Petroleum	42-gallon barrels per day	BP Petroleum Development of Norway	Ulaf fields	155,000
Do.	do.	A/S Norske Shell	Draugen field	90,000
Do.	do.	Esso Norge A/S (Exxon Mobil Corp., 100%)	Slagen Refinery at Slagentangen	6,000
Do.	do.	Statoil Mongstad A/S (Statoil ASA, 100%)	Mongstad Refinery	12,000
Pyrite		Folldal Verk A/S (Norsulfid A/S, 100%)	Mine at Hjerkinn	10
Quartzite		Elkem Tana (Elkem A/S, 100%)	Mine at Tana	540
Do.		Elkem Marnes (Elkem A/S, 100%)	Mine at Sandhornoy	200
Do.		Vatnet Kvarts A/S	Mine at Nordland	150
Do.		Snekkevik Kvartsbrudd	Mine at Kragero	110
Silicon metal		Lilleby Metall A/S (FESIL ASA, 100%)	Plant at Trondheim	9
Do.		FESIL ASA	Plant at Holla	50
Steel		Fundia AB (Norsk Jenverk, 50%, and Rautaruukki Group, 50%)	Plants at Christiania, Mandal Stal, and Spigerverk	600
Talc		A/S Norwegian Talc (Pluess-Staufer AG, 51%)	Mine and plant at Altermark/Knarrevik Mo i Rana, and Framfjord	90
Do.		Kvam Minerals A/S	Mine and plant at Kvam	6
Titanium, concentrate		Titania A/S (Kronos Norge A/S, 100%)	Mine at Tellnes	915
Zinc, metal		Boliden Odda A/S (Boliden AB, 100%)	Smelter at Odda	200

Do., do. Ditto.

THE MINERAL INDUSTRY OF POLAND

By Yadira Soto-Viruet

In 2012, Poland was estimated to be the world's 3d-ranked producer of rhenium, the 7th-ranked producer of silver, the 11th-ranked producer of cadmium, and the 12th-ranked producer of mined copper. Industrial minerals, such as feldspar, gypsum, lime, salt, sand and gravel, and sulfur, also were produced in significant quantities. For mineral fuels, Poland was ranked ninth among the world's leading producers of coal but depended on imports to meet domestic demand for oil and natural gas (Apodaca, 2013; Crangle, 2013; Dolley, 2013; Edelstein, 2013; George, 2013; Kostick, 2013; Miller, 2013; Polyak, 2013; Tanner, 2013; Tolcin, 2013; World Coal Association, 2013).

Minerals in the National Economy

In 2012, Poland's real gross domestic product (GDP) increased by 1.9% compared with that of 2011. Mining and quarrying accounted for about 2.2% of the total GDP. In 2010 (the latest year for which data were available), the total value of mineral commodity production was estimated to be $19.7 billion.[1] Mineral fuels, metals and industrial minerals made up about 49%, 31%, and 20%, respectively, of the value of Poland's mineral commodity production. Hard coal (bituminous) accounted for 76% of the total value of mineral fuel production, followed by lignite (12%), natural gas (8%) and crude oil (4%). Copper production accounted for 76% of the value of metals production, and cement production accounted for 36% of the value of industrial minerals production. As of yearend 2012, about 174,500 were employed in the mining and quarrying industry (Dmochowska, 2013, p. 483; Galos, Ney, and Smakowski, 2013, p. 11; International Monetary Fund, 2013, p. 59).

Government Policies and Programs

The new Polish Geological and Mining Law of June 9, 2011, became effective on January 1, 2012. The new Tax on Extraction of Certain Minerals became effective on March 2, 2012. The new tax establishes a tax base and imposes a tax rate on extraction of copper and silver. The tax base is based on the amount of copper and silver contained in the produced concentrate from the extracted copper ore or, if the taxpayer does not produce copper concentrate, the tax is based on the amount of copper and silver contained in the extracted copper ore. The tax rates are established separately for 1 metric ton (t) of copper and for 1 kilogram (kg) of silver. The tax rates are calculated per month on the basis of an average market price of copper and silver on stock exchanges in London and an average exchange rate from U.S. dollars to Polish zlotys. The new tax is targeted at the country's sole copper and silver producer, KGHM Polska Miedz S.A. (KGHM) (Ministry of Finance of the Republic of Poland, 2013).

The trade, distribution, and storage of gaseous fuels are regulated by the Act on Reserves of Crude Oil, Petroleum Products and Natural Gas dated February 16, 2007, and the Polish Energy Law of April 10, 1997. In March, the Government announced its new Privatisation Plan for the Years 2012–2013. The document includes plans for privatizing leading mining and energy companies, which are discussed in the Structure of the Mineral Industry section below (Ministry of the Treasury of the Republic of Poland, 2012a, p. 14, 16; Polskie Gornictwo Naftowe i Gazownictwo S.A., 2013a, p. 53).

Production

In 2012, cold-rolled steel production increased by 68% to 1.4 million metric tons (Mt) from 807,000 t in 2011; pipe steel production increased by 45% to 592,000 t from 408,000 t; moulding sands production, by 40% to 2.9 Mt; secondary copper smelter production, by 33% to 108,900 t; gold metal production, by 30% to 916 kg; refractory quartzite production, by 14% to 53,200 t; and crude petroleum production, by 9% to 5 million barrels (Mbbl). Rock salt production decreased by 37% to 782,000 t from 1.2 Mt; cadmium metal production, by 30% to 370 t from 526 t; estimated secondary refined aluminum production, by 21% to 11,000 t from 14,000 t; hydraulic cement production, by 16% to 15.9 Mt; estimated primary refined lead production, by 14% to 19,000 t from 22,000 t (revised); clinker cement production, by 13% to 11.8 Mt; and fire clay production, by 13% to 119,000 t. Data on mineral production are in table 1.

Structure of the Mineral Industry

The vast majority of companies in the mineral industry in Poland were privately owned. The Polish Government (through ownership of shares by the Polish Ministry of the Treasury) also owned shares in a small number of significant producers of mineral products. In March, the Ministry of the Treasury announced its new privatization plan for 2012 and 2013. The Ministry of the Treasury's privatization plan addresses the Ministry's strategic objectives regarding modernization of the economy and formation of better conditions for Poland's economic growth, the support of public policies, and the development of the capital markets. The plan indicates that the Ministry of the Treasury plans to privatize its shareholding in 85% of its supervised companies, which are included in the privatization plan. The privatization plan includes the energy production companies ENEA S.A, ENERGA S.A., PGE Polska Groupa Energetyczna S.A., Zespol Elektrowni "Patnow-Adamow-Konin" S.A., and Zespol Elektrowni Wodnych Niedzica S.A., and the mining companies Jastrzebska Spolka Weglowa S.A., Katowicki Holding Weglowy S.A., Kompania Weglowa S.A., Kopalnia Wegla Brunatnego "Adamow" S.A, Kopalnia Wegla Brunatnego "Konin" w Kleczewie S.A., and

---

[1]Where necessary, values have been converted from Polish zlotys (PLN) to U.S. dollars (US$) at an annual average exchange rate of PLN3.24=US$1.00 for 2012.

Lubelski Wegiel Bogdanka S.A. (Ministry of the Treasury of the Republic of Poland, 2012b, p. 14, 16).

As of 2012, one of the Government's most important holdings was a 31.79% stake in KGHM. No other shareholder owned more than 5% of the company's shares. The Government also owned 53.19% of Grupa LOTOS S.A., which owned oil and natural gas and refined petroleum production facilities, and 72.41% of Polskie Gornictwo Naftowe i Gazownictwo S.A. (PGNiG), which was the leading producer of oil and natural gas in Poland. Table 2 is a list of major mineral industry facilities.

Mineral Trade

In 2012, copper and copper alloys exports were valued at about $2.6 billion. The country's major export trade partners for copper and copper alloys were, in order of value, Germany (which received 42% of Poland's copper and copper alloys exports), China (22%), Italy (14%), and France (8%). Crude petroleum imports were valued at about $19.6 billion, and Poland's leading import partner for crude petroleum was Russia (96%). As in previous years, crude petroleum was the leading import commodity in terms of value, and refined copper was the leading export commodity (Central Statistical Office of Poland, 2013b).

Poland's exports to the United States were valued at about $4.6 billion in 2012 compared with $4.4 billion in 2011. Of this amount, fuel oil accounted for about $205 million; copper accounted for about $4.6 million; and zinc accounted for about $175,000. Imports from the United States were valued at about $3.3 billion in 2012 compared with $3.1 billion in 2011; these included $46 million in drilling and oilfield equipment; $24 million in iron and steel mill products; and $4 million in aluminum and alumina (U.S. Census Bureau, 2013a, b).

Commodity Review

Metals

Copper.—As of 2010, Poland had reported reserves of about 1.75 billion metric tons (Gt) containing about 34 Mt of copper, and about 80% of these reserves was in deposits currently operated by KGHM. KGHM was Poland's only producer of mined copper and primary copper metal, and it operated three copper refineries (the Glogow I, the Glogow II, and the Legnica refineries) and three mines (the Lubin, the Polkowice-Sieroszowice, and the Rudna Mines). In 2012, the company produced 565,800 t of electrolytically refined copper, 479,250 t of copper content of ore, and 427,100 t of copper content of concentrate (Galos, Ney, and Smakowski, 2013, p. 163).

The average copper content of ore extracted by KGHM in Poland decreased to 1.59% in 2012 from 1.61% in 2011. In recent years, copper content of ore has been steadily decreasing and has required the extraction of more ore to maintain current levels of copper mine output. KGHM also used increasing amounts of purchased copper scrap, blister, and imported concentrates to maintain refined copper production. KGHM continued to work towards achieving the development strategy's goal of increasing the company's resource base and production capacity. In 2012, KGHM reported that concessions for the extraction of copper in the Lubin-Malomice, Polkowice, Rudna, and Sieroszowice deposits would expire by yearend 2013 and that for the Radwanice-Wschod deposit would expire by 2015. These concessions were issued by the Ministry of the Environment between 1993 and 2004. The company expected to submit the appropriate documentation, which included environmental impact reports, geologic studies, deposit development plans, signed agreements with the municipalities of Lubin and Rudna, and permits, by 2013 in order for the Government to grant the mining concessions. KGHM expected to obtain the mining concessions for the extraction of copper for a period of 50 years (KGHM Polska Miedz S.A., 2013, p. 115–117).

Gold.—Gold production in Poland was entirely a result of gold obtained as a byproduct of copper ore processing at KGHM's refineries. In 2012, KGHM reported that gold metal production increased by 30%, which was attributed to a higher gold content in purchased copper-bearing materials (KGHM Polska Miedz S.A., 2013, p. 116).

Iron and Steel.—Poland had not produced iron ore since 1990 and was dependent on imported iron ore and concentrates for domestic pig iron production. In 2012, iron ores and concentrates imports increased by 10% to 6.5 Mt compared with 5.9 Mt in 2011. About 65% of these imports came from Ukraine, and about 21% came from Russia. All the imported iron ore and concentrates was used for pig iron production at ArcelorMittal Poland S.A.'s iron and steel plants at Dabrowa Gornicza and Krakow, which were the only pig iron producers in Poland (Central Statistical Office of Poland, 2013b; Galos, Ney, and Smakowski, 2013, p. 274–275).

According to the Polish Steel Association, Poland's real consumption of steel products decreased by 1% in 2012. The decrease was attributed to a reduction in building construction and automobile production in the second half of 2012. The country's apparent consumption of finished steel products in 2012 reached 10.4 Mt, which was about a 5.4% decrease compared with that of 2011. Poland's exports of steel products totaled about 5.6 Mt in 2012, which was an increase of about 11% compared with those of 2011. The major export partners for steel products were Germany and the Czech Republic. Imports of steel products increased to 8.1 Mt or by 3% compared with that of 2011, and the major import partners for steel products were Russia and Turkey. In 2012, domestic steel production capacity utilization decreased by 4% (Polish Steel Association, 2013, p. 18, 21–24).

Cognor S.A. held a 100% interest in Ferrostal Labedy Sp. z o.o. and Huta Stali Jakosciowych S.A. These companies had a combined crude steel production capacity of 636,000 t. In 2012, crude steel production from Ferrostal Labedy and Huta Stali Jakosciowych decreased by 13% to 462,874 t compared with 534,608 t in 2011. The decrease in production was attributed to low demand for steel in the country and in the rest of Europe (Cognor S.A., 2013, p. 12).

Lead and Zinc.—In November, Stalprodukt S.A. acquired an 86.92% share in Zaklady Gorniczo-Hutnicze (ZGH) "Boleslaw" S.A., which was Poland's only producer of lead and zinc ore and the leading producer of refined zinc. In September 2011, the Ministry of the Treasury had posted an invitation for bids for the purchase of shares of ZGH Boleslaw, which is located in the Bukowno region and had an estimated capacity of about 110,000 metric tons per year (t/yr) of zinc and 30,000 t/yr of lead (Ministry of the Treasury of the Republic of Poland, 2011; Thomson Reuters, 2012; Stalprodukt S.A., 2013a, p. 14; 2013b).

Industrial Minerals

Cement.—HeidelbergCement Group of Germany through its subsidiary Gorazdze Cement S.A. operated one cement plant in Gorazdze and one grinding plant in Katowice. In March, HeidelbergCement completed the construction of a new cement mill in the Gorazdze cement plant with a production capacity of about 1.4 million metric tons per year (Mt/yr) of cement. In 2011, the company also increased its clinker production capacity to 4.0 Mt/yr from 3.1 Mt/yr. In 2012, HeidelbergCement's total cement production capacity in the country increased to 5.6 Mt/yr (HeidelbergCement Group, 2012; 2013, p. 109).

Mineral Fuels

Coal.—Bituminous coal production increased by 4% to 79.8 Mt compared with 76.4 Mt in 2011, and brown coal and lignite production increased by 2% to 64.2 Mt. As of 2009, bituminous coal reserves were estimated to be 16.9 billion metric tons (Gt) and brown coal and lignite reserves were estimated to be 14.9 Gt. In 2012, Poland was the world's sixth-ranked producer of brown coal and lignite. As of 2012, bituminous coal and brown coal and lignite were Poland's major mineral fuels for electricity and accounted for about 80% of domestic electricity production. About 53% of the bituminous coal was used for electricity production in 2012. The output of brown coal and lignite was used almost entirely at thermal powerplants (Central Statistical Office of Poland, 2013a, p. 42–43, 50; European Association for Coal and Lignite, 2013; World Coal Association, 2013).

Natural Gas and Petroleum.—In Poland, crude petroleum and natural gas were produced by the following three companies: FX Energy Inc. of United States, LOTOS Petrobaltic S.A., and PGNiG. These companies had a combined estimated crude oil production capacity of about 5.6 Mbbl and an estimated natural gas production capacity of about 4.4 billion cubic meters. In 2012, Poland produced about 5.0 Mbbl of crude oil and about 5.8 billion cubic meters of natural gas but was dependent on imports for the majority of its supplies. During the year, the country imported about 183 Mbbl of crude oil, of which 96% was from Russia (Central Statistical Office of Poland, 2013b).

Grupa LOTOS through its subsidiary LOTOS Petrobaltic operated concessions for oil and gas within the Polish Economic Zone (PEZ) in the Baltic Sea Shelf. Grupa LOTOS held seven offshore exploration licenses in the eastern part of the PEZ, which included the Gaz Polnoc, Gaz Poludine, Gotlandia, Leba, Rozewie, Sambia E, and Sambia W licenses. The company also held four oil and gas production licenses on fields B3, B4, B6, and B8, respectively. In 2012, about 1.4 Mbbl of crude oil and about 30 million cubic meters of natural gas were produced from the B3 and B8 (under development) gasfields. In October, LOTOS Petrobaltic signed a joint-venture agreement with CalEnergy Resources Poland Sp. z o.o. for the development of the B4 and B6 gasfields, which had estimated resources of about 4 billion cubic meters of natural gas. Under the agreement, LOTOS Petrobaltic and CalEnergy Resources would hold 51% and 49% ownership, respectively. The joint venture planned to complete preconstruction work and a seismic acquisition program by 2013 and a preliminary field development design by 2014. As of December 31, the company's crude oil reserves and resources in the Baltic Sea Shelf were estimated to be about 32 Mbbl and 10 Mbbl, respectively. Natural gas reserves were estimated to be 0.48 billion cubic meters, and resources were estimated to be 6.46 billion cubic meters (Grupa LOTOS S.A., 2012; 2013a, p. 67, 90; 2013b).

PGNiG was Poland's leading producer of oil and gas, and, in 2012, its Barnowko-Mostno-Buszewo field accounted for about 75% of the total crude oil production in the country. In 2012, the company produced about 3.6 Mbbl of crude oil and about 4.3 billion cubic meters of natural gas. In October, PGNiG began operations at the Lubiatow oilfield and the Miedzychod gasfield, which were part of the Lubiatow-Miedzychod-Grotow (LMG) project. The LMG project is located near Gorzow Wielkoposki and had estimated recoverable reserves of about 53 Mbbl of crude oil and about 7.3 billion cubic meters of natural gas. The project included the construction of a central facility, which would provide collection, separation, treatment, and a dispatch terminal in Wierzbno. The crude oil would be transported to be processed at German refineries through the Druzhba pipeline, and the gas would be transported to the Grodzisk unit. The company expected initial production of about 2.1 Mbbl of crude oil and 100 million cubic meters of natural gas. PGNiG expected that production at the LMG field would double the company's domestic crude oil output (Polskie Gornictwo Naftowe i Gazownictwo S.A., 2012; 2013a, p. 8, 14, 33, 40, 41; 2013b).

As of 2012, PGNiG reported that the Ministry of the Environment had awarded more than 111 exploration licenses for shale gas deposits. At least 15 companies were engaged in shale gas exploration; these included Aurelian Oil and Gas plc. of the United Kingdom; BNK Petroleum Inc. of Canada; Eni S.p.A. of Italy; PGNiG; and the U.S. companies Chevron Corp., Exxon Mobil Corp., and Marathon Oil Corp. PGNiG held 15 exploration licenses in the country, which cover an area of about 12,800 square kilometers. In July, five state-owned companies (ENEA S.A., KGHM, PGNiG, PGE Polska Grupa Energetyczna, and Tauron Polska Energia S.A.) signed a cooperation agreement for shale gas exploration and extraction at PGNiG's Wejherowo concession, which is located in northern Poland. Exploration works were underway at the Kochanowo, the Cestkowo, and the Tepcz sites within the Wejherowo license area. PGNiG would be the operator during the exploration and appraisal phase (Ministry of the Treasury of the Republic of Poland, 2012b; Polish Shale Gas, 2013a–c; Polskie Gornictwo Naftowe i Gazownictwo S.A., 2013a, p. 13, 32, 39).

Outlook

The Government of Poland forecasted an increase in the GDP of 1.3% in 2013 (International Monetary Fund, 2013, p. 59). Continuing recovery from the economic crisis and an accompanying increase in domestic demand could lead to increased mineral commodity production, but much of the growth in demand for mineral products could depend on demand by Poland's trade partners, especially those in the EU. Most likely, coal and copper will remain the leading mineral products in terms of production value, and dependence on imports of mineral fuels will remain one of Poland's biggest challenges. In the longer run, plans to develop oil and natural gas and shale gas deposits are expected to strengthen the mineral fuel sector. Also, Government ownership of mineral producing companies is expected to decline as privatization efforts continue.

References Cited

Apodaca, L.E., 2013, Sulfur: U.S. Geological Survey Mineral Commodity Summaries 2013, p. 158–159.

Central Statistical Office of Poland, 2013a, Energy statistics 2011, 2012: Central Statistical Office of Poland, 290 p. (Accessed December 5, 2013, at http://www.stat.gov.pl/cps/rde/xbcr/gus/EE_energy_statistics_2011-2012.pdf.)

Central Statistical Office of Poland, 2013b, Foreign trade turnover by main commodities 2012: Central Statistical Office of Poland, August 30. (Accessed September 30, 2013, at http://stat.gov.pl/en/topics/prices-trade/trade/foreign-trade-turnover-by-main-commodities,4,12.html.)

Cognor S.A., 2013, 2012 annual report and management discussion and analysis: Cognor S.A., March 21, 15 p. (Accessed December 5, 2013, at http://www.cognor.eu/foto/domains/3/static/Cognor Group 2012 Annual Report and Management Discussion and Analysis.pdf.)

Crangle, R.D., 2013, Gypsum: U.S. Geological Survey Mineral Commodity Summaries 2013, p. 70–71.

Dmochowska, Halina, ed., 2013, Concise statistical yearbook of Poland 2013: Warsaw, Poland, Central Statistical Office of Poland [Glowny Urzad Statystyczny], June, 732 p.

Dolley, T.P., 2013, Sand and gravel (industrial): U.S. Geological Survey Mineral Commodity Summaries 2013, p. 138–139.

Edelstein, D.L., 2013, Copper: U.S. Geological Survey Mineral Commodity Summaries 2013, p. 48–49.

European Association for Coal and Lignite, 2013, Poland: European Association for Coal and Lignite. (Accessed December 5, 2013, at http://www.euracoal.org/pages/layout1sp.php?idpage=76.)

Galos, Krzysztof, Ney, Roman, and Smakowski, Tadeusz, eds., 2013, Minerals yearbook of Poland 2010: Krakow, Poland, Department of Mineral Policy, Mineral and Energy Economy Research Institute, Polish Academy of Sciences, 553 p.

George, M.W., 2013, Silver: U.S. Geological Survey Mineral Commodity Summaries 2013, p. 146–147.

Grupa LOTOS S.A., 2012, LOTOS and CalEnergy to jointly develop gas fields in the Baltic Sea: Gdansk, Poland, Grupa LOTOS S.A. press release (Accessed December 3, 2013, at http://www.lotos.pl/en/1031/p,717,n,3702/lotos_group/our_companies/lotos_petrobaltic/news/lotos_and_calenergy_to_jointly_develop_gas_fields_in_the_baltic_sea.)

Grupa LOTOS S.A., 2013a, Integrated annual report 2012: Gdansk, Poland, Grupa LOTOS S.A., 462 p. (Accessed December 3, 2013, at http://raportroczny.lotos.pl/assets/dokumenty/Integrated-Annual-Report-LOTOS-2012.pdf.)

Grupa LOTOS S.A., 2013b, LOTOS and CalEnergy working together on Baltic gas deposits: Gdansk, Poland, Grupa LOTOS S.A. press release, 462 p. (Accessed December 3, 2013, at http://www.lotos.pl/en/1031/p,695,n,3822/lotos_group/press_centre/news/lotos_and_calenergy_working_together_on_baltic_gas_deposits.)

HeidelbergCement Group, 2012, HeidelbergCement expands cement capacity in Poland—Largest cement mill in Europe commissioned: Heidelberg, Germany, HeidelbergCement Group press release, March 28. (Accessed December 6, 2013, at http://www.heidelbergcement.com/global/en/company/press_media/archive/press_releases_2012/2012-03-28.htm.)

HeidelbergCement Group, 2013, Annual report 2012: Heidelberg, Germany, HeidelbergCement Group, 256 p. (Accessed December 6, 2013, at http://www.heidelbergcement.com/NR/rdonlyres/ABE4A8AA-5288-472F-B00C-16F9E47DD0AA/0/AnnualReport_2012_Web.pdf.)

International Monetary Fund, 2013, World economic outlook: Washington, DC, International Monetary Fund, October, 188 p. (Accessed December 13, 2013, at http://www.imf.org/external/pubs/ft/weo/2013/02/pdf/text.pdf.)

KGHM Polska Miedz S.A., 2013, Annual report 2012: Lubin, Poland, KGHM Polska Miedz S.A., March, 163 p. (Accessed December 9, 2013, at http://www.kghm.pl/_files/Report/report_2012.pdf.)

Kostick, D.S., 2013, Salt: U.S. Geological Survey Mineral Commodity Summaries 2013, p. 134–135.

Miller, M.M., 2013, Lime: U.S. Geological Survey Mineral Commodity Summaries 2013, p. 92–93.

Ministry of Finance of the Republic of Poland, 2013, Tax on extraction of certain minerals: Ministry of Finance of the Republic of Poland, January 25. (Accessed December 11, 2013, at http://www.finanse.mf.gov.pl/en/other-taxes/tax-on-extraction-of-certain-minerals.)

Ministry of the Treasury of the Republic of Poland, 2011, Invitation to negotiations regarding the purchase of shares of Zaklady Gorniczo-Hutnicze "Boleslaw" S.A. with its registered office in Bukowno: Ministry of the Treasury of the Republic of Poland, September 27. (Accessed December 6, 2013, at http://msp.gov.pl/portal/en/43/2677/Invitation_to_negotiations_regarding_the_purchase_of_shares_of_Zaklady_GorniczoH.html.)

Ministry of the Treasury of the Republic of Poland, 2012a, Polish state companies PGNiG, KGHM, Tauron, PGE and ENEA have signed cooperation agreement on shale gas exploration and extraction—What does it mean for Polish energy sector and interested firms?: Ministry of the Treasury of the Republic of Poland, July 30.

Ministry of the Treasury of the Republic of Poland, 2012b, Privatisation plan for the years 2012–2013: Ministry of the Treasury of the Republic of Poland, 29 p. (Accessed December 9, 2013, at http://www.msp.gov.pl/en/privatisation-plan/privatisation-plan-for/3182, Privatisation-Plan-for-the-years-2012-2013.html.)

Polish Shale Gas, 2013a, Cooperation: Polish Shale Gas. (Accessed December 4, 2013, at http://polishshalegas.pl/en/shales-in-poland/cooperation.)

Polish Shale Gas, 2013b, Other licensees: Polish Shale Gas. (Accessed December 4, 2013, at http://polishshalegas.pl/en/shales-in-poland/other-licensees.)

Polish Shale Gas, 2013c, PGNiG licenses: Polish Shale Gas. (Accessed December 4, 2013, at http://polishshalegas.pl/en/shales-in-poland/pgnig-licenses.)

Polish Steel Association, 2013, Polish steel industry 2013: Polish Steel Association, 70 p. (Accessed December 6, 2013, at http://www.hiph.org/ANALIZY_RAPORTY/pliki/PPS_2013.pdf.)

Polskie Gornictwo Naftowe i Gazownictwo S.A., 2012, PGNiG launches the Lubiatow-Miedzychod-Grotow production facility: Warsaw, Poland, Polskie Gornictwo Naftowe i Gazownictwo S.A. press release, October 18. (Accessed December 3, 2013, at http://www.pgnig.pl/pgnig/com/8387?r%2Cnews%2CpageNumber=5&r%2Cnews%2CdateTo=&r%2Cnews%2CdateFrom=&r%2Cnews%2CnewsId=41154.)

Polskie Gornictwo Naftowe i Gazownictwo S.A., 2013a, Annual report 2012: Warsaw, Poland, Polskie Gornictwo Naftowe i Gazownictwo S.A., 121 p. (Accessed December 3, 2013, at http://www.pgnig.pl/reports/annualreport2012/download/PGNiG_AR_2012_WEB.pdf.)

Polskie Gornictwo Naftowe i Gazownictwo S.A., 2013b, LMG, Poland's largest oil and gas extraction facility starts production: Warsaw, Poland, Polskie Gornictwo Naftowe i Gazownictwo S.A. press release, March 25. (Accessed December 3, 2013, at http://www.pgnig.pl/pgnig/com/8387?r%2Cnews%2CpageNumber=3&r%2Cnews%2CdateTo=&r%2Cnews%2CdateFrom=&r%2Cnews%2CnewsId=48463.)

Polyak, D.E., 2013, Rhenium: U.S. Geological Survey Mineral Commodity Summaries 2013, p. 130–132.

Stalprodukt S.A., 2013a, Report of the Stalprodukt S.A. supervisory board for the period from 1 January 2012 to 31 December 2012: Bochnia, Poland, Stalprodukt S.A., 20 p. (Accessed December 6, 2013, at http://www.stalprodukt.com.pl/pub/File/lad_korporacyjny_EN/Sprawozdanie%20Rady%20Nadzorczej_2012_English.pdf.)

Stalprodukt S.A., 2013b, Shares and shareholders: Bochnia, Poland, Stalprodukt S.A. (Accessed December 6, 2013, at http://www.stalprodukt.com.pl/akcje_akcjonariusze.)

Tanner, A.O., 2013, Feldspar: U.S. Geological Survey Mineral Commodity Summaries 2013, p. 54–55.

Thomson Reuters, 2012, Stalprodukt SA updates on acquisition of Zakalady Gorniczo-Hutnicze Boleslaw SA: Thomson Reuters, November 7. (Accessed December 6, 2013, at http://www.reuters.com/finance/stocks/STPEUR.PAp/key-developments/article/2638729.)

Tolcin, A.C., 2013, Cadmium: U.S. Geological Survey Mineral Commodity Summaries 2013, p. 36–37.

U.S. Census Bureau, 2013a, U.S. exports to Poland from 2003 to 2012 by 5-digit end-use code: U.S. Census Bureau. (Accessed October 22, 2013, at http://www.census.gov/foreign-trade/statistics/product/enduse/exports/c4550.html.)

U.S. Census Bureau, 2013b, U.S. imports from Poland from 2003 to 2012 by 5-digit end-use code: U.S. Census Bureau. (Accessed October 22, 2013, at http://www.census.gov/foreign-trade/statistics/product/enduse/imports/c4550.html.)

World Coal Association, 2013, Coal statistics: World Coal Association, August. (Accessed September 27, 2013, at http://www.worldcoal.org/resources/coal-statistics.)

TABLE 1
POLAND: PRODUCTION OF MINERAL COMMODITIES[1]

(Thousand metric tons unless otherwise specified)

Commodity[2]		2008	2009	2010	2011	2012
METALS						
Aluminum, metal:						
Primary	metric tons	47,543	--	--	--	--
Secondary[e, 3]	do.	18,000	16,900	16,000	14,000	11,000
Total[e]	do.	65,500	16,900	16,000	14,000	11,000
Cadmium, metal, primary	do.	603	534	451	526	370
Copper:						
Ore:						
Gross weight		30,920[r]	31,253[r]	30,805[r]	31,241[r]	31,725
Cu content	metric tons	482,400	498,960	480,600	479,300	479,250
Concentrate:						
Gross weight		1,866	1,930[r]	1,842[r]	1,877[r]	1,861
Cu content	metric tons	429,400	439,000	425,400	426,700	427,100
Metal:						
Smelter:[4]						
Primary	do.	438,600	427,800	433,900[r]	449,000	470,000
Secondary	do.	43,800	68,800	94,600[r]	82,100[r]	108,900
Total	do.	482,400	496,600	528,500[r]	531,100[r]	578,900
Refined, electrolytically, primary and secondary	do.	526,808	502,491	547,100	571,000	565,800
Gold, metal	kilograms	902	814	776	704	916
Iron and steel:						
Pig iron		4,934	2,984[r]	3,638	3,975	3,944
Ferroalloys:						
Blast furnace, ferromanganese	metric tons	8,500	1,700	800[r]	800[r, e]	800[e]
Electric furnace:						
Silicomanganese	do.	25,100	--	100[r]	120[r, e]	120[e]
Ferrosilicon	do.	56,031	9,673	53,206	72,668	78,115
Other	do.	2,900	4,200	200[r]	250[r, e]	250[e]
Total ferroalloys	do.	92,531	15,573	54,306[r]	73,800[r, e]	79,300[e]
Steel, crude:						
From oxygen converters		5,225	3,235	3,995	4,424	4,333
From electric arc furnaces		4,502	3,893	4,001	4,353	4,206
Total		9,727	7,128	7,996	8,777	8,539
Finished steel products:						
Hot rolled		7,610	6,232	6,658	7,504	7,894
Cold rolled		689	558	835	807	1,353
Pipe		409	347	384	408	592
Lead:						
Mine output:						
Pb content of Pb-Zn ore	metric tons	66,400	51,500	35,300[r]	31,000	32,600
Pb content of Cu ore	do.	21,300	28,900	24,900[r]	25,000[r, e]	25,000[e]
Total	do.	87,700	80,400	60,200[r]	56,000[r, e]	57,600[e]
Concentrate, Pb content	do.	47,900	36,900	23,100[r]	22,000[e]	23,000[e]
Metal, refined, primary and secondary:[e]						
Primary	do.	42,000[r]	38,000[r]	39,000[r]	22,000[r]	19,000
Secondary	do.	66,200[r]	62,400[r]	82,000[r]	110,000[r]	121,000
Total	do.	108,200[5]	100,400[5]	121,000[r]	132,000[r]	140,000
Platinum-group metals, average content of slimes:[e, 6]						
Palladium	kilograms	15	15	15	15	15
Platinum	do.	25	25	25	25	25
Rhenium:						
Ammonium perrhenate:						
Gross weight	do.	4,900	3,500	6,709	8,650	8,700[e]
Re content of ammonium perrhenate	do.	3,400	2,400	4,656	6,000	6,000[e]
Rhenium metal in pellet form	do.	--	--	620	620[e]	620[e]
Selenium	metric tons	82	73	79	80[e]	80[e]

See footnotes at end of table.

TABLE 1—Continued
POLAND: PRODUCTION OF MINERAL COMMODITIES[1]

(Thousand metric tons unless otherwise specified)

Commodity[2]		2008	2009	2010	2011	2012
METALS—Continued						
Silver:						
Mine output, Ag content of Cu concentrate	metric tons	1,161	1,207	1,181	1,167	1,149
Metal	do.	1,221	1,221	1,175	1,280 [e]	1,290 [e]
Zinc:						
Zn content:						
Mine output	do.	136,300	116,000	88,500 [r]	89,000 [r, e]	89,000 [e]
Concentrate output	do.	132,400	115,500	92,000	92,000 [r, e]	92,000 [e]
Metal, refined, primary and secondary	do.	142,600	139,100	135,000	156,000 [e]	161,000 [e]
INDUSTRIAL MINERALS						
Cement:						
Clinker		12,443	10,659	11,768	13,630	11,807
Hydraulic		17,207	15,422	15,812	18,993	15,919
Clays and clay products:						
Bentonite:						
Crude	metric tons	3,000	3,000	2,000	3,000 [r, e]	3,000 [e]
Processed, including imported material	do.	90,412	81,354	86,000 [r]	113,800 [r]	102,100
Fire clay, crude		169 [r]	115	82	136	119
Kaolin:						
Crude		318	261	238 [r]	240 [r, e]	240 [e]
Beneficiated		155	136	125	125 [e]	125 [e]
Diatomite	metric tons	1,000	700	500 [r]	500 [r, e]	500 [e]
Feldspar:						
Run of mine	do.	599,100	445,500	513,700 [r]	510,000 [r, e]	510,000 [e]
Processed, including imported material	do.	643,700	478,000	485,000	485,000 [e]	485,000 [e]
Gypsum and anhydrite						
Natural:						
Gypsum rock		1,283	1,125 [r]	1,012 [r]	1,300 [r, e]	1,300 [e]
Anhydrite		198	152 [r]	167 [r]	200 [r, e]	200 [e]
Total		1,481	1,277	1,179	1,500 [r, e]	1,500 [e]
Synthetic gypsum		1,596	2,076	2,389	2,400 [e]	2,400 [e]
Grand total		3,077	3,353	3,568 [r]	3,900 [r, e]	3,900 [e]
Lime, hydrated and quicklime		1,952	1,704	1,799	2,036	1,799
Magnesite:						
Crude	metric tons	105,577	85,258	108,809	129,166	129,641
Concentrate	do.	60,000	47,000	63,000	60,000 [e]	60,000 [e]
Nitrogen, N content of ammonia[e]		2,000 [r]	1,600 [r]	1,700 [r]	1,900 [r]	2,100
Quartz, quartzite, and quartz schist, marketable:						
Quartz and quartz crystal	metric tons	6,500	5,100	5,600	6,100	5,400
Quartzite, refractory	do.	72,500	25,700	34,200	46,500	53,200
Quartz schist	do.	800 [r]	700 [r]	700 [r]	700 [r, e]	700 [e]
Salt:						
Rock		618	999	1,236 [r]	1,234	782
Evaporated salt		622	299	411	415	402
Other (brine and desalination of mine waste water)		2,783	2,533	2,464	2,633	2,732
Total		4,023	3,831	4,111 [r]	4,282	3,916
Sand and gravel:						
Aggregates (construction sand and gravel), natural, mine output		149,312	141,114 [r]	157,236 [r]	160,000 [r, e]	160,000 [e]
Filling sand	thousand cubic meters	5,500	5,000	4,900	5,000 [e]	5,000 [e]
Foundry sand		806 [r]	720 [r]	920 [r]	920 [r, e]	920 [e]
Lime-sand brick production sand	thousand cubic meters	834	560	615 [r]	620 [r, e]	620 [e]
Moulding sand		1,244	1,701	1,817	2,096 [r]	2,934
Silica sand (glass sand), marketable		2,398	2,127	2,458	2,570	2,354
Sodium compounds:						
Carbonate (soda ash), 98%		1,190	890	1,010	1,061	1,116
Caustic soda (96% NaOH)		409	381	285	361	388

See footnotes at end of table.

TABLE 1—Continued
POLAND: PRODUCTION OF MINERAL COMMODITIES[1]

(Thousand metric tons unless otherwise specified)

Commodity[2]		2008	2009	2010	2011	2012
INDUSTRIAL MINERALS—Continued						
Stone, mine output:						
Dimension stone		3,426	3,836	4,598	4,600 [e]	4,600 [e]
Dolomite		2,206	1,834	1,821	1,880	1,830
Limestone:						
For lime production		16,110	14,881	17,588	17,500 [e]	17,500 [e]
For non-lime end use		30,778	28,883	35,528 [r]	35,500 [e]	35,500 [e]
Road stone		300	260	169 [r]	170 [r, e]	170 [e]
Sulfur:						
Native, Frasch	metric tons	762,000	263,000	517,000	657,000	676,000
Byproduct:						
From natural gas	do.	21,300	24,800	24,900	25,000 [e]	25,000 [e]
From oil refineries and coking plants	do.	201,000	190,000	225,000 [r]	200,000 [e]	200,000 [e]
From metallurgy	do.	294,000	257,000	253,000 [r]	250,000 [e]	250,000 [e]
Other	do.	400	500	500	500 [e]	500 [e]
Total	do.	516,700	472,300 [r]	503,400 [r]	476,000 [r, e]	476,000 [e]
Grand total	do.	1,278,700	735,300 [r]	1,020,400 [r]	1,130,000 [r, e]	1,150,000 [e]
MINERAL FUELS AND RELATED MATERIALS						
Carbon black	metric tons	36,300	27,800	34,700 [r]	35,000 [e]	35,000 [e]
Coal:						
Bituminous		84,345	78,064	76,728	76,448 [r]	79,855
Brown coal and lignite		59,668	57,108	56,510	62,841	64,280
Total		144,013	135,172	133,238	139,289 [r]	144,135
Coke		9,761	7,091	9,738	9,377	8,891
Gas:						
Natural	million cubic meters	5,382	5,537	5,666 [r]	5,825	5,782
Coke oven gas, manufactured	do.	4,207	3,076	4,239	4,200 [e]	4,200 [e]
Peat, fuel and agricultural		632	594	672	746	759
Petroleum:						
Crude[7]	thousand 42-gallon barrels	5,600	5,100	5,100	4,600	5,000
Refinery products[8]	do.	144,000	150,000	158,000	160,000	165,000

[e]Estimated; estimated data are rounded to no more than three significant digits; may not add to totals shown. [r]Revised. do. Ditto. -- Zero.

[1]Table includes data available through November 26, 2013.

[2]In addition to commodities listed, beneficiated barite, cobalt, gold content of copper concentrate, nickel, sulfate, and town gas are thought to have been produced, but available information is inadequate to make reliable estimates of output.

[3]Based on official Polish Government estimates.

[4]Copper smelter production is based on production at KGHM Polska Miedz S.A. Additional smelter production may have taken place at the Institute of Non-Ferrous Metals at Gliwice, but this production was not marketable and was produced only for research purposes.

[5]Reported figure.

[6]Estimates based on reported platinum- and palladium-bearing final (residual) slimes and their average Pt and Pd content from electrolytic copper refining.

[7]Figures were converted to barrels from production in metric tons, which was reported as the following: 2008—754,907; 2009—686,992; 2010—686,487; 2011—616,525 (revised); and 2012—673,582.

[8]Figures were converted to barrels from production in metric tons, which was reported as the following: 2008—19,631,450; 2009—20,499,407; 2010—21,557,363; 2011—21,770,253 (revised); and 2011—22,497,390.

TABLE 2
POLAND: STRUCTURE OF THE MINERAL INDUSTRY IN 2012[1]

(Thousand metric tons unless otherwise specified)

Commodity	Major operating companies and major equity owners	Location of main facilities	Annual capacity
Aluminum, secondary	Huta Aluminium Konin (Impexmetal S.A., 95.52%)	Konin	NA.
Do.	Boryszew S.A. Branch Modern Products Aluminium Skawina (Grupa Boryszew, 100%)	Skawina	NA.
Do.	Grupa KETY S.A.	Kety	NA.
Do.	Nicromet	Bestwinka	NA.
Do.	Alumetal S.A.	Kety	NA.
Do.	POLST Sp. z o.o.	Walbrzych	NA.
Bentonite	Zaklady Gornico-Metalowe "Zebiec" S.A.	Starachowice	80.[e]
Cadmium, refined metric tons	Huta Cynku "Miasteczko Slaskie" S.A.	Miasteczko Slaskie Smelter	540.
Cement	Gorazdze Cement S.A. (HeidelbergCement AG, 100%)	Gorazdze	4,000 clinker, 5,600 cement.
Do.	Grupa Ozarow S.A. (CRH plc., 100%)	Plants at Ozarow and Rejowiec	2,800 clinker,[e] 3,250 cement.[e]
Do.	Cemex Polska Sp. Z o.o. (CEMEX S.A.B de C.V., 100%)	Plants at Chelm and Rudniki	2,300 clinker,[e] 3,000 cement.[e]
Do.	Cementownia Warta S.A. (Polen Zement Beteiligungsgesellschaft GmbH)	Dzialoszyn	1,500 clinker,[e] 2,000 cement.[e]
Do.	Lafarge Cement S.A.	Plants at Malogoszcz and Piechcin	3,000 clinker,[e] 5,700 cement.
Do.	Cementownia Nowiny Sp. z o.o.	Warta	1,100 clinker,[e] 1,600 cement.
Do.	Cementownia "Nowa Huta" S.A.	Krakow	300 clinker,[e] 500 cement.[e]
Do.	Cementownia "Odra" S.A.	Opole	400 clinker,[e] 800 cement.[e]
Cement, aluminous	Gorka Cement Sp. z o.o.	Trzebinia	70 clinker,[e] 70 cement.[e]
Coal:			
Bituminous	Includes:	Of which:	90,000.[e, 2]
	100% Government owned:		
	Kompania Weglowa S.A.	Upper Silesia (16 mines)	
	Katowicki Holding Weglowy S.A.	Upper Silesia (5 mines)	
	Poludniowy Koncern Weglowy S.A.	Upper Silesia (2 mines)	
	KWK Kazimierz-Juliusz Sp. z o. o.	Upper Silesia (1 mine)	
	Jastrzebska Spolka Weglowa S.A. (Government, 65.74%)	Upper Silesia (6 mines)	
	Lubelski Wegiel "Bogdanka S.A." (Government, 5%)	Bogdanka, east of Leczna, eastern Poland (1 mine)	
	SILTECH Sp. z o. o.	Upper Silesia (1 mine)	
Brown coal and lignite	Includes:	Of which:	75,000.[e, 2]
	PGE KWB Belchatow S.A. [PGE Polish Energy Group Plc. (Government, 69.29%)]	Belchatow, south of Lodz (2 open pit mines)	
	PGE KWB Turow S.A. (PGE Polish Energy Group Plc. (Government, 69.29%)]	Bogatynia, at the southwest corner of Poland (1 mine)	
	Kopalnia Wegla Brunatnego "Konin" w Kleczewie S.A. (Government, 100%)	Kleczew (4 open pit mines)	
	Kopalnia Wegla Brunatnego "Adamow" S.A. (Government, 100%)	Turek (3 open pit mines)	
	KWB Sieniawa Sp. z o.o.	Sieniawa (1 mine)	
Coke	Includes:	Of which:	9,700.[2]
	Zaklady Koksownicze Zdzieszowice (ArcelorMittal Poland S.A., 100%)	Upper Silesia (Zdzieszowice)	
	Koksownia Przyjazn S.A.	Upper Silesia (Dabrowa Gornicze)	
	Kombinat Koksochemiczny Zabrze S.A.	Upper Silesia (Cokeries at Jadwiga, Radlin, and Debiensko)	
	ArcelorMittal Poland S.A.	Upper Silesia (Krakow)	
	ISD Huta Czestochowa Sp. z o.o.	Upper Silesia (Czestochowa)	
	Zaklady Koksownicze "Victoria" S.A.	Upper Silesia (Walbrzychu)	
	CARBO-KOKS Sp. z.o.o.	Upper Silesia (Bytom)	

See footnotes at end of table.

TABLE 2—Continued
POLAND: STRUCTURE OF THE MINERAL INDUSTRY IN 2012[1]

(Thousand metric tons unless otherwise specified)

Commodity	Major operating companies and major equity owners	Location of main facilities	Annual capacity
Copper:			
Ore, gross weight (averaged 1.61% Cu)	KGHM Polska Miedz S.A. (Government, 31.79%)	Lubin Mine, Lubin-Glogow District	7,000.
Do.	do.	Polkowice-Sieroszowice Mine, Lubin-Glogow District	11,000.
Do.	do.	Rudna Mine, Lubin-Glogow District	12,000.
Concentrate, gross weight (averaged 22.8% Cu)	KGHM Polska Miedz S.A. (Government, 31.79%)	Lubin beneficiation plant, Lubin-Glogow District	800.
Do.	do.	Polkowice beneficiation plant, Lubin-Glogow District	900.
Do.	do.	Rudna beneficiation plant, Lubin-Glogow District	1,500.
Metal, refined	do.	Refineries at Glogow I, Glogow II, and Legnica	540.
Feldspar	Strzeblowskie Kopalnie Surowcow Mineralnych Sp. z o.o.	Sobotka, Lower Silesia, exploiting the Pagorki Zachodnie, Pagorki Wschodnie, and Strzeblow I deposits	500.
Do.	Pol-Skal Sp. z o.o.	Karpniki, southwestern region of Jelenia Gora	100.
Ferroalloys:			
Electric furnace (FeSiMn, FeMn, FeSi)	Huta Laziska S.A.	Upper Silesia at Laziska Gorne	120.[e]
Blast furnace (FeMn)	STALMAG Sp. z o.o.	Upper Silesia at Ruda Slaska	50.[e]
Gold, metal kilograms	KGHM Polska Miedz S.A. (Government, 31.79%)	Refineries at Glogow I, Glogow II, and Legnica	1,000.[e]
Gypsum and anhydrite	Includes:	Of which:	1,400.[2]
	Zaklady Przemyslu Gipsowego "Dolina Nidy" S.A.	Southeastern Poland, Gacki	
	Rigips Polska Stawiany Sp. z o.o.	Southeastern Poland, Szarbkow	
	Kopalnia Gipsu i Anhydrytu "Nowy Lad" Sp. z o.o.	Lower Silesia, mines at Niwnice and Iwiny	
Helium million cubic meters	Polskie Gornictwo Naftowe i Gazownictwo S.A. (PGNiG) (Government, 72.41%)	Western Poland, Odolanow	3.
Kaolin, crude and washed	KSM "Surmin-Kaolin" S.A.	Lower Silesia, Nowogrodziec	90.[e]
Do.	Grudzen Las Sp. z o.o.	Grudzen Las, in Lodz Voivodeship	55.[e]
Do.	Tomaszowskie Kopalnie Surowcow Mineralnych "Biala Gora" Sp. z o.o.	Smardzewice, Tomaszowski Voivodeship	30.[e]
Lead-zinc:			
Mine output	Zaklady Gorniczo-Hutnicze (ZGH) "Boleslaw" S.A. (Stalprodukt S.A., 86.92%)	Mine and concentrator at Olkusz and Pomorzany, Bukowno region	30 lead,[e] 110 zinc.[e]
Metal:			
Pb, refined	Huta Cynku Miasteczko Slaskie (HCM) S.A.	Refinery at Miasteczko Slaskie	35.
Do.	"Baterpol" Sp. z o.o. (Impexmetal S.A.)	Refinery at Katowice	30.
Do.	Orzel Bialy S.A.	Refinery at Bytom	40.
Do.	KGHM Polska Miedz S.A. (Government, 31.79%)	Smelter at Legnica	35.
Zn, refined	Huta Cynku Miasteczko Slaskie (HCM) S.A.	Imperial smelter at Miasteczko Slaskie	85.
Do.	Zaklady Metalurgiczny Silesia S.A.	Refinery at Katowice	12.
Do.	Zaklady Gorniczo-Hutnicze (ZGH) "Boleslaw" S.A. (Stalprodukt S.A., 86.92%)	Refinery at Boleslaw	75.
Lime	Includes:	Of which:	2,200.[e,2]
	Zaklady Przemyslu Wapienniczego (ZPW) Trzuskawica S.A. (CRH plc, 100%)	Plants in Sitkowka-Nowiny and Bielawy	
	Lhoist Group:		
	Lhoist Opolwap S.A.	Tarnow Opolski, Opole County	
	Lhoist Bukowa Sp. z o.o.	Bukowa, 90 kilometers north of Krakow	
	Zaklad Wapienniczy Wojcieszow Sp. z o.o.	Wojcieszow	
	Zaklady Wapiennicze Lhoist Sp. z o.o.	Gorazdze	

See footnotes at end of table.

TABLE 2—Continued
POLAND: STRUCTURE OF THE MINERAL INDUSTRY IN 2012[1]

(Thousand metric tons unless otherwise specified)

Commodity		Major operating companies and major equity owners	Location of main facilities	Annual capacity
Natural gas	million cubic meters	Polskie Gornictwo Naftowe i Gazownictwo S.A. (PGNiG) (Government, 72.41%)	Gasfields in southeastern Poland in the Carpathian Mountains, the Carpathian Foothills, and the Polish Lowlands	4,300.[e]
Do.	do.	FX Energy, Inc.	Western Poland	70.[e]
Do.	do.	LOTOS Petrobaltic S.A. [Grupa LOTOS S.A. (Government, 53.19%)]	Baltic Sea Shelf	30.[e]
Nitrogen:				
Ammonia (NH_3)		Includes:	Of which:	2,600.[e, 2]
		Zaklady Azotowe "Pulawy" S.A.	Pulawy in eastern Poland	
		Zaklady Azotowe "Kedzierzyn" S.A.	Kedzierzyn in Upper Silesia	
		Zaklady Azotowe "Anwil Wloclawek" S.A.	Wloclawek in central Poland	
		Zaklady Azotowe S.A. w Tarnowie	Tarnow in southern Poland	
		Azoty-Adipol S.A. (former Chorzow plant)	Chorzow in Upper Silesia	
		Zaklady Chemiezne "Police"	Police in northwestern Poland	
Petroleum:				
Crude	thousand 42-gallon barrels	Polskie Gornictwo Naftowe i Gazownictwo S.A. (PGNiG) (Government, 72.41%)	Oilfields in southeastern and western Poland with about 75% of production from the Barnowko-Mostno-Buszewo field near Debno	4,000.[e]
Do.	do.	LOTOS Petrobaltic S.A. [Grupa LOTOS S.A. (Government, 53.19%)]	Baltic Sea Shelf	1,500.[e]
Do.	do.	FX Energy, Inc.	Western Poland	70.[e]
Refined	do.	Petrochimia-Plock (PNK Orlen S.A.)	Plock in central Poland	115,000.
Do.	do.	Rafineria "Gdansk" (Grupa LOTOS S.A.)	Gdansk in northern Poland	50,000.
Do.	do.	Rafineria "Trzebinia" (PNK Orlen S.A.)	Trzebinia in southern Poland	3,000.
Do.	do.	Rafineria "Jedlicze" (PNK Orlen S.A.)	Jedlicze in southern Poland	1,000.
Rhenium:				
Rhenium content of ammonium perrhenate	kilograms	KGHM Ecoren S.A. [KGHM Polska Miedz S.A. (Government, 31.79%)]	Lubin	6,000.
Rhenium metal	do.	do.	do.	3,500.
Salt:		Includes:	Of which:	5,000.[e, 2]
Brine		Inowroclawskie Kopalnie Soli Solino S.A.	Mines at Gora and Mogilno in central Poland	
		Polskie Gornictwo Naftowe i Gazownictwo S.A. (PGNiG) (Government, 72.94%)	Mine at Mogilno in central Poland	
		Kopalnia Soli "Wieliczka" S.A.	Wieliczka in southern Poland, near Krakow, mining deposits at Barycz and Wieliczka	
Rock salt		Kopalnia Soli "Klodawa" S.A.	Klodawa in central Poland	
		KGHM Polska Miedz S.A. (Government, 31.79%)	Sieroszowice in southwestern Poland	
Desalination of mine waste water		Zaklad Odsalania Wod Dolowych "Debiensko" Sp. z o.o.	Czerwionka-Leszczyny, west of Debiensko	
Selenium	metric tons	KGHM Polska Miedz S.A. (Government, 31.79%)	Refinery at Glogow	90.
Silver, refined	do.	do.	Precious metals plant at the Glogow smelter	1,400.
Do.	do.	Institute of Non-ferrous Metals	Gliwice	30.[e]
Steel, crude		ArcelorMittal S.A., of which:		8,000.
		ArcelorMittal Poland S.A.	Steelworks at Dobrowa Gornicza (former Huta Katowice S.A.)	
		do.	Steelworks at Krakow (former Huta Sendzimir S.A.)	
		ArcelorMittal Warszawa Sp. z o.o.	Steelworks in Warsaw (former Huta "Lucchini-Warszawa" Sp. z o.o.)	

See footnotes at end of table.

TABLE 2—Continued
POLAND: STRUCTURE OF THE MINERAL INDUSTRY IN 2012[1]

(Thousand metric tons unless otherwise specified)

Commodity	Major operating companies and major equity owners	Location of main facilities	Annual capacity
Steel, crude—Continued	CMC Zawiercie S.A. (Commercial Metals Co.)	Steelworks at Zawiercie	1,900.
Do.	ISD Huta Czestochowa S.A. (Industrial Union of Donbass Corp.)	Steelworks at Czestochowa	1,000.[e]
Do.	Celsa Huta Ostrowiec S.A. (Celsa Group)	Steelworks at Ostrowiec-Swietokrzyski	1,000.[e]
Do.	Ferrostal Labedy Sp. z o.o. (Cognor S.A.)	Steelworks at Gliwice	375.
Do.	Huta Stali Jakosciowych S.A. (Cognor S.A.)	Steelworks at Stalowa Wola	261.
Do.	Huta Batory Sp. z o.o. (Alchemia S.A., 100%)	Steelworks at Chorzow	150.[e]
Sulfur	P.P. Kopalnie i Zaklady Chemiczne Siarki "Siarkopol"	Osiek deposit at Grzybow	800.

[e]Estimated. Do., do. Ditto. NA Not available.

[1]The data presented in this table were compiled, in large measure, from information provided in the Minerals Yeabook of Poland 2010, which was prepared and published by the Department of Mineral and Energy Economy Research Institute of Polish Academy of Sciences.

[2]Annual capacity listed is total for all deposits, mines, or companies that produce the commodity.

The Mineral Industry of Portugal

By Alfredo C. Gurmendi

In 2012, Portugal produced such mineral commodities as lithium (was the fifth-ranked producer after Australia, Chile, China, and Argentina) and tungsten (fifth after China, Russia, Canada, and Bolivia), as well as copper, feldspar, gypsum, kaolin, salt, silver, and talc (table 1; Jaskula, 2013; Shedd, 2013).

Portugal's real gross domestic product (GDP) was $251 billion in 2012 compared with revised $259 billion in 2011. The Portuguese economy had started to come out of the recession of 2008 and 2009, growing by 1.9% in 2010, but the country's GDP again decreased in 2011 (by 1.3%) and in 2012 (by 3.2%) owing to the continued economic crisis in the euro area. Portugal's economy was expected to recover gradually during 2014 through 2017, however, driven primarily by increased exports of goods and services. The sectors that contributed to the country's GDP were services (76.3%), industry (21.3%), and agriculture (2.4%). In 2012, unemployment increased to 15.7% from 12.7% in 2011 (Federation of International Trade Associations, The, 2013; Instituto Nacional de Estatística, 2013; International Monetary Fund, 2013, p. 48, 150; U.S. Central Intelligence Agency, 2013).

Minerals in the National Economy

In 2012, Portugal was one of the European Union's (EU's) leading producers of, in order of amount produced, rock salt, gypsum, kaolin, copper, zinc, tungsten, and silver. A number of gold and base-metal projects were undergoing feasibility studies; most of the activity was focused on the Portuguese zone of the Iberian Pyrite Belt (IPB). The IPB measures 60 kilometers (km) wide by 250 km long and extends from the southwestern coast of Portugal near Setubal to the Guadalquivir River near Seville, Spain. Portugal's mining and minerals activities are controlled by the Instituto Geologico e Mineiro (IGM) (Direcção Geral de Energia e Geologia, 2013; MBendi Information Services (Pty) Ltd., 2013a, b, d, e).

Portugal's mining and mineral processing industries represented about 1% of the GDP in 2012. The mineral sector employed about 32,970, or 0.6% of the labor force total of 5.495 million. Portugal's most valuable metallic mineral resources were copper, silver, tin, tungsten, and zinc. The most valuable resources of industrial minerals besides marble were lithium, pyrites, and rock salt. The country had limited energy resources and depended upon imports to supply the bulk of its energy needs (Direcção Geral de Energia e Geologia, 2013; Instituto Nacional de Estatística, 2013; MBendi Information Services (Pty) Ltd., 2013a, d, e; U.S. Central Intelligence Agency, 2013; U.S. Energy Information Administration, 2013).

Production

Portugal's industrial minerals sector was a producer of a variety of materials, including, in order of the amount produced, limestone, granite, sand, and diorite; the dimension stone and rock salt sectors continued to be important segments of the mineral industry in terms of value and trade. Portugal was one of the leading producers of mined copper, silver, tungsten, and zinc concentrates in the EU and a significant producer of gypsum, kaolin, and lime (table 1; Direcção Geral de Energia e Geologia, 2013; Instituto Nacional de Estatística, 2013).

Structure of the Mineral Industry

Lundin Mining Corp. (Lundin) of Canada owned the Neves-Corvo copper-zinc underground mine, which is located 100 km north of Faro, Portugal, in the western area of the IPB. A study looking at development of the Lombador copper-lead-zinc deposit as well as the Semblana copper deposit was well advanced. Sojitz Beralt Tin & Wolfram (Portugal) SA continued to mine tungsten at its Panasqueira Mine, which is located in Beira Baixa Province in the east-central region of Portugal (Lundin Mining Corp., 2013a, b; MBendi Information Services (Pty) Ltd., 2013d).

Lusosider Aços Planos S.A. and SN Servicos S.A. were Portugal's leading steel producers. Cimentos de Portugal, SGPS, S.A. (CIMPOR) was a regionally significant producer of cement. With the exception of copper, dimension stone, and tungsten, production of other minerals and related materials had only domestic significance. Some of the leading mineral-related companies were partially owned or controlled by the Government, and some operations were privately owned. In 2012, Portugal had only two metallic mines in operation—the Neves-Corvo Mine and the Panasqueira Mine (table 2; Cimentos de Portugal, SGPS, S.A., 2013a, p. 2; Lundin Mining Corp., 2013b; MBendi Information Services (Pty) Ltd., 2013a, d).

Mineral Trade

Portugal's exports amounted to $58.2 billion in 2012 compared with a revised $60.1 billion in 2011 and included such products as, in order of value, chemical products, machinery and tools, crude oil products, base metals, and industrial minerals. Portugal's leading export partners were Spain (22.5%), Germany (12.3%), France (11.9%), Angola (6.6%), the United Kingdom (5.3%), and the Netherlands (4.2%), among others. The main export destination was the EU, the 27 members of which received 46.7% of Portugal's exports. Portugal's imports amounted to $69.5 billion compared with a revised $77.7 billion in 2011 and included such products as, in order of value, machinery and tools, oil products, chemical products, base metals, and mineral products. Portugal's leading import partners were Spain (31.9%), Germany (11.5%), France (6.6%), Italy (5.3%), the Netherlands (4.9%), Angola (3.2%), and the United Kingdom (3.0%), among others. The main import origination point was the EU, the EU's 27 members supplied 72% of Portugal's imports (Federation of International Trade Associations, The, 2013; Instituto Nacional de Estatística, 2013;

U.S. Central Intelligence Agency, 2013; World Trade Organization, 2013).

Commodity Review

Metals

Copper, Gold, Lead, and Zinc.—Production from the Neves-Corvo Mine was 74,043 metric tons (t) of copper content in 2012 compared with 79,686 t in 2011. In 2012, the capacity of the Neves-Corvo's copper plant was about 2.5 million metric tons per year (Mt/yr) of ore and 310,320 metric tons per year (t/yr) of copper concentrate. Neves-Corvo produced 30,008 t of zinc content compared with 4,227 t in 2011. Its zinc plant capacity was about 1.0 Mt/yr of ore and 63,500 t/yr of zinc concentrate (table 1; Lundin Mining Corp., 2013a). According to the mine's owner (Lundin), as of June 30, 2013, Neves-Corvo's copper-rich ores amounted to 27.0 million metric tons (Mt) grading 2.9% copper, 0.8% zinc, 0.2% lead, and 37 grams per metric ton (g/t) silver, and the mine's zinc-rich ores amounted to 23.278 Mt grading 7.4% zinc, 1.8% lead, 0.4% copper, and 70 g/t silver (Instituto Nacional de Estatística, 2013; Lundin Mining Corp., 2013a, b; MBendi Information Services (Pty) Ltd., 2013a).

Lundin was engaging in the acquisition, exploration, development, and mining of base-metal deposits worldwide. Lundin held an interest in the Aljustrel project, which is a potential zinc-lead-silver mine located in the IPB area in southern Portugal, which is known to host numerous multimillion-ton base-metal deposits. The Aljustrel project hosts five known volcanogenic massive sulfide (VMS) deposits; the VMS deposits are a significant source of copper and zinc. The Aljustrel project's final feasibility study estimated total reserves to be 13.8 Mt at average grades of 5.5% zinc, 1.8% lead, and 63 g/t silver (Bloomberg Businessweek, 2013; Direcção Geral de Energia e Geologia, 2013; MBendi Information Services (Pty) Ltd., 2013e).

Tungsten.—Sojitz Beralt Tin & Wolfram's Panasqueira tungsten mine was one of the EU's leading producers of tungsten concentrates. Production from the Panasqueira Mine was 763 t in concentrate (W content) in 2012 compared with a revised 819 t in 2011. The Panasqueira Mine had the capacity to produce 1,000 t/yr of tungsten in concentrate. According to Beralt, the mine had proven and probable reserves of 1.4 Mt grading 0.233% WO_3, additional indicated resources of 3.3 Mt grading 0.263% WO_3, and inferred resources of 1.6 Mt grading 0.224% WO_3. The main end-use application for tungsten was in the manufacture of cemented carbides (60%), steel and alloys (20%), electrical and electronics products (12%), and catalysts and pigments (8%). Despite lower tungsten prices on the world market, production was continuing at the mine because it was producing concentrate on a long-term contract basis. The planned expansion of the facility was postponed because of the decrease in the tungsten market price, however (table 1; Direcção Geral de Energia e Geologia, 2013; MBendi Information Services (Pty) Ltd., 2013b, d).

Industrial Minerals

Cement.—In 2012, Portugal produced about the same estimated amount of cement as in 2011 (7.2 Mt). CIMPOR continued to be Portugal's leading cement producer and the second-ranked cement producer on the Iberian Peninsula after Cemex España S.A. In addition to cement, CIMPOR also produced aggregates, dry mortars, and precast concrete products. In line with the world economy, 2012 was a year of transition for Portugal's cement industry, and its domestic cement sales decreased to 3.4 Mt in 2012 from 3.7 Mt in 2011. The development of Portugal's infrastructure was expected to increase the demand for CIMPOR's products in the foreseeable future to a potential output of 9 Mt/yr (table 1; Cimentos de Portugal, SGPS, S.A., 2013a, p. 6–7; 2013b, p. 2).

Salt.—Rock salt was the leading industrial mineral produced in Portugal. The production of rock salt totaled 520,284 t in 2012 compared with 631,295 t in 2011 (table 1; Direcção Geral de Energia e Geologia, 2013).

Mineral Fuels and Other Sources of Energy

Petroleum, Natural Gas, and Coal.—In 2012, Portugal continued to rely on imported energy resources, such as petroleum imports of about 230,000 barrels per day (bbl/d), natural gas (5.2 billion cubic meters per year), and coal (3.6 Mt/yr). The country's leading domestic energy resource was hydropower, which is an unreliable source of power because it depends on rainfall. The Portugal Government had two crude oil refineries, which were located in the coastal cities of Porto and Sines. Argus Resources Ltd. of the United Kingdom built the petroleum refinery at Sines, which is located 90 km south of Lisbon; the refinery had a production capacity of 250,000 bbl/d and cost about $5 billion to build. Government-owned Petróleos de Portugal (Petrogal) operated the Porto and the Sines refineries, which had a combined capacity of 305,000 bbl/d. Petrogal was planning to invest about $2 billion to upgrade the country's refining processes during 2013 to 2014. The political and legal issues surrounding the EU-Russia energy relationship continued to be under review owing to concerns about the reliability of the energy supply from Russia. Production data for mineral fuels and refined products are shown in table 1 (MBendi Information Services (Pty) Ltd., 2013c; U.S. Energy Information Administration, 2013).

Renewable Energy.—Owing to Portugal's high dependence on imported energy sources, the country was emphasizing solar, wave, and wind power investment. Portugal was planning to invest about $11 billion in renewable energy projects in the near future, of which $2.5 billion would be for building the infrastructure for wind power. The financial crisis and the economic distress in many EU countries, however, were expected to have a negative effect on the wind power industry, and the increasing scarcity of finance could be a challenge to decreasing Portugal's dependence on mineral fuel imports. In 2012, the wind power production capacity in Portugal increased to 4,525 megawatts (MW) from a revised 4,379 MW in 2011, or by about 3.3%. In 2012, the wind power production capacity of the EU-27 countries amounted to 106,040 MW. The leading European countries with wind power installations were Germany (31,308 MW) and Spain (22,796 MW) followed by the United Kingdom (8,445 MW), Italy (8,144 MW), France (7,564 MW), Portugal (4,525 MW), Denmark (4,162 MW), Sweden (3,745 MW), and Poland (2,497 MW) (BP p.l.c., 2013, p. 39; European Wind Energy Association, The, 2013, p. 4–5; U.S. Energy Information Administration, 2013).

Outlook

Portugal continues to be the EU's principal producer of copper, gypsum, lithium, kaolin, rock salt, and tungsten. The structure of the Portuguese mineral industry could change in the near future, however, owing to significant mineral exploration being conducted by several foreign companies, particularly for base and precious metals, kaolin, pyrites, and tin. Feasibility studies for potential precious and base-metal projects were underway in the Portuguese zone of the IPB, which was the prime area for exploration activity. Owing in part to the financial crisis and the economic distress in many EU countries, Portugal is considering increasing investments in alternative energy sources, such as hydropower, solar, wave, wind, and other renewable energy sources to make the country less dependent on imported energy. The Government is also considering making improvements in the efficiency and performance of alternative energy sources by introducing new technologies. If the financial crisis in the EU improves, then Portugal's dependence on fuel mineral imports could be decreased in the medium term (MBendi Information Services (Pty) Ltd., 2013d, e; European Wind Energy Association, The, 2013, p. 5, 11, 13).

References Cited

Bloomberg Businessweek, 2013, Company overview of EuroZinc Mining Corp: Bloomberg Businessweek. (Accessed February 24, 2014, at http://investing.businessweek.com/research/stocks/privatesnapshot.asp?privcapId=882942.)

BP p.l.c., 2013, BP statistical review of world energy: London, United Kingdom, BP p.l.c, June, 48 p. (Accessed September 15, 2013, at http://www.bp.com/content/dam/bp/pdf/statistical-review/statistical_review_of_world_energy_2013.pdf.)

Cimentos de Portugal, SGPS, S.A., 2013a, CIMPOR—Anúncio de resultados de 2012: Cimentos de Portugal, SGPS, S.A., 14 p. (Accessed September 26, 2013, at http://www.cimpor.pt/cache/bin/Imagens/CIMPORAnuncio_de_Resultados_2012-18470.pdf.)

Cimentos de Portugal, SGPS, S.A., 2013b, CIMPOR—Novos ativos no perímetro da CIMPOR, processo de reorganização e permuta de ativos: Cimentos de Portugal, SGPS, S.A., 5 p. (Accessed September 26, 2013, at http://www.cimpor.pt/cache/binImagens/NovosAtivos-18014.pdf.)

Direcção Geral de Energia e Geologia, 2013, Divisão para a Pesquisa e Exploração de Petróleo: Direcção Geral de Energia e Geologia. (Accessed September 24, 2013, at http://www.dgge.pt/.)

European Wind Energy Association, The, 2013, Wind in power—2012 European statistics: Brussels, Belgium, The European Wind Energy Association, February, 14 p. (Accessed September 27, 2013, at http://www.ewea.org/fileadmin/files/library/publications/statistics/Wind_in_power_annual_statistics_2012.pdf.)

Federation of International Trade Associations, The, 2013, Portugal introduction: The Federation of International Trade Associations. (Accessed September 25, 2013, at http://fita.org/countries/portugal.html.)

Instituto Nacional de Estatística, 2013, Estatísticas do comércio internacional 2012: Instituto Nacional de Estatística. (Accessed September 25, 2013, at http://www.ine.pt/xportal/xmain?xpid=INE&xpgid=ine_publicacoes&PUBLICACOESpub_boui=153375654&PUBLICACOESmodo=2.)

International Monetary Fund, 2013, World economic outlook: International Monetary Fund, April, 204 p. (Accessed September 15, 2013, at http://www.imf.org/external/pubs/ft/weo/2013/01/pdf/text.pdf.)

Jaskula, B.W., 2013, Lithium: U.S. Geological Survey Mineral Commodity Summaries 2013, p. 94–95.

Lundin Mining Corp., 2013a, Neves Corvo Mine—Portugal—Copper zinc mine: Lundin Mining Corp., June, 1 p. (Accessed September 26, 2013, at http://www.lundinmining.com/i/pdf/Mine-Summary_Neves-Corvo.pdf.)

Lundin Mining Corp., 2013b, Neves–Corvo, Portugal: Lundin Mining Corp. (Accessed September 26, 2013, at http://www.lundinmining.com/s/Neves-Corvo.asp.)

MBendi Information Services (Pty) Ltd., 2013a, Copper mining in Portugal—Overview: MBendi Information Services (Pty) Ltd. (Accessed September 24, 2013, at http://www.mbendi.com/indy/ming/cppr/eu/po/p0005.htm.)

MBendi Information Services (Pty) Ltd, 2013b, Mining in Portugal—Overview: MBendi Information Services (Pty) Ltd. (Accessed September 24, 2013, at http://www.mbendi.com/indy/ming/eu/po/p0005.htm.)

MBendi Information Services (Pty) Ltd, 2013c, Oil and gas in Portugal—Exploration & production: MBendi Information Services (Pty) Ltd. (Accessed September 24, 2013, at http://www.mbendi.com/indy/oilg/eu/po/p0005.htm.)

MBendi Information Services (Pty) Ltd, 2013d, Panasqueira—Tungsten mine in Portugal: MBendi Information Services (Pty) Ltd. (Accessed September 24, 2013, at http://www.mbendi.com/facility/f15e.htm.)

MBendi Information Services (Pty) Ltd, 2013e, Zinc and lead mining in Portugal—Overview: MBendi Information Services (Pty) Ltd. (Accessed September 24, 2013, at http://www.mbendi.com/indy/ming/ldzc/eu/po/p0005.htm.)

Shedd, K.B., 2013, Tungsten: U.S. Geological Survey Mineral Commodity Summaries 2013, p. 176–177.

U.S. Central Intelligence Agency, 2013, Portugal, in The world factbook: U.S. Central Intelligence Agency. (Accessed September 15, 2013, at https://www.cia.gov/library/publications/the-world-factbook/geos/po.html.)

U.S. Energy Information Administration, 2013, Portugal energy profile: U.S. Energy Information Administration. (Accessed September 15, 2013, at http://www.eia.gov/countries/country-data.cfm?fips=PO.)

World Trade Organization, 2013, Trade profiles, Portugal: World Trade Organization. (Accessed September 26, 2013, at http://stat.wto.org/CountryProfiles/PT_e.htm.)

TABLE 1
PORTUGAL: PRODUCTION OF MINERAL COMMODITIES[1]

(Metric tons unless otherwise specified)

Commodity[2]		2008	2009	2010	2011	2012[p]
METALS						
Aluminum, secondary[e]	thousand metric tons	18	18	18	18	18
Arsenic, white[e]		15	15	15	15	15
Beryl, concentrate, gross weight[e]		5	5	5	5	5
Copper, mine output, Cu content		89,504	86,500	74,426	79,686	74,043
Iron and steel:						
Iron ore and concentrate, manganiferous:[e]						
Gross weight		14,000	14,000	14,000	14,000	14,000
Fe content		10,000	10,000	10,000	10,000	10,000
Metal:[e]						
Pig iron	thousand metric tons	100	100	100	100	100
Steel:						
Crude	do.	1,630 [3]	1,587 [3]	1,351 [3]	1,400	1,400
Hot-rolled	do.	800	800	800	800	800
Lead, refined, secondary[e]		3,000	3,000	3,000	3,000	3,000
Manganese, Mn content of iron ore[e]		300	300	300	300	300
Silver, mine output, Ag content	kilograms	28,800	22,450	23,710	28,380	27,244
Tin, mine output, Sn content		29	34	22	39	42
Tungsten mine output, W content		982	823	799 [r]	819	763
Zinc, mine output, Zn content		39,224	501	6,421	4,227	30,008
INDUSTRIAL MINERALS						
Barite		171	1,078	15	--	--
Calcium carbonate[e]		100,000	100,000	100,000	100,000	100,000
Cement, hydraulic	thousand metric tons	6,650	6,900	7,200 [4]	7,200 [4]	7,200
Clays, kaolin[5]		231,346	274,925	273,890	322,091 [r]	317,489
Feldspar		157,539	157,476 [r]	121,827	114,600 [r]	109,273
Gypsum and anhydrite		372,731	335,189	336,755	337,272	321,988
Lime, hydrated and quicklime[e]		100,000 [r]	70,000 [r]	60,000 [r]	60,000	60,000
Lithium minerals, pegmatite (1.5% Li)		34,888	37,359	40,109	37,534	20,698
Nitrogen, N content of ammonia[e]		244,000	244,000	244,000	244,000	244,000
Pyrite and pyrrhotite, including cuprous, gross weight[e]		8,000	8,000	8,000	8,000	8,000
Salt, rock		606,545	594,578	618,961	631,295	520,284
Sand	thousand metric tons	NA	9,585	7,933	7,209 [r]	NA
Sodium compounds, n.e.s.:[e, 6]						
Soda ash		150,000	150,000	150,000	150,000	150,000
Sulfate		50,000	50,000	50,000	50,000	50,000
Stone:						
Basalt		NA	326,730	240,150	361,414 [r]	NA
Calcareous:						
Dolomite	thousand metric tons	NA	144	257	195 [r]	NA
Limestone, marl, calcite	do.	NA	43,277	33,756	30,477 [r]	NA
Marble	do.	578	572	94	125 [r]	NA
Gabbro	do.	100	100	693	94 [r]	NA
Granite, ornamental	do.	877	934	21,436	21,758 [r]	NA
Graywacke	do.	NA	NA	NA	526 [r]	NA
Quartz	do.	9	35	31	29	38
Quartzite	do.	NA	NA	45	53 [r]	NA
Schist	do.	NA	679	83	70 [r]	NA
Slate	do.	38	20	NA	NA	NA
Sulfur, byproduct, all sources[e]		25,000	25,000	25,000	25,000	25,000
Talc		11,220	11,567	11,981	15,462	15,131
MINERAL FUELS AND RELATED MATERIALS						
Coke, metallurgical[e]	thousand metric tons	300	300	300	300	300
Gas, manufactured[e]	thousand cubic meters	125	125	125	125	125
Petroleum[7]	thousand 42-gallon barrels	2,730	1,728	1,723	1,725	1,730 [e]

See footnotes at end of table.

TABLE 1—Continued
PORTUGAL: PRODUCTION OF MINERAL COMMODITIES[1]

(Metric tons unless otherwise specified)

Commodity[2]		2008	2009	2010	2011	2012[p]
MINERAL FUELS AND RELATED MATERIALS—Continued						
Petroleum refinery products:[e]						
Liquefied petroleum gas	thousand 42-gallon barrels	4,444 [8]	4,450	4,450	4,450	4,450
Gasoline	do.	17,805 [8]	18,000	18,000	18,000	18,000
Kerosene and jet fuel	do.	6,508 [8]	6,500	6,500	6,500	6,500
Distillate fuel oil	do.	34,846 [8]	35,000	35,000	35,000	35,000
Residual fuel oil	do.	19,099 [8]	19,000	19,000	19,000	19,000
Unspecified	do.	15,709 [8]	16,000	16,000	16,000	16,000
Refinery fuel and losses	do.	3,800	3,800	3,800	3,800	3,800
Total	do.	102,211 [8]	103,000	103,000	103,000	103,000

[e]Estimated; estimated data are rounded to no more than three significant digits; may not add to totals shown. [p]Preliminary. [r]Revised. do. Ditto. NA Not available. -- Zero.

[1]Table includes data available through August 21, 2013.

[2]In addition to the commodities listed, Portugal produced refractory clay, crushed granite, ophite, and syenite, but information is inadequate to make reliable estimates of output.

[3]Reported by Worldsteel Association 2011 and 2012.

[4]Reported by Cimentos de Portugal, SGPS, S.A. (CIMPOR).

[5]Includes washed and unwashed kaolin.

[6]Not elsewhere specified.

[7]Reported figure. Source: U.S. Energy Information Administration, 2008 through 2012.

[8]Reported figure.

Source: USGS Minerals Questionnaires, Portugal, 2010 through 2012.

TABLE 2
PORTUGAL: STRUCTURE OF THE MINERAL INDUSTRY IN 2012

(Thousand metric tons unless otherwise specified)

Commodity		Major operating companies and major equity owners	Location of main facilities	Annual capacity
Calcium carbonate		Omya Mineral Portuguesa Lda. (Salmon & Cia Lda.)	Mine and plant at Fatima	100
Cement		Cimentos de Portugal, SGPS, S.A. (CIMPOR) (Government, 100%)	Plants (3) at Alhandra, Loule, and Souselas	12,000
Copper, concentrate		Lundin Mining Corp.	Neves-Corvo Mine near Castro Verde	300
Do.		do.	Lombador Mine near Castro Verde	20
Diatomite		Sociedade Anglo-Portugesa de Diatomite Lda.	Mines at Obidos and Rolica	150
Feldspar		A.J. da Fonseca Lda.	Seixigal Quarry, Chaves	10
Ferroalloys		Electrometalúrgia S.A.R.L.	Plant at Setubal	100
Kaolin		Saibrais Arelas e Caulinos S.A. (Denain Anzin Mineraux S.A.)	Mines at Casal dos Bracais and Mosteiros	175
Petroleum, refined	42-gallon barrels per day	Petróleos de Portugal (Petrogal) (Government, 100%)	Refineries at Porto and Sines	305,000
Do.	do.	Argus Resources Ltd. (private 100%)	Refinery at Sines	250,000
Pyrite		Pirites Alentejanas S.A. (EuroZinc Mining Corp.)	Mine at Aljustrel, plant at Setubal	100
Steel, crude		SN Servicos S.A. (Metalúrgica Galaica S.A., 100%)	Steelworks at Maia and Seixal	600
Do.		Lusosider Aços Planos S.A. (Corus Group, 50%, and Sollac S.A., 50%)	Rolling mill at Seixal	800
Tin		Sojitz Beralt Tin & Wolfram (Portugal) SA	Panasqueira Mine and plant at Barroca	42
Tungsten, concentrate	metric tons	do.	do.	1,000
Uranium	do.	Empresa Nacional de Uranio S.A. (Government, 100%)	Mines at Guargia, plant at Urgeirica	150
Zinc, concentrate	do.	Lundin Mining Corp.	Neves-Corvo Mine near Castro Verde	150,000
Do.	do.	do.	Lombador Mine near Castro Verde	NA
Zinc, refined	do.	RMC Quimigal S.A.R.L.	Electrolytic plant at Barreiro	12

Do., do. Ditto. NA Not available.

The Mineral Industry of Romania

By Alberto Alexander Perez

Romania's mineral production was not significant in terms of world production, and the mineral industry was dependent on imports of mineral ores and concentrates to produce refined metals. In 2012, Romania's gross domestic product (GDP) increased by 0.3% to $277.9 billion. The services sector accounted for 59.5% of the GDP; the agricultural sector, 31.1%; and the industrial sector, 21.1%. The Government did not report the percentage of Romania's GDP that was from mining, quarrying, and metal production activities (U.S. Central Intelligence Agency, 2013).

Minerals in the National Economy

Romania gained European Union (EU) membership on January 1, 2007. As a consequence of its EU membership, much of Romania's metallic ore mining activities for copper, iron ore, lead, and zinc stopped or declined rapidly because the country's mining and mineral processing facilities were not in conformance with EU regulations. To meet EU standards, production facilities would need to be modernized and new facilities built. At the same time, membership in the EU did not increase the possibility in the short term for the mineral industry to benefit from access to the EU market because of the world economic recession, which affected the European and the Romanian mineral industries greatly.

The Economic Ministry of Romania released a report in 2012 titled "Mineral Industry Strategy 2012–2035," which outlines a series of objectives and goals that the Government seeks to achieve. Among them are (a) ensuring the sustainable development of Romania's mineral resources; (b) the harmonization of the national interest and investment capital while meeting the mentioned sustainability requirements; and (c) reducing the dependence on imported primary energy resources and raw minerals and improving the transparency of the mineral industry (Ministerul Economiei Romania, 2012, p. 1).

As economic conditions improve, the Romanian mineral industry could become a significant supplier to the EU mineral commodities market, as Romania was one of the few countries in Europe where metal ore mining was still ongoing. Romania produced principally alumina, primary and secondary aluminum, cement, coal, copper ore, iron ore, lead, steel, and zinc (table 1).

Production

In 2012, the levels of metal mining and processing were mixed; the production of alumina decreased by 13.5% and that of primary and secondary aluminum decreased by 3.8% and 45.4%, respectively. Other metallic mineral commodities for which production decreased compared with that of 2011 included primary refined lead, for which output decreased by 79.2%; silicomanganese, 46.8%; steel (crude), 10.3%; and pig iron, 9.1%. Metallic mineral commodities for which production increased included mined copper, which increased by 23%, and zinc metal, by 36.4%. Notable increases in industrial mineral production included that of feldspar, by 140%; cement, 3%; and lime, 1.7%. The production of sulfuric acid decreased by 83.3%; caustic soda, by 59.3%; sand and gravel, by 49.7%; nitrogen, by 28.1%; gypsum, by 14.9%; and salt, by 12% (table 1).

Structure of the Mineral Industry

Since joining the EU in 2007, Romania had been in transition from a majority Government-owned mineral industry to an industry where the majority of firms are privately owned. Foreign direct investment in the mineral industry was significant; ArcelorMittal of Luxembourg, HeidelbergCement AG of Germany, Holcim Ltd. of Switzerland, Lafarge S.A. of France, and Vimetco N.V. of the Netherlands were some of the major foreign firms investing in Romania's mineral sector.

Alum S.A. (Alum), which was owned by Vimetco, was the only alumina refinery in Romania. The company had completed a modernization of the refinery's manufacturing process in 2009. The refinery's production averaged about 500,000 metric tons per year (t/yr) (Alum S.A., 2014).

Vimetco also owned Alro S.A., which was one of the principal producers of primary aluminum in Europe. Alro produced 252,000 metric tons (t) of primary aluminum in 2012.

Carpatcement Holding S.A., which was owned by HeidelbergCement, was the principal cement manufacturer in Romania; its three plants had an estimated total production capacity of 6.3 million metric tons (Mt). Holcim and Lafarge were other important cement manufacturers in Romania.

The Government-owned copper ore company S.C. Cupru Min S.A. owned the Rosia Poieni Mine, which was the largest copper mine, by production tonnage, in Romania. The company reported that the mine had an estimated capacity of 9 million metric tons per year (Mt/yr) of ore.

Commodity Review

Metals

Bauxite and Alumina and Aluminum.—Alum reported in December that it had invested more than $10 million in modernizing and upgrading its slag dump of Tulcea, which was a project that the company started in 2009. Because of this upgrade, the National Dam Committee of Romania renewed the authorization for the company's mud pond operation until 2016. Alum indicated that the modernization program included performing a partial shutdown of the former dump for red slag and building a system for dry slag deposits. Alum also reported that the company had built a new transport and depositing system for dry stacking of the slag, modernized the wetting system, consolidated all the dikes, and completed the pluvial water channel. The company projected that all construction work for capturing the meteoric waters from upstream of the dump and diverting them by means of a hydro-technical system (in accordance with the design approved by the National Dam Committee) would be completed by the end of the year (Vimetco N.V., 2012; Alum S.A., 2014).

Copper.—In April, the Government of Romania announced that it had backed out of an agreement to sell its leading copper mine to Roman Copper Corp. of Canada. Roman had won a tender to buy the Rosia Poieni Mine in Abrud from S.C. Cupra Min for $262 million in March 2012. The Government reported that after 10 days of talks, however, the two parties could not agree on the terms of the deal, which was part of a privatization plan agreed to by international lenders and designed to raise nearly $2 billion in 2012. According to S.C. Cupru Min, the Rosia Poieni Mine had estimated reserves of 900,000 t of copper, which was about 60% of Romania's copper reserves (Thomson Reuters, 2012).

Gold.—In 2012, Gabriel Resources Ltd. of Canada reported proven reserves of 5.9 million troy ounces [184,000 kilograms (kg)] of gold and 32.6 million troy ounces (1,010,000 kg) of silver, and probable reserves of 4.2 million troy ounces (131,000 kg) of gold and 15.0 million troy ounces (about 467,000 kg) of silver at its Rosia Montana project. The company estimated that the project could produce an annual average of 511,000 troy ounces (15,900 kg) of gold during a 16-year mine life, which would make Romania a significant European gold producer. The company also reported that 62.45% of the people consulted in a referendum vote were in favor of resuming the mining operations at the Rosia Montana project, which had ceased in 2006. The referendum had been initiated by 35 local mayors and was conducted on December 9, 2012. The referendum was advisory in nature and did not have the power to enforce or bind the Government to any particular action (Gabriel Resources Ltd., 2012a, b).

Carpathian Gold Inc. of Canada, which was a junior mine developer, said that it would count about 7.2 million troy ounces (224,000 kg) of gold and 635,000 t of copper in its final prefeasibility study of the Rovina Valley project. Rovina's measured and indicated resources were currently 406 Mt at grades of 0.55 gram per metric ton (g/t) gold and 0.16% copper (Keen, 2012).

Industrial Minerals

Cement.—In 2012, reported cement production in Romania increased compared with that of the previous year, continuing a trend of increased production but still below the 5-year high of 11 Mt in 2008 (table 1). The long-term decrease was possibly owing to weak demand in the domestic construction sector and to weak foreign demand owing to the protracted recession that countries in the region had been subject to since 2009. Romania's cement production capacity had increased in 2010 (the latest year for which data were available) by 2.8 Mt/yr to 16.9 Mt/yr owing to investments made in existing plants. Holcim (Romania) S.A. had increased the capacity of its plants in 2010 by 1 Mt/yr, Carpatcement Holding had increased the capacity of its plant at Bicaz by 1.4 Mt/yr, and Lafarge Ciment S.A. had increased the capacity of its plants by 0.4 Mt/yr (HeidelbergCement AG, 2010, p. 67; Holcim Ltd., 2010, p. 187; Lafarge S.A., 2010, p. 30).

Mineral Fuels

Petroleum.—In July, Romania's Zeta Petroleum plc. announced that a new energy and gas law (law No.123/2012) had been published in the Romanian official Gazette No. 485, dated July 16, 2012. The main objective of this law is to put into Romanian law the provisions of the European Commission's third energy package concerning rules for the internal market in natural gas. The new law provides a calendar for the elimination of regulated prices for end customers. These regulated prices ended on December 1, 2012, for nonhousehold customers and were to end on July 1, 2013, for household clients (Zeta Petroleum plc., 2012).

Outlook

Romania's mineral production will likely increase as the resumption of mining activity and privatizations take place according to the new mineral strategy that the Government has adopted. The level of output will mostly be driven by the international demand for the country's mineral products, particularly demand from EU countries. The modernization of facilities to meet EU standards is likely to be of the highest priority for the mineral industry (Ministerul Economiei Romania, 2012).

References Cited

Alum S.A., 2014, Welcome to Alum S.A.: Alum S.A. (Accessed October 10, 2014, at http://www.alum.ro/en.)

Gabriel Resources Ltd., 2012a, Reserves and resources: Toronto, Ontario, Canada, Gabriel Resources Ltd. (Accessed November 9, 2013, at http://www.gabrielresources.com/site/reserves.aspx.)

Gabriel Resources Ltd., 2012b, Referendum confirmed overwhelmingly in favour of mining in Rosia Montana: Gabriel Resources Ltd. press release, December 12, 2 p. (Accessed August 9, 2013, at http://www.gabrielresources.com/documents/GBU_Referendum_Results_000.pdf.)

HeidelbergCement AG, 2010, Annual report 2009: Heidelberg, Germany, HeidelbergCement AG, 160 p.

Holcim Ltd., 2010, Annual report 2009: Zurich, Switzerland, Holcim Ltd., 202 p.

Keen, Kip, 2012, Carpathian doubles gold-copper resource bound for Rovina prefeasibility study: Mineweb.com, July 17. (Accessed August 13, 2013, at http://www.mineweb.com/mineweb/content/en/mineweb-europe-and-middle-east?oid=155250&sn=Detail.)

Lafarge S.A., 2010, 2009 annual report: Paris, France, Lafarge S.A., 263 p.

Ministerul Economiei Romania, 2012, Strategia Industriei Miniere pentru perioada 2012–2035, Forma actualizata la data de 31.05.2012: Ministerul Economiei Romania, May 31, 31 p. (Accessed August 13, 2013, at http://www.minind.ro/resurse_minerale/Strategia_Industriei_Miniere_2012_2035.pdf.)

Thomson Reuters, 2012, Romania backs out of mine sale, privatization hits snag: Thomson Reuters, April 7. (Accessed August 9, 2013, at http://ca.reuters.com/article/businessNews/idCABRE83605Y20120407?sp=true.)

U.S. Central Intelligence Agency, 2013, Romania, in The world factbook: U.S. Central Intelligence Agency. (Accessed August 6, 2013, at https://www.cia.gov/library/publications/the-world-factbook/geos/ro.html.)

Vimetco N.V., 2012, Alum Tulcea invested USD 10 million to upgrade its slag dump red mud lake: Rotterdam, Netherlands, Vimetco N.V., December 12. (Accessed August 9, 2013, at http://www.alum.ro/en/article/alum-tulcea-invested-usd-10-million-upgrade-its-slag-dump-red-mud-lake.)

Zeta Petroleum plc., 2012, New energy and gas law; liberalises Romanian domestic gas market: Zeta Petroleum plc., July 26, 2 p. (Accessed August 9, 2013, at http://www.zetapetroleum.com/files/files/70_120726_RomanianOfficialGazette_1_3.pdf.)

TABLE 1
ROMANIA: PRODUCTION OF MINERAL COMMODITIES[1]

(Metric tons unless otherwise specified)

Commodity[2]		2008	2009	2010	2011	2012
METALS						
Alumina, calcined, gross weight		--	44,000	450,000 [e]	520,000 [e]	450,000
Aluminum:						
Primary		264,752	201,000	240,000 [e]	262,000 [e]	252,000 [e]
Secondary		12,149	10,544	18,282	23,970 [r]	13,089
Total		276,901	211,544	258,000 [e]	285,000 [r]	265,000 [e]
Copper:						
Mine output, Cu content of concentrate[e]		2,000	1,000	5,000	6,500	8,000
Metal, refined:						
Primary		12,000	3,000	3,000	--	--
Secondary[e]		3,000	1,000	1,000	--	--
Total[e]		15,000	4,000	4,000	--	--
Gold, mine output, Au content[e]	kilograms	400	400	400	--	--
Iron and steel:						
Metal:						
Pig iron	thousand metric tons	2,945	1,575	1,726	1,595	1,450
Ferroalloys, electric furnace:						
Ferrochromium[e]		6,000	15,000	14,000	--	--
Silicomanganese[e]		10,000	--	20,000	31,000	16,500
Total[e]		16,000	15,000	34,000	31,000	16,500
Steel, crude	thousand metric tons	5,035	2,761	3,724	3,811	3,417
Finished products:[e]						
Pipes and tubes	do.	850	450	678	799	789
Rolled products	do.	4,500	2,800	3,762 [3]	4,061 [3]	3,472 [3]
Lead, refined:						
Primary		34,000	9,000	11,000	6,500	1,350
Secondary[e]		5,000	3,000	3,000	3,000	3,000
Total[e]		39,000	12,000	14,000	9,500	4,300
Manganese, Mn content of ore:[e]						
Gross weight	thousand metric tons	50	15	--	--	--
Mn content	do.	10	3	--	1 [r]	--
Silver, mine output, Ag content[e]		18	18	--	--	--
Zinc metal, refined, primary and secondary		62,000 [e]	300	200	220	300
INDUSTRIAL MINERALS						
Cement, hydraulic	thousand metric tons	11,000 [e]	7,800 [e]	7,000	7,846	8,082
Clays:						
Bentonite, marketable		16,643	13,756	20,000 [e]	18,008	17,942
Kaolin, marketable		3,166	1,000 [e]	500	--	--
Feldspar[e]		25,000	14,000	5,500 [r]	2,500	6,000
Fluorspar[e]		15,000	15,000	15,000	--	--
Graphite		-- [e]	20,000 [e]	7,000	--	--
Gypsum	thousand metric tons	885	600	600 [e]	834 [r]	710
Lime[e]	do.	2,000	1,600	1,700	1,679 [3]	1,708 [3]
Nitrogen, N content of ammonia[e]	do.	1,300	40	80	160	115
Salt:[e]						
Rock	thousand metric tons	50	40	40	40	40
Other	do.	2,400	2,000	2,400	2,500	2,200
Total	do.	2,450	2,040	2,440 [r]	2,540 [r]	2,240
Sand and gravel[e]	do.	5,000	3,000	2,700	5,873 [r,3]	2,952 [3]
Sodium compounds:						
Caustic soda[e]	do.	650	300	300	540 [r]	220
Soda ash, manufactured, 100% Na_2CO_3 basis	do.	450 [e]	400 [e]	350	420	430
Sulfuric acid[e]		--	500	383	2,100	350
Talc[e]		1,700	500	307	100	--

See footnotes at end of table.

TABLE 1—Continued
ROMANIA: PRODUCTION OF MINERAL COMMODITIES[1]

(Metric tons unless otherwise specified)

Commodity[2]		2008	2009	2010	2011	2012
MINERAL FUELS AND RELATED MATERIALS						
Coal, brown and lignite[e]	thousand metric tons	35,000	30,000	30,000	35,000	33,500
Coke, metallurgical	do.	1,080	320	--	--	--
Gas, natural, gross:						
Associated	million cubic meters	1,000 [e]	1,000 [e]	1,161	1,166 [r]	1,150
Nonassociated	do.	10,400 [e]	9,860	9,694	9,733 [r]	9,783
Total	do.	11,400 [e]	10,860	10,855	10,901 [r]	10,933
Petroleum:						
Crude[4]	thousand 42-gallon barrels	36,000	33,700	33,000 [e]	31,000 [e]	31,000 [e]
Refinery products[5]	do.	112,000	97,300	95,000 [e]	79,000 [e]	78,000 [e]
Uranium, U_3O_8 content[e]		91	88	88	88	88

[e]Estimated; estimated data are rounded to no more than three significant digits; may not add to totals shown. [r]Revised. do. Ditto. -- Zero.

[1]Table includes data available through August 2, 2013.

[2]In addition to the commodities listed, a variety of construction materials are produced; antimony, asbestos, bismuth, and pyrites may have been produced; and molybdenum may have been produced as a byproduct of copper from 1988 onward; however, available information is inadequate to make reliable estimates of output.

[3]Reported figure.

[4]Figures converted to barrels from production in metric tons, which was reported as the following: 2007—5,086,000; 2008—4,798,000; 2009—4,494,000; 2010—4,490,000 (estimated); 2011—4,000,000 (estimated); and 2012—4,000,000 (estimated).

[5]Figures converted to barrels from production in metric tons, which was reported as the following: 2007—13,648,000; 2008—13,974,000; 2009—12,165,000; 2010—12,000,000 (estimated); 2011—10,000,000 (estimated) and 2012—9,800,000 (estimated).

TABLE 2
ROMANIA: STRUCTURE OF THE MINERAL INDUSTRY IN 2012

(Thousand metric tons unless otherwise specified)

Commodity	Major operating companies and major equity owners	Location of main facilities	Annual capacity
Alumina	Alum S.A. (Vimetco N.V., 99.4%)	Plant at Tulcea, Danube Delta	600
Aluminum:			
Primary	Alro S.A. (Vimetco N.V., 88%)	Slatina, 120 kilometers west of Bucharest	265
Secondary	Neferal S.A. (member of Metanef Group)	Bucharest	NA
Cement	Holcim (Romania) S.A. (Holcim Ltd., 100%)	Plants at Alesd, Campulung, and Turda	5,700
Do.	Carpatcement Holding S.A. (HeidelbergCement AG, 99%)	Plant at Bicaz, northeastern Romania	3,000
Do.	do.	Plant at Fieni, 90 kilometers northwest of Bucharest	1,650
Do.	do.	Plant at Deva, western Romania	1,650
Do.	Lafarge Ciment S.A. (Lafarge S.A., 99%)	Plants at Hoghiz and Medgidia	4,900
Coal:			
Bituminous	Compania Nationala a Huilei-Petrosani (Government)	7 mines located near Petrosani	3,500 [e]
Lignite	Societatea National a Lignitului Oltenia (Government)	Tismana I-II, Garla-Rovinari Est, and Pinoasa opencast mines at Rovinari	8,000 [e]
Do.	do.	Rosia, Pesteana Nord, and Pesteana Sud-Udari opencast mines at Balteni	6,900 [e]
Do.	do.	Udari underground mine at Udari	300 [e]
Do.	do.	Jilt Sud and Jilt Nord opencast mines at Matasari	7,000 [e]
Do.	do.	Dragotesti underground mine at Matasari	600 [e]
Do.	do.	Lupoaia and Rosiuta opencast mines at Motru	5,000 [e]
Do.	do.	Plostina, Horasti, and Lupoaia underground mines at Motru	1,600 [e]
Do.	do.	Seciuri, Oltet, Berbesti-Vest, and Panga opencast mines near Berbesti	2,000 [e]
Do.	do.	Albeni underground mine at Bolbocesti	555 [e]
Do.	do.	Husnicioara-Vest opencast mine near Drobeta Turnu Severin	2,500 [e]
Do.	do.	Zegujani underground mine about 18 kilometers northeast of Drobeta Turnu Severin	600 [e]
Do.	Societatea National a Carbunelui Ploiesti (Government)	8 mines located near Campulung, Baraolt City, Sarmasag, Popesti Commune, Comanesti Commune, Filipestii de Padure Commune, Sotanga Commune, and Borsec City	3,000 [e]
Do.	SC Complexul Energetic Craiova SA (Ministry of Economy, 73%)	Prigaria Mine	1,000 [e]
Coke	ArcelorMittal Galati (ArcelorMittal, 100%)	Galati, north of Brail	2,100
Copper, ore	S.C. Cupru Min S.A., REMIN S.A., Compania Nationala Minvest, and Moldomin S.A. (Government)	Borsa Balan, Rosia Poieni Mine; Moldova Noua	9,000
Ferroalloys, ferrochromium	S.C. Feral s.r.l.	Complex at Tulcea	NA
Iron ore	Compania Nationala Minvest (Government)	Mining complex at Hunedoara, west-central Romania	1,320
Do.	do.	Resita Mining Complex, southwestern Romania	660
Do.	do.	Napoca-Cluj Mining Complex, northwestern Romania on the Somesul River	990
Lead:			
In ore	Compania Nationala REMIN S.A. (Government)	Baia Mare Mine, near Ukrainian and Hungarian borders	12
Do.	Compania Nationala Minvest (Government)	Vetel Mine, near Deva	5
Metal	Neferal S.A. (member of Metanef Group)	Bucharest	25

See footnotes at end of table.

TABLE 2—Continued
ROMANIA: STRUCTURE OF THE MINERAL INDUSTRY IN 2012

(Thousand metric tons unless otherwise specified)

Commodity		Major operating companies and major equity owners	Location of main facilities	Annual capacity
Natural gas	million cubic meters	SNGN Romgaz S.A. (Romanian Ministry of Economy, 85.01%, and Fondul Proprietatea, 14.99%)	Operated more than 150 reservoirs in Moldova, Muntenia, and Transylvania regions	NA
Do.	do.	S.C. OMV Petrom S.A. [OMV AG, 51.01%; Ministry of Economy, 20.64%; Property Fund S.A. (Government), 20.11%]	Approximately 250 commercial oil and gas fields located in southern and western Romania and offshore in the Black Sea	NA
Petroleum:				
Crude		do.	do.	NA
Refined		do.	Arpechim refinery, just south of Ploiesti	3,500
Do.		do.	Petrobrazi refinery, just south of Ploiesti	4,500
Do.		Rompetrol Rafinarie S.A. (Rompetrol Group)	Refinery at Navodari	4,500
Do.		Vega Ploiesti Refinery (Rompetrol Group)	Refinery, just north of Ploiesti	500
Do.		S.C. RAFO S.A.	Refinery at Onesti	3,000 [e]
Do.		S.C. Petrotel S.A. (OAO Lukoil)	Refinery, just east of Ploiesti	2,400
Do.		Astra Refinery	Refinery in Ploiesti	NA
Do.		Refinaria Petrolsub Suplacu de Barcau	Refinery at Suplacu de Barcau	NA
Do.		Darmanesti refinery	Refinery at Darmanesti, eastern Romania	NA
Steel		ArcelorMittal Galati (ArcelorMittal, 100%)	Galati, north of Brail	4,500 [e]
Do.		ArcelorMittal Hunedoara (ArcelorMittal, 100%)	Hunedoara, west-central Romania, near Calan	NA
Do.		S.C. Donasid S.A. (Tenaris S.A., 99%)	Calarasi, southeastern Romania	470
Do.		S.C. TMK-Resita S.A. (OAO TMK, 100%)	Resita, southwestern Romania	450
Do.		Mechel Targoviste (OAO Mechel, 87%)	Targoviste	630 [e]
Do.		Mechel Campia Turzii SA (OAO Mechel, 87%)	Campia Turzii, northwestern Romania	300 [e]
Do.		Ductil Steel (OAO Mechel, 100%)	Otelu Rosu, southwestern Romania	400 [e]
Uranium		Compania Nationala a Uraniului (Government)	Suceava County	NA
Zinc, ore		Compania Nationala REMIN S.A. (Government)	Vetel Mine, near Deva	45

[e]Estimated. Do. do. Ditto. NA Not available.

THE MINERAL INDUSTRY OF RUSSIA

By Elena Safirova

In 2012, Russia was ranked among the world's leading producers or was a leading regional producer of such mineral commodities as aluminum, antimony, arsenic, asbestos, barite, bauxite, boron, cadmium, cement, coal, cobalt, copper, diamond, diatomite, fluorspar, gallium, germanium, gold, graphite, gypsum, indium, iodine, iron ore, lead, lime, magnesium compounds and metals, mica (flake, scrap, and sheet), molybdenum, natural gas, nickel, nitrogen, palladium, peat, petroleum, phosphate rock, pig iron, platinum, potash, rhenium, selenium, silicon, silver, steel, sulfur, tellurium, titanium sponge, tungsten, uranium, vanadium, and vermiculite (Apodaca, 2013a–c; Bedinger, 2013; BP p.l.c., 2013; Bray, 2013a, b; Carlin, 2013; Corathers, 2013; Crangle, 2013a–c; Edelstein, 2013a, b; Fenton, 2013; George, 2013a–d; Guberman, 2013a, b; Jasinski, 2013a, b; Jaskula, 2013; Kramer, 2013a, b; Kuck, 2013; Loferski, 2013; Miller, 2013a–c; Olson, 2013a–c; Polyak, 2013a–d; Shedd, 2013a, b; Tanner, 2013; Tolcin, 2013a, b; Tuck, 2013; van Oss, 2013; Virta, 2013; Willett, 2013).

Minerals in the National Economy

In 2012, the growth rate of the real gross domestic product (GDP) of Russia was 3.4%, which was a decrease compared with the 4.3% GDP growth rate in 2011; in 2012, the nominal GDP increased to 62,599 billion rubles ($1,931 billion[1]). Industrial production contributed 36.0% to the total GDP, and the industrial sector accounted for 27.4% of the country's overall employment. Mining and quarrying contributed 5.8 billion rubles ($179 million), or 10.9%, to the total value added in the economy in 2012, and the total value of output from mining and quarrying in current prices was 8.81 billion rubles ($272 million), or 14.1% of the GDP (Federal'naya Sluzhba Gosudarstvennoy Statistiki, 2013; U.S. Central Intelligence Agency, 2013).

During 2012, the value of mining and quarrying production increased by 1.1%. Mining and quarrying of fuel and energy products increased at a faster rate (by 1.2%) than did the mining and quarrying of nonenergy minerals, which increased at a rate of 0.9%. Among the nonmining industrial sectors, the value of metallurgy and the production of finished metal products increased by 4.5%; the value of the production of coke and petroleum products increased by 2.2%; that of chemicals, by 1.3%; and that of other nonmetal mineral products, 5.6%. In 2012, 14.3% of all investment in the economy was directed to the mining and quarrying industry, of which 89.5% was invested in the mining and quarrying of fuel and energy products (Federal'naya Sluzhba Gosudarstvennoy Statistiki, 2013).

Government Policies and Programs

In 2012, the Government of Russia used legislative and other methods to stimulate sectors of the economy that were considered particularly important for the Russian economy. In November, the State Duma (Parliament) had adopted a law that reduced export tariff rates by 10% for the companies that produce hard-to-recover petroleum. The reduced rates would apply to ultra-heavy oil and bituminous oil as well as to petroleum produced from deposits that are either completely or partially located within the borders of Sakha Republic (Yakutiya), Irkutskaya Oblast', Krasnoyarskiy Kray, and the Nenetskiy Avtonomnyy Okrug; on the Yamal Peninsula; on the Russian sector of Caspian Sea; within the borders of an internal sea; or on the continental shelf of the Russian Federation. The law went into effect on January 1, 2013. By March 1, 2013, the Government was expected to issue specific regulations and rules related to new tariff rates. In particular, the Government was expected to establish monitoring of the average petroleum prices on the world markets. The Ministry of Economic Development estimated that the reduced tariff regime for the hard-to-recover crude oil would increase petroleum production in Russia by between 40 and 100 million metric tons per year (Mt/yr) (Mineral.ru, 2012f).

In August, the Government changed the criteria that determine whether a particular natural resource is considered strategic; this change in policy has implications for foreign companies, which face restrictions with respect to investment in Russian strategic resources. In the case of gold mining, a deposit was previously considered strategic if its resources exceeded 50 metric tons (t). According to the new definition, a gold deposit is considered strategic if its resources exceed 250 t. The analysts expected this change to attract new foreign companies to Russia. The new rules are also likely to be of benefit to Russian companies with foreign investors, such as OAO Polymetal and OAO Polyus Zoloto. Similar rules for other strategic resources were also relaxed. Those new rules allow the share of foreign capital in Russian enterprises to reach 25% instead of the previous 10% (Mineral.ru, 2012i; MinerJob.ru, 2012v).

In July, the President of Russia brought a new bill to the State Duma for ratification. The bill states that for all resources of Federal importance, development and mining licenses can be issued only through an auction. In the previous version of the Mining Code, licenses for such resources could be issued as a result of an auction or a tender. The auctions, however, have at least two advantages compared with tenders. First, they are more transparent and are less likely to be disputed, and, second, they are likely to increase the Government revenue obtained for the license. At the same time, compared with auctions, tenders may better meet the multifaceted objectives that a Government agency might have when issuing a mining license. For example, in addition to the bidding price, projects may differ according to the particular features of the proposed

---

[1]Where necessary, values have been converted from Russian rubles (RUB) to U.S. dollars (US$) at an annual average exchange rate of RUB32.41=US$1.00 for 2012 and RUB30.63=US$1.00 for 2011. All values are nominal, at current prices, unless otherwise stated.

mine construction, the company's contribution to environmental goals or community development, and so forth. The introduction of the new bill appears to have been at least partially a reaction by the Government to the outcomes of two high-profile tenders for nickel-containing resources, both of which resulted in legal disputes brought about by the party that lost the tender. At yearend, it was not clear if the auction amendment would be ratified (Mineral.ru, 2012w).

Production

In 2012, Russia's production of mineral commodities was largely stable and demonstrated modest growth compared with that of 2011. Production of diatomite increased by 112%; that of titanium sponge, by 71%; tin ore, by an estimated 33%; boron, by an estimated 25%; vanadium metal, by 15.5%; ferronickel (high nickel), by 15.3%; antimony, by 15%; anthracite coal, by 14%; phosphate rock, by an estimated 13.6%; ferromanganese, by 13%; bituminous coal, by 11%; and platinum, by 10.6%. At the same time, production of fluorspar decreased by 61%; that of bismuth, by 20%; and potash, by 14.4%. Production data for these and other mineral commodities are in table 1.

Structure of the Mineral Industry

At the end of 2012, Russia had 17,300 enterprises engaged in mining and quarrying, which was a 0.58% increase compared with the number of enterprises active in mining and quarrying in 2011. Of these enterprises, 7,100 were engaged in extracting fuel minerals and the other 10,200 were engaged in mining nonfuel minerals. Out of all mining and quarrying enterprises, only about 200 were owned by the central and municipal governments, 15,300 were owned by Russian citizens, and about 400 were either owned by foreign companies or were jointly owned by domestic and foreign entities. Data on the capacity and ownership of selected mineral operations are in table 2 (Federal'naya Sluzhba Gosudarstvennoy Statistiki, 2013).

Mineral Trade

In 2012, the total value of Russian exports was $529.1 billion, which was a 2.7% increase compared with the revised value of exports in 2011. The value of Russian imports also increased in 2012, to $335.8 billion, or by 5.4%. In 2012, Russia had a positive trade balance of $193.3 billion (Federal'naya Sluzhba Gosudarstvennoy Statistiki, 2013).

The main export categories for Russia were chemicals, manufactured goods, metals, natural gas, petroleum and petroleum products, and wood and wood products. Mineral products made up 71.4% of the total value of Russian exports, and crude oil alone contributed 34.4% to the total value of exports. Petroleum refinery products accounted for another 19.7%; natural gas, 11.8%; and ferrous metals, 4.3%. Among ferrous metals and products made out of them, the leading categories were semifinished products made from carbon steel (34.8%) and flat-rolled iron and steel (25.9%). Other mineral products that contributed significant amounts to Russia's export revenue were bituminous coal (2.5%), aluminum (1.2%), complex mineral fertilizers (0.76%), nickel (0.71%), nitrogen fertilizers (0.69%), and ferrous ores and concentrates (0.47%).

The major export partners of Russia in 2012 were the Netherlands (which received 14.6% of Russia's exports), China and Germany (6.8% each), Italy (6.2%), Ukraine (5.2%), Turkey (5.2%), Belarus (4.7%), Poland (3.8%), and the United States (2.5%) (Federal'naya Sluzhba Gosudarstvennoy Statistiki, 2013).

In 2011, Russia imported $7,629 million worth of products made of ferrous metals (which constituted 2.3% of total imports) and $6,403 million worth of ferrous metals (1.9%). The major import partners of Russia were China (which supplied 16.5% of Russia's imports), Germany (12.2%), Ukraine (5.7%), Japan (5.0%), the United States (4.9%), France (4.1%), Italy (4.0%), Belarus (3.9%), and the Republic of Korea (3.3%) (Federal'naya Sluzhba Gosudarstvennoy Statistiki, 2013).

Commodity Review

Metals

Antimony.—In 2012, Russia produced an estimated 7,300 t of antimony in concentrate, which was a 15% increase compared with the production level in 2011. Most antimony in the country was mined by GeoProMining, Ltd. (GPM). GPM was a private mining company and operated mines in Russia and Armenia. In Russia, GPM was mining and processing gold and antimony at its two gold-antimony mines (Sarylakh and Sentachan) and was processing the ore at the Sarylakh processing plant located in the Yakutsk region. The operations were located in Oimyakon Ulus—the coldest place in Russia—and GPM was running seasonal production because of the lack of reliable infrastructure. The mine was in operation from November through April when the winter road was established and maintained along the frozen rivers. During spring and summer, the Sentachan Mine was accessible only by helicopter. GPM held two exploration licenses in Yakutiya and was planning to restructure its production to eventually move away from the seasonal scheme of production (MinerJob.ru, 2012b, c).

Another antimony producer in Russia, the Zabaykal'skiy mining and beneficiation complex (GOK) (ZabGOK) was located in Zabaykal'skiy Kray. In the beginning of 2012, the company announced that the beneficiation plant was expected to restart operations sometime during the year. The plant's capacity was estimated to be 360,000 metric tons per year (t/yr) of antimony ore. ZabGOK had a stockpile of 60,000 t of antimony ore and was expecting to receive more ore from the mine at the Bulykta-Solntsevskoye deposit, which was located in Sakhalinskaya Oblast'. OOO NefteChimMash, which was majority owner of ZabGOK, was intending to create the largest antimony enterprise in Russia by 2015; the enterprise would have the capacity to produce 15,000 t/yr of flotation antimony concentrate and 5,000 t/yr of antimony metal. Until the metal plant is built, the antimony concentrate would continue to be shipped to the Kadamjay Antimony Complex in Kyrgyzstan. In addition to antimony, ZabGOK was planning to produce beryllium, lithium, niobium, rare-earth metals (REMs), and tantalum (Mineral.ru, 2012j).

Bauxite and Alumina and Aluminum.—All Russian production of alumina, bauxite, and primary aluminum was controlled by United Company RUSAL (RUSAL), which was the world's leading producer of alumina and aluminum.

RUSAL operated 15 aluminum smelters, which were located in four countries: Russia (12 plants), Nigeria (1 plant), Sweden (1 plant), and Ukraine (1 plant).[2] Globally, RUSAL operated 11 alumina refineries, 8 bauxite mines, 4 plants for producing aluminum foil, 3 plants for producing aluminum powder, and 2 plants for producing secondary aluminum. In 2012, RUSAL produced 4.17 Mt of aluminum, 7.48 Mt of alumina, and 12.37 Mt of bauxite at its facilities worldwide (MinerJob.ru, 2012m; United Company RUSAL, 2013a–c).

In Russia, RUSAL had 12 aluminum plants in operation. The leading 6 of the 12 were Krasnoyarskiy AZ, which produced 1.00 Mt of primary aluminum; Bratskiy AZ (995,000 t); Sayanogorskiy AZ (541,000 t); Irkutskiy AZ (413,000 t); Khakasskiy AZ (295,000 t); and Novokuznetskiy AZ (291,000 t). In addition, RUSAL was planning to open two more aluminum plants—Boguchanskiy AZ in Krasnoyarskiy Kray, which would have an annual capacity of 600,000 t/yr of aluminum, and Tayshetskiy AZ in Irkutskaya Oblast', which would have an annual capacity of 750,000 t/yr. Boguchanskiy AZ was expected to be commissioned in 2013 and to reach full production capacity in 2014. The opening of the Tayshetskiy AZ plant was delayed from the initial target of 2013 to sometime between 2014 and 2015 because of the low price of aluminum on the world market. By the time of completion, the total investment in Tayshetskiy AZ was expected to reach about $1.77 billion (Mineral.ru, 2012o, r).

Because of the low aluminum prices in 2012, RUSAL devoted much effort to cutting costs and reducing production. The general plan adopted by RUSAL was to move aluminum production, which requires large amounts of electricity, to the company's Eastern Division, where hydropower plants produce cheap energy, from the Western Division, where energy is more expensive. In August, the company decided to reduce primary aluminum production at Bogoslovskiy AZ, Nadvoitskiy AZ, Novokuznetskiy AZ, and Volkhovskiy AZ by a combined 275,000 t/yr before 2018. The first 150,000-t/yr reduction in output was expected to be achieved in 2012, mostly by decreasing production at Bogoslovskiy AZ, where high electricity prices made that plant's electrolysis-based production inefficient. Bogoslovskiy AZ was expected to continue to produce alumina for use in the Siberian plants, however. The Government of Sverdlovskaya Oblast' and RUSAL agreed that RUSAL would continue to invest in energy-efficient technologies and would not fire workers; however, workers at the Bogoslovskiy AZ who did not think that the plant had a future started quitting their jobs in October and began looking for other opportunities (Mineral.ru, 2012d, v; MinerJob.ru, 2012p).

Another strategy implemented by RUSAL was to increase the production of aluminum ferroalloys and flat ingots, instead of aluminum metal, at the Eastern Division plants. Flat ingots were in high demand by the packaging materials industry and the automobile industry. In 2012, RUSAL invested a total of $5 million into technological changes related to increasing its production of ferroalloys (Mineral.ru, 2012h).

Cobalt.—In 2012, Russia mined 6,300 t of cobalt, which was a 3.3% increase compared with the output of mined cobalt in 2011. In January, OJSC MMC Norilsk Nickel (Nornickel) announced that it was planning to invest about 2 billion rubles ($61.7 million) into its own production of electrolytic cobalt, which would enable it to export cobalt metal rather than just mined cobalt concentrate. Kolskaya Gorno-Metallurgicheskaya Kompaniya (Kolskaya GMK), which was a part of Nornickel, previously had facilities in place to produce cobalt concentrate, but the concentrate requires additional processing to produce cobalt metal. Kolskaya GMK had had in place an experimental-scale production line of processed cobalt since 2007. With the construction of the new production line, Nornickel would have the facilities in place to perform the full cycle of mining and processing of cobalt. The new cobalt plant was expected to be constructed at the Monchegorskaya Industrial Area and was to be commissioned in 2014 (MinerJob.ru, 2012h).

Copper.—In 2012, Russia's production of copper in concentrate increased by 3.1% to 883,000 t compared with the 2011 production level, but the output of refined copper decreased by 4.4% to 844,400 t. Russia had three leading vertically integrated copper-producing companies—Nornickel, OAO Ural'skaya Gorno-Metallurgicheskaya Kompaniya (UGMK), and ZAO Russkaya Mednaya Kompaniya (RMK). In 2012, UGMK was the leading producer of copper, with output of 389,900 t (a 1.9% increase compared with its production in 2011). UGMK was in the process of reconstructing some of its facilities and was expecting to increase its production to between 450,000 and 500,000 t/yr by 2018. Nornickel, on the other hand, decreased its production in 2012 by 4% to 363,800 t (RIA-Analitika, 2013, p. 28–29).

RMK was building two new plants in Chelyabinskaya Oblast' in the South Urals—the Tominskiy GOK and the Mikheevskiy GOK. The Tominskiy GOK had an initial planned capacity of 14 Mt/yr of copper ore, which was expected to increase to 28 Mt/yr by the end of the second stage of mine development. RMK was planning to invest a total of 22.2 billion rubles (about $685 million) in the new plant and to produce copper concentrate and copper cathodes. The Tominskiy GOK was expected to be commissioned in December 2014. The Mikheevskiy GOK was expected to become the largest newly constructed mining plant in the country in the post-Soviet times. Its projected capacity was 18 Mt/yr of copper ore, and the total budgeted investment was $787 million. The Mikheevskoye deposit was extensively studied between 1984 and 1987, and its resources were estimated to be 400 Mt of ore; in addition to copper, the deposit contains gold, molybdenum, rhenium, and silver. The Mikheevskiy GOK was expected to be commissioned in late 2013. Together, the Mikheevskiy GOK and the Tominskiy GOK were expected to provide RMK with mined copper for 25 years (MinerJob.ru, 2012g, n, u).

Metalloinvest Holding and the State Corporation Gostecknologii continued to prepare the Udokan deposit in Zabaykal'skiy Kray for mining. The Udokan deposit is one of the largest copper deposits in the world. According to the Russian-style estimates—which use geologic data, but do not account for the profitability of extraction—the deposit's resources are 1,375 Mt

---

[2]As of 2012, United Company RUSAL was engaged in a dispute with the Government of Ukraine about the ownership of the Zaporozhye plant.

grading 1.56% copper and 9.6 grams per metric ton (g/t) silver. According to the 2010 Joint Ore Reserves Committee (JORC) Code estimates, the probable resources for open pit mining were evaluated to be 795 Mt of ore with an average grade of 1.24% copper, which corresponds to 9.86 Mt of copper. The sum of the measured, indicated, and inferred resources was estimated to be 2,700 Mt grading 0.95% copper with a total copper content of 25.7 Mt. According to the license rules, in 2012 and 2013, the companies were expected to build the mine and the mine infrastructure and, in 2014, to reach the target of mining 12 Mt/yr of ore. Then, under the license rules, the new GOK must reach its projected mining capacity of 36 Mt/yr of ore, production of 474,000 t/yr of copper cathodes, and production of 277 t/yr of silver by 2016. According to different estimates, the total project cost was between $5 billion and $8 billion. The Udokan deposit was thought to be the third largest undeveloped copper deposit in the world after the Oyu Tolgoi deposit in Mongolia and the Pebble deposit in the United States (Baikal Mining Co. LLC, 2011; Mineral.ru, 2012b; MinerJob.ru, 2012k, l).

Gold.—In 2012, gold production in Russia increased by 8.3% to a record of 226,300 kilograms (kg); the gold produced from mine output increased to 217,800 kg, or by about 9.1%, compared with that of 2011. Russia was the fourth-ranked gold producer in the world after China, Australia, and the United States. According to the Union of Gold Producers (a Russian trade group), the country would continue to increase gold production by, on average, between 3% and 5% per year through 2020. As of 2012, Russia had 26 large gold mining companies, which together produced about 80% of all gold; the rest of gold production in Russia was performed by about 400 gold mining companies (Mineral.ru, 2012e; 2013c).

The top 10 gold mining companies in Russia remained the same in 2012 as in 2011. The leading gold producer in Russia, OAO Polyus Gold, produced 48.8 t, which was an increase of 13.9% compared with its output in 2011. Petropavlovsk plc produced 22,100 kg, which was an increase of 13.0% compared with its output in 2011, and Polymetal increased its output by 39.0% to 15,200 kg. ZAO Chukotskaya Mining and Geological Co. (Chukotskaya GGK), which was a part of Kinross Gold Corp. of Canada, decreased its production by 7.8% to 14,500 kg. The decrease was related to reduced gold content in the ores of the Kupol deposit, which the company was mining. It was expected that the Chukotskaya GGK would increase its overall production in the future as new mines at four deposits—the Dvoynoye, the Kekura, the Mayskoye, and the Valunistoye—become operational. Other gold companies in the top 10 included Nordgold N.V. (the gold-mining arm of Severstal of Russia), which produced 10,200 kg of gold; Highland Gold Mining Ltd. (HGM) (6,700 kg); OOO Yuzhuralzoloto (6,500 kg); OOO Vysochayshiy (GV Gold) (5,200 kg); OOO Sovrudnik (3,900 kg); and OAO Susumanzoloto (3,700 kg). The 11th-ranked producer, OAO Seligdar, produced 2,980 kg in 2012 and was planning to become one of the top 10 gold producers in the country by 2015. Seligdar, which operated eight mines and plants in Altayskiy Kray, Buryatiya, and the Sakha Republic (Yakutiya), was opening a new heap-leaching plant at the Podgolechnoye deposit in Sakha Republic (Yakutiya).

Overall, Seligdar had mining licenses for 17 hard-rock deposits and 2 alluvial deposits, of which 7 were mined in 2012 (MinerJob.ru, 2012d).

In 2012, gold was mined in 24 regions of the country, and 14 of these regions produced more than 1 t of gold. The leading producing regions were Krasnoyarskiy Kray (44,000 kg), Amurskaya Oblast' (29,270 kg), Sakha Republic (Yakutiya) (21,220 kg), Magadanskaya Oblast' (19,660 kg), Irkutskaya Oblast' (19,000 kg), Chukotskiy Avtonomnyy Okrug (17,980 kg), and Khabarovskiy Kray (13,000 kg) (RIA-Analitika, 2013, p. 11–15).

By the end of 2013, two new mining and beneficiation complexes (GOKs) in Magadanskaya Oblast' were expected to become operational. The two deposits were located within 12 km of each other and both had large resources. The Natalkinskiy GOK was being built at the Natalka deposit, which was thought to be the third largest gold deposit in the world after the Grasberg deposit in Indonesia and the Muruntau deposit in Uzbekistan (Expert.ru, 2014). Its proven and probable reserves were evaluated to be 1,270 t of gold in a deposit grading 1.13 g/t. Polyus Gold (the leading gold producer in Russia) was building a GOK at which the initial annual ore-processing capacity was projected to be 10 Mt/yr, and the gold production capacity was expected to reach 15 t/yr. By 2020, when the GOK reaches its full projected capacity, the plant would process 40 Mt/yr of ore and produce 50 t/yr of gold. To implement the second stage of the GOK's construction, however, Magadanskaya Oblast' needed to commission a new hydropowerplant, Ust-Srednekanskaya GES, which would be owned and operated by OAO RusHydro. Polyus Gold was working with the Magadanskaya Oblast' administration to speed up the powerplant's construction (MinerJob.ru, 2012z).

Another GOK was being built at the Pavlik deposit. According to the Russian-style resource measurement system, as of January 2010, Pavlik's resources in the C1+C2 categories had a grade of 2.5 g/t gold and contained 100.2 t of gold. According to JORC Code methodology, the deposit's reserves were evaluated to be 150.9 t of contained gold. The initial capacity of the new GOK would be 3 Mt/yr of ore and between 5 and 6 t/yr of gold. It was expected that, at full capacity, the GOK would be producing 40 t/yr of gold. By 2014, the new GOK was projected to employ 600 workers (MinerJob.ru, 2012j).

Nickel.—Russia was the world's second-ranked nickel mining country in 2012 after the Philippines, and it produced 11.5% of the world's mined nickel. Nornickel was the country's leading nickel producer and the world's leading nickel mining company. Other significant nickel producers in Russia included OAO Ufaleynickel and OAO Yuzhuralnickel (Kuck, 2013; OJSC MMC Norilsk Nickel, 2013).

Nornickel's operations in Russia were located on the Kola Peninsula in the northwest of the country and in the Norilsk region on the Taymyr Peninsula in Eastern Siberia. Nornickel also owned assets in Australia, Botswana, Finland, and South Africa. In 2012, Nornickel invested 17.7 billion rubles ($546 million) in production development and was planning to invest a total of 120 billion rubles ($3.7 billion) through 2016. In 2012, the company planned to increase the amount of ore mined in its Zapolyarnyi division to 18 Mt from 16.5 Mt by 2016; however, it was not clear if Nornickel would be able to realize those plans. In 2012, because of lower prices on the world market, the company was considering reducing its investments or even halting the processing of ores mined on the Kola Peninsula by 2015 if the nickel prices remain low (Mineral.ru, 2012m; MinerJob.ru, 2012f, y; OJSC MMC Norilsk Nickel, 2013).

Other Russian nickel producers also were affected by financial difficulties because of decreased nickel prices. Yuzhuralnickel (part of OAO Mechel) reduced production investments by 10%, halted production for part of the year, and sent workers on forced leave with partial pay. In December, Yuzhuralnickel made a decision to stop production for an extended period of time to minimize losses. Ufaleynickel, on the other hand, was trying to avoid bankruptcy by promoting efficiency and modernization. The company was able to reduce the cost of coke used in production by 40% and to reduce the per-unit nickel cost by 14%. It was also able to sign a new sales contract for 100 t of granulated nickel with ALPICOM S.A. of Switzerland. All those measures saved the company from involuntary delays in production in 2012, but it was not clear whether Ufaleynickel would be able to avoid stopping production in 2013 (MinerJob.ru, 2012q, w, x).

For Nornickel (the leading producer in the world), 2012 was a difficult year not only because of low nickel prices, but because it lost two attractive tenders for development of nickel deposits in Russia. The first loss took place in May, when Voronezhskaya Oblast' announced the winner of the tender for two polymetallic deposits—the Elanskoe deposit and the Elkinskoe deposit. The total cost of mining the deposits was estimated to be 50 billion rubles ($1.7 billion). Although Nornickel had already conducted exploration and other preliminary works at the deposits, Voronezhskaya Oblast' announced that the winner of the tender was OOO Mednogorskiy Copper & Sulfur Complex, which was a part of UGMK. Nornickel initially attempted to dispute the Voronezh decision in court but eventually accepted the decision (Mineral.ru, 2012k, l).

The other tender was for exploration and mining of the copper and nickel ores in the southern part of the Norilsk-1 deposit. In June, it was announced that the exploration and development license for Norilsk-1 had been won by Russian Platinum Co. (the application was sent in the name of OAO AS Amur, which was a part of Russian Platinum). Complications with respect to the Norilsk-1 deposit included that the Zapolyarnyi division of Nornickel already had been mining parts of Norilsk-1 and over the years had developed its own transportation and production infrastructure. Nornickel disputed the decision about Norilsk-1 in court. At yearend, a final court decision had not yet been reached and it was not clear if Nornickel would have a chance to win the license. In addition to court disputes, the Nornickel leadership was proposing changes to the system of distributing exploration and mining licenses in Russia. In particular, Nornickel suggested that the Federal Agency for Subsoil Use (Rosnedra) and local government agencies use auctions rather than tenders because auctions will tend to favor the highest bidders and make the process of choosing the winners less subjective (Mineral.ru, 2012t).

Platinum-Group Metals.—In 2012, Russia produced 82,400 kg of palladium and 30,200 kg of platinum. The country was the world leader in palladium production and was ranked second in the world in platinum mine production. Compared with that of 2011, palladium output decreased by 2%, and platinum production increased by 10.6%. The leading platinum-group metal (PGM) producer in Russia was Nornickel, whose Zapolyarnyi division was mining three large PGM deposits in Krasnoyarskiy Kray—Norilsk-1, Oktyabr'skoye, and Talnahskoye. Another division within Nornickel, Kolskaya GMK, was mining several deposits in Murmanskaya Oblast'—Kotsel'vaara-Kammikivi, Semiletka, Zapolyarnoye, and Zhdanovskoye. Altogether, Nornickel produced practically all the palladium and about three-quarters of the platinum output in Russia (Mineral.ru, 2012s).

Another significant producer of PGMs in Russia was Russian Platinum Co. In the past several years, Russian Platinum acquired mining rights for several prominent PGM deposits. In 2007, Russian Platinum acquired OAO AS Amur, which was mining the Kondyor deposit, which was the largest placer deposit of platinum in Khabarovskiy Kray. In 2011, Russian Platinum acquired Chernogorskaya Gornorudnaya Kompaniya (ChGRK), which had a license for the exploration, development, and mining of the Chernogorskoye deposit of nonferrous and precious metals in Krasnoyarskiy Kray. In the summer, Russian Platinum won the tender for the southern part of the Norilsk-1 deposit, but Nornickel (the main competitor) was contesting the tender result in court (MinerJob.ru, 2012i; MinerJob.ru, 2012o).

As of 2012, ChGRK was continuing with the building of a mining and a beneficiation plant at the Chernogorskoye deposit. The company planned to start mining in 2013 and would stockpile the ore until 2015, when construction and commissioning of the beneficiation plant was scheduled to be completed. ChGRK expected to reach its production capacity of between 13,000 and 16,000 t/yr of copper, 500 kilograms per year (kg/yr) of gold, between 7,000 and 9,000 t/yr of nickel, between 12,000 and 13,000 kg/yr of palladium, and 6,000 kg/yr of platinum by 2016. The total cost of the project was estimated to be $1.2 billion (Mineral.ru, 2012c).

Tungsten.—In 2012, Russia produced 3,025 t of tungsten in concentrate, which was an 8.7% reduction compared with 2011 production. Russia was ranked a distant second after China in tungsten production. Tungsten was mined in five hard-rock deposits and one alluvial deposit. The largest mine in terms of production amount, Vostok-2, which was operated by the OAO Primorskiy GOK, produced about one-half of the total tungsten output. Other significant producers were AS Quartz, which was mining the Bom-Gorkhon deposit in Zabaykal'skiy Kray; KGUP Primteploenergo, which was mining the Lermontovskoye deposit in Primorskiy Kray; ZAO Novoorlovskiy GOK, which was mining the Spokoyninskoye deposit in Zabaykal'skiy Kray; and ZAO Zakamensk, which was mining the Ruchey Inkur alluvial deposit and the Barun-Narynskoye technogenic deposit in Buryatiya Republic. A technogenic deposit refers to the accumulated tailings from a previous mine operation (Ministry of Natural Resources and Ecology of the Russian Federation, 2013).

In April, the government of the Republic of Kabardino-Balkariya announced that a new mining and metallurgical complex was planned to be built at the Tyrnyauz tungsten and molybdenum deposit in Kabardino-Balkariya. The complex had a planned capacity of 1 Mt/yr of ore and was to be constructed during a 5-year period from 2013 to 2017. The proven resources of the deposit were about 30 Mt, and the new enterprise was expected to create 1,000 jobs in the region. OAO Kabardino-Balkarskaya Tungsten-Molybdenum Co. (which was majority owned by the government of Kabardino-Balkariya) was planning to invest a total of 7 billion rubles (about $216 million) in this project and was expecting to recover the investment within 8 years of beginning full-scale production. The Tyrnyauz deposit was mined previously, but operations had been stopped in 2001 because of the poor market conditions (MinerJob.ru, 2012e).

Industrial Minerals

Potash.—OAO Uralkali was the world's second-ranked producer of potash. In 2012, Uralkali reduced its production of potash by 14.4% to 5.56 Mt in K_2O equivalent (or 9.12 Mt in potassium chloride equivalent). The Uralkali mines were operating, on average, at 80% of capacity throughout the year, and the company was able to complete the expansion of the Berezniki-4 Mine. After the expansion was complete, the total capacity of Uralkali increased to 13 Mt/yr from 11.5 Mt/yr of potassium chloride (OAO Uralkali, 2014).

Although in 2012 Uralkali was the only company mining potash in Russia, several other projects were underway. OOO Verkhnekamskaya Potash Co., which was a part of OAO Akron, continued preparations for development of potash production at the Talitskiy sector of the Verkhnekamskoye potash deposit. The new mining complex was expected to be commissioned in 2016 and to reach full capacity by 2018. The planned capacity of the new complex was 2 Mt/yr (Mineral.ru, 2012a; MinerJob.ru, 2012a, s).

Another company, OAO MHK EuroChem, and its subsidiary OOO EuroChem VolgaKaliy, were continuing to build a mine at the Gremyachinskoye potash deposit. The companies were using a freezing method for creating the mine shaft and, as of January 2012, the shaft was 400 meters (m) deep; the planned depth was 1,147 m. EuroChem was planning to produce the first 150,000 t of potassium chloride in 2014 (Mineral.ru, 2012g).

Rare Earths.—Rare-earth metals (REMs) were being produced in Russia in limited amounts in 2012. The majority of ores containing rare-earth elements on Russian territory were mined by OAO Apatit at apatite-nepheline deposits in Khibines in Murmanskaya Oblast'. The rare-earth elements were extracted only from the loparite ores mined by the Lovozerskiy GOK, however. The ore was processed at the Karnasurtskaya beneficiation plant into intermediate loparite concentrate and then into marketable concentrate, which contained 95% loparite. The Solikamskiy Magnesium Plant (SMZ) then processed the loparite concentrate and extracted rare earths. In 2012, SMZ produced about 2,400 t/yr of rare-earth oxides, primarily from the cerium group.

In November, the President of Sakha Republic (Yakutiya) announced an auction for the Tomtor deposit of REMs. The ores of the Burannyi section of the Tomtor deposit contain, on average, from 9% to 12% REMs; they are, in fact, a natural concentrate. The ores also contain up to 7% niobium, as well as scandium and yttrium. According to estimates made by scientists from the Russian Academy of Sciences, the concentration of REMs and rare metals in Tomtor is so high that the value of 1 kilogram of ore, at current prices, is about $10. Development of Tomtor would require a large amount of infrastructure to be built from scratch, however (MinerJob.ru, 2012r, t).

In October, the Government of Zabaykal'skiy Kray announced that ZabGOK in Zabaykal'skiy Kray would be reopened for the mining and processing of rare metals and REMs. It was expected that the project would be financed by the Fund for Development of the Far East and the Baykal Region, by OAO Atomredmetzoloto (ARMZ), and possibly by German investors. According to experts, the plant would need an investment of 200 million rubles ($6.17 million) to restart production. ZabGOK already had a contract in place for antimony processing. The government of Zabaykal'skiy Kray offered ZabGOK 100 million rubles ($3.08 million) in loan guarantees to restart production (Mineral.ru, 2012j).

Mineral Fuels and Related Materials

Coal.—In 2012, coal production in Russia increased by 8.9% to 366 Mt. According to the BP Statistical Review of World Energy, by the end of 2012, Russia had coal resources of 157,000 Mt, which was second only to those of the United States. Those resources are 430 times the current annual production levels. About 50% of Russia's coal resources were in anthracite and bituminous coal types, and the rest of the resources were in lignite. About 20% of the resources were in coking coal, and the rest were in thermal types of coal. The coal resources are spread across the country unevenly: the majority of resources are located in Eastern Russia, whereas most demand is concentrated in the European part of Russia where coal resources are limited. A large share of the resources is found in parts of the country where the climate is cold and the infrastructure is underdeveloped, which makes coal mining more costly. The largest coalfield in Russia is the Kuznetskiy bituminous field in Kemerovskaya Oblast', which contains about 25% of Russia's coal resources and about 60% of its coking coal resources. The Kansko-Achinskiy lignite field located in Krasnoyarskiy Kray contains about 40% of the country's resources. The coals of the Kansko-Achinskiy field are of high quality (usually do not require beneficiation) and are located in easily accessible strata, which are between 25 and 80 m thick. Other Siberian fields include the Ulughemskiy field in Tyva Republic, the Irkutskiy bituminous field in Irkutskaya Oblast', the Minusinskiy bituminous field in the Republic of Khakasiya, and the Yuzhno-Yakutskiy bituminous field and the Lenskiy field, both in Sakha Republic (Yakutiya). The resources of each of those Siberian fields do not exceed 5% of the total Russian resources. The largest coalfield in the European part of Russia is the Donetskiy coalfield located in Rostovskaya Oblast', which contains about 3.6% of all Russian resources; about 75% of those resources are anthracite coal. The Donetskiy coalfield is largely depleted, and its resources do not have a significant potential to be increased in the future. Finally, the Pechorskiy bituminous coalfield located in Komi Republic and in the Nenetskiy Avtonomnyy Okrug contains about 2.8% of the Russia's resources, and about one-half of its resources are coking coal. The resources of the Pechorskiy field are unlikely to be increased in the future (Mineral.ru, 2011, 2012n; BP p.l.c., 2013).

The coal industry in Russia was mostly privately owned, and joint-stock companies (often consolidated into large holdings) dominated the industry. About 80% of coal was mined by 12 major companies; those companies included both "proper" coal mining corporations and metallurgical holdings companies, which included divisions specializing on coking coal mining. Siberian Coal Energy Co. (SUEK) was the largest, in terms of annual production, coal producer in Russia. In 2012, it produced 97.5 Mt of thermal coal, or about 30% of the entire Russian coal output. OAO UK Kuzbassrazrezugol (part of UGMK), which mined coking coal deposits in Kuznetskiy coalfield, was the second-ranked coal producer; its output accounted for about 15% of Russia's coal production. Other leading producers included OAO KhK SDS-Ugol, which operated mines and pits in the Kuznetskiy field, and OAO Mechel, which mined coking coal in the Kuznetskiy and the Yuzhno-Yakutskiy fields.

In February 2011, Russia adopted a new program for development of the coal industry through 2030. According to forecasts by the Ministry of Energy, annual coal production could increase to about 450 Mt by 2030. The total cost of the coal program for the Government would be 3.7 billion rubles (about $115 million) (Mineral.ru, 2012n).

During the past two decades, domestic coal consumption was reduced by 50%, and coal producers targeted primarily the export market. One of the main reasons for this reduction in domestic consumption of coal was the coal industry's weak position with respect to the natural gas industry; the domestic prices of gas were regulated (and kept artificially low) by the Government, and at those prices, coal producers were unable to compete with natural gas producers. By 2030, when natural gas prices were planned no longer to be regulated, annual domestic consumption of coal was projected to increase by 100% to 220 Mt. The Ministry of Energy projected that Russia would construct more than 100 new coal enterprises within the next 20 years. Because most of the new coal mines were to be located in Siberia and the Far East, the Government considered that its main role would be to assist coal producers by providing better and less expensive infrastructure facilities, such as ports and railroads. As of 2011, transportation costs accounted for about 40% of the delivered cost of coal (Mineral.ru, 2011).

Petroleum.—In 2012, Russia produced 497 Mt (3,615 million barrels) of crude oil (which was a 1% increase compared with the production level in 2011), and it was one of the leading petroleum producers in the world. The estimates of various research organizations and agencies, however, disagreed on the ranking. According to the U.S. Energy Information Administration, Russia was the third-ranked producer of liquid fuels after the United States and Saudi Arabia; according to the BP Statistical Review of World Energy, Russia was the second-ranked petroleum producer after Saudi Arabia, and according to the International Energy Agency and the Organization of Petroleum Exporting Countries, Russia was the leading petroleum producer. In addition to the crude oil,

Russia's output of gas condensate was 21 Mt, for a total output of 518 Mt of petroleum and gas condensate. According to the BP Statistical Review of World Energy, at the end of 2012, Russia's proven reserves of petroleum were 11,900 Mt (or 87,200 million barrels), which constituted 5.2% of the world's proven reserves of petroleum. In 2012, Russia exported 239.6 Mt of petroleum (or 48.2% of its total output), which was a 1% reduction compared with its 2011 petroleum exports (BP p.l.c., 2013; Mineral.ru, 2013a, b).

Among Russia's regions, the leader in petroleum production was Khanty-Mansiyskiy Avtonomnyy Okrug (HMAO), which accounted for about 52% of the country's production. Other leading producers were Yamalo-Nenetskiy Avtonomnyy Okrug (YNAO) (7.3%), Tatarstan Republic (6.6%), Orenburgskaya Oblast' (4.6%), and Krasnoyarskiy Kray (3.7%). Production by HMAO was continuing to decrease—in 2012 alone, HMAO reduced petroleum production by 4.2 Mt and, in the past 5 years, production decreased to 258 Mt in 2012 from 275 Mt in 2008 (RIA-Rating, 2013).

The main factor in the increase in overall Russian production in 2012 was an acceleration of production in deposits in Eastern Siberia. In its turn, a significant factor in stimulating production in Eastern Siberia was the opening of the eastern branch of the Eastern Siberia-Pacific Ocean oil pipeline. Another factor was the introduction of new technologies at old fields in Western Siberia and Povolzhye that could help in the extraction of hard-to-recover crude oil deposits (RIA-Rating, 2013).

In 2012, most petroleum production in Russia was carried out by nine vertically-integrated oil and gas (VIOG) companies, which together included about 150 extracting units. The nine companies accounted for about 90% of the country's petroleum production, and about two-thirds of the country's petroleum production was controlled by just four companies (in order of output volume): OAO NK Rosneft', the OAO LUKOIL group, OAO TNK–BP Holding, and OAO Surgutneftegaz. In 2012, Rosneft produced 125.8 Mt (25.3% of total Russian production), LUKOIL produced 84.2 Mt (16.9%), TNK–BP produced 79.2 Mt (14.7%), and Surgutneftegaz produced 61.4 Mt (12.4%). In 2012, Rosneft announced its merger with TNK–BP; if the merger goes through, the joint company would become one of the top five oil companies in the world. The buyout of TNK–BP from its current shareholders (BP p.l.c. of the United Kingdom and a consortium of private Russian shareholders) would cost Rosneft a total of $55 billion (Mineral.ru, 2012q).

In addition to traditional methods of petroleum extraction, companies operating in Western Siberia adopted hydraulic fracturing techniques to produce hard-to-recover crude oil. TNK–BP developed an advanced technology of multistage hydraulic fracturing to increase effectiveness and reduce costs when developing mature deposits. The key element of the technology is using special equipment to clear the shaft of the drilling hole after each of the six consecutive hydraulic fracturing stages. The company applied the multistage hydraulic fracturing at 25 holes of the Samotlor field in 2012, and was planning to apply the procedure at least 50 more times in 2013. OOO LUKOIL-West Siberia (one of the LUKOIL companies) applied a somewhat different multistage hydraulic fracturing technique in a horizontal shaft of a drilling hole. The procedure includes a method of hydraulic jet perforation using flexible coil tubing. The procedure eliminates the need to build a hydraulic fracturing column and reduces the preparation time for the drilling hole and potentially increases the well's productivity. At five test holes, the daily output was doubled. LUKOIL announced that it was planning to use this new technique at other company wells (Mineral.ru, 2012x, y).

In 2012, Russia had 28 large crude oil refineries and about 200 refineries of average and small size. The total refining capacity of Russia's refineries was about 290 Mt/yr of petroleum. More than 90% of the total refining capacity in Russia belonged to VIOG companies. In 2012, Russian companies refined 271 Mt of petroleum. Of this amount, Rosneft refined 50.9 Mt of petroleum at its seven refineries; LUKOIL refined 44.4 Mt; Gazprom Neft' refined 31.6 Mt; and TNK–BP refined 27.8 Mt. In 2011, 12 petroleum companies signed an agreement with the Federal Antimonopoly Service in which the oil companies agreed to modernize their refineries by 2015 so that they can supply the domestic market with enough gasoline that satisfies Euro-5 standards to meet domestic demand. Beginning in 2013, Russia was planning to ban the sale of automotive gasolines for which the environmental requirements are below the Euro-3 standard. In 2012, several companies were reportedly modernizing their refineries. The Ryazanskiy refinery (owned by TNK–BP) increased the percentage of its production of Euro-5 gasoline to 30% and was planning to increase this percentage to 46% during 2013. Slavneft'- YaNOS announced its decision not to produce any gasoline below the Euro-5 standard starting from July 2012. One of the reasons to accelerate the transition to cleaner gasolines was the decision of the Government to restructure the excise taxes on automobile fuels to encourage refineries to make the switch faster than they would do otherwise (Mineral.ru, 2012p, u).

Outlook

Russia has large reserves of a variety of mineral commodities and most likely will continue to be one of the world's leading mineral producers. In 2012, the country's mineral sector demonstrated several interesting trends. First, the country's leading mineral producers, such as Nornickel, RUSAL, and Uralkali, encountered serious difficulties related mostly to the reductions in the world prices of their products. Even under such conditions, however, they appeared more financially stable than their smaller counterparts in Russia. It was not yet clear whether those leading companies had adapted to the most recent economic downturn, but 2012 demonstrated that such downturns can lead the Russian mineral and metallurgical industries to improve the efficiency of their operations. Second, Russia, as a country with comprehensive mineral production and vast resources, is starting to use Government regulation more often to stimulate the production of minerals that are considered strategic yet are not steadily produced in the market economy, such as rare earths. Finally, Russian regulators appear to have started taking environmental concerns more seriously than they did in previous years, which is likely to affect the image of the Russian mineral industry, both domestically and abroad. If these trends continue for the next several years, Russia's mineral sector is likely to become more resilient to volatile prices of minerals, technology changes, and the cyclic nature of the economy.

References Cited

Apodaca, L.E., 2013a, Nitrogen (fixed)—Ammonia: U.S. Geological Survey Mineral Commodity Summaries 2013, p. 112–113.
Apodaca, L.E., 2013b, Peat: U.S. Geological Survey Mineral Commodity Summaries 2013, p. 114–115.
Apodaca, L.E., 2013c, Sulfur: U.S. Geological Survey Mineral Commodity Summaries 2013, p. 158–159.
Baikal Mining Co. LLC, 2011, Udokan project: Baikal Mining Co. LLC, September. (Accessed December 3, 2014, at http://www.cerbanet.org/intranet/Documents/22_Sept_Investing_in_Russia/Presentation/2. Direct Investments Opportunities and Success Stories %E2%80%93 How Do They Do It/Panel2_Andrei_Varichev_Metalloinvest.pdf.)
Bedinger, G.M., 2013, Titanium and titanium dioxide: U.S. Geological Survey Mineral Commodity Summaries 2013, p. 172–173.
BP p.l.c., 2013, BP statistical review of world energy: BP p.l.c., June, 48 p. (Accessed April 22, 2014, at http://www.bp.com/content/dam/bp/pdf/statistical-review/statistical_review_of_world_energy_2013.pdf.)
Bray, E.L., 2013a, Aluminum: U.S. Geological Survey Mineral Commodity Summaries 2013, p. 16–17.
Bray, E.L., 2013b, Bauxite and alumina: U.S. Geological Survey Mineral Commodity Summaries 2013, p. 26–27.
Carlin, J.F., Jr., 2013, Antimony: U.S. Geological Survey Mineral Commodity Summaries 2013, p. 18–19.
Corathers, L.A., 2013, Silicon: U.S. Geological Survey Mineral Commodity Summaries 2013, p. 144–145.
Crangle, R.D., Jr., 2013a, Boron: U.S. Geological Survey Mineral Commodity Summaries 2013, p. 32–33.
Crangle, R.D., Jr., 2013b, Diatomite: U.S. Geological Survey Mineral Commodity Summaries 2013, p. 52–53.
Crangle, R.D., Jr., 2013c, Gypsum: U.S. Geological Survey Mineral Commodity Summaries 2013, p. 70–71.
Edelstein, D.L., 2013a, Arsenic: U.S. Geological Survey Mineral Commodity Summaries 2013, p. 20–21.
Edelstein, D.L., 2013b, Copper: U.S. Geological Survey Mineral Commodity Summaries 2013, p. 48–49.
Expert.ru, 2014, Natalkinskoye zolotorudnoye mestorozhdenietretye v mire po velichine zapasov [Natalkinskoye gold deposit—The third in the world by the size of resources]: Expert.ru. (Accessed December 3, 2014, at http://expert.ru/ratings/table_276747.)
Federal'naya Sluzhba Gosudarstvennoy Statistiki [Federal State Statistical Service], 2013, Rossiyskiy Statisticheskiy Yezhegodnik [Statistical yearbook 2012]: Federal'naya Sluzhba Gosudarstvennoy Statistiki. (Accessed April 22, 2014, at http://www.gks.ru/bgd/regl/b13_13/Main.htm.)
Fenton, M.D., 2013, Iron and steel: U.S. Geological Survey Mineral Commodity Summaries 2013, p. 78–79.
George, M.W., 2013a, Gold: U.S. Geological Survey Mineral Commodity Summaries 2013, p. 66–67.
George, M.W., 2013b, Selenium: U.S. Geological Survey Mineral Commodity Summaries 2013, p. 142–143.
George, M.W., 2013c, Silver: U.S. Geological Survey Mineral Commodity Summaries 2013, p. 146–147.
George, M.W., 2013d, Tellurium: U.S. Geological Survey Mineral Commodity Summaries 2013, p. 164–165.
Guberman, D.E., 2013a, Germanium: U.S. Geological Survey Mineral Commodity Summaries 2013, p. 64–65.
Guberman, D.E., 2013b, Lead: U.S. Geological Survey Mineral Commodity Summaries 2013, p. 90–91.
Jasinski, S.M., 2013a, Phosphate rock: U.S. Geological Survey Mineral Commodity Summaries 2013, p. 118–119.
Jasinski, S.M., 2013b, Potash: U.S. Geological Survey Mineral Commodity Summaries 2013, p. 122–123.
Jaskula, B.W., 2013, Gallium: U.S. Geological Survey Mineral Commodity Summaries 2013, p. 58–59.
Kramer, D.A., 2013a, Magnesium compounds: U.S. Geological Survey Mineral Commodity Summaries 2013, p. 96–97.
Kramer, D.A., 2013b, Magnesium metal: U.S. Geological Survey Mineral Commodity Summaries 2013, p. 98–99.
Kuck, P.H., 2013, Nickel: U.S. Geological Survey Mineral Commodity Summaries 2013, p. 108–109.

Loferski, P.J., 2013, Platinum-group metals: U.S. Geological Survey Mineral Commodity Summaries 2013, p. 120–121.

Miller, M.M., 2013a, Barite: U.S. Geological Survey Mineral Commodity Summaries 2013, p. 24–25.

Miller, M.M., 2013b, Fluorspar: U.S. Geological Survey Mineral Commodity Summaries 2013, p. 56–57.

Miller, M.M., 2013c, Lime: U.S. Geological Survey Mineral Commodity Summaries 2013, p. 92–93.

Mineral.ru, 2011, K 2030 g. vnutrennee potreblenie uglya v Rossii vozrastet bole chem vdvoe [By 2030, domestic consumption of coal in Russia would more than double]: Mineral.ru, May 6. (Accessed February 5, 2014, at http://www.mineral.ru/News/44711.html.)

Mineral.ru, 2012a, "Akron" privlek sredstva dlya stroitel'stva Talitskogo kaliynogo GOKa [Akron obtained funding for constructing the Talitskiy potassium GOK]: Mineral.ru, November 7. (Accessed April 22, 2014, at http://www.mineral.ru/News/50686.html.)

Mineral.ru, 2012b, BGK letom 2012 g. nachnet programmu promyshlennykh ispytaniy rudy na Udokane [In summer 2012, BGK will start a program of industrial tests of the ore at Udokan]: Mineral.ru, November 23. (Accessed April 22, 2014, at http://www.mineral.ru/News/48614.html.)

Mineral.ru, 2012c, ChGRK pristupit k dobyche s 2013 g. [ChGRK will start mining in 2013]: Mineral.ru, February 13. (Accessed April 22, 2014, at http://www.mineral.ru/News/50892.html.)

Mineral.ru, 2012d, Division "Vostok" OK "Rusal" nameren uvelichit' v 2012 g. proizvodstvo alyuminiya na 1,7% [In 2012, OK RUSAL's Vostok division intends to increase aluminum production by 1.7%]: Mineral.ru, February 13. (Accessed April 22, 2014, at http://www.mineral.ru/News/47610.html.)

Mineral.ru, 2012e, Dobycha zolota v Rossii do 2020 goda budet rasti na 3–5% v god [Through 2020, Russia's gold production will increase by 3–5% per year]: Mineral.ru, December 8. (Accessed April 22, 2014, at http://www.mineral.ru/News/51090.html.)

Mineral.ru, 2012f, Gosudarstvennaya Duma prinyala zakon o l'gotax dlya mestorozhdeniy trudnoizvlekaemoy nefti [The State Duma adopted the law about reduced tariffs for the deposits of hard-to-extract crude oil]: Mineral.ru, November 26. (Accessed April 22, 2014, at http://www.mineral.ru/News/50929.html.)

Mineral.ru, 2012g, Khloristyi kaliy na Gremyachinskom mestorozhdenii nachnut dobyvat' b 2014 g. [Gremyachinskoye deposit will start producing potassium in 2014]: Mineral.ru, January 11. (Accessed April 22, 2014, at http://www.mineral.ru/News/47208.html.)

Mineral.ru, 2012h, Krupneyshiy alyuminievyi division "Rusala" uvelichit proizvodstvo splavov na 70% [RUSAL's largest aluminum division will increase ferroalloy production by 70%]: Mineral.ru, February 20. (Accessed April 22, 2014, at http://www.mineral.ru/News/47696.html.)

Mineral.ru, 2012i, MPR ocen'yu zhdet zakona, uvelichivayushego dolyu inostrantsev v strategicheskih proektah [The Ministry of Natural Resources expects that the law increasing the share of foreigners in strategic projects will be adopted in the fall]: Mineral.ru, July 31. (Accessed April 22, 2014, at http://www.mineral.ru/News/49511.html.)

Mineral.ru, 2012j, Na baze Pervomayskogo GOK'a budet sozdan metallurgicheskiy kombinat [A new metallurgical complex will be created at the site of the Pervomayskiy GOK]: Mineral.ru, October 9. (Accessed April 22, 2014, at http://www.mineral.ru/News/50393.html.)

Mineral.ru, 2012k, Nazvan pobeditel' konkursa na razrabotku medno-nikelevyh mestorozhdeniy v Voronezhskoy oblasti [The winner of the tender for development of copper and nickel deposits in Voronezhskaya oblast' is announced]: Mineral.ru, May 23. (Accessed April 22, 2014, at http://www.mineral.ru/News/48694.html.)

Mineral.ru, 2012l, Nornickel otkazyvaetsya ot bor'by za Voronezhskiye mestorozhdeniya [Nornickel stops fight for Voronezh deposits]: Mineral.ru, July 3. (Accessed April 22, 2014, at http://www.mineral.ru/News/49172.html.)

Mineral.ru, 2012m, OAO "GMK Norilskiy Nickel"—krupneyshiy mirovoy proizvoditel' nikelya i palladiya—nameneno investirovat' v razvitiye mineral'no-syr'evogo kompleksa do 2016 g. okolo 120 mlrd. rub. [OAO GMK Norilsk Nickel—The leading world producer of nickel and palladium—Intends to invest about 120 million rubles in its resource base through 2016]: Mineral.ru, August 25. (Accessed April 22, 2014, at http://www.mineral.ru/News/49818.html.)

Mineral.ru, 2012n, Ob'em programmy "Razvitiye ugol'noy promyshlennosti RF do 2030 g." sostavit 3,7 trln rub [The coal industry development through 2030 program will cost 3.7 trillion rubles]: Mineral.ru, January 24. (Accessed April 22, 2014, at http://www.mineral.ru/News/47364.html.)

Mineral.ru, 2012o, OK "RUSAL" nameren nachat' proizvodstvo alyuminiya na Tayshetskom zavode [UC RUSAL is planning to start aluminum production at the Tayshetskiy Aluminum plant]: Mineral.ru, February 14. (Accessed April 22, 2014, at http://www.mineral.ru/News/47611.html.)

Mineral.ru, 2012p, RNPK uvelichil dolyu topliva standarta Evro-5 v proizvodstve do 30% [Ryazan petroleum refinery increased the production share of Euro-5 compliant fuel to 30%]: Mineral.ru, January 12. (Accessed April 22, 2014, at http://www.mineral.ru/News/50992.html.)

Mineral.ru, 2012q, "Rosneft'" zaplatit za 100% aktsiy TNK–BP 55 milliardov dollarov [Rosneft will pay $55 billion for 100% share of TNK–BP]: Mineral.ru, October 23. (Accessed April 22, 2014, at http://www.mineral.ru/News/50569.html.)

Mineral.ru, 2012r, "RUSAL" planiruet zapustit' Taysheyskiy alyuminievyi zavod ne ran'she 2014-2015 [RUSAL is planning to start operations at the Tayshetskiy aluminum plant no earlier than 2014–2015]: Mineral.ru, September 17. (Accessed April 22, 2014, at http://www.mineral.ru/News/50113.html.)

Mineral.ru, 2012s, Russkaya Platina vlozhit 78 mlrd. Rub. v Norilsk-1 [Russian Platinum will invest 78 billion rubles in Norilsk-1]: Mineral.ru, July 2. (Accessed April 22, 2014, at http://www.mineral.ru/News/49158.html.)

Mineral.ru, 2012t, "Russkaya Platina vyigrala yug Noril'ska-1 [Russian Platinum won the south portion of Norilsk-1]: Mineral.ru, June 22. (Accessed April 22, 2014, at http://www.mineral.ru/News/49054.html.)

Mineral.ru, 2012u, S iyulya 2012 goda vse vypuskaemoye YaNOSom motornoye toplivo budet sootvetstvovat' standartu Evro-5 [From July 1, 2012, all motor fuels produced by YaNOS will be Euro-5 compliant]: Mineral.ru, June 17. (Accessed April 22, 2014, at http://www.mineral.ru/News/48988.html.)

Mineral.ru, 2012v, Sovet directorov OK "Rusal" odobril dolgosrochnuyu programmu poetapnogo zakrytiya neeffektivnykh mishnostey [The Board of Directors of UC RUSAL approved the long-term program of gradual closure of inefficient facilities]: Mineral.ru, August 29. (Accessed April 22, 2014, at http://www.mineral.ru/News/49865.html.)

Mineral.ru, 2012w, Strategicheskiye uchastki dolzhny rasprerdelyat'sya na auktsionah [Strategic deposits should be distributed using auctions]: Mineral.ru, July 12. (Accessed April 22, 2014, at http://www.mineral.ru/News/49300.html.)

Mineral.ru, 2012x, TNK-BP vnedrila usovershenstvovannuyu tekhnologiyu gidrorazryva plasta dlya povysheniya effektivnosti bureniya v Zapadnoy Sibiri [TNK–BP introduced an improved technology of hydraulic fracturing to improve the effectiveness of drilling in West Siberia]: Mineral.ru, August 17. (Accessed April 22, 2014, at http://www.mineral.ru/News/49717.html.)

Mineral.ru, 2012y, V OOO "LUKOIL—Zapadnaya Sibir' uspeshno osvoili technologiyu mnogostadiynogo gidrorazryva plasta [OOO LUKOIL—Western Siberia successfully applied a technology of multistage hydraulic fracturing]: Mineral.ru, November 12. (Accessed April 22, 2014, at http://www.mineral.ru/News/50743.html.)

Mineral.ru, 2013a, Dobycha nefti v RF v 2012 godu vyrosla na 1,3%, gaza snizilas' na 2,3% [In 2012, petroleum production in RF increased by 1.3%, and gas production decreased by 2.3%]: Mineral.ru, January 10. (Accessed April 22, 2014, at http://www.mineral.ru/News/51386.html.)

Mineral.ru, 2013b, Eksport nefti iz Rossii v 2012 godu sokratilsya na 1% [In 2012, petroleum exports from Russia were reduced by 1%]: Mineral.ru, January 12. (Accessed April 22, 2014, at http://www.mineral.ru/News/51421.html.)

Mineral.ru, 2013c, V 2012 g. v Rossii bylo dobyto 226 t zolota [In 2012, Russia produced 226 tons of gold]: Mineral.ru, February 22. (Accessed April 22, 2014, at http://www.mineral.ru/News/51953.html.)

MinerJob.ru, 2012a, "Belgorkhimprom" podgotovilo TEO stroitel'stva Talitskogo GOKa v Permskom Krae [Belgorkhimprom prepared technical and economic blueprints for constructing the Talitskiy GOK in Permskiy Kray]: MinerJob.ru, March 6. (Accessed April 22, 2014, at http://www.minerjob.ru/viewnew.php?id=20963.)

MinerJob.ru, 2012b, GeoProMining otkryla sezon dobychi 2012-2-13 gg. na Sentachane [GeoProMining has opened the 2012–2013 production season at Sentachan]: MinerJob.ru, November 14. (Accessed April 22, 2014, at http://www.minerjob.ru/viewnew.php?id=22483.)

MinerJob.ru, 2012c, GeoProMining zavershaet pervyi etap rekonstruktsii Magadanskogo otdeleniya "Sarylakh-Sur'ma" [GeoProMining is finishing up the first reconstruction stage of Sarylakh-Antimony Magadan division]: MinerJob.ru, October 3. (Accessed April 22, 2014, at http://www.minerjob.ru/viewnew.php?id=21954.)

MinerJob.ru, 2012d, K 2015 godu AOA "Seligdar" rasschityvaet voyti v desyatku liderov zolotodobychi Rossii [By 2015, Seligdar intends to join the top 10 Russian gold producers]: MinerJob.ru, August 21. (Accessed April 22, 2014, at http://www.minerjob.ru/viewnew.php?id=21686.)

MinerJob.ru, 2012e, Kabardino-Balkariya vosstanovit Tyrnauzskiy GOK [Kabardino-Balkariya will restore the Tyrnauzskiy GOK]: MinerJob.ru, September 14. (Accessed April 22, 2014, at http://www.minerjob.ru/viewnew.php?id=21856.)

MinerJob.ru, 2012f, Na Kol'skoy GMK pererabotka rudy mozhet stat' nerentabel'noy [Kol'skaya GMK may start processing ore at a loss]: MinerJob.ru, July 6. (Accessed April 22, 2014, at http://www.minerjob.ru/viewnew.php?id=21400.)

MinerJob.ru, 2012g, Na Mikheevskom GOKe nachato stroitel'stvo obogatitel'noy fabriki [Mikheevskiy GOK started building a beneficiation plant]: MinerJob.ru, May 30. (Accessed April 22, 2014, at http://www.minerjob.ru/viewnew.php?id=21277.)

MinerJob.ru, 2012h, Nornickel sozdaet pervoye v RF polnoye proizvodstvo kobal'ta [Nornickel is creating the first full production chain of cobalt in RF]: MinerJob.ru, January 20. (Accessed April 22, 2014, at http://www.minerjob.ru/viewnew.php?id=20161.)

MinerJob.ru, 2012i, Novye gorizonty "Russkiy Platiny" [New horizons for Russian Platinum]: MinerJob.ru, October 15. (Accessed April 22, 2014, at http://www.minerjob.ru/viewnew.php?id=22071.)

MinerJob.ru, 2012j, Osvoeniye Kolymskogo mestorozhdeniya "Pavlik" nachnetsya po planu [Development of the Pavlik deposit on Kolyma will start as planned]: MinerJob.ru, July 27. (Accessed April 22, 2014, at http://www.minerjob.ru/viewnew.php?id=21545.)

MinerJob.ru, 2012k, Phantom Udokana [Phantom of Udokan]: MinerJob.ru, May 21. (Accessed April 22, 2014, at http://www.minerjob.ru/viewnew.php?id=21244.)

MinerJob.ru, 2012l, Rosnedra sohranili srok vvoda Udokanskogo mestorozhdeniya v Ekspluatatsiyu [Rosnedra kept the date for the Udokan deposit's start of operations]: MinerJob.ru, April 3. (Accessed April 22, 2014, at http://www.minerjob.ru/viewnew.php?id=21092.)

MinerJob.ru, 2012m, Rusal podvodit itogi proizvoditel'noy deyatel'nosti kompanii v 2012 godu [RUSAL is announcing its production results for 2012]: MinerJob.ru, February 8. (Accessed April 22, 2014, at http://www.minerjob.ru/viewnew.php?id=23399.)

MinerJob.ru, 2012n, Russkaya mednaya kompaniya zapustit Tominskiy GOK v 2014 godu [Russian Copper Co. will start operations at the Tominskiy GOK in 2014]: MinerJob.ru, June 18. (Accessed April 22, 2014, at http://www.minerjob.ru/viewnew.php?id=21339.)

MinerJob.ru, 2012o, "Russkaya Platina" planiruet v tekushem godu sokhranit' dobychu platiny na urovne 2011 goda [Russian Platinum plans to keep its current production level the same as in 2011]: MinerJob.ru, October 4. (Accessed April 22, 2014, at http://www.minerjob.ru/viewnew.php?id=21959.)

MinerJob.ru, 2012p, S BAZa uvol'nyayutsya rabochie, kotorye ne veryat v spaseniye zavoda [Workers who don't believe that the plant can be saved quit BAZ]: MinerJob.ru, October 30. (Accessed April 22, 2014, at http://www.minerjob.ru/viewnew.php?id=22266.)

MinerJob.ru, 2012q, Ufaleynickel pytaetsya spastis' ot bankrotstva svoimi silami [Ufaleynickel is trying to avoid bankruptcy using its own resources]: MinerJob.ru, July 9. (Accessed April 22, 2014, at http://www.minerjob.ru/viewnew.php?id=21404.)

MinerJob.ru, 2012r, V 2013 vozmozhno budet proveden auktsion po Tomtorskomu mestorozhdeniyu redkozemel'nykh metallov v Yakutii [In 2013, an auction of the Tomtor REM deposit will likely take place in Yakutiya]: MinerJob.ru, November 21. (Accessed April 22, 2014, at http://www.minerjob.ru/viewnew.php?id=22577.)

MinerJob.ru, 2012s, "Verkhnekamskaya kaliynaya kompaniya k kontsu 2012 goda udvoit kolichestvo rabochih mest na Talitskom uchastke v Permskom Krae [By the end of 2012, the Verkhnekamskaya potash company will double the number of workers at the Talitskiy section in Permskiy Kray]: MinerJob.ru, October 12. (Accessed April 22, 2014, at http://www.minerjob.ru/viewnew.php?id=22047.)

MinerJob.ru, 2012t, V Sibiri nashli redkozemel'nye zapasy na 250 mlrd dollarov [Rare earth resources with a value of $250 billion are found in Siberia]: MinerJob.ru, October 1. (Accessed April 22, 2014, at http://www.minerjob.ru/viewnew.php?id=21931.)

MinerJob.ru, 2012u, V Sosnovskom rayone odobrili stroitel'stvo Tominskogo GOKa [The construction of a new GOK was approved in Sosnovskiy Rayon]: MinerJob.ru, June 14. (Accessed April 22, 2014, at http://www.minerjob.ru/viewnew.php?id=21324.)

MinerJob.ru, 2012v, Vlasti uprostyat dustup k zolotym nedram dlya inostrantsev [The Government will make Russian gold resources more accessible for foreigners]: MinerJob.ru, August 9. (Accessed April 22, 2014, at http://www.minerjob.ru/viewnew.php?id=21615.)

MinerJob.ru, 2012w, "Yuzhuralnickel konserviruet proizvodstvo na dlitel'nyi srok [Yuzhuralnickel is freezing production long term]: MinerJob.ru, December 19. (Accessed April 22, 2014, at http://www.minerjob.ru/viewnew.php?id=23009.)

MinerJob.ru, 2012x, "Yuzhuralnickel" prosit gospodderzhki. Tseny na nickel' sdelali kombinat nerentabel'nym [Yuzhuralnickel asks for Government support—The nickel prices are causing its losses]: MinerJob.ru, September 25. (Accessed April 22, 2014, at http://www.minerjob.ru/viewnew.php?id=21889.)

MinerJob.ru, 2012y, Zapolyarnyi Filial Nornickelya planiruet k 2016 godu uvelichit' dobychu rudy na 13% do 18 mln tonn [By 2016, Nornickel's Zapolyarnyi division plans to increase ore production by 13% to 18 million tons]: MinerJob.ru, August 20. (Accessed April 22, 2014, at http://www.minerjob.ru/viewnew.php?id=21688.)

MinerJob.ru, 2012z, ZRK "Pavlik" planiruet v kontse 2013 goda vvesti v stroy GOK na Kolyme moshnost'yu 5-6 tonn zolota v god [In 2013, ZRK Pavlik is planning to start operations of a GOK with a capacity of 5–6 tons per year of gold]: MinerJob.ru, October 9. (Accessed April 22, 2014, at http://www.minerjob.ru/viewnew.php?id=22013.)

Ministry of Natural Resources and Ecology of the Russian Federation, 2013, Wolfram [Tungsten], chap. of Gosudarstvennyi doklad o sostoyanii i ispol'zovanii mineral'no-syr'evyh resursov Rossiyskoy federatsii v 2011 gody [State report on conditions and use of mineral resources in the Russian Federation in 2011]: Moscow, Russia, Mineral Center, p. 259–263. (Accessed April 22, 2014, at http://mineral.ru/Facts/russia/161/528/3_14_w.pdf.)

OAO Uralkali, 2014, About: OAO Uralkali. (Accessed April 22, 2014, at http://www.uralkali.com/about.)

OJSC MMC Norilsk Nickel, 2013, Fact sheet: OJSC MMC Norilsk Nickel. (Accessed February 5, 2014, at http://www.nornik.ru/en/investor/fact.)

Olson, D.W., 2013a, Diamond (industrial): U.S. Geological Survey Mineral Commodity Summaries 2013, p. 50–51.

Olson, D.W., 2013b, Gemstones: U.S. Geological Survey Mineral Commodity Summaries 2013, p. 62–63.

Olson, D.W., 2013c, Graphite (natural): U.S. Geological Survey Mineral Commodity Summaries 2013, p. 68–69.

Polyak, D.E., 2013a, Iodine: U.S. Geological Survey Mineral Commodity Summaries 2013, p. 76–77.

Polyak, D.E., 2013b, Molybdenum: U.S. Geological Survey Mineral Commodity Summaries 2013, p. 106–107.

Polyak, D.E., 2013c, Rhenium: U.S. Geological Survey Mineral Commodity Summaries 2013, p. 130–131.

Polyak, D.E., 2013d, Vanadium: U.S. Geological Survey Mineral Commodity Summaries 2013, p. 178–179.

RIA-Analitika, 2013, Metallurgiya: Tendentsii i prognozy, vypusk # 9, Itogi 2012 goda [Metallurgy—Trends and forecasts—Issue #9—Results of 2012]: RIA Novosti, 35 p. (Accessed February 5, 2014, at http://vid1.rian.ru/ig/ratings/met_9.pdf.)

RIA-Rating, 2013, Neftegazovaya i neftedobyvayushaya promyshlennost': Tendentsii i prognozy, vypusk # 9, Itogi 2012 goda [Oil and gas and oil refining industry— Trends and forecasts—Issue #9—Results of 2012]: RIA Novosti, 46 p. (Accessed February 5, 2014, at http://vid1.rian.ru/ig/ratings/oil9.pdf.)

Shedd, K.B., 2013a, Cobalt: U.S. Geological Survey Mineral Commodity Summaries 2013, p. 46–47.

Shedd, K.B., 2013b, Tungsten: U.S. Geological Survey Mineral Commodity Summaries 2013, p. 176–177.

Tanner, A.O., 2013, Vermiculite: U.S. Geological Survey Mineral Commodity Summaries 2013, p. 180–181.

Tolcin, A.C., 2013a, Cadmium: U.S. Geological Survey Mineral Commodity Summaries 2013, p. 36–37.

Tolcin, A.C., 2013b, Indium: U.S. Geological Survey Mineral Commodity Summaries 2013, p. 74–75.

Tuck, C.A., 2013, Iron ore: U.S. Geological Survey Mineral Commodity Summaries 2013, p. 84–85.

United Company RUSAL, 2013a, Annual report 2012: United Company RUSAL, 219 p. (Accessed February 5, 2014, at http://www.rusal.ru/upload/uf/6fb/EWF101_smaller size.pdf.)

United Company RUSAL, 2013b, Key facts: United Company RUSAL. (Accessed February 5, 2014, at http://www.rusal.ru/about/facts.aspx.)

United Company RUSAL, 2013c, Key figures: United Company RUSAL. (Accessed February 5, 2014, at http://www.rusal.ru/investors/kpi.aspx.)

U.S. Central Intelligence Agency, 2013, Russia, in The world factbook: U.S. Central Intelligence Agency, July 5. (Accessed February 5, 2014, at https://www.cia.gov/library/publications/the-world-factbook/geos/rs.html.)

van Oss, H.G., 2013, Cement: U.S. Geological Survey Mineral Commodity Summaries 2013, p. 38–39.

Virta, R.L., 2013, Asbestos: U.S. Geological Survey Mineral Commodity Summaries 2013, p. 22–23.

Willett, J.C., 2013, Mica (natural): U.S. Geological Survey Mineral Commodity Summaries 2013, p. 104–105.

TABLE 1
RUSSIA: PRODUCTION OF MINERAL COMMODITIES[1]

(Metric tons unless otherwise specified)

Commodity[2]		2008	2009	2010	2011	2012
METALS						
Aluminum:						
Ore and concentrate:						
Alumina	thousand metric tons	3,112	2,794	2,930 [r]	2,825 [r]	2,719
Bauxite		5,675,000	5,775,000	5,688,000 [r]	5,943,000 [r]	5,700,000
Nepheline concentrate, 25% to 30%		4,760,000	500,000 [r]	1,000,000 [r]	997,000 [r]	1,056,700
Metal, smelter, primary		4,190,000	3,815,000	3,947,000	3,992,000 [r]	3,924,000
Antimony, mine output, recoverable Sb content[e]		3,500	3,500	6,040 [r]	6,348	7,300
Arsenic, white[e]		1,500	1,500	1,500	1,500	1,500
Bismuth:[e]						
Mine output, Bi content		70	65	50	50	40
Metal, refined		13	12	10	10	8
Cadmium, metal, smelter		580 [e]	581	733 [r]	800 [r]	850 [e]
Chromium, chrome ore, marketable		747,000 [r]	347,000 [r]	699,000 [r]	662,000 [r]	670,000 [e]
Cobalt:[e]						
Mine output, recoverable Co content		6,200	6,100	6,200	6,100 [r]	6,300
Metal, refined		2,500	2,352 [3]	2,460	2,337 [r,3]	2,186 [3]
Copper:						
Ore, recoverable Cu content[e]		750,000	676,000	703,000	856,200 [3]	883,000
Metal:						
Blister, smelter:[e]						
Primary		630,000	580,000	590,000	596,490 [3]	621,200 [3]
Secondary		235,000	220,000	240,000	242,640 [3]	253,800 [3]
Total		865,000	800,000	830,000	839,130 [3]	875,000 [3]
Refined:						
Primary		610,000	612,000	656,000	663,200	635,000
Secondary		250,000	250,000	218,000	220,400	209,400
Total		860,000	862,000	874,000	883,600	844,400
Gallium[e]		11	11	11	11	10
Gold:						
Mine output, Au content	kilograms	172,031	192,832	189,000	199,650	217,800
Secondary recovery	do.	8,140	12,404	12,600	9,334	8,500
Indium[e]		10	4	4	5	5
Iron and steel:						
Iron ore:						
Gross weight		99,900,000	92,000,000	95,900,000	104,000,000	104,000,000
Fe content, 55% to 63%[e]		57,800,000	53,200,000	56,600,000	61,400,000	61,400,000
Metal:						
Pig iron		48,300,000	43,930,000	48,000,000	48,000,000	50,500,000
Direct-reduced iron[e]		4,560,000	4,670,000	4,700,000	4,900,000	5,200,000
Ferroalloys:[e]						
Blast furnace:						
Ferromanganese		110,000	88,000	171,600 [3]	146,000	165,000
Ferrophosphorus		3,500	3,000	3,600	3,600	3,600
Spiegeleisen		7,000	6,500	5,500	6,000	6,000

See footnotes at end of table.

TABLE 1—Continued
RUSSIA: PRODUCTION OF MINERAL COMMODITIES[1]

(Metric tons unless otherwise specified)

Commodity[2]		2008	2009	2010	2011	2012
METALS—Continued						
Iron and steel—Continued:						
Metal—Continued:						
Ferroalloys—Continued:[e]						
Electric furnace:						
Ferrochromium		490,000	378,000 [3]	414,288 [3]	501,700 [3]	477,600 [3]
Ferrochromiumsilicon		4,000	3,500	4,200	4,200	4,100
Ferronickel, gross weight:[3, 4]						
High-nickel		17,971	17,489	19,763	20,200	23,300 [e]
Other		13,440	14,040	13,165	13,800 [r]	13,000 [e]
Ferroniobium (ferrocolumbium)		500 [r]	500 [r]	700 [r]	700 [r]	700
Ferrosilicon		850,000	745,000	916,000	1,030,000	1,042,000 [3]
Ferrovanadium		12,000	8,029 [3]	13,507 [3]	13,500	12,500
Silicomanganese		40,000	98,700	147,900 [3]	150,000	160,000
Silicon metal		54,000	23,900	48,700	52,000	52,000
Ferrotitanium		--	--	4,000	4,000	4,000
Other		22,000	22,000 [r]	18,000	18,000	18,000
Total, ferroalloys		1,620,000 [r]	1,410,000 [r]	1,780,000 [r]	1,960,000 [r]	1,980,000
Steel:						
Crude		68,700,000	59,800,000	66,800,000	68,100,000	70,400,000
Finished, rolled		56,564,000	51,900,000	55,000,000 [r]	56,500,000 [r]	59,000,000
Pipe		7,772,000	6,655,000	9,149,000	10,017,000	9,656,000
Lead:[e]						
Mine output, recoverable Pb content		60,000	70,000	97,000	94,500	92,700
Metal, refined, primary and secondary		80,000	73,000	89,000	86,700	85,100
Magnesium:[e]						
Magnesite		1,200,000	1,000,000	1,200,000	1,200,000	1,300,000
Metal, including secondary		37,000	29,000	29,000	29,000	31,000
Manganese ore, marketable:[e]						
Gross weight		45,000	45,000	45,000	120,000	130,000
Mn content		9,200	9,200	9,200	30,000	32,500
Mercury[e]		50	50	50	50	50
Molybdenum, in concentrate		4,061 [r]	4,562 [r]	4,495 [r]	4,843 [r]	4,800 [e]
Nickel:						
Marketable mine production, Ni content:						
Laterite ore		36,804	32,298	41,184	34,000	32,000
Sulfide concentrate		229,765	229,493	228,093	231,000 [r]	223,000
Total		266,569	261,791	269,277	265,000 [r]	255,000
Matte, for export, primarily to China		--	--	660	700 [e]	700 [e]
Nickel products:						
Metal		258,800 [r]	255,000 [r]	262,400 [r]	264,900 [r]	255,000
Chemicals[e]		2,900	2,700	2,900	2,900	2,900
Total		261,700 [r]	257,700 [r]	265,300 [r]	267,800 [r]	257,900
Niobium (columbium)[e]	kilograms	150	150	150	150	150
Platinum-group metals:						
Platinum	do.	27,000 [r]	25,900 [r]	25,700 [r]	27,300 [r]	30,200
Palladium	do.	84,000 [r]	83,200 [r]	84,700	84,100 [r]	82,400
Other	do.	12,500	11,900	12,000	12,000	12,000
Total	do.	123,500 [r]	121,000 [r]	122,400 [r]	123,400 [r]	124,600
Rhenium[e]	do.	500 [r]	500 [r]	500 [r]	500 [r]	500
Selenium[e]	do.	130,000	140,000	140,000	140,000	145,000
Silicon[e]		1,000,000 [r]	1,000,000 [r]	1,000,000 [r]	1,031,000 [r,3]	1,043,000 [3]
Silver:						
Mine output, Ag content	kilograms	1,400,000 [r]	1,590,000 [r]	1,545,000 [r]	1,543,000 [r]	1,679,000
Secondary recovery	do.	265	228	408	393	400

See footnotes at end of table.

TABLE 1—Continued
RUSSIA: PRODUCTION OF MINERAL COMMODITIES[1]

(Metric tons unless otherwise specified)

Commodity[2]		2008	2009	2010	2011	2012
METALS—Continued						
Tin:[e]						
Mine output, recoverable Sn content		400 [r, 3]	127 [r, 3]	144 [r, 3]	75 [r, 3]	100
Metal, smelter:						
Primary		1,425 [r, 3]	1,129 [r, 3]	1,081 [r, 3]	526 [r, 3]	500
Secondary		300	300	300	200 [r]	200
Total		1,700 [r]	1,400 [r]	1,400 [r]	700 [r]	700
Titanium sponge		34,730 [r]	22,600 [r]	26,500 [r]	24,600 [r]	42,000
Tungsten, concentrate, W content		3,163	2,665	2,785	3,314	3,025
Vanadium, metal[e]		14,500	14,500	15,000	12,860 [r, 3]	14,856 [3]
Zinc:						
Mine output, recoverable Zn content		174,000	241,700	186,900	176,300	179,800
Metal, smelter, primary and secondary		262,700	227,000	248,600	255,600	260,000 [e]
Zirconium, baddeleyite concentrate, averaging 98% ZrO$_2$		7,094	8,249	9,308	8,914	9,000 [e]
INDUSTRIAL MINERALS						
Asbestos, grades I through VI		1,017,000	1,000,000 [e]	995,174	1,031,880	1,050,000 [e]
Barite[e]		63,000	63,000	60,000	63,000	63,000
Boron[e]	thousand metric tons	400	300 [r]	200 [r]	200 [r]	250
Cement, hydraulic		53,548,000	44,266,000	50,400,000	56,200,000	61,700,000
Clays:						
Bentonite		500,000	500,000	500,000	500,000	500,000
Kaolin concentrate		107,500	90,300	105,000	120,000	120,000 [e]
Diamond:[e]						
Gem	carats	21,925,000 [3]	17,791,400 [3]	17,800,000	20,140,000	19,900,000
Industrial	do.	15,000,000	15,000,000	15,000,000	15,000,000	15,000,000
Synthetic	do.	80,000,000	80,000,000	80,000,000	80,000,000	80,000,000
Total	do.	117,000,000	113,000,000	113,000,000	115,000,000	115,000,000
Diatomite		28,000	30,000	32,000	33,000	70,000
Feldspar[e]		45,000	45,000	45,000	45,000	45,000
Fluorspar, concentrate, 55% to 96.4% CaF$_2$		177,000 [r, 3]	114,000 [r, 3]	100,000 [r]	258,000 [r, 3]	100,000 [e]
Germanium[e]		2	2	5 [r, 3]	5	5
Graphite		14,000	14,000	14,000	14,000	14,000
Gypsum[e]		3,600,000	2,900,000 [3]	2,900,000	4,960,000 [r, 3]	5,000,000
Iodine[e]		300,000	250,000 [r]	230,000 [r]	210,000 [r]	200,000
Lime, industrial and construction[e]		8,200,000	7,000,000	9,500,000	10,100,000	10,800,000
Limestone		7,420,000	7,000,000 [e]	7,000,000 [e]	7,000,000 [e]	7,200,000
Mica[e]		100,000	100,000	100,000	100,000	100,000
Nitrogen, N content of ammonia		10,425,000	10,441,000	10,400,000	10,400,000	10,300,000
Perlite[e]	cubic meters	200,000	200,000	200,000	200,000	200,000
Phosphate rock:[e]						
Gross weight		10,400,000	9,500,000	11,000,000	11,000,000	12,500,000
P$_2$O$_5$ content		3,800,000	3,500,000	4,000,000	4,000,000	4,500,000
Potash, marketable, K$_2$O equivalent		5,992,400	3,727,000	6,283,000	6,498,000	5,563,000
Rare earths, total rare-earth oxides		3,400	2,600	2,300	2,500	2,400 [e]
Salt, all types		1,800,000	1,600,000	1,800,000	1,800,000	1,850,000
Soda ash		2,800,000 [e]	2,322,000	2,670,000	2,822,000	2,807,000
Sulfur:[e]						
Native		50,000	50,000	50,000	50,000	50,000
Pyrites		200,000	200,000	200,000	200,000	200,000
Byproduct:						
Metallurgy		100,000 [r]	100,000 [r]	100,000 [r]	200,000 [r]	300,000
Natural gas		6,100,000 [3]	6,000,000	6,000,000	6,000,000	6,000,000
Petroleum		500,000	600,000	600,000	600,000	700,000
Total		6,950,000 [r]	6,950,000 [r]	6,950,000 [r]	7,050,000 [r]	7,250,000

See footnotes at end of table.

TABLE 1—Continued
RUSSIA: PRODUCTION OF MINERAL COMMODITIES[1]

(Metric tons unless otherwise specified)

Commodity[2]		2008	2009	2010	2011	2012
INDUSTRIAL MINERALS—Continued						
Sulfur, sulfuric acid		9,106,000	8,600,000	10,200,000	10,700,000	11,000,000
Talc[e]		160,000	160,000	160,000	160,000	160,000
Vermiculite[e]		25,000	25,000	25,000	25,000	25,000
MINERAL FUELS AND RELATED MATERIALS						
Coal:						
Anthracite	thousand metric tons	6,383	7,100	8,700	10,000	11,400
Bituminous	do.	216,049	200,982	236,100 [r]	249,100 [r]	276,500
Lignite	do.	82,485	69,011	76,800 [r]	76,900 [r]	78,100
Total	do.	304,917	277,093	321,600 [r]	336,000 [r]	366,000
Coke, metallurgical, 6% moisture content	do.	32,082	24,200	26,800	26,800 [r]	26,900
Natural gas, marketed	million cubic meters	663,000	583,610	651,000	671,000	655,000
Peat, horticultural and fuel uses[e]		1,200,000	1,200,000	1,258,000 [r, 3]	1,337,000 [r, 3]	1,400,000
Petroleum:						
Crude:						
In gravimetric units		488,105,000	479,000,000	486,000,000	492,000,000	497,000,000
In volumetric units[e]	thousand 42-gallon barrels	3,550,000	3,590,000	3,530,000	3,578,000 [3]	3,615,000 [3]
Refinery products:						
In gravimetric units		237,000,000	237,000,000	250,000,000	258,000,000	271,000,000
In volumetric units[e]	thousand 42-gallon barrels	1,910,000	1,910,000	2,010,000	2,080,600 [3]	2,185,500 [3]
Uranium:						
U content		3,521	3,564	3,562	2,993	2,862
U_3O_8 content		4,152	4,203	4,200	3,502	3,348 [3]

[e]Estimated; estimated data are rounded to no more than three significant digits; may not add to totals shown. [r]Revised. do. Ditto. -- Zero.

[1]Table includes data available through February 28, 2014.

[2]In addition to the commodities listed, Russia produces a number of other mineral commodities, which include lithium, oil shale, scandium, tantalum, titanium ore, and vanadium ore, but available information is inadequate to make reliable estimates of output.

[3]Reported figure.

[4]Excludes nickel-chromium remelt alloy produced from scrap. The remelt alloy typically has a nickel content of 20% to 50%.

TABLE 2
RUSSIA: STRUCTURE OF THE MINERAL INDUSTRY IN 2012[1]

(Metric tons unless otherwise specified)

Commodity	Major operating companies, main facilities, or deposits	Location or deposit names	Annual capacity[e]
Alumina	Achinsk (United Company RUSAL)	Achinsk in East Siberia	900,000
Do.	Bogoslovsk (United Company RUSAL)	Krasnotur'insk	1,050,000
Do.	Boksitogorsk (United Company RUSAL)	Leningradskaya Oblast'	200,000
Do.	Pikalyovo (United Company RUSAL)	Pikalyovo	300,000
Do.	Uralsk (United Company RUSAL)	Kamensk-Uralskiy	700,000
Aluminum, primary smelters	Bogoslovskiy AZ (United Company RUSAL)	Krasnotur'insk	175,000
Do.	Bratskiy AZ (United Company RUSAL)	Bratsk	1,000,000
Do.	Irkutskiy AZ (United Company RUSAL)	Irkutskaya Oblast'	420,000
Do.	Kandalakskiy AZ (United Company RUSAL)	Kola Peninsula	75,000
Do.	Khakasskiy AZ (United Company RUSAL)	Khakasiya	300,000
Do.	Krasnoyarskiy AZ (United Company RUSAL)	Krasnoyarskiy Kray	1,000,000
Do.	Nadvoitskiy AZ (United Company RUSAL)	Nadvoitsy, Kareliya Republic	75,000
Do.	Novokuznetskiy AZ (United Company RUSAL)	Novokuznetsk	300,000
Do.	Sayanogorskiy AZ (United Company RUSAL)	Sayanogorsk	550,000
Do.	Uralskiy AZ (United Company RUSAL)	Kamensk-Uralskiy	150,000
Do.	Volgogradskiy AZ (United Company RUSAL)	Volgogradskaya Oblast'	175,000
Do.	Volkhovskiy AZ (United Company RUSAL)	Volkhov, east of St. Petersburg	20,000
Amber	Kaliningrad Amber enterprise (Kaliningrad regional authorities and Alrosa Co. Ltd.)	Kaliningrad Oblast'	250
Antimony:			
Sb content of concentrate	GeoProMining, Ltd. (GPM)	Sarylakh deposit, Ust'-Nera region, Sakha Republic (Yakutiya)	8,000 [2]
Do.	do.	Sentachan deposit, Northeastern Sakha Republic (Yakutiya)	NA
Do.	Zabaykal'skiy GOK (ZabGOK) (OOO NefteChimMash)	Zabaykal'skiy Kray	360,000
Compounds and metals	Ryazsvetmet plant	Ryazanskaya Oblast'	NA
Apatite, concentrate	Khibiny apatite association (OAO Apatit)	Kola Peninsula	15,000,000
Do.	Kovdor iron ore mining association	do.	700,000
Asbestos	Bazenovskoye chrysotile deposit	Sverdlovskaya Oblast'	NA
Do.	Molodeznoye deposit	Zabaykal'sk (Chita) Oblast'	NA
Do.	"Orenburg Minerals" Co., Kiembaevskoye chrysotile deposit	Orenburgskaya Oblast'	500,000
Do.	"Tuvaasbest" plant, Ak-Dovurakskoye chrysotile deposit	Tyva Republic	250,000
Do.	"Uralasbest" mining and clarification plant	Central Urals	1,100,000
Barite	Salarinskiy mining and beneficiation complex	Kvartsitovaya Sopka deposit	100,000
Bauxite	OAO Sevuralboksitruda (United Company RUSAL)	Severoural'sk region	NA
Do.	South-Urals mining company (United Company RUSAL)	South Urals	NA
Do.	Severnaya Onega Mine (United Company RUSAL)	Northwest region	800,000
Do.	Komi Aluminum (United Company RUSAL)	Sredne-Timanskiy	3,000,000
Boron, boric acid	Bor Association	Primorskiy Kray	140,000
Do.	Amur River complex	Russian Far East	8,000
Do.	Alga River chemical complex	do.	12,000
Chromite	Saranov complex	Saranovskiy	200,000
Coal	Donetskiy (east) basin	Rostovskaya Oblast'	30,000,000
Do.	Irkutskiy basin	Irkutskaya Oblast'	NA
Do.	Kansko-Achinskiy basin	Eastern Siberia	50,000,000
Do. thousand metric tons	Kuznetskiy basin (Kuzbass)	Western Siberia	160,000
Do.	Lenskiy basin	Sakha Republic (Yakutiya)	NA
Do.	Minusinskiy field	Khakasiya Republic	NA
Do.	Moscovskiy basin	Moscow region	15,000,000
Do.	Neryungri basin	Sakha Republic (Yakutiya)	15,000,000
Do.	Pechorskiy basin	Komi Republic	30,000,000

See footnotes at end of table.

TABLE 2—Continued
RUSSIA: STRUCTURE OF THE MINERAL INDUSTRY IN 2012[1]

(Metric tons unless otherwise specified)

Commodity		Major operating companies, main facilities, or deposits	Location or deposit names	Annual capacity[e]
Coal—Continued		South Yakutiya basin	Sakha Republic (Yakutiya)	17,000,000
Do.		Ulughemskiy basin	Tyva Republic	NA
Do.		Yuzhno-Yakutskiy basin	Sakha Republic (Yakutiya)	NA
Cobalt		OJSC MMC Norilsk Nickel (Nornickel)	Norilsk, Kola Peninsula	4,000
Do.		Rezh and Yuzhuralnikel enterprises	South Urals	2,100
Do.		Ufaleynikel Co.	Chelyabinskaya Oblast', Urals	4,000
Do.		Khovu-Aksynskoe (nickel-cobalt) deposit	Khovu-Aksy, Tyva Republic	NA
Copper:				
Cu in concentrate		OJSC MMC Norilsk Nickel (Nornickel)	Norilsk region, Kola Peninsula	500,000
Do.		ZAO Russkaya Mednaya Kompaniya (RMK)	Urals	70,000
Do.		Metalloinvest Holding	Udokan, Zabaykal'skiy Kray	NA
Do.		OAO Ural'skaya Gorno-Metallurgicheskaya Kompaniya (UGMK)	do.	230,000
Metal, refined		OJSC MMC Norilsk Nickel (Nornickel)	Norilsk region, Kola Peninsula	450,000
Do.		ZAO Russkaya Mednaya Kompaniya (RMK)	Urals	170,000
Do.		OAO Ural'skaya Gorno-Metallurgicheskaya Kompaniya (UGMK)	do.	360,000
Diamond, gem and industrial	thousand carats	Almazy Rossii-Sakha Joint Stock Co. (Alrosa Co. Ltd.) enterprises: Udachnyy mining and beneficiation complex	Sakha Republic (Yakutiya) mines: Zarnitsa and Udachnyy	NA
Do.	do.	Mirny mining and beneficiation complex	Mir and International	NA
Do.	do.	Aikhal mining and beneficiation complex	Aikhal and Komsomol'skiy	NA
Do.	do.	Anabaraskiy mining and beneficiation complex	Alluvial mines	NA
Do.	do.	Nyurbinskiy mining and beneficiation complex	Nyurbinskiy and Botuobinskiy	NA
Do.	do.	Lomonosov	Arkhangel'skaya Oblast'	NA
Feldspar		Kheto-Lanbino and Lupikko deposits	Kareliya Republic	NA
Ferroalloys		ChEMK Industrial Group enterprises:		
Do.		Chelyabinsk electrometallurgical plant	Chelyabinskaya Oblast'	450,000
Do.		Kuznetsk ferroalloys plant	Novokuznetsk	400,000
Do.		Chusovoy iron and steel plant	Permskiy Kray	NA
Do.		Klyuchevsk ferroalloy plant	Dvurechensk	160,000
Do.		Kosaya Gora iron works	Kosaya, Gora	200,000
Do.		Lipetsk iron and steel works	Lipetskaya Oblast'	NA
Do.		Serov ferroalloy plant [a subsidiary of Eurasian Natural Resources PLC (ENRC)]	Sverdlovskaya Oblast'	NA
Ferronickel		Ufaleynikel Co.	Chelyabinskaya Oblast', Urals	5,000
Ferrovanadium		Vanadii-Tulachermet (Evraz Group)	Tula, North Caucasus	NA
Fluorspar		Abagaytuy deposit	Transbaikal	NA
Do.		Usugli Mine	do.	NA
Do.		Kyakhtinsky deposit	do.	NA
Do.		Kalanguy mining complex	Zabaykal'skiy Kray	NA
Do.		Yaroslavsky mining and beneficiation complex	Pogranichnoye and Vosnesenskoye deposits, Primorskiy Kray	NA
Gallium		Achinsk (United Company RUSAL)	Achinsk in Eastern Siberia	15 [2]
Do.		OOO Galliy	NA	NA
Do.		Novosibirsk tin complex	Novosibirsk	NA
Do.		Pikalevo (United Company RUSAL)	Pikalevo	NA
Germanium, metal and products		Federal State Unitary Enterprise Germanium	Kranoyarsk	7
Gold	kilograms	Mining companies: ZAO Amur a/s	Mining regions: Khabarovskiy Kray	5,500
Do.	do.	OAO Buryatzoloto	Buryatiya Republic	5,000
Do.	do.	ZAO Chukotskaya Mining and Geological Co. (Chukotskaya GGK)	Chukotskiy Avtonomnyy Okrug	15,000
Do.	do.	OOO Mining and Geological Co. (GRK) Aldanzoloto	Sakha Republic (Yakutiya)	4,000
Do.	do.	Highland Gold Mining Ltd. (HGM)	Khabarovskiy and Zabaykal'skiy Kray	NA

See footnotes at end of table.

TABLE 2—Continued
RUSSIA: STRUCTURE OF THE MINERAL INDUSTRY IN 2012[1]

(Metric tons unless otherwise specified)

Commodity		Major operating companies, main facilities, or deposits	Location or deposit names	Annual capacity[e]
Gold—Continued		Mining companies—Continued:	Mining regions—Continued:	
Do.	kilograms	Kinross Gold Corp.	Chukotskiy Avtonomnyy Okrug	NA
Do.	do.	LT-Resurs, ZAO	Irkutskaya Oblast'	2,700
Do.	do.	OOO Neryungri-Metallik	Sakha Republic (Yakutiya)	1,500
Do.	do.	OOO Nirungan	do.	1,100
Do.	do.	OAO Omchak	Magadanskaya Oblast'	3,000
Do.	do.	OAO Omolonskaya ZRK	do.	5,000
Do.	do.	ZAO Omsukchanskaya GGK	do.	3,000
Do.	do.	Oyna, a/s	Tyva Republic	1,500
Do.	do.	Petropavlovsk plc	Petropavlovsk	23,000
Do.	do.	OAO Pokrovskiy Mine	Amurskaya Oblast'	6,000
Do.	do.	OAO Polimetal	Magadanskaya and Sverdlovskaya Oblast's, Khabarovskiy Kray	7,500
Do.	do.	Polyarnaya, a/s	Chukotskiy Avtonomnyy Okrug	1,000
Do.	do.	OAO Polyus Gold	Krasnoyarskiy Kray	50,000
Do.	do.	OOO Priisk Drazhnyy	do.	1,200
Do.	do.	OAO Priisk Solov'yevskiy	Amurskaya Oblast'	1,500
Do.	do.	OOO Ros-DV	Khabarovskiy Kray	1,100
Do.	do.	OOO Russdragmet	Khabarovskiy Kray, Zabaykal'skiy Kray	6,000
Do.	do.	OAO Seligdar	Sakha Republic (Yakutiya)	3,000
Do.	do.	Severstal Nordgold NV	Russia, Kazakhstan, and West Africa	10,200
Do.	do.	OOO Sovrudnik	Krasnoyarskiy Kray	3,900
Do.	do.	OAO Susumanzoloto	Magadanskaya Oblast'	3,000
Do.	do.	OAO Uralelktomed'	Sverdlovskaya Oblast'	1,400
Do.	do.	Vitim, a/s	Irkutskaya Oblast'	2,900
Do.	do.	Vostok, a/s	Khabarovskiy Kray	1,100
Do.	do.	OOO Vysochayshiy (GV Gold)	Irkutskaya Oblast' and Sakha Republic (Yakutiya)	5,500
Do.	do.	OOO Yuzhuralzoloto	Chelyabinskaya Oblast'	6,500
Do.	do.	Zapadnaya, a/s	Krasnoyarskiy Kray	1,900
Do.	do.	ZAO Zolotaya, ZDK	Khakasiya Republic	1,200
Indium:				
Primary		Chelyabinsk zinc plant	Chelyabinskaya Oblast'	6
Secondary		Elektrozink plant	Vladikavkaz, North Caucasus	6
Iron ore		Kursk Magnetic Anomaly (KMA) region, which contains the following enterprises:	Locations:	50,000,000 [2]
		Lebedi and Stoilo	Gubkin	
		Mikhaylovka	Zheleznogorsk	
Do.		Northwest region, which contains the following enterprises:	Locations:	22,000,000 [2]
		Kostomuksha	Kostomuksha	
		Kovdor	Kola Peninsula	
		Olenegorsk	Olenegorsk	
Do.		Siberia region, which contains the following enterprises:	Locations:	18,000,000 [2]
		East:		
		Korshunovo	Zheleznogorsk	
		Rudnogorsk	Rudnogorsk	
		West:		
		Abakan	Abaza	
		Sheregesh	Sheregesh	
		Tashtagol	Tashtagol	
		Teya	Vershina Tel	

See footnotes at end of table.

TABLE 2—Continued
RUSSIA: STRUCTURE OF THE MINERAL INDUSTRY IN 2012[1]

(Metric tons unless otherwise specified)

Commodity	Major operating companies, main facilities, or deposits	Location or deposit names	Annual capacity[e]
Iron ore—Continued	Urals region, which contains the following enterprises:	Locations:	22,000,000 [2]
	Akkermanovka	Novotroitsk	
	Bakal	Bakal	
	Goroblagodat	Kushva	
	Kachkanar	Kachkanar	
	Magnitogorsk	Magnitogorsk	
	Peshchanka	Rudnichnyy	
Lead, metal	Dalpolymetal lead smelter	Rudnaya in Primorskiy Kray	20,000
Do.	Elektrozink lead smelter [Ural Mining and Metallurgical Co. (UMMC)]	Vladikavkaz, North Caucasus	40,000
Lead-zinc, recoverable content of ore:			
Lead, recoverable Pb content of ore	Altay mining-beneficiation complex	Altay Kray, Southern Siberia	2,000
Do.	Dalpolymetal mining-beneficiation complex	Primorskiy Kray	20,000
Do.	Nerchinsk polymetallic complex	Zabaykal'skiy Kray	7,000
Do.	Sadon lead-zinc complex	North Ossetia	5,000
Do.	Salair mining-beneficiation complex	Kemerovskaya Oblast'	2,000
Zinc, recoverable Zn content of ore	Altay mining-beneficiation complex	Altay Kray, Southern Siberia	1,000
Do.	Dalpolymetal mining-beneficiation complex	Primorskiy Kray	25,000
Do.	Nerchinsk polymetallic complex	Zabaykal'skiy Kray	12,500
Do.	Sadon lead-zinc complex	Severnaya Osetiya	14,000
Do.	Salair mining-beneficiation complex	Kemerovskaya Oblast'	10,500
Limestone	Mazulsky Mine (United Company Rusal)	Goryachegorsk massif, Eastern Siberia	NA
Lithium and its compounds	JSC Novosibirsk Chemical Plant (TVEL Corp.)	Novosibirsk	NA
Do.	JSC Chemical-Metallurgical Plant (TVEL Corp.)	Kransnoyarsk	NA
Magnesite	Karagayskiy open pit (Magnezit Group) and Magnezitovaya underground mine (Magnezit Group)	Sakha group of deposits (Chelyabinskaya Oblast')	3,800,000 [2]
Magnesium, metal (for sale)	Avisma plant	Berezniki	35,000
Do.	Solikamsk plant (Uralkaliy)	Permskiy Kray	30,000
Mica	Emel'dzhak deposit, Aldan Shield	Sakha Republic (Yakutiya)	NA
Do.	Lopatova Guba mica pit, Northern Kareliya	Kareliya Republic	NA
Do.	Kovdor phlogopite Mine (Mica Mine; Slyuda Mine; Kovdorslyuda Shaft)	Kola Peninsula, Murmanskaya Oblast'	NA
Do.	Irkutsk complex (JSC "Vostoksluda")	Mam deposit, Irkutskaya Oblast'	NA
Molybdenum	Dzhida tungsten-molybdenum mine	West Transbaikal	NA
Do.	Sorsk molybdenum mining enterprise	Khakasiya Republic	NA
Do.	Tyrnyauz tungsten-molybdenum mine [OAO Kabardino-Balkarskaya Tungsten-Molybdenum Co. (Government of Kabardino-Balkarskaya Republic)]	Republic of Kabardino-Balkariya, North Caucasus	NA
Do.	Shakhtaminskoye molybdenum mining enterprise	Zabaykal'skiy Kray	NA
Natural gas million cubic meters	Komi Republic	Komi Republic	8,000
Do. do.	Norilsk area	Norilsk region and Kola Peninsula	5,500
Do. do.	North Caucasus	North Caucasus	6,000
Do. do.	Sakhalin	Russian Far East	2,000
Do. do.	Tomsk Oblast	Western Siberia	500
Do. do.	Tyumen Oblast, of which:	do.	575,000 [2]
Do. do.	Medvezhye field	do.	(75,000)
Do. do.	Urengoy field	do.	(300,000)
Do. do.	Vyrngapur field	do.	(17,000)
Do. do.	Yamburg field	do.	(170,000)
Do. do.	Bovanenko field	Yamal Peninsula	NA
Do. do.	Pestsovoyy field	Ob-Taz Gulf area	NA
Do. do.	Zapolyarnyy field	do.	NA
Do. do.	Schtokmanov field	Barents Sea	NA

See footnotes at end of table.

TABLE 2—Continued
RUSSIA: STRUCTURE OF THE MINERAL INDUSTRY IN 2012[1]

(Metric tons unless otherwise specified)

Commodity		Major operating companies, main facilities, or deposits	Location or deposit names	Annual capacity[e]
Natural gas— Continued	million cubic meters	Urals	Ural'skiye Gory	45,000
Do.	do.	Volga	Vologodskaya Oblast'	6,000
Do.	do.	Yakut-Sakha	Sakha Republic (Yakutiya)	1,500
Nepheline syenite		Apatite complex	Kola Peninsula	1,500,000
Do.		Kiya-Shaltyr Mine (United Company RUSAL)	Goryachegorsk massif, Eastern Siberia	NA
Nickel:				
Ni in ore		OJSC MMC Norilsk Nickel (Nornickel)	Kola Peninsula and Norilsk region	300,000
Do.		OAO Ufaleynikel [Koks Company of Industrial Metallurgical Holding]	Chelyabinskaya Oblast', Urals	17,000
Do.		OAO Yuzhuralnikel [OAO Mechel]	South Urals	3,000
Do.				
Metal:				
Smelting		OJSC MMC Norilsk Nickel (Nornickel)	Norilsk region, Kola Peninsula	160,000
Do.		do.	Pechenga	50,000
Do.		do.	Monchegorsk	50,000
Refining		do.	do.	140,000
Do.		do.	Norilsk region, Kola Peninsula	100,000
Ni products and Ni in FeNi		Enterprises: ZAO Rezhnickel [Ural Mining and Metallurgical Co. (UMMC)]	Location: South Urals	65,000 [2]
		OAO Ufaleynikel [Koks Industrial Metallurgical Holding Co.]	do.	
		Yuzhuralnikel [Mechel OAO]	do.	
Niobium (columbium)		Karnarsurt mining enterprise (AO Sevredmet)	Lovozerskoye deposit, Kola Peninsula	12,000
Oil shale		Leningradslanets Association	Slantsy, Leningradskaya Oblast'	5,000,000
Petroleum		Bashneft'	Bashkortostan Republic	12,000,000
Do.		Gazprom Neft'	Deposits throughout Russia	50,000,000
Do.		OAO Lukoil	West Siberian deposits: Kechimovskoye Nivagalskoye Urals deposits Volga Region Timen Pechora deposit: Yuzhnaya Khylchuya Komi Republic deposits: Kyrtayelskoye Pashshorskoye Perevoznoye	100,000,000 [2]
Do.		OAO Novatek	Western Siberia	5,000,000
Do.		OAO NK Rosneft'	Deposits throughout Russia	120,000,000
Do.		Russneft'	Central and Western Siberia, Urals and Volga regions	15,000,000
Do.		Slavneft'	Western Siberia and Krasnoyarskiy Kray	20,000,000
Do.		OAO Surgutneftegas	Khanty-Mansiyskiy Avtonomnyiy Okrug (HMAO)	60,000,000
Do.		Tatneft'	Deposits: Romashkinskoye Novo-Elkhovskoye Bavlinskoye Bondyuzskoye Pervomayskoye Sabandchinskoye	30,000,000 [2]

See footnotes at end of table.

TABLE 2—Continued
RUSSIA: STRUCTURE OF THE MINERAL INDUSTRY IN 2012[1]

(Metric tons unless otherwise specified)

Commodity	Major operating companies, main facilities, or deposits	Location or deposit names	Annual capacity[e]
Petroleum—Continued	OAO TNK–BP Holding	Deposits:	80,000,000 [2]
		Kamennoye	
		Kovyatka	
		Russkoye	
		Suzunskoye	
		Tagulskoye	
		Uvat	
		Verkhnechonsk	
Phosphate rock	Kingisepp complex (OAO Fosforit)	Leningradskaya Oblast'	3,500,000
Do.	Lopatino and Yegorevsk deposits	Moscow Oblast'	NA
Do.	Polpinskoye deposit	Bryanskaya Oblast'	NA
Do.	Verkhnekamsk deposit	Urals	NA
Phosphate rock, apatite concentrate	OAO Apatit (Phosagro)	Kola Peninsula	12,000,000
Do.	Kovdor iron mining complex	do.	700,000
Platinum-group metals:			
Ore, PGM content	OJSC MMC Norilsk Nickel	Norilsk region, Kola Peninsula	150
Do.	AO Koryakgeoldobycha, Amur Prospectors	Placer deposits (mostly platinum), Urals; Siberia; Russian Far East	10
Do.	Lopatino and Yegorevsk deposits	Moscow Oblast'	NA
Do.	Polpinskoye deposit	Bryanskaya Oblast'	NA
Do.	Verkhnekamsk deposit	Ural'skiye Gory	NA
Do.	OAO AS Amur (Russian Platinum Co.)	Placer deposits (mostly platinum), Urals; Siberia; Russian Far East	10
Metals	Krasnoyarsk Nonferrous Metals Plant (Krastsvetmet)	Krasnoyarskiy Kray	NA
Do.	Ekaterinburgskiy plant (EZOTsM)	Sverdlovskaya Oblast'	NA
Do.	Priobsk plant (OJSC Gazprom Neft)	Khanty-Mansiyskiy Avtonomnyi Okrug (HMAO)	NA
Potash, K_2O equivalent	OAO Uralkali	Verkhnekamskoye deposit	8,000,000
Do.	OAO Silvinit[3]	Solikamsk-Berezniki regions, Urals	NA
Do.	OAO Akron	Novgorod	NA
Rare earths	OAO Apatit	Lovozerskoe deposit, Kola Peninsula	NA
Salt	AO Bassol'	Lake Baskunchak in Astrakhanskaya Oblast'	2,500,000
Do.	Dus-Dagskoe deposit	Dus-Dag Mountains	25,000
Silver	Dukat Mine	Magadanskaya Oblast'	1,000
Do.	Kinross Gold Corp.	Chukotskiy Avtonomnyy Okrug	NA
Soda ash	Achinsk plant	Eastern Siberia	595
Do.	Berezniki plant	Ural'skiye Gory	1,080
Do.	Pikalevo plant	Leningradskaya Oblast'	200
Do.	Sterlitamak plant	Bashkortostan Republic	2,135
Do.	Volkhov plant	Leningradskaya Oblast'	20
Steel, crude	OAO Amurmetal	Komsomol'sk-na-Amure	1,600,000
Do.	JSC Asha Metallurgical Plant	Chelyabinskaya Oblast'	450,000
Do.	Beloretsk Iron and Steel Works	Bashkirskoye	380,000
Do.	Chusovskoy Iron and Steel Works	Permskiy Kray	570,000
Do.	JSC Electrostal Metallurgical Plant	Moscow	314,000
Do.	Gorkovskoy Metallurgichesky Zavod	Nizhegorodskaya Oblast'	78,000
Do.	Gur'yevsk Steel Works	Kemerovskaya Oblast'	160,000
Do.	Karaganda	Karagandinskaya Oblast'	6,300,000
Do.	Kuznetsk Steel Works	Kemerovskaya Oblast'	4,700,000
Do.	Lys'va Metallurgical Plant	Permskiy Kray	350,000
Do.	OAO Magnitogorsk mining and metallurgical complex (MMK)	Chelyabinskaya Oblast'	16,200,000

See footnotes at end of table.

TABLE 2—Continued
RUSSIA: STRUCTURE OF THE MINERAL INDUSTRY IN 2012[1]

(Metric tons unless otherwise specified)

Commodity	Major operating companies, main facilities, or deposits	Location or deposit names	Annual capacity[e]
Steel, crude—Continued	OAO Mechel (Mechel)	Chelyabinskaya Oblast'	7,000,000
Do.	Nizhniy Sergi Steel Works	Sverdlovskaya Oblast'	300,000
Do.	Nizhniy Tagil mining and metallurgical complex (NTMK) (Evraz Group)	do.	8,000,000
Do.	Nosta JSC (JSC Orsk-Kahlilovo Iron and Steel Works)	Novotroitsk, Orenburgskaya Oblast'	4,600,000
Do.	Novolipetskiy mining and metallurgical complex (NLMK)	Lipetskaya Oblast'	9,900,000
Do.	Novosibirsk Steel Works (Novosibprokat)	Novosibirskaya Oblast'	1,100,000
Do.	CJSC Omutninsk Metallurgical Plant	Kirovskaya Oblast'	210,000
Do.	Oskol Electric Steel Works (OEMK)	Staryi Oskol	2,500,000
Do.	Petrovsk-Zabaykal'skiy Steel Works	Petrovsk-Zabaykal'skiy	426,000
Do.	Revdinskiy Steel and Wire Production Works	Sverdlovskaya Oblast'	281,000
Do.	Salda Steel Works	do.	1,900
Do.	Serov Steel Works	do.	1,000,000
Do.	Serp i Molot (Moscow Metallurgical Works)	Moskovskaya Oblast'	70,000
Do.	Severskiy Tube Works	Polevskoy, Sverdlovskaya Oblast'	825,000
Do.	OAO Severstal	Vologodskaya Oblast'	14,000,000
Do.	Sibelektrostal Metallurgical Works	Krasnoyarskiy Kray	110,000
Do.	Sulinskiy Steel Works (Staks)	Rostovskaya Oblast'	280,000
Do.	Taganrog Iron and Steel Works (Tagmet)	do.	925,000
Do.	OAO Tulachermet	Tul'skaya Oblast'	18,400
Do.	Viz-Stal (Verkh-Isetsk Steel Works)	Sverdlovskaya Oblast'	132,000
Do.	Volgograd Steel Works (Red October)	Volgogradskaya Oblast'	2,000,000
Do.	Vyksa Steel Works	Nizhegorodskaya Oblast'	540,000
Do.	Zapadno-Sibirskiy mining and metallurgical complex (ZSMK) (Evraz Group)	Kemerovskaya Oblast'	6,900,000
Do.	Zlatoust Iron and Steel Works	Zlatoust, Chelyabinskaya Oblast'	1,200,000
Talc	Onotsk deposit	Irkutskaya Oblast'	NA
Do.	Kirgiteysk deposit	Krasnoyarskiy Kray	NA
Do.	Miass deposit	Chelyabinskaya Oblast'	NA
Do.	Shabrovsk deposit	Sverdlovskaya Oblast'	NA
Tantalum, ore	Facilities: Zabaykalskiy mining and beneficiation complex NA	Deposits: Etykinskoye deposit Lovozerskoye deposit, Kola Peninsula	10 [2]
Tellurium	OJSC MMC Norilsk Nickel	NA	5
Do.	Ural Mining and Metallurgical Co. (UMMC)	Urals	35
Tin:	Novosibirsk mining and beneficiation complexes:	Locations:	
Ore	Khinganskoye olovo (Jewish Autonomous District)	Khabarovskiy Kray	11 [4]
Do.	Tin Ore Co.	Solnechnyi deposit, Khabarovskiy Kray	NA
Do.	Pravourmiyskoye	Khabarovskiy Kray	NA
Do.	Deputatskiy (Sakhaolovo)	Sakha Republic (Yakutiya)	NA [4]
Do.	Vostokolovo	Russian Far East	NA [4]
Do.	Iultin mining and beneficiation complex	Magadanskaya Oblast'	NA [4]
Do.	Khrustalnyy mining and beneficiation complex	Primorskiy Kray	NA [4]
Do.	Pevek mining and beneficiation complex	Magadanskaya Oblast'	NA [4]
Metal	Novosibirsk Processing Plant Ltd.	Novosibirskaya Oblast'	NA [4]
Titanium:			
Ore	OOO Lovozerskiy GOK	Murmanskaya Oblast	NA
Do.	OAO Apatit	Kykisvumchorrskoye and Yuksporskoye deposits	NA
Do.	OAO TGOK Ilmenit	Tyuganskoye deposit	NA
Do.	OOO Olekminskiy Rudnik	Kuranakhskoye deposit	NA

See footnotes at end of table.

TABLE 2—Continued
RUSSIA: STRUCTURE OF THE MINERAL INDUSTRY IN 2012[1]

(Metric tons unless otherwise specified)

Commodity	Major operating companies, main facilities, or deposits	Location or deposit names	Annual capacity[e]
Titanium—Continued:			
Metal	Moscow plant	Moscow	NA
Do.	Podol'sk plant	Podol'sk	NA
Do.	OAO Corp. VSMPO-Avisma	Bereznikovskiy Complex, Permskiy Kray	NA
Sponge	do.	do.	40,000
Do.	Solikamskiy Magnium Plant (SMZ)	Solikamsk, Permskiy Kray	NA
Tungsten:			
Concentrates, W content	AS Quartz	Bom-Gorkhom deposit, West Transbaikal, Zabaykal'skiy Kray	NA
Do.	ZAO Novoorlovskiy GOK	Spokoyninskoye deposit, Zabaykal'skiy Kray	NA
Do.	KGUP Primteploenergo	Lermontovskoye deposit, Primorskiy Kray	NA
Do.	OAO Primorsky GOK	Vostok-2 deposit	NA
Do.	ZAO Zakamensk	Ruchey Inkur deposit, Barun-Narynskoye deposit	NA
Do.	Tyrnyauz tungsten-molybdenum mine [OAO Kabardino-Balkarskaya Tungsten-Molybdenum Co. (Government of Kabardino-Balkarskaya Republic)]	Republic of Kabardino-Balkariya, North Caucasus	NA
Metal	Gidrometallurg plant	do.	NA
Uranium, U content	Uranium Holding OAO Atomredmetzoloto (ARMZ) ZAO Dalur mining enterprise OAO Khiagda mining enterprise Priargunsky mining and chemical enterprise	Locations: Kurganskaya Oblast' Buryatiya Republic Krasnokamensk, Zabaykal'skiy Kray	3,500
Vanadium:			
Ore	Kachkanar iron mining complex	Ural'skiye Gory	NA
Metal	Chusovoy and Nizhniy Tagil plants	do.	17,000
Pentoxide	Vanadii-Tulachermet	Tul'skaya Oblast', North Caucasus	NA
Zinc:			
Zn content of copper-zinc ore	Bashkir copper-zinc complex	Sibai, Southern Urals	5,000
Do.	Buribai copper-zinc mining complex	Buribai, Southern Urals	1,500
Do.	Gai copper-zinc mining and beneficiation complex	Gai, Southern Urals	25,000
Do.	Kirovgrad copper enterprise	Kirovgrad, Central Urals	1,200
Do.	Sredneuralsk copper complex	Revda, Central Urals	5,000
Do.	Uchali copper-zinc mining and beneficiation complex	Uchalinskiy Rayon, Southern Urals	90,000
Metal	Chelyabinsk electrolytic zinc plant	Chelyabinskaya Oblast'	200,000
Do.	Elektrozink plant [Ural Mining and Metallurgical Co. (UMMC)]	Vladikavkaz, North Caucasus	90,000
Do.	Uralelektromed plant [Ural Mining and Metallurgical Co. (UMMC)]	Verkhnaya Pyshma	17,000
Zirconium:			
Baddeleyite concentrate	Kovdor iron ore mining and beneficiation complex	Kola Peninsula	3,500
Metal	Chepetsky metallurgical plant (TVEL Corp.)	Glazov, Udmurtiya Republic	NA

[e]Estimated; estimated data are rounded to no more than three significant digits. Do., do. Ditto. NA Not available.

[1]Many location names have changed since the breakup of the Soviet Union. Many enterprises, however, are still named or commonly referred to based on the former location name, which accounts for discrepancies in the names of enterprises and that of locations.

[2]Capacity estimates are totals for all enterprises that produce that commodity.

[3]Merged with Uralkali in February 2011.

[4]Not in operation as of 2012.

The Mineral Industry of Serbia

By Yadira Soto-Viruet

Serbia's mineral industry was dominated by copper, iron and steel, and refined petroleum products. Other mineral and mineral-based commodities produced in the country included cement, coal, gold, lead, natural gas, nitrogen, salt, and selenium.

Minerals in the National Economy

In 2012, Serbia's gross domestic product (GDP) decreased by 1.7%. In 2011 (the most recent year for which data were available), mining and quarrying made up about 1.9% of the GDP. The value of exports of mining and quarrying products in 2011 was about $92 million, and the value of imports of mining and quarrying products was about $2.5 billion. About 87% of crude petroleum and 90% of natural gas imports were from Russia. The country also imported about 98% of its iron ores and concentrates from Ukraine (Statistical Office of the Republic of Serbia, 2012, p. 121, 289–290, 298–299; 2013, p. 1, 4).

In 2012, Serbia's exports to the United States were valued at about $145 million compared with about $134 million in 2011; these included about $47,000 in copper. Imports from the United States were valued at about $128 million in 2012 compared with $126 million in 2011; these included about $3.1 million in drilling and oilfield equipment and about $1.0 million in excavating machinery (U.S. Census Bureau, 2013a, b).

Production

In 2012, primary refined copper production increased by 28% to 32,229 metric tons (t) from 25,251 t (revised) in 2011; estimated primary metal copper production increased by 27% to 32,500 t from 25,500 t (revised); copper content of concentrate production increased by 23% to 34,400 t from 28,000 t (revised); and estimated crude petroleum production increased by 10% to 8,340,000 barrels (bbl) from 7,570,000 bbl. Crude steel production decreased by 74% to 345,000 t from 1.3 million metric tons (Mt); pig iron production, by 75% to 312,000 t from 1.2 Mt; and salt production, by 29% to 16,506 t from 23,144 t. Data on mineral production are in table 1.

Structure of the Mineral Industry

Cement was produced by Cementara Kosjeric a.d., Holcim (Srbija) d.o.o., and Lafarge Beocinska Fabrika Cementa. These companies had a combined cement production capacity of 4.1 million metric tons per year (Mt/yr). Rudarsko Topionicki Bazen Bor (RTB Bor) was the only producer of copper, gold, and silver in the country. Table 2 is a list of major mineral industry facilities.

At least five companies were engaged in gold and copper exploration in Serbia. These included Canadian companies Avala Resources Ltd., Dunav Resources Ltd., Euromax Resources Ltd., and Reservoir Minerals Inc., and Orogen Gold plc of the United Kingdom.

Commodity Review

Metals

Copper.—In May, state-owned RTB Bor announced the rehabilitation and reopening of the Cerovo Mine after an investment of about $23 million. The open pit mine, which was closed in 2002, is located 15 kilometers (km) northwest of Bor and had estimated reserves of about 150 Mt at an average grade of 0.35% copper. The company envisioned that initial output when production resumes at Cerovo would be 2.5 Mt/yr of ore and would increase to 5.5 Mt/yr after 2015. In 2011, RTB Bor awarded the contract for the modernization of its existing copper smelting complex to SCN-Lavalin Group Inc. of Canada, at a cost of about $211 million. The project, which was to include a flash furnace and a sulfuric acid plant, was scheduled to be completed by yearend 2013. The new smelter was expected to produce 80,000 metric tons per year (t/yr) of anodes (SCN-Lavalin Group Inc., 2011; Outotec Oyj, 2012; Rudarsko Topionicki Bazen Bor, 2012a, b; Savic, 2012).

Gold.—Avala Resources through its subsidiary Avala Resources d.o.o. held copper and gold exploration licenses for the 1,050-square-kilometer Timok Gold Project, which is located about 270 km southeast of the capital city of Belgrade. In January 2013, a technical report and mineral resource estimates were completed by AMC Consultants Ltd. of the United Kingdom. Total inferred mineral resources at the Bigar Hill, the Korkan, and the Kraku Pester deposits were estimated to be 48.7 Mt at an average grade of 1.5 grams per metric ton (g/t) gold using a cutoff grade of 0.6 g/t gold. A preliminary economic assessment was expected to be completed in mid-2013 (AMC Consultants Ltd., 2013, p. 1, 8; Avala Resources Ltd., 2013).

Iron and Steel.—On January 31, United States Steel Corp. of Pennsylvania sold its 2.2-Mt/yr-capacity steel plant in Smederevo (the only steel plant in Serbia) to the Government of Serbia. After the sale, the plant's name was changed to Zelezara Smederevo d.o.o. The Government was seeking to prevent any job losses and intended to find a new strategic partner for state-owned Zelezara Smederevo. In June, the company announced that it would temporarily suspend production at its second blast furnace in July, stating that the company needed to find a strategic partner. In October, the Government, through the Serbian Ministry of Finance and Economy, opened a second public invitation for a new strategic partner and for further developments of the Zelezara Smederevo plant. The public invitation was scheduled to close on February 28, 2013 (Ministry of Finance and Economy, 2012a, b; United States Steel Corp., 2012, p. 4; Zelezara Smederevo d.o.o., 2012).

Industrial Minerals

Lithium.—Rio Tinto plc of the United Kingdom held 100% interest in the Jadar lithium-borate project, which is located about 100 km from the capital city of Belgrade in western Serbia. By yearend, total mineral resources at Jadar were estimated to be 118 Mt at an average grade of 1.6% lithium oxide. A prefeasibity study was underway. The reported inferred resources represented only those in the lowest of the three vertically stacked zones of the deposit. The company estimated that, if developed, the project could begin production by 2016 (Rio Tinto Minerals, 2011, 2013; Rio Tinto plc 2013, p. 27, 55, 57).

Mineral Fuels

Natural Gas and Petroleum.—Naftna Industrija Srbije a.d. (NIS), which was a joint venture between JSC Gazprom Neft of Russia (56.15%) and the Government (29.87%), was the sole producer of natural gas, petroleum, and refined petroleum products in Serbia. In November, the company completed a modernization project at its Pancevo refinery. The project included a new light hydrocracking and motor fuel hydrotreatment facility. From 2013 onward, the Pancevo refinery would produce only EURO-5 standard gasoline and diesel with sulfur content of less than 10 parts per million (in accordance with European Union environmental requirements). NIS expected eventually to increase the refinery's production capacity to about 640,000 t/yr of standard gasoline and 1.5 Mt/yr of diesel (JSC Gazprom Neft, 2012; Naftna Industrija Srbije a.d., 2012a, b).

Outlook

Serbia's GDP growth is projected to reach 2% in 2013 based on the planned increase in exports, mostly in the agriculture and automobile sectors (International Monetary Fund, 2012, p. 194; European Bank for Reconstruction and Development, 2013). The reopening of the Cerovo Mine, the expansion plans at RTB Bor, and the modernization and expansion at the NIS refinery are expected to strengthen the mineral industry in the short run. In the longer run, new projects in the nonfuel mineral sector, such as possible developments of copper, gold, and lithium deposits, are likely to provide significant revenue to the Government and to the GDP. Plans to rehabilitate the country's infrastructure (including railways, roads, water supply, and wastewater treatment plants) are all likely to attract increased foreign direct investment in the mineral sector and to increase interest in nonfuel mineral prospecting.

References Cited

AMC Consultants Ltd., 2013, Timok Gold Project Serbia—Technical report and mineral resources estimates: London, United Kingdom, AMC Consultants Ltd. technical report, January 9, 161 p. (Accessed July 18, 2013, at http://www.avalaresources.com/i/pdf/TIMOK_GOLD_PROJECT-NI-43-101-JAN-9-2013.pdf.)

Avala Resources Ltd., 2013, Summary of total mineral resources: Longueuil, Quebec, Canada, Avala Resources Ltd. (Accessed July 18, 2013, at http://www.avalaresources.com/s/Resources.asp.)

European Bank for Reconstruction and Development, 2013, Serbia: European Bank for Reconstruction and Development, April, 2 p. (Accessed July 18, 2013, at http://www.ebrd.com/downloads/research/factsheets/serbia.pdf.)

International Monetary Fund, 2012, World economic outlook: Washington, DC, International Monetary Fund, October, 228 p. (Accessed July 22, 2013, at http://www.imf.org/external/pubs/ft/weo/2012/02/pdf/text.pdf.)

JSC Gazprom Neft, 2012, Gazprom Neft completes refinery modernization project at Pancevo: St. Petersburg, Russia, JSC Gazprom Neft press release, November 22. (Accessed July 15, 2013, at http://www.gazprom-neft.com/press-center/news/741930/.)

Ministry of Finance and Economy, 2012a, Amendments number 2 to the public invitation: Belgrade, Serbia, Ministry of Finance and Economy, [undated], 1 p. (Accessed October 29, 2013, at http://www.mfin.gov.rs/UserFiles/File/dokumenti/2012/Amendments No_ 2 of the Public Invitation for the selection of a Strategic Partner for a company engaged in the production and processing of steel Zelezara Smederevo.pdf.)

Ministry of Finance and Economy, 2012b, Selection of a strategic partner for "Zelezara Smederevo": Belgrade, Serbia, Ministry of Finance and Economy, October 4, 2 p. (Accessed October 29, 2013, at http://www.mfin.gov.rs/UserFiles/File/dokumenti/2012/Public Invitation Zelezara Smederevo2.pdf.)

Naftna Industrija Srbije a.d., 2012a, Gazprom Neft completes refinery modernization project at Pancevo: Naftna Industrija Srbije a.d. press release, November, 2. (Accessed July 15, 2013, at http://www.nis.rs/prescentar/vesti/modernizovana-rafinerija-pancevo?lang=en.)

Naftna Industrija Srbije a.d., 2012b, The company "Gazprom Neft" fulfilled investment obligations from the purchase agreement of 51% of NIS's shares: Naftna Industrija Srbije a.d. press release, April 24. (Accessed July 15, 2013, at http://www.nis.rs/prescentar/vesti/gasprom-neft-ispunjene-obaveze-ugovor?lang=en.)

Outotec Oyj, 2012, New flash smelter for RTB-Bor in central Serbia (under construction): Outotec Oyj smelting news, December 19, no. 1, 2 p. (Accessed July 22, 2013, at http://www.outotec.com/imagevaultfiles/id_919/cf_2/2012_issue1_new_flash_smelter_to_rtbbor_in_central.pdf.)

Rio Tinto Minerals, 2011, Rio Tinto Minerals and Loznica officials sign cooperation agreement in support of Jadar lithium-borate project: Belgrade, Serbia, Rio Tinto Minerals press release, July 1, 1 p. (Accessed July 5, 2013, at http://www.riotintominerals.com/documents/RTM-NewsRelease-JadarMOU-1July2011-FINAL.pdf.)

Rio Tinto Minerals, 2013, Our history: Rio Tinto Minerals. (Accessed July 8, 2013, at http://www.riotintominerals.com/ENG/ourbusiness/1885_history_of_the_talc_business.asp.)

Rio Tinto plc, 2013, 2012 annual report: London, United Kingdom, Rio Tinto plc, 232 p. (Accessed July 5, 2013, at http://www.riotinto.com/reportingcentre2012/pdfs/rio_tinto_2012_annual_report.pdf.)

Rudarsko Topionicki Bazen Bor, 2012a, Iz uspensne u godinu velike zavrsnice: Rudarsko Topionicki Bazen Bor press release, December 29. (Accessed July 22, 2013, at http://rtb.rs/novogodisnja-pres-konferencija-generalnog-direktora-rtb-a-bor/.)

Rudarsko Topionicki Bazen Bor, 2012b, Potekla ruda iz Cerova: Rudarsko Topionicki Bazen Bor press release, May 18. (Accessed July 22, 2013, at http://rtb.rs/juce-zavrsene-poslednje-provere-u-obnovljenom-rudniku/.)

Savic, Misha, 2012, RTB Bor reopens Cerovo Mine to increase copper production: Bloomberg L.P., May 9. (Accessed July 22, 2013, at http://www.bloomberg.com/news/2012-05-09/rtb-bor-reopens-cerovo-mine-to-increase-copper-production.html.)

SCN-Lavalin Group Inc., 2011, SCN-Lavalin awarded copper smelter modernization project in Serbia: Montreal, Quebec, Canada, SCN-Lavalin Group Inc. press release, January 18. (Accessed August 30, 2013, at http://www.snclavalin.com/news.php?lang=en&id=1301.)

Statistical Office of the Republic of Serbia, 2012, Statistical yearbook of Serbia 2012: Statistical Office of the Republic of Serbia, October, 484 p. (Accessed July 15, 2013, at http://pod2.stat.gov.rs/ObjavljenePublikacije/G2012/pdf/G20122007.pdf.)

Statistical Office of the Republic of Serbia, 2013, Quarterly gross domestic product in the Republic of Serbia, 4th quarter 2012: Statistical Office of the Republic of Serbia, March 30, 4 p. (Accessed July 15, 2013, at http://webrzs.stat.gov.rs/WebSite/repository/documents/00/00/97/34/NR40_073_srb+eng.pdf.)

United States Steel Corp., 2013, 2012 annual report and Form 10–K: United States Steel Corp., 171 p. (Accessed July 10, 2013, at http://www.ussteel.com/uss/wcm/connect/3d39a02d-688a-401c-a70a-8c2db248e2ef/USS-2012-Annual-Report.pdf?MOD=AJPERES&CACHEID=3d39a02d-688a-401c-a70a-8c2db248e2ef.)

U.S. Census Bureau, 2013a, U.S. exports to Serbia from 2003 to 2012 by 5-digit end-use code: U.S. Census Bureau. (Accessed July 10, 2013, at http://www.census.gov/foreign-trade/statistics/product/enduse/exports/c4801.html.)

U.S. Census Bureau, 2013b, U.S. imports from Serbia from 2003 to 2012 by 5-digit end-use code: U.S. Census Bureau. (Accessed July 10, 2013, at http://www.census.gov/foreign-trade/statistics/product/enduse/imports/c4801.html.)

Zelezara Smederevo d.o.o., 2012, Temporary production stoppage in Zelezara Smederevo: Zelezara Smederevo d.o.o., June 26. (Accessed August 19, 2013, at http://www.zelsd.rs/index.php?link=en/news-view/1241/temporary-production-stoppage-in-zelezara-smederevo.)

TABLE 1

SERBIA: PRODUCTION OF MINERAL COMMODITIES[1]

(Metric tons unless otherwise specified)

Commodity[2]		2008	2009	2010	2011	2012
METALS						
Aluminum ingot, including alloys, secondary		1,882	1,789	1,739	-- [r]	--
Copper:						
Mine and concentrator output:						
Ore:						
Gross weight	thousand metric tons	7,746	9,896	10,665	12,216 [r]	14,346
Cu content of ore		21,900	24,400	28,400	33,300 [r]	41,531
Concentrate, Cu content		20,800 [r]	23,400 [r]	24,600 [r]	28,000 [r]	34,400
Metal:						
Blister and anodes, primary[e]		19,000 [r]	19,000 [r]	21,500 [r]	25,500 [r]	32,500
Refined:						
Primary		18,550 [r]	18,875 [r]	21,240 [r]	25,251 [r]	32,229
Secondary		2,641 [r]	1,186 [r]	963 [r]	3,198 [r]	2,473
Total		21,191 [r]	20,061 [r]	22,203 [r]	28,449 [r]	34,702
Gold, refined	kilograms	712	452	356	1,032 [r]	900
Iron and steel, metal:						
Pig iron		1,582,000	1,008,000 [r]	1,265,000	1,226,000	312,000
Crude steel		1,662,000	1,097,000	1,254,000	1,324,000	345,000
Semimanufactures		2,393,430	1,556,017	2,127,960	1,885,019 [r]	480,000 [e]
Lead:						
Ore:						
Gross weight (Pb-Zn ore)		202,000	225,000	219,000	412,000 [r]	420,000 [e]
Pb content[e]		1,600	1,800	1,800	1,800	1,800
Metal, including alloys, secondary, refined[e]		929 [3]	1,000	1,000	1,000	1,000
Platinum-group metals:						
Palladium	kilograms	70	38	22	4 [r]	22
Platinum	do.	--	12	--	6 [r]	3
Selenium	do.	16,827	19,075	10,592	12,947 [r]	13,200
Silver, mine output, Ag content	do.	2,300 [e]	2,500 [e]	4,820	4,750 [r]	5,224
INDUSTRIAL MINERALS						
Cement	thousand metric tons	2,843	2,232	2,130	2,095	1,831
Lime	do.	292	251	239	274	239
Nitrogen, N content of ammonia[e]		47,000	53,000	84,000	132,000 [r]	130,000
Salt, all sources		30,115	28,783	30,816	23,144	16,506
Sand and gravel, excluding glass sand[e]	thousand cubic meters	8,670	5,790	6,950	6,532 [r]	6,167
MINERAL FUELS AND RELATED MATERIALS						
Coal:						
Bituminous	thousand metric tons	66	69	108	142 [r]	141
Brown coal and lignite	do.	38,519	38,828	38,490	40,636 [r]	38,587
Total	do.	38,585	38,897	38,598	40,778 [r]	38,728
Natural gas, gross production	million cubic meters	282	283	427	616	672
Petroleum:[4]						
Crude	thousand 42-gallon barrels	4,720	4,920	6,420	7,570 [e]	8,340 [e]
Refinery products[e]	do.	18,700	16,700	16,000	13,200 [r]	13,100 [r]

[e]Estimated; estimated data are rounded to no more than three significant digits. [r]Revised. do. Ditto. -- Zero.

[1]Table includes data available through November 7, 2013.

[2]In addition to the commodities listed, crude gypsum, secondary magnesium metal, zinc, and other mineral commodities may have been produced, but available information is inadequate to make reliable estimates of output.

[3]Reported figure.

[4]Data were converted to barrels from thousand metric tons, and were reported as follows: for crude production, 2008—636; 2009—663; 2010—865; 2011—1,020; and 2012—1,124; for petroleum products, 2008—2,462; 2009—2,227; 2010—2,130; 2011—1,760 (revised); and 2012—1,740 (estimated).

TABLE 2
SERBIA: STRUCTURE OF THE MINERAL INDUSTRY IN 2012

(Thousand metric tons unless otherwise specified)

Commodity	Major operating companies and major equity holders	Location of main facilities	Annual capacity
Antimony:			
Ores and concentrates	Farmakom M.B.	Mines and mills near Zajaca	NA
Metal	do.	Smelter at Zajaca	NA
Cement	Beocinska Fabrika Cementa (Lafarge S.A., 100%)	Plant at Beocin	2,000
Do.	Holcim (Srbija) a.d. (Holcim Ltd., 100%)	Plant at Popovac	1,400
Do.	Cementara Kosjeric a.d. (Titan Group, 100%)	Plant at Kosjeric	750
Coal:			
Bituminous	JP PEU Resavica	Ibarski Rudnici Mines near Baljevac and Vrska Cuka Mines	70 [e]
Brown coal	do.	Underground mines near Resavica, Bogdinac, Bogovina, Krepoljin, and Stavalj	400 [e]
Lignite	MB Kolubara Ltd. (Electric Power Industry of Serbia)	Opencast mines: Field B, Field D, Veliki Crljeni, and Tamnava West near Vreoci	31,000
Do.	TPPs-OCMs Kostolac Ltd. (Electric Power Industry of Serbia)	Opencast mine at Drmno near Kostolac	8,500
Do.	JP PEU Resavica	Underground mine at Lubnica	60 [e]
Copper:			
Mine production, Cu content of concentrate	Rudarsko Topionicki Bazen Bor (RTB Bor) (Government owned)	Mine and mill "Jama" at Bor	8
Do.	do.	Mine and mill at Majdanpek	65
Do.	do.	Mine and mill at Veliki Krivelj	35
Metal	do.	Smelter at Bor	170
Do.	do.	Electrolytic refinery at Bor	170
Lead, metal, secondary	Farmakom M.B.	Smelter at Zajaca	NA
Lead-zinc ore	Contango d.o.o.	Mine and mill at Rudnik	250 [e]
Do.	NA	Grot Mine near Vranje	300 [e]
Do.	Farmakom M.B.	Mines at Rajiceva Gora, Ravnaja, and Veliki Cip	350 [e]
Do.	Mineco Group	Mine at Veliki Majdan, near Ljubovija	60 [e]
Lime	Jelen Do a.d. (Nexe Grupa)	Plant in Jelen Dol, west of Cacak	90
Do.	Zelezara Smederevo d.o.o.[1]	Plant at Kucevo	NA
Do.	Ravnaja AD	Plant at Mali Zvornik	NA
Magnesite, concentrate	Magnohrom d.o.o.	Mines near Kraljevo	NA
Magnesium:			
Mine (byproduct of dolomite mining)	MG Serbien d.o.o.	Bela Stena, near Baljevac	NA
Metal:			
Primary	do.	do.	8,500
Secondary	do.	do.	12,000
Natural gas million cubic meters	Naftna Industrija Srbije a.d. (NIS) (JSC Gazprom Neft, 56.15%, and Government, 29.87%)	Gas is produced from wells located throughout northern Serbia	620 [e]
Petroleum:			
Crude thousand 42-gallon barrels per day	do.	Crude petroleum produced mainly in northeastern Serbia	21 [e]
Refined	do.	Refinery at Pancevo	4,800
Do.	do.	Refinery at Novi Sad	2,500
Pig iron	Zelezara Smederevo d.o.o.[1]	Two blast furnaces at Smederevo	NA
Steel, crude	do.	Plant at Smederevo	2,400
Zinc, metal	Hemijska Industrija Zorka	Electrolytic plant at Sabac	30

[e]Estimated. Do., do. Ditto. NA Not available.

[1]The lime plant at Kucevo and the pig iron and steel plant at Smederevo were sold to the Serbian Government by United States Steel Corp. of Pennsylvania on January 31, 2012.
After the sale, the company's name was changed to Zelezara Smederevo d.o.o.

The Mineral Industry of Slovakia

By Harold R. Newman

In 2012, Slovakia continued to produce a modest range of mineral products but was not a significant world producer of mineral commodities. Slovakia was dependent on foreign imports to meet most of its domestic demand for minerals. In 2012, the real growth rate of Slovakia's gross domestic product (GDP) was 2.0% compared with 3.2% in 2011. This decrease was primarily the result of the reduced demand for Slovakia's exports. Services amounted to 59.2% of the GDP and industrial production amounted to 37% of the GDP. Mining and quarrying of minerals made up about 0.6% of the GDP. Slovakia's exports were valued at $81 billion and included machinery and electrical equipment (35.9% of the total), base metals (11.3%), and chemical and mineral products (8.1%). Slovakia's main export partners were Germany (22.4%), the Czech Republic (14.6%), Poland (8.6%), and Hungary (7.8%). Slovakia's imports of $76 billion included mineral products (13%), base metals (9%), and chemicals (8%). Slovakia's main import partners were Germany (18.5%), the Czech Republic (17.9%), Russia (9.9%), and Austria (7.7%) (U.S. Central Intelligence Agency, 2013).

The State Geological Institute of the Slovak Republic is a Government organization supervised by the Ministry of Environment. In accordance with the Geological Act (Act No. 569/2007) and the Mining CT (Act No. 44/1988), the State Geological Institute is responsible for basic and regional geologic research, geologic mapping, the compilation and publishing of general geologic maps as well as specialized and thematic geologic maps; research and evaluation of mineral resource deposits; metallogenetic investigation and modeling of the deposits; mineral exploration; and geophysical works. The Ministry of Economy is responsible for mineral resource development and the issuance of mining permits. A total of 150 exploration licenses and 621 mineral deposits were registered with the State Geological Institute as of yearend 2011 (State Geological Institute of the Slovak Republic, 2012).

New Government energy regulations came into effect on September 1, 2012. The new regulations change the rules for the energy sector and implement relevant European Union (EU) directives. The principal aim of the new Energy Act is to implement the latest EU directives and regulations, which are aimed at achieving an internal energy market in Europe, increasing competition among market participants, and strengthening the rights of consumers. Also, the Energy Act facilitates stronger market liberalization (Power Engineering, 2012).

Production

In 2012, aluminum and steel were two of Slovakia's more valuable metal products. Some industrial minerals, including cement, dolomite, lime, and magnesite, were produced. Brown coal, including lignite, was also produced. The production levels of most of the country's mineral commodities in 2012 remained steady. Production of construction materials and industrial minerals was sufficient to meet domestic demand (table 1).

Structure of the Mineral Industry

Table 2 lists the major mineral companies that were operating in Slovakia in 2012 and their respective mine and (or) plant locations and capacities. No significant changes in ownership took place in 2012.

Commodity Review

Metals

Antimony, Copper, and Silver.—In 2012, Global Minerals Ltd. of Canada was focused on developing its Strieborna antimony, copper, and silver deposit in Roznava in eastern Slovakia. Global Minerals' licenses covered about 135 square kilometers in the Strieborna project area. The Strieborna deposit is a high-grade vein-type deposit that occurs within a mineralized structure that is 1.2 kilometers (km) long and 600 meters (m) deep with an average thickness of 3.4 m. The mineralization is characterized by antimony, copper, and silver-bearing minerals, mainly tetrahedrite. According to a National Instrument (NI) 43–101 estimate, the Strieborna deposit contains an estimated measured and indicated resource of 2.3 million metric tons (Mt) with average grades of 0.85% antimony, 1.2% copper, and 266 grams per metric ton (g/t) silver. Copper and silver were the principal metallic minerals of economic interest. The project's existing infrastructure included electric power, railway access, and paved highways, and the local workforce was experienced in underground mining operations (Global Minerals Ltd., 2012a).

Global Minerals announced that the dewatering and rehabilitation of the Strieborna Mine had reached Level 6, which is located about 180 m below the surface, thereby allowing Global Minerals to re-establish underground drill stations, begin bulk metallurgical sampling, and, ultimately, prepare the deposit for mining. Dewatering was to continue to Level 8 (located about 280 m below the surface), which would allow inspection of the Maria vertical shaft. The Maria vertical shaft was intended to provide access between Level 6 and Level 13 located about 525 m below the surface. Global Minerals continued with other mine development activities, primarily metallurgical testing, process flow sheet design, and mine planning with the goal of completing a preliminary economic assessment by 2014 (Global Minerals Ltd., 2012b).

Gold.—The activities of United Kingdom-based EMED Mining Public Ltd. in eastern Europe were focused on exploring for gold deposits in Slovakia. EMED Mining's principal asset was the Biely Vrch gold project located within the Detva license. EMED Mining stated that a preliminary Joint Ore Reserves Committee (JORC) Code-compliant assessment indicated the probable economic viability for development of the project. The Biely Vrch deposit was estimated to contain indicated resources of 17.7 Mt grading an average of 0.81 g/t gold and containing 14,340 kg of gold and inferred resources of 24 Mt grading an average of 0.77 g/t gold and containing 18,480 kg of gold (EMED Mining Ltd., 2012).

Ortac Resources plc of the United Kingdom's Sturec deposit is located 17 km west of Banska Bystrica. Total estimated resources were 25 Mt grading 1.44 g/t gold and 11.2 g/t silver. The main part of the Sturec zone is the Schramen vein, which is about 100 m wide along a 500-m strike section. The deposit accounts for about 90% of the gold contained in the JORC Code-compliant estimated measured and indicated mineral resources. Schramen, which is a massive-to-sheeted quartz vein, strikes almost due north and dips steeply to the east and thins at depth (Ortac Resources plc, 2012).

Mineral Fuels and Related Materials

Slovakia does not have significant indigenous primary energy reserves. Although mineral fuel resources were thought to be abundant, the majority of these resources were not exploited in 2012. Economic resources of mineral fuels were limited to brown coal and uranium. The country did not have significant reserves of hard coal, natural gas, and petroleum, and demand for these commodities was satisfied mainly by imports from Russia. Estimated exploitable coal reserves were about 100 Mt; estimated natural gas reserves, about 10 billion cubic meters; and estimated petroleum reserves, were about 2 Mt. A deposit of hard coal located in eastern Slovakia was considered by the European Association for Coal and Lignite to be insignificant and not exploitable (European Association for Coal and Lignite, 2012).

Uranium.—European Uranium Resources Ltd. (the name was changed from Tournigan Energy Ltd. in 2011) of Canada's Kuriskova deposit is located about 10 km northeast of Kosice. The deposit had indicated resources of 2.3 Mt at a grade of 0.555% U_3O_8 (commonly called uranium oxide) with a U_3O_8 content of 12,900 t (28.5 million pounds) and additional inferred resources of 3.1 Mt at 0.185% U_3O_8 with a U_3O_8 content of 5,760 t (12.7 million pounds) using a cutoff grade of 0.05% U.

European Uranium continued its efforts to define the structure, which would allow uranium production from Kuriskova license area. The project was thought to be able to be developed as an underground mine and processing facility from which uranium could be extracted using conventional alkaline (nonacid) processing (European Uranium Resources Ltd., 2012).

Outlook

No major increases in the production of mineral commodities are expected in Slovakia in the foreseeable future. Decreased coal production, however, is expected to take place during the next few decades as reserves are depleted. The country will likely continue to import the majority of its metallic ores and concentrates and to depend on imported mineral fuels for its domestic consumption.

References Cited

EMED Mining Ltd., 2012, Slovakia: EMED Mining Ltd. (Accessed November 2, 2013, at http://www.emed-mining.com/projects/slovakia.)

European Association for Coal and Lignite, 2012, Slovakia: European Association for Coal and Lignite. (Accessed November 2, 2013, at http://www.eurocoal.be/pages/layout1sp.php?idpage=79.)

European Uranium Resources Ltd., 2012, European Uranium drills double expected grade at Kuriskova deposit, Slovakia: European Uranium Resources Ltd. news release, June 26. (Accessed August 12, 2014, at http://www.euresources.com/s/news.asp?ReportID=533239.)

Global Minerals Ltd., 2012a, About Strieborná: Global Minerals Ltd. (Accessed November 2, 2013, at http://www.globalminerals.com/projects/about-strieborna/.)

Global Minerals Ltd., 2012b, Global Minerals successfully reaches working levels at Strieborná: Global Minerals Ltd., July 19. (Accessed November 2, 2013, at http://www.globalminerals.com/2012/07/global-minerals-successfully-reaches-working-levels-at-strieborna/.)

Ortac Resources plc, 2012, The Sturec deposit represents a great opportunity for a responsible investment to create sustainable and profitable results: Ortac Resources plc. (Accessed November 2, 2013, at http://www.ortacresources.com/operations-detail/660443-turec-deposit.)

Power Engineering, 2012, Slovakia—New energy law enacted: Power Engineering. (Accessed September 15, 2012, at http://www.power-eng.com/com/news/2012/08/13/slovakia-new-energy-law-enacted.html.)

State Geological Institute of the Slovak Republic, 2012, Department of Geological Exploration: State Geological Institute of the Slovak Republic. (Accessed November 1, 2013, at http://www.geology.sk/test/en/node/318.)

U.S. Central Intelligence Agency, 2013, Slovakia, in The world factbook: U.S. Central Intelligence Agency. (Accessed November 1, 2013, at https://www.cia.gov/library/publications/the-world-factbook/geos/lo.html.)

TABLE 1
SLOVAKIA: PRODUCTION OF MINERAL COMMODITIES[1]

(Metric tons unless otherwise specified)

Commodity[2]		2008	2009	2010	2011	2012[e]
METALS						
Aluminum ingot, primary		162,995	149,604	162,997 [r]	162,840	180,671 [3]
Copper:						
Anode, primary		27,100	34,000	46,250 [e]	48,806 [r]	41,713 [3]
Smelter, secondary		10,000 [r]	9,560 [r]	9,015 [r]	9,014 [r]	8,236 [3]
Gold, Au content of concentrate	kilograms	90 [e]	346	534 [e]	398 [r]	546 [3]
Iron and steel:						
Iron ore:[4]						
Gross weight	thousand metric tons	392	--	--	-- [r]	--
Metal content[e]	do.	130	--	--	-- [r]	--
Concentrate, gross weight	do.	182	--	--	-- [r]	--
Pig iron	do.	3,529	3,019	3,649	3,346	3,519 [3]
Ferroalloys:						
Ferromanganese		61,194	21,000	35,449 [r]	18,180 [r]	12,862 [3]
Ferrosilicomanganese		59,940	32,000	34,960 [r]	25,023 [r]	50,089 [3]
Ferrosilicon		8,622	10,844	26,419	31,845	36,869 [3]
Silver	kilograms	200	250	300	330	441 [3]
Steel:						
Crude	thousand metric tons	4,478	3,747	4,580 [e]	4,236	4,403 [3]
Semimanufactures	do.	4,477	3,740	4,567	4,223 [r]	4,391 [3]
INDUSTRIAL MINERALS						
Asbestos		200	200	200	--	--
Barite:[e]						
Mine output		20,000 [3]	30,000 [3]	17,000	18,000	21,000
Concentrate		12,950 [3]	8,000 [3]	13,000	14,000	15,000
Basalt[e]		63,000 [3]	101,000 [3]	60,000	60,000	60,000
Cement, hydraulic	thousand metric tons	4,157	3,021	2,888	3,219	2,915 [3]
Clays:						
Bentonite		145,000	109,000	130,521	119,323	129,930 [3]
Ceramic[e]		47,000	47,000	47,000	40,000	40,000
Kaolin[e]		44,000 [3]	44,000	44,000	4,000	3,000
Refractory[e]		12,000 [3]	12,000	12,000	--	--
Feldspar[e]		10,000 [3]	10,000	10,000	--	--
Gypsum and anhydrite, crude		152,000	131,000	87,000 [e]	88,000 [e]	85,000
Lime, hydrated and quicklime	thousand metric tons	1,082	867	986 [r]	971	903 [3]
Magnesite, concentrate		807,000	800,000	800,000 [e]	751,700 [r]	618,400 [3]
Nitrogen, N content of ammonia[e]		260,000	260,000	493,000 [r]	486,689 [r,3]	485,518 [3]
Perlite		25,000	25,000	25,000 [e]	23,000	24,000
Salt		99,000	38,000	38,000 [e]	-- [r]	--
Sand and gravel	thousand metric tons	9,300 [e]	8,500 [e]	6,932 [r]	6,479	4,238 [3]
Silica sand (foundry and glass sands)	do.	619	620	620 [e]	600 [e]	600
Stone:						
Dolomite	do.	1,249	908	895	952	1,000
Limestone and other calcareous stones for cement	do.	4,992	5,099	4,952	5,630	5,228 [3]
Crushed stone	do.	18,500 [e]	10,571	11,904	9,855	8,065 [3]
MINERAL FUELS AND RELATED MATERIALS						
Coal, brown	thousand metric tons	2,423	2,572	2,378	2,376	2,292 [3]
Coke, unspecified		1,737	1,573	1,570 [e]	1,500 [e]	1,500
Natural gas	million cubic meters	111	110	110 [e]	106 [r]	110
Petroleum:						
Crude[e]	thousand 42-gallon barrels	100	130	130	125	105 [3]
Refinery products:						
Fuel oil, distillate	do.	21,863	21,718	21,389	21,000 [e]	21,000
Fuel oil, residual	do.	3,066	3,759	3,577	3,600 [e]	3,600
Gasoline, motor	do.	3,759	2,738	2,811	3,614	3,395 [3]
Jet fuel	do.	694	548	465	365	365
Liquid petroleum gas	do.	1,168	949	1,022	1,000 [e]	1,000
Other products[e]	do.	10,000	10,293 [3]	10,326 [3]	10,000	10,000
Total[e]		40,600 [r]	40,005 [r,3]	39,590 [r,3]	39,600 [r]	39,400

See footnotes at end of table.

TABLE 1—Continued
SLOVAKIA: PRODUCTION OF MINERAL COMMODITIES[1]

[e]Estimated; estimated data are rounded to no more than three significant digits. [r]Revised. do. Ditto. -- Zero.
[1]Table includes data available through October 31, 2013.
[2]In addition to the commodities listed, a small amount of silver occurs in concentrate produced by gold ore processing at the Banska Hodrusa deposit.
[3]Reported figure.
[4]Production ceased after 2008.

TABLE 2
SLOVAKIA: STRUCTURE OF THE MINERAL INDUSTRY IN 2012

(Thousand metric tons unless otherwise specified)

Commodity	Major operating companies and major equity owners[1]	Location of main facilities	Annual capacity
Aluminum	Slovalco, a.s. (Norsk Hydro ASA, 55.3%, and ZSNP SCO, a.s., 44.7%)	Ziar nad Hronom, central Slovakia	165
Brown coal and lignite	Hornonitranske Bane Prievidza, a.s. (HBP)	Mines at Cigel, Handlova, and Novaky	2,200
Do.	Bana Dolina, a.s.	Mine east of V'lky Krtis, southern Slovakia	150
Do.	Bana Cary, a.s.	Mine at Cary, western Slovakia	500 [e]
Cement	Povazska Cementaren, a.s.	Ladce	NA
Do.	Cemmac a.s. (Asamer & Hufnagl Baustoff Holding Wien GmbH, 82.72%)	Horne Srnie	NA
Do.	Vychodoslovenske staebne hmoty a.s.	Turna	1,300
Do.	Holcim (Slovensko), a.s.	Rohoznik	2,200
Coke	U.S. Steel Kosice, s.r.o.	Kosice, eastern Slovakia	NA
Copper, smelter, secondary	Kovohuty, a.s. (Umcor Holding GmbH)	Krompachy, central Slovakia	90
Ferroalloys	Oravske Feroziliatinarske Zavody (OFZ), a.s. (ArcelorMittal S.A.)	Istebne	170
Gold in concentrate	Slovenska Banska, s.r.o.	Hodrusa-Hamre	NA
Iron:			
Ore	SIDERIT, s.r.o. Nizna Slana	Nizna Slana, central Slovakia	600 [e]
Concentrate	do.	do.	400 [e]
Magnesite	SMZ, a.s. Jelsava	Jelsava, eastern Slovakia	370 [e]
Do.	Slovenske Magnezitove zavody a.s.	Lubenik, central Slovakia	NA
Do.	GE.NE.S., a.s.	Mutnik, near Hnusta in central Slovakia	NA
Natural gas million cubic meters	NAFTA, a.s.	Oilfields and natural gas fields in western and eastern Slovakia	NA
Do. do.	ENGAS, s.r.o.	Brno	NA
Petroleum:			
Crude	NAFTA, a.s.	Oilfields and natural gas fields in western and eastern Slovakia	NA
Refinery	SLOVNAFT, a.s. (MOL Plc., 98.5%)	Bratislava	6,000
Do.	Petrochema, a.s.	Dubova	150
Pig iron	U.S. Steel Kosice, s.r.o.	Kosice, eastern Slovakia	4,500
Salt	Solivary, a.s. Presov (Garantovana Group)	Presov, eastern Slovakia	NA
Steel, crude	U.S. Steel Kosice, s.r.o.	Kosice, eastern Slovakia	4,900
Do.	Zeleziarne Podbrezova, a.s.	Podbrezova	600 [e]
Zeolites	Zeocem, a.s.	Quarry near Nižný Hrabovec and processing plant near Bystre	NA
Do.	VSK Pro-Zeo Ltd.	Humenne	NA

[e]Estimated, Do., do. Ditto. NA Not available.
[1]Abbreviations used for types of companies in this table include the following: a.s., joint stock company; s.r.o., limited company.

THE MINERAL INDUSTRY OF SLOVENIA

By Harold R. Newman

Slovenia's output of mineral commodities was not significant on either a world or a regional scale. The country continued to be an importer of most mineral commodities, including mineral fuels, ferrous and nonferrous ores and metals, and other mining and quarrying products. In 2012, Slovenia's main export partners were Germany (20%), Italy (12%), Austria (7.9%), and Croatia (6.2%). The country's main import partners were Italy (16.5%), Germany (16.3%), Austria (10.4%), and Croatia (4.8%). The United States was the leading non-European trading partner. In 2012, Slovenia's gross domestic product (GDP) based on purchasing power parity was $58.9 billion compared with the $58.3 billion in 2011. Mining and quarrying accounted for only 0.4% of the GDP, and the manufacture of basic metals accounted for 0.5% of the GDP (U.S. Central Intelligence Agency 2013).

In 2012, the United States exported goods valued at $308 million to Slovenia, including fuel oil valued at $57.5 million, metallurgical-grade coal ($53.6 million), other coal and fuels ($18.3 million), and other petroleum products ($3.7 million) (U.S. Census Bureau, 2012a). The United States imported goods valued at $562 million from Slovenia, including semifinished iron and steel mill products valued at $49.4 million, aluminum and bauxite ($17.4 million), finished metal shapes ($7.7 million), and advanced iron and steel manufactures ($2.2 million) (U.S. Census Bureau, 2012b).

The two primary laws that govern energy and natural resources in Slovenia are the Energy Act of 2008 and the Mining Act of 2011, respectively. The Mining Act regulates the exploration for, exploitation of, and management of mineral resources; mine closures; and health and safety issues in mining-related works. Mineral resources are defined by the Government as nonrenewable natural resources that are owned and licensed by the Government and are directly or indirectly economic to exploit. Energy is regulated by the Energy Act of 2008 and applies to legal entities and persons that perform a variety of activities in the energy field (International Law Office, 2012).

The Geological Survey of Slovenia (GeoZS) is a research institute established by the Government. Geologic data are fundamental for decisionmaking in the mineral sector. The main purpose of the GeoZS is to provide this geologic expertise and information about the geology of Slovenia. The GeoZS has the following six research departments: Environmental Geology and Geochemistry; Geological Maps; Groundwater; Mineralogy and Petrology; Mineral Resources; and Paleontology, Sedimentary, and Stratigraphy (Geological Survey of Slovakia, 2012).

Production

The production of Slovenia's mineral commodities depended on the demand for them, and the world financial crisis continued to affect the demand for these mineral commodities. No base-metal or precious-metal mining was conducted in Slovenia in 2012. Production of bentonite, coal, crude steel, and natural gas decreased whereas production of petroleum increased (table 1).

Structure of the Mineral Industry

The major mineral industry facilities that were operating in Slovenia in 2012 and their respective locations and capacities are listed in table 2.

Commodity Review

Metals

Aluminum.—Talum Aluminium Ltd. was formed on January 1, 2011, from two former units of Talum d.d. Kidricevo (DE and DE Anode Analysis) after the parent company was reorganized. In 2012, Talum Aluminum produced primary aluminum along with billets, castings, and discs using raw materials imported from Birac AD of Bosnia and Herzegovina. The company produced all major types of aluminum alloys using alloying elements of copper, magnesium, manganese, and silicon. Talum sold its output in Slovenia through a network of warehousing and distribution centers across the country (Emerging Markets Information Service, 2012).

Iron and Steel.—Slovenska Industria Jekla d.d. (SIJ) [Slovenian Steel Group] was composed of 14 metallurgical enterprises. The two principal enterprises were Acroni Jesenice d.o.o. and Metal Ravne d.o.o. Acroni was the leading steel company in Slovenia, and its electric arc furnace was capable of melting 85 metric tons per hour (t/hr) of steel using iron- and nickel-containing scrap as the principal source of raw materials. Acroni's main products were flat steel products and construction steel plates. Metal Ravne manufactured long-steel products, including high-speed steel, construction steel, and tool steel products (Industrial Metallurgical Holding Management Co., 2012).

Industrial Minerals

Cement.—The Salonit Anhovo d.o.o. cement works was the leading cement producer in Slovenia. The company also produced lime; products made from concrete, such as pipe and stone; and ready-mix concrete. Salonit invested significantly in the modernization of its entire production plant. The purpose of the €75 million ($103 million[1]) modernization project, which was completed in 2011, was to increase production capacity, reduce energy consumption per production unit, and update environmental protection measures. Also, the project laid the basis for increased use of alternative fuels in the future (Wietersdorfer Gruppe, 2012).

---

[1]Where necessary, values have been converted from euro area euros (€) to U.S. dollars (US$) at an average annual exchange rate of €0.72=US$1.00.

Mineral Fuels and Related Materials

Coal.—Resources of brown coal, including lignite, were estimated to be 1,174 million metric tons (Mt), and reserves were estimated to be about 144 Mt. Brown coal was mined from the Rudnik Trbovlje Hrastnik (RTH) Mine and used by the TET Trbovlje powerplant. Lignite coal was mined from two deposits located at Trbovlje and Velenje, respectively. The Velenje Mine, which has coal layers estimated to be 16 meters thick, was one of the leading coal mines in Europe. A major part of the Velenje Mine's output was used at the 779-megawatt Šoštanj powerplant, which was owned by Termoelektrarna Šoštanj (European Association of Coal and Lignite, 2012).

Natural Gas.—In midyear 2012, the Governments of Russia and Slovenia met to discuss the status of Russia-Slovenia cooperation in the energy sector and the terms and conditions for Russian gas supplies to Slovenia. A Slovenian company, South Stream Slovenia LLC, was established to deal with spatial planning, environmental impact assessment, and basic design work on the South Stream natural gas pipeline. Shareholding in the company was split on a parity basis between Gazprom OAO of Russia and Geoplin Plinovodi d.o.o. of Slovenia. Construction of the South Stream pipeline started at yearend 2012, and first supplies were expected to be transported through the pipeline by yearend 2015. Throughput of the pipeline was projected to reach 63 billion cubic meters per year (OAO Gazprom, 2012).

Outlook

Production of minerals and mineral products in Slovenia will likely continue to be modest. As a result, Slovenia is expected to remain dependent on mineral imports to satisfy its domestic requirements. The production of mineral commodities, mainly aluminum and crude steel, is expected to depend mainly on demand from external trade partners. Exploration efforts for uranium have shown some positive results, and exploration is expected to continue.

References Cited

Emerging Markets Information Service, 2012, Talum: Emerging Markets Information Service. (Accessed November 30, 2013, at http://www.securities.com/Public/company-profile/SI/Talum_en_1538991.html.)

European Association for Coal and Lignite, 2012, Slovenia: European Association for Coal and Lignite. (Accessed October 29, 2013, at http://www.euracoal.be/pages/layout1sp.php?idpage=80.)

Geological Survey of Slovenia, 2012, Our purpose: Geological Survey of Slovenia. (Accessed October 29, 2013, at http://www.geo-zs.si/podrocje.aspx?langid=1033.)

Industrial Metallurgical Holding Management Co., 2012, Slovenian Steel Group: Industrial Metallurgical Holding Management Co. (Accessed November 26, 2013, at http://www.metholding.ru/en/enterprises/slov/.)

International Law Office, 2012, Energy and natural resources—Slovenia—New Mining Act enters into force: International Law Office. (Accessed October 29, 2013, at http://www.internationallawoffice.com/newsletters/detail.aspx?g=f84e7c91-544e-4b41-86bc-fe91bac12133.)

OAO Gazprom, 2012, Gazprom and Slovenia setting up joint project company for South Stream: OAO Gazprom. (Accessed December 31, 2013, at http://www.gazprom.com/press/news/2012/may/article136482./)

U.S. Census Bureau, 2012a, U.S. exports to Slovenia by 5-digit end-use code: U.S. Census Bureau. (Accessed December 30, 2013, at http://www.census.gov/foreign-trade/statistics/product/enduse/exports/c4792.html.)

U.S. Census Bureau, 2012b, U.S. imports from Slovenia by 5-digit end-use code: U.S. Census Bureau. (Accessed December 30, 2013, at http://www.census.gov/foreign-trade/statistics/product/enduse/imports/c4792.html.)

U.S. Central Intelligence Agency, 2013, Slovenia, in The world factbook: U.S. Central Intelligence Agency. (Accessed October 29, 2013, at http://www.cia.gov/library/publications/the-world-factbook/geos/si.html.)

Wietersdorfer Gruppe, 2012, Salonit cement: Wietersdorfer Gruppe. (Accessed November 26, 2013, at http://www.wietersdorfer.com/Salonit.697.0.html?&L=1.)

TABLE 1
SLOVENIA: PRODUCTION OF MINERAL COMMODITIES[1]

(Metric tons unless otherwise specified)

Commodity[2]		2008	2009	2010	2011	2012[e]
METALS						
Aluminum:						
Primary		83,328	35,148	40,177	41,000 [r]	40,000
Secondary[e]		20,000	10,000	17,000	16,000	15,000
Iron and steel, metal:						
Crude steel from electric furnaces		642,000	430,000	606,000	648,000 [r]	632,000 [3]
Semimanufactures		692,320	462,700	656,943	706,133	700,864 [3]
Lead, refined, secondary[e]		15,000	14,000	14,000	12,000	10,000
INDUSTRIAL MINERALS						
Cement[e]	thousand metric tons	1,000	1,082 [r]	799 [r]	620 [r]	1,000
Clays:[e]						
Bentonite		130	130	135	168 [3]	98 [3]
Ceramic clay, crude		32,200	9,478	12,279	10,103 [3]	12,000
Salt, all sources		535	2,924	590 [r]	4,291	5,684 [3]
Sand and gravel	thousand metric tons	19,171	15,720	12,965	10,691 [r]	7,612 [3]
Silica sands (quartz and quartzite)[4]		353,983	326,636	253,866 [r]	230,908 [r]	300,000
Stone, excluding quartz and quartzite, crude:						
Aggregate	thousand metric tons	19,489	16,611	14,495	11,527	12,000
Chert		21,648	16,695	16,114	18,907	20,000
Crushed, dolomite and limestone	thousand metric tons	14,982	13,609	12,073 [r]	9,627 [r]	7,557 [3]
Dimension[e]		15,000	15,000	14,000 [r]	16,000 [r]	15,000
MINERAL FUELS AND RELATED MATERIALS						
Coal:						
Brown coal	thousand metric tons	489	511	419	453	354 [3]
Lignite	do.	4,032	3,918	4,011	4,066	3,967 [3]
Total	do.	4,521	4,429	4,430	4,519	4,321 [3]
Natural gas	thousand cubic meters	2,610	2,575	6,675	2,489	1,616 [3]
Petroleum, crude	42-gallon barrels	1,170	994	1,561	1,762	2,093 [3]

[e]Estimated; estimated data are rounded to no more than three significant digits; may not add to totals shown. [r]Revised. do. Ditto.
[1]Table includes data available through October 31, 2013.
[2]In addition to the commodities listed, common clay and lime also were produced, but available information is inadequate to make reliable estimates of output.
[3]Reported figure.
[4]In previous years, this commodity category was listed as "Quartz, quartzite, glass sand."

TABLE 2
SLOVENIA: STRUCTURE OF THE MINERAL INDUSTRY IN 2012

(Thousand metric tons)

Commodity	Major operating companies and major equity owners	Location of main facilities	Annual capacity
Aluminum, primary and secondary	Talum Aluminium Ltd. (Elektro Slovenija d.o.o., 80%)	Smelter at Kidricevo	120 [e]
Cement	Salonit Anhovo d.o.o. (Wietersdorfer Gruppe, 100%)	Plant at Anhovo	1,100 [e]
Do.	Lafarge Cement d.d. (Lafarge S.A., 56%, and European Bank for Reconstruction and Development, 44%)	Plant at Trbovlje	600
Coal:			
Brown	Rudnik Trbovlje-Hrastnik d.o.o. (RTH)	Mine Rudnik Trbovlje Hrastnik near Trbovlje	NA
Lignite	Premogovnik Velenje, d.d.	Mine at Velenje	NA
Ferroalloys, (ferromanganese, ferrosilicomanganese, and ferrosilicon)	OFZ a.s. (ArcelorMittal S.A., 100%)	Plant at Istebné	NA
Lead, metal, secondary	MPI-Reciklaza Metalurgija, plastika in inženiring d.o.o.	Refinery at Zerjav	35
Salt	SOLINE Pridelava soli d.o.o.	Salt pans at Secovlje and Strunjan	2
Steel, crude	Acroni Jesenice d.o.o. (OAO Koks., 55.4%, and Government, 25%)	Plant at Acroni	400
Do.	Metal Ravne d.o.o. (OAO Koks, 55.4%, and Government, 25%)	Plant at Ravne	130 [e]
Do.	Štore Steel d.o.o. (Unior d.d., 54%)	Plant at Store	185 [e]

[e]Estimated. Do. Ditto. NA Not available.

THE MINERAL INDUSTRY OF SPAIN

By Alfredo C. Gurmendi

In 2012, Spain was a significant European producer of industrial mineral commodities, such as gypsum (third after China and Iran), sand and gravel (industrial) (fifth after the United States, Italy, Germany, and Australia), and fluorspar (sixth after China, Mexico, Mongolia, South Africa, and Russia), among others (table 1; Crangle, 2013; Dolley, 2013; Miller, 2013).

Spain encompasses almost 90% of the Iberian Peninsula, which is considered still to be the most mineralized zone in the European Union (EU), as it includes the volcanic massive sulfide (VMS) deposits of the Iberian Pyrite Belt (IPB). The IPB is more than 240 kilometers (km) in length and extends from Sevilla in southwestern Spain to south of Lisbon, Portugal; it hosts massive-sulfide and associated stockwork deposits that date from the Late Devonian to Middle Carboniferous periods. It varies in thickness from a few meters to several hundred meters and consists of a sequence of bimodal volcanics and associated pyroclastic and tuffaceous rocks. Within the IPB, at least 80 VMS deposits are thought to occur. About 1.7 billion metric tons of sulfides have been produced from the IPB. The leading polymetallic deposits include the Aguablanca nickel-copper mine, the Aguas Teñidas copper-lead-zinc mine, Las Cruces copper mine, the Masa Valverde copper-lead-zinc deposit, Los Santos tungsten mine, the Rio Tinto copper mine (which is the largest copper mine in Spain), the Slave gold project, and others, such as the Aljustrel, the Aznalcollar, the Neves-Corvo, the Scotiel, the Tharsis, and La Zarza Mines (Gibbons and Moreno, 2002, p. 473–510; Bastida and others, 2010; Cambridge Mineral Resources plc, 2013a, b; Lannin, 2013).

Spain has a long history of mining and has attracted interest from many large gold and base-metal mining companies. Several factors have contributed to this interest, including the highly prospective geology of the IPB and that of the Rio Narcea Belt, plus the gold discoveries at the Carlés and El Valle deposits at Boinas, Asturias, in northern Spain, and the Masa Valverde polymetallic volcanic-hosted massive-sulfide deposit at Andalucia in southeastern Spain. International mineral investment interest has also been encouraged by Spain's transparent legislative framework and positive fiscal environment for the extraction of natural resources, its well-developed infrastructure and skilled workforce, its long mining tradition and past success in exploration and mine development, and the availability of nonrefundable Government grants for both exploration and mine development (Cambridge Mineral Resources plc, 2013a, b; EMED Mining Public Ltd., 2013).

In 2012, Spain's gross domestic product (GDP) based on purchasing power parity decreased by 1.4% following an increase of 0.4% in 2011. The GDP in 2012 was $1.515 trillion compared with a revised $1.475 trillion in 2011, and the rate of inflation was 2.4% in 2012 compared with 3.1% in 2011. A new law to ensure budgetary stability and financial sustainability was applied for the first time in 2012. Spain had a population of about 47.4 million in 2012. The total labor force was 23.05 million, of which services accounted for 71.7%; industry, 24.1%; and agriculture, 4.2%. In 2012, the mining and mineral processing industries contributed 0.8% of Spain's GDP and employed about 1% (55,600) of the industry total of 5.6 million (Sociedad Geológica de España, 2012, p. 9–11; Banco de España, 2013a, p. 27, 62–63; 2013b, p. 4–5; Federation of International Trade Associations, 2013a, b; Instituto Nacional de Estadística, 2013a; 2013b, p. 51).

Minerals in the National Economy

In 2012, Spain's most valuable mineral products included, in order of value, alumina, cement, coal, steel, gold, zinc, and copper. Spain was the fifth-ranked economy in the EU and, in spite of weak market confidence throughout the EU and reduction in foreign investment inflows in the second half of 2011 and early 2012, the country continued to attract the interest of many major world mining companies, which invested in prospecting and exploration for base metals, gold, and uranium (Banco de España, 2013a, p. 23–25; Instituto Nacional de Estadística, 2013b, p. 20, 24, 26).

Government Policies and Programs

Minerals are owned by the Government under an arrangement known as the Regalía Principal. The Mining Law of July 21, 1973, and the Hydrocarbon Law of October 7, 1998, continued to control Spain's mineral industry. The Dirección General de Política Energética y Minas implements these laws. Law 20 of June 5, 2006, modified the Finance Regime of the Sociedad Estatal de Participaciones Industriales (SEPI), which is a Government-owned holding company that has mining as one sector in its portfolio. In the mineral sector, SEPI owned 100% of Hulleras del Norte, S.A. (HUNOSA), which produced coal; 60% of Enusa Industrias Avanzadas, S.A. (ENUSA), which produced nuclear energy; 20% of Red Eléctrica Corporación, S.A. (RECSA), which produced electricity; and 5% of Enagás, S.A. (ENAGAS), which produced natural gas. El Instituto Geológico y Minero de España (IGME) is the principal Government science organization that provides assistance in the fields of geology and mining to the private and public sectors through the production of maps and scientific publications (Sociedad Estatal de Participaciones Industriales, 2013, p. 13–14, 19–20; Instituto Geológico y Minero de España, 2013).

Production

Production data for selected mineral commodities are in table 1. Spain continued to be a leading EU producer of natural sodium sulfate, slate, and strontium minerals and a regionally significant processor of domestic and imported raw materials. In 2012, the sources of Spain's domestic energy production were nuclear (47.1%), biomass (18.9%), wind and solar power combined (14.1%), hydroelectric (10.5%), coal (8.8%), petroleum (0.4%), and natural gas (0.2%). Spain's energy consumption sources were petroleum (44.1%), natural gas (19.5%), coal (13.3%), renewable energies (10.3%), nuclear energy (9.6%), and hydroelectricity (3.2%) (BP p.l.c, 2013, p. 41; Instituto Nacional de Estadística, 2013c; International Energy Agency, 2013, p. 13, 23; Ministerio de Industria, Energía y Turismo, 2013).

Structure of the Mineral Industry

Data on the capacity and ownership of selected mineral operations are in table 2. The mineral industry was made up of a mix of Government-owned companies, joint ventures of public and private-sector companies, and privately owned companies. Spain's accession to the EU in January 1986 required the country to conform to EU guidelines. Spain followed the U.S.-EU mutual recognition agreements in its application of nontariff regulations and conformity assessment procedures (Banco de España, 2013a, p. 31, 37; 2013b, p. 20).

Cambridge Mineral Resources plc (CMR) of the United Kingdom, which owned the Lomero-Poyatos auriferous polymetallic massive-sulfide deposit, was one of the leading mining companies in Spain. Another of the leading mining companies, Ormonde Mining Plc of Ireland, was developing mining projects in Spain, including the Barruecopardo tungsten project, which is located in Salamanca Province, Castilla y Leon, in northwestern Spain. Another leading mining company, Orvana Minerals Corp. (OMC) of Canada, through its wholly owned subsidiary Kinbauri Gold Corp. (Kinbauri), owned and operated El Valle-Boinas/Carlés gold mines in the Rio Narcea gold belt in northwestern Spain in Oviedo Province, Asturias. The Asturias airport and the port city of Aviles are located approximately 40 km northeast of the property. Alcoa Inc. of the United States had six production centers across Spain. Alcoa was the leading alumina and aluminum producer in Spain, the sole producer of aluminum oxide (alumina) and primary aluminum, and a manufacturer of rolled products made from those materials (Alcoa Inc., 2013b; Ormonde Mining plc, 2013; Orvana Minerals Corp., 2013).

Mineral Trade

Spain's total exports amounted to $292.0 billion and total imports amounted to $328.0 billion in 2012 compared with a revised $305.0 billion in exports and $369.2 billion in imports, respectively, in 2011. In 2012, Spain's leading export partners were France (16.8%), Germany (10.8%), Italy (7.7%), Portugal (7.1%), and the United Kingdom (6.5%). Its leading import partners were Germany (11.8%), France (11.5%), Italy (6.7%), China (5.6%), the Netherlands (5.4%), and the United Kingdom (4.1%). The share of foreign trade in Spain's GDP was about 42.9% in 2012 compared with 45.7% in 2011. Mineral fuels and derivatives accounted for almost 5% of total exports and almost 15% of total imports; iron and steel, 3.3% of total exports and 4.3% of total imports; industrial minerals, 2.6% of total exports and 3% of total imports; base metals, 2.3% of total exports and 2% of total imports; and aluminum, 1.2% of total exports and 1% of total imports. Spain's foreign direct investment (FDI) reached about $27.8 billion in 2012 compared with more than $26.8 billion in 2011, which represented an increase of more than 3.7%. The increase in FDI led to increased exports for Spain in 2012 (Federation of International Trade Associations, 2013a–c; Instituto Nacional de Estadística, 2013a, b; U.S. Central Intelligence Agency, 2013).

Commodity Review

Metals

Bauxite and Alumina and Aluminum.—Alcoa Inc., which was one of the world's leading alumina (aluminum oxide) producers, owned and operated the San Ciprian industrial complex on the east coast of Spain in Lugo Province, Galicia. In 2012, the San Ciprian facility produced 1.5 million metric tons (Mt) of aluminum oxide, calcinated alumina, and hydrate from bauxite. Bauxite for the facility was supplied mainly by Alcoa's mines in Brazil and Guinea. San Ciprian also produced 250,000 metric tons (t) of primary aluminum as ingots for casting, billets for extrusion, and sheets for lamination. The production of alumina (aluminum oxide) was used for the production of primary aluminum and chemical aluminum oxides for the ceramics and chemical sectors. About 70% of San Ciprian's alumina output was supplied to Alcoa's Aviles, Coruña, and San Ciprian aluminum smelters in Spain, and the remaining output was sold largely as hydrated alumina to European chemical manufacturers. San Ciprian's location allows commodity-grade alumina (aluminum oxide) to be sold within the EU without the high tariffs imposed on non-European suppliers. Alumina, alumina chemicals, alumina hydrates, and primary aluminum were produced by Aluminio/Alúmina Española, S.A. (AESA) in San Ciprian for domestic consumption and for export; primary aluminum was produced by Alcoa Inespal S.A. (AISA) at its 93,000-metric-ton-per-year (t/yr)-capacity smelter in Aviles Province, Asturias, and its 87,000-t/yr-capacity smelter in Coruña Province, Galicia, for domestic consumption and export (Alcoa Inc., 2013a, p. 12, 14; 2013b; Alumina Ltd., 2013).

Copper, Gold, Lead, Silver, and Zinc.—In 2012, Spain produced 75,057 t of copper content compared with 74,246 t in 2011, which was a slight increase of almost 1.1%. Gold output was 3,600 kilograms (kg) compared with a revised 3,550 kg in 2011. Output of secondary refined lead (125,000 t), silver (3,500 kg) and primary and secondary refined zinc (490,000 t) also remained at about the same levels as output in 2011. First Quantum Minerals Ltd. of Canada became one of the world's leading copper producers with its acquisition of Inmet Mining Corp. of Canada in March 2013. First Quantum owned Las Cruces copper mine, which is located on the eastern edge of the IPB about 15 km northwest of Sevilla Province in the Andalucia Region of Spain. As of December 31, 2012, Las Cruces had estimated proven and probable copper reserves of 14.1 Mt grading 5.44% copper. Las Cruces' designed production capacity was 72,000 t/yr of copper cathode as an end product and 1.0 Mt of copper ore for the period 2009 to 2022. In 2012,

Las Cruces produced 67,700 t of copper cathode compared with 42,000 t in 2011. First Quantum had a geographically diversified portfolio of development and operation projects and a strategic plan to produce more than 1.3 million metric tons per year (Mt/yr) of copper within 5 years; thus, in the near future, First Quantum could become a leading European copper producer and one of the top five copper producers in the world (First Quantum Minerals Ltd., 2013).

Cambridge Mineral Resources plc's (CMR's) Lomero-Poyatos auriferous polymetallic massive-sulfide deposit occurs within Devonian-age intermediate volcaniclastic rocks of the IPB. The deposit is located 30 km west of the Río Tinto Mine and 8 km west of the Aguas Teñidas Mine. CMR owned 100% of its local subsidiary Recursos Metállicos SA (RMSA), which held the mining licenses for the Lomero-Poyatos deposit, including the right to production. The licenses were valid for 45 years. The Lomero-Poyatos deposit was reported to contain reserves of 3.71 Mt at grades of, in order of value, 3.26 grams per metric ton (g/t) gold, 27.9 g/t silver, 0.87% copper, 1.57% lead, and 1.16% zinc at a 1.5 g/t gold-equivalent cutoff. CMR's wholly owned Spanish subsidiary Cambridge Minería España SL was the successful applicant for mineral exploration permits on the Masa Valverde polymetallic volcanic-hosted massive-sulfide deposit located in the central part of the IPB. The Masa Valverde cupriferous stockwork deposit contains a massive resource estimated to be more than 75 Mt at grades of, in order of value, 0.43 g/t gold, 22.4 g/t silver, 0.76% copper, 0.38% lead, and 1.28% zinc at a 1.0 g/t gold-equivalent cutoff. CMR was seeking significant joint-venture partners to develop both projects (Bastida and others, 2010; Cambridge Mineral Resources plc, 2013a, b).

In Spain, Ormonde's permits covered mineral prospects and occurrences with potential for high-grade vein-hosted deposits and large, bulk-tonnage low-grade deposits. Ormonde was exploring the Salamanca and the Zamora prospects for gold mineralization associated with granites. La Zarza deposit contains significant copper-gold-zinc resources and was also considered to be a potential source of silver; it is located in Huelva Province, Andalucia, in southwestern Spain in the IPB mining district. La Zarza deposit was estimated to contain 9.9 Mt of ore grading 3% zinc, 1% copper, 1% lead, 38.9 g/t silver, and 1.6 g/t gold; its inferred resources were 1.3 Mt grading 1.9% copper (Ormonde Mining plc, 2013).

Orvana's El Valle-Boinás/Carlés copper and gold skarn deposits had estimated proven and probable reserves of 7.7 Mt grading 3.30 g/t gold and 0.52% copper and inferred resources of 8.4 Mt grading 4.88 g/t gold and 0.39% copper. In February 2012, Orvana outlined a 10-year mine life and expected to produce an average of 2.3 t/yr (73,000 troy ounces per year) of gold and 2,570 t/yr of copper (Orvana Minerals Corp., 2013).

Asturiana de Zinc S.A. was owned by Xstrata plc of Switzerland. Xstrata's operations in Spain included the San Juan de Nieva zinc smelter and the Arnao zinc semis plant in Asturias and the Hinojedo roasting plant in Cantabria. Asturiana's core business was the refining and production of zinc metal, mainly zinc ingots. According to Asturiana, the San Juan de Nieva plant, which had the capacity to produce 510,000 t/yr of zinc metal, was the leading zinc smelter in the world and also one of the world's lowest cost operations (Asturiana de Zinc S.A., 2013; Xstrata plc, 2013, p. 123, 127).

Minas de Aguas Teñidas S.A. (MATSA), which was a wholly owned Spanish subsidiary of Iberian Minerals Corp. of Switzerland, owned 100% of the Aguas Teñidas Mine. The mine is based on one of an east-west striking chain of VMS deposits on the northernmost limb of the IPB. According to MATSA, the Aguas Teñidas' geology is composed primarily of heavily tectonic volcano-sedimentary rocks with cross-cutting shear zones. The deposit is made up of four main mineralization types: polymetallic lead-zinc rock, massive cupriferous, barren pyrite, and a cupriferous stockwork (Iberian Minerals Corp., 2013).

Iron and Steel.—According to Worldsteel Association, Spain produced almost 3.6 Mt of iron and 16.1 Mt of crude steel in 2012 compared with 3.5 Mt and 15.6 Mt, respectively, in 2011. Compañía Española de Laminación S.A. (Celsa) produced about 2.5 Mt of steel in 2012. Celsa also produced corrugated and smooth round rods, rolled wire, flat rods, squares, angle rods, structural sections, and electro-welded mesh (Compañía Española de Laminación, S.L., 2013; Worldsteel Association, 2013a, b). Corporación Gerdau Sidenor S.A. (Sidenor), which was a leading producer of special steels in Spain, was planning to start producing stainless steel by modernizing the electric arc furnace at its works in Basauri in northern Spain (Corporación Gerdau Sidenor S.A., 2013, p. 3, 14).

Nickel.—Spain produced 6,300 t of nickel content in concentrate, which is about the same amount that it produced in 2011. Lundin Mining Corp. of Canada's Aguablanca nickel-copper deposit (which is located in the Province of Badajoz, Extremadura Region, in southwestern Spain) consisted of an open pit and a processing facility with a production capacity of 1.9 Mt/yr of ore and 8,000 t/yr of nickel content, respectively. The initial open pit mine life was estimated to be 5 years based on mineral reserves of 14 Mt of ore grading 0.6% nickel, 0.5% copper, 0.47 g/t platinum-group metals, and 0.13 g/t gold. According to Lundin, mine instabilities reoccurred in the south wall of the open pit during the third quarter of 2012. Mining operations continued in the north side of the pit, which resulted in the production of 2,398 t of nickel and 2,260 t of copper in concentrate during the year. Lundin initiated a study on the future configuration of the open pit and anticipated having the study completed and the results available during the second quarter of 2013 (Lundin Mining Corp., 2013).

Tungsten.—According to Ormonde, the Barruecopardo tungsten project in western Spain is one of the premier undeveloped tungsten projects in the world outside of China. The company completed a feasibility study on February 12, 2012. The Barruecopardo project contains 27.4 Mt of tungsten reserves grading 0.26% WO_3. The development of an open pit would be based on measured and indicated mineral resources of 8.7 Mt at a grade of 0.32% WO_3. The planned production rate of 1.1 Mt/yr of ore was expected to produce about 227,000 t/yr of tungsten metal by late 2013 (Ormonde Mining plc, 2013).

Industrial Minerals

Cement.—In 2012, Spain's estimated cement output was 20.0 Mt compared with 22.2 Mt in 2011. According to the Spanish Cement Association, cement consumption decreased owing to the almost total paralysis of the construction industry and budget cuts in the public works sector. The decrease in demand accelerated in the first 9 months of 2012. For the year, domestic consumption decreased by 34.6% to 10.6 Mt and production decreased by 28.1% to 17.6 Mt compared with consumption and production, respectively, in 2011. Tudela Veguin S.A. of La Robla in southern Spain was one of the country's leading producers of clinker; Tudela's plant produced clinker at a rate of 2,600 metric tons per day (International Cement Review, 2013; Oficemen, 2013).

Fluorspar.—Spain's estimated fluorspar output was 128,090 t in 2012 compared with a revised 117,333 t in 2011. MINERSA, which was the EU's leading fluorspar producer, operated three fluorite deposits in Asturias in the north of Spain. MINERSA's production capacity was 150,000 t/yr of fluorspar, mainly acid grade, but also metallurgic and ceramic grades. MINERSA's fluorspar operations were located close to the deepwater Port of Aviles (Minerales y Productos Derivados S.A., 2013).

Potash.—In 2012, Spain produced an estimated 436,000 t of potash compared with 436,026 t in 2011. Iberpotash S.A., which manufactured and distributed potash and fertilizer as a subsidiary of Israel Chemicals Ltd. of Israel (also known as ICL Fertilizers Europe), was a leading producer of potash in the EU. In 2012, ICL announced the first stage of an efficiency plan for Iberpotash, which would include increasing the company's potash granulating capacity to meet increased demand for granulated potash, as well as construction of a plant at Suria to produce 1.5 Mt of vacuum salt for the food and chemical industries. Execution of this plan was expected to be completed in early 2014 at a cost of about $260 million (160 million euros).[1] The second project, which had not yet been approved, would expand Suria's potash production capacity to 1.1 Mt of potash; the new capacity would include 630,000 t of granulated potash and 50,000 t ornamental potash (Iberpotash S.A., 2013).

Sepiolite.—In 2012, Spain's sepiolite output was at about the same level as that of 2011 (567,000 t). Grupo Tolsa S.A. (Tolsa), which was based in Toledo, had reserves of sepiolite in the River Tajo Basin that were thought to be about 20 Mt. Tolsa was the discoverer of sepiolite, which is a light mineral with a high capacity to absorb and retain water and which can be used to decrease rain runoff. Sepiolite is used in a variety of products and applications (Grupo Tolsa S.A., 2013).

Mineral Fuels and Other Sources of Energy

Spain has limited energy resources; thus, the country was strongly dependent upon imports of energy. Spain had no major oilfields, one natural gas field located offshore, and coal mines that contained mainly low-quality coal. In 2012, proved reserves of petroleum were estimated to be 150 million barrels; proved reserves of natural gas were estimated to be 2.5 billion cubic meters; and proved reserves of coal were estimated to be 530 Mt. Spain's petroleum refinery capacity was more than 1.5 million barrels per day (Mbbl/d) (BP p.l.c., 2013, p. 16, 30; Instituto Nacional de Estadística, 2013b, p. 26, 48; U.S. Energy Information Administration, 2013).

Coal.—In 2012, estimated proved reserves of anthracite and bituminous coal amounted to 200 Mt, and subbituminous coal and lignite amounted to 330 Mt, for a total of 530 Mt. Spain's coal production decreased to about 6.3 Mt in 2012 from about 6.6 Mt in 2011. Coal continued to be Spain's most plentiful indigenous energy source; however, no production of lignite was reported during 2008 through 2012. Spain's coal consumption amounted to 23.1 Mt compared with 19.5 Mt in 2011 (BP p.l.c., 2013, p. 30; Instituto Nacional de Estadística, 2013c; Ministerio de Industria, Energía y Turismo, 2013; U.S. Energy Information Administration, 2013). Private companies produced most of the coal in the country, although the leading producer of bituminous coal was the HUNOSA Group, which was owned by the Government through the SEPI (Ministerio de Industria, Energía y Turismo, 2013).

Crude Oil.—Spain's annual production of crude oil was about 323,000 barrels (bbl) in 2012 compared with 234,000 bbl in 2011. Spain imported 1.180 Mbbl/d, including from the Middle East (Saudi Arabia, Iran, Iraq, and others), 37.2%; Africa (Algeria, Libya, Nigeria, and others), 27.2%; Europe (the United Kingdom, Russia, and others), 17.2%; Latin America (Mexico, Venezuela, and others), 16.7%; and other countries, 1.7% (Instituto Nacional de Estadística, 2013c; Ministerio de Industria, Energía y Turismo, 2013; U.S. Energy Information Administration, 2013).

Natural Gas.—Spain's natural gas production was about the same level as that of 2011 (44 million cubic meters). Most of Spain's natural gas production came from one offshore field, Poseidon, which was operated by Repsol YPF S.A. of Argentina. The country's natural gas imports amounted to about 34.2 billion cubic meters and were from Algeria (37.8%), Nigeria (19.7%), Qatar (13.0%), Norway (8.2%), Trinidad and Tobago (6.9%), Egypt (6.5%), and other countries (7.9%) (BP p.l.c., 2013, p. 28; Instituto Nacional de Estadística, 2013c; Ministerio de Industria, Energía y Turismo, 2013; U.S. Energy Information Administration, 2013).

Renewable Energy.—In Europe, Germany and Spain continued to attract the majority of investments in wind power. In 2012, the wind power production capacity of the two countries totaled 54,104 megawatts (MW), which was more than 49% of the EU's total capacity of 109,581 MW. The leading European countries with wind power installations were Germany (31,308 MW) and Spain (22,796 MW) followed by the United Kingdom (8,445 MW), Italy (8,144 MW), France (7,564 MW), Portugal (4,525 MW), Denmark (4,162 MW), Sweden (3,745 MW), Poland (2,497 MW), the Netherlands (2,391 MW), Romania (1,905 MW), Greece (1,749 MW), Ireland (1,738 MW), Austria (1,378 MW), and Belgium (1,375 MW). In 2012, Spain's additional installed capacity was 1,122 MW compared with 1,050 MW in 2011, which was an increase of almost 7% (European Wind Energy Association, 2013, p. 4–5).

---

[1]Where necessary, values have been converted from euro area euros (€) to U.S. dollars (US$) at an annual average exchange rate of €0.809=US$1.00.

Outlook

In spite of the double dip recessions (second half of 2008 through late 2009 and in late 2011 through 2012) in the Spanish economy and a crisis in the EU market that created a substantive decrease of foreign investment in early 2012, Spain continued to be a significant producer of such mineral commodities as, in terms of quantity, coal, iron ore, pyrites, copper, lead, zinc, and potash. Spain's economic downturn appears to have increased Government and community support for mining projects, however, particularly in the Iberian Peninsula. The IPB continues to be of interest to domestic and foreign mining companies and is a prime target for mineral exploration because of past discoveries of large VMS deposits. The Andalucian government was considering awarding an exploration license to EMED by mid-2013. EMED was planning to reopen the Río Tinto copper mine near Sevilla in late 2013 and expected to produce about 12 Mt/yr of copper ore by 2014 and to increase the mine's output to 15 Mt/yr by 2015. Spain's output of mineral fuels is not sufficient to satisfy domestic demand, and the country continued to be a large-scale importer of fuel minerals. Owing to the country's strong dependence on imported energy sources, the Government is expected to support and direct its attention toward renewable energy investments, such as biofuels, and geothermal, solar, and wind energy, while at the same time increasing its total installed capacity to meet increasing demand. In 2012, more renewable-energy-generating capacity was installed in Spain; renewable energy represented 25% of total new installed capacity. Nevertheless, Spain's current power sector continues to be dependent on coal, crude oil, and nuclear power (BP p.l.c., 2013, p. 41; EMED Mining Public Ltd., 2013; European Wind Energy Association, 2013, p. 7–8).

References Cited

Alcoa Inc., 2013a, 2012 annual report and Form 10–K: Alcoa Inc., 200 p. (Accessed August 12, 2013, at http://www.alcoa.com/global/en/investment/pdfs/2012_Annual_Report.pdf.)

Alcoa Inc., 2013b, Alcoa in Spain: Alcoa Inc. (Accessed August 7, 2013, at http://www.alcoa.com/spain/en/info_page/home.asp.)

Alumina Ltd., 2013, Global operations, Spain—San Ciprian alumina refinery: Alumina Ltd. (Accessed August 16, 2013, at http://www.aluminalimited.com/european-african-operations/.)

Asturiana de Zinc S.A., 2013, Asturiana de Zinc—Principales aplicaciones del zinc: Asturiana de Zinc S.A. (Accessed August 20, 2013, at http://www.azsa.es/ES/CalidadyProductos/Paginas/PrincipalesAplicacionesdelZinc.aspx.)

Banco de España, 2013a, Annual report 2012: Banco de España, 120 p. (Accessed July 27, 2013, at http://www.bde.es/f/webbde/SES/Secciones/Publicaciones/PublicacionesAnuales/InformesAnuales/12/Files/inf2012e.pdf.)

Banco de España, 2013b, Economic bulletin 2011: Banco de España, 62 p. (Accessed July 27, 2013, at http://www.bde.es/f/webbde/SES/Secciones/Publicaciones/InformesBoletinesRevistas/BoletinEconomico/13/Jul/Files/indicae.pdf/.)

Bastida, F., Aller, J., Pulgar, J.A., Toimil, N.C., Fernández, F.J., Bobillo-Ares, N.C., and Menéndez, C.O., 2010, Folding in orogens—A case study in the northern Iberian Variscan Belt: Geological Journal, v. 45, p. 597–622. (Accessed December 12, 2012, at http://onlinelibrary.wiley.com/doi/10.1002/gj.1199/abstract.)

BP p.l.c., 2013, BP statistical review of world energy: London, United Kingdom, BP p.l.c, June, 48 p. (Accessed August 7, 2013, at http://www.bp.com/content/dam/bp/pdf/statistical-review/statistical_review_of_world_energy_2013.pdf.)

Cambridge Mineral Resources plc, 2013a, Acquisition of large base-metal resource at Masa Valverde, Spain: London, United Kingdom, Cambridge Mineral Resources plc. (Accessed July 24, 2013, at http://www.cambmin.co.uk/?i=home.)

Cambridge Mineral Resources plc, 2013b, Spain, Lomero Poyatos project: London, United Kingdom, Cambridge Mineral Resources plc. (Accessed July 24, 2013, at http://www.cambmin.co.uk/?i=projects&s=spa.)

Compañía Española de Laminación, S.L., 2013, Manufacturing process—Celsa Barcelona, España: Compañía Española de Laminación, S.L. (Accessed August 20, 2013, at http://www.celsa.com/Celsa.mvc/Presentacion.)

Corporación Gerdau Sidenor S.A., 2013, Special steels in Spain: Corporación Gerdau Sidenor S.A. (Accessed August 20, 2013, at http://www.gerdau.es/archivos/descargas/Catalogo_Aceros_Especiales_GSE_-Ingles-_2013_05_22_16_34_42.pdf.)

Crangle, R.D., Jr., 2013, Gypsum: U.S. Geological Survey Mineral Commodity Summaries 2013, p. 70–71.

Dolley, T.P., 2013, Sand and gravel (industrial): U.S. Geological Survey Mineral Commodity Summaries 2013, p. 138–139.

EMED Mining Public Ltd. 2013, New exploration license awarded: EMED Mining Public Ltd. (Accessed July 24, 2013, at http://finance.yahoo.com/news/emed-mining-public-limited-exploration-074805046.html.)

European Wind Energy Association, 2013, Wind in power—2012 European statistics: Brussels, Belgium, European Wind Energy Association, February, 14 p. (Accessed August 22, 2013, at http://www.ewea.org/fileadmin/files/library/publications/statistics/Wind_in_power_annual_statistics_2012.pdf.)

Federation of International Trade Associations, 2013a, Spain economic and political outline—Economic indicators: Federation of International Trade Associations. (Accessed August 2, 2013, at http://fita.org/countries/economic_and_political_outline_17.html.)

Federation of International Trade Associations, 2013b, Spain information: Federation of International Trade Associations. (Accessed August 2, 2013, at http://fita.org/countries/spain.html.)

Federation of International Trade Associations, 2013c, Spain investing—FDI in figures: Federation of International Trade Associations. (Accessed August 2, 2013, at http://fita.org/countries/investing_17.html.)

First Quantum Minerals Ltd., 2013, Our company: First Quantum Minerals Ltd. (Accessed August 19, 2013, at http://www.first-quantum.com/Investors-Centre/Corporate-Presentations/2013/default.aspx.)

Gibbons, Wes, and Moreno, Teresa, eds., 2002, The geology of Spain: Bath, United Kingdom, The Geological Society, 609 p.

Grupo Tolsa S.A., 2013, Industrial products—Sepiolite: Grupo Tolsa S.A. (Accessed August 21, 2013, at http://www.tolsa.com/index.php?seccion=23&contenido=125&padre=25&idioma=1.)

Iberian Minerals Corp., 2013, Our projects—Aguas Teñidas Mine: Iberian Minerals Corp. (Accessed August 20, 2013, at http://iberianminerals.com/English/Our-Projects/Spain/default.aspx.)

Iberpotash S.A., 2013, ICL to beef up efficiency plans for Iberpotash in Spain: Iberpotash S.A. (Accessed August 21, 2013, at http://www.commodityonline.com/news/icl-to-beef-up-efficiency-plans-for-Iberpotash-in-Spain-38197-3-38198.html.)

Instituto Geológico y Minero de España, 2013, Acceso a la Información Geocientífica del IGME: Instituto Geológico y Minero de España. (Accessed August 5, 2013, at http://www.igme.es/internet/default.asp.)

Instituto Nacional de Estadística, 2013a, Gross domestic product at market prices: Instituto Nacional de Estadística. (Accessed August 6, 2013, at http://www.ine.es/jaxi_XML/tabla.do.)

Instituto Nacional de Estadística, 2013b, Informe anual 2012: Instituto Nacional de Estadística, 59 p. (Accessed August 6, 2013, at http://www.ine.es/ine/planine/informe_anual_2012.pdf.)

Instituto Nacional de Estadística, 2013c, Other results about energy production, consumption, and foreign trade: Instituto Nacional de Estadística. (Accessed August 7, 2013, at http://www.ine.es/jaxi/menu.do?type=pcaxis&path=/t04/a082/a1998/&file=pcaxis.)

International Cement Review, 2013, Spain—Fall in demand accelerates: International Cement Review, December, p. 7.

International Energy Agency, 2013, Key world energy statistics 2012: International Energy Agency, 82 p. (Accessed August 5, 2013, at http://www.iea.org/publications/freepublications/publication/kwes.pdf.)

Lannin, Sue, 2013, Mining company looking to reopen iconic Rio Tinto Mine in Spain: ABC News, 2013 (Australian Broadcast-Windows Internet Explorer) (Accessed August 12, 2013, at http://www.abc.net.au/news/2013-08-12/mining-company-looking-to-reopen-iconic-rio-tinto/4880000.)

Lundin Mining Corp., 2013, Aguablanca nickel/copper mine, Spain: Lundin Mining Corp. (Accessed August 21, 2013, at http://www.lundinmining.com/s/Aguablanca.asp.)

Miller, M.M., 2013, Fluorspar: U.S. Geological Survey Mineral Commodity Summaries 2013, p. 56–57.

Minerales y Productos Derivados S.A., 2013, MINERSA—Operaciones de fluorita: Asturias, Spain, Minerales y Productos Derivados S.A. (Accessed August 21, 2013, at http://www.grupominersa.es/minersa/gencontent.aspx/actividades-139/operaciones-de-fluorita-155.)

Ministerio de Industria, Energía y Turismo, 2013, Consulta de estadísticas mineras—Minería y explosivos—Energía: Ministerio de Industria, Energía y Turismo. (Accessed August 7, 2013, at http://www.minetur.gob.es/energia/mineria/Estadistica/Paginas/Consulta.aspx.)

Oficemen, 2013, Annual report—Agrupación de fabricantes de cemento en España: Oficemen. (Accessed August 21, 2013, at http://www.oficemen.com/noticia.asp?id_rep=1479.)

Ormonde Mining plc, 2013, Projects overview: Ormonde Mining plc. (Accessed August 7, 2013, at http://www.ormondemining.com/en/projects/projects_overview.)

Orvana Minerals Corp., 2013, El Valle-Boinás/Carlés gold-copper project: Orvana Minerals Corp. (Accessed August 9, 2013, at http://www.orvana.com/projects/el-valle/index.html.)

Sociedad Estatal de Participaciones Industriales, 2013, SEPI ESTRATEGIAS: Sociedad Estatal de Participaciones, v.2006, no.18.)

Sociedad Geológica de España, 2012, VIII Congreso Geológico de España 2012—Oviedo: Salamanca, Spain, Sociedad Geológica de España, 205 p.

U.S. Central Intelligence Agency, 2013, Spain: U.S. Central Intelligence Agency, January 31. (Accessed February 25, 2014, at https://www.cia.gov/library/publications/the-world-factbook/geos/sp.html.)

U.S. Energy Information Administration, 2013, Spain—Overview data for Spain: U.S. Energy Information Administration. (Accessed August 5, 2013, at http://www.eia.gov/countries/country-data.cfm?fips=SP.)

Worldsteel Association, 2013a, Iron production 2012: Worldsteel Association. (Accessed August 20, 2013, at http://www.worldsteel.org/statistics/statistics-archive/2012-iron-production.html.)

Worldsteel Association, 2013b, Steel production 2012: Worldsteel Association. (Accessed August 20, 2013, at http://www.worldsteel.org/statistics/statistics-archive/2012-steel-production.html.)

Xstrata plc, 2013, Annual report 2012: Xstrata plc, 152 p. (Accessed August 20, 2013, at http://www.glencorexstrata.com/assets/Uploads/2012-Annual-Report-FINAL.pdf.)

TABLE 1
SPAIN: PRODUCTION OF MINERAL COMMODITIES[1]

(Metric tons unless otherwise specified)

Commodity[2]		2008	2009	2010	2011	2012[p]
METALS						
Aluminum:						
Alumina[e, 3]		1,500,000 [4]	1,500,000	1,500,000	1,500,000	1,500,000
Metal:						
Primary		408,000 [4]	408,000	408,000	408,000	408,000
Secondary		243,000	243,000	243,000	243,000	243,000
Total		651,000	651,000	651,000	651,000	651,000
Copper:						
Mine output, Cu content		7,067	12,587	46,333 [5]	74,246 [5]	75,057 [5]
Metal:[e]						
Blister:						
Primary		259,900 [6]	260,000	260,000	260,000	260,000
Secondary		10,000	10,000	10,000	10,000	10,000
Total		269,900 [6]	270,000	270,000	270,000	270,000
Refined:						
Primary		255,000	255,000	255,000	255,000	255,000
Secondary		35,000	35,000	35,000	35,000	35,000
Total		290,000	290,000	290,000	290,000	290,000
Gold, mine output, Au content	kilograms	3,400	3,450 [r]	3,500 [r]	3,550 [r]	3,600
Iron and steel, metal:						
Pig iron[e]	thousand metric tons	4,200	4,200	3,572 [5]	3,540 [5]	3,570 [5]
Steel:						
Crude[7]	do.	18,600	14,400	16,343 [5]	15,591 [5]	15,600
Hot rolled	do.	15,000	15,000	15,000	15,000	15,000
Lead, metal, refined, secondary		125,000	125,000	125,000	125,000	125,000
Nickel, Ni content of concentrate		8,136	8,035	5,402 [5]	6,296 [5]	6,300
Silver, mine output, Ag content	kilograms	3,450 [r]	3,500 [r]	3,500 [r, e]	3,505 [r]	3,500 [e]
Tungsten, mine output[8]		150	200	229	497	542
Zinc, metal, primary and secondary		456,050	500,776	505,000	489,000 [5]	490,000

See footnotes at end of table.

TABLE 1—Continued
SPAIN: PRODUCTION OF MINERAL COMMODITIES[1]

(Metric tons unless otherwise specified)

Commodity[2]		2008	2009	2010	2011	2012[p]
INDUSTRIAL MINERALS						
Barite, $BaSO_4$		11,100	2,814	2,050 [5]	NA	NA
Cement, hydraulic	thousand metric tons	42,088	29,505	26,217	22,200	20,000
Clays:						
Bentonite		154,534	140,000	157,001 [5]	110,721 [5]	110,750
Kaolin, washed		355,739	300,000	310,993 [5]	302,580 [5]	303,000
Diatomite and tripoli[e]		50,000	29,194 [5]	64,346 [5]	50,000	50,000
Feldspar		690,256	597,496 [5]	691,894 [5]	580,000	580,000 [e]
Fluorspar, CaF_2 content:						
Acid-grade		127,300	111,810	126,730 [5]	109,284 [5]	120,000
Ceramic-grade		15,930	6,485	1,824 [5]	2,639 [5]	2,640
Metallurgical-grade		5,506	4,238 [5]	3,787 [5]	5,410 [5]	5,450
Total		148,736	122,533 [5]	132,341 [5]	117,333 [5]	128,090
Gypsum and anhydrite, crude[e]	thousand metric tons	15,000	11,500	6,990 [5]	7,400 [r]	7,100
Lime, hydrated and quicklime[e]	do.	2,000	2,000	2,200	2,200	2,200
Magnesite, calcined		187,626	163,930	195,893 [5]	200,000 [5]	200,000
Mica		4,254	4,000	4,034 [5]	3,775 [5]	3,775
Potash, K_2O equivalent[e]		435,000	481,455 [5]	418,778 [5]	436,026 [5]	436,000
Pumice		600,000	436,542 [r,5]	432,364 [r,5]	430,500	430,500
Salt:						
Rock, including byproduct from potash works		2,850	2,850	3,116 [r,5]	3,200 [6]	3,200
Marine and other	thousand metric tons	1,291	1,439 [r,5]	1,234 [r,5]	1,171 [6]	1,185
Sand and gravel, silica sand[9]	do.	134,000	134,000	170,000 [r,5]	130,000	130,000
Sepiolite, meerschaum		707,950	573,937 [r,5]	557,862 [r,5]	566,970 [6]	567,000
Sodium compounds, n.e.s., sulfate, natural:[e,10]						
Glauberite, Na_2SO_4 content		9,500,000	1,121,784 [r,5]	1,216,787 [r,5]	1,200,000 [e]	1,200,000
Thenardite, Na_2SO_4 content		165,000	166,362 [5]	156,776 [5]	160,000	160,000
Stone:[e]						
Basalt	thousand metric tons	5,000	2,703 [5]	2,252 [5]	3,000	3,000
Chalk	do.	1,000	744 [5]	679 [5]	700	700
Dolomite	do.	15,000	13,843 [5]	10,431 [5]	12,000	12,000
Granite, ornamental	do.	12,500	20,964 [5]	12,464 [5]	12,500	12,500
Limestone	do.	270,000	195,138 [5]	134,864 [5]	170,000	170,000
Marble, ornamental	do.	2,600	1,669 [5]	1,209 [5]	1,600	1,600
Marl	do.	10,000	9,000 [5]	8,057 [5]	8,000	8,000
Ophite	do.	4,000	5,181 [5]	3,446 [5]	4,000	4,000
Phonolite	do.	1,800	1,011 [5]	1,018 [5]	1,200	1,200
Porphyry	do.	1,100	2,682 [5]	1,548 [5]	1,550	1,550
Quartz	do.	1,100	789 [5]	1,129 [5]	1,100	1,100
Quartzite	do.	2,800	4,685 [5]	2,500 [5]	2,500	2,500
Sandstone	do.	3,400	2,905 [5]	2,219 [5]	2,400	2,400
Slate	do.	1,200	416 [5]	357 [5]	380	380
Other	do.	8,000	8,418 [5]	8,000 [5]	8,000	8,000
Strontium minerals, Sr_2O_4 content		188,000 [e]	57,466 [5]	83,035 [5]	97,102 [5]	97,100
Sulfur, byproduct:[e]						
Metallurgy	thousand metric tons	500	536	539	539	540
Petroleum	do.	136 [r]	136 [r]	136 [r]	136 [r]	140
Coal (lignite) gasification	do.	1	1	1	1	1
Total	do.	637 [r]	673 [r]	676 [r]	676 [r]	681
Talc and steatite		100,000	47,218 [6]	51,897 [6]	52,000	52,000

See footnotes at end of table.

TABLE 1—Continued

SPAIN: PRODUCTION OF MINERAL COMMODITIES[1]

(Metric tons unless otherwise specified)

Commodity[2]		2008	2009	2010	2011	2012[p]
MINERAL FUELS AND RELATED MATERIALS						
Coal, marketable:						
Anthracite	thousand metric tons	7,238	7,700	5,990 [5]	4,265 [5]	4,073
Bituminous	do.	2,890	2,892	2,444 [5]	2,358 [5]	2,252
Total	do.	10,128	10,592	8,434 [5]	6,623 [5]	6,325
Gas, natural						
Produced	thousand cubic meters	46,354	32,280	58,425 [5]	43,888 [5]	44,000
Peat		60,000	58,678 [5]	64,962 [5]	65,000	65,000
Petroleum:						
Crude	thousand 42-gallon barrels	298	250	284 [5]	234 [5]	323
Refinery products:[e]						
Liquefied petroleum gas	do.	34,240 [6]	34,200	34,200	34,200	34,100
Naphtha	do.	25,401 [6]	25,400	25,400	25,400	25,300
Gasoline, motor	do.	89,006 [6]	89,000	89,000	89,000	88,700
Jet fuel	do.	21,540 [6]	21,500	21,500	21,500	21,400
Kerosene	do.	16,257 [6]	16,300	16,300	16,300	16,200
Distillate fuel oil	do.	113,740 [6]	114,000	114,000	114,000	113,500
Residual fuel oil	do.	58,220 [6]	58,200	58,200	58,200	58,000
Other	do.	81,284 [6]	81,300	81,300	81,300	81,000
Refinery fuel and losses	do.	26,417 [6]	26,400	26,400	26,400	26,300
Total	do.	466,105 [6]	466,000 [r]	466,000 [r]	466,000 [r]	465,000

[e]Estimated; estimated data are rounded to no more than three significant digits; may not add to total shown. [p]Preliminary. [r]Revised. do. Ditto. NA Not available.

[1]Table includes data available through October 18, 2013.

[2]In addition to the mineral commodities listed, Spain had produced attapulgite, ferroalloys, germanium oxide, other clays, pigment ocher and red iron oxide, sand and gravel (industrial), soda ash and sulfate (manufactured), coke (metallurgical), and natural gas (marketed).

[3]Reflects aluminum hydrate.

[4]Reported figure. Source: Alcoa Inc., 2012 Annual Report.

[5]Reported figure. Source: Minerals Questionnaires for 2011 and 2012.

[6]Reported figure.

[7]Reported figure. Source: Worldsteel Association, December 2012.

[8]Reported figure. Source: Heemskirk, Almonty, December 2012.

[9]Includes sand obtained as a byproduct of feldspar and kaolin production.

[10]Not elsewhere specified.

Sources: Industria y Minería, 2012. Ministerio de Industria, Turismo y Comercio—Secretaría General de Energía, 2012. Instituto Geológico y Minero de España, 2012.

TABLE 2
SPAIN: STRUCTURE OF THE MINERAL INDUSTRY IN 2012

(Thousand metric tons unless otherwise specified)

Commodity	Major operating companies and major equity owners	Location of main facilities	Annual capacity
Alumina	Aluminio/Alúmina Española S.A. (AESA) (Alcoa Inc., 100%)	Alumina plant at San Ciprian, Lugo	1,500
Aluminum	do.	Electrolytic plant at San Ciprian, Lugo	228
Do.	Alcoa Inespal S.A. (AISA) (Alcoa Inc., 100%)	Electrolytic plant at Aviles	93
Do.	do.	Electrolytic plant at La Coruña	87
Barite	Minerales y Productos Derivados S.A. (MINERSA)	Mine and plant at Vera, Almería	100
Bentonite	Süd-Cheme España SL	Mine and plant at Yuncos, Toledo	150
Cement	Ashland S.A.	Puerto de Sagunton, Valencia	2,000
Do.	do.	Villaluenga de la Sagra, Toledo	2,000
Do.	do.	3 other plants	2,000
Do.	35 other companies	49 other plants	38,000
Coal:			
Anthracite	Antracitas Gaiztarro S.A.	Mines at María and Paulina	2,000
Do.	do.	Mines near Oviedo	2,000
Do.	Antracitas del Bierzo S.A.	Mines near Leon	1,000
Bituminous	Hulleras del Norte S.A. (HUNOSA)	Various mines and plant	3,300
Do.	Hulleras Vasco Leonesa S.A.	Santa Lucía Mine, León	2,000
Do.	Minas de Figaredo S.A.	Mines near Oviedo	1,000
Do.	Nacional de Carbon del Sur (Encasur)	Rampa 3 and San José Mines, Cordoba	200
Lignite	Empresa Nacional de Electricidad S.A. (Endesa)	As Pontes Mine, and Andorra Mine, La Coruña	15,000
Copper, metal, content	Atlantic Copper S.A. (Freeport MacMoRan Copper & Gold Inc., 100%)	Refinery at Huelva	270
Do.	do.	Electrolytic refinery at Huelva	105
Do.	Industrias Reunidas de Cobre	Smelter at Asua-Bilbao	30
Do.	Elmet SL	Smelter and electrolytic refinery at Berango, Vizcaya	60
Do.	Atlantic Copper S.A. (Freeport MacMoRan Copper & Gold Inc., 100%)	Mines and plant at Arientero near Santiago de Compostela	12
Do.	do.	Alfredo underground mine in Río Tinto area	30
Do.	Inmet Mining Corp., 100%	Open pit mines in Sevilla, Andalucía, Spain	210
Do.	do.	Cathode Electrowinning at Las Cruces in Sevilla	72
Do.	Minas de Rio Tinto S.A.	Cerro Colorado open pit mine	20
Do.	Río Narcea Gold Mines, Ltd. (Lundin Mining Corp., 100%)	Aguablanca Mine, Extremadura	7
Dunite	Pasek España S.A.	Mines and plant at Landoy, Ortigueira	1,500
Fluorspar	Minerales y Productos Derivados S.A. (MINERSA)	Plant at Torre, Asturias	150
Do.	do.	Underground mines at Emilio, Jaimina, and Moscona, Asturias	420
Gold kilograms	Río Narcea Gold Mines, Ltd.	El Valle and Carlés Mines, Asturias	3,750
Lead, metal, content	Española del Zinc S.A.	Refinery at Cartagena, Murcia	50
Do.	Compañia La Cruz, Minas y Fundaciones de Plomo S.A.	Smelter at Lineares, Jaen	40
Do.	do.	Refinery at Lineares, Jaen	40
Do.	Tudor S.A.	Secondary smelter at Saragoza	16
Do	Ferroaleaciones Españolas, S.A.	Secondary smelter at Medina del Campo	12
Do	Derivados de Minerales y Metales	Secondary smelter at Barcelona	5
Do	Sociedad Minera y Metalúrgica de Peñarroya de España S.A. (Peñarroya, France, 90%)	Opencast mine at Montos de Los Azules	25
Do.	Exploración Minera Internacional España S.A. (EXMINESA)	Underground mine at Rubiales, Lugo	16
Magnesite	Magnesitas Navarras S.A.	Mine at Eugui, plant at Zubiri	600
Do.	Magnesitas de Rubián S.A.	Plant at Monte Castel	70
Do.	SA Reverte	Plant at Zaragoza	443
Nickel, metal, content	Río Narcea Gold Mines, Ltd. (Lundin Mining Corp., 100%)	Aguablanca Mine, Extremadura	8

See footnotes at end of table.

TABLE 2—Continued
SPAIN: STRUCTURE OF THE MINERAL INDUSTRY IN 2012

(Thousand metric tons unless otherwise specified)

Commodity		Major operating companies and major equity owners	Location of main facilities	Annual capacity
Petroleum:				
Crude	42-gallon barrels per day	Chevron S.A.	Oilfield at Casablanca	300
Refined	do.	Repsol YPF, S.A.	Refinery at Escombreras	200,000
Do.	do.	do.	Refinery at Puertollano	14,000
Do.	do.	do.	Refinery at Tarragona	260,000
Do.	do.	Refinería de Petróleos del Norte S.A. (Petronor)	Refinery at Somorrostro	240,000
Do.	do.	Compañía Española de Petróleos S.A. (Cepsa) [Total SA, 51.17%, and International Petroleum Investment Co. (IPIC), 48.83%]	Refinery at Gibraltar-San Roque	88,000
				35,000
Do.	do.	do.	Refinery at La Rabida	37,000
Do.	do.	do.	Refinery at Tenerife	35,000
Do.	do.	Petróleos del Mediterraneo S.A. (Petromed)	Refinery at Castellon de la Plana	120,000
Do.	do.	BP p.l.c., 100%	Refinery at Castellon, Iberia	45,000
Do.	do.	Compañía Ibérica Refinadora de Petróleos S.A. (Petroliber)	Refinery at La Coruña	140,000
Potash, ore		Iberpotash S.A. (ICL Fertilizers Europe)	Mines and plants at Suria near Barcelona	1,100
Pyrite		Compañía Española de Mines de Tharsis	Mines and plants at Tharsis and Zarza (closed)	1,300
Do.		do.	Plant at Huelva	600
Sepiolite		Grupo Tolsa S.A.	Mine and plant at Vicalvaro near Madrid	1,000
Do.		Silicatos-Anglo-Ingleses S.A.	Mine and plant at Villecas near Madrid	200
Silver	metric tons	Polar Minin Oy (Dragon Mining NL, 50%, and Ormonde Mining plc, 50%)	Valiña silver project, Lugo Province	4
Sodium sulfate		Crimidesa S.A.	Mine and plant at Cerezo de Rio, Burgos	600
Steel		Aceralia Corporación Siderúrgica (Arbed S.A., 35%)	Plants at Aviles, Gijon, Sagunto, and Sestao	8,000
Do.		Compañia Española de Laminacion S.L. (Celsa), 100%	Plant at Barcelona	2,600
Do.		Corporación Gerdau Sidenor S.A. (Sidenor) (Gerdau Group, 50%, and Santander Group, 50%)	Plant at Basauri	2,500
Strontium		Solvay Minerales S.A.	Mines and plant at Escuzar, Granada	85
Do.		Bruno S.A.	Mine and plant at Montevives, Granada	50
Uranium, U_3O_8	metric tons	Empresa Nacional del Uranio (Enusa) (Government, 100%)	Mines and plant near Ciudad Real	500
Zinc, metal, content		Asturiana de Zinc S.A. (Azsa) (Xstrata plc, 100%)	Electrolytic zinc plant at San Juan de Nieva Castillon	510
Do.		Española del Zinc S.A.	Electrolytic plant at Cartagena	50
Do.		Exploración Minera International España S.A. (EXMINESA)	Underground mine at Rubiales, Lugo	500
Do.		Sociedad Minera y Metalúrgica de Penarroya-Espana S.A.	Mines and plants at Montos de los Azules y Sierra de Lujar, San Agustín	200

Do., do. Ditto.

The Mineral Industry of Sweden

By Alberto Alexander Perez

Sweden is located on part of the Fennoscandian Shield, an area of Precambrian crystalline and metamorphic rocks. Common rocks of the shield and surrounding platform are gneiss, granite, granodiorite, sandstone, and marble. Glacial till covers about 75% of the landscape (Geological Survey of Sweden, 2014a).

In 2012, Sweden was among the most active mining countries in Europe. Sweden was the leading producer of iron ore in the European Union (EU), and it was one of the EU's leading producers of copper, gold, lead, silver, and zinc. Most notably, Sweden has alum shale-hosted uranium-molybdenum-vanadium deposits and Kiruna-type iron deposits in the north (Eilu, 2011, p. 14; Geological Survey of Sweden, 2014b).

Although the country has abundant hydroelectric power, it also relied on nuclear power produced by 10 active reactors for 42.7% of its electricity. The real gross domestic product (GDP) based on purchasing power parity was $399.4 billion, which was 1.2% more than in 2011. Sweden had reached an advanced state of industrialization; as such, the largest portion of its GDP was from its services sector. Industry accounted for 27.4% of the country's GDP (U.S. Central Intelligence Agency, 2012; International Atomic Energy Agency, 2014).

Minerals in the National Economy

According to the Mining Inspectorate of Sweden [a part of the Geological Survey of Sweden (SGU)], growth in the Swedish mineral sector is crucial for creating employment in Sweden, particularly in those regions of Sweden where mines are located. The Inspectorate has also said that mining is vital to the development of the mining equipment industry, which is an important part of the Swedish industrial sector, regardless of where in the country it is situated. In 2010 (the latest year for which data were available), Sweden was the 17th-ranked country in the world in terms of the value of production of its mineral industry, which totaled $3.96 billion, or 0.9% of the total value of world mineral production. Mineral production accounted for about 0.9% of Sweden's GDP, and the value of exports generated by the mineral industry was about 5.3% (International Council on Mining and Metals, 2012; Geological Survey of Sweden, 2014c).

Government Policies and Programs

The Mining Inspectorate is the Government office responsible for issuing permits for exploration and mining. It also arbitrates on matters relating to the Minerals Act of 1991, which is the law that governs the mineral industry in Sweden. The SGU is the Swedish Government's authority on matters relating to geology and minerals management, both nationally and at the EU level. The SGU monitors the developments in the minerals markets at the Swedish level and internationally, and also publishes statistics on the production of mineral commodities (including aggregates and peat) and other commodities in Sweden and in the global market (Geological Survey of Sweden, 2014a).

Production

Sweden was the leading producer of iron ore in the EU and the 13th-ranked producer of iron ore in the world (Tuck, 2013). The country also produced copper, gold, lead, silver, and zinc, and extracted industrial minerals, including feldspar and limestone. In 2012, production of primary aluminum, iron ore, mined lead, and silver metal all increased whereas production of mined copper and crude steel decreased (table 1).

Structure of the Mineral Industry

The Swedish mineral industry was composed mostly of privately owned companies, and it operated on a free-market basis. The Government was the major equity owner of Luossavaara-Kiirunavaara AB's (LKAB's) iron ore operation, and had significant ownership in the Svenskt Stal AB steel operation.

Boliden AB was a leading Swedish mining and mineral processing company; it had operations in Sweden and abroad. The company principally produced copper, gold, lead, and silver. Boliden's main mines were the Aitik Mine and the Kankberg Mine.

Nordkalk AB was a leading international producer of limestone, crushed and ground limestone, concentrated calcite, quicklime, and slaked lime as well as dolomite and wollastonite, which Nordkalk extracted as a byproduct of mining for limestone. Nordkalk had operations in 30 locations in nine countries as well as mines in five countries. In Sweden, Nordkalk's limestone operations were located in Storugns. Table 2 is a list of Sweden's major mineral industry facilities in 2012.

The Canadian company Lundin Mining Corp. had significant operations in Sweden. The company produced lead, silver, and zinc from its Ammeberg Mine.

HeidelbergCement AG of Germany owned Cementa AB, which had three cement plants in Sweden. The plants were located at Degerhamn, Skivode, and Slite and had a combined production capacity of about 3.4 million metric tons per year (Mt/yr) (table 2).

Mineral Trade

In 2012, significant mineral commodity exports to Sweden from the United States included, in order of value, nuclear fuel material ($116.03 million), metallurgical-grade coal ($87.12 million), and coal and other mineral fuels ($45.26 million). Significant mineral commodity imports from Sweden to the United States included semifinished iron and steel products ($575.77 million); petroleum products ($341.20 million); fuel oil ($262.59 million); iron and steel products, except those of advanced manufacture ($99.45 million); and unmanufactured steelmaking and ferroalloying materials ($80.13 million) (U.S. Census Bureau, 2013a, b).

The principal products that the Swedish industry produced in 2012 were iron and steel, motor vehicles, precision instruments, processed food, wood pulp, and paper products. Sweden's main export partners were Norway (which received 10.4% of Sweden's exports, in terms of value), Germany (10.3%), the United Kingdom (8.1%), Denmark and Finland (6.7% each), the United States (5.5%), the Netherlands (5.2%), Belgium (5%), and France (4.8%). Its main import partners were Germany (which supplied 17.4% of Sweden's imports, in terms of value), Denmark (8.5%), Norway (8.4%), the United Kingdom (6.5%), the Netherlands (6.4%), Russia (5.6%), Finland (5.1%), China (4.9%), and France (4.2%) (U.S. Central Intelligence Agency, 2012).

Commodity Review

Metals

Aluminum.—Kubikenborg Aluminium AB (KUBAL), which was a wholly owned subsidiary of United Company RUSAL of Russia, was the only major aluminum producer in the country. KUBAL increased its output in 2012 by 16% compared with that of the previous year. As was the case in 2011, increased output was in response to an increase in demand for its products and the ability of KUBAL to produce them at a relatively low cost (United Company RUSAL, 2013, p. 29).

Copper.—Boliden's Aitik Mine, which is located in northern Sweden, was Boliden's and Sweden's largest copper mine in terms of production quantity. According to Boliden, Aitik's copper grade is low (about 0.22%), but the open pit mine uses a combination of large-scale extraction and high levels of automation to ensure high levels of productivity, which compensates for the low grades yielded by the mine. The Aitik Mine had been under expansion for the past several years, and the expansion was scheduled to be completed by 2014. Once completed, the mine would have an ore production capacity of 36 Mt/yr. In 2012, the mine produced 34.3 million metric tons (Mt) of milled ore (an increase of about 8.7% compared with that of 2011) at grades of about 0.22% copper, 2.50 grams per metric ton (g/t) silver, and 0.11 g/t gold (Boliden AB, 2013a, p. 17, 19, 97).

Boliden's Rönnskär smelter was a leading facility (in terms of tonnage produced) for the recycling of copper and precious metals in Sweden. The main products were copper, gold, lead, and zinc clinker. The smelter produced 844,000 metric tons (t) of concentrates and secondary materials and 214,000 t of copper cathodes in 2012 (Boliden AB, 2013b).

Gold.—Dragon Mining Ltd. of Australia and Elgin Mining Inc. of Canada owned gold mines in the Skelleftea mining district. Dragon Mining's Svartliden Mine is located 700 kilometers (km) north of Stockholm, and Elgin's Bjorkdal Mine is located 750 km north of Stockholm. In 2012, the Svartliden production center produced 382,104 t of mined ore at a grade of 3.33 g/t gold and a gold recovery rate of 90.9%, which resulted in output of more than 1,000 kilograms (kg) of gold. In February, Gold-Ore Resources Ltd. announced an updated measured and indicated mineral resource estimate for the Bjorkdal Mine's open pit and underground mine of 30,295 kg of gold. The Skelleftea mining district where the Bjorkdal and the Svartliden Mines are located had been the focus of exploration for gold-rich polymetallic deposits since the mid-1920s (Gold-Ore Resources Ltd., 2012; Dragon Mining Ltd., 2013).

Boliden was the other main producer of gold in Sweden. Its polymetallic mines had an estimated production capacity of about 2,000 kilograms per year of gold. Its major gold mining operations were the Aitik Mine, which was principally a copper-producing mine, and the operations at the Boliden and the Garpenberg sites (table 2; Boliden AB, 2013a, p. 19).

Iron and Steel.—LKAB's Kiruna Mine was one of the world's largest underground iron ore mine in terms of volume; it has an ore body that is 4 km long and 80 meters wide and reaches to a depth of about 2 km. LKAB announced that it had been granted an environmental permit for a new open pit mine located at Gruvberget. This would be LKAB's first new iron ore mine in 50 years. Production at the new Gruvberget Mine was expected to be 2 Mt/yr. The ore body contains both hematite and magnetite (Luossavaara-Kiirunavaara AB, 2012a, b).

Industrial Minerals

Limestone.—SMA Mineral AB produced principally limestone and lime; however, the company also produced dolomite, magnesium hydroxide, and magnesium oxide. The company had operations in Boda and Rattvik. Svenska Kyanite AB, which was a fully owned subsidiary of SMA Mineral, produced kyanite in Halskoberg (SMA Mineral AB, 2014).

Rare Earths.—LKAB was investigating the tailings ponds at the Kiruna and the Malmberget operations; these ponds were thought to contain large quantities of rare-earth elements (REEs) bound in the phosphate mineral apatite, which is considered an impurity in iron ore. LKAB planned to conduct a study to determine the conditions for the recovery of apatite and REEs from the tailings in the ponds. Test drilling results indicated the occurrence of different REEs in the apatite. Estimates showed that there was enough apatite in the tailings ponds for the production of 400,000 metric tons per year of apatite concentrates for a period of 14 years. Startup of production was not expected before 2015 (Steel Orbis, 2011).

Mineral Fuels and Other Sources of Energy

Renewable Energy.—In 2011 (the latest year for which data were available), Sweden had the largest share of renewable energy in the EU. About 40% of Swedish energy consumption was covered by renewable energy sources. The Government had set its target at 49% use of renewable energy by 2020. In comparison, renewable energy was projected to cover only about 20% of the whole EU's energy consumption by 2020 (Nordic Energy Solutions, 2012).

Outlook

Mining, although a small part of the country's GDP, is expected to remain important to Sweden's economy. Iron ore production is expected to increase and reach 50 Mt/yr within 10 years. Sweden has substantial base-metal, gold, and iron ore deposits that are expected to continue to attract investors in the near future. Foreign companies are likely to continue to explore actively in Sweden for base metals, diamond, and, particularly, gold. The Government is expected to continue to support the production and use of renewable energy in electricity, heating, cooling, and transport.

References Cited

Boliden AB, 2013a, Annual report 2012: Boliden AB, 129 p. (Accessed November 14, 2013, at http://www.boliden.com/Documents/Press/Publications/Boliden_AR12_ENG.pdf.)

Boliden AB, 2013b, Rönnskär: Boliden AB, 2 p. (Accessed November 15, 2013, at http://www.boliden.com/Documents/Press/Publications/Fact sheets/facts-ronnskar-en.pdf.)

Dragon Mining Ltd., 2013, Annual report 2012: Dragon Mining Ltd., 102 p. (Accessed November 19, 2013, http://www.dragon-mining.com.au/sites/default/files/dragon_ar2012_final_web_version_0.pdf.)

Eilu, Pasi, 2011, Metallic mineral resources of Fennoscandia, *in* Nenonen, Keijo, and Nurmi, P.A., eds., Geoscience for society—125th anniversary volume: Espoo, Finland, Geological Survey of Finland Special Paper 49, p. 13–21.

Geological Survey of Sweden, 2014a, Geology of Sweden: Geological Survey of Sweden. (Accessed October 13, 2014, at http://www.sgu.se/en/geology-of-sweden/.)

Geological Survey of Sweden, 2014b, Minerals of Sweden: Geological Survey of Sweden. (Accessed October 13, 2014, at http://www.sgu.se/en/mineral-resources/minerals-of-sweden/.)

Geological Survey of Sweden, 2014c, Why legislation on minerals?: Geological Survey of Sweden. (Accessed October 13, 2014, at http://www.sgu.se/en/mining-inspectorate/legislation/why-legislation-on-minerals/.)

Gold-Ore Resources Ltd., 2012, Gold-Ore reports increase in gold resource estimates at Bjorkdal gold mine: Gold-Ore Resources Ltd. (Accessed November 22, 2013, at http://www.businesswire.com/news/home/20120305005916/en/Gold-Ore-Reports-Increase-Gold-Resource-Estimates-Bjorkdal.)

International Atomic Energy Agency, 2014, Sweden: International Atomic Energy Agency. (Accessed October 13, 2014, at http://www.iaea.org/PRIS/CountryStatistics/CountryDetails.aspx?current=SE.)

International Council on Mining and Metals, 2012, The role of mining in national economies: International Council on Mining and Metals, 20 p. (Accessed October 13, 2014, at http://www.icmm.com/document/4440.)

Luossavaara-Kiirunavaara AB, 2012a, Gruvberget: Luossavaara-Kiirunavaara AB. (Accessed November 15, 2013, at http://www.lkab.com/en/Future/Urban-Transformations/Why/What-is-Iron-Ore/The-Ore-in-Svappavaara1/Gruvberget/.)

Luossavaara-Kiirunavaara AB, 2012b, LKAB gets the go-ahead for new iron ore mine in Sweden: Luossavaara-Kiirunavaara AB. (Accessed November 15, 2013, at http://www.lkab.com/en/About-us/Overview/Operations-Areas/Kiruna/.)

Nordic Energy Solutions, 2012, Sweden's energy system: Nordic Energy Solutions. (Accessed October 12, 2014, at http://www.nordicenergysolutions.org/performance-policy/sweden/swedens-energy-system/.)

SMA Mineral AB, 2014, Applications: SMA Mineral AB. (Accessed October, 13, 2014, at http://www.smamineral.com/Applications.aspx.)

Steel Orbis, 2011, LKAB seeks to produce rare earth elements in Sweden: Steel Orbis, January 14. (Accessed September 19, 2012, at http://www.steelorbis.com/steel-news/latest-news/lkab-seeks-to-produce-rare-earth-elements-in-sweden-577451.htm.)

Tuck, C.A., 2013, Iron ore: U.S. Geological Survey Mineral Commodity Summaries 2013, p. 84–85.

United Company RUSAL, 2013, RUSAL annual report 2012: United Company RUSAL, 236 p. (Accessed October 11, 2012, at http://www.rusal.ru/upload/uf/f44/UC_RUSAL_Annual_Report_2012_eng.pdf.)

U.S. Census Bureau, 2013a, U.S. exports to Sweden by 5-digit end-use code: U.S. Census Bureau. (Accessed November 18, 2013, at http://www.census.gov/foreign-trade/statistics/product/enduse/exports/c4010.html.)

U.S. Census Bureau, 2013b, U.S. imports from Sweden by 5-digit end-use code: U.S. Census Bureau. (Accessed November 18, 2013, at http://www.census.gov/foreign-trade/statistics/product/enduse/imports/c4010.html.)

U.S. Central Intelligence Agency, 2012, Sweden, *in* The world factbook:, U.S. Central Intelligence Agency. (Accessed October 28, 2012, at https://www.cia.gov/library/publications/the-world-factbook/geos/sw.html.)

TABLE 1
SWEDEN: PRODUCTION OF MINERAL COMMODITIES[1]

(Metric tons unless otherwise specified)

Commodity[2]		2008	2009	2010	2011	2012[e]
METALS						
Aluminum, metal:[e]						
Primary		81,546 [3]	69,708 [3]	93,000	111,000	129,000
Secondary		32,000	30,000	30,000	30,000	30,000
Total		113,546 [3]	99,708 [3]	123,000	141,000	159,000
Copper:						
Mine output, Cu content		57,688 [3]	55,400 [3]	76,500	83,000	82,422 [3]
Metal:						
Smelter:[e]						
Primary		204,204 [3]	125,398 [3]	137,000	155,000	214,000
Secondary		67,795 [3]	65,000	42,000	44,000	46,000
Total		271,999 [3]	190,000 [3]	179,000	199,000	260,000
Refined:						
Primary		227,774	205,759	150,497	179,316	179,000
Secondary[e]		25,000	25,000	40,000	40,000	40,000
Total[e]		253,000	231,000	190,000	219,000	219,000
Gold:						
Mine output, Au content	kilograms	4,900	5,461	6,242	5,935	6,015 [3]
Metal, primary and secondary[4]	do.	13,425 [r]	13,282	12,450	10,600	12,532 [3]
Iron and steel:						
Iron ore concentrate and pellets:						
Gross weight	thousand metric tons	27,713	20,389	27,917	22,968	26,039 [3]
Fe content (60%)	do.	16,628	12,233	16,750	15,159 [5]	17,186 [3,5]
Metal:						
Pig iron and sponge iron	do.	3,583	1,966	3,447	3,240	5,253 [3]
Ferroalloys, ferrochromium[e]		117,053 [3]	31,345 [3]	32,000	32,000	32,000
Steel, crude	thousand metric tons	5,196	2,805	4,844	4,866	4,326 [3]
Lead:						
Mine output, Pb content		65,100 [e]	69,300	67,700	61,999	63,551 [3]
Metal, refined:[e]						
Primary		56,800	55,000	56,000	52,400	62,000
Secondary		42,600	42,000	40,000	41,000	44,000
Total		99,400	97,000	96,000	93,400	106,000
Nickel, metal, secondary[e]		50	50	50	--	--
Silver:						
Mine output, Ag content	kilograms	293,100	288,600	302,100	238,030	309,337 [3]
Metal, primary	do.	429,637	481,223	385,684	415,066	447,759 [3]
Zinc, mine output, Zn content		188,048	192,538	198,687	190,251	188,300 [3]
INDUSTRIAL MINERALS						
Cement, hydraulic[e]	thousand metric tons	2,900	2,950	2,900	2,900	3,000
Feldspar, salable, crude and ground[e]		42,000	44,000	44,000	30,000 [r,3]	27,000
Lime[e]	thousand metric tons	600	600	700	960 [6]	960 [6]
Quartz and quartzite	do.	151 [r]	56 [r]	85 [r]	163 [r]	101 [3]

See footnotes at end of table.

TABLE 1—Continued
SWEDEN: PRODUCTION OF MINERAL COMMODITIES[1]

(Metric tons unless otherwise specified)

Commodity[2]		2008	2009	2010	2011	2012[e]
INDUSTRIAL MINERALS—Continued						
Stone:						
Dimension:						
Mostly unfinished	thousand metric tons	170	170	180	NA	NA
Granite	do.	132	132	124	92 [r]	79 [3]
Limestone	do.	32	32	43	23 [r]	21 [3]
Other	do.	6	6	6	67 [r]	82 [3]
Crushed:[e]						
Dolomite	do.	450	450	450	483 [r,3]	429
Limestone	do.	8,980	8,980	8,980	7,317 [r,3]	7,385
Sandstone	do.	20	20	20	NA	629
Undifferentiated	do.	30,000	30,000	30,000	NA	101
Talc, soapstone[e]		4,000 [r]	4,000 [r]	4,000 [r,3]	3,000 [r]	--
MINERAL FUELS AND RELATED MATERIALS						
Coke, metallurgical	thousand metric tons	1,177	987	1,197	1,190 [e]	1,200
Peat:						
Agricultural use	thousand cubic meters	1,434	1,198	1,250	1,611	1,000
Fuel	do.	2,135	2,143	2,213	2,139	1,800
Petroleum, refinery products:						
Liquefied petroleum gas	thousand 42-gallon barrels	3,886	3,248	3,978 [r]	3,900 [r,e]	3,900
Gasoline, motor	do.	38,444	38,070	32,740 [r]	33,000 [r,e]	33,000
Jet fuel	do.	1,933	1,679	1,424 [r]	1,400 [r,e]	1,400
Distillate fuel oil	do.	59,450	57,232	56,393 [r]	56,400 [r,e]	56,400
Residual fuel oil	do.	29,826	28,543	33,252 [r]	33,200 [r,e]	33,200
Other	do.	--	22,119	25,331 [r]	25,300 [r,e]	25,300
Total	do.	133,539	150,891	153,118 [r]	153,000 [r,e]	153,000

[e]Estimated; estimated data are rounded to no more than three significant digits; may not add to totals shown. [r]Revised. do. Ditto. NA Not available.
-- Zero.
[1]Table includes data available through November 18, 2013.
[2]In addition to the commodities listed, Sweden produced synthetic diamond, manufactured fertilizer, manufactured gas, granite, limestone, molybdenum, selenium, slate, steel semimanufactures, and sulfur, but available information was inadequate to make reliable estimates of output.
[3]Reported figure.
[4]Series was updated to include metal production from ores and electronics scrap recycling.
[5]Iron content reported to be 66%.
[6]Quicklime; estimate based on volume sold.

TABLE 2
SWEDEN: STRUCTURE OF THE MINERAL INDUSTRY IN 2012

(Thousand metric tons unless otherwise specified)

Commodity		Major operating companies and major equity owners	Location of main facilities	Annual capacity
Aluminum		Kubikenborg Aluminium AB (KUBAL) (United Company RUSAL, 100%)	Smelter at Sundsvall	125
Cement		Cementa AB (HeidelbergCement AG, 100%)	Plants at Degerhamn, Skivode, and Slite	3,400
Copper:				
Ore, copper content		Boliden AB	Mines at Aitik, Garpenberg, Kankberg, Kristineberg, Maurlinden, Maurlinden Ostra, and Renstrom	NA
Metal		do.	Smelter and refinery at Ronnskar	240
Feldspar		Berglings Malm & Mineral AB (Omya GmbH)	Mines at Beckegruvan, Hojderna, and Limbergsbo	50
Do.		Silbelco Nordic AS	Mines at Forshammar	30
Ferroalloys		Vargon Alloys AB (Yildrim Group, 100%)	Plant at Vargon	255
Gold:				
Ore, gold content	kilograms	Dragon Mining Ltd.	Svartliden Mine, Skelleftea District	300
Do.	do.	Elgin Mining Inc.	Bjorkdal Mine, Skelleftea District	1,200
Do.	do.	Boliden AB	Mines at Aitik, Akerberg, Kankberg, Kristineberg, and Renstrom	2,000
Metal	do.	do.	Smelter and refinery at Ronnskar	15,000
Iron and steel		Svenskt Stal AB (Government, 48%)	Steelworks at Lulea and Oxelosund	3,900
Iron ore		Luossavaara-Kiirunavaara AB (LKAB) (Government, 98%)	Mines at Kiruna and Malmberget	32,500
Do.		Northland Resources S.A.	Mine at Kauniavaara	15,000
Kyanite		Svenska Kyanite AB (Svenska Mineral AB, 100%)	Quarry at Halskoberg	10
Lead:				
Ore, lead content		Boliden AB	Mines at Garpenberg and Renstrom	100
Do.		Lovisagruvan AB	Lovisa Mine	3
Do.		Lundin Mining Corp.	Zinkgruvan Mine at Ammeberg	20
Metal		Boliden AB	Smelter and refinery at Ronnskar	30
Do.		do.	Smelter at Bergsoe	50
Lime		Svenska Minerals AB	Plants at Rattvik and Boda	250
Limestone		Kalproduction Storugns AB (Rettig Group, 100%)	Mines at Gotland Island	3,000
Do.		NordKalk AB	Storugns	3,200
Marble	cubic meters	Borghamnsten AB	Quarry at Askersund	15,000
Petroleum, refined	42-gallon barrels per day	Preem AB (Corral Petroleum Holdings AB, 100%	Refinery at Lysekil and Goteborg	210,000
Do.	do.	St1 Group Oy	do.	82,000
Do.	do.	AB Nynas Petroleum	Refineries at Gothenburg and Nynashamn	50,000
Silver, metal	kilograms	Boliden AB	Smelter and refinery at Ronnskar	408,000
Do.	do.	Lundin Mining Corp.	Zinkgruvan Mine at Ammeberg	25,000
Zinc, ore, zinc content		Boliden AB	Mines at Garpenberg, Laisvall, Langdal, and Renstrom	112
Do.		Lovisagruvan AB	Lovisa Mine	3
Do.		Lundin Mining Corp.	Zinkgruvan Mine at Ammeberg	78

Do., do. Ditto. NA not available.

THE MINERAL INDUSTRY OF SWITZERLAND

By Harold R. Newman

Switzerland has a very limited amount of mineral resources. The mineral sector, exclusive of industrial minerals for construction, played a minor role in Switzerland's economy and remained mostly inactive in 2012. Reserves of the small deposits of cobalt-nickel, gold, iron ore, and silver ores that had existed in the past were mostly depleted, and new mining activities were discouraged for environmental reasons. Metal processing, which was limited mostly to secondary aluminum, secondary lead, and steel production, and metal refining, which was limited mostly to precious metals, depended on imported raw materials and scrap (Encyclopedia.com, 2012).

The economic policy of the Government was based on free trade and featured low import duties and minimal import quotas. Although not a member of the European Union (EU), Switzerland was a member of the European Free Trade Association, and trade in general, including of minerals, continued to be of major importance to the economy in 2012 (U.S. Department of State, 2012).

Switzerland was the world's 17th-ranked exporter in 2012; the value of its exports was $333 billion. The principal recipients of Switzerland's exports were Germany (19.8%), the United States (11.1%), Italy (7.2%), France (7.1%), the United Kingdom (5.4%) and China (4.4%). As the world's 19th-ranked importer, Switzerland imported $287.7 billion worth of goods in 2012. The principal suppliers were Germany (29.7%), Italy (10.2%), France (8.4%), the United States (5.6%), and China (5.5%). Switzerland served as a major diamond exchange and was involved in the cutting and polishing of diamond. In 2011 (the latest year for which data were available), Switzerland exported polished diamond valued at $1.1 billion (Antwerp Facets Online, 2012; U.S. Central Intelligence Agency, 2012).

In 2012, U.S. exports to Switzerland totaled $26.4 billion and U.S. imports from Switzerland totaled $25.7 billion. U.S. exports to Switzerland included, in order of value, nonmonetary gold ($14.2 billion), coal and other fuels ($72 million), fuel oil ($48 million), iron and steel products ($12 million), and nuclear fuel materials ($3 million). U.S. imports from Switzerland included, in order of value, iron and steel products ($62 million), iron and steel manufactures ($57 million), fuel oil ($43 million), steelmaking and ferroalloying materials ($23 million), and other petroleum products ($6 million) (U.S. Census Bureau, 2012a, b).

Production

Industrial minerals produced by mining and processing included cement, gypsum, and lime, which were used domestically in construction. The country's rolled aluminum production was mainly for export to the automotive industry and its salt production was for domestic consumption and export. Estimated crude steel production remained at about the same level as in 2011 whereas salt production increased. Estimated secondary lead production decreased. Data on mineral production are in table 1.

Structure of the Mineral Industry

The Swiss mineral industry was owned privately or by regional governments (Cantons). Federal regulatory control of mineral resources is administered by the Government; the 26 Cantons grant mining and processing licenses and directly operate electricity generating facilities, gas utilities, and water resource facilities. Regulations for the mineral industry are vested in the Federal Council (U.S. Central Intelligence Agency, 2012). Table 2 is a list of major mineral industry facilities, including their locations and capacities.

Commodity Review

Metals

Aluminum.—Novelis Switzerland S.A. (a subsidiary of Hindalco Industries Ltd. of India) was a world leader in the production of aluminum rolled products in terms of both production and technology, and it was the leading rolled-products producer in Europe. The Novelis complex at Sierre had ingot-casting, hot- and cold-rolling, and heat-treatment capability and modern laser-blanking lines. The complex was also the location of Novelis's Novelis Fusion™ process for producing multialloy rolled aluminum with different combinations of aluminum core properties. Novelis's particular specialty was providing flat-rolled products for the automotive sector (PRNewswire, 2012).

Copper.—Schmelzmetall AG was a leading manufacturer of copper-based high-performance alloys in the EU. Schmelzmetall's HOVADUR® alloys were manufactured from raw materials that were smelted and cast in inductively heated vacuum furnaces and further processed into semifinished products and finished parts for use in various types of applications, from aerospace components to casting moulds (Schmelzmetall AG, 2012).

Gold.—In 2012, Switzerland did not mine gold; however, it was home to a number of refineries. These included Argor-Heraeus S.A.'s refinery at Mendrisio, Cendres+Métaux S.A.'s refinery at Biel-Bienne, Produits Artistiques de Métaux Précieux S.A.'s (PAMP) refinery at Castel San Pietro, and Valcambi S.A.'s refinery at Balerna (table 2).

Argor-Heraeus was a leading international gold refiner and bar manufacturer. Production took place at its facilities in Mendrisio in southeastern Switzerland and ranged from gold refining to the manufacture of gold bars and other precious-metals products. Argor-Heraeus processed precious metals by refining, melting, assaying, and minting the metals, and making them into semifinished products. A process of electrolysis was used to refine the gold and silver to a degree of fineness of 999.9 parts per thousand for gold and 999.0 parts per thousand for silver. Platinum-group metals (PGMs) underwent a chemical method of refining to attain a degree of fineness of 999.5 parts per thousand (Argor-Heraeus S.A., 2012).

Cendres+Métaux was a small foundry located in Biel-Bienne that produced semifinished and finished products for the dental, electronics, and jewelry industries. Cendres+Métaux specialized in the recovery of gold and other precious metals, including PGMs and silver. The metals were refined to meet a standard of 999.9 parts per thousand (Cendres+Métaux S.A., 2012).

PAMP was a leading gold, PGM, and silver refinery based in Castel San Pietro. PAMP handled a complete range of precious metals in bars of various shapes, weights, purities, and sizes and, in 2012, was the only precious metals refinery in Switzerland to simultaneously hold the International Organization for Standardization (ISO) 9001, ISO 14001, ISO 17025, and Occupational Health and Safety Management Systems (OHSMS) 18001 accreditations (Produits Artistiques Métaux Précieux, 2012).

Valcambi S.A.'s precious metals refinery was one of the leading such facilities in the world. Valcambi offered a range of services, from assaying and refining through the manufacturing of cast and minted bars and the development of semifinished products. Apart from gold, Valcambi also refined PGMs and silver to produce bars of various weights (Valcambi S.A., 2012).

Iron and Steel.—Scrap steel is a valuable source of material in a country short of raw materials. In Switzerland, about 1 million metric tons per year (Mt/yr) of scrap steel was collected through a network of collection points, scrap processors, and traders. Stahl Gerlafingen AG was a major consumer of this material and operated a modern high-efficiency electric arc furnace (EAF) for melting scrap steel at its plant at Gerlafingen. The company was the leading supplier of reinforced-steel products in Switzerland (Stahl Gerlafingen AG, 2012).

In 2012, steel was also produced from recycled iron scrap of mainly Swiss origin at Swiss Steel AG. The scrap was melted in an 80-metric-ton (t) EAF and transferred to a ladle furnace for alloy and micro-alloy treatment and adjustment. Steel billets that were 11 meters long by 150 millimeters wide were produced in a continuous-casting machine that formed steel billets, which were then rolled into bars or wire rod (Swiss Steel AG, 2012).

Industrial Minerals

Cement.—Holcim (Schweiz) AG was the leading cement, concrete, and gravel producer in Switzerland and, as a subsidiary of the Holcim Group, was part of one of the world's leading suppliers of cement and aggregates, as well as a supplier of asphalt and ready-mix concrete. Holcim operated seven cement plants and grinding stations and had a production capacity of 4.3 Mt/yr [Holcim (Schweiz) AG, 2012].

Salt.—Saline de Bex S.A. was active in the extraction and production of salt for the Canton of Vaud. Although most of the salt produced was destined for winter road maintenance, about 6,000 t was for water treatment, 3 t was for human consumption, and about 2.5 t was for animal feed (Saline de Bex S.A., 2012).

Mineral Fuels and Other Sources of Energy

In 2012, Switzerland did not produce natural gas; instead, natural gas was imported to meet a part of the country's energy requirement. Switzerland's energy needs were met mainly by hydropower, which supplied about 56% of the country's electric power requirements from 550 hydroelectric plants that generated about 35.8 terawatthours per year of electricity; nuclear power, which supplied about 40% of the country's electric power requirements; and other sources (including natural gas imports), which supplied about 4% of the country's electric power requirements (Swissinfo, 2012).

Nuclear Energy.—Switzerland has five nuclear reactors that supplied 40% of the country's electricity requirements. In 2011, the Government resolved not to replace any reactors and to phase out nuclear power by 2034. This decision was thought to be a reaction to Japan's Fukushima-Daiichi nuclear powerplant accident in March 2011. This move would be a total reversal of the 2007 energy policy that focused on nuclear energy efficiency and renewable resources and called for aging nuclear units to be replaced in due course of time with new ones. Given life spans of 50 years, the first Swiss reactor to shut down could be the Beznau 1 plant in 2019, followed by the Beznau 2 plant in 2020 and the Mühleberg plant around 2022. The larger units that could close would be the 985-megawatt (MW)-capacity plant in Gösgen in 2029 and the 1,165-MW-capacity plant in Leibstadt in 2034. According to the Government's Department of the Environment, Transport, Energy, and Communications, the decision to close the nuclear powerplants could cost the Government about $33 billion up to 2050 (World Nuclear News, 2012).

Petroleum.—Tamoil (Suisse) S.A.'s Collombey refinery, which was one of two petroleum refineries in Switzerland, is located in the Canton of Valais about 100 kilometers (km) from Geneva. The refinery produced about 2.7 Mt/yr of petroleum products from crude petroleum brought to Collombey from the Port of Genoa, Italy, through a 340-km-long pipeline. Products produced at the refinery included diesel, heating oil (light and heavy), kerosene, liquefied petroleum gas, unleaded 98 octane gasoline, and unleaded 95 octane gasoline. Tamoil has 90 storage tanks with a combined capacity of 795,000 cubic meters for crude petroleum and products [Tamoil (Suisse) S.A., 2012].

The Switzerland-based Vitrol Group announced that Varo Holding SA, which was a joint venture between the Vitrol Group and the AtlasInvest Group, had entered into an agreement to purchase the former Petroplus Refining Cressier S.A. refinery at Cressier and the related Swiss marketing and logistics assets. These assets included Petroplus Tankstorage AG, Oléoduc du Jura Neuchâtelois, and Société Française du Pipeline du Jura. Varo Holding stated that it intended to restart the refinery and continue refining operations. The plant was an integrated atmosphere-vacuum distillation visbreaking and thermal-cracking refinery with a nameplate capacity of 68,000 barrels per day (TankTerminals.com, 2012).

Outlook

The outlook for Switzerland's mineral industry is for little change. Metal mining is not likely to be initiated, and industrial minerals are expected to be produced according to local demand. Limited exploration for natural gas and petroleum in the future is possible.

References Cited

Antwerp Facets Online, 2012, Antwerp balance sheet: Antwerp Facets Online. (Accessed February 22, 2012, at http://www.antwerpfacetsonline.be/nc/articles/single/amtwerps-november-imports-and-exports-indicate-higher-per-carat-prices.)

Argor-Heraeus S.A., 2012, Refining goal—Highest purity and fastest turnaround: Argor-Heraeus S.A. (Accessed August 9, 2013, at http://www.argor.com/index.php/eng/OUR-OPERATIONS/REFINING.)

Cendres+Métaux S.A., 2012, Welcome to the refining division: Cendres+Métaux S.A. (Accessed August 9, 2013, at http://www.cmsa.ch/en/refining/Pages/default.aspx.)

Encyclopedia.com, 2012, Mining—Switzerland: Encyclopedia.com, p. 13. (Accessed August 9, 2013, at http://www.encyclopedia.com/topic/Switzerland.aspx.)

Holcim (Schweiz) AG, 2012, Global company with Swiss roots: Holcim (Schweiz) AG. (Accessed August 10, 2013, at http://www.holcim.ch/uber-holcim/portraet/zahlen-und-fakten.html.)

PRNewswire, 2012, Novelis starts European production of multi-alloy sheet ingot: PRNewswire, August 9. (Accessed August 9, 2013, at http://www.prnewswire.co.uk/news-releases/novelis-starts-european-production-of-multi-alloy-sheet-ingot.)

Produits Artistiques Métaux Précieux, 2012, About PAMP: Produits Artistiques Métaux Précieux. (Accessed August 10, 2013, at http://www.pamp.ch/aboutus.)

Saline de Bex S.A., 2012, The salt work in numbers: Saline de Bex S.A. (Accessed August 9, 2013, at http://www.selbex.com/en/saline-de-bex/the-saltwork-in-numbers.html.)

Schmelzmetall AG, 2012, Leading manufacturer of copper-based high-performance materials: Schmelzmetall AG. (Accessed September 18, 2013, at http://www.kunststoff-sshweiz.ch/html/schmelzmetall_ag_englisch.html.)

Stahl Gerlafingen AG, 2012, Portrait: Stahl Gerlafingen AG. (Accessed August 9, 2013, at http://www.stahl-gerlafingen.com/en/Portrait/tabid/908/language/de-CH/Default.aspx) Business/Fromscraptosteel/tabid/911/language/de-CH/default.aspx.)

Swissinfo, 2012, Hydropower: Swissinfo. (Accessed August 10, 2013, at http://www.swissinfo.ch/eng/science_technology/Swiss_Alps_proposed-as-powerhouse_of_Europe.html?cid=32762310.)

Swiss Steel AG, 2012, From scrap to high-grade steel: Swiss Steel AG. (Accessed August 10, 2013, at http://www.swiss-steel.com/en/process/.)

Tamoil (Suisse) S.A., 2012, Collombey: Tamoil (Suisse) S.A. (Accessed August 9, 2013, at http://www.tamoil.com/Tamoil+World/Activities+and+Products/Downstream/Refining/Collombey/.)

TankTerminals.com, 2012, Vitrol to buy Swiss refinery: TankTerminals.com. (Accessed August 10, 2013, at http://www.tanktermimals.com/news_detail.php?id=1834.)

U.S. Census Bureau, 2012a, U.S. exports to Switzerland by 5-digit end-use code 2003–2012: U.S. Census Bureau. (Accessed August 9, 2013, at http://www.census.gov/foreign-trade/statistics/product/enduse/exports/c4419.html.)

U.S. Census Bureau, 2012b, U.S. imports from Switzerland by 5-digit end-use code 2003–2012: U.S. Census Bureau. (Accessed August 9, 2013, at http://www.census.gov/foreign-trade/statistics/product/enduse/imports/c4419.html.)

U.S. Central Intelligence Agency, 2012, Switzerland, in The world factbook: U.S. Central Intelligence Agency. (Accessed August 9, 2013, at https://www.cia.gov/library/publications/the-world-factbook/geos/sz.html.)

U.S. Department of State, 2012, U.S. relations with Switzerland: U.S. Department of State. (Accessed August 9, 2013, at http://www.state.gov/r/pa/ei/bgn/3431.htm.)

Valcambi S.A., 2012, About Valcambi: Valcambi S.A. (Accessed August 9, 2013, at http://www.valcambigold.com/AboutValcambi.aspx.)

World Nuclear News, 2012, Switzerland's challenging energy policy: World Nuclear News. (Accessed August 10, 2013, at http://www.world-nuclear-news.org/EE-switzerlands_challenging_energy_policy-0307124.html.)

TABLE 1
SWITZERLAND: ESTIMATED PRODUCTION OF MINERAL COMMODITIES[1, 2]

(Thousand metric tons unless otherwise specified)

Commodity[3]		2008	2009	2010	2011	2012
METALS						
Aluminum, secondary	metric tons	50	30	25	25	25
Iron and steel, metal:						
Crude steel		1,257 [4]	984 [4]	1,330 [4]	1,350	1,400
Semimanufactures		700	600	700	700	700
Lead, refined, secondary	metric tons	8,000	5,000	5,000	3,000	2,500
INDUSTRIAL MINERALS						
Cement, hydraulic		4,000	4,000	4,000	4,750 [r, 4]	4,467 [4]
Gypsum		300	300	300	350 [r, 4]	320 [4]
Lime		90	80	80	85	85
Salt		535	435 [4]	500	501 [r, 4]	528 [4]
Sulfur, from petroleum refining	metric tons	3,000	3,000	3,000	3,000	3,000
MINERAL FUELS AND RELATED MATERIALS						
Petroleum refinery products:						
Liquefied petroleum gas	thousand 42-gallon barrels	2,774 [4]	2,373 [4]	2,000	1,840 [r, 4]	1,960 [4]
Gasoline	do.	11,534 [4]	12,045 [4]	12,000	12,630 [r, 4]	10,280 [4]
Distillate fuel oil	do.	17,301 [4]	17,739 [4]	20,000	22,090 [r, 4]	17,030 [4]
Residual fuel oil	do.	3,833 [4]	2,482 [4]	3,000	3,440 [r, 4]	2,750 [4]
Other	do.	2,190 [4]	2,373 [4]	3,000	3,060 [r, 4]	3,790 [4]
Total	do.	37,632 [4]	37,012 [4]	40,000	43,060 [4]	35,810 [4]

[r]Revised. do. Ditto.

[1]Estimated data are rounded to no more than three significant digits; may not add to totals shown.

[2]Table includes data available through August 31, 2013.

[3]In addition to the commodities listed, a variety of crude construction materials (common clay, sand and gravel, and stone) were thought to be produced, but output was not reported, and available information is inadequate to make estimates of output.

[4]Reported figure.

TABLE 2
SWITZERLAND: STRUCTURE OF THE MINERAL INDUSTRY IN 2012

(Thousand metric tons unless otherwise specified)

Commodity		Major operating companies and major equity owners	Location of main facilities	Annual capacity
Aluminum		Novelis Switzerland S.A. (Hindalco Industries Ltd., 100%)	Plant at Sierre	130
Cement		Holcim (Schweiz) AG (Holcim Group, 100%)	Plants at three locations	4,300
Copper, alloy	metric tons	Schmelzmetall AG	Refinery at Gurtnellen	2,400
Gold, refined	kilograms	Argor-Heraeus S.A.	Refinery at Mendrisio	350,000
Do.	do.	Cendres+Métaux S.A.	Refinery at Biel-Bienne	NA
Do.	do.	Metalor Group	Refinery at Neuchatel	270,000
Do.	do.	Produits Artistiques de Métaux Précieux S.A. (MKS Finance SA, 100%)	Refinery at Castel San Pietro	450,000
Do.	do.	Valcambi S.A. (European Gold Refineries Holding S.A., 100%)	Refinery at Balerna	350,000
Lead, secondary		Metallum Group	Smelter at Pratteln	32
Petroleum, refinery	barrels per day	Tamoil (Suisse) S.A. (Colony Capital LLC, 65%, and Government of Libya, 35%)	Refinery at Collombey	72,000
Do.	do.	Varo Holdings S.A. (AtlasInvest Group, 50%, and Vitrol Group, 50%.)	Refinery at Cressier	68,000
Platinum-group metals	kilograms	Produits Artistiques de Métaux Précieux S.A. (MKS Finance SA, 100%)	Refinery at Castel San Pietro	30,000
Salt		United Swiss Salt Works (25 Cantons, except Vaud, 100%)	Saline plants at Riburg and Schweizerhalle	500
Do.		Saline de Bex S.A. (Canton of Vaud, 100%)	Saline mine and plant at Bex	50
Steel		Stahl Gerlafingen AG (Schmolz and Bickenbach AG, 100%)	Plant at Gerlafingen	720
Do.		Swiss Steel AG	Plant at Emmenbrucke	300

Do., do. Ditto. NA Not available.

The Mineral Industry of Tajikistan

By Elena Safirova

Tajikistan reportedly has more than 600 mineral deposits containing such minerals as anthracite coal, antimony, bismuth, boron, copper, gemstones, gold, iron ore, lead, manganese, molybdenum, natural gas, nickel, petroleum, phosphor, salt, silver, strontium, tin, tungsten, uranium, and zinc. Mining that was successfully developed in the 1980s had declined during the 1990s because of economic and political difficulties in the country. In the past decade, Tajikistan was expending significant effort to increase its mineral production (Tajik Gateway, 2013).

Tajikistan had significant hydropower resources. The hydropower plants with the largest electricity generating capacity were the Nurek plant [which had a capacity of 3,000 megawatts (MW)], the Sangtuda plant (670 MW), and the Baipaza plant (600 MW). Another hydropower plant—the Rogunskaya plant on the Vaksh River—was under construction. In the absence of significant reserves of oil and gas, Tajikistan relied on imported hydrocarbons for its industrial production; most of the imported hydrocarbons were from Russia, Turkmenistan, and Uzbekistan (European Bank for Reconstruction and Development, 2010).

Minerals in the National Economy

In 2012, Tajikistan's gross domestic product (GDP) was reported to be $7.6 billion and real GDP growth was 7.5%.[1] Industrial production accounted for 22.8% of the GDP; in nominal terms, industrial production increased by 10.4% compared with that of 2011. Production of mineral products constituted a significant portion of the national economy. Mining and quarrying made up 12.7% of the value of industrial production, and the rate of growth of the mining and quarrying sector was 24.0% (Agency on Statistics Under the President of the Republic of Tajikistan, 2012, 2013; U.S. Central Intelligence Agency, 2013).

In 2012, the country's revenue from exports amounted to $1.36 billion, which was far less than the $3.78 billion that it spent on imports. The main categories of exported commodities were, in order of the contribution to export revenue, basic metals, which accounted for 40.9% of total exports; mineral products (22.4%); textiles and products made out of them (19.5%); and vehicles, machinery, and equipment (6.2%). A single export category (aluminum) contributed 39.5% of the total export revenue. Overall, Tajikistan's export revenue became more diversified compared with that of 2011; the share of aluminum was reduced to 39.5% from 54.6% in 2011, and the share of cotton increased to 16.5% from 15.7% in 2011. The main export partners were, in order of export value, Turkey (which received 36.3% of Tajikistan's exports), Afghanistan (14.1%), China (13.3%), Russia (7.9%), Kazakhstan (7.5%), Switzerland (6.7%), and Iran (4.9%). The main import categories were, in order of export value, vehicles, machinery and equipment (26.6%); mineral products (20.2%); chemical products (12.6%); plants and products made out of them (10.4%); basic metals and products made out of them (6.3%); and timber and products made out of wood (5.4%). The main import partners were, in order of the value of the imports, Russia (which supplied 25.4% of Tajikistan's imports), Kazakhstan (16.0%), China (12.9%), the United States (5.1%), Turkmenistan (4.1%), Iran (4.0%), and Lithuania (3.8%) (Agency on Statistics Under the President of the Republic of Tajikistan, 2012, 2013; U.S. Central Intelligence Agency, 2013).

Production

In 2012, Tajikistan decreased production of natural gas by 40.6%; cement, by 22.4%; and aluminum, by 2%. Coal production increased by 73.9%; that of gold, by 7.2%; mercury output, by an estimated 6.7%, and lead output, by an estimated 5%. These and other production data are in table 1.

Structure of the Mineral Industry

Table 2 is a list of major mineral industry facilities.

Commodity Review

Metals

Aluminum.—The aluminum smelter Tajik Aluminum Co. (TALCO) was Tajikistan's only aluminum producer; it had the capacity to produce 517,000 metric tons per year (t/yr). In 2012, the smelter produced 272,500 metric tons (t), which was a 2.0% decrease compared with the output in 2011. The decrease in production was attributed to TALCO's need for modernization as well as to production interruptions related to the shortages of natural gas in the country (Ergasheva, 2013a).

Gold.—In 2012, Tajikistan produced 2,401 kilograms (kg) of gold, which was an increase of 7.2% compared with the output in 2011. In 2012, the enterprises that produced gold in Tajikistan were the Tajik-Chinese joint venture SP Zerafshan, the Tajik-British joint venture Darvaz, the Tajik-Canadian joint venture Aprelevka, Government-owned Tilloi Tochik [Tajik Gold], the Tajik-Kyrgyz joint venture OOO Takom Gold, and the private Tajik company Arteli Odina (Ergasheva, 2013b).

The leading gold producer in Tajikistan, SP Zerafshan, produced 1,511 kg of gold, which was 182 kg more than it produced in 2011. The company mined several deposits in Sughd Oblast' in the northern part of the country, including the Tarror deposit, the Jilau deposit, the Khirskhona deposit, and the Olympiyskoye deposit. In 2012 the Zijin Mining Group reevaluated the company's gold resources and increased them to 184 t of gold (MinerJob.ru, 2012d; BizTass, 2013).

---

[1]Where necessary, values have been converted from Tajikistani somoni (TJS) to U.S. dollars (US$) at an annual average exchange rate of TJS4.76=US$1.00 for 2012.

The country's second-ranked gold enterprise, Aprelevka, produced 442 kg of gold, which was 3 kg more than it produced in 2011; it also produced about 1.5 t of silver. The company's plan called for it to increase gold output to 1 t/yr and silver output, to 3 t/yr by 2015. Aprelevka was planning to invest $100 million in enterprise development. As of 2012, the company employed 850 workers, and it was planning to increase the number of employees to between 1,200 and 1,300 workers by 2015 (MinerJob.ru, 2012j).

In November 2011, the Government of Tajikistan issued a gold mining license to Kryso Resources Ltd. of the United Kingdom to produce gold at the Pakrut deposit. The company was planning to start operations at the end of 2013 and expected to be able to produce about 2 t/yr of gold and silver as a byproduct during the first several years of mining. According to Kryso's initial estimates, the Pakrut deposit had gold reserves of 111.3 t contained in ore grading up to 4.43 grams per metric ton (g/t) gold, which is considered to be a very high grade. Kryso Resources had spent about $20 million on prospecting work in the area and was planning to spend another $25 million on plant construction; the company's planned total investment in Pakrut was estimated to be $108 million (MinerJob.ru, 2012a, b; Tjkrus4all.ru, 2012).

In February, the Government announced that it had begun accepting applications for participation in an international tender for development of two gold deposits—the Chore deposit and the Konchoch deposit, both of which are located in Ayninskiy Rayon of Sughd Oblast'. The tender was expected to be conducted during 2012. According to the tender rules, priority would be given to projects that would produce more final products within the borders of the country. The Chore deposit is characterized by gold-arsenic ores; the average grade is estimated to be between 2.8 and 8.2 g/t gold. The Konchoch deposit contains antimony, mercury, gold, and silver; previously, Konchoch was considered to be purely a mercury deposit. Two other deposits—the Ikar tungsten deposit and the Gumas nickel-copper-platinum deposit, both of which are located in the Gorno-Badakhanshkaya Autonomous Region—were also expected to be put up for tender (MinerJob.ru, 2012e).

Silver.—In 2012, Tajikistan produced 1,767 kg of silver, which was practically unchanged compared with production in 2011. In 2010, the Government announced an international tender for the right to develop the Koni Mansur Kalon [Big Konimansur, also known as Konimansuri] polymetallic deposit. The deposit had been prospected in the 1970s and was one of the largest silver deposits in the world. It contained about 1 Mt of ore containing 49 g/t silver, 0.49% lead, and 0.38% zinc. Total resources of silver at the Koni Mansur Kalon were reported as 70,000 t. As of October 2012, the only major contender was an international consortium led by Kazzinc of Kazakhstan. Other participants in the consortium were Glencore International plc of Switzerland, Konimansur of Kazakhstan, and the ore refinery OAO Adrasman of Tajikistan. The Government expected that the total proposed investment would amount to between $3 billion and $4 billion (KyrTag.kg, 2012; MinerJob.ru, 2012g, i).

In November, the Nukrafon company, which planned to specialize in silver production, opened in the city of Istiklol in the Sogd region of northern Tajikistan. The company would operate in Kanchol Valley. According to press reports, the enterprise would be working with a Swiss mining company and planned to hire 200 employees (MinerJob.ru, 2012c).

Mineral Fuels

Coal.—In 2011, Tajikistan produced about 412,000 t of coal, which was an increase of 73.9% compared with production in 2011. Tajikistan had significant reserves of coal amounting to 36 deposits containing, according to some sources, 3,600 Mt of coal; proven resources were estimated to be 714 Mt (Naumova and Chorshanbiyev, 2012). Tajikistan had a variety of coal types, from lignite and bituminous to the highest grades of anthracite. In 2012, 10 coal deposits were mined by 14 companies; 6 of the companies were Government owned. The leading producers were OAO Anguisht, SP Anzob, and UP Fon-Yagnob. UP Fon-Yagnob, which was located in Zarafshan Valley, produced 38.8% of the country's total output. On May 1, the Government introduced a ban on coal exports. The reason for the ban was the shortage of natural gas, which had effectively led to reduced output or even complete stoppage of several enterprises. For example, TALCO and Tajikcement were unable to work at full capacity, and SP Azot was idle throughout the year. The Government was encouraging enterprises to undertake renovations to be able to use coal as their primary fuel source. As of the end of 2012, 154 industrial enterprises in Tajikistan were using coal as their main source of energy (MinerJob.ru, 2012f, h; 2013; Naumova and Chorshanbiyev, 2012).

Oil and Natural Gas.—In July, Tethys Petroleum Ltd. of the United Kingdom announced that it had discovered large resources of oil and gas in Tajikistan. The company reported estimated gross unrisked mean recoverable resources of 27.5 billion barrels of oil equivalent, of which about 69% was natural gas and 31% was oil and condensate. The company was planning to refine those estimates using seismic data. In December, Tethis signed an agreement with Total S.A. of France and China National Oil and Gas Exploration and Development Corp. (CNODC), a division of the leading Chinese oil and gas producer China National Petroleum Corp. (CNPC), about joint geologic work in Tajikistan. Tethys was planning to start production in the second part of 2014 (Rosbalt.ru, 2012; Tethys Petroleum Ltd., 2012; Pressa.tj, 2013).

Outlook

Tajikistan has significant undeveloped mineral resources, including a large number of metals, rare-earth minerals, and uranium. During the past decade, the country started to intensify its efforts to revive its mineral industry and increase its mineral production, which had decreased significantly or ceased during the 1990s after the breakup of the Soviet Union. In particular, the Government was holding mining tenders and was trying to attract foreign investment into its mineral industry. Given those efforts, it is likely that Tajikistan will be able to increase its output of gold and silver in the next few years and perhaps will continue to expand the production of coal and cement. The country, however, currently does not produce enough energy to support its industrial sector. The recent trend to provide energy for most of its industry by burning coal (rather than by using imported natural gas, as it has done in the past) may lead to additional long-term environmental problems. On the other hand, the recent discovery of large reserves of hydrocarbons, if further confirmed and eventually developed and produced, could help to address the country's energy shortages and, perhaps, give a boost to the national economy.

References Cited

Agency on Statistics Under the President of the Republic of Tajikistan, 2012, Tajikistan v Tsifrah [Tajikistan in figures]: Agency on Statistics Under the President of the Republic of Tajikistan, 169 p. (Accessed June 28, 2013, at http://www.stat.tj/en/im/2342f4d3bcc13e5b6247c13f8e1fe06c_1311130128.pdf.)

Agency on Statistics Under the President of the Republic of Tajikistan, 2013, Sotsial'no-ekonomichesoe polozhenie respubliki Tadzhikistan [Social and economic situation of the Republic of Tajikistan]: Agency on Statistics Under the President of the Republic of Tajikistan, 130 p. (Accessed June 28, 2013, at http://www.stat.tj/ru/img/a6069090cb7edbe5efb67aec241e9816_1359030405.pdf.)

BizTass, 2013, Zijin planiruet ezhegodno vypuskat' bole 6 tonn zolota v Tadzhikistane i Kirgizii [Zijin is planning to produce more than 6 tons of gold in Tajikistan and Kyrgyzstan]: BizTass.ru, January 11. (Accessed June 28, 2013, at http://www.biztass.ru/news/id/53691.)

Ergasheva, Zarina, 2013a, Proizvodstvo alyuminiya v Tadzhikistane sokratilos' [Aluminum production in Tajikistan decreased]: News.tj, January 29. (Accessed June 28, 2013, at http://news.tj/ru/news/proizvodstvo-alyuminiya-v-tadzhikistane-sokratilos.)

Ergasheva, Zarina, 2013b, Tadzhikskie gornorudnye predpriyatiya uvelichili dobychu zolota i serebra [Tajik mining companies increased production of gold and silver]: TopTJ, April 17. (Accessed June 28, 2013, at http://www.toptj.com/News/2013/04/17/tadzhikskie_gornorudnye_predpriyatiya_uvelichili_dobychu_zolota_i_serebra.)

European Bank for Reconstruction and Development, 2010, Tajikistan—Country profile: European Bank for Reconstruction and Development. (Accessed June 28, 2013, at http://ebrdrenewables.com/sites/renew/countries/Tajikistan/profile.aspx.)

KyrTag.kg, 2012, Krupneyshee mestorozhdeniye serebra v mire "Bol'shoy Koni Mansur" mozhet dostat'sya 'Kaztsinku' [Kazzinc might get the world's largest silver deposit Konimansur]: KyrTag.kg, February 13. (Accessed June 28, 2013, at http://www.kyrtag.kg/?q=ru/news/16353.)

MinerJob.ru, 2012a, Britanskaya kompaniya do kontsa 2013 goda zavershit stroitel'stvo zavoda na zolotom mestorozhdenii "Pokrud" v Tadzhikistane [The British company will finish plant construction at the Pokrud gold deposit in Tajikistan by the end of 2013]: MinerJob.ru, December 5. (Accessed June 28, 2013, at http://minerjob.ru/viewnew.php?id=22824.)

MinerJob.ru, 2012b, Britantsy budut dobyvat' zoloto v Tadzhikistane [The British will mine gold in Tajikistan]: MinerJob.ru, January 12. (Accessed June 28, 2013, at http://minerjob.ru/viewnew.php?id=20790.)

MincrJob.ru, 2012c, Na severe Tadzhikistana zarabotalo novoe predpriyatie po dobyche serebra [A new silver-producing enterprise started operations in the north of Tajikistan]: MinerJob.ru, November 20. (Accessed June 28, 2013, at http://minerjob.ru/viewnew.php?id=22569.)

MinerJob.ru, 2012d, SP "Zarafshon" uvelichivaet dobychu zolota I sokrashaet proizvodstvo serebra [SP Zarafshon is increasing gold production and reducing silver output]: MinerJob.ru, October 31. (Accessed June 28, 2013, at http://minerjob.ru/viewnew.php?id=22286.)

MinerJob.ru, 2012e, Tadzhikistan ob'yavil tender na mestorozhdeniya zolota [Tajikistan announced a tender for gold deposits]: MinerJob.ru, February 24. (Accessed June 28, 2013, at http://minerjob.ru/viewnew.php?id=20931.)

MinerJob.ru, 2012f, Tadzhikistan s pervogo maya vvel zapret na eksport uglya, cootvetstvuyushee postanovlenie prinyato pravitel'stvom [Starting from May 1, Tajikistan introduced a ban on coal exports, a relevant decree is issued by the Government]: MinerJob.ru, May 1. (Accessed June 28, 2013, at http://minerjob.ru/viewnew.php?id=21192.)

MinerJob.ru, 2012g, Tender po mestorozhdeniyu serebra na severe Tadzhikistana "Bol'shoy Konimansur" podoshel k svoemu zavershayushemu etapu [A tender for Big Konimansur silver deposit in northern Tajikistan comes to its final step]: MinerJob.ru, October 10. (Accessed June 28, 2013, at http://minerjob.ru/viewnew.php?id=22030.)

MinerJob.ru, 2012h, V Tadzhikistane dobycha uglya vyrosla na 110 tys. tonn [Coal production in Tajikistan increased by 110 thousand tons]: MinerJob.ru, September 10. (Accessed June 28, 2013, at http://minerjob.ru/viewnew.php?id=21815.)

MinerJob.ru, 2012i, V Tadzhikistane eshe ne opredelilis' s pretendentom dlya osvoeniya mestorozhdeniya "Bol'shoy Konimansur" [In Tajikistan, the contender for the development of Big Konimansur is not determined yet]: MinerJob.ru, July 24. (Accessed June 28, 2013, at http://minerjob.ru/viewnew.php?id=21537.)

MinerJob.ru, 2012j, Zolotodobyvayushee predpriyatie "Aprelevka" v Tadzhikistane namereno uvelichit' proizvodstvo zolota v dva raza [Gold mining company in Tajikistan Aprelevka intends to double its gold production]: MinerJob.ru, October 29. (Accessed June 28, 2013, at http://minerjob.ru/viewnew.php?id=22221.)

MinerJob.ru, 2013, V 2012 godu v Tadzhikistane bylo proizvedeno 412 tys. ton uglya [In 2012, Tajikistan produced 412 tons of coal]: MinerJob.ru, January 15. (Accessed June 28, 2013, at http://minerjob.ru/viewnew.php?id=23182.)

Naumova, Victoria, and Chorshanbiyev, Payrav, 2012, S Nachala 2012 goda v tadzhikistane dobyto svyshe 33 tys. ton uglya [More than 33 thousand tons of coal have been produced in Tajikistan since the beginning of 2012]: News.tj, May 5. (Accessed June 28, 2013, at http://news.tj/ru/print/121223.)

Pressa.tj, 2013, Tadzhikistan v 2014 godu, vozmozhno, stanet energonezavisimym, ili obeshaniya kanadsko-britanskoy kompanii Tethys Petroleum [In 2014, Tajikistan will probably become energy independent, or The promises of the Canadian-British Tethys Petroleum]: Pressa.tj, January 22. (Accessed June 28, 2013, at http://www.pressa.tj/news/tadzhikistan-v-2014-godu-vozmozhno-stanet-energonezavisimym-ili-obeshchaniya-kanadsko.)

Rosbalt.ru, 2012, "Sverkhgigantskiye" zapasy nefti v Tadzhikistane izmenyat situatsiyu v Sredney Azii ['Supergigantic' oil resources will change the situation in Central Asia]: Rosbalt.ru, August. (Accessed June 28, 2013, at http://www.rosbalt.ru/exussr/2012/08/10/1021409.html.)

Tajik Gateway, 2013, Poleznye Iskopaemye [Mineral Resources]: Tajik-Gateway.org. (Accessed June 28, 2013, at http://www.tajik-gateway.org/index.phtml?lang=ru&id=163.)

Tethys Petroleum Ltd., 2012, Tethys Petroleum Limited—Tajikistan resource upgrade 27.5 billion BOE: Tethys Petroleum Ltd., July 19. (Accessed June 28, 2013, at http://www.tethyspetroleum.com/tethys/newscontent.action?articleId=2341092.)

Tjkrus4all.ru, 2012, Novyi zavod po dobyche zolota poyavitsya v Tadzhikistane v kontse 2013 goda [By the end of 2013, a new gold plant will be built in Tajikistan]: Tjkrus4all.ru, December 5. (Accessed June 28, 2013, at http://tjk.rus4all.ru/news/20121205/723638888.html.)

U.S. Central Intelligence Agency, 2013, Tajikistan, in The world factbook: U.S. Central Intelligence Agency. (Accessed June 28, 2013, at https://www.cia.gov/library/publications/the-world-factbook/geos/ti.html.)

TABLE 1
TAJIKISTAN: PRODUCTION OF MINERAL COMMODITIES[1]

(Metric tons unless otherwise specified)

Commodity[2]		2008	2009	2010	2011	2012
METALS						
Aluminum, primary		339,450	359,385	348,850	278,000	272,500
Antimony ore, Sb content[e]		2,000	2,000	2,000	4,500 [r]	4,700
Gold	kilograms	1,672	1,361	2,049	2,240	2,401
Lead ore, Pb content[e]		800	800	800	800	840
Mercury, Hg content[e]		30	30	30	30	32
Silver, Ag content	kilograms	3,110	1,268	2,652	1,764 [r]	1,767
INDUSTRIAL MINERALS						
Cement, hydraulic		190,400	195,000	288,200	299,400 [r]	232,400
Gypsum		8,500	26,400	14,600	16,000 [e]	16,000 [e]
Nitrogen, N content of ammonia		23,000	--	--	--	--
Salt[e]		52,000	52,000	52,000	27,000 [r]	27,954 [3]
MINERAL FUELS AND RELATED MATERIALS						
Coal, bituminous and lignite		198,500	178,300	199,700	236,800	411,789
Natural gas	thousand cubic meters	16,100	19,900	22,800	18,800	11,170
Petroleum, crude:						
In gravimetric units		25,900	26,200	27,000	28,700	29,918
In volumetric units	42-gallon barrels	85,900	79,600	78,500	83,440	86,762

[e]Estimated; estimated data are rounded to no more than three significant digits. [r]Revised. -- Zero.

[1]Table includes data available through June 27, 2013.

[2]In addition to the commodities listed, Tajikistan had produced a number of other mineral commodities in the past, but available information is inadequate to determine if production was still taking place.

[3]Reported figure.

TABLE 2
TAJIKISTAN: STRUCTURE OF THE MINERAL INDUSTRY IN 2012[1]

(Metric tons unless otherwise specified)

Commodity	Major operating companies and major equity owners	Location of main facilities	Annual capacity[e]
Aluminum	TALCO aluminum smelter [formerly the Tajikistan Aluminum Smelter (TadAZ)]	Tursunzade	517,000
Antimony, ore	Anzob mining-beneficiation complex	Dzhizhikrutskoye Sb-Hg deposit	700,000
Antimony, metal	Isfara hydrometallurgical plant	Isfara	500
Arsenic	Mosrif deposit	NA	NA
Bismuth	Isfara hydrometallurgical plant	Isfara	500
Do.	Leninabad mining-beneficiation complex	Yuzhno-Yangikanskiy deposit	25
Bismuth, copper, fluorspar, gold, silver, zinc (ore processing)	Adrasman mining-beneficiation complex	Kanimansurskoye deposit	650,000
Boron	Yakarkharskoye deposit	Badakhshan region	NA
Cement	OAO Tajikcement	Dushanbe	1,000,000
Coal	UP Fon-Yagnob	Pyandzh region	50,000
Do.	Isfara hydrometallurgical plant	Isfara	300,000
Do.	OAO Anguisht	Shurab region	NA
Do.	Shurab brown coal deposit	do.	NA
Do.	SP Anzob	do.	NA
Copper-lead-zinc	Leninabad mining-beneficiation complex	Yuzhno-Yangikanskiy deposit	2,500
Dolomite	Yavan electrochemical complex	Pashkharvoskoye deposit	NA
Fluorspar, concentrate	Takob mining-beneficiation complex	Takob and Krasnye Kholmy deposits	60,000 [2]

See footnotes at end of table.

TABLE 2—Continued
TAJIKISTAN: STRUCTURE OF THE MINERAL INDUSTRY IN 2012[1]

(Metric tons unless otherwise specified)

Commodity		Major operating companies and major equity owners	Location of main facilities	Annual capacity[e]
Gold, in ore	kilograms	Aprelevka joint venture	Aprelevka deposit	200
Do.	do.	Arteli Odina	NA	NA
Do.	do.	Darvaz joint venture	Yak-Suyskoye deposit, Khatlonskaya region	2,000
Do.	do.	OOO Takom Gold	NA	NA
Do	do.	Tilloi Tochik	NA	NA
Do.	do.	Zerafshan Gold Co.	Jilau, Khirskhona, Olympiyskoye, and Taror deposits, Sughd Oblast'	2,500
Gold, ore processing	do.	Kansayskaya factory	Aprelevka, Burgunda, Kyzyl-Chek, and Shkol'noye deposits	165,000 [2]
Do.	do.	Vostokredmet refinery	Qizfaquz	NA
Lead-zinc		Adrasman mining-beneficiation complex	NA	NA
Do.		Kansayskoye mining complex	Karamazor region	NA
Do.		Takaeliyskiy metallurgical complex	NA	NA
Limestone		Dushanbe cement complex	Kharangonskoye deposit	NA
Loam		do.	Varzobskoye Ushchel'ye deposit	NA
Marble		Dal'yan Bolo deposit	Ganchinskiy region	NA
Do.		Dashtak deposit	Darvaz region	NA
Do.		Jilikul deposit	Pendzhikentskiy region	NA
Mercury		Anzob mining-beneficiation complex	Dzhizhikrutskoye deposit	150
Natural gas and petroleum:				
Natural gas	thousand cubic meters	Sixteen oil-gas deposits under exploration, which include Ayritanskoye, Madaniyatskoye, and Ravatskoye	Fergana depression	200,000 [2]
Petroleum		Beshtentyakskoye, Kichik-Belskoye, Shaambary, and Uzunkhorskoye deposits	Southern Tajik depression	200,000 [2]
Salt		Ashtskiy plant	Kamyshkurganskoye deposit	NA
Do.		Khoja-Sartez, Samanchi, and Tanabchi deposits	NA	NA
Do.		Voseyskiy plant	Khodzha-Muminskoye deposit	NA
Do.		Yavan electrochemical complex	Tut-Bulakskoye deposit	NA
Silver	kilograms	Zerafshan Gold Co.	Jilau and Taror deposits, Sughd Oblast'	NA
Do.	do.	Aprelevka joint venture	Aprelevka deposit	NA
Do.	do.	Nukrafon Co.	Soghd region	NA
Strontium, ore		Chaltash, Chilkutan, and Davgir deposits	Khatlon region	180,000
Tin-tungsten		Tafkon deposit	NA	NA
Tungsten ore		Maykhura deposit	Central Tajikistan	150,000

[e]Estimated; estimated data are rounded to no more than three significant digits. Do., do. Ditto. NA Not available.

[1]Many location names have changed since the breakup of the Soviet Union. Many enterprises, however, are still named or commonly referred to based on the former location name, which accounts for discrepancies in the names of enterprises and that of locations.

[2]Capacity estimates are totals for all enterprises that produce that commodity.

The Mineral Industry of Turkmenistan

By Elena Safirova

Turkmenistan has a wide variety of mineral deposits, but the most important from an economic perspective are its oil and gas resources. Turkmenistan had the world's fourth-largest natural gas reserves, after Russia, Iran, and Qatar, with proven reserves of 24.3 trillion cubic meters. Turkmenistan's total oil reserves as of September 2012 were estimated by the Government to be 20.8 billion metric tons (Gt), including 11 Gt located in the Turkmen sector of the Caspian Sea. According to the Turkmengeologiya State Concern, Turkmenistan has 38 petroleum deposits, but, according to BP Statistical Review of World Energy, the country had only about 100 million metric tons (Mt) of proven reserves (BP p.l.c., 2012; U.S. Energy Information Administration, 2012; Easttime.ru, 2013b; U.S. Central Intelligence Agency, 2013; Ustimenko, 2013).

Among the nonfuel minerals produced in Turkmenistan are bentonite, bischofite, bromine, epsomite, gypsum, iodine, kaolin, lime, quartz sands, salt, sodium sulfate, and sulfur. According to the Government, Turkmenistan has more than 160 deposits of solid minerals and significant resources of "hydrominerals"—in particular, iodine-bromine brines; surface brines of Kara-Bogaz-Gol Bay; and brines of oil, gas, and sulfur deposits (Ministerstvo Ekonomicheskogo Razvitiya Rossiyskoy Federatsii, 2012).

Production

Detailed production data and other information regarding mineral production for most mineral commodities except natural gas and oil have not been available for a number of years. The State Committee on Statistics of Turkmenistan reported only production growth rates for most of the economic categories that it tracks, including those for construction materials, metallurgy, mineral fertilizers, and mineral products. Production estimates in table 1 are based on past levels of production and occasional data reports published in mass media.

Minerals in the National Economy

Turkmenistan's gross domestic product (GDP) in 2012 was estimated to be $33.5 billion, and real GDP increased by an estimated 8.0%. Although the country's growth rate was lower than in 2011, when real GDP increased by 14.7%, Turkmenistan was still among the 10 fastest growing economies in the world. According to estimates by the U.S. Central Intelligence Agency, industrial production contributed 54.4% of Turkmenistan's GDP. The oil and gas sector accounted for 75.4% of industrial production, including 51.4% for natural gas, 12.9% for oil refining, and 11.1% for crude oil production (U.S. Energy Information Administration, 2012; Fergananews, 2013; State Committee on Statistics of Turkmenistan, 2013; U.S. Central Intelligence Agency, 2013).

The country's exports in 2012 amounted to an estimated $16.2 billion, whereas the imports were valued at $10.5 billion. The major export commodities of Turkmenistan were cotton, crude oil, natural gas, petrochemicals, and textiles. The main export partners of Turkmenistan in 2012 were China, Italy, and Turkey. Turkmenistan's main imported commodities were chemicals, foodstuffs, and machinery and equipment. Turkmenistan's major import partners were China, Russia, Turkey, and the United Arab Emirates (UAE). The United States imported from Turkmenistan $62.5 million worth of petroleum and exported $92 million worth of goods, including $9.3 million worth of drilling and oilfield equipment (U.S. Census Bureau, 2013a, b; U.S. Central Intelligence Agency, 2013).

Structure of the Mineral Industry

Table 2 is a list of major mineral industry facilities.

Commodity Review

Industrial Minerals

Bromine and Iodine.—For many years, bromine and iodine were produced at the Cheleken and the Nebitdag plants. In April, it was announced that two new plants would be built in Balkan Welayaty. One of them would be located next to the city of Balkanabat and was planned to have a capacity of 2,400 metric tons per year (t/yr) of bromine and 250 t/yr of iodine. Another one, to be built in the city of Khazar, would have a capacity of 4,500 t/yr of bromine and 300 t/yr of iodine. In addition to construction of new plants, the existing plants were expected to be modernized and to increase production as well (Turkmeninform.com, 2012).

Potash.—In January 2010, the state enterprise OAO Belgorkhimprom of Belarus and the Turkmenkhimia concern, which was owned by the Government of Turkmenistan, signed a contract according to which the Belarusian company would build Turkmenistan's first mining and beneficiation complex (GOK), called the Garlyk GOK, within a 5-year period. The Garlyk GOK, which would be located in Turkmenistan's Lebap Welayaty, would produce potash fertilizers and have a production capacity of 1.4 million metric tons per year (Mt/yr). The total cost of the project, which was specified in the contract, was $1 million, and the GOK was expected to start operations in January 2015 (MinerJob.ru, 2013).

In September 2012, the Turkmen Government noted that the construction of the Garlyk complex was slower than expected, and the main reason was insufficient number of workers. To address the problem, 800 construction workers and 200 specialists were recruited from Belarus, in addition to the about 1,200 workers already working on the construction sites. As of December, about 50 structures were at various degrees of completion. The Government of Belarus considered the construction of the Garlyk GOK in Turkmenistan a project of national importance for Belarus, because good working relations with Turkmenistan, which has large gas reserves, were likely to provide insurance should any problems with Belarus's energy supply arise in the future. Accordingly, the Government of Belarus provided incentives for the Belarusian businesses involved in construction as well as to Belarusian suppliers of materials used for the construction. In particular, starting from December 2012, the Belarusian Government reduced to zero the value-added tax on all materials used for construction in Turkmenistan and on other projects in Belarus related to the construction of the Garlyk GOK. In addition to Belneftekhim, OAO Trest Shalhtospetsstroy, OAP Promtechmontazh, and other construction companies from Belarus were involved in construction (MinerJob, 2012a, b; 2013).

Mineral Fuels

Natural Gas.—In 2012, Turkmenistan produced 69 billion cubic meters of natural gas, about 80% of which was exported. The 2012 production level constituted a 4.2% increase compared with the level of production in 2011. The Dauletabad field, which is located in the Amu Darya basin in the southeast, was one of Turkmenistan's largest and oldest gas-producing fields and produced most of the country's gas (Casfactor.com, 2012; Fergananews.com, 2013).

Turkmenistan's largest natural gas fields are located in the eastern part of the country in Mary Welayaty. In November 2011, the President of Turkmenistan issued a decree to rename several natural gas deposits, including Minara, Osman, South Yolotan, and Yashlar, into one gasfield, Galkynysh (which means "revival" in Turkmen); the total resources of Galkynysh were estimated to be 26.2 trillion cubic meters of natural gas, according to Gaffney, Cline and Associates of the United Kingdom. The preparation for gas production at Galkynysh started in December 2009, and production was expected to begin in 2013. The preparation included drilling wells, construction of sulfur-removing plants with a total capacity of 30 billion cubic meters of natural gas per year, and construction of pipelines. The four major international companies involved in Galkynysh construction activities were China National Petroleum Corp. of China, Hyundai Motor Co. and LG Chem Ltd. of the Republic of Korea, and Petrofac Ltd., of the United Kingdom; the total cost of the service contracts with these companies was $10 billion. It was expected that the natural gas from Galkynysh would be exported to China, Iran, and Russia (Neftegaz.ru, 2012; Easttime.ru, 2013a).

In the past 5 years, Turkmenistan concentrated its efforts on developing new export markets for its natural gas. In addition to the two Central Asia–Center (CAC) pipelines, which connected Turkmenistan with Russia, and the Korpezh-Kurt Kui pipeline, which linked Turkmenistan and Iran, Turkmenistan was constructing several new gas pipelines with the goal of bypassing transit through Russia. A second pipeline connecting Turkmenistan to Iran (the Dauletabad-Khangiran pipeline) was initiated in the beginning of 2010; when the second phase of construction is completed, the $550 million pipeline was expected to have capacity of 12 billion cubic meters per year. The Central Asia-China Pipeline (CACP) connected Turkmenistan's eastern fields through Uzbekistan to western China. The pipeline began operations at the end of 2009 and had an initial capacity of 30 billion cubic meters per year; the pipeline's capacity was expected to increase to about 60 billion cubic meters per year by 2015. The construction of the East-West pipeline within Turkmenistan's borders was initiated in May 2010. The pipeline would connect Turkmenistan's southeastern gasfields to the Caspian Sea and serve as a potential transit link to Europe using routes along the Caspian Sea. The 766-kilometer (km) pipeline would connect Turkmenistan's natural gas deposits in the west of the country with those in the east and would permit greater flexibility in transporting natural gas both within the country and for export. The pipeline's capacity was expected to be 30 billion cubic meters, and the construction was planned to be completed by 2015. As of the end of 2012, the first 40 km of the pipeline were completed (Easttime.ru, 2012; Rossiyskoye Energeticheskoye Agenstvo, 2012; U.S. Energy Information Administration, 2012).

The Turkmenistan-Afghanistan-Pakistan-India (TAPI) pipeline was a trans-Afghanistan pipeline intended to reach markets in Pakistan and India. The pipeline's proposed capacity was 35 billion cubic meters per year, and it would be more than 1,500 km long. During 2012, Turkmenistan signed gas export agreements with Afghanistan, India, and Pakistan and was continuing work on assembling resources to start construction (U.S. Energy Information Administration, 2012).

The Trans-Caspian Gas pipeline (TCGP) was proposed to connect Turkmenbashi City, Turkmenistan, and Baku, Azerbaijan. The pipeline would bypass both Russia and Iran and would connect either to the proposed Nabucco pipeline between Turkey and Austria or the Trans-Adriatic pipeline. The TCGP was proposed to have a capacity of 30 billion cubic meters per year and would run across the floor of the Caspian Sea at an estimated cost of $5 billion; however, the disputes with Azerbaijan concerning Caspian seabed jurisdiction could significantly undermine the project's viability (U.S. Energy Information Administration, 2012).

Petroleum.—In 2012, Turkmenistan produced about 11 Mt (80 million barrels) of petroleum, which was an increase of 11.3% compared with that of 2011. The majority of oil produced in Turkmenistan was extracted by foreign companies working in the county under production-sharing agreements. According to such contracts, the profits were split between the company and the Government of Turkmenistan (usually at 50-50 shares). The companies involved were from Canada, China, Germany, Italy, Malaysia, Russia, the UAE, and the United Kingdom. The leading companies working in the Turkmen Caspian Sea were Petronas of Malaysia and Dragon Oil Ltd. of the UAE (Nefterynok.info, 2012; Fergananews, 2013).

Turkmenistan had two oil refineries—one in Turkmenbashi and one in Seidi—which had a combined annual capacity of 12 Mt/yr. The refineries' main products were compressed gas, diesel fuel, gasoline, kerosene, polypropylene, and machine oils. Turkmenistan was planning to increase its oil refining capacity to 15 Mt/yr by 2015 and to 30 Mt/yr by 2030. To achieve these goals, it planned to modernize its existing refineries and to build three new refineries. One of the new refineries would be built in the city of Ekrem on the Caspian Sea in western Turkmenistan by JGC Corp. of Japan; construction was expected to begin in 2015. The new refinery was planned to have an annual capacity of 3 Mt/yr that would later be increased to 5 Mt/yr (Gasanov, 2012; Gurt, 2012).

Outlook

For the next few years, Turkmenistan is expected to continue expanding its production of hydrocarbons. The Galkynysh gasfield, which is expected to come online in 2013, is likely to give a boost to the country's natural gas production. Although Turkmenistan's existing oilfields are likely to be less bountiful, in the short and medium term they are expected to provide sufficient resources to keep up with domestic demand. The Government's plan to further develop the national oil refining sector is likely to provide higher returns for Turkmenistan's petroleum resources because exporting refined petroleum products is usually more profitable than exporting crude oil.

In the past decade, Turkmenistan's economy had exceptionally fast growth, fueled mostly by the export of hydrocarbons, and it is likely to continue this fast-growth trend in the next 5 to 10 years. It is possible that in the next decade the country will focus on diversification of its economy, with further development of manufacturing. One sign of these developments is Turkmenistan's investment in powerplants. The country was planning to build eight more powerplants in 2013 to 2016 that would be powered by natural gas, and thus to double electricity production in the country. It is possible that a part of the additional energy would be exported and that the other part would be used for the needs of local industry, including, perhaps, the metals and industrial minerals sector (Easttime.ru, 2013c).

References Cited

BP p.l.c., 2012, BP statistical review of world energy: BP p.l.c., June, 45 p. (Accessed June 11, 2013, at http://www.bp.com/content/dam/bp/pdf/Statistical-Review-2012/statistical_review_of_world_energy_2012.pdf.)

Casfactor.com, 2012, V Turkmenistane vyrosla dobycha nefti i gaza [Gas and oil production in Turkmenistan increased]: Casfactor.com, February 6. (Accessed July 11, 2013, at http://www.casfactor.com/rus/news/2548.html.)

Easttime.ru, 2012, Turkmenistan samostoyatel'no stroit gazoprovod vostok-zapad [Turkmenistan is building east-west gas pipeline by itself]: Easttime.ru, April 12. (Accessed July 11, 2013, at http://easttime.ru/news/turkmenistan/turkmenistan-samostoyatelno-stroit-gazoprovod-vostok-zapad.)

Easttime.ru, 2013a, Perspektivy mestorozhdeniya Galkynysh v Turkmenistane [Perspectives of Turkmenistan's Galkynysh deposit]: Easttime.ru, May 24. (Accessed July 11, 2013, at http://easttime.ru/news/turkmenistan/2013/05/24/perspektivy-mestorozhdeniya-galkynysh-v-turkmenistane.)

Easttime.ru, 2013b, Pochetnoye 4-e mesto po zapasam gaza teper' u Turkmenistana. BP podtverzhdaet [Turkmenistan now has a prestigious fourth place by its gas reserves, BP confirms]: Easttime.ru, July 17. (Accessed July 11, 2013, at http://easttime.ru/news/turkmenistan/pochetnoe-4-e-mesto-po-zapasam-gaza-teper-u-turkmenistana-vr-podtverzhdaet.)

Easttime.ru, 2013c, Turkmeniya planiruyet intensivnoye razvitiye energetiki [Turkmenistan plans intensive energy development]: Easttime.ru, April 14. (Accessed July 11, 2013, at http://www.easttime.ru/news/turkmenistan/turkmeniya-planiruet-intensivnoe-razvitie-energetiki.)

Fergananews.com, 2013, V Turkmenistane v 2012 godu dobyto 11 mln ton nefti i 69 mlrd kubometrov gaza [In 2012, Turkmenistan produced 11 million tons of petroleum and 69 billion cubic meters of gas]: Fergananews.com, March 27. (Accessed July 11, 2013, at http://www.fergananews.com/news/20413.)

Gasanov, G., 2012, NPZ v Turkmenistane pererabotali okolo 4,8 mln. ton nefti [Refineries in Turkmenistan processed about 4.8 million tons of oil]: Trend.az, October 15. (Accessed July 11, 2013, at http://www.trend.az/regions/casia/turkmenistan/2076776.html.)

Gurt, Marat, 2012, Pravitel'stvo Turkmenistana zayavilo, chto v 2015 godu ono nachnet stroit' novyi neftepererabatyvayushiy zavod [Turmenistan's Government announced that it would start construction of a new refinery in 2015]: InoZpress.kg, September 14. (Accessed July 11, 2013, at http://inozpress.kg/news/view/id/37229.)

MinerJob.ru, 2012a, "Belneftekhim" aktiviziroval raboty po stroitel'stvu Garlykskogo gorno-obogatitel'nogo kombinata [Belneftekhim speeded up construction of Garlyk mining and beneficiation complex]: MinerJob.ru, December 20. (Accessed July 11, 2013, at http://www.minerjob.ru/viewnew.php?id=23023.)

MinerJob.ru, 2012b, Pravitel'stvu Belarusi nuzhny dobrovol'tsy dlya raboty v Turkmenistane [Belarus' Government needs people to work in Turkmenistan]: MinerJob.ru, November 1. (Accessed July 11, 2013, at http://www.minerjob.ru/viewnew.php?id=22296.)

MinerJob.ru, 2013, Stroiteli Garlykskogo GOKa v Turkmenistane poluchat nalogovye l'goty [The builders of Turkmenistan's Garlyk GOK will get tax breaks]: MinerJob.ru, January 9. (Accessed July 11, 2013 at http://www.minerjob.ru/viewnew.php?id=23106.)

Ministerstvo Ekonomicheskogo Razvitiya Rossiyskoy Federatsii, 2012, Osnovnye ekonomicheskiye pokazateli i otsenki sostoyaniya otrasley ekonomiki i perspektivy ikh razvitiya [Main economic indicators and evaluation of current situation and perspectives of economic sectors]: Ministerstvo Ekonomicheskogo Razvitiya Rossiyskoy Federatsii. (Accessed July 11, 2013, at http://www.ved.gov.ru/exportcountries/tm/about_tm/eco_tm/.)

Neftegaz.ru, 2012, Dobycha gaza na krupneyshem mestorozhdenii Turkmenistana Galkynysh nachnetsya v 2013 g. [Gas production on Turkmenistan's largest deposit will start in 2013]: Neftegaz.ru, November 16. (Accessed July 11, 2013, at http://neftegaz.ru/news/view/105676/.)

Nefterynok.info, 2012, V 2012 godu 65% dobychi nefti Turkmenistana pridetsya na dolyu Kaspiyskih mestorozhdeniy [In 2012, 65% of Turkmen oil production will be from Caspian deposits]: Nefterynok.info, December 10. (Accessed July 11, 2013, at http://www.nefterynok.info/news.phtml?news_id=8524.)

Rossiyskoye Energeticheskoye Agenstvo, 2012, TEK Turkmenistana [Energy complex of Turkmenistan]: Rossiyskoye Energeticheskoye Agenstvo, August 24. (Accessed July 11, 2013, at http://rosenergo.gov.ru/upload/medialibrary/41d/Turkmenistan.pdf.)

State Committee on Statistics of Turkmenistan, 2013, Osnovnye social'no-ekonomicheskie pokazateli [Main socio-economic indicators]: State Committee on Statistics of Turkmenistan. (Accessed July 11, 2013, via http://www.stat.gov.tm/.)

Turkmeninform.com, 2012, V Turkmenistane obnovyat proizvodstvo broma [Turkmenistan will modernize its bromine production]: Turkmeninform.com, April 23. (Accessed July 11, 2013, at http://www.turkmeninform.com/ru/news/20120423/05646.html.)

U.S. Census Bureau, 2013a, U.S. exports to Turkmenistan by 5-digit end-use code 2003–2012: U.S. Census Bureau. (Accessed July 11, 2013, at http://www.census.gov/foreign-trade/statistics/product/enduse/exports/c4643.html.)

U.S. Census Bureau, 2013b, U.S. imports to Turkmenistan by 5-digit end-use code 2003–2012: U.S. Census Bureau. (Accessed July 11, 2013, at http://www.census.gov/foreign-trade/statistics/product/enduse/imports/c4643.html.)

U.S. Central Intelligence Agency, 2013, Turkmenistan, in The world factbook: U.S. Central Intelligence Agency, May 15. (Accessed July 11, 2013, at https://www.cia.gov/library/publications/the-world-factbook/geos/tx.html.)

U.S. Energy Information Administration, 2012, Turkmenistan: U.S. Energy Information Administration country analysis brief, January. (Accessed July 11, 2013, at http://www.eia.gov/countries/cab.cfm?fips=TX.)

Ustimenko, Artem, 2013, Turkmenistan 'primeryaetsya' k 'bol'shoy nefti [Turkmenistan is getting close to big oil]: Oilnews.kz. (Accessed July 11, 2013, at http://oilnews.kz/1/analitika/turkmenistan-primeryaetsya-k-bolshoj-nefti/.)

TABLE 1
TURKMENISTAN: ESTIMATED PRODUCTION OF MINERAL COMMODITIES[1, 2]

(Metric tons unless otherwise specified)

Commodity[3]		2008	2009	2010	2011	2012
METALS						
Rolled steel		--	10,000	40,000	50,000	120,000
INDUSTRIAL MINERALS						
Bentonite		50,000	50,000	50,000	50,000	53,000
Bentonite powder		250	250	250	250	255
Bischofite		100	100	100	100	105
Bromine		400 [r]	420 [r]	445 [r]	460 [r]	480
Cement		1,025,000 [4]	1,100,000 [4]	1,140,000 [4]	1,500,000 [4]	1,900,000
Ferrous bromide, 74% Br		85	85	85	85	95
Gypsum		100,000	100,000	100,000	100,000	105,000
Iodine	kilograms	468,400 [r, 4]	470,000 [r]	470,000 [r]	470,000 [r]	480,000
Lime		16,000	16,000	16,000	16,000	18,000
Nitrogen, N content of ammonia		270,000	270,000	270,000	270,000	280,000
Salt		215,000	215,000	215,000	215,000	220,000
Sodium sulfate		60,000	60,000	60,000	60,000	62,000
Sulfur		9,000	9,000	9,000	9,000	10,000
MINERAL FUELS AND RELATED MATERIALS						
Natural gas[4]	million cubic meters	70,501	38,000	44,270	66,200	69,000
Petroleum:						
Crude:						
In gravimetric units		9,678,000 [4]	8,850,000	9,097,800 [4]	9,882,300 [4]	11,000,000 [4]
In volumetric units	42-gallon barrels	70,400,000	64,300,000	66,100,000	71,800,000	79,915,000 [4]
Refinery products:						
In gravimetric units		7,300,000 [4]	7,600,000	7,752,000 [4]	7,900,000	8,000,000
In volumetric units	42-gallon barrels	58,700,000	61,100,000	62,322,000 [4]	63,200,000	64,300,000

[r] Revised. -- Zero.

[1] Estimated data are rounded to no more than three significant digits.

[2] Table includes data available through July 8, 2013.

[3] In addition to the commodities listed, barite, bench gravel, coal, dolomite, epsomite, and kaolin are thought to be produced, but available information is inadequate to make reliable estimates of output.

[4] Reported figure.

TABLE 2
TURKMENISTAN: STRUCTURE OF THE MINERAL INDUSTRY IN 2012[1]

(Metric tons unless otherwise specified)

Commodity		Major operating companies, main facilities, or deposits[2]	Location or deposit names	Annual capacity
Ammonia		Maryzoat Association	Mary Welayaty	400,000
Argillite	cubic meters	Keramzit plant	Yagmanskoye deposit	200,000
Barite-witherite		Arpaklenskiy mining enterprise	Arpaklen deposit	10,000
Do.		NA	Kumytash deposit and other deposits	NA
Bench gravel and loam:				
Bench gravel		Bezmeinskiy deposit	Near Ashgabat	1,200,000
Loam		do.	do.	12,000
Bischofite, epsomite, Caspian Sea salt, Glauber's salt		Karabogazsulfate Association	Kara-Bogaz-Gol Lagoon, off the Caspian Sea	NA
Bromine		Cheleken plant	Cheleken region	4,740
Do.		Nebitdag plant	Nebitdag region	2,370
Cement		Bakharlinskiy cement plant	Bakharly	1,000,000
Do.		Kelyata cement plant	Kelyata	1,000,000
Do.		Jebel cement plant	Jebel	1,000,000
Clays:				
Bentonite		Oglanly Mine	Oglanly region	100,000
Kaolin		Ashkhabad glass plant	Kyzylkainskoye deposit	80,000 [e]
Do.		Tuarkyrskoye deposit	250 kilometers southeast of Turkmenbashi	NA
Coal		do.	do.	NA
Dolomite		Ashkhabad glass plant	Kelyatinskoye deposit	6,000 [e]
Gypsum		IA Turkmenmineral	Mukry, Tagorin deposits	300,000
Do.		Wastes from Gaurdak sulfur deposit	Gaurdak, Gora	400,000
Do.		Krasnovodsk Aylagy (anhydride) deposit	9 kilometers east of Turkmenbashi	160,000
Iodine		Cheleken plant	Cheleken region	355
Do.		Nebitdag plant	Nebitdag region	255 [e]
Limestone		Deposits:		NA
		Gaurdak	4 kilometers northeast of Gaurdak	
		Kara-Dzhumalakskoye	60 kilometers from Gaurdak	
Limestone, for facing materials		NA	Charshanginskoye, Gaurdakskoye, Geok-Tepinskoye, Kaylyu, Krasnovodsk Aylagy (tuff and granite), and Tyuzmergenskoye deposits	NA
Do.	cubic meters	Tagarinskoye deposit	8 kilometers from Gaurdak	1,000 [e]
Limestone, for aggregates	do.	Aeroport deposit	21 kilometers northeast of Turkmenbashi	2,000
Do.	do.	Bekdashskoye deposit	200 kilometers north of Turkmenbashi	5,000
Do.	do.	Dostluksoye deposit	230 kilometers southeast of Turkmenbashi	2,000
Do.	do.	Mukrinskoye deposit	60 kilometers southwest of Gaurdak	25,000
Natural gas	million cubic meters	Achakskoye, Dauletabad, Doviet-Denmez (Donmez), Gygyrlinskoye, Ioltan (South Yolotan-Osman), North and South Naipskiye, Shatlyk, and Yashlar deposits	Onshore in eastern and southwestern parts of the country and offshore in the Caspian Sea; Amu-Darya and Murgab basins; Dashoguzskiy, Lebapskiy, Maryyskiy deposits	90,000 [e, 3]
Ozokerite		Cheleken mining enterprise	Cheleken region	NA

See footnotes in the end of the table.

TABLE 2—Continued
TURKMENISTAN: STRUCTURE OF THE MINERAL INDUSTRY IN 2012[1]

(Metric tons unless otherwise specified)

Commodity		Major operating companies, main facilities, or deposits[2]	Location or deposit names	Annual capacity
Petroleum:				
Crude	thousand metric tons	Barsa-Gelmesskoye, Burunskoye, Cheleken, Gograndagskoye, Ioltan (South Yolotan-Osman), Kamyshldzhinskoye, Korturtepinskoye, Kum Dag, Kuydzhikskoye, Okaremskoye, and Yashlar deposits	Centered in Caspian plain in west Turkmenistan and in offshore oilfields to the west of the Cheleken Peninsula in the Caspian Sea	11,000 e, 3
Refined	do.	Refineries:		12,000 [3]
		Seidi oil refinery	Lebap Welayaty	
		Turkmenbashi complex of oil refineries	Turkmenbashi	
Potash (sylvinite, carnallite)		Karlyuk deposit (experimental mine closed in 1998)	25 kilometers from Gaurdak	NA
Do.		Karabil'skoye deposit	17 kilometers south of Gaurdak	NA
Quartz sand		Annauskoye, Babadurmazskoye, Bakhardenskoye, and Kelyatinskoye deposits	NA	NA
Rock salt		Gaurdak deposit	8 kilometers from Gaurdak	15,000 e
Do.		Khodzhaguymaskoye deposit	4 kilometers west of Gaurdak	NA
Do.		Kugitangskoye deposit	75 kilometers from Gaurdak	2,000 e
Do.		Uzun-Kudukskoye deposit	20 kilometers from Gaurdak	2,000 e
Salt		Kuulinskoye	40 kilometers north of Turkmenbashi	650,000 e
Sand and gravel	cubic meters	Dushaksoye deposit	NA	1,150,000
Do.	do.	Kala-I-Morskoye deposit	NA	925,000
Do.	do.	Kernayskoye deposit	NA	36,000
Do.	do.	Kubatayskoye deposit	NA	740,000
Do.	do.	Ufrinskoye deposit	NA	900,000
Sodium sulfate		Karabogazsulfate Association	Bekdash, Kara-Bogaz-Gol Lagoon (off the Caspian Sea)	400,000
Steel, rolled		Turkmen metallurgical plant	Near Ashgabat	160,000
Do.		Shakhtaminskoye deposit	do.	NA
Do.		IA Turkmenmineral	Gora deposit	340,000
Do.		Kugitangskoye deposit	75 kilometers from Gaurdak	NA

eEstimated; estimated data are rounded to no more than three significant digits. Do., do. Ditto. NA Not available.

[1]Many location names have changed since the breakup of the Soviet Union. Many enterprises, however, are still named or commonly referred to based on the former location name, which accounts for discrepancies in the names of enterprises and that of locations.

[2]The majority of companies are owned by the Government.

[3]Capacity estimates are totals for all enterprises that produce that commodity.

THE MINERAL INDUSTRY OF UKRAINE

By Elena Safirova

Ukraine was among the world's leading producers of a number of minerals. It was one of the world's top four producers of gallium, the 4th-ranked producer of rutile (accounting for 7% of world output), the 6th-ranked producer of titanium sponge (3% of world output) and of iron ore (3% of world output), the 7th-ranked producer of manganese ore (2.4% of world output), the 9th-ranked producer of steel (2.1% of world output), and the 11th-ranked producer of ilmenite (2.3% of world output). The country had significant coal reserves but was dependent on imports to satisfy most of its petroleum and natural gas demand. Ukraine was also an important transit country for natural gas and petroleum from Central Asia and Russia to Europe (Corathers, 2013; Fenton, 2013; Gambogi, 2013a, b; Jaskula, 2013; Tuck, 2013).

Minerals in the National Economy

In 2012, Ukraine's real gross domestic product (GDP) increased by 0.3% compared with that of 2011. The nominal GDP in 2012 amounted to $174.6 million. During the year, overall annual industrial output was reduced by 4.5% compared with that in 2011; the share of industrial production in the country's GDP was 58%. The State Statistics Committee of Ukraine reported that, in 2012, mining and quarrying activities accounted for about 14.1% of all industrial production, and manufacturing, for 57.1%. The share of metallurgical production in overall industrial production was 15.1%, out of which production of ferroalloys, pig iron, steel accounted for 13.0 percentage points and production of nonferrous and precious metals accounted for 0.6 percentage points. Out of the 14.1% that constituted the total share of mining and quarrying in manufacturing production, mining of metallic ores accounted for 6.4 percentage points, coal mining accounted for 4.4 percentage points, and production of crude oil and natural gas accounted for 2.4 percentage points (State Statistics Service of Ukraine, 2013b; 2014).

Government Policies and Programs

In August 2012, Ukraine's Government proposed to simplify the process of issuing special licenses for the use of subsoil, including mining licenses. Special licenses are usually required for mineral resources that have been determined to be of national importance. According to the existing rules, the special licenses are issued by the national Government following special auctions or tenders, but an applicant has to obtain the approval of the corresponding local elected body, usually at the oblast level, before being allowed to participate in an auction. The corresponding elected bodies customarily consider such petitions during their biannual sessions. This procedure significantly slows the Government's ability to conduct auctions, which serve as an important source of revenue for the Government. To simplify the process, the proposal suggests delegating the issuance of approvals to local governments, which can process the approvals more quickly. To implement those changes, the Ukraine Parliament would need to approve relevant amendments to the Mining Code and the Code on Regional and Local Administrations (MinerJob.ru, 2012b).

Beginning on February 1, 2012, the value-added tax was lifted on lead-acid accumulators and lead-containing scrap imported into Ukraine. The tax had already been lifted on imports of many other types of nonferrous metal scrap, such as aluminum, bismuth, cadmium, chromium, cobalt, copper, germanium, magnesium, manganese, molybdenum, nickel, tantalum, thallium, tin, titanium, vanadium, and zirconium. Among ferrous metals, scrap of alloyed steel, pig iron, and stainless steel containing nickel were also not subject to the value-added tax (Mineral.ru, 2012o).

Production

Production of many mineral commodities was down in 2012. The output of refinery products decreased by 49%; that of primary aluminum, by 40%; gypsum, by 35%; and secondary aluminum, by an estimated 31%. The output of several other mineral commodities also decreased significantly, including that of secondary copper, which decreased by an estimated 25%; zirconium concentrates, by 23%; ferrosilicon, by 21%; kaolin, by 20%; feldspar, by 16%; gallium, by an estimated 15%, and steel pipe, by 15%. On the other hand, output of anthracite coal increased by 48%; that of other ferroalloys, by 23%; and manganese ore and concentrates, by 22%. These and other production data are in table 1.

Mineral Trade

The total value of Ukraine's exports increased to about $68.8 billion in 2012 from $68.4 billion in 2011. The value of exports was equal to about 39% of Ukraine's GDP in 2012. Ukraine's leading export category in terms of value was ferrous metals, and in 2012, exports of ferrous metals were valued at $15.3 billion and made up 22.3% of the total value of exports; exports of mineral fuels and petroleum products were valued at $3.7 billion and made up 5.3% of the total value of exports. Another $3.3 billion (4.8% of the total value of exports) was contributed by exports of cinder, ores, and slag. The value of exports of mineral products and metals made up about 33.4% of the value of total exports. The main export partners of Ukraine were Russia (which received 25.6% of Ukraine's exports); Turkey (5.4%); Egypt (4.2%); Poland (3.7%); Italy and Kazakhstan (3.6% each); and India (3.3%) (State Statistics Service of Ukraine, 2013a, 2014).

The total value of Ukraine's imports was about $84.7 billion in 2012 and $82.6 billion in 2011. The leading imported commodities were mineral fuels and refined petroleum products, which made up about 32.5% of the value of total imports in 2012.

Natural gas was the leading individual imported product, in terms of value, and accounted for 16.6% of the value of total imports. The country's main import partners in 2012 were Russia (which supplied 32.4% of Ukraine's imports); China (9.3%); Germany (8.0%); Belarus (6.0%); Poland (4.2%); the United States (3.4%); and Italy (2.6%) (State Statistics Service of Ukraine, 2013a, 2014).

Commodity Review

Metals

Aluminum.—On April 26, 2011, United Company RUSAL (RUSAL) of Russia halted primary aluminum production at the Zaporozhye smelter (ZAlK), which was Ukraine's only producer of primary aluminum. According to RUSAL, the main reason for halting production at the smelter was the high cost of electricity used in primary aluminum production, which was as high as $0.08 per kilowatthour compared with about $0.03 per kilowatthour in a typical smelter located in Europe and Russia. The company was planning to continue to produce rolled aluminum at the ZAlK plant using the ingots shipped from other RUSAL facilities (Kommersant Ukraina, 2011; Stasovskaya, 2012).

The Government of Ukraine had expressed dissatisfaction with RUSAL's operations in the country. In 2000, Ukraine's alumina producer, the Nikolaevskiy alumina plant (NGZ), had been privatized and sold to RUSAL. As a part of the deal, RUSAL had promised to build an additional aluminum plant in Ukraine that would have the capacity to produce at least 100,000 metric tons per year (t/yr) of aluminum. In 2004, because of low aluminum prices, RUSAL asked the Government to replace the requirement to build a new plant with a requirement to modernize the NGZ plant extensively and to increase alumina production to 1.6 million metric tons per year (Mt/yr). In 2007, the ZAlK plant became a part of RUSAL when the former RUSAL, Siberian-Urals Aluminum Co. (SUAL), and the alumina holdings of Glencore International AG were merged into the new United Company RUSAL. Before the merger, SUAL had owned 68.01% of the shares in ZAlK, which it acquired in 2004 from AvtoVAZ-Invest. Since 2008, annual production of primary aluminum at the smelter had gradually decreased to about 25,000 metric tons (t) in 2011 from 113,000 t in 2008. As a result of these developments, by yearend 2011, Ukraine, instead of having two aluminum plants, was about to be left with no plants at all (Vzglyad, 2012).

In the beginning of 2012, Ukraine's General Prosecutor's office filed a lawsuit against RUSAL claiming that RUSAL had violated the conditions of the ZAlK sale, in particular the requirement that ZAlK's $75 million loan be refinanced by the investor. On March 24, the Kyev Economic Court pronounced the initial sale of ZAlK to be invalid and ordered that ZAlK be returned to its original owner, the Ukraine State Property Fund, which had privatized ZAlK in 2001. RUSAL appealed the decision and, on May 23, the Kyev Economic Court of Appeals confirmed the decision of the lower court. RUSAL announced that it was planning to appeal this decision in the Ukraine Court of Cassation and was not planning to return the plant. RUSAL offered to restart production at ZAlK if the Government of Ukraine agreed to sign a long-term agreement about preferential electricity tariffs for aluminum production in the country, which would render production at ZAlK profitable. It was expected that the court battle between RUSAL and the Government might last for years while production at ZAlK stays at very low levels (Vzglyad, 2012).

Ferroalloys.—Production of ferromanganese, ferrosilicon, and silicomanganese decreased significantly in 2012 as Ukrainian producers continued to encounter high production costs and reduced market demand for their products. The Nikopol ferroalloys plant produced 103,000 t of ferromanganese and 554,800 t of silicomanganese. The Stakhanov ferroalloys plant produced 119,400 t of ferrosilicon and 49,400 t of silicomanganese. The Zaporozhye ferroalloys plant produced 54,100 t of ferromanganese and 129,900 t of silicomanganese (Metallosnabzhenie i sbyt, 2013).

In September, the PrivatBank Group, which was the owner of all three ferroalloys plants as well as two manganese mining and beneficiation complexes—the Marganets mining and beneficiation complex (GOK) and the Ordzhonikidze GOK, threatened to begin massive layoffs if the Government did not provide a discount for electric power to all three plants. The Cabinet of Ministers of Ukraine appeared to agree to form an interagency working group to revise power rates for electrometallurgical plants. Some members of the Government, however, made statements indicating that the Government might file a lawsuit against PrivatBank for violation of the investment obligations agreed to when the ferroalloys plants were privatized (Unian.net, 2013a).

In October, the Government introduced a reduced price for electric power for the electrometallurgical plants whose share of electric power in the production cost was greater than 30%; those enterprises included the Nikopol, the Stakhanov, and the Zaporozhye ferroalloys plants. The reduction in power prices was supposed to be in effect from October 1, 2012, through March 1, 2013. It was expected that the price reduction for electrometallurgical plants would eventually result in increased electricity prices for other industrial electric power consumers (Unian.net, 2013a).

Gold.—Ukraine did not produce gold in 2012. According to the Prime Minister of Ukraine, the only existing gold-mining enterprise, OOO Zakarpatpolimetally, was in need of investment. When the plant stopped production in December 2006, 400 workers lost their jobs and the plant had a debt of 29 million hryvnias ($5.8 million).[1] The company had been in bankruptcy since 2008, but, as of September 2012, no court decision about the company's debt resolution had been finalized. The Muzhievskoe gold mine, which was operated by Zakarpatpolimetally, had estimated resources of about 44 t of gold, and additional resources of lead, zinc, and silver. Zakarpatpolimetally also had a beneficiation plant with a capacity of 200,000 t/yr of ore (Mineral.ru, 2012b).

Meanwhile, other companies continued to invest in other gold projects in Ukraine. Lugansk Gold Ltd. of Australia was

---

[1]Where necessary, values have been converted from Ukrainian hryvnias (UAH) to U.S. dollars (US$) at an annual average exchange rate of UAH8.12=US$1.00 for 2012 and UAH5.00=US$1.00 for 2006. All values are nominal, at current prices, unless otherwise stated.

planning to invest a total of $50 million in a gold project at the Bobrikovskoye gold sulfide deposit in Luhans'ka Oblast'. The project included construction of a beneficiation plant that would have the capacity to process up to 500,000 t/yr of ore. The project was expected to have a life of between 10 and 15 years and to produce about 2,000 kilograms per year of gold. The Bobrikovskoye deposit was grading between 1 and 16 grams per metric ton (g/t) gold and 42.5 g/t silver. According to the Joint Ore Reserves Committee (JORC) Code, its gold reserves were evaluated to be 1.012 million troy ounces [31,500 kilograms (kg)], out of which about 500,000 troy ounces (15,600 kg) was classified as measured and probable, and silver reserves were evaluated to be between 12.25 million troy ounces (381,000 kg) and 14.5 million troy ounces (451,000 kg) (Mineral.ru, 2012e, f).

The state enterprise Severgeologiya obtained a license for additional exploration and mining of two gold deposits in Dnipropetrovs'ka Oblast'. The two were the Balka Zolotaya deposit, which graded 6.2 g/t gold, and the Balka Shirokaya deposit, which graded between 4 and 7 g/t gold. The resources of the deposits were estimated to be 50 and 130 t of gold, respectively (MinerJob.ru, 2012i).

Manganese.—Ukraine had two producers of mined manganese—the Marganets GOK and the Ordzhonikidze GOK, both of which were located in Dnipropetrovs'ka Oblast' and owned by PrivatBank. The Marganets GOK produced manganese concentrate, and the Ordzhonikidze GOK produced both manganese concentrate and manganese agglomerate. The Marganets GOK was the only manganese ore producer in Ukraine that had an underground mine; it produced about 80% of its ore from underground mining and the other 20% was from open pit mining. The Ordzhonikidze GOK produced all its ore from open pit mining. In 2012, the Marganets GOK produced 696,100 t of manganese concentrate, which was a 7% decrease compared with its output in 2011. The Ordzhonikidze GOK, which was the leading manganese ore producer in Ukraine based on capacity, had production interruptions throughout 2011 and 2012; it did not produce manganese concentrate between September and December of 2011, and it did not produce any manganese agglomerate from January to April 2012. As a result, in 2012, the manganese concentrate production at the Ordzhonikidze GOK increased by 53.9% compared with that in 2011, whereas manganese agglomerate production decreased by 60% (MinerJob.ru, 2012e).

Hubei Changyang Hongxin Industrial Group, Inc. (HCHIG), which was the leading producer of electrolytic manganese in China, expressed interest in obtaining a development license for the Velikotokmakskoye manganese deposit in Zaporiz'ka Oblast'. The company created a plan for development and exploitation of the deposit that would span 20 years and cost a total of $1 billion. According to this plan, the maximum capacity of the mine would be 5 Mt/yr of manganese ore and 200,000 t/yr of electrolytic manganese. This was not the first attempt to develop the Velokotokmakskoye deposit. In the 1980s, the Government of the Soviet Union had started construction of the Tavricheskiy GOK; activity at the only mine at the Tavrickeskiy GOK was suspended in 1995 because of the enterprise's losses, however, and the mine had not been in operation since then (MinerJob.ru, 2012d).

Titanium.—The titanium industry in Ukraine consisted of ilmenite and rutile concentrate production, titanium sponge production at the Government-owned Zaporozhye Titanium & Magnesium Complex (ZTMK), and titanium ingot production by a small number of producers, including OOO Antares, OOO Fiko, and ZTMK, which had a combined capacity to produce about 12,000 t/yr of titanium ingots. Titanium dioxide pigment was produced by Crimea Titan CJSC and OAO Sumykhimprom. Ukraine did not have the ability to produce titanium metal products used in the aerospace industry and other industries that require more technically advanced titanium metal products (Metall Ukrainy, 2010, p. 61).

Velta LLC was planning to open the second stage of its mining and beneficiation complex at the Birzulovkoe ilmenite deposit at the end of 2012. The first stage officially started operations in December 2011. The total investment in the new complex was approximately $90 million, out of which $80 million was used to fund the development of the first stage, and the rest was invested in the second stage. Once the second stage of the construction is completed, the complex was projected to have an annual capacity of 240,000 t/yr of titanium ore (Kabash, 2011; Mineral.ru, 2012p).

In January 2012, the Cabinet of Ministers made a decision to privatize ZTMK, which was the leading producer of titanium sponge in Ukraine. Later, it was announced that a potential investor was willing to invest $700 million into the enterprise. In December, it was announced that Tolexis Trading Ltd. of Cyprus (a part of the DF Group) had won the tender and would receive 49% of the shares of ZTMK. According to the rules of the tender, the winner was to invest at least $100 million in the company to provide ZTMK with ilmenite from a domestic source, which would be used to pay off company debts, including delayed payments of wages, taxes, and energy costs. The DF Group was likely to select Velta as its ilmenite supplier (MinerJob.ru, 2012h).

In July, a world leading titanium producer, VSMPO-Avisma of Russia, acquired 75% of the shares of Limpeza Ltd. of Cyprus, which owned the Demurinskiy GOK. The Demurinskiy GOK had a license to mine the Volchanskiy deposit, which contains alluvial titanium and zirconium. The GOK was projected to be able to reach its capacity of 50,000 t/yr of ilmenite concentrate, 13,000 t/yr of rutile concentrate, and 3,000 t/yr of zircon in the next 3 to 5 years. VSMPO-Avisma reportedly invested $30 million into the GOK and was planning to build a beneficiation plant (MinerJob.ru, 2012f, g, j).

Mineral Fuels and Related Materials

Coal.—According to BP p.l.c. of the United Kingdom, proven reserves of coal in Ukraine were 33.9 billion metric tons, or about 4.0% of the world's reserves. According to those estimates, Ukraine's coal reserves are 390 times larger than the country's current annual production. Ukraine has three coalfields—the Dneprovskiy, the Donetskiy, and the Lvovsko-Volynskiy fields. The largest of these fields is the

Donetskiy field, which contains 87% of the proven reserves (Baker Tilly, 2013).

In 2012, Ukraine produced 85.7 million metric tons (Mt) of coal, which was a 4.64% increase compared with its coal output in 2011. The country was the fourth-ranked coal producer in Europe after Russia, Germany, and Poland. The output from the mines located in Donets'ka Oblast', totaled 36.46 Mt of coal in 2012, which was an 8.8% increase compared with production in 2011. The mines located in Luhan'ska Oblast' produced 26.94 Mt of coal (which was a 1.2% decrease); Dnipropetrovs'ka Oblast', 17 Mt (a 10.3% increase); Lvivs'ka Oblast', 1.96 Mt (a 16.8% decrease); and Volyns'ka Oblast', 392,600 t (a 28.6% decrease) (Mineral.ru, 2012l; 2013).

Despite the increases in production, according to the Government estimates, about 80% of all coal mines in Ukraine operated at a loss in 2012. The coal mined from the Donetskiy field, for example, contained large amounts of sulfur and other undesirable constituents. In addition, as of 2012, a total of about 400,000 t of unsold coal was being kept in coal storage facilities. Consequently, the Government was facing significant problems with the country's coal industry. On one hand, increases in coal production could potentially serve as part of the strategy to increase the country's energy independence and reduce its natural gas imports from Russia. On the other hand, Ukrainian coal was unable to compete with coal from Germany and Poland either in terms of price or quality (MinerJob.ru, 2012a).

To overcome those problems, the Government developed two potential solutions. One of them was privatization of Government-owned coal mines. Beginning in 2013, several specified mines could be privatized for a symbolic price of 1 hryvnia ($0.12) in exchange for taking on investment obligations with respect to the privatized mines. Another approach to saving Ukraine's coal industry was to artificially generate domestic demand for coal. In particular, the Government mandated that heating plants in the country were to switch from natural gas, which was imported from Russia, to domestically produced coal as their energy source. According to the Ministry of Energy, this switch would allow Ukraine to reduce domestic consumption of natural gas by 6 billion cubic meters per year. On the other hand, many residents and local leaders voiced concern that the switch to coal would significantly increase pollution and worsen the environmental situation in many cities (Mineral.ru, 2012g, h; MinerJob.ru, 2012c).

Natural Gas.—In 2012, Ukraine was expanding its natural gas production from shale and from the Black Sea shelf. After long and largely fruitless negotiations with Russia about the price of natural gas imported from Russia, the Government of Ukraine decided to develop a set of measures to reduce gas consumption and simultaneously increase natural gas production domestically. The Government was hoping to reach production levels of between 20 billion and 30 billion cubic meters per year of shale gas within the next 10 years (Mineral.ru, 2012k, m).

In May, the Government conducted tenders for the Yuzovskoye and the Olesskoye shale gas deposits, which were won by Royal Dutch Shell plc of the Netherlands and Chevron Corp. of the United States, respectively. According to some estimates, the total resources of the Yuzovskoye deposit and the Olesskoye deposit are about 2 trillion cubic meters and 1.5 trillion cubic meters, respectively. Ukraine was hoping to start commercial production of shale gas in 2017. In August, the Government held a tender for the Skifskaya oil and gas field, which is located in the Black Sea shelf; the winner of the tender was Exxon Mobil Corp. of the United States. According to preliminary data, the Skifskaya field has the potential to produce between 3 billion and 4 billion cubic meters per year of natural gas. Ukraine was also planning to announce a tender for the Slobozhanskaya field. Annual gas production at the Slobozhanskaya field was estimated to be between 6 billion and 8 billion cubic meters per year (Mineral.ru, 2012c, j, n).

In September, Ukrainian State Company Chernomorneftegaz started industrial exploitation of the Odesskoye gas deposit located on the Black Sea shelf. The total resources of the deposit were estimated to be 22 billion cubic meters of natural gas. Overall, Chernomorneftegaz was planning to triple natural gas production from the Black Sea shelf to 3 billion cubic meters per year of natural gas by 2015 from 1.056 billion cubic meters per year in 2011. The company was planning to add three more shelf gas projects to the four already in operation—the Arkhangel'skoye, the Bezymyannoye, the Golitsinskoye, and the Odesskoye. It was expected that the demand for natural gas on the Crimean Peninsula, where annual natural gas consumption was about 2 billion cubic meters, would be fully met by the shelf gas produced in the Black Sea (Mineral.ru, 2012a, q).

Petroleum.—Ukraine's production of petroleum refinery products decreased by 49% in 2012 to 4.57 Mt of petroleum. The total annual capacity of all refineries in Ukraine was 43.65 Mt; therefore, overall, the country's refineries produced only at 10.5% of their capacity. In 2012, Ukraine had seven refineries, only two of which were operating. The leading Ukrainian refinery, the Kremenchug refinery, processed about 3 Mt of petroleum in 2012, whereas its capacity was 18.6 Mt/yr. The other working refinery—the Shebelinsk refinery—which had facilities to process both petroleum and natural gas, had the capacity to process about 1 Mt/yr of petroleum, but in 2012 it refined only 660,000 t (Somov, 2013).

Three more refineries—the Drogobychsk refinery, the Kherson refinery, and the Nadvoryansk refinery—had obsolete equipment and were incapable of producing petroleum products to Ukraine's national quality standards. Modernization of these refineries would require significant investments, but the current owners were not interested in making such investments. The Odessa refinery was shut down in 2010 because it had difficulty in obtaining crude oil supplies after the directional flow of oil in the Odessa-Brody pipeline was reversed. Previously, the pipeline had shipped Russian oil from Brody to Odessa, but beginning in 2011, it began shipping Caspian Sea oil from Odessa to Brody. Finally, the Lisichansk refinery was shut down in March 2012 because of financial losses. The current owner of the Lisichansk refinery, TNK-BP, was reportedly looking for a buyer for the refinery. The Lisichansk refinery had the capacity to refine 8 Mt/yr of petroleum (Mineral.ru, 2012d).

Uranium.— Uranium ore was mined in Ukraine from underground mines by the state-owned company Vostochny GOK, and it was processed into concentrate at the company's hydrometallurgical plant at Zheltye Vody. The concentrate was then sent to Russia to be processed by OAO TVEL of Russia into nuclear fuel for use in Ukraine; the remaining nuclear fuel required for Ukraine's nuclear powerplants was purchased from TVEL. In 2012, domestic uranium production accounted for 32% of all uranium used in Ukraine's nuclear powerplants. In 2012, the Vostochny GOK increased its production of uranium concentrate by 8% (compared with its output in 2011) to 960.2 t (Unian.net, 2013b).

In September 2010, TVEL was awarded a contract for the construction of a nuclear fuel assembly plant in Ukraine, and, in 2011, a Russian-Ukrainian joint venture between TVEL and GK Nuclear Fuel of Ukraine completed its registration procedures. GK Nuclear Fuel had a 50% plus one share in the new enterprise, and TVEL had 50% minus one share. The fuel assembly plant would allow Ukraine to execute the final stage of nuclear fuel preparation, but it would still be necessary for Ukraine to ship its uranium concentrate out of the country for the intermediate process of uranium conversion and enrichment. The construction of the plant was expected to begin in 2013 (Mineral.ru, 2012i).

In July 2011, the Vostochny GOK began uranium ore production from the Novokonstantinovskoye deposit in Kirovohrads'ka Oblast'. In June 2012, the Government of Ukraine approved the construction plan for a new uranium processing plant at the Novokonstantinovskoye deposit with a total cost of about $800 million. The plant would have the capacity to process 1.5 Mt/yr of uranium ore and to produce 500,000 t/yr of uranium oxides. The construction of the first stage of the project was planned to be completed in 42 months, and the entire project was expected to be built within 96 months. By 2014, Ukraine was planning to increase its uranium ore production to 1.88 Mt/yr (Mineral.ru, 2012i).

Outlook

Ukraine mining and metallurgy sectors had significant setbacks in the past few years. Ukrainian aluminum production was essentially halted because of the high electricity prices; ferroalloys plants had been granted temporary electricity price discounts provided by the Government, but they had no clear way forward in the absence of these subsidies; and coal mines and petroleum refineries were outdated and required significant investments to become competitive or even to break even. In terms of energy production, Ukraine was trying to increase its domestic production of petroleum and natural gas, ramp up coal and uranium production, and reduce its reliance on imported natural gas from Russia.

Going forward, Ukraine is likely to remain one of the leading world producers of manganese ore, titanium ores, and titanium sponge. Remaining competitive in metallurgy may prove to be difficult because of high energy requirements, a need of new investments, and the often differing interests of plant owners and the Government. It remains to be seen if the Government and the owners of privately owned industrial facilities will be able to reach compromises and if the country will be able to attract new investments to move the mineral and metallurgical industries of the country forward.

References Cited

Baker Tilly, 2013, Ugol' Ukrainy [Ukrainian coal]: Baker Tilly International, July, 13 p. (Accessed March 26, 2014, at http://www.bakertilly.ua/media/Baker Tilly - Report_coal_industry_rus.pdf.)

Corathers, L.A., 2013, Manganese: U.S. Geological Survey Mineral Commodity Summaries 2013, p. 100–101.

Fenton, M.D., 2013, Iron and steel: U.S. Geological Survey Mineral Commodity Summaries 2013, p. 78–79.

Gambogi, Joseph, 2013a, Titanium and titanium dioxide: U.S. Geological Survey Mineral Commodity Summaries 2013, p. 172–173.

Gambogi, Joseph, 2013b, Titanium mineral concentrates: U.S. Geological Survey Mineral Commodity Summaries 2013, p. 174–175.

Jaskula, B.W., 2013, Gallium: U.S. Geological Survey Mineral Commodity Summaries 2013, p. 58–59.

Kabash, Natalya, 2011, "Velta" uvelichit dobychu il'menita [Velta will increase ilmenite production]: Elzvestiya.com, November 11. (Accessed March 26, 2014, at http://markets.eizvestia.com/full/velta-uvelichit-dobychu-ilmenita.)

Kommersant Ukraina, 2011, Spasite nashy chushki [Save our ingots]: Kommersant Ukraine, April 29. (Accessed March 26, 2014, at http://www.kommersant.ua/doc/1631718.)

Metall Ukrainy [Metal of Ukraine], 2010, Rynok titana [The titanium market]: Metall Ukrainy, no. 16, August 16–31, 70 p.

Metallosnabzhenie i sbyt, 2013, V 2012 godu Ukraina sokratila proizvodstvo ferrosplavov I margantsevogo syr'ya [In 2012, Ukraine reduced production of ferroalloys and mined manganese]: Metallosnabzhenie i sbyt, January 15. (Accessed March 26, 2014, at http://www.metalinfo.ru/ru/news/61083.)

Mineral.ru, 2012a, "Chernomorneftegaz" planiruet utroit' dobychu gaza na shel'fe [Chernomorneftegaz is planning to triple gas production from the shelf]: Mineral.ru, May 10. (Accessed March 26, 2014, at http://www.mineral.ru/News/47880.html.)

Mineral.ru, 2012b, Dobycha zolota na Ukraine prekrashena [Gold mining in Ukraine stopped]: Mineral.ru, March 7. (Accessed March 26, 2014, at http://www.mineral.ru/News/47880.html.)

Mineral.ru, 2012c, Kompaniya Shell i "Ukrgazdobycha" nachali bureniye pervoy poiskovoy skvazhiny v Khar'kovskoy oblasti [Shell and Ukrgazdobycha have started drilling the first search hole in Khar'kovskaya Oblast]: Mineral.ru, October 28. (Accessed March 26, 2014, at http://www.mineral.ru/News/50623.html.)

Mineral.ru, 2012d, Lisichanskiy NPZ smenil sobstvennika [Lisichanskiy refinery changed owners]: Mineral.ru, October 2. (Accessed March 26, 2014, at http://www.mineral.ru/News/50301.html.)

Mineral.ru, 2012e, Lugansk Gold investiruet v zolotodobychu 50 mln dollarov [Lugansk Gold invests $50 million in gold mining]: Mineral.ru, January 25. (Accessed March 26, 2014, at http://www.mineral.ru/News/47377.html.)

Mineral.ru, 2012f, Na Ukraine budut dobyvat' bol'she zolota [Ukraine will produce more gold]: Mineral.ru, March 31. (Accessed March 26, 2014, at http://www.mineral.ru/News/48127.html.)

Mineral.ru, 2012g, Perekhod na ugol' na Ukraine uhudshit ekologiyu [The switch to coal in Ukraine will worsen the environmental situation]: Mineral.ru, May 4. (Accessed March 26, 2014, at http://www.mineral.ru/News/48511.html.)

Mineral.ru, 2012h, Perekhod ukrainskih kitel'nykh na ugol' privedet k ekologicheskoy katastrofe [The switch of the heating plants to coal will lead to environmental catastrophe]: Mineral.ru, January 18. (Accessed March 26, 2014, at http://www.mineral.ru/News/47287.html.)

Mineral.ru, 2012i, Pravitel'stvo Ukrainy predpochlo fabrikatsiyu yadernogo topliva s TVELom sobstvennoy dobyche urana [Ukrainian Government preferred nuclear fuel production in partnership with TVEL to its own uranium mining]: Mineral.ru, June 27. (Accessed March 26, 2014, at http://www.mineral.ru/News/49102.html.)

Mineral.ru, 2012j, Ukraina opredelila gruppu kompaniy vo glave s "ExxonMobil" pobeditelem konkursa na razrabotku Skifskogo uchastka shel'fa Chernogo morya [Ukraine selected a consortium of companies headed by ExxonMobil as the winner of the tender for the Skifskoe sector of the Black Sea]: Mineral.ru, August 15. (Accessed March 26, 2014, at http://www.mineral.ru/News/49693.html.)

Mineral.ru, 2012k, Ukraina planiriet skoro dobyvat' 20–30 mlrd kub. m.slantsevogo gaza v god [Ukraine is planning to produce 20–30 billion cubic meters per year of shale gas soon]: Mineral.ru, December 2. (Accessed March 26, 2014, at http://www.mineral.ru/News/51004.html.)

Mineral.ru, 2012l, Ukraina planiruet v 2012 g. narastit' dobychu uglya do 84 mln t. [In 2012, Ukraine is planning to increase its coal production to 84 million tons]: Mineral.ru, January 16. (Accessed March 26, 2014, at http://www.mineral.ru/News/47257.html.)

Mineral.ru, 2012m, Ukraina predostavila plan trehkratnogo uvelicheniya dobychi gaza na shel'feChernogo I Azovskogo morey k 2015 g. [Ukraine presented a plan of a threefold increase of shelf gas production in the Black and the Azov Seas by 2015]: Mineral.ru, October 3. (Accessed March 26, 2014, at http://www.mineral.ru/News/50315.html.)

Mineral.ru, 2012n, Ukraina pristupaet k dobyche slantsevogo gaza [Ukraine is starting shale gas production]: Mineral.ru, June 18. (Accessed March 26, 2014, at http://www.mineral.ru/News/48999.html.)

Mineral.ru, 2012o, Ukraina vremenno otmenila NDS na import svintsovogo loma [Ukraine temporarily removed VAT on lead scrap imports]: Mineral.ru, January 25. (Accessed March 26, 2014, at http://www.mineral.ru/News/47384.html.)

Mineral.ru, 2012p, "Velta" zapustit vtoruyu ochered' titanovogo GOKa [Velta will start operations at the second stage of the titanium mining and beneficiation plant]: Mineral.ru, February 15. (Accessed March 26, 2014, at http://www.mineral.ru/News/47646.html.)

Mineral.ru, 2012q, Yanukovich zapustil v ekspluatatsiyu mestorozhdeniye gaza v Chernom more [Yanukovich opened operation at a new gas deposit in the Black Sea]: Mineral.ru, September 6. (Accessed March 26, 2014, at http://www.mineral.ru/News/49981.html.)

Mineral.ru, 2013, Ukraina za god narastila dobychu uglya na 4,8% [Ukraine increased coal mining by 4.8% in one year]: Mineral.ru, January 15. (Accessed March 26, 2014, at http://www.mineral.ru/News/51452.html.)

MinerJob.ru, 2012a, Donetskiy basis: Nuzhno li spasat' ukrainskuyu ugledobychu ot kollapsa [Problem in Donetsk—Should the coal mining industry be saved from collapse?]: MinerJob.ru, May 10. (Accessed March 26, 2014, at http://minerjob.ru/viewnew.php?id=21204.)

MinerJob.ru, 2012b, Dostup k nedram uprostyat v Ukraine [In Ukraine, access to subsoil will be simplified]: MinerJob.ru, August 11. (Accessed March 26, 2014, at http://minerjob.ru/viewnew.php?id=21607.)

MinerJob.ru, 2012c, Gosudarstvennye shakhty Ukrainy privatiziruyutsya za odnu grivnu [Ukrainian state-owned mines are privatized for 1 hryvnia]: MinerJob.ru, June 25. (Accessed March 26, 2014, at http://minerjob.ru/viewnew.php?id=21365.)

MinerJob.ru, 2012d, Kitayskaya HCHIG khochet dobyvat' margantsevuyu rudu v Ukraine [Chinese HCHIG wants to mine manganese in Ukraine]: MinerJob.ru, August 10. (Accessed March 26, 2014, at http://minerjob.ru/viewnew.php?id=21588.)

MinerJob.ru, 2012e, Ordzhonikidzevskiy complex skoree vsego budet rabotat' zimoy? [Will Ordzhonikidze GOK most likely work in winter?]: MinerJob.ru, September 28. (Accessed March 26, 2014, at http://minerjob.ru/viewnew.php?id=21920.)

MinerJob.ru, 2012f, Rossiyskaya titanovaya korporatsiya investirovala $30 mln. v razvitie svoego ukrainskogo GOKa [Russian titanium corporation invested $30 million in development of its Ukrainian GOK]: MinerJob.ru, November 29. (Accessed March 26, 2014, at http://minerjob.ru/viewnew.php?id=22747.)

MinerJob.ru, 2012g, Titanovaya korporatsiya "VSMPO-Avisma" obzavelas' cyr'evym aktivom [Titanium corporation VSMPO-Avisma acquired its own source of raw materials]: MinerJob.ru, July 16. (Accessed March 26, 2014, at http://minerjob.ru/viewnew.php?id=21429.)

MinerJob.ru, 2012h, Ukraina otdala oligarkhu Firtashu Zaporozhskiy titano-magnievyi kombinat [Ukraine gave its Zaporozhye Titanium and Magnesium Complex to Firtash]: MinerJob.ru, December 21. (Accessed March 26, 2014, at http://minerjob.ru/viewnew.php?id=23035.)

MinerJob.ru, 2012i, Ukrainskoe zoloto khotyat pribrat' v odni ruki [Some want to consolidate Ukrainian gold]: MinerJob.ru, September 27. (Accessed March 26, 2014, at http://minerjob.ru/viewnew.php?id=21909.)

MinerJob.ru, 2012j, "VSMPO-Avisma" kupil GOK na Ukraine [VSMPO-Avisma bought a GOK in Ukraine]: MinerJob.ru, July 9. (Accessed March 26, 2014, at http://forum.minerjob.ru/viewnew.php?id=21401.)

Somov, Kirill, 2013, Ukrainskaya neftepererabotka: Nizhe nekuda [Ukrainian petroleum refining—Can't get lower]: LB.ua, March 27. (Accessed March 26, 2014, at http://economics.lb.ua/business/2013/03/27/194249_ukrainskaya_neftepererabotka_nizhe.html.)

Stasovskaya, Svetlana, 2012, Tsvetmet Ukrainy: Aluminiy "splavili" [Non-ferrous metals of Ukraine—Aluminum has been sent away]: UGMK.info, October 24. (Accessed March 26, 2014, at http://ugmk.info/art/cvetmet-ukrainy-aljuminij-splavili/0.html.)

State Statistics Service of Ukraine, 2013a, Commodity pattern of foreign trade of Ukraine, 2012: State Statistics Service of Ukraine. (Accessed December 5, 2013, at http://www.ukrstat.gov.ua/.)

State Statistics Service of Ukraine, 2013b, Statistical yearbook of Ukraine for 2012: State Statistics Service of Ukraine. (Accessed December 5, 2013, at http://www.ukrstat.gov.ua/.)

State Statistics Service of Ukraine, 2014, Gross domestic product [1990–2012]: State Statistics Service of Ukraine. (Accessed December 12, 2014, at http://www.ukrstat.gov.ua/.)

Tuck, C.A., 2013, Iron ore: U.S. Geological Survey Mineral Commodity Summaries 2013, p. 84–85.

Unian.net, 2013a, Reshenie o l'gotnykh tarifah dlya elektrometallurgicheskih predpriyatiy soglasovano [The decision about reduced electric rates for electrometallurgical plants is agreed upon]: Unian.net, March 1. (Accessed March 26, 2013, at http://economics.unian.net/industry/757545-reshenie-o-lgotnyih-tarifah-dlya-elektrometallurgicheskih-predpriyatiy-soglasovano.html.)

Unian.net, 2013b, Vostochnyi GOK v 2012 godu uvelichil proizvodstvo kontsentrata urana na 8% [In 2012, Vostochnyi GOK increased uranium concentrate production by 8%]: Unian.net, March 4. (Accessed March 26, 2013, at http://economics.unian.net/industry/758433-vostochnyiy-gok-v-2012-godu-uvelichil-proizvodstvo-kontsentrata-urana-na-8.html.)

Vzglyad, 2012, Natsionalizatsiya po sudu [Nationalization by the court of law]: Vzglyad, March 24. (Accessed March 26, 2013, at http://www.vz.ru/economy/2012/3/24/570691.print.html.)

TABLE 1
UKRAINE: PRODUCTION OF MINERAL COMMODITIES[1]

(Metric tons unless otherwise specified)

Commodity[2]		2008	2009	2010	2011	2012
METALS						
Alumina		1,673,000	1,524,000	1,534,000	1,601,000	1,429,000
Aluminum:						
Primary		113,000	50,000	25,000	24,830 r	14,829
Secondary[e]		130,000	130,000	130,000	130,000	90,000
Total[e]		243,000	180,000	155,000	155,000 r	105,000
Copper, metal, secondary[e]		20,000	20,000	20,000	20,000	15,000
Gallium[e]		13	13	13	13	11
Germanium[e]	kilograms	1,032 [3]	690 [3]	700	700	700
Iron and steel:						
Iron ore, marketable ore and concentrate:						
Gross weight		72,688,000	66,476,000	78,170,700	80,580,800	81,966,400
Fe content[e]		40,000,000	36,600,000	43,000,000	44,300,000	45,100,000
Metal:						
Pig iron		30,982,000	25,682,900	27,361,000	28,881,100	28,513,500
Ferroalloys, electric furnace:						
Ferromanganese		362,400	129,400	280,100	180,500	157,100
Ferronickel:[e]						
Gross weight		89,825 [3]	61,449 [3]	62,000	62,000	62,000
Ni content		16,224 [3]	12,392 [3]	12,400	12,400	12,400
Ferrosilicon		152,800	150,300	195,500	150,900	119,400
Silicomanganese		894,900	741,900	940,400	843,500	734,200
Other[e]		23,000	23,900	28,500	28,500	35,000
Total[e]		1,520,000	1,110,000	1,510,000	1,270,000	1,110,000
Steel:						
Crude		37,279,000	29,855,000	33,559,000	35,332,000	32,394,000
Finished products:						
Rolled		20,493,000	16,097,600	17,549,300	19,511,000	18,457,300
Pipe		2,542,000	1,742,000	1,928,400	2,371,800	2,014,000
Lead, refined, secondary[e]		7,000	7,000	7,000	13,500 r	13,700
Magnesium metal[e]		2,000	2,000	2,000	2,000	--
Manganese, marketable ore and concentrate:						
Gross weight		1,446,600 [4]	932,000 [4]	1,589,300 [4]	971,500	1,189,240
Mn content[e]		492,000	317,000	540,000	330,000	396,000
Manganese, metal		8,585	14,330	16,137	16,100 [e]	14,575
Nickel, laterite ore[e]		8,000	--	--	--	--
Titanium:						
Ilmenite concentrate:[e]						
Gross weight		520,000	500,000	500,000	260,700 r, 3	246,800 [3]
TiO$_2$ content, 59%		306,000	295,000	295,000	153,800 r, 3	145,640 [3]
Rutile concentrate, 95% TiO$_2$[e]		60,000	60,000	60,000	60,000	58,000
Metal, sponge[e]		9,930	6,830	7,400	9,000	8,500
Zirconium concentrates[e]		36,000	31,000	30,000	26,000	20,000
INDUSTRIAL MINERALS						
Bromine[e]		4,416 [3]	4,121 [3]	4,100	4,100	4,100
Cement, hydraulic		14,918,400	9,495,700	9,456,500	10,515,300	9,801,100
Clays:						
Ball clay[e]		650,000	600,000	600,000	600,000	600,000
Bentonite[e]		200,000	195,000	185,000	211,000	210,000
Kaolin	thousand metric tons	1,457	764	1,085	1,317	1,050
Kaolinitic clays	do.	318	354	306	575	580

See footnotes at end of table.

TABLE 1—Continued
UKRAINE: PRODUCTION OF MINERAL COMMODITIES[1]

(Metric tons unless otherwise specified)

Commodity[2]		2008	2009	2010	2011	2012
INDUSTRIAL MINERALS—Continued						
Feldspar		83,420	84,757	146,000	179,000	150,000 [e]
Graphite[e]		5,800	5,500	6,000	6,000	5,800
Gypsum		1,158,000	711,000	679,000	676,000	436,200
Lime	thousand metric tons	5,128	4,101	4,220	4,487	4,196
Limestone	do.	26,700	18,000	20,600	22,800	20,387
Nitrogen, N content of ammonia[e]	do.	4,000	2,500	3,400	4,300	4,160
Salt		4,425,000	5,395,000	4,908,000	5,938,000	6,189,446
Soda ash		977,800	680,000	706,700	700,000 [e]	720,000
Sulfuric acid	thousand metric tons	1,479	890	1,296	1,537	1,376
Sulfur, native[e]		135,000	120,000	130,000	130,000	120,000
Vermiculite[e]		65,000	55,000	60,000	60,000	60,000
MINERAL FUELS AND RELATED MATERIALS						
Coal, raw:[e]						
Anthracite	thousand metric tons	14,000	13,000	14,000	14,059 [r, 3]	20,763 [3]
Bituminous	do.	63,400	59,000	61,000	67,600 [r, 3]	64,690 [3]
Lignite	do.	200	200	200	200	200
Total	do.	77,600	72,200	75,200	81,900 [r]	85,700
Marketable	do.	59,312 [3]	54,820 [3]	54,444 [3]	62,700 [3]	66,700
Coke		19,543,000	17,424,000	18,599,700	19,599,100	18,939,100
Natural gas[5]	thousand cubic meters	21,467,000	21,545,000	20,458,000	19,934,900	19,318,300
Peat:						
Fuel use		358,000	449,000	321,000 [r]	301,000 [r]	263,000
Horticultural use		200,000 [e]	242,000	138,000 [r]	129,000 [r]	116,000
Total		558,000 [e]	691,000	459,000 [r]	430,000 [r]	379,000
Petroleum:						
Crude and gas condensate[6]	42-gallon barrels	30,300,000	28,500,000	25,400,000	24,000,000	24,110,000
Refinery products[7]	do.	83,700,000	85,700,000	80,300,000	69,000,000	35,508,900
Uranium, mine output:[e]						
U content		830	830	850	890	960
U_3O_8 content		980	980	1,000	1,050	1,130

[e]Estimated; estimated data are rounded to no more than three significant digits; may not add to totals shown. [r]Revised. do. Ditto. -- Zero.

[1]Table includes data available through February 20, 2014.

[2]In addition to the commodities listed, other mineral commodities may be produced, but available information was inadequate to make reliable estimates of output.

[3]Reported figure.

[4]Includes secondary production.

[5]The data series for natural gas production is based on natural gas production as reported by the State Statistics Service of Ukraine and includes associated petroleum gas production.

[6]Figures were converted to barrels from metric tons, which were reported as follows: 2008—4,168,300; 2009—3,916,600; 2010—3,493,400; 2011—3,297,800; and 2012—3,316,500.

[7]Figures were converted to barrels from metric tons, which were reported as follows: 2008—10,717,000; 2009—10,947,000; 2010—10,333,000; 2011—8,787,000; and 2012—4,570,000.

TABLE 2
UKRAINE: STRUCTURE OF THE MINERAL INDUSTRY IN 2012

(Metric tons unless otherwise specified)

Commodity	Major operating companies and major equity owners[1,2]	Location or deposit names	Annual capacity[e]
Alumina and aluminum:			
Alumina	Nikolaevskiy alumina refinery [United Company RUSAL (RUSAL)]	20 kilometers south of Mykolaiv	1,601,000
Do.	Zaporozhye refinery [United Company RUSAL (RUSAL)]	Zaporizhia	275,000
Aluminum, primary	Zaporozhye smelter [United Company RUSAL (RUSAL)]	do.	114,000
Coal	About 150 active surface and underground mines, including:	About 95% of coal produced in Donets'ka, Dnipropetrovs'ka, and Luhans'ka Oblasts	90,000,000 [3]
	Donbass Fuel and Energy Co. (DTEK) (System Capital Management, 100%):		
	DTEK Pavlogradugol	10 mines in Dnipropetrovs'ka and Donets'ka Oblasts	
	DTEK Komsomolets Donbassa Mine	Kirovskoe, Donets'ka Oblast'	
	DTEK Dobropolyeugol	5 mines near Dobropillya, Donets'ka Oblast'	
	DTEK Sverdlovanthracite	5 coal mines and 3 processing plants in Luhans'ka Oblast'	
	DTEK Rovenkyanthracite	6 mines and 3 processing plants in Luhans'ka Oblast'	
	Krasnoarmeiskaya-Zapadnaya No. 1	1 mine at Krasnoarmeisk, Donets'ka Oblast'	
	JSC Krasnodon Coal Co. (Metinvest B.V.)	7 mines and 2 processing plants in Luhans'ka Oblast'	
	Smaller producers	Donets'ka, Dnipropetrovs'ka, Luhans'ka, Lvivs'ka, and Volyns'ka Oblasts	
Coke	Evraz Group:	Dnipropetrovs'ka Oblast':	3,000,000
	OAO Dneprkoks coke plant	Dnipropetrovsk	
	OAO Bagliykoks coke plant	Dniprodzerzhinsk	
	OAO Dneprodzerzhinsk coke plant	do.	
Do.	Metinvest B.V.:		
	JSC Avdiivka coke plant	Avdeyevka, Donets'ka Oblast'	4,000,000
Do.	JSC Azovstal Iron and Steel Works	Mariupol, Donets'ka Oblast'	3,182,000
Do.	OJSC ArcelorMittal Kryviy Rih	Kryviy Rih, Dnipropetrovs'ka Oblast'	3,300,000
Do.	JSC Donetskkoks (Metinvest B.V., 24.5%, and OJSC Ilyich Iron and Steel Works, 12.96%)	Donetsk, Donets'ka Oblast'	390,000
Do.	Yenakievo coke plant	Yenakievo, Donets'ka Oblast'	NA
Do.	OAO Zaporozhkoks (JSC Zaporizhstal, 42%, and Metinvest B.V., 25%)	Zaporizhia	NA
Do.	Makeevka coke plant	Makeevka, Donets'ka Oblast'	NA
Do.	OAO Yasinovskiy coke plant	do.	NA
Do.	OAO Alchevsk coke plant [Industrial Union of Donbass (ISD Corp.)]	Alchevsk, Luhans'ka Oblast'	3,700,000
Do.	Horlivka coke plant	Horlivka, Donets'ka Oblast'	440,000
Do.	Kharkov coke plant	Kharkiv	225,000
Ferroalloys:			
Ferromanganese	Zaporozhye ferroalloys plant (PrivatBank Group)	Zaporizhia	100,000
Do.	Nikopol ferroalloys plant (PrivatBank Group and EastOne Group)	Nikopol	300,000
Do.	Stakhanov ferroalloys plant (PrivatBank Group)	Luhans'ka Oblast'	NA

See footnotes at end of table.

TABLE 2—Continued
UKRAINE: STRUCTURE OF THE MINERAL INDUSTRY IN 2012

(Metric tons unless otherwise specified)

Commodity	Major operating companies and major equity owners[1, 2]	Location or deposit names	Annual capacity[e]
Ferroalloys—Continued:			
Ferromanganese, blast furnace	Konstantinovka Iron and Steel Works	Konstyantynivka, Donets'ka Oblast'	NA [4]
Do.	Kramatorskiy ferroalloys plant	Kramatorsk, Donets'ka Oblast'	NA
Ferronickel	Pobuzhskiy ferronickel plant	Pobuzhye, Kirovohrads'ka Oblast'	100,000
Ferrosilicon	Stakhanov ferroalloys plant (PrivatBank Group)	Luhans'ka Oblast'	120,000
Do.	Zaporozhye ferroalloys plant (PrivatBank Group)	Zaporizhia	100,000
Silicomanganese	Stakhanov ferroalloys plant (PrivatBank Group)	Luhans'ka Oblast'	50,000
Do.	Zaporozhye ferroalloys plant (PrivatBank Group)	Zaporizhia	250,000
Do.	Nikopol ferroalloys plant (PrivatBank Group and EastOne Group)	Nikopol	900,000
Gallium	Nikolaev alumina refinery [United Company RUSAL (RUSAL)]	20 kilometers south of Mykolaiv	13
Germanium	Zaporozhye titanium-magnesium plant	Zaporizhia	19,000
Graphite	Zavalyevskiy graphite complex	Zavalyevskiy deposit	NA
Iron ore:			
Underground mining	Krivorozhskiy Iron Ore Complex (Metinvest B.V., 50%, and PrivatBank Group, 50%)	4 mines, in Kryvorizkiy iron ore basin	7,000,000
Do.	Sukha Balka (Evraz Group)	2 mines in Dnipropetrovs'ka Oblast'	3,100,000
Do.	PJSC ArcelorMittal Kryviy Rih	2 mines at Kryviy Rih	1,500,000
Do.	Zaporozhye Iron Ore Complex	Eksplutatsionnay Mine in Zaporiz'ka Oblast'	4,500,000
Do.	JSC Central Iron Ore Enrichment Works (Metinvest B.V.)	1 mine in Dnipropetrovs'ka Oblast'	2,200,000
Open pit mining	do.	3 mines in Dnipropetrovs'ka Oblast'	12,000,000
Do.	JSC Northern Iron Ore Enrichment Works (Metinvest B.V.)	2 mines in Dnipropetrovs'ka Oblast'	30,000,000
Do.	JSC Ingulets Iron Ore Enrichment Works (Metinvest B.V.)	Ingulets mine south of Kryviy Rih	35,000,000
Do.	Yuzhniy GOK (Evraz Holding, 50%, and Smart Holding, 50%)	Mine at Kryviy Rih	22,000,000
Do.	PJSC ArcelorMittal Kryviy Rih	2 mines at Kryviy Rih	26,550,000
Do.	Poltava GOK (Ferrexpo Plc.)	Gorishne-Plavninskoye and Lavrikovskoye (GPL) Mine 15 kilometers east of Kremenchug	30,000,000
Lead, secondary	CJSC Svinets	Kostyantynivka	20,000
Magnesium metal	Magnii concern	Kalush	22,000
Manganese:			
Ore, marketable	Ordzhonikidze GOK (PrivatBank Group)	Ordzhonikidze	700,000
Do.	Marganets GOK (PrivatBank Group)	Marhanets	NA
Metal	Zaporozhye ferroalloys plant (PrivatBank Group)	Zaporizhia	NA
Mercury	OOO Nikitrtyt	Horlivka, Donets'ka Oblast'	300
Natural gas	Yuzovskoye deposit (Royal Dutch Shell plc)	Kharkiv and Donets'ka Oblasts	NA
Do.	Olesskoye deposit (Chevron Corp.)	Lvivs'ka and Ivano-Frankovs'ka Oblasts	NA
Nickel, Ni content in FeNi	Pobuzhskiy GOK (comprises three open pit mines and the Pobuzhskiy ferronickel plant)	Pobuzhye, Kirovohrads'ka Oblast	20,000
Petroleum, refined 42-gallon barrels	Kherson oil refining plant	Kherson	NA
Do. do.	Odessa refinery (OAO Lukoil)	Odessa	23,000,000 [5]
Do. do.	Lisichansk refinery (TNK-BP)	Lisichansk	62,000,000
Do. do.	Halychyna refinery (Ukraine Oil Co.)	Drohobych, Lvivs'ka Oblast'	28,600,000
Do. do.	Kremenchug refinery (CJSC Ukrtatnafta)	Kremenchug	150,000,000
Do. do.	JSC Naftokhimik Prykarpattya	Nadvirna, Ivano-Frankivs'ka Oblast'	18,400,000
Do. do.	Shebelinka refinery	Shebelinka, Kharkivs'ka Oblast'	NA
Steel, crude	Industrial Union of Donbass Corp. (ISD Corp.):		
	OJSC Alchevsk Iron and Steel Works	Alchevsk, Luhans'ka Oblast'	5,200,000
Do.	Dneprovskiy Metallurgical Plant "Dzerzhinsky"	Dniprodzerzhinsk	3,850,000

See footnotes at end of table.

TABLE 2—Continued
UKRAINE: STRUCTURE OF THE MINERAL INDUSTRY IN 2012

(Metric tons unless otherwise specified)

Commodity	Major operating companies and major equity owners[1, 2]	Location or deposit names	Annual capacity[e]
Steel, crude—Continued	OJSC ArcelorMittal Kryviy Rih	Kryviy Rih, Dnipropetrovs'ka Oblast'	7,400,000
Do.	Metinvest B.V.:		
	JSC Azovstal Iron and Steel Works	Mariupol, Donets'ka Oblast'	6,200,000
Do.	JSC Yenakiieve Iron and Steel Works	Yenakievo, Donets'ka Oblast'	2,700,000
Do.	OJSC Ilyich Iron and Steel Works	Mariupol, Donets'ka Oblast'	6,000,000
Do.	Dnepropetrovsk Metals Plant "Petrovskovo" (DMZP) (Evraz Group S.A., 96.77%)	Dnipropetrovsk	1,360,000
Do.	JSC Zaporizhstal (Metinvest B.V., 24.9%) (Mechel OAO)[6]	Zaporizhia	4,350,000
Do.	Kramatorskiy Metal Plant "Kuibiysheva"	Kramatorsk, Donets'ka Oblast'	NA
Do.	Donetskstal	Donetsk	NA
Do.	Donetsk electrometallurgical plant	do.	1,000,000 [6]
Do.	Dneprospetsstal	Zaporizhia	918,000
Do.	OOO Elektrostal	Kurakhovo, Donets'ka Oblast'	NA
Do.	JSC Energomashspetsstal (OJSC Atomenergomash)	Kramatorsk, Donets'ka Oblast'	NA
Do.	PJSC Azovelectrostal (JSC Azovmash)	Mariupol, Donets'ka Oblast'	500,000
Titanium:			
Concentrate:			
Ilmenite	Irshansk GOK [Leased from the Government by Crimea Titan CJSC (Government, 50% plus one share, and OstChem GmbH, 50% minus one share)]	Irshansk, 50 kilometers north of Zhytomyr	400
Do.	OOO Valki-Ilmenit (OstChem GmbH, 75%)	do.	70
Do.	Mezhdurechensk GOK (OstChem GmbH, 75%)	Zhytomyrs'ka Oblast'	84
Do.	Velta LLC	Korobchino, Novomirgorod district, Kirovograds'ka Oblast'	185 [7]
Do.	Volnogorsk state mining-metals complex [Leased from the Government by Crimea Titan CJSC (Ukraine Government, 50% plus one share, and OstChem GmbH, 50% minus one share)]	Volnogorsk, 70 kilometers west of Dnipropetrovsk	200
Do.	Demurinskiy GOK (Limpeza Ltd. of Cyprus 25%, and VSMPO-Avisma of Russia, 75%)	Dnipropetrovs'ka Oblast'	NA
Rutile	do.	do.	65
Sponge	Zaporozhye Titanium & Magnesium Complex (ZTMK) (Government, 51%, and Tolexis Trading Ltd., 49%)	Zaporizhia	NA
Ingots	OOO Antares	Kyev	NA
Do.	OOO Fico	do.	NA
Do.	Zaporozhye Titanium & Magnesium Complex (ZTMK) (Government, 51%, and Tolexis Trading Ltd., 49%)	Zaporizhia	NA
Titanium dioxide pigment	Crimea Titan CJSC	Crimea	NA
Do.	OAO Sumykhimprom	Sumy	NA
Uranium:			
Ore thousand metric tons	Vostochny GOK (Government)	Ingulskaya Mine at Kirovohrad	450
Do. do.	do.	Smolinskaya Mine at Smolino	600
Do. do.	do.	Novokonstantinovskoye deposit in Kirovohrads'ka Oblast'	100
Concentrate	do.	Hydrometallurgical concentration plant at Zheltye Vody	1,000
Zinc, secondary	Ukrzinc plant	Kostyantynivka	25,000
Do.	CJSC Svinets	do.	30,000

See footnotes at end of table.

TABLE 2—Continued
UKRAINE: STRUCTURE OF THE MINERAL INDUSTRY IN 2012

(Metric tons unless otherwise specified)

Commodity	Major operating companies and major equity owners[1,2]	Location or deposit names	Annual capacity[e]
Zirconium:			
Concentrate	Volnogorsk state mining-metals complex [Leased from the Government by Crimea Titan CJSC (Ukraine Government, 50% plus one share, and OstChem Gmbh, 50% minus one share)]	Volnogorsk, 70 kilometers west of Dnipropetrovsk	35
Metal and compounds	State Research and Production Enterprise "Zirconium"	Dniprodzerzhinsk	NA

[e]Estimated; estimated data are rounded to no more than three significant digits. Do., do. Ditto. NA Not available.

[1]Inconsistencies in enterprise and location names may appear in this table because both Ukrainian and Russian spellings were used for transliterations. English versions of company names are used as given by official company sources (Web sites, press releases, and so forth). Ukrainian versions of location names are used wherever possible.

[2]GOK is the abbreviation for gorno-obogotitelniy kombinat, which translates as "mining and beneficiation complex."

[3]Capacity estimates are totals for all enterprises that produce that commodity.

[4]Konstantinovka Iron and Steel Works stopped production of blast furnace ferromanganese in 2008.

[5]The Odessa refinery stopped production in the fourth quarter of 2010. Production could restart in the future if business conditions improve.

[6]In December 2011, Mechel OAO of Russia purchased 100% of the shares of the Donetsk electrometallurgical plant.

[7]Velta LLC began production of ilmenite concentrate in December 2011, but its first deliveries of commercial concentrate were not made until April 2012.

THE MINERAL INDUSTRY OF THE UNITED KINGDOM

By Alberto Alexander Perez

In 2012, the United Kingdom's gross domestic product (GDP) was $2.443 trillion in real terms at official exchange rates. This was an increase of 0.2% compared with that of 2011. Within the European Union (EU), the United Kingdom's economy was ranked second after Germany's in terms of GDP based on purchasing power parity. The country's heavy industry was composed of companies that produced automotive and aviation products, chemicals, and machine tools. These industries relied on many imported metal ores, concentrates, and refined metals, as well as on imported industrial minerals and mineral fuels. The mineral fuels sector, which included coal, natural gas, and petroleum, was a significant part of the United Kingdom's mineral industry (U.S. Central Intelligence Agency, 2014).

In 2011 (the latest year for which data were available), the country accounted for about 2.4% of the world's refined nickel production (including nickel content of chemicals). In 2012, the United Kingdom produced 2.1% of the world's crude salt output and 1.26% of the world's potash output (Jasinski, 2013; Kostick, 2013; Kuck, 2013, p. 51.21).

Minerals in the National Economy

The United Kingdom's mineral sector served domestic economic needs, but the country's mining and processing companies also played an important role in global mineral prospecting, development, and production, and in mineral commodity trade. The London Metal Exchange remained the world's leading central market for nonferrous metals.

Government Policies and Programs

The 1971 Minerals Act, as amended, is the statute that governs the development and exploitation of mineral deposits. Minerals, as defined in Section 209 of the Act, include all minerals and materials in or under the land of a kind ordinarily worked for removal by underground or surface workings; they do not, however, include peat cut for purposes other than for sale. Mineral development is specifically addressed in the Town and Country Planning (Minerals) Regulations, 1971, and the Town and Country Planning (Minerals) Act, 1981. Mineral rights to mineral fuels, such as coal, petroleum, and uranium, belong to the state. The Coal Authority is authorized to license open pit and underground mines to the private sector subject to restrictions on their size and the payment of a royalty on the amount of coal produced.

Most other mineral rights in England, Scotland, and Wales are privately owned with the exceptions of gold and silver, which are vested in the Royal Family. A different situation regarding mineral rights applies to Northern Ireland where, under the Mineral Development Act (Northern Ireland), 1969, the rights to work minerals and to license others to do so are vested in the state.

The Government of the United Kingdom had ratified the Kyoto Protocol. The European Commission, however, decided that the requirements under the Kyoto Protocol would be met by the EU as a whole rather than as individual signatories, and each member state was subsequently given a different emissions target by the European Commission (British Geological Survey, 2010).

Production

In 2012, the production of primary aluminum decreased by an estimated 71.8%, and the production of both crude steel and pig iron increased slightly. In the industrial minerals sector, production remained at about the same level as in 2011. Natural gas production decreased by an estimated 13.6%, and crude petroleum production decreased by 12.99%, which continued the United Kingdom's trend of decreasing production of these commodities (table 1).

Structure of the Mineral Industry

Domestic and foreign-owned corporations produced minerals and mineral-based commodities. Table 2 is a list of major mineral industry facilities.

Mineral Trade

The United Kingdom was a net importer of coal, iron and steel, natural gas, crude petroleum, and petroleum products in 2011 (the latest year for which data were available). It exported metal manufactures valued at $7.78 billion.[1] The value of the country's iron and steel and nonferrous metals exports was $9.42 billion and $13.4 billion, respectively. The United Kingdom became a net importer of natural gas in 2004, of crude petroleum in 2005, and of refined petroleum products in 2006 after many previous years of self-sufficiency (Office for National Statistics, 2012, p. 12, 35).

Commodity Review

Metals

Aluminum.—In 2012, primary aluminum production in the United Kingdom decreased to 60,000 metric tons (t) from 213,000 t owing to the closure in March of Rio Tinto Alcan Ltd.'s smelter, which was located in Lynemouth. Rio Tinto Alcan indicated that the high cost of power and new regulations imposed on the site were the main reasons for the permanent shutdown. The plant had a stated capacity of 182,000 metric tons per year (Rio Tinto plc, 2011; 2012, p. 17).

---

[1]Where necessary, values have been converted from the British pound sterling (£) to U.S. dollars (US$) at the average exchange rate of £0.637=US$1.00 for 2012.

Gold.—The number of licenses for exploration and development of gold mines in the United Kingdom was 24. The number of leases remained constant at four. The more active areas of exploration were located in Cononish in Perthshire County, Scotland, and in County Armagh and in Omagh, County Tyrone, Northern Ireland. In Scotland, Scotgold Resources plc of Australia had licenses from Mines Royal (the license-granting authority in the United Kingdom) for the areas around Glen Lyon, Glen Orchy, and Inverliever, and the company owned the gold and silver assets of the Cononish deposit near Tyndrum (British Geological Survey, 2011, p. 49–50).

In Northern Ireland, the Omagh (formerly Cavanacaw) deposit located 10 kilometers southwest of the town of Omagh was owned by Omagh Minerals Ltd., which was a wholly owned subsidiary of Galantas Gold Corp. of Canada. The deposit had a proven and probable reserve of 367,310 t grading 7.52 grams per metric ton gold across a width of 4.43 meters within the designated open pit operation. Galantas had been granted exploration licenses to the west and north of its existing license and currently held licenses for an area totaling 460 square kilometers. Conroy Diamonds and Gold plc was exploring in the Clontibret district, which is located on the border of Northern Ireland and the Republic of Ireland near Co. Monaghan (British Geological Survey, 2011, p. 49–50).

Industrial Minerals

The United Kingdom remained a significant producer of such minerals as barite and calcareous material for cement, clays, and fluorspar.

Barite.—The United Kingdom's barite production was dominated by output from M–I Drilling Fluids (UK) Ltd., which operated the underground Foss Mine located near Aberfeldy in Perthshire County, Scotland. The production of this mine accounted for most of the barite production in the United Kingdom; the remaining output of barite was from the Southern Pennine Orefield, where barite was derived as a byproduct of fluorspar mining. Barite production from the Foss Mine decreased by 6% in 2011 (British Geological Survey, 2011, p. 25; 2012, p. 25).

Fluorspar.—Glebe Mines Ltd. was the United Kingdom's only domestic producer of fluorspar (calcium fluorite), and it supplied the country's two fluorochemical producers with acid-grade fluorspar. Glebe's operations were based on surface extraction and processing in the Southern Pennine Orefield. Glebe operated the Cavendish Mill near Stoney Middleton to produce acid-grade fluorspar together with byproduct barite and lead concentrate. Glebe Mines' ore reserves were estimated to be about 1.2 million metric tons. The British Geological Survey reported no production of fluorspar for 2011 and 2012 (British Geological Survey, 2011, p. 48; 2012, p. 43).

Mineral Fuels

Coal.—In 2011, the number of Coal Authority licenses for opencast sites in production totaled 35 and included 19 in Scotland, 9 in England, and 7 in Wales. Scottish Coal Co. Ltd. was the leading opencast coal mining company in the United Kingdom and the second-ranked net coal producer (British Geological Survey, 2011, p. 33).

The generation of electricity accounted for the majority of the country's total coal consumption. About one-third of all electricity generated in the United Kingdom was supplied by coal (British Geological Survey, 2011, p. 33).

Natural Gas and Petroleum.—In May, Maersk Oil UK Ltd. (Maersk) signed an agreement with Noble Energy Inc. for the purchase of 30% of its interest in the Maersk-operated Dumbarton and Lochranza fields. Maersk reported that this agreement also included control of the Global Producer III floating production storage and offloading installation in the central North Sea. With this investment, Maersk would hold a 100% interest in the Dumbarton and the Lochranza fields. Maersk stated that it had paid Noble Energy Inc. $127 million for the assets. The Dumbarton and the Lochranza fields produced a combined output of about 20,000 barrels per day of oil equivalent (Maersk Oil Ltd., 2012).

The Buzzard oilfield in the Outer Moray Firth was the most prolific oilfield on the United Kingdom Continental Shelf (UKCS) in 2012. The North Sea holds Europe's largest natural gas and petroleum reserves. At the end of 2013, the United Kingdom's estimated proven crude oil reserves totaled 3.1 billion barrels, which was the largest within the EU; the reserves were located mostly offshore on the UKCS. Most of the country's production came from basins located east of Scotland in the central North Sea. The northern North Sea, east of the Shetland Islands, also contains considerable reserves, and smaller deposits are located in the North Atlantic Ocean. Besides these offshore assets, the country had the Wytch Farm field, which was the largest onshore oilfield in Europe (British Geological Survey, 2011, p. 75; U.S. Energy Information Administration, 2014).

Outlook

The United Kingdom's mineral production remained concentrated in the industrial minerals and energy sectors. Overall, its manufacturing industry is likely to continue to import the raw minerals that it needs for the foreseeable future. The country is expected to continue to be a leading European producer of crude oil and refined products, but the decline in domestic production is likely to continue. Exploration for gold and mixed sulfide ores will likely continue, particularly in Northern Ireland. Production of steel will likely remain stable and continue to reflect international trends in demand.

References Cited

British Geological Survey, 2010, Minerals UK, Legislation and policy: London, United Kingdom, British Geological Survey. (Accessed December 8, 2012, at http://www.bgs.ac.uk/mineralsuk/planning/legislation/home.html.)
British Geological Survey, 2011, United Kingdom minerals yearbook 2010: Keyworth, Nottingham, United Kingdom, British Geological Survey, 108 p.
British Geological Survey, 2012, United Kingdom minerals yearbook 2011: Keyworth, Nottingham, United Kingdom, British Geological Survey, 89 p.
Jasinski, S.M., 2013, Potash: U.S. Geological Survey Mineral Commodity Summaries 2013, p. 122–123.
Kostick, D.S., 2013, Salt: U.S. Geological Survey Mineral Commodity Summaries 2013, p. 134–135.

Kuck, P.H., 2013, Nickel, *in* Metals and minerals: U.S. Geological Survey Minerals Yearbook 2011, v. I, p. 51.1–51.31.

Maersk Oil Ltd., 2012, Maersk Oil buys 30% interest in Dumbarton and Lochranza fields in UK North Sea: Maersk Oil Ltd. (Accessed January 5, 2014, at http://www.maerskoil.com/Media/Newsroom/Pages/MaerskOilbuys30interestinDumbartonandLochranzafieldsinUKNorthSea.aspx.)

Office for National Statistics [United Kingdom], 2012, Monthly review of external trade statistics: Office for National Statistics, December, 77 p. (Accessed December 8, 2012, at http://www.ons.gov.uk/ons/rel/uktrade/monthly-review-of-external-trade-statistics/december-2011/bdl-monthly-review-of-external-trade-statistics-december-2011.pdf.)

Rio Tinto plc, 2011, Rio Tinto Alcan announces intention to close Lynemouth aluminium smelter: London, United Kingdom, Rio Tinto plc media release, November 16, 1 p. (Accessed January 5, 2014, at http://www.riotinto.com/documents/PR932g_Rio_Tinto_Alcan_announces_intention_to_close_Lynemouth_aluminium_smelter.pdf.)

Rio Tinto plc, 2012, Second quarter 2012 operations review: London, United Kingdom, Rio Tinto plc media release, July 17, 25 p. (Accessed January 5, 2014, at http://www.riotinto.com/documents/PR969g_Second_quarter_2012_operations_review.pdf.)

U.S. Central Intelligence Agency, 2014, United Kingdom, *in* The world factbook: U.S. Central Intelligence Agency. (Accessed January 5, 2014, at https://www.cia.gov/library/publications/the-world-factbook/geos/uk.html.)

U.S. Energy Information Administration, 2014, United Kingdom: U.S. Energy Information Administration country analysis brief, May, 13 p. (Accessed January 5, 2014, at http://www.eia.gov/countries/cab.cfm?fips=UK.)

TABLE 1

UNITED KINGDOM: PRODUCTION OF MINERAL COMMODITIES[1]

(Metric tons unless otherwise specified)

Commodity		2008	2009	2010	2011	2012
METALS						
Aluminum, metal:						
Primary		326,000	253,000	186,000	213,000 [e]	60,000
Secondary		205,200	288,397	311,741	301,250	300,000 [e]
Total		531,200	541,397	497,741	514,000 [e]	360,000
Iron and steel:						
Iron ore and concentrate, manganiferous:						
Gross weight		100	--	--	--	--
Fe content, 54% Fe		54	--	--	--	--
Metal:						
Pig iron	thousand metric tons	10,137	7,671	7,233	6,625	7,183
Steel:						
Crude	do.	13,500	10,079	9,709	9,478	9,579
Hot-rolled	do.	9,517	7,091	8,395	7,963	7,042
Lead:						
Mine output, Pb content[e]		300	300	300	300	100
Metal:						
Smelter, bullion from imported concentrate[e]		36,000	36,000	36,000	36,000	36,000
Refined:[e]						
Primary[2]		139,000 [3]	135,000	150,000	150,000	157,000
Secondary[4]		144,000 [3]	144,000	144,000	144,000	155,000
Total		283,000 [3]	279,000	294,000	294,000	312,000
Nickel, metal[5]		41,000	38,700	38,000	37,400 [r]	34,300
INDUSTRIAL MINERALS						
Barite[6]		43,000	36,000 [e]	33,000	31,000	30,000
Cement, hydraulic[e]	thousand metric tons	10,071 [3]	7,800 [r]	7,900 [r]	8,500 [r]	7,900
Clays:						
Fire clay[e]	do.	180	129	150	100	150
Kaolin, china clay[7]	do.	1,355	1,060 [e]	900	1,000	1,150
Ball clay and pottery clay[e, 8]	do.	1,020	727	1,000	1,000	750
Other, including shale	do.	8,459	5,310	4,721	5,483	5,800
Feldspar, china stone[e]		500	400	500	500	--
Fluorspar, all grades[e, 9]		37,000	19,000	26,000	--	--
Gypsum and anhydrite[e]	thousand metric tons	1,700	1,700	1,700	1,700	1,700
Lime, hydrated and quicklime[e]	do.	1,500	1,500	1,500	1,500	1,500
Nitrogen, N content of ammonia[e]	do.	1,100	1,100	1,100	1,100	1,100
Potash, KCL product		673,000	673,000 [e]	700,000	770,000	770,000 [e]

See footnotes at end of table.

TABLE 1—Continued
UNITED KINGDOM: PRODUCTION OF MINERAL COMMODITIES[1]

(Metric tons unless otherwise specified)

Commodity		2008	2009	2010	2011	2012
INDUSTRIAL MINERALS—Continued						
Salt:[e]						
Rock	thousand metric tons	2,000	2,000	2,000	2,000	2,200
From brine	do.	1,000	1,000	1,000	1,000	1,000
In brine, sold or used as such	do.	2,800	2,800	2,800	2,800	2,800
Sand and gravel, common sand and gravel	do.	85,473	65,800	66,800	62,000	61,000
Sodium compounds, carbonate, n.e.s.[e, 10]	do.	1,000	1,000	1,000	1,000	1,000
Stone:						
Chalk	do.	5,874	4,047	3,626	3,996 [r]	3,600
Dolomite	do.	5,509	3,164	4,540	4,490 [r]	4,400
Igneous rock	do.	53,489	44,618	44,876	44,490 [r]	40,000
Limestone	do.	74,143	60,111	60,207	57,930 [r]	57,000
Sandstone	do.	12,255	12,335	11,556	12,477 [r]	11,000
Slate, including fill	do.	1,058	683	695	763 [r]	750
Total	do.	152,328	124,958	125,500	124,146 [r]	116,750
Talc, soapstone, pyrophyllite		2,000	3,000	3,000	4,000	4,000
Titanium, titanium dioxide[e]	thousand metric tons	200	200	200	200	200
MINERAL FUELS AND RELATED MATERIALS						
Coal, anthracite and bituminous	thousand metric tons	17,912	18,054	18,159	18,492	16,788
Coke:[e]						
Metallurgical	do.	4,000	4,000	4,000	4,000	4,000
Breeze, all types	do.	250	250	250	250	250
Gas, natural, marketable[11]	billion cubic meters	74	68	59	66 [r]	57 [e]
Peat[e]	cubic meters	760	887	1,004	825 [r]	800 [e]
Petroleum:						
Crude[12]	thousand 42-gallon barrels	507,850	484,643	430,791	422,568	368,139
Refinery products	do.	612,632	578,014	565,422	565,000 [e]	565,000 [e]

[e]Estimated; estimated data are rounded to no more than three significant digits; may not add to totals shown. [r]Revised. do. Ditto. -- Zero.
[1]Table includes data available through January 5, 2014.
[2]Produced entirely from imported bullion and includes the lead content of alloys.
[3]Reported figure.
[4]Includes a small quantity of primary lead from domestic concentrate.
[5]Refined nickel.
[6]Includes witherite.
[7]Sales, dry weight.
[8]Salable product.
[9]Proportions of grades not available; probably about two-thirds acid grade.
[10]Not elsewhere specified.
[11]Methane, excluding gas flared or reinjected.
[12]Excludes gases and condensates.

TABLE 2
UNITED KINGDOM: STRUCTURE OF THE MINERAL INDUSTRY IN 2012

(Thousand metric tons unless otherwise specified)

Commodity	Major operating companies and major equity owners	Location of main facilities[1]	Annual capacity
Aluminum:			
Primary	Rio Tinto Alcan Ltd.	Lynemouth Smelter, Northumberland County, England (closed in March 2012)	182
Do.	do.	Locchaber Smelter, Fort William County, Scotland	41
Do.	Anglesey Aluminium Metal Ltd. (Rio Tinto Corp., 51%, and Kaiser Aluminum and Chemical Corp., 49%)	Holyhead, Gwynedd County, Wales	144
Secondary	Hydro Aluminium Deeside Ltd. (Hydro Aluminium AS)	Wrexham, Clwyd County, Wales	60
Do.	Cohen Alloys Ltd.	Glasgow, Scotland	NA
Do.	Coleshill Aluminium Ltd.	Coleshill, Warwickshire, England	NA
Do.	Dolgarrog Aluminium Ltd.	Dolgarrog, Conwy, Gwynedd County, Wales	9
Barite	M–I Drilling Fluids (UK) Ltd.	Foss Mine, near Aberfeldy, Perthshire County, Scotland	50
Do.	Glebe Mines Ltd. (on care-and-maintenance status by beginning of 2011)	Arthurton West, Bow Rake, High Rake, and Watersaw Mines, Southern Pennine Orefield, Derbyshire County, England	15
Celestite	Bristol Minerals Co. Ltd.	Yate, Avon County, England	30
Cement	Lafarge Cement UK, Ltd. (Lafarge Group)	Aberthaw plant, East Aberthaw, Barry, South Glamorgan County, Wales	500
Do.	do.	Barnstone plant, near Langar, Nottinghamshire County, England	--[2]
Do.	do.	Cauldon plant, near Leek, Staffordshire County, England	1,000
Do.	do.	Cookstown plant, Cookstown, County Tyrone, Northern Ireland	500
Do.	do.	Dunbar plant, Dunbar, East Lothian, Scotland	1,000
Do.	do.	Hope plant, Hope Valley, Derbyshire County, England	1,300
Do.	do.	Northfleet plant, Northfleet, Kent County, England	1,000
Do.	do.	Westbury plant Westbury, Wiltshire County, England	700
Do.	Castle Cement Ltd. (HeidelbergCement AG, 100%)	Ketton plant, Rutland County, near Stamford, Lincolnshire County, England	1,400
Do.	do.	Padeswood plant, Mold, Flintshire County, Wales	1,400
Do.	do.	Ribblesdale plant, Clitheroe, Lancashire County, England	1,400
Do.	CEMEX UK Operations, Ltd. (CEMEX, S.A.B. de C.V., 100%)	Rugby plant, Rugby, Warwickshire County, England	1,800
Do.	do.	Barrington plant, Barrington, Cambridgeshire County, England	300
Do.	do.	South Ferriby plant, North Lincolnshire County, England	800
Do.	Tarmac Buxton Lime and Cement Industries Ltd.	Tunstead plant, Buxton, Derbyshire County, England	800
Clay:			
Ball clay	WBB Minerals (S.C.R.-Sibelco NV)	Various operations in northern and southern Devon County, England	500
Do.	Imerys Group	Operations in Bovey and Wareham Basins, Dorset County, England	300
China clay (kaolin)	do.	Mines and plants in Cornwall and Devon Counties, England	3,000
Do.	WBB Minerals (S.C.R.-Sibelco NV)	Mines and plants in Cornwall County, England	1,000

See footnotes at end of table.

TABLE 2—Continued
UNITED KINGDOM: STRUCTURE OF THE MINERAL INDUSTRY IN 2012

(Thousand metric tons unless otherwise specified)

Commodity		Major operating companies and major equity owners	Location of main facilities[1]	Annual capacity
Coal:				
Underground mines		UK Coal plc	Operations in England include the Daw Mill Colliery, Warwickshire County; the Kellingley Colliery, North Yorkshire County; the Maltby Colliery, Rotherham, Yorkshire County; the Thoresby Colliery, Nottinghamshire County; the Welbeck Colliery, Nottinghamshire County	30,000
Do.		Goitre Tower Colliery Ltd.	Tower Colliery, Hirwaun, Mid Glamorgan County, Wales	500
Surface pits		Scottish Coal Company Ltd.	Operations in Scotland include the Broken Cross pit near Douglas, South Lanarkshire County; Chalmerston pit, Dalmellington, East Ayrshire County; Chapelhill pit, South Lanarkshire County; Glentaggart pit, near Douglas, South Lanarkshire; Newbigging Farm pit, near Howgate, Midlothian County; Powharnal pit, near Muirkirk, East Ayrshire County; St. Ninians (Greenbank) pit, northeast of Dunfermline, Fife	4,000
Do.		ATH Resources PLC	Operations in Scotland include the Grievehill, the Laigh Glenmuir, and the Skares road pits in Ayrshire County; Glenmuckloch pit, Dumfries and Galloway County	1,600
Do.		Celtic Energy Ltd.	Margam pit, near Bridgend, Mid Glamorgan County, Wales	1,000
Do.		do.	Nant Helen Extension pit, Abercraf, West Glamorgan, Wales	400
Do.		do.	Selar pit, Glynneath, West Glamorgan, Wales	400
Do.		Energybuild Ltd.	Nant-y-Mynydd pit, Neath, West Glamorgan, Wales	130
Do.		H.J. Banks Mining (Banks Group)	Dehli pit, Stannington, Northumberland County, England	NA
Fluorspar		Glebe Mines Ltd.	Mill at Stoney Middleton, mines in Derbyshire County, England	60
Gold	kilograms	Galantas Gold Corp.	Omagh Mine, near Omagh, County Tyrone, Northern Ireland	900 [3]
Gypsum		British Gypsum Ltd.	Several mines and quarries in England, which include the Barrow Mine, Barrow upon Soar, southeast of Loughborough, Leicestershire County; the Brightling Mine, Robertsbridge, East Sussex County; the Birkshead Mine, Kirby Thore, near Penrith, Cumbria County; the Fauld Mine, Tutbury, near Burton on Trent, Staffordshire County; the Kilvington Quarry, Staunton in the Vale, Kilvington, Nottinghamshire County; the Marbleegis Mine, East Leake, northeast of Loughborough, Leicestershire County; the Newbiggin Mine, Newbiggin, near Kirby Thore, Cumbria County	3,500
Lead:				
Primary		Britannia Refined Metals Ltd. (Xstrata plc)	Northfleet, Kent County, England	180
Secondary		Britannia Recycling Ltd. (Xstrata plc)	Wakefield, West Yorkshire County, England	20
Do.		H.J. Enthoven Ltd. (Quexco Inc, 100%)	Darley Dale, Derbyshire County, England	75
Natural gas	billion cubic meters per year	Numerous domestic and international oil companies	North Sea gasfields	100

See footnotes at end of table.

TABLE 2—Continued
UNITED KINGDOM: STRUCTURE OF THE MINERAL INDUSTRY IN 2012

(Thousand metric tons unless otherwise specified)

Commodity		Major operating companies and major equity owners	Location of main facilities[1]	Annual capacity
Nickel, refined		INCO Europe Ltd. (CVRD INCO Ltd.)	Clydach Refinery, near Swansea, West Glamorgan County, Wales	30
Nitrogen, N content of ammonia		Terra Nitrogen Ltd.	Billingham, Durham County, England, and Severnside, near Bristol, Avon County, England	550
Do.		GrowHow UK Ltd. (Kemira GroHow Oyj)	Ince, Lancashire County, England	400
Petroleum:				
Crude	million 42-gallon barrels per day	Numerous domestic and international oil companies, which include Apache North Sea Ltd., BG Group, BHP Billiton Ltd., BP p.l.c., Challenger Minerals Inc., Chevron Corp., ConocoPhillips Corp., Dana Petroleum plc, Eni S.p.A.., Exxon Mobil Corp., Hess Corp., Lundin Britain Ltd., Maersk Oil UK Ltd., Marathon Oil U.K. Ltd., Midmar Energy Onshore Ltd., Nexen Petroleum Inc., Noble Energy (Europe) Ltd., Oilexco Inc., Perenco UK Ltd., Petro-Canada UK Ltd., Premier Oil plc, Royal Dutch Shell plc, Statoil (U.K.) Ltd.,Talisman Ltd., Total S.A., and Tullow Oil (U.K.) Ltd.	North Sea oilfields	2
Refined	million 42-gallon barrels	Exxon Mobil Corp.	Fawley refinery, Southampton, Hampshire County, England	120
Do.	do.	Royal Dutch Shell plc	Stanlow manufacturing complex, Ellesmere Port, Cheshire County, England	100
Do.	do.	ConocoPhillips Co.	Humber refinery, South Killingholme, North Lincolnshire County, England	90
Do.	do.	Total S.A.	Lindsey refinery, Killingholme, North Lincolnshire County, England	85
Do.	do.	Chevron Corp.	Pembroke refinery, Pembroke, Dyfed County, Wales	82
Do.	do.	Ineos Group	Grangemouth refinery, Grangemouth, Stirling County, Scotland	80
Do.	do.	BP p.l.c.	Coryton refinery, Stanford-le-Hope, Essex County, England	80
Do.	do.	Petroplus Holdings AG	Teesside refinery, Middlesborough, Cleveland County, England	43
Do.	do.	Total S.A., 70%, and Murco Petroleum Ltd., 30%	Milford Haven, Dyfed County, Wales	40
Do.	do.	Eastham Refinery Ltd. (Shell UK Ltd., 50%, and AB Nynas Ltd., 50%)	Eastham refinery, Ellesmere Port, Cheshire County, England	9
Do.	do.	AB Nynas Ltd.	Dundee refinery, Dundee, Scotland	4
Platinum-group metals		Johnson Matthey plc	Refineries at Enfield (London) and Royston, Hertfordshire County, England	NA
Do.		Vale Acton (Vale Group)	Acton refinery, London, England	NA
Potash		Cleveland Potash Ltd. (Israel Chemicals Ltd., 100%)	Boulby Mine, Yorkshire County, England	1,000
Salt:				
Road		do.	do.	600
Rock		British Salt Ltd.	Middlewich, Cheshire County, England	800
Do.		Irish Salt Mining and Exploration Co. Ltd.	Kilroot Mine, Carrick Fergus, Northern Ireland	500
Sand and gravel		Hanson plc (HeidelbergCement AG, 100%)	Various offshore and onshore locations	NA
Silica sand		WBB Minerals (S.C.R.-Sibelco NV)	Various operations in Cheshire, Humberside, and Norfolk Counties, England	5,000
Do.		Hanson plc	Various locations	NA

See footnotes at end of table.

TABLE 2—Continued
UNITED KINGDOM: STRUCTURE OF THE MINERAL INDUSTRY IN 2012

(Thousand metric tons unless otherwise specified)

Commodity	Major operating companies and major equity owners	Location of main facilities[1]	Annual capacity
Slate, natural	Alfred McAlpine Slate Ltd. (Welsh Slate Ltd.)	Operations in Wales include the Penrhyn quarry, Bethesda, Conwy County; the Pen Yr Orsedd quarry, Nantlle, Gwynedd County; quarries at Blaenau Ffestiniog and Cwt y Bugail, Gwynedd County	1,000
Do.	Greaves Welsh Slate Company Ltd.	Llechwedd Slate Mines, Blaenau Ffestiniog, Gwynedd County, Wales	NA
Soda ash	Brunner Mond Group (Tata Chemicals Ltd.)	Northwich, Cheshire County, England	900
Steel	Tata Steel Europe (Tata Steel Group)	Scunthorpe Works, Scunthorte, Lincolnshire County, England	4,500
Do.	Tata Steel Europe Teesside Cast Products (Tata Steel Group)	Teesside Works, Redcar, Cleveland County, England	3,900
Do.	Tata Steel Europe Strip Products UK (Tata Steel Group)	Port Talbot works, Port Talbot, West Glamorgan, Wales	3,750
Do.	Tata Steel Europe Engineering Steels (Tata Steel Group)	Rotherham Works, Rotherham, South Yorkshire County, England	1,200
Do.	do.	Stocksbridge Works near Sheffield, South Yorkshire County, England	NA [4]
Do.	Tata Steel Europe Special Profiles (Tata Steel Group)	Skinningrove, Carlin How, near Saltburn-by-the-Sea, Cleveland County, England	NA
Do.	Celsa Manufacturing Ltd. (Grupo Celsa, 100%)	Tremorfa Works, Cardiff, South Glamorgan County, Wales	850
Stone, crushed	Hanson plc	90 quarries in various locations	70,000
Tin, ore	Celeste Copper Corp.	South Crofty Mine, Cornwall County, England	400

Do., do. Ditto. NA Not available. -- Zero.

[1]Location names may include historic, postal, or preserved counties instead of current regional governments, such as cities, county boroughs, or unitary authorities.
[2]Grinding plant only. Kilns closed in May 2006.
[3]Under construction.
[4]Remelt facilities.

The Mineral Industry of Uzbekistan

By Elena Safirova

Uzbekistan has substantial natural resources, which include more than 1,800 known mineral deposits. The two minerals produced in the country in significant amounts were gold and uranium. In addition, Uzbekistan was one of the leading world producers of kaolin, molybdenum, nitrogen, rhenium, oil and natural gas, and sulfur. Other valuable minerals produced included copper, gypsum, silver, tungsten, and zinc. Many other mineral commodities (such as iron ore and lithium) had been identified that were not being mined. In previous decades, mineral production was limited by the country's inefficient infrastructure, remote location with respect to world markets, and tight regulatory environment that attracted only a small amount of foreign investment. In the past several years, however, the country had made significant efforts to increase its mineral production, including through expansion of copper and gold production facilities, construction of new potash and tungsten plants, and development of shale oil and gas condensate deposits (Apodaca, 2013a, b; George, 2013; Polyak, 2013a, b; U.S. Central Intelligence Agency, 2013; U.S. Energy Information Administration, 2013; Virta, 2013).

Minerals in the National Economy

In 2012, Uzbekistan's real gross domestic product (GDP) increased by 8.2%; the nominal GDP was 96,589.8 billion soums ($48.7 billion).[1] The value of exports was reported to be to $14.3 billion, which was a decrease of 5.1% compared with that of 2011. The main mineral export commodities were ferrous and nonferrous metals, gold, mineral fertilizers, and oil and gas. The country's main export partners were China (which received 18.5% of Uzbekistan's exports, by value), Kazakhstan (14.6%), Turkey (13.8%), Russia (12.8%), Ukraine (12.5%), and Bangladesh (8.9%). The value of imports increased to $12.0 billion, or by 11.4% compared with that of 2011. The main mineral import commodities were chemicals and ferrous and nonferrous metals. The major import partners were Russia (which supplied 20.6% of Uzbekistan's imports, by value), China (16.5%), the Republic of Korea (16.3%), Kazakhstan (12.8%), Germany (4.6%), and Turkey (4.2%) (State Committee of the Republic of Uzbekistan on Statistics, 2013).

In 2012, the share of industrial production in the GDP was 52.6%. The main industries (as a percentage of the value produced by all industries) were the fuel industry (18.0%), machine building and metal processing (17.5%), food processing (13.1%), textile manufacturing (12.9%), nonferrous mining and metallurgy (10.0%), electric power production (7.7%), and the chemical sector and construction material manufacturing (5.5% each) (State Committee of the Republic of Uzbekistan on Statistics, 2013; U.S. Central Intelligence Agency, 2013).

Production

In 2012, production of tungsten increased by 173%; that of zinc, by 11.3%; and potash, by 11.1%. Uranium production increased by an estimated 8.0%; copper, by 4.5%; and gypsum, by an estimated 4.2%. On the other hand, crude petroleum production decreased by 12.1% and molybdenum mine output decreased by 6.2%. These and other production data are in table 1.

Structure of the Mineral Industry

Table 2 is a list of major mineral industry facilities.

Commodity Review

Metals

Copper.—The only producer of copper in Uzbekistan was the Almalyk mining and metallurgical complex (Almalyk GMK), which was located in Toshkent Province (Toshkent Viloyati). Two large copper porphyry deposits, the Kalmakyr and the Sary-Cheku deposits, were the complex's sources of copper. An additional copper deposit, Dal'neye, was on reserve. The mineral deposits of Toshkent Viloyati are highly complex and contain more than 170 types of minerals. In addition to copper, the Almalyk GMK mined and processed lead-zinc-barite ores from the Uch-Kulach deposit located in Jizzax Viloyati and the Khandiza polymetallic deposit located in Qashqadaryo Viloyati (Almalyk Mining-Metallurgical Complex, 2013).

In December, the Almalyk GMK completed reconstruction and expansion of its Kalmakyr Mine. The reconstruction started in 2009 and included the purchase of mining and transportation equipment, reconstruction of railroad tracks at the mine, and overburden removal at some sections of the mine. Because of the reconstruction, the mine capacity was expected to increase to 31.5 million metric tons per year (Mt/yr) of ore from 27 Mt/yr. The project was financed by the Fund for Reconstruction and Development of Uzbekistan (FRRU), which provided a $63 million loan; AKIB Ipotekabank, which provided a $20 million loan; and the Almalyk GMK, which used $39.4 million of its own funds (MinerJob.ru, 2013b).

In July 2011, Rio Tinto announced plans to start copper exploration of the Gava property in Namangan Viloyati and announced that it was prepared to invest up to $100 million in the project if the results of exploration were promising. In December, the State Geology and Mineral Resources Committee (Goskomgeo) granted Rio Tinto Ltd. of the United Kingdom a 5-year license for exploration of copper deposits in Namangan Viloyati. Rio Tinto was planning to start exploration in the beginning of 2013 and to invest $1 million within 1 year (MinerJob.ru, 2012b).

---

[1]Where necessary, Uzbekistani soums (UZS) were converted to U.S. dollars (US$) at the average annual rate of 1,983UZS=US$1.00.

The Almalyk GMK was planning to start mining its reserve Dal'neye deposit, which was located in Toshkent Viloyati, by the end of 2014. The company was planning to complete a technical assessment of the project by mid-2013 and would then seek Government approval for the project. The Almalyk GMK expected to start mine construction in 2014 and to complete the first stage of mine development within 5 years, when it expected to be mining 10 Mt/yr of ore. The total cost of the first stage was estimated to be $330 million. After the completion of the second stage (the terms of which were not disclosed), the company was planning to mine 35 Mt/yr of ore. The resources of Dal'neye were expected eventually to replace the partially depleted Kalmakyr and Sary-Cheku deposits (Mineral.ru, 2012g).

Gold.—Uzbekistan's significant reserves of gold were estimated to total 5,300 metric tons (t). According to Goskomgeo, the country had 33 primary gold deposits. The main gold producers of the country were two Government-owned mining and metallurgical complexes—the Almalyk GMK and the Navoi mining and metallurgical complex (Navoi GMK). The Muruntau deposit in the Central Qizilqum region was thought to be unique in the world because of the high quality of its ores and the relatively low extraction costs (because the ore was accessible by open pit mining). Another prospective gold deposit, the Tamdybulak, is located 25 kilometers north of Muruntau (Almalyk Mining-Metallurgical Complex, 2013; Navoi Mining and Metallurgical Combinat, 2013).

The Navoi GMK's share of total gold production in Uzbekistan was about 80%; it had control of 13 gold deposits, most of which were either already being mined or were planned to be developed in the near future. Production at the Navoi GMK was conducted at four metallurgical plants located in Navoi, Uchkuduk, Zamaritan, and Zarafshan. In 2012, the Navoi GMK was planning to complete construction of a new gold mining complex that would use bioleaching (BIOX) technology. The new complex was being built at the hydrometallurgical plant in Uchkuduk, and the total cost of the project was $210 million. The complex was expected to process 5 Mt/yr of ore and to produce 20 metric tons per year (t/yr) of gold when it reaches its full capacity (Mineral.ru, 2012a).

In August, the Navoi GMK announced that it was planning to invest $28 million to modernize its heap-leaching facilities between 2012 and 2015. The project was to be financed by the Navoi GMK's own funds and an $8.3 million loan from the FRRU. At Muruntau, heap-leaching technology was first used in 1996 by the Zarafshan-Newmont company, which was a joint venture between two Uzbek partners—Goskomgeo and the Navoi GMK—and Newmont Mining Co. of the United States. In 2007, Goskomgeo decided to liquidate the joint venture, and its equipment was transferred to the Navoi GMK. After the transfer, the heap-leach plant, which was originally a part of Newmont, continued to operate, but was in need of modernization. The modernization currently planned by the Navoi GMK was expected to increase the ore throughput by 20%; it would take 2.5 years to complete and was projected to extend the life of the mine to at least 2025 (MinerJob.ru, 2012c).

In December, the Navoi GMK announced that, because of a recent series of improvements, the annual capacity of the Zamaritan plant had been increased to 1.4 Mt/yr of ore from 1.1 Mt/yr. The improvements included opening new mining horizons at the underground Guzhumsai Mine and starting up new equipment for gravity-based fine gold extraction. The total cost of improvements was $330 million, and the production capacity of the Zamaritan plant was expected to reach 1.8 Mt/yr of ore by 2015 (MinerJob.ru, 2012a).

Also in December, the Navoi GMK opened a new conveyor system at the southeastern part of its Muruntau Mine. The new conveyor was 540 meters (m) long and was 2 m wide. The new system was expected to reduce the costs of transporting ore within the mine as well as reduce the transportation time and distance traveled by the mined materials. In 2013, the transportation capacity of the new system was expected to reach 6 million cubic meters per year (MinerJob.ru, 2012d).

In 2012, the Almalyk GMK started construction of three new gold mines; all the deposits are located in Toshkent Viloyati, and the total cost of construction was expected to reach $132 million. The construction of the three mines was expected to increase the Almalyk GMK's gold production by between 25% and 30%. The first, the Samarchuk Mine, was to be constructed at the Kyzyl-Alma deposit. The new mine was expected to have an annual capacity of 200,000 t/yr of ore. Construction of the Samarchuk Mine would cost $74 million, and it was to be financed by the FRRU ($14.2 million), Uzbek banks ($24.8 million), and the Almalyk GMK's own funds ($35 million). The mine was planned to go online in 2014 (Forbes.kz, 2013; Regnum.ru, 2013).

The second mine under construction was a new underground mine at the Kairagach deposit. The total project cost was $30.6 million, and the mine was expected to be completed in 2 years. The project was being financed by Uzbek banks, which loaned a total of $13.2 million; the FRRU, which provided $6.7 million in loans; and the Almalyk GMK's own funds. The total production capacity of the new mine was expected to be 80,000 t/yr of ore. To stimulate construction of the Kairagach Mine, the Government reduced the customs duties on all imported vehicles, materials, and equipment used in the project through November 2014. The third mine under construction was an open pit mine at the Uzun portion of the Kochbulak gold deposit. The total cost of the project was expected to be $15 million, and the work was scheduled to be completed by the end of 2013. Because of the existing mining operations at the Kochbulak deposit, the Uzun mine construction works were able to use existing infrastructure and therefore reduce construction costs (Mineral.ru, 2012b, 2013; Forbes.kz, 2013; MinerJob.ru, 2013a).

Tungsten.—In 2012, OAO Uzbek Refractory and Heatproof Metals Complex (UzKTZhM) produced 130.8 t of metallic tungsten, which was a 173% increase compared with the 2011 production level. UzKTZhM produced tungsten on a tolling scheme from raw materials imported from Russia. The complex was the leading producer of metal tungsten in the Commonwealth of Independent States and, in addition to tungsten, it also produced molybdenum and alloys of refractory metals. The complex was originally built in 1956 and had 14 major plants and 12 auxiliary production lines. As of 2012, however, UzKTZhM was operating well below its capacity; tungsten production, for example, was only at about 50% of full capacity level (Regnum.ru, 2012a; UZDaily.uz, 2013).

A newly formed Uzbekistan-Korea Tungsten Co. was planning to start mining tungsten at the Sautbai deposit, which is located in Navoiy Viloyati. Uzbekistan-Korea Tungsten was a joint venture between Goskomgeo of Uzbekistan and Shindong Resources Co. Ltd. of the Republic of Korea. According to the announced schedule, the partners were expected to finalize the economic assessment of the project and to start ground work by the end of 2013. The project included construction of a mining and beneficiation complex with a projected capacity of 1,500 t/yr of highly beneficiated tungsten concentrate. Goskomgeo reported that the Sautbai deposit contains 4 million metric tons (Mt) of ore that included 19,900 t of tungsten trioxide. The Government expected that the Almalyk GMK and UzKTZhM would eventually join the joint venture to increase the effectiveness of the project (Mineral.ru, 2012d; Uzreport.uz, 2012b).

Industrial Minerals

Potash.—In the summer of 2010, State Joint Stock Co. Uzkimyosanoat opened the new Dekhkanabad potash fertilizer mining and beneficiation complex in Qashqadaryo Viloyati on the border with Turkmenistan. The mine was built by ZAO Zapadno-Ural'skiy Mashinostroitel'nyy Kontsern (ZUMK) of Russia, and CITIC Pacific Ltd. of Hong Kong constructed the beneficiation plant. The mine construction was financed by FRRU ($61.9 million), Eximbank of China ($41.7 million), and Uzkimyosanoat's own funds. The cost of mine construction was $56 million, and the construction cost for the beneficiation plant was $43.9 million. The total cost of the complex was $123.7 million, and the annual capacity of the complex was expected to reach 200,000 t/yr of potassium chloride (Gazeta.uz, 2010; Newchemistry.ru, 2012).

The Dekhkanabad complex is built at the Tubegatan deposit of potash salts. The total resources of the deposit are estimated to be 400.2 Mt of "ore" containing 36.8% potassium chloride. The Uzbek portion of the deposit's resources is estimated to be about 200 Mt. In 2011, the plant produced 180,000 t of potassium chloride, and, in 2012, it reached its production capacity of 200,000 t/yr, most of which was exported. In May, ZUMK and CITIC Pacific started working on the second stage of construction, which was expected to cost a total of $254.7 million. According to the company's plan, after the second stage is completed, the plant's capacity would be tripled to 600,000 t/yr, of which 350,000 t/yr was expected to be exported. The construction of the second stage was expected to be finished by the end of 2013 (Gazeta.uz, 2010; RCCnews.ru, 2012).

Sulfuric Acid.—In 2012, Uzbekistan produced 1.27 Mt of sulfuric acid and was planning to double its production by 2015. Two modernization projects were underway—a new plant within the Almalyk GMK copper smelter and a new plant at the Navoi GMK. The new plant at the Almalyk GMK was planned to have a capacity of 500,000 t/yr at a total cost of $80.2 million. The Almalyk GMK expected to invest $21.2 million of its own funds and to obtain a $30 million loan from the FRRU and a $29 million loan from AKIB Ipotekabank. The plant at the Almalyk GMK was to be completed by the end of 2013 (Trend.az, 2012).

The new plant at the Navoi GMK was planned to have a capacity of 650,000 t/yr and was expected to cost a total of $94 million, of which $64 million would come from the Navoi GMK's own funds and the other $30 million was to be financed by the FRRU. The new plant was expected to start operations in 2014 (Uzreport.uz, 2012a).

Mineral Fuels and Related Materials

Coal.—Uzbekistan's identified resources of coal were estimated to be 1,900 Mt, including 46.3 Mt of bituminous coal. Undiscovered resources of coal in Uzbekistan were estimated to total an additional 323 Mt. Significant bituminous coal resources are concentrated in the south of the country, in particular, in Qashqadaryo and Surxondaryo Viloyatis; resources of lignite are concentrated in Fergana, Navoiy, and Toshkent Viloyatis, as well as the Karakalpak Autonomous Republic. As of 2012, coal mining was conducted at three main deposits—the Angren lignite deposit and the Baysun and the Shargun bituminous coal deposits. In 2012, the four companies that produced coal in Uzbekistan were, in the order of the tonnage produced, OAO Uzbekugol, OAO Apartak, OAO Shargunkumir, and OAO Erostigaz (Mineral.ru, 2012e).

In recent years, the Government had been trying to increase coal production with the goal of exporting some of the domestically produced hydrocarbons that had previously been used within Uzbekistan. In particular, the Government planned to increase lignite production to 6.4 Mt/yr by 2014 and 16.3 Mt by 2021, and to increase bituminous coal production to 900,000 t by 2021. The Government was expecting that direct coal consumption by residents would increase to 2.4 Mt by 2020, or by 2.9 times that of 2012. By then, the share of coal in the country's total domestic energy consumption would increase to 12% from 3.9% in 2012 (Mineral.ru, 2012c, h).

According to Government plans, Government-owned OAO Uzbekugol, which was the leading producer of solid fuels in the country, was developing the Angren lignite deposit. The company had adopted a program of modernization that would consist of six projects which would cost a total of about $500 million. The first two stages of the project were to be completed by 2015, which would increase the production capacity to 11.5 Mt by 2018. OAO Apartak, which was also developing the Angren deposit, was planning to modernize its lignite mine operations by improving the mine's infrastructure and logistics. The project would include construction of a new technological complex to improve coal sorting and loading into railroad cars. The Apartak modernization project was expected to cost about $80 million and to increase the company's annual lignite production to 1.82 Mt by 2016 (12uz.com, 2012).

OAO Shargunkumir, which held a mining license for the Baysun and the Shargun deposits in Surxondaryo Viloyati, was also planning to modernize its production facilities by investing $92.2 million in upgrades. The modernization program would include acquisition of new equipment for processing, sorting, and loading coal and the construction of a new $15 million beneficiation plant that would have a capacity of 600,000 t/yr. In March, Uzbekugol became one of the shareholders of

Shargunkumir by purchasing some shares on the secondary market that used to be owned by foreign shareholders. The amount of shares and the purchasing price were not disclosed. Prior to the transaction, the Government owned 51% of the shares, the company workers owned 7.43% of the shares, and the other 41.57% of the shares belonged to a consortium of companies, including M.Metal & Co. Ltd., SAB Energy, and Shadella Inc., all of the United Kingdom (UZDaily.uz, 2012).

Natural Gas and Petroleum.—Uzbekistan had significant hydrocarbon resources and was one of only a few countries in the region that were not dependent on a foreign supply of energy. The country had 171 discovered oil and natural gas fields, 51 of which produced oil and 17 of which produced gas condensate. Because of aging production equipment, however, oil production at existing facilities had been decreasing since 2003, and the currently producing fields were being rapidly depleted. In 2012, Uzbekistan's production of liquid hydrocarbons decreased by 12.1% to 3.165 Mt—1.561 Mt of petroleum and 1.604 Mt of gas condensate (Nefttrans.ru, 2013).

In May, National Holding Company Uzbekneftegaz announced that it planned to build an oil and gas shale processing complex at the Sangruntau deposit in Navoiy Viloyati. The shale project was a joint project with Japan Gas Corp. (JGC) of Japan. According to preliminary plans, the new complex would produce up to 1 Mt/yr of petroleum products. According to expert estimates, probable resources of oil and gas shale amount to 47 billion metric tons (Gt), and most of the resources are located at depths of up to 600 m. The resources of the Aktau, the Baisun, the Jam, the Kulbeshkak, the Sangruntau, the Uchkyr, and the Urtabulak deposits were included in that estimate. In addition to hydrocarbon resources, the deposits reportedly contain a wide spectrum of nonferrous and rare metals (Azizov, 2012).

In November, it was announced that Uzbekneftegaz would start development of natural gas condensate deposits at the border with Turkmenistan. The $320 million project would involve development of seven deposits, the largest of which were the Girsan, the Samantepe, and the Taylyak deposits. The deposits were discovered in the 1990s, but no attempts have been made to develop them until recently. It was expected that once the deposits were all producing (by 2017), the total production from all seven of them would amount to 3.5 billion cubic meters per year of natural gas (Mineral.ru, 2012f).

Uranium.—The Navoi GMK was the only enterprise in the country that conducted mining, beneficiation, and export of uranium as uranium oxide (U_3O_8). The Navoi GMK had three mining units and Hydrometallurgical Plant #1 (GMZ–1) that were involved in uranium production. The primary method of uranium mining used at Navoi was in situ leaching (ISL). This technology made profitable extraction of uranium from sandstone-type deposits with low uranium content possible. Navoi GMK's uranium resources consisted of 20 deposits and 10 additional prospective areas. According to Goskomgeo, explored and evaluated resources of uranium in Uzbekistan amounted to 185,800 t, of which 138,800 t was of sandstone type and the other 47,000 t was of black shale type. Based on the proven and probable resources, the Navoi GMK was expecting to continue uranium mining for the next 40 years (Navoi Mining and Metallurgical Combinat, 2013).

In 2012, the Navoi GMK invested $230.5 million into modernization of existing uranium processing facilities and was planning to invest $55.2 million in two new uranium mines. The first mine at the Aulbek deposit began operations in May; mine construction was continuing, however, and the second stage was expected to be completed in 2013. The total cost of construction was expected to be $20.9 million, of which $8.9 million would be spent in 2012. Construction of another new mine at the North Kanimekh deposit started in 2012 and was expected to be finished in 2013; the total cost of the project was estimated to be $34.3 million. The ores of the two new mines had higher carbonate content and were located deeper underground than were existing mines operated by the Navoi GMK; it was likely that mining of these mines would require new technologies to make uranium extraction cost effective (Podrobno.uz, 2012; Regnum.ru, 2012b).

Outlook

In the past several years, Uzbekistan has intensified its efforts to grow the country's industry, including manufacturing and, especially, automobile production, chemical production, and machine building. In 2012, the share of the country's GDP produced by industrial enterprises was greater than 50%. Increased industrial production and higher living standards in the country are expected eventually to increase the demand for energy goods. Uzbekistan will likely seek to increase its production and export of hydrocarbons during the next decade by expanding its pipelines and modernizing its production facilities and infrastructure. The Government is also likely to continue to form partnerships with Asian and Russian firms to help achieve this objective. It also plans to increase coal production for domestic heating and electricity production significantly.

The country is likely to increase its production of copper, gold, and uranium. In the past several years, Uzbekistan has made concerted efforts to modernize its Almalyk and Navoi GMKs and to ramp up their production. Barring unforeseen events in the world economy, therefore, Uzbekistan's mineral production is expected to continue to increase in the next several years.

References Cited

12uz.com, 2012, V Uzbekistane namereny uvelichit' dobychu uglya k 2021 godu v 4 raza [By 2021, Uzbekistan intends to quadruple coal production]: 12uz.com, September 4. (Accessed November 20, 2013, at http://www.12uz.com/news/show/economy/11175/.)

Almalyk Mining-Metallurgical Complex, 2013, Home page: Almalyk Mining-Metallurgical Complex. (Accessed November 20, 2013, at http://www.agmk.uz/index.php?option=com_contact&view=contact&id=10&Itemid=80&lang=en.)

Apodaca, L.E., 2013a, Nitrogen (fixed)—Ammonia: U.S. Geological Survey Mineral Commodity Summaries 2013, p. 112–113.

Apodaca, L.E., 2013b, Sulfur: U.S. Geological Survey Mineral Commodity Summaries 2013, p. 158–159.

Azizov, D., 2012, Uzbekistan v 2012-2013 godu planiruet sozdat' proizvodstvo po pererabotke goryuchikh slantsev [In 2012-2013 Uzbekistan plans to build facilities for processing of oil and gas shale]: Trend.az, May 17. (Accessed November 20, 2013, at http://www.trend.az/print/2028009.html.)

Forbes.kz, 2013, Almalykskiy GMK stroit zolotoy rudnik stoimost'yu $74 mln v Uzbekistane [Almalyk GMK is building a $74 million gold mine in Uzbekistan]: Forbes.kz, February 5. (Accessed November 20, 2013, at http://forbes.kz/news/2013/02/05/newsid_18044.)

Gazeta.uz, 2010, Kaliy na vyrost [Potash for future growth]: Gazeta.uz, January 7. (Accessed November 20, 2013, at http://www.gazeta.uz/2010/01/07/plant.)

George, M.W., 2013, Gold: U.S. Geological Survey Mineral Commodity Summaries 2013, p. 66–67.

Mineral.ru, 2012a, NGMK v 2012 g. zavershit stroitel'stvo kompleksa BIOX na GMZ-3 [In 2012, NGMK will finish construction of a BIOX complex at GMZ–3]: Mineral.ru, March 13. (Accessed November 20, 2013, at http://www.mineral.ru/News/47913.html.)

Mineral.ru, 2012b, OAO "AGMK" pristupil k modernizatsii rudnika na Kochbulake [OAO AGMK started modernization of the Kochbulak Mine]: Mineral.ru, February 7. (Accessed November 20, 2013, at http://www.mineral.ru/News/47546.html.)

Mineral.ru, 2012c, Potreblenie uglya naseleniem Uzbekistana vozrastet k 2020 g. v 2,9 raza – do 2,4 mln t [By 2020, coal consumption by Uzbekistan's residents will increase 2.9 times—to 2.4 million tons]: Mineral.ru, October 7. (Accessed November 20, 2013, at http://www.mineral.ru/News/50354.html.)

Mineral.ru, 2012d, SP Uzbekistan-Korea Tungsten v 2013 g. pristupit k dobyche vol'frama [In 2013, a joint venture, Uzbekistan-Korea Tungsten, will start tungsten production]: Mineral.ru, November 12. (Accessed November 20, 2013, at http://www.mineral.ru/News/50731.html.)

Mineral.ru, 2012e, Uzbekistan nameren uvelichit' dobychu uglya k 2021 godu v 4 raza – do 17,2 mln t [By 2021, Uzbekistan intends to quadruple its coal production to 17.2 million tons]: Mineral.ru, September 7. (Accessed November 20, 2013, at http://www.mineral.ru/News/49995.html.)

Mineral.ru, 2012f, Uzbekistan v nachale 2013 g. nachnet razrabotku semi gazokondensatnykh mestorozhdeniy [In the beginning of 2013, Uzbekistan will start development of seven condensate deposits]: Mineral.ru, November 1. (Accessed November 20, 2013, at http://www.mineral.ru/News/50676.html.)

Mineral.ru, 2012g, Uzbekskiy AGMK v 2014 g. nachnet razrabotku novogo mestorozhdeniya [In 2014, Uzbek AGMK will start development of a new deposit]: Mineral.ru, November 26. (Accessed November 20, 2013, at http://www.mineral.ru/News/50930.html.)

Mineral.ru, 2012h, "Uzbekugol' k 2016 g. uvelichit dobychu uglya v dva raza [By 2016, Uzbekugol will double coal production]: Mineral.ru, May 29. (Accessed November 20, 2013, at http://www.mineral.ru/News/48762.html.)

Mineral.ru, 2013, Uzbekskiy AGMK k 2015 g. postroit rudnik na Kayragache [By 2015, Uzbek AGMK will build a mine at Kairagach]: Mineral.ru, March 4. (Accessed November 20, 2013, at http://www.mineral.ru/News/52072.html.)

MinerJob.ru, 2012a, NGMK na tret' uvelichil zolotodobychu [NGMK increased gold production by one-third]: MinerJob.ru, December 19. (Accessed November 20, 2013, at http://minerjob.ru/viewnew.php?id=23006.)

MinerJob.ru, 2012b, Rio Tinto Ltd poluchila litsenziyu na razvedku mednykh mestorozhdeniy Uzbekistana [Rio Tinto Ltd. obtained an exploration license for copper deposits in Uzbekistan]: MinerJob.ru, December 24. (Accessed November 20, 2013, at http://minerjob.ru/viewnew.php?id=23055.)

MinerJob.ru, 2012c, V 2012-2015 godakh Navoiskiy gorno-metallurgicheskiy kombinat investiruet $28 mln v modernizatsiyu proizvodstva kuchnogo vyshelachivaniya zolota [In 2012–2015, Navoi GMK will invest $28 million in modernization of gold heap-leaching process]: MinerJob.ru, August 9. (Accessed November 20, 2013, at http://minerjob.ru/viewnew.php?id=21618.)

MinerJob.ru, 2012d, V NGMK zapustili novyy konveyernyi kompleks tsiklichno-potochnoy tekhnologii "Yugo-bostok-poroda" [NGMK opened a new conveyor complex using cyclic and continuous technology "South-East-ore"]: MinerJob.ru, December 17. (Accessed November 20, 2013, at http://minerjob.ru/viewnew.php?id=22967.)

MinerJob.ru, 2013a, AGMK postroit rudnik na mestorozhdenii "Kayragach" [AGMK will build a mine at Kairagach deposit]: MinerJob.ru, March 4. (Accessed November 20, 2013, at http://minerjob.ru/viewnew.php?id=23635.)

MinerJob.ru, 2013b, AGMK zavershil rasshirenie rudnika "Kal'makyr" [AGMK finished Kalmakyr Mine expansion]: MinerJob.ru, January 31. (Accessed November 20, 2013, at http://minerjob.ru/viewnew.php?id=23357.)

Navoi Mining and Metallurgical Combinat, 2013, Home page: Navoi Mining and Metallurgical Combinat. (Accessed November 20, 2013, at http://www.nkmk.uz/.)

Nefttrans.ru, 2013, Uzbekistan postroit vtoruyu ochered' Bukharskogo NPZ [Uzbekistan will build the second part of the Bukhara refinery]: Nefttrans.ru, September 4. (Accessed November 20, 2013, at http://www.nefttrans.ru/news/uzbekistan-postroit-vtoruyu-ochered-bukharskogo-npz-.html.)

Newchemistry.ru, 2012, Uzbekistan nachal proizvodstvo kaliynykh udobreniy [Uzbekistan started production of potash fertilizers]: Newchemistry.ru. (Accessed November 20, 2013, at http://newchemistry.ru/letter.php?n_id=9225.)

Podrobno.uz, 2012, Navoiskiy GMK nachal dobychu uarna na mestorozhdenii Aul'bek [Navoi GMK started uranium production from the Aulbek deposit]: Podrobno.uz, May 21. (Accessed November 20, 2013, at http://podrobno.uz/cat/economic/Navoiski-GMK-nachal-dobichu-urana-na-mestorojdenii-Aulbek/.)

Polyak, D.E., 2013a, Molybdenum: U.S. Geological Survey Mineral Commodity Summaries 2013, p. 106–107.

Polyak, D.E., 2013b, Rhenium: U.S. Geological Survey Mineral Commodity Summaries 2013, p. 130–131.

RCCnews.ru, 2012, Uzbekistan narashivaet proizvodstvo kaliynykh udobreniy [Uzbekistan is increasing its potash fertilizer production]: RCCnews.ru, October 31. (Accessed November 20, 2013, at http://rccnews.ru/ru/news/fertilizers/85583/.)

Regnum.ru, 2012a, Proizvodstvo vol'frama v Uzbekistane v 2012 godu planiruetsya uvelichit pochti v dva raza [In 2012, tungsten production in Uzbekistan is planned to almost double]: Regnum.ru, August 15. (Accessed November 20, 2013, at http://www.regnum.ru/news/1561585.html.)

Regnum.ru, 2012b, Uzbekistan: NGMK planiruet v 2012 godu uvelichit' dobychu urana v 1,7 raz [Uzbekistan: In 2012, NGMK is planning to increase uranium production 1.7 times]: Regnum.ru, March 22. (Accessed November 20, 2013, at http://www.regnum.ru/news/1512782.html.)

Regnum.ru, 2013, AGMK Uzbekistana pristupilo k stroitel'stvu rudnika po dobyche zolota v Tashkentskoi oblasti [Uzbekistan's AGMK started building a gold mine in Toshkent Oblast']: Regnum.ru, February 5. (Accessed November 20, 2013, at http://www.regnum.ru/news/1621488.html.)

State Committee of the Republic of Uzbekistan on Statistics, 2013, Quarterly reports, basic economic and social indicators: State Committee of the Republic of Uzbekistan on Statistics. (Accessed November 20, 2013, at http://stat.uz/en/reports/165/.)

Trend.az, 2012, Uzbekistan vkladyvaet bolee 174 mln dollarov dlya uvelicheniya proizvodstva sernoy kisloty [Uzbekistan is investing more than $174 million to increase sulfuric acid production]: Trend.az, March 15. (Accessed November 20, 2013, at http://www.trend.az/print/2003896.html.)

U.S. Central Intelligence Agency, 2013, Uzbekistan, in The world factbook: U.S. Central Intelligence Agency, April 18. (Accessed November 20, 2013, at https://www.cia.gov/library/publications/the-world-factbook/geos/uz.html.)

U.S. Energy Information Administration, 2013, Country analysis note, International energy data and analysis for Uzbekistan: U.S. Energy Information Administration, August 22. (Accessed November 20, 2013, http://www.eia.gov/countries/country-data.cfm?fips=UZ.)

UZDaily.uz, 2012, OAO "Uzbekugol" stalo aktsionerom OAO "Shargun'kumir" [OAO Uzbekugol became a sharcholder of OAO Shargunkumir]: UZDaily.uz, March 6. (Accessed November 20, 2013, at http://www.uzdaily.uz/articles-id-10145.htm.)

UZDaily.uz, 2013, Uzbekistan v 2012 godu uvelichil proizvodstvo metallicheskogo vol'frama [In 2012, Uzbekistan increased production of metallic tungsten]: UZDaily.uz, February 4. (Accessed November 20, 2013, at http://www.uzdaily.uz/articles-id-14341.htm.)

Uzreport.uz, 2012a, K 2015 godu Uzbekistan vdvoe uvelichit proizvodstvo sernoi kisloty [By 2015, Uzbekistan will double its sulfuric acid production]: Uzreport.uz, March 13. (Accessed November 20, 2013, at http://news.uzreport.uz/news_4_r_95136.html.)

Uzreport.uz, 2012b, OAO "UzKTZhM" planiruet dovesti v 2012 godu ob'em proizvodstva produktsii do 40,8 mlrd sumov [In 2012, OAO UzKTZhM is planning to increase production to 40.8 billion soums]: Uzreport.uz, July 4. (Accessed November 20, 2013, at http://news.uzreport.uz/news_4_r_98108.html.)

Virta, R.L., 2013, Clays: U.S. Geological Survey Mineral Commodity Summaries 2013, p. 44–45.

TABLE 1
UZBEKISTAN: PRODUCTION OF MINERAL COMMODITIES[1]

(Metric tons unless otherwise specified)

Commodity[2]		2008	2009	2010	2011	2012
METALS						
Aluminum, secondary[e]		NA [r]	NA [r]	NA [r]	NA [r]	NA
Copper:						
Mine output, Cu content		95,000	95,000	90,000	91,500 [3]	95,600
Metal:[e]						
Blister		92,000	92,000	92,000	92,000	93,000
Refined		71,000	80,000	90,000	91,500 [3]	95,600
Gold[e]	kilograms	85,000	90,000	90,000	91,000	93,000
Molybdenum, mine output, Mo content[e]		500	500	500	557 [r, 3]	522 [3]
Rhenium[e]	kilograms	4,800	4,800	4,800	5,400	5,400
Silver, mine output	do.	74,648	52,876	59,097	60,000	60,000
Steel:						
Crude		685,700	716,400	731,373	733,400 [r, 3]	736,300
Rolled		640,000	670,000	691,910	709,900 [3]	710,500
Tungsten		--	--	--	48	131
Zinc, metal, smelter, primary		70,445	40,000	40,000 [e]	54,900 [r, 3]	61,100
INDUSTRIAL MINERALS						
Cement[e]		6,600,000	6,850,000	6,800,000 [r]	6,698,000 [3]	6,800,000
Clays:[e]						
Bentonite		20,000 [r]	20,000 [r]	20,000 [r]	25,000	25,000
Kaolin		5,500,000	5,500,000	5,500,000	7,000,000	7,000,000
Feldspar		NA [r]	NA [r]	NA [r]	NA [r]	NA
Fluorspar		NA [r]	NA [r]	NA [r]	NA [r]	NA
Graphite		NA [r]	NA [r]	NA [r]	NA [r]	NA
Gypsum[e]		50,000 [r]	48,400 [r, 3]	44,000 [r, 3]	48,000 [r]	50,000
Iodine		NA [r]	NA [r]	NA [r]	NA [r]	NA
Nitrogen, N content of ammonia[e]		1,000,000	1,000,000	1,344,029 [r, 3]	1,294,300 [3]	1,300,000
Phosphate rock:[e]						
Gross weight		600,000	600,000	800,000	800,000	800,000
P_2O_5 content		140,000	140,000	187,000	187,000	187,000
Potash, K_2O equivalent		--	--	33	110	122
Sulfur:[e]						
Byproduct:						
Metallurgy		170,000	170,000	170,000	170,000	170,000
Natural gas and petroleum		350,000	350,000	350,000	350,000	370,000
Total		520,000	520,000	520,000	520,000	540,000
Sulfuric acid		600,000	1,023,800	1,192,600	1,200,000	1,270,000
MINERAL FUELS AND RELATED MATERIALS						
Coal:						
Bituminous		198,000	101,000	198,000	244,000	252,900
Lignite		3,092,000	3,553,000	3,102,000	3,600,000	3,600,000
Total		3,290,000	3,654,000	3,300,000	3,844,000 [3]	3,852,900
Natural gas, dry	million cubic meters	67,593	65,000	65,937	63,036	62,911
Petroleum:						
Crude:[4]						
In gravimetric units		2,533,000	2,331,000	1,866,000	3,600,000	3,165,000
In volumetric units[e]	42-gallon barrels	18,400,000	16,900,000	13,600,000	26,200,000	23,000,000
Petroleum refinery products:						
In gravimetric units		4,117,000	4,117,000	3,296,000	5,000,000	5,000,000
In volumetric units	42-gallon barrels	33,100,000	33,100,000	26,480,000	40,165,000	40,165,000
Uranium:						
U content		2,338	2,429	2,400	2,500	2,700 [e]
U_3O_8 content		2,757	2,865	2,830	2,950	3,190

See footnotes at end of table

TABLE 1—Continued
UZBEKISTAN: PRODUCTION OF MINERAL COMMODITIES[1]

eEstimated; estimated data are rounded to no more than three significant digits; may not add to totals shown. rRevised. do. Ditto. -- Zero. NA Not available.

[1]Table includes data available through November 1, 2013.

[2]In addition to the commodities listed, Uzbekistan is thought to produce a number of other mineral commodities, including cesium, caustic soda, iron ore, lead, lithium, manganese, rubidium, selenium, tellurium, tungsten, and vermiculite, but available information is not adequate to estimate production.

[3]Reported figure.

[4]Includes gas condensate.

TABLE 2
UZBEKISTAN: STRUCTURE OF THE MINERAL INDUSTRY IN 2012[1,2]

(Metric tons unless otherwise specified)

Commodity		Major operating companies, main facilities, or deposits	Location or deposit names	Annual capacity[e]
Bismuth		Ustarassay deposit (depleted)	Chotqol and Kuraminskiy Khrebet regions	NA
Cement		OAO Kyzylkumcement	Navoi City	3,150,000
Do.		OAO Akhangarcement	Sirdaryo Viloyati	1,740,000
Do.		OAO Kuvasaycement	Farg'ona Viloyati	1,100,000
Cesium, lithium, rubidium		Shava-Say deposit	NA	NA
Clays:				
Bentonite		Arab-Dasht and Khaudag deposits	NA	NA
Kaolin		Angren deposit	Angren region	8,000,000
Coal:				
Lignite		OAO Uzbekugol and OAO Apartak	Angren deposit, Toshkent Viloyati	4,500,000
Bituminous		OAO Shargunkumir and OAO Erostigaz	Baysun and Shargun deposits, Surxondaryo Viloyati	700,000 [3]
Copper:				
Mine output, Cu content		Almalyk mining and metallurgical complex (Almalyk GMK)	Dal'neye, Kalmakyr, and Sary-Cheku deposits	100,000 [3]
Concentrate		Almalyk polymetallic beneficiation plant	Qashqadaryo Viloyati	5
Metal		Almalyk refinery	Olmaliq	130,000
Diamond		Karashok and Kok-Say deposits	Navoiy Vilolyati	NA
Feldspar		Karichasayskoye and other deposits	Deposits in Samarqand Viloyati, Toshkent Viloyati, and Qoraqalpog'iston Respublikasi	120,000 [3]
Fertilizers		Ammophos production association	Olmaliq	NA
Do.		Azot production association	Farg'ona area	NA
Do.		Elektrokhimprom production association	Chirchiq	NA
Do.		Kokand superphosphate plant	Qo'qon	NA
Do.		Naviazot production association	Navoiy Viloyati	NA
Do.		Samarkand chemicals plant	Samarqand Viloyati	NA
Fluorspar		Agata-Chibargata, Aurakhmat, Kengutan, Kyzylbaur, Naugarzan, and Nugisken deposits	East of Toshkent Viloyati	150,000
Do.		Syrpatash deposit	Namangan Viloyati	NA
Gold	kilograms	Various facilities and deposits, which include:	Of which:	93,000 [3]
		Adzhi-Bugutty, Amantaytau, Balpantau, Bulutkan, Donguz-Tau, Muruntau, and Taurbay deposits	Central Qizilqum region	
		Navoi mining and metallurgical complex (Navoi GMK) (Uzbekistan State Committee for Geology and Mineral Resources) Navoi, Uchkuduk, Zamaritan, and Zarafshan gold refineries	Muruntau deposit and 12 others	
		Kochbulak and Kyzyl-Al'ma-Say deposits	Toshkent Viloyati	
		Almalyk mining and metallurgical complex (Almalyk GMK)	Dalneye, Kalmakyr, and Sary-Cheku deposits	
Graphite		Tadzhi-Kazgan deposit	Navoiy Viloyati	NA
Iron ore		Syurenata deposit	Toshkent Viloyati	NA

See footnotes at end of table.

TABLE 2—Continued
UZBEKISTAN: STRUCTURE OF THE MINERAL INDUSTRY IN 2012[1,2]

(Metric tons unless otherwise specified)

Commodity	Major operating companies, main facilities, or deposits	Location or deposit names	Annual capacity[e]
Lead, mine output, Pb content	Almalyk mining and metallurgical complex (Almalyk GMK)	Uch-Kulach deposit in Jizzax Viloyati	40,000 [3]
Manganese	Dautashskoye deposit	Qashqadaryo Viloyati	40,000
Molybdenum:			
Mine output, Mo content	Almalyk mining and metallurgical complex (Almalyk GMK); Kalmakyr and Sary-Cheku deposits	Toshkent Viloyati	900 [3]
Metal	Uzbek refinery and hard metals plant	Chirchiq	NA
Natural gas million cubic meters	Gazli, Kandym, Khauzak, Kokdumalak, Pamuk, and Shurtan-Say deposits (major)	Amu-Dar'ya Basin; Muborak region	70,000 [3]
Do.	Itera/Lukoil (Russia), Uzbekneftegaz JSC	Kan-Dam field	NA
Natural gas condensate	Trinity Energy	Ustyurt Platosi region	NA
Natural gas liquids million cubic meters	Mubarek gas processing plant	Muborak region	28,000
Do.	Shurtan gas-chemical complex	Shurtan-Say deposit, Qashqadaryo Viloyati	137,000
Petroleum:			
Crude	Kokdumalak and Mingbulak deposits (major)	NA	9,000,000 [3]
Refinery products	Fergana oil refinery	Farg'ona area	8,800,000
Do.	Bukhara oil refinery	Buxoro area	2,500,000
Phosphate rock	Kyzyl Kum complex	Dzheroy-Sardarin Moroccan type; Karaktay, Severnyy, and Dzhetymtau deposits	NA
Polyethylene	Shurtan gas-chemical complex	Shurtan-Say deposit, Qashqadaryo Viloyati	125,000
Potash	Dekhkanabad potash fertilizer plant	Tubegatan Mine, Qashqadaryo Viloyati	200,000
Rhenium	Almalyk mining and metallurgical complex (Almalyk GMK)	Toshkent Viloyati	NA
Selenium	do.	do.	NA
Silver	do.	do.	NA
Do.	Kosmanachi, Okzhetpes, and Vysokovoltnoye deposits	Namangan Viloyati	NA
Steel, crude	Bekabad steel mill	Bekobod region	1,100,000
Sulfur	Almalyk mining and metallurgical complex (Almalyk GMK)	Dalneye, Kalmakyr, and Sary-Cheku deposits	NA
Do.	Mubarek gas processing plant complex	Muborak region	2,000,000
Tellurium	Almalyk mining and metallurgical complex (Almalyk GMK)	Toshkent Viloyati	NA
Tungsten:	Deposits:	Locations:	1,200 [3]
Mine output, W content	Koytash deposit	Northeastern Uzbekistan	
	Ingichka and Lyangar deposits	Zirabulak Mountains	
	Ugat deposit	Northern Uzbekistan	
Mine output, WO$_3$ content (0.49%)	Sautbay wolframite deposit	Qizilqum region	NA
Metal	Uzbek refractory and hard metals complex (UzKTZhM)	Chirchiq, Toshkent Viloyati	NA
Uranium, U content	Navoi mining and metallurgical complex (Navoi GMK)	Central Qizilqum region	3,000
Vermiculite cubic meters	Tebin-Bulak deposit	NA	25,000
Zinc:			
Mine output, Zn content	Almalyk mining and metallurgical complex (Almalyk GMK)	Khandiza and Uch-Kulach deposits	NA
Concentrate	Almalyk polymetallic beneficiation plant	Qashqadaryo Viloyati	60,000
Metal	do.	do.	80,000

[e]Estimated; estimated data are rounded to no more than three significant digits. Do., do. Ditto. NA Not available.
[1]Table includes data and information available through October 15, 2013.
[2]Many location names have changed since the breakup of the Soviet Union. Many enterprises, however, are still named or commonly referred to based on the former location name, which accounts for discrepancies in the names of enterprises and that of locations.
[3]Capacity estimates are totals for all enterprises that produce that commodity.

VINCENNES UNIVERSITY LIBRARY